Lecture Notes in Artificial Intelligence 10464

Subseries of Lecture Notes in Computer Science

LNAI Series Editors

Randy Goebel
  University of Alberta, Edmonton, Canada
Yuzuru Tanaka
  Hokkaido University, Sapporo, Japan
Wolfgang Wahlster
  DFKI and Saarland University, Saarbrücken, Germany

LNAI Founding Series Editor

Joerg Siekmann
  DFKI and Saarland University, Saarbrücken, Germany

More information about this series at http://www.springer.com/series/1244

YongAn Huang · Hao Wu
Honghai Liu · Zhouping Yin (Eds.)

Intelligent Robotics and Applications

10th International Conference, ICIRA 2017
Wuhan, China, August 16–18, 2017
Proceedings, Part III

 Springer

Editors
YongAn Huang
School of Mechanical Science
 and Engineering
Huazhong University of Science
 and Technology
Wuhan
China

Hao Wu
School of Mechanical Science
 and Engineering
Huazhong University of Science
 and Technology
Wuhan
China

Honghai Liu
Institute of Industrial Research
University of Portsmouth
Portsmouth
UK

Zhouping Yin
School of Mechanical Science
 and Engineering
Huazhong University of Science
 and Technology
Wuhan
China

ISSN 0302-9743   ISSN 1611-3349 (electronic)
Lecture Notes in Artificial Intelligence
ISBN 978-3-319-65297-9   ISBN 978-3-319-65298-6 (eBook)
DOI 10.1007/978-3-319-65298-6

Library of Congress Control Number: 2017948191

LNCS Sublibrary: SL7 – Artificial Intelligence

Printed on acid-free paper

This Springer imprint is published by Springer Nature
The registered company is Springer International Publishing AG
The registered company address is: Gewerbestrasse 11, 6330 Cham, Switzerland

Preface

The International Conference on Intelligent Robotics and Applications (ICIRA 2017) was held at Huazhong University of Science and Technology (HUST), Wuhan, China, during August 16–18, 2017. ICIRA 2017 was the 10th event of the conference series, which focuses on: (a) fundamental robotics research, including a wide spectrum ranging from the first industrial manipulator to Mars rovers, and from surgery robotics to cognitive robotics, etc.; and (b) industrial and real-world applications of robotics, which are the force driving the research further.

This volume of *Lecture Notes in Computer Science* contains the papers that were presented at ICIRA 2017. The regular papers in this volume were selected from more than 350 submissions covering various topics on scientific methods and industrial applications for intelligent robotics, such as soft and liquid-metal robotics, rehabilitation robotics, robotic dynamics and control, robot vision and application, robotic structure design and mechanism, robot learning, bio-inspired robotics, human–machine interaction, space robotics, mobile robotics, intelligent manufacturing and metrology, benchmarking and measuring service robots, real-world applications, and so on. Papers describing original works on abstractions, algorithms, theories, methodologies, and case studies are also included in this volume. Each submission was reviewed by at least two Program Committee members, with the assistance of external referees. The authors of the papers and the plenary and invited speakers come from the following countries and areas: Australia, Austria, China, Cyprus, Germany, Hong Kong, Japan, Korea, Singapore, Spain, Switzerland, UK, and USA.

We wish to thank all who made this conference possible: the authors of the submissions, the external referees (listed in the proceedings) for their scrupulous work, the six invited speakers for their excellent talks, the Advisory Committee for their guidance and advice, and the Program Committee and the Organizing Committee members for their rigorous and efficient work. Sincere thanks also go to the editors of the *Lecture Notes in Computer Science* series and Springer for their help in publishing this volume in a timely manner.

In addition, we greatly appreciate the following organizations for their support:

National Natural Science Foundation of China
School of Mechanical Science and Engineering, HUST
State Key Laboratory of Digital Manufacturing Equipment and Technology, China
State Key Laboratory of Robotic Technology and System, China
State Key Laboratory of Mechanical System and Vibration, China
State Key Laboratory of Robotics, China

August 2017

YongAn Huang
Hao Wu
Honghai Liu
Zhouping Yin

Organization

Honorary Chair

Youlun Xiong   HUST, China

General Chair

Han Ding   HUST, China

General Co-chairs

Naoyuki Kubota	Tokyo Metropolitan University, Japan
Kok-Meng Lee	Georgia Institute of Technology, USA
Xiangyang Zhu	Shanghai Jiao Tong University, China

Program Co-chairs

YongAn Huang	HUST, China
Honghai Liu	University of Portsmouth, UK
Jinggang Yi	Rutgers University, USA

Advisory Committee Chairs

Jorge Angeles	McGill University, Canada
Tamio Arai	University of Tokyo, Japan
Hegao Cai	Harbin Institute of Technology, China
Xiang Chen	Windsor University, Canada
Toshio Fukuda	Nagoya University, Japan
Huosheng Hu	University of Essex, UK
Sabina Jesehke	RWTH Aachen University, Germany
Yinan Lai	National Natural Science Foundation of China, China
Jangmyung Lee	Pusan National University, Korea
Ming Li	National Natural Science Foundation of China, China
Peter Luh	University of Connecticut, USA
Zhongqin Lin	Shanghai Jiao Tong University, China
Xinyu Shao	HUST, China
Xiaobo Tan	Michigan State University, USA
Guobiao Wang	National Natural Science Foundation of China, China
Michael Wang	The Hong Kong University of Science and Technology, SAR China
Yang Wang	Georgia Institute of Technology, USA

| Huayong Yang | Zhejiang University, China |
| Haibin Yu | Chinese Academy of Science, China |

Organizing Committee Chairs

Feng Gao	Shanghai Jiao Tong University, China
Lei Ren	The University of Manchester, UK
Chunyi Su	Concordia University, Canada
Jeremy L. Wyatt	University of Birmingham, UK
Caihua Xiong	HUST, China
Jie Zhao	Harbin Institute of Technology, China

Organizing Committee Co-chairs

Tian Huang	Tianjin University, China
Youfu Li	City University of Hong Kong, SAR China
Hong Liu	Harbin Institute of Technology, China
Xuesong Mei	Xi'an Jiaotong University, China
Tianmiao Wang	Beihang University, China

Local Chairs

Kun Bai	HUST, China
Bo Tao	HUST, China
Hao Wu	HUST, China
Zhigang Wu	HUST, China
Wenlong Li	HUST, China

Technical Theme Committee

| Gary Feng | City University of Hong Kong, SAR China |
| Ming Xie | Nanyang Technological University, Singapore |

Financial Chair

| Huan Zeng | HUST, China |

Registration Chair

| Jingrong Ge | HUST, China |

General Secretariat

| Hao Wu | HUST, China |

Sponsoring Organizations

National Natural Science Foundation of China (NSFC), China
Huazhong University of Science and Technology (HUST), China
School of Mechanical Science and Engineering, HUST, China
University of Portsmouth, UK
State Key Laboratory of Digital Manufacturing Equipment and Technology, HUST,
   China
State Key Laboratory of Robotic Technology and System, Harbin Institute
   of Technology, China
State Key Laboratory of Mechanical System and Vibration, Shanghai Jiao Tong
   University, China
State Key Laboratory of Robotics, Shenyang Institute of Automation, China

Contents – Part III

Sensors and Actuators

Mobile Robotics and Path Planning

Virtual Reality and Artificial Intelligence

Aerial and Space Robotics

Mechatronics and Intelligent Manufacturing

Sensors and Actuators

Modeling of Digital Twin Workshop
Based on Perception Data

Qi Zhang[1,2(✉)], Xiaomei Zhang[1,2], Wenjun Xu[1,2],
Aiming Liu[1,3], Zude Zhou[1,3], and Duc Truong Pham[4]

[1] School of Information Engineering, Wuhan University of Technology, Wuhan 430070, China
zhangqiwhut@126.com, may125z@126.com, liuaiming@cbmi.com.cn,
{xuwenjun,zudezhou}@whut.edu.cn
[2] Hubei Key Laboratory of Broadband Wireless Communication and Sensor Networks,
Wuhan University of Technology, Wuhan 430070, China
[3] Key Laboratory of Fiber Optic Sensing Technology and Information Processing,
Ministry of Education, Wuhan University of Technology, Wuhan 430070, China
[4] Department of Mechanical Engineering, School of Engineering,
University of Birmingham, Birmingham, B15 2TT, UK
d.t.pham@bham.ac.uk

Abstract. In recent years, the new generation of information technology has been widely applied in manufacturing domain. Building intelligent workshop and achieving intelligent manufacturing have become the purposes of industry development. The current workshop service systems just fulfill the mapping between physical and digital layer, while have not completed the interconnection and interaction between physical world and information world. It is the emergence of digital twin that become one of the solution to this bottleneck. In this paper, a system framework of digital twin workshop is proposed. In the light of the framework, a model of digital twin workshop based on physical perception data is established. This model is divided into three parts, which including workshop physical model, digital model based on ontology, and virtual model. Moreover, the operation mechanism of digital twin workshop is described among the three models. Finally, a three-dimensional model for a production line is built, and the connection between digital and virtual layer is established and demonstrated.

Keywords: Intelligent manufacturing · Digital twin workshop · Modeling · Perception data

1 Introduction

As the important pillar industry of our country, the workshop manufacturing promotes the development of the national economy and society. However with the much more fierce global competition and the increasing social demand, the workshop manufacturing encounters enormous pressure, including the complicated assembly technology and the fast-speed upgrading. In addition, the requirement of operators' professional level is rising constantly. During manufacturing, it is significant to facilitate the sustainable

© Springer International Publishing AG 2017
Y. Huang et al. (Eds.): ICIRA 2017, Part III, LNAI 10464, pp. 3–14, 2017.
DOI: 10.1007/978-3-319-65298-6_1

development of enterprises, in view of the relationship among economic, environment and the social.

In the past, the collection and processing of multiple data in workshop production mainly relied on manual management. The transmission and storage of information had low real-time performance and bad efficiency according to the paper medium. Meanwhile, the material configuration, production schedule and the process monitoring were depended on human experience and feeling, with low accuracy. With the transformation of the pattern of economic development and the adjustment of industrial structure, the connection between the manufacturing and information industry is more and more closely. The emergence of information technology has greatly increased the production level of workshop, making it possible to collect data in time, draw up plans dynamically, and monitor status automatically. However the data between workshop physical space and information space lack interaction and interoperability, far from implementing fully intelligence.

Digital twin, as one of the approaches to achieving the interconnection between physical world and digital world, has attracted broad attention [1]. Digital twin not only refers to product digitization, but also incorporates the digitization of factory itself, technical process and equipment [2]. It can help enterprises to simulate and test before putting into production, as well as synchronous optimize the whole process, so as to find how running conditions affect performance, improving the productivity [3].

The digital twin workshop was studied in this paper. Section 2 introduced current researches about the digital twin and the workshop modeling. The system framework of digital twin workshop was posed in Sect. 3. In addition, the digital twin workshop model based on physical data was established and described in detail as well. Section 4 gave a simulation using Flexsim. Finally, the conclusion and outlook were drawn in Sect. 5.

2   Related Works

Digital Twin was first introduced at a presentation about Product Lifecycle Management in a University of Michigan in 2002 [4]. Afterwards, the concept of digital twin has been applied in some fields. The U.S. defense department utilized Digital Twin for the maintenance and security guarantee of the aerospace vehicle [5]. And digital twin specimens were modeled and simulated in Abaqus/Explicit, to resolve crack path ambiguity in the SFC geometry [6]. By using Digital Twin, it integrated realistic simulation and vehicle management system with abundant data together to enable safety and reliability [7]. In addition, a way was proposed to the industrial IOT life cycle management and optimization through the Digital Twin [8]. Besides, Siemens hammered at integrate manufacturing process system in information space, achieving the digitization of the whole process from product design to execution [9].

Serviceability, intelligence and sustainable had become the development trend of manufacturing workshop [10]. It is very important for manufacturing system to interact between realistic model of current process status and real environment [11]. Digital twin is one of the best ways to achieve the communication and interaction between physical space and information space [2]. Based on the digital twin technology, the concept of digital twin workshop was proposed and it was composed of physical workshop, cyber workshop, and workshop service system and workshop digital twin data [1].

A Digital Twin model included three parts including real physical entity, virtual model and process information as a bond linking real and virtual world [2]. In physical modeling respect, the allocation of resources in manufacturing execution system was introduced, and the description model of the shop manufacturing task and equipment resources was analyzed [12]. A multidimensional information model of manufacturing capability was put forward, furthermore, a description language framework was proposed too [13]. In digital modeling respect, a shared concept can be described definitely and normally by building an ontology model [14]. An ontology model of manufacturing equipment capability was promoted, what's more, a mapping relationship between the attributes and real-time data was established [15].

In this paper, a system framework of digital twin workshop was posed. In addition, a digital twin workshop model based on physical data was established and described in detail as well. Finally, a simulation was given using Flexsim.

3 Modeling of Digital Twin Workshop Based on Perception Data

3.1 System Framework of Digital Twin Workshop

Digital twin, as an effective way to realize the communication and interaction between physical and information world, received extensive concern recently. The features are as follows. Firstly, it is the completely mapping of the physical system. Secondly, this conceptual model is dynamic and existing throughout the lifecycle of the system. Thirdly, it can visualize and simulate complex systems, optimizing the physical system constantly.

Applying the digital twin technology to workshop manufacturing may be a good approach to improving the intelligence, foreseeability and initiative of the plant. The digital twin workshop is a new operational mode. It can meet the demand of the development of intelligent manufacturing, through the integration and fusion of status data, model data, instruction data, and through the two-way real mapping and iterative optimization between physical plant, digital plant and 3D virtual plant. The system framework is shown as Fig. 1.

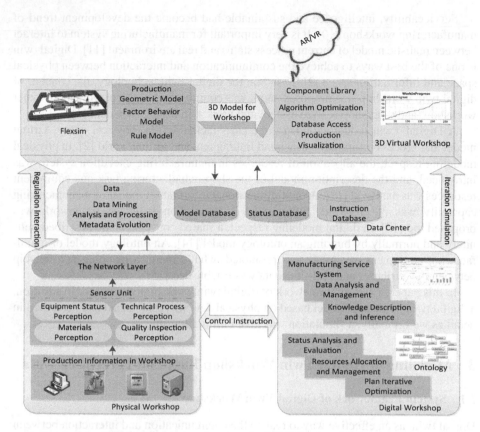

Fig. 1. The system framework of digital twin workshop

3.2 Modeling of Digital Twin Workshop Based on Perception Data

A model of digital twin workshop is established based on the framework proposed above. The model is composed of three modules, including plant physical model, digital model and 3D virtual model. The operation of system model relies on the interaction and iterative optimization among those three modules. And the three modules link together throughout the full life cycle, achieving the two-way mapping and interoperability by data interaction. The model is shown as Fig. 2.

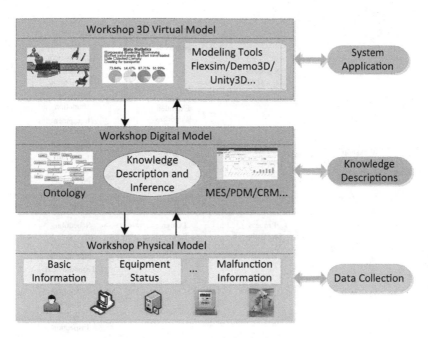

Fig. 2. The modeling of digital twin workshop

The Physical Information Model of Workshop. The plant is a complex, varied and dynamic production environment. In a narrow sense, manufacturing workshop is a physical production space which is made up of five elements including material, people, equipment, environment, and knowledge. In a broad sense, it is a production system where the staff utilizes existing or outsourcing materials, realizing production tasks by physical production unit and processing the raw materials or semi-finished products into the finished products.

According to the discussion above, the workshop physical elements are summarized. Meanwhile, the physical data in workshop is redundant and abundant, as shown in Fig. 3, which can be analyzed from resources, tasks and process dimensions.

According to the analysis of the manufacturing resources and the physical data in workshop, a physical information model for manufacturing workshop was established as shown in Fig. 4.

On the physical workshop side, the multi-sourced heterogeneous data is collected and transmitted by intelligent sensor and communication equipment, including element attribute, status, technological process, perturbation and so on. Those data will be incorporated into the digital plant and virtual plant through the transmission network. In addition, relevant physical elements will make timely and correctly response, in the light of instructions from digital plant and adjustment from virtual plant. It is very important to establish a mapping relation between attributes and real-time data, which is divided into direct mapping and indirect mapping. The connection between sensory data and model property makes it possible to realize the interaction between the workshop physical model and the digital model.

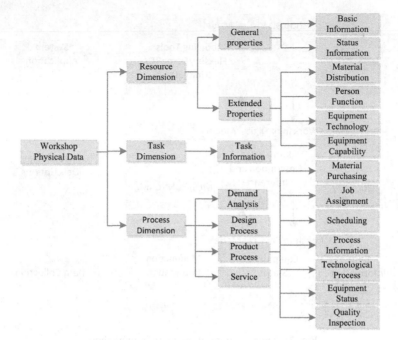

Fig. 3. Physical data for workshop

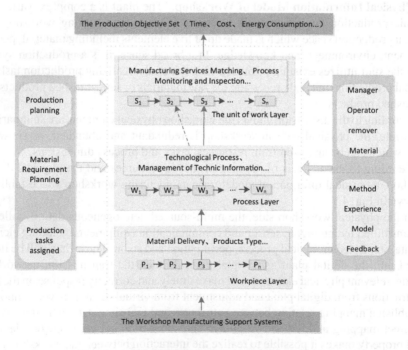

Fig. 4. A physical information model for workshop

The Digital Model of Workshop Based on Multi-dimensional Ontology. Ontology is the formal specification of conceptual models. We utilized ontology to depict the model which was established above. The formal description of workshop is as follow. The ontology of workshop manufacturing system includes resource information ontology, task information ontology and process information ontology.

$$PlantManufacturing$$
$$= (ResourceInformation, TaskInformation, ProcessInformation) \tag{1}$$

The following definition describes the resource information of the workshop, which can be divided into general properties and extended properties in order to meet the demand of different resources. And they all are subclasses of ResourceInformation.

$$ResourceInformation =$$
$$(BasicInformation, StateInformation, TechnicalPara, MaterialDelivery \ldots) \tag{2}$$

The formal description of manufacturing task ontology is as follow, which is about the information of manufacturing task.

$$TaskInformation =$$
$$(Basicproperty, ManufacturingObjects, ProductionDemand, Others) \tag{3}$$

The formal description of manufacturing process is as follow, which mainly describes dynamic parameters of workshop production. Each class has its own subclasses, such as ManufacturingProcess, which has subclasses including scheduling, technological process, equipment status, etc.

$$ProcessInformation$$
$$= (DemandAnalysis, DesignProcess, ManufacturingProcess, Service) \tag{4}$$

On the digital workshop side, the manufacturing service system should be established for the knowledge description, data management, state evaluation, generation of plan and instructions, etc. Meanwhile, the ontology model built above can be stored in database using Jena, so as to be easily obtained when analyzing production status or building workshop virtual model. The production plan is based on digital twin data, and simulated constantly to iterative optimization in the virtual plant. Besides, the instructions will be sent to the physical plant to process production.

The Three-dimensional Virtual Model of Workshop. A three-dimensional virtual model of workshop is required and the connection should be established for the interaction and communication between physical and virtual model. The three-dimensional model can be built via Flexsim, which can not only simulate but also optimize the production system with the abundant modeling units in 3D virtual reality environment. Then, the attributes of virtual model can be obtained from database. Meanwhile the feedback also can be deposit in database on account of the ODBC interface. Thus the linkage between physical and virtual model of workshop is created.

On the virtual workshop side, the virtual plant, as the real mapping to the physical one, can not only visualize products, but also simulate complex systems. When conflicts and perturbations occur in physical plant, the virtual model can test timely and feedback regulations to the real production line synchronously.

4   Case Study and Analysis

The model of digital twin workshop is composed of physical, digital and virtual model. And the perception data drives the operation of system. In the process of the modeling of digital twin workshop, after the collection of multi-source heterogeneous data from intelligent manufacturing environment, it is significant to give a knowledge description using ontology. In this paper, Protégé, a general ontology editor, was used to realize the ontology structure. Figure 5 shows the ontology of workshop manufacturing system. And the workshop can be defined as a class which possesses resource information, task information and process information. The ontology of manufacturing resource is shown in Fig. 6. Besides, the Figs. 7 and 8 show the manufacturing task ontology and the manufacturing process ontology.

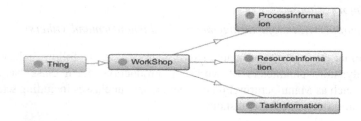

Fig. 5. Ontology of workshop

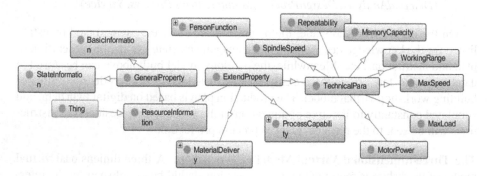

Fig. 6. Ontology of manufacturing resource

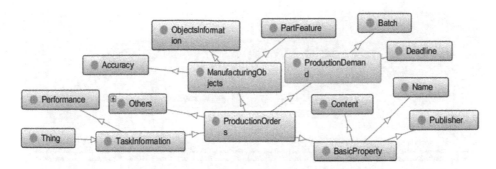

Fig. 7. Ontology of manufacturing task

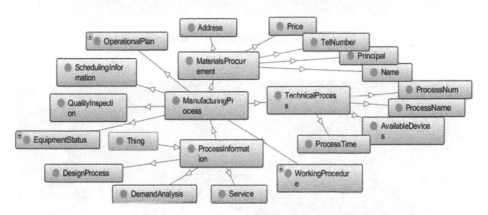

Fig. 8. Ontology of manufacturing process

For the connection between digital and virtual model, a 3D workshop model was established. Figure 9 gives a virtual model specimen of production line built by Flexsim. The parts and pallets enter into the system from two different sources, and the parts loop through the processor, synthesizer and robots to complete different operations. And the Fig. 10 shows the basic status information in the manufacturing process.

The interaction between the physical and virtual plant mainly relies on the database connection. There are three steps:

(1) A database named workshop data was built in MySQL, and several tables were created for saving the information of physical and virtual models.
(2) A data source was established in windows system, which possessed the MySQL ODBC Driver.
(3) Open the connection and load the physical information into the virtual model, which was saved in table processtimes before, as shown in Fig. 11(a). Then close the connection. Figure 11(b) gives some codes.

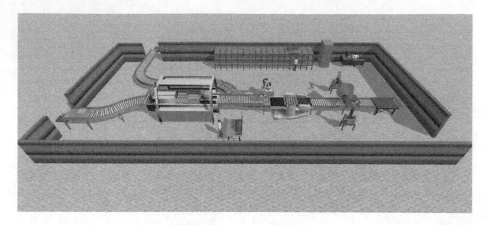

Fig. 9. The 3D virtual model of workshop

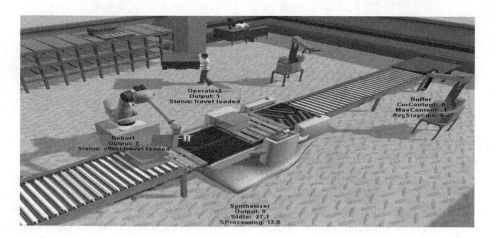

Fig. 10. The status shown besides the materials

Taking the process time as an example, the process time of each machine was read from database. Running the virtual model for a period of time, parts of model data would be saved into database, as shown in Fig. 12, and the first column was the process time of virtual model, which was the same as the data given before. Furthermore, the information statistics of virtual model were displayed dynamically, as shown in Fig. 13, which help analyze and do the correct adjustments in time.

id	Proc1	Proc2	Proc3	Proc4
1	2.1	2.4	2.6	2.9
2	3.1	3.4	3.6	3.9
3	4.1	4.4	4.6	4.9
4	5.1	5.4	5.6	5.9

```
1 /**Custom Code*/
2 treenode current = ownerobject(c);
3 treenode item = parnode(1);
4
5 dbopen("workshopdata","processtimes",1);
6 double proctime=dbgettablenum(getitemtype(item),2);
7 dbclose();
8 return proctime;
```

Fig. 11. (a) The process time saved in data table. (b) Database connection

id	Proctime	Genertime	Time
7	5.4	17.274	43.004
8	5.9	21.142	50.326
9	2.9	27.613	55.663
10	3.4	29.606	61.217
11	4.6	35.041	68.475

Fig. 12. The information of virtual model

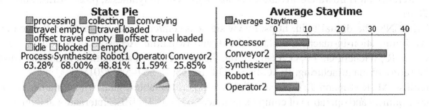

Fig. 13. Information statistics of virtual model

5 Conclusion and Future Work

In this paper, a framework of digital twin workshop has been proposed. In the light of that, the model of digital twin workshop based on physical data was established, which included physical model, digital description model based on ontology and also a simulation model built by Flexsim. And a linkage between the physical and digital model has been proposed. Finally, the simulation of a production line was realized and the linkage between the physical and virtual model was built. In the future, to realize the digital twins, a dynamic linkage among the three models should be present to make the interaction between the virtual and the real world come true, furthermore, a fully functional workshop service system and the vivid three-dimensional scenes combining with Augmented Reality technology should be considered conscientiously.

Acknowledgment. This research is supported by National Natural Science Foundation of China (Grant Nos. 51305319 and 51475343), the International Science & Technology Cooperation Program, Hubei Technological Innovation Special Fund (Grant No. 2016AHB005), and the Fundamental Research Funds for the Central Universities (Grant No. 2017III5XZ).

References

1. Tao, F., Zhang, M., Cheng, J., et al.: Digital twin workshop: a new paradigm for future workshop. Comput. Integr. Manuf. Syst. **23**(1), 1–9 (2017). (in Chinese)
2. APRISO Digital twin: manufacturing excellence through virtual factory replication. http://www.apriso.com. Accessed 06 May 2014
3. Boschert, S., Rosen, R.: Digital twin—the simulation aspect. In: Hehenberger, P., Bradley, D. (eds.) Mechatronic Futures. Springer, Cham (2016)
4. Grieves, M., Vickers, J.: Digital twin: mitigating unpredictable, undesirable emergent behavior in complex systems. In: Kahlen, J., Flumerfelt, S., Alves, A. (eds.) Transdisciplinary Perspectives on Complex Systems. Springer, Cham (2017)
5. Tuegel, E., Ingraffea, A., Eason, T., et al.: Reengineering aircraft structural life prediction using a digital twin. Int. J. Aerosp. Eng. **2011**, 1687–5966 (2011)
6. Hochhalter, J.: On the effects of modeling as-manufactured geometry: toward digital twin. Int. J. Aerosp. Eng. **2014**(439278), 1–10 (2014)
7. Glaessgen, E., Stargel, D.: The digital twin paradigm for future NASA and U.S. air force vehicles. In: Proceedings of the 53rd Structures Dynamics and Materials Congerence, pp. 1–14. AIAA, Reston (2012)
8. Canedo, A.: Industrial IoT lifecycle via digital twins. In: 11th Ieee/acm/ifip International Conference on Hardware/Software Codesign and System Synthesis, p. 29. IEEE, Pittsburgh (2016)
9. SIEMENS The digital twin. https://www.siemens.com/customer-magazine/en/home/industry/digitalization-in-machine-building/the-digital-twin.html. Accessed 17 Nov 2015
10. Rosen, R., Wichert, G., Lo, G., et al.: About the importance of autonomy and digital twins for the future of manufacturing. IFAC-PapersOnLine **48**(3), 567–572 (2015)
11. Schluse, M., Rossmann, J.: From simulation to experimentable digital twins: simulation-based development and operation of complex technical systems. In: IEEE International Symposium on Systems Engineering, pp. 1–6. IEEE, Edinburgh (2016)
12. Du, L., Fang, Y., He, Y.: Manufacturing resource optimization deployment for manufactuing execution system. In: Second International Symposium on Intelligent Information Technology Application, pp. 234–238. IEEE, Shanghai (2008)
13. Luo, Y., Lin, Z., Fei, T., et al.: Key technologies of manufacturing capability modeling in cloud manufacturing mode. Comput. Integr. Manuf. Syst. **18**(7), 1357–1367 (2012)
14. Studer, R., Benjamins, V., Fensel, D.: Knowledge engineering: principles and methods. Data Knowl. Eng. **25**(1–2), 161–197 (1998)
15. Xu, W., Yu, J., Zhou, Z., et al.: Dynamic modeling of manufacturing equipment capability using condition Information in cloud manufacturing. J. Manuf. Sci. Eng. **137**(4), 1–14 (2015)

A Stable Factor Approach
of Input-Output-Based Sliding-Mode Control
for Piezoelectric Actuators with Non-minimum
Phase Property

Haifeng Ma, Jianhua Wu, and Zhenhua Xiong$^{(\boxtimes)}$

The State Key Laboratory of Mechanical System and Vibration,
School of Mechanical Engineering,
Shanghai Jiao Tong University, Shanghai 200240, China
mexiong@sjtu.edu.cn

Abstract. This paper presents a new stable factor approach of input-output-based discrete-time sliding-mode control (IODSMC-SF), which dedicates to piezoelectric actuators (PEAs) with non-minimum phase (NMP) property. This control approach is developed based on a linear discrete-time input-output nominal model. A stable factor, which ensures stable and accurate motion control for PEAs with NMP nature, is designed, analyzed and introduced into the controller. One unique feature of the proposed controller lies in that it ensures stable and precision motion control for PEAs with NMP property. The construction of either a hysteresis model or a state observer is not needed. Moreover, the proposed controller releases the burden on parameter selection since only a stable factor is needed to stabilize the NMP system and this factor can be obtained by optimization algorithm. Experimental results with a piezoelectric actuator are presented to demonstrate the effectiveness of the proposed controller.

Keywords: Piezoelectric actuator · Precision motion control · Discrete-time Sliding-mode control · Non-minimum phase

1 Introduction

Piezoelectric actuators (PEAs) have been widely employed in many fields. However, the nonlinear effects existing in PEAs, including hysteresis nonlinearity and creep, can greatly degrade the motion accuracy [1]. Thus, for precise motion control of PEAs, it is essential to suppress these nonlinear effects to negligible levels.

As already be known, the discrete-time sliding-mode control (DSMC) is a very efficient feedback control approach featuring robustness in the presence of disturbance. The majority of existing DSMC have been realized based on system state feedback [2]. For actual application, it is difficult to measure all states of a system. Hence, state observers are always required for the DSMC implementation. Nonetheless, the state observer design clearly increases the burden of control design procedure. Furthermore, an inappropriately designed state observer affects the stability of the system. Thus, for practical implementation, it is desired to eliminate the use of state observers.

© Springer International Publishing AG 2017
Y. Huang et al. (Eds.): ICIRA 2017, Part III, LNAI 10464, pp. 15–24, 2017.
DOI: 10.1007/978-3-319-65298-6_2

Many efforts have been made to release the use of state observers in DSMC strategies [3–5]. Chan [3] proposed a discrete sliding-mode controller based on input-output model in the presence of modeling uncertainty and disturbance, but the reference inputs and disturbances of this method were assumed to be varying slowly. Janardhanan et al. [4] proposed a feedback sliding-mode controller based on multirate output feedback, where the system output is sampled at a rate faster than the control input. Sha et al. [5] designed an input-output-based adaptive sliding-mode controller, where only input and output data was needed. However, this controller was dedicated to a first-order model with long dead time. Yet, these restrictions make the control schemes not applicable to a piezoelectric actuation system which always possesses a high order model along with complicated nonlinearity. An input-output-based digital sliding-mode control (IODSMC) has been proposed by Xu [6] for piezoelectric micro/nano positioning systems, which successfully suppresses the unmolded hysteresis nonlinearity and disturbance in piezoelectric micro/nano positioning systems. However, these IODSMC controllers are only suitable for minimum phase systems. Many PEAs or piezoelectric micro/nano systems have non-minimum phase (NMP) nature and this greatly restricts the application of these methods.

In this paper, a stable factor approach of input-output-based discrete-time sliding-mode control (IODSMC-SF) for PEAs, which possess NMP nature preceded by hysteresis nonlinearity. This control scheme is developed based on a linear discrete-time input-output nominal model. A stable factor, which ensures stable and accurate motion control for PEAs with NMP nature, is designed, analyzed and introduced into the controller. The contribution of this controller can be summarized as follows:

(1) The proposed controller can realize stable and accurate motion tracking control for PEAs with NMP nature by overcoming the hysteresis nonlinearity and external disturbance.

(2) The proposed controller releases the burden on parameter selection. Unlike existing methods, only a stable factor is required to stabilize PEAs with NMP nature. This factor can be obtained by optimization algorithm. Moreover, the hysteresis model and state observer are not needed for the proposed controller.

The rest of this paper is organized as follows. The experiment setup of a PEA device and its model, which has NMP property, are presented in Sect. 2. The IODSMC-SF scheme is developed in Sect. 3. The stability analysis for NMP systems and optimized selection of the stable factor are presented. For experimental verification, in Sect. 4, the IODSMC-SF is implemented on the PEA device. Finally, Sect. 5 concludes this research.

2 Experimental Setup and PEA Model

The experimental setup used in this study is a commercially available PEA system as shown in Fig. 1 The PEA system consists of an amplifier module (model: E505 from PI in Germany), a dSPACE1103, a piezoelectric actuator (model: P-843.30 from PI in Germany) with an inbuilt position sensor, a signal processing unit (model: E509 from PI in Germany), and a host PC. The dSPACE-DS1103 controller board equipped with

16-bit digital-to-analog converters and 16-bit analog-to-digital converters is adopted to generate control codes and obtain the displacement information. The sampling frequency of the dSPACE1103 is set to be 20 kHz in present work.

Fig. 1. Experimental platform.

Figure 2 shows the block diagram of the PEA model. The hysteresis nonlinearity, dynamic and creep are represented by blocks of H, G_M, and F, respectively. $u(t)$, $f(t)$, $y_o(t)$, and $y(t)$ represent the input voltage, internal actuation force, output displacement without creep and the final displacement of PEA, respectively.

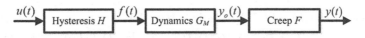

Fig. 2. PEA model.

In this paper, only the dynamic G_M is considered, while the hysteresis nonlinearity H and the creep F are taken as part of the disturbance. A generalized discrete-time dynamic model of G_M is established for PEAs as follow

$$x(k) = \sum_{i=1}^{n} a_i x(k-i) + \sum_{i=1}^{m} b_i u(k-i) + d(k) \tag{1}$$

where a_i and b_i are model coefficients. $x(k)$ and $u(k)$ represent the output position and input voltage of PEAs, respectively. In addition, $d(k)$ describes the lumped effect of unmodeled hysteresis nonlinear, residual dynamic, creep and external perturbations. If the plant Eq. (1) is NMP, then all or partial zeros of the plant are out of unit disk.

Experiments are performed to identify the parameters of Eq. (1), the plant transfer function can be obtained.

$$x(k) = 0.211x(k-1) + 0.153x(k-2) + 0.188x(k-3) - 0.009u(k-1)$$
$$+ 0.064u(k-2) - 0.039u(k-3) \tag{2}$$

The system has an unstable zero at 6.129, which is outside the unit disk, i.e., the PEA is a NMP system.

3 The Proposed Controller Design

3.1 Design of IODSMC-SF

The dynamic model Eq. (1) can be further expressed as

$$x(k+1) = A(z^{-1})x(k) + B(z^{-1})u(k) + d(k) \tag{3}$$

where $A(z^{-1})$ and $B(z^{-1})$ are polynomials in the unit-delay operator q^{-1} defined as

$$A(z^{-1}) = a_1 + a_2 z^{-1} + \cdots + a_n z^{-n+1}, \quad B(z^{-1}) = b_1 + b_2 z^{-1} + \cdots + b_m z^{-m+1} \tag{4}$$

The position error is represented as $e(k) = x(k) - r(k)$, where $r(k)$ is the desired position trajectory, an incremental type sliding function can be written as

$$s(k) = s(k-1) + F(z^{-1})[x(k) - r(k)] \tag{5}$$

where $F(z^{-1}) = 1 + f_1 z^{-1} + \ldots + f_n z^{-n}$. The controller is designed based on the reaching law $s(k + 1) = 0$. Substituting Eq. (3) into $s(k + 1) = 0$, the following deduction can be got

$$\begin{aligned} s(k+1) &= s(k) + F(z^{-1})[x(k+1) - r(k+1)] \\ &= s(k) + F(z^{-1})[A(z^{-1})x(k) + B(z^{-1})u(k) + d(k) - r(k+1)] \\ &= 0 \end{aligned} \tag{6}$$

The equivalent control $u_{eq}(k)$ is the solution to Eq. (6) by adding the stable factor $Q(z^{-1})$

$$u^{eq}(k) = G(z^{-1})[-F(z^{-1})A(z^{-1})x(k) + F(z^{-1})r(k+1) - s(k) - F(z^{-1})d(k)] \tag{7}$$

where $G(z^{-1}) = 1/(F(z^{-1}) B(z^{-1}) + Q(z^{-1}))$.

From Eq. (7) we can see that the equivalent control only requires knowledge of measured output position, so the state observer is not needed. The unknown disturbance term $d(k)$ is obtained by its one-step delayed estimation $d(k - 1)$.

Since there exists unmodeled hysteresis nonlinearity, modeling uncertainty, and external disturbances, the switching control $u_{sw}(k)$ is needed here.

$$K_s \mathrm{sat}\{s(k)\} = \begin{cases} K_s \mathrm{sign}\{s(k)\}, & \text{if } |s(k)| > \varepsilon \\ K_s s(k)/\varepsilon, & \text{if } |s(k)| \leq \varepsilon \end{cases} \tag{8}$$

where K_s is a positive control gain and sat$\{s(k)\}$ is the saturation function. The positive constant ε, which ensures that $s(k)$ is always bounded by $\pm\ \varepsilon$, represents the boundary layer thickness. The overall control action can be obtained

$$
u(k) = (F(z^{-1})B(z^{-1})+Q(z^{-1}))^{-1}[-F(z^{-1})A(z^{-1})x(k)+F(z^{-1})r(k+1)-s(k) \\
-F(z^{-1})d(k-1)] - K_s\mathrm{sat}\{s(k)\} \tag{9}
$$

3.2   Stability Analysis

The stability of IODSMC-SF is analyzed in this section. Substituting the equivalent control $u_{eq}(k)$ Eqs. (7) into (3), the control law gives rise to the closed-loop response

$$
x(k+1)=A(z^{-1})x(k)+\frac{B(z^{-1})}{H(z^{-1})}[-F(z^{-1})A(z^{-1})x(k)+F(z^{-1})r(k+1) \\
-s(k)-F(z^{-1})d(k-1)]+d(k) \tag{10}
$$

where $H(z^{-1}) = F(z^{-1})B(z^{-1}) + Q(z^{-1})$. The closed-loop response can be simplified as

$$
[z - A(z^{-1}) + \frac{B(z^{-1})F(z^{-1})A(z^{-1})}{H(z^{-1})}]x(k) = \frac{B(z^{-1})}{H(z^{-1})}[F(z^{-1})r(k+1)-s(k)] \\
+ d(k) - \frac{B(z^{-1})F(z^{-1})}{H(z^{-1})}d(k-1) \tag{11}
$$

The closed-loop system behavior is governed by the roots of polynomial in Eq. (12). The system will be stable if the eigenvalues of characteristic equation locate inside the unit disk in the z-plane.

$$
z - A(z^{-1}) + \frac{B(z^{-1})F(z^{-1})A(z^{-1})}{H(z^{-1})} = 0 \tag{12}
$$

It can be concluded that the parameters $Q(z^{-1})$ and $F(z^{-1})$ are both included in the roots of the characteristic equation. So, by properly adjusting the parameters of $Q(z^{-1})$ and $F(z^{-1})$, the eigenvalues of the system could locate inside the unit disk in the z-plane, even if the system has NMP property.

3.3   Parameter Optimization

As shown in Eq. (7), the equivalent control $u_{eq}(k)$ has an additional control parameter: the stable factor $Q(z^{-1})$. A minimization optimization is designed to release the burden on factor selection.

Selection of the fitness function (or objective function) is the key point of the optimization. We can see that if $G(z^{-1})$ has suitable fitted amplitude- and phase-frequency characteristics with objective output transfer function $G_o(z^{-1}) = 1/(F(z^{-1})B(z^{-1}))$ in the desired frequency range, $u_{eq}(k)$ can remain the tracking accuracy. $G(z^{-1})$ is defined as

$$G(z^{-1}) = \frac{k(z - z_1)(z - z_2)\ldots(z - z_m)}{(p - p_1)(p - p_2)\ldots(p - p_n)} \tag{13}$$

where k, z_i, and p_i are gain, the zeros and poles of $G(z^{-1})$. In this work, the number of zeros and poles are determined by trials via experiments. So in this work, the fitness function is chosen as

$$E(w) = \sum_{k=0}^{n} [G_o(jw_k) - G(jw_k)] \tag{14}$$

where w_k denotes the total number of the frequency data. When $G(z^{-1})$ is close to the experimental data of $G_o(z^{-1})$, the identified result is thought to be a good result. Therefore, the minimize optimization problem with constraints can be formulated as follows

$$\min E(w) = \sum_{k=0}^{n} [G_{FB}(jw_k) - G_{FB+Q}(jw_k)] \tag{15}$$

s.t.

$$\begin{cases} |z_i| < 1, |p_i| < 1 \\ |\lambda_i| < 1 \end{cases} \tag{16}$$

where λ_i are the solutions of the characteristic equation Eq. (12).

4 Experimental Studies

4.1 Parameter Selection

First, sliding surface parameter $F(z^{-1}) = 1 - 0.75z^{-1} + 0.0425z^{-2}$ is chosen by trials for the IODSMC-SF controller. In the present research, a fourth-order model is employed to make a tradeoff between the model accuracy and complexity in order to demonstrate the effectiveness of the proposed scheme. The stable factor $Q(z^{-1})$ and $G(z^{-1})$ can be obtained by a minimization optimization, which is shown as below

$$G(z^{-1}) = \frac{(z + 0.1183)^4}{0.0213(z - 0.8322)(z - 0.263)(z^2 + 1.214z + 0.3729)} \tag{17}$$

Spectral analysis is conducted to obtain the frequency responses, as depicted in Fig. 3. It is obvious that the identified output transfer function $G(z^{-1})$ (dot red curve) is a suitable fitting of $G_o(z^{-1})$ behavior up to 100 Hz. So the IODSMC-SF controller can preserve the tracking accuracy in the frequency range 1–100 Hz.

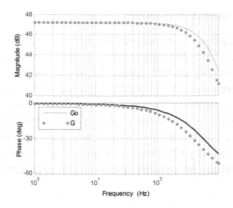

Fig. 3. Frequency response of $G_o(z^{-1})$ and $G(z^{-1})$.

The eigenvalues of the closed-loop system can be got by substituting the optimized stable factor into the constraint Eq. (12). The solutions are shown in Table 1, where all the solutions locate inside the unit disk in the z-plane.

Table 1. Solutions of the constraint equation.

	Solutions	
0.0627	−0.1659 −0.1259i	0.2537 −0.6120i
−0.0666 +0.0265i	−0.1659 +0.1259i	0.2537 +0.6120i
−0.0666 −0.0265i	0.7363 +0.0473i	0.7363 −0.0473i
−0.6961 +0.6257i	−0.6961 −0.6257i	

For comparison, the input-output-based discrete-time sliding-mode control (IODSMC) [6], and a traditional PID controller are also implemented on the model Eq. (2) to demonstrate the effectiveness of the proposed method.

Moreover, there are two control parameters both incorporated in the IODSMC-SF and IODSMC controllers: switching gain K_s and the boundary layer thickness ε. The parameters can be tuned by trials via experiments. The PID gains K_p, K_i, and K_d are also tuned by trials via experiments. Control parameters of IODSMC-SF and PID are listed in Table 2.

Table 2. Parameters of IODSMC-SF and PID.

Controller	Parameters	Value
IODSMCSF	K_s	5
	ε	10
PID	K_p	2.2
	K_i	200
	K_d	0.00001

4.2   Tracking Performance Experiment

Here, the reference trajectories are sinusoidal signals with amplitudes of 9 μm and frequencies of 10 Hz and 100 Hz, respectively. It can be found that no matter how to adjust control parameters, IODSMC cannot be stable for both sinusoidal reference inputs. The tracking errors of the IODSMC-SF controller and PID are illustrated in Fig. 4(b) and (d). The root mean square (RMS) tracking errors are listed in Table 3. Compared with the popular PID controller, the tracking errors of the IODSMC-SF are much smaller. Moreover, as the frequency increases, the performance improvement is more obvious. The reason why PID produces such a worse result is mainly attributed to its band-width limit and the inherent hysteresis nonlinearity effect of PEA.

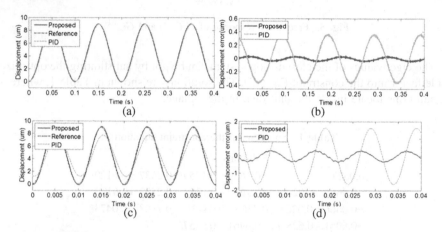

Fig. 4. (a) Sinusoidal tracking results at 10 Hz. (b) Tracking error at 10 Hz. (c) Sinusoidal tracking results at 100 Hz. (d) Tracking error at 100 Hz.

Table 3. Tracking error RMS (%) of IODSMC-SF and PID.

Input frequency	IODSMC-SF	PID
10 Hz	0.19%	2.82%
100 Hz	2.13%	12.88%

4.3   Robust Performance Experiment

The robustness of the proposed IODSMC-SF and PID is examined by internal and external disturbance. As shown in Sect. 4.2, since no hysteresis model-based feedforward controller is used, the hysteresis nonlinearity is taken as part of the internal disturbance to the feedback controller. Note that smaller tracking error of the proposed method is achieved under the influence of hysteresis nonlinearity, the proposed method shows better robustness than PID.

Step load disturbance is added into the control system to evaluate the robustness of IODSMC-SF and PID with external disturbance. The desired reference is a sinusoidal

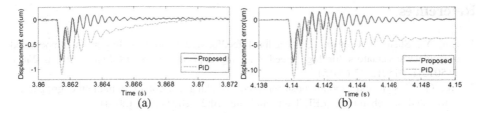

Fig. 5. Tracking error with external disturbance: (a) 5 V step load. (b) 50 V step load.

trajectory with 9 μm in amplitude and 5 Hz in frequency. The disturbance is added in the 5th second with voltages of 5 V and 50 V, respectively. The tracking errors of two situations are exhibited in Fig. 5.

As shown in Fig. 5(a), it can be concluded that both controllers are capable of precisely tracking the predefined trajectory with the 5 V step load disturbance. On the other hand, the IODSMC-SF and PID produce the settling time of 5 ms and 10 ms, respectively. As compared with PID, the IODSMC-SF renders a more rapid transient response with smaller tracking error. As shown in Fig. 5(b), it is notable that the tracking error of PID cannot converges to zero within a ±0.2 μm band and leaves large tracking error, whereas the tracking error of the proposed controller converges to zero within a ±0.2 μm band in 11 ms. The reason why PID produces such a worse result in the second test is mainly attributed to the large disturbance of 50 V step load. The experimental results demonstrate the robustness of IODSMC-SF under the influence of the internal and external disturbance. The proposed controller shows better performance than PID.

5 Conclusion

Simplification of the SMC controller design in front to hysteresis nonlinearity is a challenging task in many applications requiring precision motion control. If the controlled object is NMP, the problem becomes even more difficult. In this proper, a stable factor approach of input-output-based discrete-time sliding-mode control, which dedicates to precision motion tracking control of piezoelectric actuation systems with NMP property, was presented. The stability condition of the proposed control algorithm was analyzed, and based on the stability condition, the stable factor can be obtained by an optimization process, which released the burden on parameter selection. To illustrate the effectiveness of the proposed controller, experiments were performed and the results were compared with the IODSMC and PID. It has been shown that the proposed controller is stable for NMP systems, and shows superior performance than PID.

Acknowledgments. This research work is supported in part by the National Basic Research Program of China (2013CB035804) and National Natural Science Foundation of China (U1201244).

References

1. Liu, Y., Shan, J., Gabbert, U.: Feedback/feedforward control of hysteresis-compensated piezoelectric actuators for high-speed scanning applications. Smart Mater. Struct. **24**(1), 015012-1–015012-92 (2015)
2. Xu, Q.: Design and smooth position/force switching control of a miniature gripper for automated microhanding. IEEE Trans. Ind. Inf. **10**(2), 1023–1032 (2014)
3. Chan, C.Y.: Robust discrete quasi-sliding mode tracking controller. Automatica **31**(10), 1509–1511 (1995)
4. Janardhanan, S., Bandyopadhyay, B.: Multirate output feedback based robust quasi-sliding mode control of discrete-time system. IEEE Trans. Autom. Control **52**(3), 499–503 (2007)
5. Sha, D., Bajic, V.B.: Robust discrete adaptive input-output-based sliding mode controller. Int. J. Syst. Sci. **31**(12), 1601–1614 (2000)
6. Xu, Q.: Digital sliding-mode control of piezoelectric micropositioning system based on input-output model. IEEE Trans. Ind. Electron. **61**(10), 5517–5526 (2014)

Design of Quadrotor Unmanned Aerial Vehicle

Mofei Wu[1]($\boxtimes$), Zhigang Cheng[2], Lin Yang[2], and Lamei Xu[2]

[1] Department of Energy and Power, Wuhan University of Technology,
Wuhan 430070, China
981025897@qq.com
[2] Department of Automation, Wuhan University of Technology,
Wuhan 430070, China

Abstract. This paper designs the dynamic system and main control circuit of quadrotor unmanned aerial vehicle (UAV). Power system of the quadrotor UAV includes brushless DC motor (BDCM), matching paddle, power lithium battery and drive circuit of (BDCM). Its main control hardware circuit includes the main controller minimum system, the sensor module, the wireless communication module and the power module. The attitude algorithm and attitude control algorithm are designed. The Kalman filter algorithm is used to calculate the attitude of quadrotor UAV. The cascade PID control algorithm is used to control attitude. Finally, a quadrotor UAV prototype was trial-produced, and the tested results show that the aerial vehicle meets the design requirements.

Keywords: Quadrotor Unmanned Aerial Vehicle (UAV) · Attitude control · Cascade PID control algorithm

1 Introduction

Compared with the traditional fixed-wing aerial vehicle, the quadrotor UAV has many irreplaceable advantages, such as hovering in the air, vertical take off and landing, low-speed flight and indoor flight. Having been widely concerned in road cruising, unmanned investigation, traffic monitoring, forest fire prevention, aerial photography and so on, the quadrotor UAV has become an international research hot spots in recent years. GRASP laboratory of Pennsylvania University researches a small indoor quadrotor UAV which can be multi-aircraft co-flight in a small space, and has done the experiment bee colony technology of the quadrotor UAV. With the principle of same behavior, in the absence of leadership control of the case, the bee colony of quadrotor UAV can complete the complex behavior or task. Stanford University developed Mesicopter micro quadrotor UAV only a little more than a dollar coin, The prototype can fly at a small drop point, which the research goal is also multi-machine bee collaboration [5]. In 2006, Germany Microdrones GmbH pushed out the MD4-200, an electric quadrotor unmanned flight system, which is specially used in the field of specialization. MD4-200 is widely used in fire, security, geology, surveying and mapping, environmental protection, film and television, police and other industries. In 2010, Parrot (France) pushed out AR. Drone micro quadrotor UAV, which can be controlled by Apple's Iphone through the WIFI. In 2014, Taijiang innovation

© Springer International Publishing AG 2017
Y. Huang et al. (Eds.): ICIRA 2017, Part III, LNAI 10464, pp. 25–34, 2017.
DOI: 10.1007/978-3-319-65298-6_3

technology company produced its latest commercialized high-performance quadrotor UAV Inspire 1, which internal integrated aircraft control board with good environmental adaptability. Its aerial functions such as single shooting and continuous shooting can be realized through the plug-in three-axis stabilization platform equipped with wide-angle 12 million pixel high-definition camera. In order to solve the problem that the conventional unmanned aerial vehicle is large and can not be applied to the narrow space flight, this paper designs a quadrotor UAV to calculate its current attitude by the measurement system composed of three-axis gyroscope and three-axis accelerometer, and control its flight attitude adopting based on Euler angle feedback then by the remote control of the appropriate manual adjustment, the quadrotor UAV can in stable and flexible flight in a small complex indoor space

2 Establishment of Mathematical Model of the Quadrotor UAV

The gravity of the UAV is:

$$G = mg \tag{1}$$

The lift to each blade is:

$$F_{li} = \frac{1}{2} \rho C_l \omega_i^2 \tag{2}$$

The resistance to each blade is:

$$F_{ri} = \frac{1}{2} \rho C_r \omega_i^2 \tag{3}$$

The total lift of the UAV in the body coordinate system can be set as:

$$F_{Bl} = \begin{bmatrix} 0 \\ 0 \\ U_l \end{bmatrix} U_1 = F_{l1} + F_{l2} + F_{l3} + F_{l4} = \sum_1^4 b\omega_i^2 \tag{4}$$

Where "ω" is the rotational angular velocity of the blade.

The lift of the UAV in the ground coordinate system is:

$$F_{El} = U_l \begin{bmatrix} \cos\alpha \sin\beta \cos\gamma + \sin\alpha \sin\gamma \\ \cos\alpha \sin\beta \sin\gamma - \sin\alpha \cos\gamma \\ \cos\alpha \cos\beta \end{bmatrix} \tag{5}$$

Where 'α, β, γ' represents the roll angle, pitch angle and heading angle of the UAV respectively.

In the ground coordinate system, the displacement vector of the vehicle is $r = [x \quad y \quad z]^T$, then the linear displacement equation of the system is:

$$\begin{cases} \ddot{x} = U_l(\cos \alpha \sin \beta \cos \gamma + \sin \alpha \sin \gamma)/m \\ \ddot{y} = U_l(\cos \alpha \sin \beta \sin \gamma - \sin \alpha \cos \gamma)/m \\ \ddot{z} = U_l(\cos \alpha \cos \beta)/m - g \end{cases} \qquad (6)$$

Supposed the angular velocity as $[p \quad q \quad r]^T$, the moment of inertia matrix is:

$$I = \begin{bmatrix} I_x & I_{xy} & I_{xz} \\ I_{xy} & I_y & I_{yz} \\ I_{xz} & I_{yz} & I_z \end{bmatrix} \qquad (7)$$

So the moment matrix of the aircraft is:

$$\begin{bmatrix} M_x \\ M_y \\ M_z \end{bmatrix} = \begin{bmatrix} \dot{p}I_x - \dot{r}I_{xz} + qr(I_z - I_y) - pqI_{xz} \\ \dot{q}I_y + pr(I_x - I_z) + (p - r)I_{xz} \\ \dot{r}I_z - \dot{p}I_{xz} + pq(I_y - I_x) + qrI_{xz} \end{bmatrix} \qquad (8)$$

The angular velocity of the aircraft has the following relationship with the attitude angle:

$$\begin{bmatrix} p \\ q \\ r \end{bmatrix} = \begin{bmatrix} 1 & 0 & -\sin \beta \\ 0 & \cos \alpha & \sin \alpha \cos \beta \\ 0 & -\sin \alpha & \cos \alpha \cos \beta \end{bmatrix} \begin{bmatrix} \dot{\alpha} \\ \dot{\beta} \\ \dot{\gamma} \end{bmatrix} \qquad (9)$$

The attitude angle is very small when the UAV is stable, so it is approximated to have this relationship:

$$[p \quad q \quad r]^T = [\dot{\alpha} \quad \dot{\beta} \quad \dot{\gamma}]^T$$

Then,

$$\begin{bmatrix} \ddot{\alpha} \\ \ddot{\beta} \\ \ddot{\gamma} \end{bmatrix} = \begin{bmatrix} \left[M_x + \dot{\beta}\dot{\gamma}(I_x - I_z)\right]/I_x \\ \left[M_y + \dot{\beta}\dot{\gamma}(I_z - I_x)\right]/I_y \\ \left[M_z + \dot{\alpha}\dot{\beta}(I_x - I_y)\right]/I_z \end{bmatrix} \qquad (10)$$

Supposed four state control variables as $(U_1, U_2, U_3 U_4)$ according to the four motion states of the UAV. The coupled nonlinearity models are divided into four relatively independent control systems by the four state control variables:

$$\begin{cases} U_1 = F_{l1} + F_{l2} + F_{l3} + F_{l4} \\ U_2 = F_{l4} - F_{l2} \\ U_3 = F_{l3} - F_{l1} \\ U_4 = F_{l2} + F_{l4} - F_{l1} - F_{l3} \end{cases} \tag{11}$$

Introduced the four state variables into Eqs. (6) and (10), the system mathematical model of the aircraft is following:

$$\begin{cases} \ddot{x} = U_1(\cos \alpha \sin \beta \cos \gamma + \sin \alpha \sin \gamma)/m \\ \ddot{y} = U_1(\cos \alpha \sin \beta \sin \gamma - \sin \alpha \cos \gamma)/m \\ \ddot{z} = U_1(\cos \alpha \cos \beta)/m - g \\ \ddot{\alpha} = \left[lU_2 + \dot{\beta}\dot{\gamma}(I_x - I_z) \right]/I_x \\ \ddot{\beta} = \left[lU_3 + \dot{\beta}\dot{\gamma}(I_z - I_x) \right]/I_y \\ \ddot{\gamma} = \left[U_4 + \dot{\alpha}\dot{\beta}(I_x - I_y) \right]/I_z \end{cases} \tag{12}$$

3   Hardware Design of the Quadrotor UAV

In order to be able to mount the camera or video camera for low-altitude aerial or reconnaissance missions, the aircraft need to ensure adequate load and life time, the specific design indicators of quadrotor UAV are as follows:

(1) Rack size does not exceed 50 cm * 50 cm, height is not more than 30 cm;
(2) No-load weight is not more than 1000 g;
(3) Full load weight does not exceed 3000 g;
(4) Life time is not less than 10 min;
(5) The maximum flight speed is not less than 10 m/s;
(6) Automatic flight radius is 1800 m, remote control flight radius is 600 m;
(7) Wind resistance level 3;
(8) Control mode: automatic/remote control;

The overall structure of the miniature quadrotor UAV is shown in Fig. 1, which is divided into six parts: aircraft frame, power system, main control board of aircraft, GPS navigation module, wireless communication module and remote control receiver.

(1) The frame of the aircraft is the support structure of the UAV. The standard cross-shaped frame, the middle flight control board and the battery mounting board are sandwiched by four motor arms with a cross shape. The wheelbase of the symmetrical motor is 450 mm. The landing gear with the high of bottom 16 mm and the wide 26.5 mm, can effectively ensure that the aircraft take-off and landing is not reflected by the ground reflection of airflow.
(2) The power system of the aircraft is composed of a lithium battery, a BDCM, a driver of BDCM and four propellers. The BDCM is mounted vertically at the end of the four motor arms. The four propellers paralleling to the body four

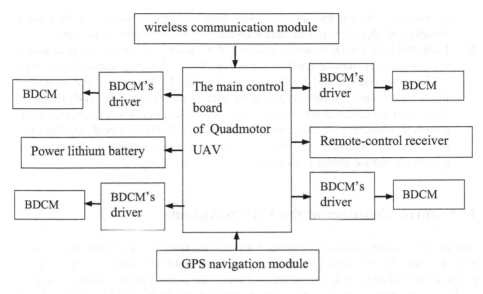

Fig. 1. Design of the aircraft structure

symmetrical positions provide lift perpendicular to the plane's plane. Power lithium battery with high-capacity and high-magnification provides the explosive force to guarantee life time and flight attitude adjustment.

(3) As the core of the whole system the main control board of the aircraft is integrated accelerometer, gyroscope and geomagnetism sensor, and adopts high-precision air pressure sensor to feedback current height information, and provides GPS navigation module, wireless communication module communication interface of the remote control receiver. The main control chip drives the integrated sensor and the external module, obtains the digital data of the sensor data in parallel and then calculates the current flight attitude and height. According to the current flight condition the flight control algorithm adjusts the lift caused by the four BDCM to achieve self-flight by controlling the flight attitude and Height.

(4) Connected to the main control board through the serial port, GPS navigation module searches signal of the measured satellite which is strong enough, and the protocol is consistent with it, then continue to track these signal. Adopting the positioning algorithm, the current location of the module's latitude and longitude, altitude, real-time speed, current time and other information are calculated by amplifying, transforming and processing the satellite signal obtained. Comparing with the local sensor data, these positioning information is sent to the main control panel through the serial to achieve long-distance flight path navigation.

(5) Wireless communication module for a pair, using 433 M frequency band communication, using single-chip and dedicated RF transceiver chip. Built-in single-chip drive RF chip to achieve wireless communication protocol, FM algorithm and software error correction algorithm, The interface of external and wireless communication module is provided by the USART serial port of the

microcontroller, and the same wireless communication module can be used to achieve complete transparent data transmission with the ground station.

(6) Using the FuSi FS-T6 matching receiver of the model remote controller as remote control receiver, remote controller can Interfere with the quadrotor UAC, especially in the take-off and landing stage when the UAV are susceptible to interference from the environment and other unfavorable factors. The 2.4G frequency band communication is used between the remote controller and receiver, and the automatic frequency modulation digital system (ADHDS) protocol can effectively reduce communication interference, improve communication distance, and effectively reduce power consumption.

4   Control Algorithm of the UAV'S Attitude

The aircraft can only obtain the current flight attitude through the attitude algorithm. In order to obtain smooth flight and make a variety of flight movements according to the given attitude change, it is also need to control the current flight attitude. When the angle occurs between the aircraft and the horizontal, the body of UAV will be affected by the component force in the horizontal direction and produce displacement, so the angular motion of the UAV must have an effect on the linear motion, and the autonomous route flight of the aircraft needs to control the line motion. In order to solve the problems above, the attitude control algorithm uses cascade PID to control the attitude angle and linear displacement at the same time. The actuator controls the four motion states of yaw, pitch, vertical and roll, and requires independent control of at least four control input variables, while ensuring low coupling of the control channels under the conditions used. Supposed our state control variables as $(U_1, U_2, U_3 U_4)$ according to the four motion states of the UAV, the coupled nonlinear model of the aircraft is divided into four relatively independent control systems. Using a linear motion and angular motion control subsystem to describe the two states(pitch and yaw movement), the linear motion subsystem is affected by the angular motion subsystem, but the angular motion subsystem is not directly affected by the linear motion subsystem. So consider the control of the attitude movement firstly, followed by the position control.

In the X-axis direction of the ground coordinate system and the pitch angle control are showed in Fig. 2. The outer loop controls the displacement of the quadrotor UAV in the X axis direction, and the inner loop controls the pitch angle of the OXZ plane of the quadrotor UAV. Because the pitch angle response fast, modal frequency band is wide, so the inner loop is pitch angle control loop; But the displacement response of the X axis direction is slow, the modal frequency band is narrow, so the outer loop is the displacement control loop in the X axis direction. The displacement control commands in the X-axis direction can be set remotely or generated in real time by the GPS navigation control system. The displacement control loop in the X-axis direction acts to keep the quadrotor UAV in the X-axis direction or in accordance with the reference trajectory. The pitch angle control loop enables the quadrotor UAV to keep horizontal on the X axis or to generate the acceleration on the X axis by the pitch angle. Through the joint action of the two loops, the output control signal is used to adjust the four motor, and the quadrotor UAV can complete the control of the displacement and angle in the X axis direction according to the control requirements.

Figure 2 Cascade PID control chart of aircraft pitch angle and X-axis displacement

The Y-axis displacement and roll angle control are similar to those in the X-axis direction, as shown in Figs. 3, 4 and 5. Since the yaw angle does not affect the displacement (height) of the UAV on the Z-axis, only the single-stage PID controller can be used to meet the control requirements. The PID control charts of the Z-axis and the heading angle of the aircraft is shown in Figs. 4 and 5.

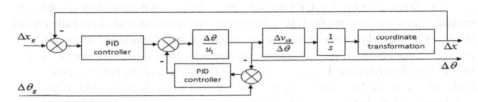

Fig. 2. Cascade PID control of aircraft pitch angle and X-axis displacement

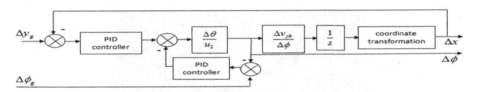

Fig. 3. Cascade PID control chart of aircraft roll angle and Y-axis displacement

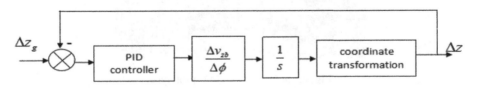

Fig. 4. PID control chart of aircraft Z axis displacement

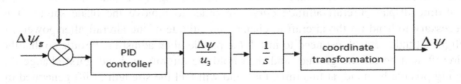

Fig. 5. PID control chart of aircraft yaw angle

5   Assembly and Debugging of the Quadmotor UAV

In order to avoid unnecessary damage during the flight test due to hardware failure, it is necessary to be careful in the assembly stage. Firstly, take the rack and install the motor and fix it with screws. Then connect the motor, the ESC and the power splitter with 18AWG cable and fix it with solder. Use a 12AWG silicone cable to connect the manifold and lithium battery, and the XT60 high-power connector to turn off the power when it is turned off. The main control board is installed in the middle of the aircraft. It should be noted that the x-axis of the sensor direction is consistent with the flight direction. Then connect the GPS module, the wireless communication module, the remote control receiver and the ESC cable to the corresponding port of the main control board. And the GPS antenna should be faced up and surrounded away from the wireless communication module antenna and remote control receiver antenna to prevent mutual Interferences. Finally, by changing the location of the battery, fine-tune the center of gravity of the aircraft to make it be centered as much as possible. With the tie and double-sided adhesive on all facilities to re-fixed, the prototype will be completed, as shown in Fig. 6.

Fig. 6. A Quadrotor UAV prototype

Flight attitude calculation test uses the ground station data display software as the host computer software to realize the wireless monitoring of the quadrotor UAV status, real-time display aircraft attitude curve. In order to remove the blade safely, it is necessary to hand up the aircraft to change the attitude of the aircraft after power on. Flight attitude control test need to install the paddle, then adjust the throttle so that the aircraft will be off the ground, and then hold the aircraft by hand. First change the flying posture by hand, at this time, the hand will feel the strength of lift generated by four blades, then adjust the PID parameters for each change posture corresponding to the lift variation. Then use the remote control to change the flight attitude, with the gradual increase in control efforts, it is significantly to feel the adjustment force of the attitude growing at this time. The corresponding action attitude angle curve through the posture calculation can be seen in the host computer. As shown in Figs. 7, 8, 9, and 10,

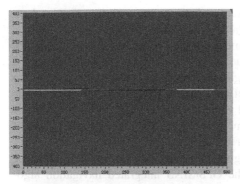

Fig. 7. The attitude angle curve of the quadrotor UAV at rest

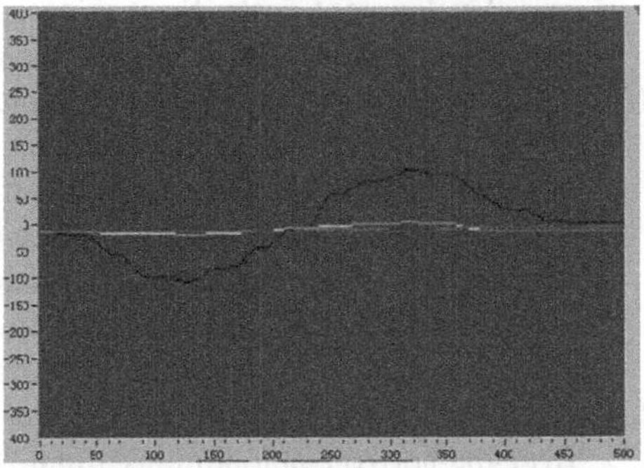

Fig. 8. The attitude angle curve of the quadrotor UAV in pitching motion

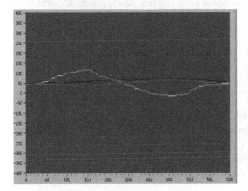

Fig. 9. The attitude angle curve of the quadrotor UAV in roll motion

Fig. 10. The attitude angle curve of the quadrotor UAV in yaw motion

the attitude angle response curve can fully reflect the current attitude changes obviously. Where the blue curve shows the pitch angle σ, the yellow curve shows the roll angle β, and the red curve shows the yaw angle γ. The ordinate represents the magnitude of the angle (°), and the abscissa indicates the time of flight(s).

6 Conclusion

The design of the quadrotor UAV can meet the design requirements, and it has certain value in the research of automatic control algorithm and autonomous navigation algorithm. However, due to the limitation of R&D funding and the shallow knowledge of aircraft-related expertise, there is a huge room for improvement. There are some prominent aspects need to be improved:

(1) The UAV need higher development costs. Due to the lack of protecting function, if high drop occurred during the test, hardware replacement will increase research

and development costs. So it is necessary to increase the anti-drop system from software and hardware to reduce the damage caused by falling.

(2) The height data of the UAV is complemented by the air pressure sensor and GPS navigation module, but the detection accuracy of two methods above is limited and get the altitude. Low-precision high-value feedback will cause the UAV to take off and landing on the impact of excessive resulting in damaging to the landing gear. Hitting the raised ground when flying at a high altitude in terms of altitude probably, so it is necessary to increase the sensor to measure the height of the ground, such as ultrasonic or infrared.

(3) Although the quadrotor UAV designed is characterized by long-distance over-the-horizon autonomous flight, the GPS/SINS integrated navigation algorithm is not mature enough, so it is necessary to introduce the higher precision navigation method to correct. For example, install a digital camera in the bottom of the body to take terrain photos, after identification processing and stored route map comparison to correct the course to achieve a higher navigation accuracy.

References

1. Bo-wen, N., Hong-xian, M., Wang, J., et al.: Research status and key technology of micro-four rotorcraft. Electro-Optic Control **14**, 113–117 (2007)
2. Ji, J., Feifei, H., He, Z.: Application of quadrotor micro-aircraft in farmland information acquisition. J. Agric. Mechanization Res. **2**, 1–4 (2013)
3. Hamel, T., Mahony, R., Lonano, R., et al.: Dynamic modeling and configuration stabilization for an X4-Flyer. In: Proceedings of the IFAC World Congress, Barcelona, Spain, pp. 336–384 (2002)
4. Jiang, B.: Small quadrotor low-altitude unmanned aerial vehicle integrated design. Zhejiang University (2013)
5. Yang, Y., Xian, B., Yin, Q.: Four rotor unmanned aerial vehicle architecture and flight control research status. China Automation Society of Control Theory Professional Committee. China Automation Society Control Theory Professional Committee C roll. China Automation Society Control Theory Professional Committee, p. 6 (2011)
6. Li, J., Li, Y.: Dynamic modeling and PID control of four rotorcraft. J. Liaoning Tech. Univ. **31**, 114–117 (2012)
7. Zhou, Q., Huang, X.: Experimental study on attitude stabilization control of quadrotor micro flight platform. J. Sens. Microsyst. **28**(5), 72–79 (2009)
8. Liu, W.: Design and experimental study of four rotor unmanned aerial vehicles. Harbin Engineering University (2011)
9. Wang, L.: Four rotor unmanned aerial vehicle control technology research. Harbin Engineering University (2012)

Dust Detection System Based on Capacitive Readout IC MS3110

Xiaoqin Tong[✉]

Wuhan University of Science and Technology City College, Wuhan, China
254013697@qq.com

Abstract. A real timework-shop dust detection system based on Single-Chip Microcomputer was designed. Aiming at complexity of manual configuration for MS3110, the concept of self-correcting was proposed and realized. The system provides RS232 communication and Bluetooth wireless communication mode. Comparing with the traditional detecting devices, it can greatly shorten the time consuming of detection as ensuring the measuring accuracy.

Keywords: Dust detection · MS3110 · Bluetooth

1 Introduction

At present, the environmental pollution has become increasingly serious, dust pollution phenomenon is particularly prominent. If people stay in an environment with a lot of dust for a long time, different degree of harm will be brought to their bodies. Dust concentration detection is the principle of controlling dust pollution and validation method of the governance effect. It is a key technology to guarantee human's health. In this paper, we design a digital dust detection system, which can detect the leakage powder weight or concentration of workshop equipment, test data can be stored and displayed through the PC terminal.

2 The Working Principle of the Detection System

Computational formula of capacitance value between two metal plates is presented as $C = \varepsilon * s/4k\pi d$. When dust exists between the plates, it will cause change to capacitance value. From the principle, we transform the measured physical quantity as: dust concentration (weight\concentration) -> capacitance value -> electrical signal.

This system adopts common capacitive readout IC MS3110 produced by the Irvine Sensor company to detect the capacitance value, which has low power consumption, high sensitivity and low noise. MS3110 in this system using modem capacitance detection methods. Oppositely phased square wave signals with same amplitude are generated in chip as carrier signal of output capacitance to realize the modulation of capacitance change. Through charge integrator, capacitance changes are converted into voltage changes. In this process, modulation signals are demodulated through sampling

© Springer International Publishing AG 2017
Y. Huang et al. (Eds.): ICIRA 2017, Part III, LNAI 10464, pp. 35–43, 2017.
DOI: 10.1007/978-3-319-65298-6_4

keeping circuit. Then, through low-pass filter and amplification, a voltage signal which is proportional to capacitor difference is obtained.

The voltage signals generated by master control chip are compared with actual weight or concentration of powder. Through curve fitting, then the physical relationship between electrical signal and concentration of powder can be studied, and we can detect weight or concentration of the powder.

2.1 Internal Parameter Configuration of MS3110

MS3110 has high flexibility that it can realize interior EEPROM data read and write by programming to configure register data. In this system, we configure internal registers without internal EEPROM function. According to the data sheet of MS3110, when offering the clock signal CLK to MS3110, we write a 64-bit data to its registers at the same time. For configuration of each register we use the reference value MS3110 data sheet provides.

Here, MS3110 internal parameters configuration is divided into two processes as explained below:

- When the system has been started, first step of system initialization is required. At this point, the external capacitance under test CS1IN and CS2IN are not connected to circuit. Parameters needs to be configured are presented below: the compensating capacitor value CS1, CS2 are set to zero, current value of pin TESTSEL, voltage value on pin V2P25 and the square wave frequency of pin C2SIN.
- After the first step has done, the external capacitance under test CS1IN and CS2IN are connected to the circuit, At the second step, parameters need to be configured are feedback-capacitor Cf value, compensating capacitor value CS1 and CS2, reference voltage value Vref and gain value Gain.

2.2 Principle of Hardware Parameters Automatic Correction of MS3110

To make the MS3110 sensors to achieve the best state rang during initialization, there are sixty data bits required to be set for the configuration of the internal parameters at the same time. Manual adjustment is very troublesome. The innovation of this paper is introducing the automatic correction (later known as the "self-correcting") concept. It's important to note that internal parameters correction of MS3110 belongs to hardware correction which would be distinguished from software correction. The purpose of hardware correction is to make the system working in the best condition as soon as possible. If hardware correction can't meet with the system's requirements, it shall combine software correction to make improvements.

2.3 Current Correction of Pin TESTSEL

From MS3110data sheet, the current ranges from 8uA to10uA in normal operation. There may be three states when correcting:

- If the measured values are within the normal working current range, this parameter does not need to correct;
- If the measured values are near the theoretically calculated adjustable dynamic range, error will be defined as endurable; To rise or fall parameter values via configuration bits, making it comply with the dynamic range of normal work;
- If the measured value deviates from the theoretical calculated adjustable dynamic range too far, then measured value will be defined as uncorrectable or considered as meaningless (or hardware failures).

2.4 Clock Division Correction of Oscillator

MS3110 provides an 800 kHz inner clock oscillator. After signal generated by oscillator is divided frequency by eight, it is applied to shift registers and some analog circuits. There one very important point is: the frequency division signal is used as the carrier signal of the output capacitance to implement modulation of capacitor changes. As one of the basic requirements, constant clock is essential to chip MS3310.

By MS3110 data sheet, this signal after frequency division can be tested by pinCS2IN. The configuration of frequency parameter is shown in Table 1.

Table 1. MS3110 crystal vibration control register

D2	D1	D0	Frequency
0	0	0	Nominal
0	0	1	+15%
0	1	0	+24%
0	1	1	+33%
1	0	0	Nominal
1	0	1	−35%
1	1	0	−47%
1	1	1	−81%

When MS3110 works normally, the frequency of pin CS2IN must keep between 100 kHz ± 5 kHz. Changing D1, D2 and D0 can change frequency parameter. By changing these configuration bits to implement the self-rectification by software.

MS3110 Initial Output Voltage Correction

Known by MS3110 data book, that the system output voltage and input power satisfy the following formula:

$$Vo = GAIN * V2P25 * 1.14 * (CS2T - CS1T)/CF + VREF.$$

When external under-test capacitances were switched on in the system, CS1 and CS2 which were set to 0 and VREF was configured to 0.5 V by software. According to the formula above, system initial output voltage Vo = VREF, if output voltage was automatically detected as 0.45 V by software, then VREF will be modified to 0.45 V, and it will participate in the following-up measurement operations. In the actual correction, though capacitance CS1T is configured to 0 when CS2T is invariant, CS2T inevitably has certain amount of electricity charge due to interference of environment, so it will affect the initial voltage. So far, we still need to configure the feedback capacitance value CF, making the initial voltage as closer to 0.5 V as possible. The actual value of VREF was obtained by averaging the testing values of repeated measurements.

3   System Hardware Circuit Design

The system uses STM32F103VBT6 as system processor, and adopts the character type LCD 1602 which is based on HD44780 controller to display the system running state and real-time detecting data. The whole system is divided into seven parts such as power circuit, main control circuit, sensor circuit, AD sampling circuit, communication circuit, display module circuit and so on. The hardware circuit structure is shown in Fig. 1.

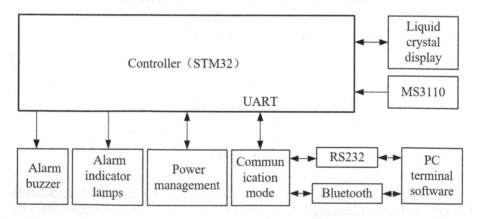

Fig. 1. General structure of detection system

3.1 Ms3110 Sensor Circuit

System uses two-way relays connecting external under-test capacitance CS1IN, CS2IN, to disconnect and access the under-test capacitance during system initialization, which as shown in Fig. 2.

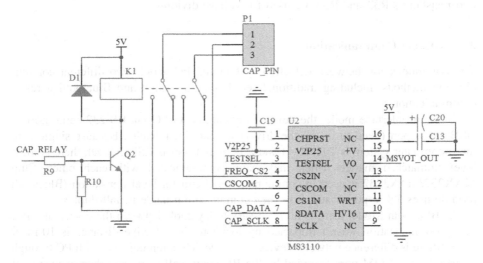

Fig. 2. MS3110 sensor circuit

TESTSEL Pin Current Test Circuit

Because TESTSEL pin current is one of the preconditions for sensor chipMS3110 to work properly, the system must monitor the TESTSEL Pin current during the initialization. To test whether current is within the accepted error range [8uA, 10uA]. To measure the pin current, the current signal is translated to a voltage signal, which then is amplified to a suitable scope. Since the op-amp output error is subject to offset voltage Vos, and the bias current Ib, so if you want to achieve micro current measurement, the following conditions must be satisfied:

- The op-amp offset current Ib < measured current Is;
- Input impedance R1 ≫Rf(feedback resistance);
- High CMRR, Low Offset Voltage, Small Zero drift.

We choose MAX4238 chip here, and all the indexes can meet the design requirements. For some of its indicators: Vos = 20uV; Ib = 1 Pa; CMRR is 130Db. This circuit adopts two stage amplifier circuit. For example, the first stage translates the pin current to 50 ∼ 100 mV voltage output, which is amplified to 20 times by the second stage.

MS3110 CS2IN Pin Frequency Test Circuit

Because the signal waveform of pin CS2IN of MS3110 sensor is not a standard square wave signal in the actual detection, which brings difficulty to frequency detection.

Therefore, the Schmitt trigger consists of NE555 is used to shape the signal to a standard square wave signal, then it is send to the controller.

When the controller detects the frequency, it mainly testing pulse rising edge or pulse width. The power supply of the controller is 3.3 V. To avoid the output voltage of the Schmitt trigger being greater than the withstand voltage of the controller pins, two resistances R32 and R33 are used for voltage dividing.

3.2 System Communication

The communication between detection system and PC adopts two different communication methods, including traditional serial communication and Bluetooth wireless communication.

In serial interface mode, the device is connected to PC via RS-232 serial port or USB serial port (need to install the driver) for data interchange. Because single chip microcomputer system is TTL level and a PC serial port is RS232 level, there is logic level unmatched matters when they have communicated with each other. Chip MAX232/MAX3232 is used in communication circuit for level translation (Bluetooth module uses 3.3 V power supply). The communication mode is half-duplex.

In Bluetooth Mode, there is no necessary to lay cable between the device and PC, their communication wasn't impacted by location, the effective distance is 10 m or more (there is difference between devices). The device communicates with PC through a virtual serial COM port provided by the Bluetooth soft-ware, and then exchange of wireless values can be realized.

This system adopts the Bluetooth module BF10 to communicate with the existing embedded system by a serial port. The Bluetooth is configured as follows: (Note: Only suitable for the PC equipped with Bluetooth adapter)

– The Bluetooth module BF10 was provided with 3.3 V power supply. The module BF10 is used as a slave module of MCU
– Starting the dust testing equipment and install the driver software of the Bluetooth driver module, then query the Bluetooth module around. After the matching code is configured. The connection between Bluetooth module and PC is established with virtual serial ports.

4 Software Design

This system software design of the main process is as follows:

– First the controller corrects the initialization parameters of the MS3110, make it work in the best state;
– After system initialization is completed, we begin dust semaphore (voltage values) sampling and process the sampled data;
– Then sending the data to the PC via a serial port or Bluetooth for da-ta display, storage and analysis.

The overall flowchart is shown in Fig. 3 below:

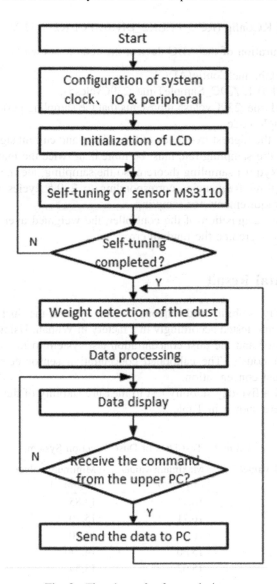

Fig. 3. Flowchart of software design

5 AD Sampling and Filtering

Controller STM32F103VBT6 includes a built-in 12-bit ADC converter. Before AD sampling, the system and ADC clock, ADC hardware need to be configured; System master clock is configured to 72 M. Due to the max ADC clock must not exceed 14 M, here we take 6 frequency division, reducing the ADC clock to 12 M.

- For the ADC clock configuration, the article gave some detailed program codes as follow (Here only the core code is given):

RCC_ADCCLKConfig (RCC_PCLK2_Div6);//PCLK2 = 72 M

– For the configuration of the ADC device, the code is shown below:

ADC_RegularChannelConfig(ADC1,
ADC_Channel_0,1, ADC_SampleTime_239Cycles5);
ADC_SampleTime_239Cycles5 represents that the sampling periods is 239.5 clock cycles in this sample code.

In this system, the highest frequency of the voltage and current signal do not exceed 8 kHz by testing, the sampling rate must be more than twice the maximum frequency according to the Nyquist sampling theorem, so the sampling rate must be greater than 16 kHz, the sampling frequencies corresponding to 239.5 cycles is about 50 kHz, which meets the requirements for sampling.

During the data acquisition of the controller, the weighted average filtering algorithm was used to overcome the random error.

6   Experimental Result

To verify feasibility, stability and reliability of the system, the dust detection system has been adjusted and tested accordingly in a factory in Wuhan. Using a calibrated dust concentration sensor and the dust concentration test system to measure the dust concentration simultaneously. The calibrated concentration sensor determined the theoretical value of dust concentration.

We conducted a five-day stability test to measure stability of the detection system, the test results were shown in Table 2.

Table 2.   Test Data of Dust Detection System.

Day	Theoretical value/ $(mg \cdot m^{-3})$	Maximum value/ $(mg \cdot m^{-3})$	Minimum value/ $(mg \cdot m^{-3})$	Deviation/ $(mg \cdot m^{-3})$
1	15.25	16.32	13.85	$-1.07 \sim 1.40$
2	15.90	16.71	15.70	$-0.93 \sim 0.41$
3	15.50	16.30	14.80	$-1.35 \sim 0.70$
4	14.70	15.60	14. 90	$-1.24 \sim 0.53$
5	14.20	15.02	13.59	$-0.82 \sim 0.61$

As we can see from Table 2, the theoretical value of dust concentration is stable at about 15 mg/m^{-3}. The system has continuously detected for five days, and maximum and minimum values are obtained, the deviation was stable at ± 1.5 mg/m^{-3}. The error can be controlled at $\pm 10\%$, it shows that the detection system has good stability.

To test the detection accuracy of this system, we sampled the dust concentration and recorded the data of the dust detection system, taking the arithmetic mean values as the measuring results. The results were compared with the result of the calibrated dust concentration sensor. The data of the detection system for dust concentration test is shown in Table 3.

Table 3. Test Data of Dust Detection System.

Nom.	Sampling time/min	Dust mass concentration/ $(mg \cdot m^{-3})$	Detection value/ $(mg \cdot m^{-3})$	Error/ %
1	10	7.90	8.31	−5
2	10	13.1	13.56	−4
3	10	18.8	19.81	−5
4	10	26.6	25.65	+4
5	10	21.3	20.9	+2
6	10	31.7	29.74	+6
7	10	37.5	35.8	+5
8	10	40.3	37.9	+6
9	10	42.5	41.32	+3

According to the results analysis from Table 3, the test results of the dust detection system were basically consistent with the test result of calibration. Field tests results has proved that the detection system is operable and practical, which can measure the concentration of dust in the factory, and the relative error was less than 10%.

7 Conclusion

The system is simple structure, high precision, and real-time transmission. It has great engineering application value. To use this system better, we can raise the AD digits to increase the resolution of the sampling; Using multi channels sensor at the same time, achieving the indoor distributed detection.

References

1. Ping, X.: Differential capacitance detection method based on MS3110 re-search and its electromagnetic compatibility. Instrument Technique and Sensor (2008)
2. Mingjie, L., Yunfeng, L., et al.: MEMS accelerometer based on capacitive readout IC MS3110. J. Chin. Inertial Technol. **18**, 236–239 (2010)
3. Weihua, L., Wei, Y.: The coal dust on-line monitoring system based on image analysis. Electron Measur. Technol. (2010)
4. Bin, S., Chuanlin, J., Yuanyong, C.: Design of vehicle air purification devices based on optical dust sensor. Electron. Tech. (2013)
5. Hongguang, G., Fei, W.: Preliminary study on PM2.5 standard of the coal mine dust control. Coal Mine Saf. (2015)

Design and Modeling of a Compact Rotary Series Elastic Actuator for an Elbow Rehabilitation Robot

Qiang Zhang, Benyan Xu, Zhao Guo$^{(\boxtimes)}$, and Xiaohui Xiao$^{(\boxtimes)}$

Wuhan University, Wuhan 430072, China
{zhangqiang007,Benyan.Xu,guozhao,xhxiao}@whu.edu.cn

Abstract. Rehabilitation robot has direct physical interaction with human body, in which the adaptability to interaction, safety and robustness is of great significance. In this paper, a compact rotary series elastic actuator (SEA) is proposed to develop an elbow rehabilitation robot for assisting stroke victims with upper limb impairments perform activities of daily living (ADLs). The compliant SEA ensures inherent safety and improves torque control at the elbow joint of this rehabilitation robot. After modeling of the rotary stiffness and dynamics of the SEA, a PD feedback plus feedforward control architecture is introduced. A test bench has been designed to experimentally characterize the performance of the proposed compliant actuator with controller. It shows an excellent torque tracking performance at low motion frequency, which can satisfy the elbow rehabilitation training requirement. These preliminary results can be readily extended to a full upper limb exoskeleton-type rehabilitation robot actuated by SEA without much difficulty.

Keywords: Series elastic actuator (SEA) · PD feedback control · Torque control · Elbow rehabilitation robot

1 Introduction

When human beings suffer from stroke and other neurological diseases, the difficulty of their limb motion will influence activities of daily living (ADLs). According to the World Health Organization, for every year 15 million people suffer from stroke and 5 million of them remain permanently disabled [1]. Thus, the development of effective rehabilitation devices has attracted much attention, whose merit is the ability to intelligently interact with users. The actuators applied to traditional rehabilitation robots are preferred to be as stiff as possible to make precise position movements or trajectory tracking control easier (faster systems with high bandwidth) [2, 3].

Recent years, researchers have proposed SEA concept, and the elastic element gives SEAs several unique properties compared to rigid ones, including low output impedance, increased peak power output, and passive mechanical energy storage [4–7]. Many different types of SEAs have been developed for rehabilitation devices. Ragonesi et al. [8] propose rotary powered exoskeleton primarily on commercially available off-the-shelf components, compression springs as the compliant element, and power transmission through a bevel gear. In [9], the authors use linear springs coupled to rotary

© Springer International Publishing AG 2017
Y. Huang et al. (Eds.): ICIRA 2017, Part III, LNAI 10464, pp. 44–56, 2017.
DOI: 10.1007/978-3-319-65298-6_5

shafts and place the springs between the motor and external load to achieve compact rotary SEA. In [10, 11], the proposed SEA is built based on linear springs coupled to a ball screw which is connected with servo motor. In [12], the authors use Bowden cable connected to linear springs to achieve rotary SEA. In [13, 14], the linear springs are placed within the reduction phase, which can reduce the torque requirement on the spring compared to placing the spring at the actuator output. Although current rehabilitation devices actuated by SEAs have achieved reasonable performance through separated experiments, they still face some limitation. For example, in [15] the SEA is fixed onto the knee and ankle joint of rehabilitation robot directly, which increases the inertia of the joint and affects its performance.

Apart from the mechanical development of SEAs, the control approach is also gaining much attention these years. The basic purpose of these controllers is to guarantee the SEA's output torque as desired. For instance, the output force can be observed by measuring change of resistance [4], or by measuring the spring deflection and based on Hooke's law [16–18]. Pratt et al. [19] propose that the position feedback can be used as the innermost control structure for torque control. This kind of control method has been adopted by other researchers, considering torque control as a position control problem. In recent years, new controllers include PID control as well as take the dynamics of mechanical system to improve response frequency of torque control [20, 21]. In [13] the authors combine PID, model-based, and disturbance observer (DOB) together to realize impressive torque tracking performance.

Existing rehabilitation devices fix stiff actuator or compliant actuator onto joint directly. One drawback is that they are too heavy, bulky, and stationary. To address this issue, in this paper, we present a novel compact rotary SEA with Bowden cable to develop a light weight, low profile, and potable elbow rehabilitation robot. Unlike the former SEA design method, we propose a new structure configuration for shorter length and better output torque measurement. The description and mathematic model of this proposed SEA is presented in Sect. 2. In Sect. 3 the controller including PD feedback term and feedforward term is described as well as the numerical parameters selection of the PD. The physical prototype experiments results are presented to evaluate the proposed SEA in Sect. 4. Finally, Sect. 5 presents conclusions and future work.

2 Description and Modeling

2.1 Description of the Compact Rotary SEA

Existing SEAs usually put elastic elements between servo motor and load, such as linear translational or torsional springs. In general, servo motor shell of SEA is stationary, and the output side of SEA is away from servo motor, which brings the difficulty of spring deflection measurement and increases the whole length of SEA, as shown in Fig. 1(a). In this paper, we propose a new structural arrangement of rotary SEA, which is shown in Fig. 1(b). The definition of parameters in Fig. 1 will be given in modeling and control section. Figure 2 shows the CAD prototype of the SEA. Our design consists of a servo motor with gear reducer, a rotary encoder, a four-spoke component, eight linear springs, deep groove ball bearing, output pulley, inner and outer sleeve.

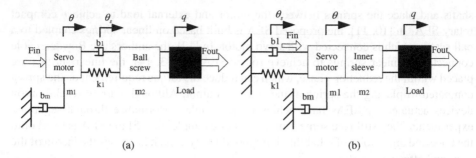

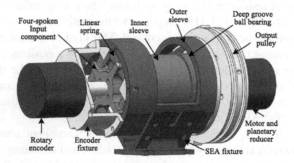

Fig. 1. Structural configuration of compliant actuator, (a) The general case model; (b) The modified model.

Fig. 2. The CAD prototype of rotary SEA design.

The power source comes from DC servo motor connected with planetary gear reducer, whose output shaft is fixed with a four-spoke component. Between the four-spoke and stationary outer sleeve, there are eight linear translational springs, and each spring shares same stiffness and size. The shell of servo motor and reducer is fixed with the inner sleeve, and the inner sleeve is also fixed with output pulley, which means they will rotate together when SEA works. Deep groove ball bearings are used between inner and outer sleeve to guarantee smooth rotation motion. In Fig. 2, define left side as front and right side as end, so the proposed SEA design can be regarded as a kinematic chain in series. For the output torque measurement, a rotary encoder installed in the carriage is to measure the rotary deflection angle of linear springs. When SEA works, the load acts on the output pulley, which leads to two parameters, reducer output position θ and pulley output position q. The output torque of SEA can be calculated through springs' compression.

This structural configuration is beneficial to install encoder for measuring the deflection of springs. The proposed actuator is designed to be able to provide up to 10 N·m assistive torque for elbow joint. A Maxon DC servo motor (20 W 24 V) connected with planetary reducer is used for the design due to its lightweight (0.4 kg) and low movement inertia. The linear springs have spring constant of 13.75 N/mm and a working stroke of 10 mm. They can provide an output force of 125 N before fully compressed.

The incremental rotary encoder fixed on the outer sleeve has a resolution of 2500 lines/rev. And the total mass of the actuator is 0.95 kg.

2.2 Stiffness Modeling of the Compact Rotary SEA

Based on above mechanical description of SEA, the approximate SEA stiffness model was presented in our previous work [22]. In this paper, we propose the precise SEA stiffness model. As shown in Fig. 3, the eight linear translational springs inserted in the SEA experience a pre-contraction equal to half of the maximum acceptable deflection. Here the angle deflection θ_s is designed changing from minus 10 degrees to 10 degrees.

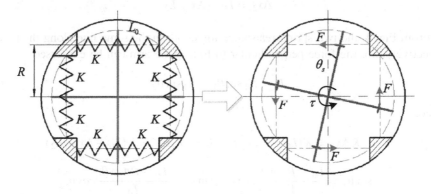

Fig. 3. Compression of springs related to module deflection.

The precise spring compression diagram is presented in Fig. 4, where L_0 is the length of pre-contraction and Δx is the compression. Then the initial spring length can be expressed as:

$$L = L_0 + \Delta x \tag{1}$$

When the four-spoke experiences a deflection of θ_s, based on Cosine theorem we can obtain the length of two springs:

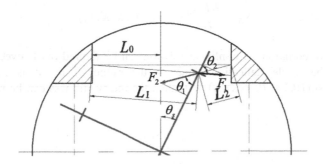

Fig. 4. Precise compression model of springs.

$$L_1 = \sqrt{\left(\frac{R}{\cos\theta_s} - R\right)^2 + (L_0 + R\tan\theta_s)^2 - 2\left(\frac{R}{\cos\theta_s} - R\right)(L_0 + R\tan\theta_s)\sin\theta_s}$$

$$L_2 = \sqrt{\left(\frac{R}{\cos\theta_s} - R\right)^2 + (L_0 - R\tan\theta_s)^2 + 2\left(\frac{R}{\cos\theta_s} - R\right)(L_0 - R\tan\theta_s)\sin\theta_s}$$

$$(2)$$

Thus, the compression of the two springs now is written as:

$$\Delta x_1 = L_0 + \Delta x - L_1$$
$$\Delta x_2 = L_0 + \Delta x - L_2$$

$$(3)$$

From Fig. 4, the two spring resistance forces are in the same line along their axis, respectively. So, the force perpendicular to four-spoke arm can be expressed as:

$$F = F_2 \sin\theta_1 - F_1 \sin\theta_2 \qquad (4)$$

Where

$$F_1 = K\Delta x_1 = K(L_0 + \Delta x - L_1), F_2 = K\Delta x_2 = K(L_0 + \Delta x - L_2) \qquad (4)$$

$$\sin\theta_1 = \frac{L_0 - R\tan\theta_s}{L_2}\cos\theta_s, \sin\theta_2 = \frac{L_0 + R\tan\theta_s}{L_1}\cos\theta_s \qquad (6)$$

In conclusion, the combined torque on the compliant actuator considering the axial forces of four pairs of springs is:

$$T_{total} = 4FR = 4KR\left[(L_0 + \Delta x - L_2)\frac{L_0 - R\tan\theta_s}{L_2}\cos\theta_s - (L_0 + \Delta x - L_1)\frac{L_0 + R\tan\theta_s}{L_1}\cos\theta_s\right]$$

$$(7)$$

By directly differentiating the torque equation, the rotary stiffness of the four-spoke component which is due to the axial deflection of the springs can be expressed as:

$$K_A = \frac{\partial T_{total}}{\partial\theta_s} = f(L_0, \Delta x, R, K, \theta_s) \qquad (8)$$

From above equations, a nonlinear relation can be figured out between the output torque and the angle deflection. Due to the range of θ_s, it is assumed that $\tan(\theta_s) \approx \theta_s, \sin(\theta_s) \approx \theta_s, \cos(\theta_s) \approx 1$, then the rotary stiffness can be regarded as a constant:

$$K_A = 8KR^2 \qquad (9)$$

2.3   Dynamic Modeling of the Compact Rotary SEA

The mechanical implementation of the compact rotary SEA is shown in Figs. 2 and 3. As we mentioned before, the whole system can be regarded as a kinematic chain in series, and the whole mechanical system's mathematical equation can be expressed as:

$$J\ddot{q} + D_q\dot{q} + \tau_{out} = \tau_e(\theta_s)$$
$$B\ddot{\theta} + D_\theta\dot{\theta} + \tau_e(\theta_s) = \tau \qquad (10)$$
$$K_A(\theta - q) = \tau_e(\theta_s)$$

where q and θ are respectively the pulley angle position and gear reducer shaft angle position, and $\theta_s = \theta - q$. J and B are the inertias of inner sleeve with power source and servo motor, respectively. D_q is the viscous friction coefficient of the inner sleeve, while D_θ is the viscous friction coefficient of servo motor. τ_{out} is the output torque of pulley. τ and $\tau_e(\theta_s)$ are output torque of servo motor and elastic element, respectively. K_A is the rotary stiffness coefficient of our proposed SEA.

Based on the Eq. (10), assuming a fixed load end, which means $q = 0$, and $\theta_s = \theta$, the Newton's second law of the whole system is given by:

$$\ddot{\tau}_{out} = \frac{K_A}{B}\tau - \frac{D_\theta}{B}\dot{\tau}_{out} - \frac{K_A}{B}\tau_{out} \qquad (11)$$

The system model above does not include friction and disturbance terms, so in this paper we consider these two terms as follows:

$$\ddot{\tau}_{out} = \frac{K_A}{B}\tau - \frac{D_\theta}{B}\dot{\tau}_{out} - \frac{K_A}{B}\tau_{out} + [\Delta f(\dot{\theta}_s) + d_1] \qquad (12)$$

where $\Delta f\left(\dot{\theta}_s\right)$ represents the friction and d_1 represents the disturbance. The output torque of servo motor has a relationship with motor current i, which can be expressed as $\tau = \beta i$. β is the product of torque current coefficient and reducer ratio.

3   Torque Control Design for SEA

In this paper, the control objective is to get well performance of torque tracking, which means that the output torque should follow the desired torque trajectory as closely as possible. A torque controller consisting of PD feedback and feedforward term is proposed here. The diagram is presented in Fig. 5.

First, we design the model-based feedback control. Here, define the torque error $e_1(t) = \tau_d(t) - \tau_{out}(t)$, the derivative error is given by $e_2(t) = \dot{\tau}_d(t) - \dot{\tau}_{out}(t)$. So, we define a state as $e = [e_1, e_2]^T$, and the state equation is given by:

$$\dot{e} = A_1 e - B_1 i + B_1\left[\frac{B}{\beta K_A}\ddot{\tau}_d + \frac{D_\theta}{\beta K_A}\dot{\tau}_d + \frac{1}{\beta}\tau_d + \Delta f(\dot{\theta}_s) + d_1\right] \qquad (13)$$

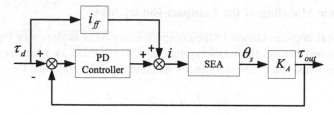

Fig. 5. Torque control method of rotary SEA.

Where

$$A_1 = \begin{bmatrix} 0 & 1 \\ -\frac{K_A}{B} & -\frac{D_\theta}{B} \end{bmatrix}, B_1 = \begin{bmatrix} 0 \\ \frac{\beta K_A}{B} \end{bmatrix}, \Delta f(\dot{\theta}_s) = \mu_1 \mathrm{sgn}(\dot{\theta}_s) \tag{14}$$

Considering the state e, it is natural to think that a proportional-derivative (PD) controller should be employed, that is:

$$i(t) = K_p(\tau_d - \tau_{out}) + K_d(\dot{\tau}_d - \dot{\tau}_{out}) \tag{15}$$

where K_p and K_d are the PD parameters which should be chosen elaborately. In this design, the dominant linear model is given by:

$$\dot{e} = A_1 e - B_1 i \tag{16}$$

so, we will set the PD control parameters based on this linear model. Combining the Eqs. (14) and (16), we can get:

$$\dot{e} = \begin{bmatrix} 0 & 1 \\ -\frac{K_A}{B} - \frac{\beta K_A}{B} K_p & -\frac{D_\theta}{B} - \frac{\beta K_A}{B} K_d \end{bmatrix} e \tag{17}$$

Then, the system characteristic equation of closed-loop can be described as:

$$s^2 + \left(\frac{D_\theta}{B} + \frac{\beta K_A}{B} K_d\right)s + \frac{K_A}{B} + \frac{\beta K_A}{B} K_p = 0 \tag{18}$$

Here, the pole assignment method is used to design the PD parameters.

From the above analysis, note that the friction and disturbance term are not considered in PD feedback control design. Thus, we set two compensation terms for friction and disturbance in the feedforward term. Although the term $\Delta f(\dot{\theta}_s)$ is nonlinear, it can be estimated by off-line experiments. In Eq. (14), μ_1 should be tested before the control design. The friction compensation can be written as:

$$i_r = \mu_1 \mathrm{sgn}(\dot{\theta}_s) \tag{19}$$

For the term d_1, it is assumed to be bounded by a value, which is $|d_1| \leq \alpha_1$.

Combining Eqs. (15) and (19), the proposed torque control is given by:

$$i = K_p(\tau_d - \tau_{out}) + K_d(\dot{\tau}_d - \dot{\tau}_{out}) + i_{ff} \tag{20}$$

Where

$$i_{ff} = \frac{B}{\beta K_A}\ddot{\tau}_d + \frac{D_\theta}{\beta K_A}\dot{\tau}_d + \frac{1}{\beta}\tau_d + \mu_1 \mathrm{sgn}(\dot{\theta}_s) + \alpha_1 \tag{21}$$

4 Experimental Results

Once our compact rotary SEA is controlled by using the proposed controller, it is desired to verify the following performance objectives: (1) Bandwidth of the output torque feedback system; (2) Torque fidelity.

4.1 Experimental Setup and Model Identification

To analyze the performance of the proposed SEA, the test bench based on the design principles and design solutions has been built, which is shown in Fig. 6.

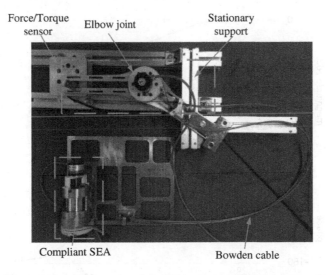

Fig. 6. The physical prototype of the SEA test bench and worn a subject.

The physical parameters of the compliant actuator prototype are list in Table 1, where the original number and the equivalent values are presented for dynamic modeling and analysis purpose. The equivalent values are obtained from open-loop step current response, which is identified in MATLAB [23]. Based on the model

Table 1. Parameters of the physical actuator prototype.

Hardware parameters	Values
Four-spoke arm length R	20 mm
Spring length of pre-contraction L_0	11 mm
Spring contraction Δx	11 mm
Inertia J	4.89×10^{-5} kg.m^2
Inertia B	2.65×10^{-5} kg.m^2
Damping coefficient D_θ	0.0005
Damping coefficient D_q	0.0001
Torque current constant β	10.775 N·m/A

identification between i and θ_s, we can get the SEA closed-loop transfer function as follows:

$$G(s) = \frac{7.988 \times 10^5}{s^2 + 1.362 \times 10^4 s + 7.988 \times 10^5} \tag{22}$$

Closed-loop output torque control bandwidth is defined as the transfer function between the desired torque and the actual torque, which is shown in Fig. 7. The measurement can present the working frequency of the compliant actuator.

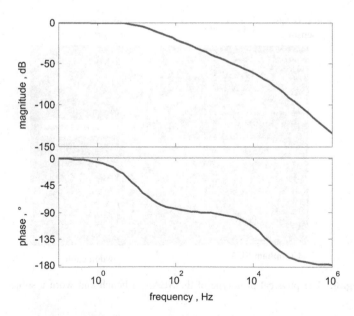

Fig. 7. The closed-loop Bode diagram of rotary SEA.

From Fig. 7, it is observed that the bandwidth of closed-loop system is around 10 Hz at the −3 dB. As is known to us, the frequency of human elbow joint motion is usually within 4 Hz [12]. Thus, we can make sure that the bandwidth of this proposed SEA is enough for assisting human elbow joint motion.

4.2 Performance of Output Torque Fidelity

Based on the linear relationship between output torque and springs deflection angle, the output position θ_s could be transformed to the output torque. In this section, first we will verify the closed-loop step response with proposed controller, which is presented in Fig. 8. The rise time is 0.17 s, and there is no obvious overshoot.

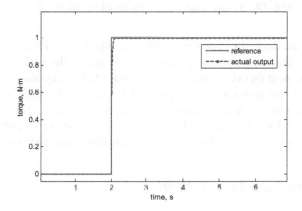

Fig. 8. Step torque response of SEA with controller.

Then with the same controller's parameters, we test different frequency sinusoidal torque signals to examine how the SEA can perform an accurate torque tracking ability. This will involve the frequency of 0.5 Hz and 1.0 Hz, whose experimental results are shown in Figs. 9 and 10.

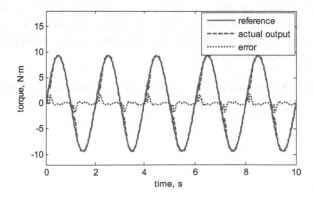

Fig. 9. Torque tracking of sinusoidal signal at 0.5 Hz.

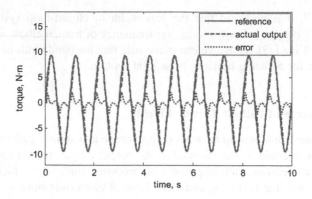

Fig. 10. Torque tracking of sinusoidal signal at 1.0 Hz.

In our tests, the desired torque signals have a range from −10 N·m to 10 N·m. The test results show that torque tracking have periodic error. The peak value of the error occurs on the moment that translational springs reach half acceptable compression, and it increases with the frequency increasing. To evaluate the tracking performance, average absolute error is used in different tests. With the frequency increasing, the average value is 0.3698 N·m and 0.4014 N·m. The tracking results show that at low frequency, the tracking error is less than 5%, indicating very high force fidelity.

5   Conclusions and Future Work

In this paper, based on the concept of SEA, we design a new compact rotary SEA for an initial Bowden cable driven elbow rehabilitation robot. After modeling the stiffness and dynamics of the compliant actuator, we propose the PD feedback plus feedforward controller, with which the SEA achieved good torque control fidelity and working bandwidth. The preliminary experimental results of the SEA prototype validate the design concept and demonstrate excellent torque tracking performance at low motion frequency, which can satisfy the elbow rehabilitation training requirement.

In the future, based on the preliminary experimental results of the compliant actuator, we will figure out the Coulomb friction analytical model of Bowden cable and achieve impedance control and torque fidelity with friction compensation on the designed elbow rehabilitation robot.

Acknowledgment. This research is sponsored by National Natural Science Foundation of China (NSFC, Grant No. 51605339).

References

1. Mackay, J., Mensah, G.A., Mendis, S., et al.: The Atlas of Heart Disease and Stroke. World Health Organization, Geneva (2004)
2. Suin, K., Bae, J.: Force-mode control of rotary series elastic actuators in a lower extremity exoskeleton using model-inverse time delay control (MiTDC). In: IEEE/RSJ International Conference on Intelligent Robots and Systems, pp. 3836–3841. IEEE, Daejeon, Korea (2016)
3. Ham, R., Sugar, T.G., Vanderborght, B., et al.: Compliant actuator designs. IEEE Robot. Autom. Mag. **16**(3), 81–94 (2009)
4. Zhang, Q., Teng, L., Wang, Y., Xie, T., Xiao, X.: A study of flexible energy-saving joint for biped robots considering sagittal plane motion. In: Liu, H., Kubota, N., Zhu, X., Dillmann, R., Zhou, D. (eds.) ICIRA 2015. LNCS, vol. 9245, pp. 333–344. Springer, Cham (2015). doi:10.1007/978-3-319-22876-1_29
5. Pratt, G., Williamson, M.: Series elastic actuators. In: IEEE/RSJ International Conference on Intelligent Robots and Systems, pp. 399–406. IEEE, Pittsburgh, USA (1995)
6. Arumugom, S., Muthuraman, S., Ponselvan, V.: Modeling and application of series elastic actuators for force control multi legged robots. J. Comput. **1**(1), 26–33 (2009)
7. Zhang, Q., Xiao, X., Wang, Y.: Compliant joint for biped robot considering energy consumption optimization. J. Cent. South Univ. **46**(11), 4070–4076 (2015). (In Chinese)
8. Ragonesi, D., Agrawal, S., Sample, W., et al.: Series elastic actuator control of a powered exoskeleton. In: IEEE International Conference on Engineering in Medicine and Biology, pp. 3515–3518. IEEE, Boston, USA (2011)
9. Hutter, M., Remy, C., Hoepflinger, M., et al.: ScarlETH: design and control of a planar running robot. In: IEEE/RSJ International Conference on Intelligent Robots and Systems, pp. 562–567. IEEE, San Francisco, USA (2011)
10. Robinson, D.W., Pratt, J.E., Paluska, D.J., et al.: Series elastic actuator development for a biomimetic walking robot. In: IEEE/ASME International Conference on Advanced Intelligent Mechatronics, pp. 561–568. IEEE, Atlanta, USA (1999)
11. Pratt, J.E., Krupp, B.T., Morse, C.J., et al.: The RoboKnee: an exoskeleton for enhancing strength and endurance during walking. In: IEEE International Conference on Robotics and Automation, pp. 2430–2435. IEEE, New Orleans, USA (2004)
12. Veneman, J.F., Ekkelenkamp, R., Kruidhof, R., et al.: A series elastic- and Bowden-cable-based actuation system for use as torque actuator in exoskeleton-type robots. Int. J. Robot. Res. **25**(3), 261–281 (2006)
13. Kong, K., Bae, J., Tomizuka, M.: A compact rotary series elastic actuator for human assistive systems. IEEE/ASME Trans. Mechatron. **17**(2), 288–297 (2012)
14. Taylor, M.D.: A compact series elastic actuator for bipedal robots with human-like dynamic performance. Master's thesis, Robotics Institute of Carnegie Mellon University, Pittsburgh, USA (2011)
15. Yu, H., Huang, S., Chen, G., et al.: Control design of a novel compliant actuator for rehabilitation robots. Mechatronics **23**(8), 1072–1083 (2013)
16. Kong, K., Bae, J., Tomizuka, M.: A compact rotary series elastic actuator for knee joint assistive system. In: IEEE International Conference on Robotics and Automation, pp. 2940–2945. IEEE, Anchorage, USA (2010)
17. Guo, Z., Yu, H., Pang, Y., et al.: Design and control of a novel compliant differential shape memory alloy actuator. Sens. Actuators, A **225**(3), 71–80 (2015)
18. Yu, H., Chen, G., Huang, S., et al.: Human-robot interaction control of rehabilitation robots with series elastic actuators. IEEE Trans. Rob. **31**(5), 1089–1100 (2015)

19. Pratt, G., Willisson, P., Bolton, C., et al.: Late motor processing in low-impedance robots: impedance control of series-elastic actuators. In: America Control Conference, pp. 3245–3251. IEEE, Boston, USA (2004)
20. Hurst, J., Chestnutt, J., Rizzi, A.: The actuator with mechanically adjustable series compliance. IEEE Trans. Rob. **26**(4), 597–606 (2010)
21. Pan, Y., Guo, Z., Li, X., et al.: Output feedback adaptive neural control of a compliant differential SMA actuator. IEEE Trans. Control Syst. Technol. **PP**(99), 1–9 (2017)
22. Zhang, Q., Xiao, X., Guo, Z.: Power efficiency-based stiffness optimization of a compliant actuator for underactuated bipedal robot. In: Kubota, N., Kiguchi, K., Liu, H., Obo, T. (eds.) ICIRA 2016. LNCS, vol. 9834, pp. 186–197. Springer, Cham (2016). doi:10.1007/978-3-319-43506-0_16
23. Wang, Y., Zhang, Q., Xiao, X.: Trajectory tracking control of the bionic joint actuated by pneumatic artificial muscle based on robust modeling. ROBOT **38**(2), 248–256 (2016). (In Chinese)

A Vibro-tactile Stimulation and Vibro-signature Synchronization Device for SSSEP-Based Study

Huanpeng Ye, Tao Xie, Lin Yao, Xinjun Sheng$^{(\boxtimes)}$, and Xiangyang Zhu

State Key Laboratory of Mechanical System and Vibration,
Shanghai Jiao Tong University, 800 Dongchuan Road,
Minhang District, Shanghai, China
xjsheng@sjtu.edu.cn

Abstract. A vibro-tactile stimulation and vibro-signature synchronization device was proposed in this paper. The device was used to elicit steady-state somatosensory evoked potential (SSSEP) and synchronize the actual vibraion information with the corresponding electroencephalography (EEG) signals in temporal domain. The device provided five independent stimulation channels and could generate vibro-tactile stimulation with arbitrary waveform, amplitude and frequency. Each channel used the intended stimulation waveform to drive a linear resonant actuator (LRA). The vibro-signature of each channel could be detected by a force sensing resistor (FSR) and fed back into the EEG recording system. Four stimulation patterns with a random sequence were applied to the index finger of three healthy subjects. Results showed that SSSEP features could be evoked with different stimulation patterns and the actual vibro-signature could be synchronized with the EEG signals. Comparing the actual stimulation with the intended stimulation, the errors of carrier frequency and modulation frequency were quite small. Results indicated that this novel device stood a good chance of serving in SSSEP-based studies.

Keywords: Vibro-tactile stimulation · Steady-state somatosensory evoked potential (SSSEP) · Synchronization

1 Introduction

Repetitive tactile stimulation can elicit a brain response called steady-state somatosensory evoked potential (SSSEP), whose temporal frequency follows the driving stimulation, and may contain higher harmonics [1,2]. Since the feasibility of Brain-computer Interface (BCI) studies using SSSEP was validated by Muller-Putz and colleagues [3], SSSEP-based BCIs have already been widely developed. The possibility of classifying different finger attention within one hand using SSSEP was assessed [4,5]. SSSEP is a promising tool for a wheelchair controlling with three commands [6]. SSSEP-based hybrid BCI studies include SSSEP with tactile selective attention [7], SSSEP with motor imagery [8], and SSSEP with transient somatosensory event-related potentials (ERPs) [9].

© Springer International Publishing AG 2017
Y. Huang et al. (Eds.): ICIRA 2017, Part III, LNAI 10464, pp. 57–68, 2017.
DOI: 10.1007/978-3-319-65298-6_6

Some vibro-tactile stimulation devices are custom-built to generate appropriate stimulations, which can elicit SSSEP [2,10,11]. The combination of computer and audio amplifier is commonly used as the stimulator [3,12,13]. Generally, the stimulation actuators are the electromagnetic transducer [4,12–14], the plastic spherical shell [2,10], and the knock-out pins [9,11]. The commonly used stimulation waveforms are sinusoidal signal [9,15,16], or the combination of a sinusoidal carrier signal with a sinusoidal [2,10,13,17], or rectangular [4,12] modulation signal. The most suitable vibro-tactile stimulator for extracting patterns of SSSEPs is developed by Pokory and colleagues [14], which can generate modulated sinusoidal waveform, rectangular waveform, and other various kinds of waveform patterns. Some vibro-tactile stimulation devices are used in functional magnetic resonance imaging (fMRI) or magnetoencephalography (MEG) studies. For example, a mechanical shaft driven by a direct current motor, and a pneumatically driven inflatable plastic membrance can elicit steady-state responses in somatosensory cortex, during fMRI and MEG experiment respectively [18,19]. These devices mentioned above are all custom-built. The existing commercial stimulation devices, such as C-2 TACTOR (Engineering Acoustics Inc., Casselberry, FL, USA) and g.STIMbox (g.tec medical engineering GmbH, Schiedlberg, Austria) are usually unable to elicit a significant SSSEP [20]. Even though the commercial QS Piezo (QuaeroSys Medical Devices, Schotten, Germany) stimulator can elicit SSSEP, it only drives Braille-like pin matrix transducers [14].

The devices mentioned above only generate tactile stimulation, but we do not know the actual vibratory stimulation signature, such as waveform, amplitude and frequency. In this study, the actual output tactile stimulation signatures and electroencephalography (EEG) signals were recorded synchronously. We could observe the waveform and roughly read the amplitude and the frequency online, so as to monitor the stimulator performance in the real time. By off-line analysis of the actual vibration information and EEG signals, we could know the exact vibration amplitude and phase at any moment. In this condition, it is possible to investigate the signal trial amplitude (or phase) relationship between the vibratory stimulation and the brain response. To satisfy the synchronous requirements, the goal of this work was to develop a flexible, portable and safe multiple channel tactile stimulation as well as vibro-signature synchronization device. The device should produce vibro-tactile stimulation with different amplitude-modulated waveforms to reliably evoke SSSEPs, and should synchronize the actual vibratory stimulation output with the EEG recording. Remarkably, the acquired vibro-signatures and SSSEP signals shared a common clock. Moreover, this device could serve as vibro-tactile stimulator and actual vibro-signature detector during EEG or other similar neurophysiological measurements.

2   Materials and Methods

2.1   Device Architecture

The overall architecture of the multiple channel vibro-tactile stimulation and vibro-signature synchronization device is described in Fig. 1. The whole device

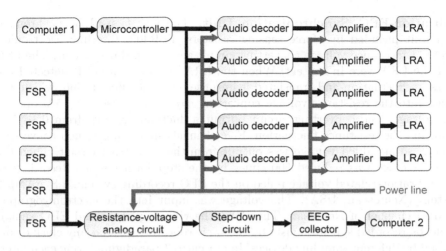

Fig. 1. Schematic diagram of the simultaneous vibro-tactile stimulation and vibro-signature synchronization device in an EEG experiment.

could be divided into two functional units, the stimulation unit and the vibration-EEG synchronization unit, with details as follows.

Stimulation Unit. The stimulation unit consisted of a microcontroller (Arduino mega2560), five audio decoders, five amplifiers and five linear resonant actuators (LRA, 10 mm, C10-100, Precision Microdrives Ltd. Typical Normalized Amplitude 1.4 G). A host computer software (E-Prime, Psychology Software Tools, Inc., USA) generated the paradigm by configuring the stimulation channel, the waveform and the duration. The microcontroller was embedded in the device. It received the real time control command from the computer via universal serial bus, and then sent switch signals to the audio decoders. Each audio decoder was powered with 5 V operating voltage and was equipped with a MicroSD memory card. Four modulated waveforms and a vibration burst waveform were stored in the memory card. Five audio trigger ports and a stop port of each audio decoder could be activated respectively by receiving signals from the microcontroller. When a trigger port of an audio decoder was activated, the corresponding channel worked and one of the predetermined waveforms was generated. The waveform from the audio decoder needed to be amplified strong enough to drive the LRA and reliably elicit SSSEP. Therefore, for each channel, an amplifier powered with 5 V operating voltage received the mentioned waveforms and then used the amplified waveforms to drive the LRA. The custom-designed stimulation pattern, frequency and amplitude in the memory card could be rewritten and the vibration amplitude could also be regulated by adjusting the potentiometer of the amplifier.

Vibration-EEG Synchronization Unit. The vibration-EEG synchronization unit included five force sensing resistors (FSR402, Interlink Electronics, USA), a resistance-voltage converting circuit and a step-down circuit. The FSR and the LRA were in contact. When the LRA vibrated, the FSR detected the pressure fluctuation and translated it into the resistance fluctuation, which was input into the resistance-voltage converting circuit, powered with 5 V operating voltage. The converting circuit output a fluctuant voltage from 0 to 5 V, following the resistance fluctuation. Five channels shared a common converting circuit, so an added voltage was output from the converting circuit. Then the added voltage was reduced 500 times by the step-down circuit. Recording of this vibration-related voltage relied on the EEG recording system, a SynAmps2 system (Neuroscan, USA). The voltage was input into the electromyography (EMG) channel of Neurosan. Therefore, the vibro-signature and EEG signals could be recorded synchronously by the same collector, without any clock errors caused by different sampling devices. In our current paradigm, during each interval for recording, only one vibration channel worked. Consequently, during each trial, the voltage which was input into Neuroscan represented only one vibro-signature rather than a superposition of five vibro-signatures. One to four independent vibro-signatures could be recorded synchronously when necessary, using the same number of resistance-voltage converting circuits, step-down circuits and EMG channels of Neurosan.

2.2   Test

Three EEG experiments were performed to verify the availability of the device. The purpose of the experiment was to test whether the vibro-tactile stimulation and vibro-signature synchronization device can reliably evoke SSSEP and synchronize the vibro-signature with EEG signals.

Subjects. Three healthy subjects participated in the experiment, all male with mean age of 24 years old. Subject s1 is left handed and subject s2 and s3 are right handed. They were informed with the whole experiment process. The experiment was approved by the Ethics Committee of Shanghai Jiao Tong University.

Stimulation Pattern. The stimulation unit stored and generated four different stimulation patterns, as Fig. 2(a) shows: (i) the combination of a sinusoidal carrier signal and an amplitude-modulated sinusoidal signal, named as sinusoidal modulation pattern, (ii) the combination of a sinusoidal carrier signal and an amplitude-modulated rectangular signal whose duty cycle was 50%, named as rectangular modulation pattern, (iii) a sinusoidal carrier signal modulated with a piecewise ascending ramp signal, named as triangular modulation pattern, and (iv) a sinusoidal carrier signal modulated with a piecewise descending ramp signal, named as inversed triangular modulation pattern. For all four stimulation patterns, the carrier frequency was set to be 205 Hz (resonant frequency of LRA) and the modulation frequency was chosen as 27 Hz throughout the experiment. The four waveforms produced by Matlab had the same maximum amplitude.

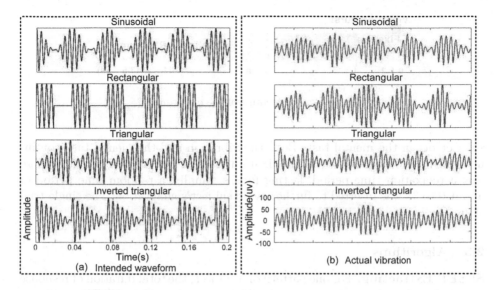

Fig. 2. Four stimulation patterns in 0.2 s. (a) The intended stimulation waveforms generated by Matlab. (b) The actual vibrations of the actuator.

EEG and Vibro-signature Recording. A SynAmps2 system was used to record EEG signals. 62 channel EEG signals were collected using a 64 channel quick-cap, and the electrodes were placed according to the extended 10/20 system. The reference electrode was located on the vertex, and the ground electrode was located on the forehead. An analog bandwidth filter of 0.5 Hz to 70 Hz and a notch filter of 50 Hz were applied to the raw signals. Signals were digitally sampled at 2500 Hz. The vibro-signature detected by LRA was recorded using the EMG channel.

Experimental Paradigm. In this work, one vibro-tactile stimulation was applied to subjects' index finger of the dominant hand, and each subject was required to perform only one mental task, namely stimulation sensation. The LRA and the FSR were tied together and adhered to the subject's index finger. The FSR was sandwiched between the LRA and the index finger. Vibrations were detected by the FSR during the experiment. Each subject sat in a comfortable armchair in an electrically shielded room throughout the experiment, with forearms and hands resting on the armrest. A total of 40 trials were performed in a run. In the first 10 trials, one of the four stimulation patterns presented and kept constant. Then in the second 10 trials, another stimulation pattern presented. And so on, per neighbouring 10 trials shared the same stimulation pattern. The 4 patterns served in a random order. At the beginning of each trial, a fixation cross appeared in the screen center. In the meantime, a vibration burst lasting 200 ms presented to attract subject's attention for the subsequent task. Then at the 2nd second, the vibration was applied to the index finger and the

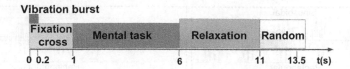

Fig. 3. Trial procedure of the experiment.

subject began the mental task. Both the vibration and the mental task continued until the disappearance of the fixation cross at the 7th second. Next there was a relaxation time period, lasting for 5 s. Finally, a random time period of about 0 to 3 s was inserted to further avoid subject's adaptation. The single-trial structure is presented in Fig. 3.

2.3   Algorithms

SSSEP Extracting. For one pattern of a subject, the total duration of response was divided into 20 sections averagely. Then for each section, SSSEP amplitude at each electrode was computed by means of Fast Fourier transform (FFT). 20 groups of FFT spectra were averaged, and then the general SSSEP amplitudes of each channel were generated.

Artifact Removing. When the LRA and the FSR were fixed on the index finger together, baselines of signals acquired by the FSR showed regular fluctuation, with a cycle of about 1 s. When the LRA and the FSR combined unit was taken off from the finger, the baseline fluctuation disappeared. Therefore, we could infer that this artifact was produced by human pulse. To avoid influence caused by the pulse fluctuation and reproduce the actual vibration signals furthest, the raw signals detected by FSR were processed using a pulse removal algorithm (see Fig. 4). Peaks and troughs of the detected raw vibration signals were lined respectively, as a result, the upper and lower envelope curves were generated. Then the values in the upper and lower envelope curves at the same time point were averaged, so the average value curve was generated. We regarded the average value curve as the baseline, namely the pulse wave. The actual vibration signals produced by the LRA was the raw vibration signals detected by the FSR minus the pulse wave.

Frequency Analysis. For all the four vibration patterns of three subjects, the carrier frequencies of the processed vibration signals were computed, with details as follows. For one pattern of a subject, a 4 s vibration was intercepted from each trial. FFT was applied to the 10 intercepted signals respectively. Then 10 FFT spectra were averaged and a general spectra was generated. To calculate errors of carrier frequency, FFT was also applied to the intended waveforms generated by Matlab. To calculate the modulation frequency, number of main peaks in the 10 intercepted vibrations mentioned above were summed for each pattern

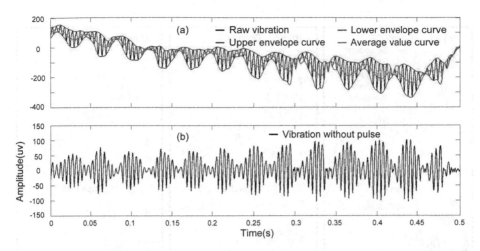

Fig. 4. Signals in pulse removal algorithm. (a) The detected raw vibration signal, together with its lower envelope curve, upper envelope curve and average value curve. (b) The vibration after removal of the pulse wave.

of a subject, and the value divided by 40 s was the modulation frequency. Each pattern of all subjects was processed as this method.

3   Results

3.1   SSSEP

According to the localizations where the maximal SSSEP amplitude appears most probably [5], FCZ and FC2 channels of subject s1 were chosen to observe SSSEP. FCZ and FC1 channels were chosen for subject s2 and s3. For all six channels, maximal SSSEP amplitudes were at 26.55 Hz, very close to the stimulation frequency 27 Hz (see Fig. 5).

3.2   Artifact Removing

After the removal of the pulse artifact, the actual vibration signals from the LRA were reconstructed further, and their baseline basically showed a horizontal linear shape. The actual vibrations of four patterns are shown in Fig. 2(b). Comparing the actual vibrations with the intended waveforms in 0.2 s, it could preliminarily demonstrate that they had extremely similar carrier frequency and modulation frequency.

3.3   Frequency

Figure 6(a–c) shows the FFT spectra of the processed vibration signals among patterns and subjects. Similar phenomenon occurred in the spectra of all

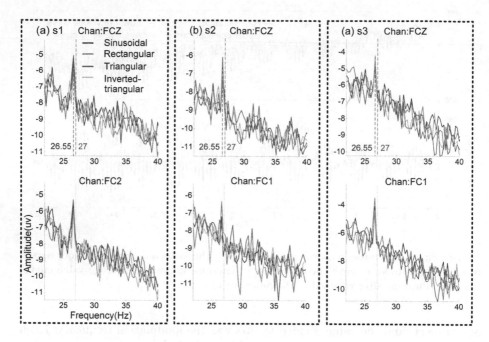

Fig. 5. SSSEP amplitude and frequency of the selected EEG channels. (a) Subject s1. (b) Subject s2. (c) Subject s3.

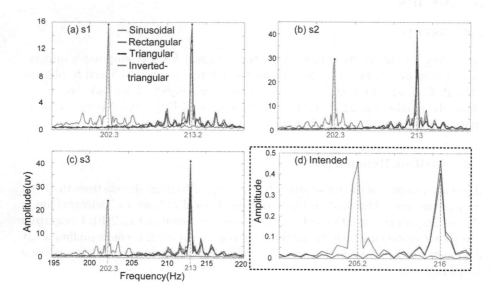

Fig. 6. (a–c) FFT spectra of the processed vibration signals of subject s1, s2 and s3. (d) FFT spectra of the intended waveforms. The spectra of the triangular modulation pattern and the inverted triangular modulation pattern coincide completely.

three subjects. For sinusoidal modulation pattern, the maximum amplitudes of spectra were localized at 202.3 Hz (subject s1, s2 and s3). For other three patterns, the maximum amplitudes were localized at 213.2 Hz (subject s1) and 213 Hz (subject s2 and s3). These frequencies were derived from the carrier frequency of actual vibrations. Figure 6(d) shows the spectra of the intended waveforms. For sinusoidal modulation pattern, the maximum amplitude of spectra was localized at 205.2 Hz. For other three patterns, the maximum amplitudes were localized at 216 Hz. We regarded the frequency corresponding to the maximum spectra amplitude as the dominant frequency to describe the carrier wave. Therefore, for all three subjects, the percentage error of carrier frequency of sinusoidal modulation pattern was 1.41%, and that of other three patterns were 1.29% (subject s1) and 1.39% (subject s2 and s3) respectively. The measured modulation frequencies of each pattern among subjects are shown in Table 1. According to the intended 27 Hz modulation frequency, percentage errors were also calculated.

Table 1. Modulation frequency of actual vibrations among subjects and patterns

Subject	Pattern	Modulation frequency (Hz)	Percentage error (%)
s1	Sinusoidal	26.925	0.28
	Rectangular	26.775	0.83
	Triangular	27	0
	Inverted triangular	27.25	0.93
s2	Sinusoidal	26.875	0.46
	Rectangular	26.775	0.83
	Triangular	27.075	0.28
	Inverted triangular	27.225	0.83
s3	Sinusoidal	26.9	0.37
	Rectangular	26.875	0.46
	Triangular	27	0
	Inverted triangular	27.2	0.74

4   Discussion

The actual vibration signals were reproduced successfully. For all patterns, from the 0.2 s intercepted actual vibrations, the number of cycles in a window could be identified clearly, thus most frequency information was reserved. The four stimulation patterns had different degrees of distortion, compared to the intended waveforms generated by Matlab. For sinusoidal modulation pattern, most amplitude information in temporal domain was preserved, whereas for other three patterns, some amplitude information missed. The distortions mainly appeared at the border between cycles. At these points in time, amplitudes of the intended waveforms changed suddenly, from the maximum to zero or on the contrary. In fact, the actual vibration signals showed transitional oscillations at these cycle borders, instead of sudden changes. An alternative explanation for this kind of distortion was the inertia of coil in the LRA, leading a transitional oscillation between two distant coil positions. Another possible explanation was the hysteresis characteristic of the FSR. In addition, at some time points where the slope of pulse wave was too high, oscillation of the raw signal detected by the FSR was not obvious. As a result, the amplitude of the actual vibration signal which was computed using the pulse removal algorithm was relatively small, compared to amplitude where the slope of pulse wave was low.

The device has served as a stimulator and a detector in EEG measurements. The vibro-tactile stimulation elicited significant EEG changes in three subjects, therefore, the feasibility of the device as a stimulator in SSSEP-based study was confirmed. By means of the EMG channel of Neurosan, synchronization between EEG and the vibro-tactile stimulation was realized. The detected vibration signals were observed online and analysed offline.

In future work, using different actuators and sensors, we would find out where the waveform distortion mainly occurred, which might be meaningful for stimulation pattern design in vibro-tactile experiments. Finer pulse removal methods would be considered so that more stable actual vibration signals would be obtained. Besides, the high level input channel of Neurosan could also serve as the vibration input channel. Obviously, our device could be modified to adapt to other similar brain response measurements, such as electrocorticography (ECoG) and stereo-electroencephalography (sEEG), relying on peripheral signals input channels of the corresponding recording devices. This device would provide conditions for vibro-tactile stimulation related neurophysiological studies, especially could satisfy their requirements of synchronization between the stimulation and the response.

5   Conclusion

We developed a novel device with functions of vibro-tactile stimulation and vibro-signature synchronization. SSSEP could be evoked by the stimulation unit which could generate different kinds of stimulation patterns. The errors between the intended stimulation frequencies and the actual measured frequencies were quite small. Using the same EEG recording system, the actual vibration signature and EEG signals could be recorded synchronously, which established a foundation for precise coupling analysis between the SSSEP and the actual vibro-signature.

Acknowledgments. This work is supported by the National Natural Science Foundation of China (Grant No. 51620105002). The authors thank all the volunteers for their participation in the study.

References

1. Snyder, A.Z.: Steady-state vibration evoked potentials: descriptions of technique and characterization of responses. Electroencephalogr. Clin. Neurophysiol. **84**(3), 257 (1992)
2. Tobimatsu, S., Zhang, Y.M., Kato, M.: Steady-state vibration somatosensory evoked potentials: physiological characteristics and tuning function. Clin. Neurophysiol. **110**(11), 1953–1958 (1999)
3. Mullerputz, G.R., Scherer, R., Neuper, C., Pfurtscheller, G.: Steady-state somatosensory evoked potentials: suitable brain signals for brain-computer interfaces? IEEE Trans. Neural Syst. Rehabil. Eng. **14**(1), 30–37 (2006)
4. Breitwieser, C., Pokorny, C., Neuper, C., Muller-Putz, G.R.: Somatosensory evoked potentials elicited by stimulating two fingers from one hand-usable for BCI?. In: International Conference of the IEEE Engineering in Medicine and Biology Society, pp. 6373–6376 (2011)
5. Pang, C.Y., Mueller, M.M.: Competitive interactions in somatosensory cortex for concurrent vibrotactile stimulation between and within hands. Biol. Psychol. **110**, 91–99 (2015)
6. Kim, K.T., Suk, H.I., Lee, S.W.: Commanding a brain-controlled wheelchair using steady-state somatosensory evoked potentials. IEEE Trans. Neural Syst. Rehabil. Eng. **PP**(99), 1 (2016). A Publication of the IEEE Engineering in Medicine & Biology Society
7. Ahn, S., Jun, S.C.: Feasibility of hybrid BCI using ERD- and SSSEP- BCI. In: International Conference on Control, Automation and Systems, pp. 2053–2056 (2012)
8. Yi, W., Qiu, S., Wang, K., Qi, H., Zhao, X., He, F., Zhou, P., Yang, J., Ming, D.: Enhancing performance of a motor imagery based brain-computer interface by incorporating electrical stimulation-induced sssep. J. Neural Eng. **14**(2), 026002 (2016)
9. Severens, M., Farquhar, J., Duysens, J., Desain, P.: A multi-signature brain-computer interface: use of transient and steady-state responses. J. Neural Eng. **10**(2), 026005 (2013)
10. Tobimatsu, S., Zhang, Y.M., Suga, R., Kato, M.: Differential temporal coding of the vibratory sense in the hand and foot in man. Clin. Neurophysiol. Official J. Int. Fed. Clin. Neurophysiol. **111**(3), 398–404 (2000)
11. Wang, H., Ge, Y., Song, A., Li, B.: The vibro-tactile stimulations experiment to verify the optimal resonance frequency of human's tactile system. In: IEEE International Conference on Information and Automation, pp. 2960–2964 (2015)
12. Breitwieser, C., Kaiser, V., Neuper, C., Muller-Putz, G.R.: Stability and distribution of steady-state somatosensory evoked potentials elicited by vibro-tactile stimulation. Med. Biol. Eng. Comput. **50**(4), 347–357 (2012)
13. Yao, L., Meng, J., Zhang, D., Sheng, X., Zhu, X.: Selective sensation based brain-computer interface via mechanical vibrotactile stimulation. PLoS ONE **8**(6), e64784 (2013)
14. Pokorny, C., Breitwieser, C., Mullerputz, G.R.: A tactile stimulation device for eeg measurements in clinical use. IEEE Trans. Biomed. Circ. Syst. **8**(3), 305 (2013)
15. Giabbiconi, C.M., Trujillo-Barreto, N.J., Gruber, T., Muller, M.M.: Sustained spatial attention to vibration is mediated in primary somatosensory cortex. Neuroimage **35**(1), 255–262 (2007)

16. Adler, J., Giabbiconi, C.M., Muller, M.M.: Shift of attention to the body location of distracters is mediated by perceptual load in sustained somatosensory attention. Biol. Psychol. **81**(2), 77–85 (2009)
17. Lin, Y., Meng, J., Zhang, D., Sheng, X.: Combining motor imagery with selective sensation toward a hybrid-modality BCI. IEEE Trans. Bio-med. Eng. **61**(8), 2304 (2014)
18. Golaszewski, S.M., Siedentopf, C.M., Baldauf, E., Koppelstaetter, F., Eisner, W., Unterrainer, J., Guendisch, G.M., Mottaghy, F.M., Felber, S.R.: Functional magnetic resonance imaging of the human sensorimotor cortex using a novel vibrotactile stimulator. Neuroimage **17**(1), 421–430 (2002)
19. Nangini, C., Ross, B., Tam, F., Graham, S.J.: Magnetoencephalographic study of vibrotactile evoked transient and steady-state responses in human somatosensory cortex. Neuroimage **33**(1), 252–262 (2006)
20. Ahn, S., Kim, K., Jun, S.C.: Steady-state somatosensory evoked potential for brain-computer interface-present and future. Front. Hum. Neurosci. **9**, 716 (2014)

Improved Indoor Positioning System Using BLE Beacons and a Compensated Gyroscope Sensor

Jae Heo and Younggoo Kwon[✉]

Department of Electronics Engineering, Konkuk University,
120 Neungdong-Ro, Gwangjin-Gu, Seoul, South Korea
{gjwo35,ygkwon}@konkuk.ac.kr

Abstract. When it comes to indoor positioning system (IPS) using smartphones, most of pedestrian dead reckoning (PDR) based IPS systems rely on the embedded sensors in a smartphone. Unfortunately, those sensors cannot avoid having noisy data due to their poor performance. Unpredictable indoor environments and a user's motion may also distort sensor data. For these reasons, an error-compensating algorithm is required. In this paper, we propose a compensated gyroscope sensor algorithm to clear away those noises. Also, we used Bluetooth-Low-Energy (BLE) beacons based positioning system. It aims for reducing the compensated gyroscope sensor's cumulative error as well as providing precise user's initial point. Finally, we propose an integrated system making cooperative relationship between BLE beacons and sensors, especially compensated gyroscope sensor. This paper also includes implementations in the hallway of a building to describe the effectiveness by showing error rates.

Keywords: Indoor positing system (IPS) · Bluetooth-low-energy (BLE) beacons · Pedestrian dead reckoning (PDR) · A compensated gyroscope sensor

1 Introduction

Most of IPSs based on PDR depend on the magnetic sensor to determine its heading orientation. The unpredictable indoor surroundings such as electronic gadgets result in the uncertainty of the magnetic sensor's data. To reduce errors caused by the magnetic sensor, researchers propose the using the combined form of magnetic sensor and gyroscope sensor at the same time [1, 2, 4]. To provide more precision, other extra systems are added. Xin Li, Jian Wang [1] proposed a fused form of PDR and BLE IPS. In this paper, it used BLE beacons for correcting PDR's errors. They provide quite reliable trajectories compared to what is not combined with. Especially, greatly magnetically unstable areas restrict using the combination of magnetic and gyroscope sensor for heading orientation. In this paper, the idea starts from the fact that gyroscope sensor offers temporarily robust angular values regardless of external surroundings. However, it has lethal drift problem in its sensory values due to the initial bias [5]. The user's trembling motions also make the drift problem worse [4]. To cope with those errors, we propose a compensated gyroscope sensor algorithm in the Sect. 3.

© Springer International Publishing AG 2017
Y. Huang et al. (Eds.): ICIRA 2017, Part III, LNAI 10464, pp. 69–76, 2017.
DOI: 10.1007/978-3-319-65298-6_7

This algorithm aims for clearing away noises caused by user's trembling motion. In this way, we can reduce the noises greatly and cope with the gyroscope sensor's drift problem temporarily.

In wireless networks, a received signal strength (RSS) fingerprint based system can be used for estimating a user's location. As indoor surroundings are well-equipped with Wi-Fi access point (AP), a number of IPSs use them for correcting location error. However, it does not provide a precise location information the place where we need for. For this reason, BLE beacons can be built in the location we want. Pavel and Filip [3] used 17 BLE beacons for IPS. They provide considerably precise location information. But such a system costs a lot for the actual usage. It also requires learning data process to create a radio-map. In this paper, we used a path-loss model of RSS in the Sect. 3.2. This model lessens the burden of making radio maps. This paper aims for a simple and inexpensive system for the actual service.

We propose a solution by using few number of BLE beacons and a compensated gyroscope sensor. we propose BLE beacon's role-estimating an initial user's location, compensating errors caused by gyroscope sensor. BLE beacons are full of help for correcting distorted errors. We also propose a compensated gyroscope sensor's role-it provides a wide range of localization coverage resulting in the less number of infrastructures such as BLE beacons. Finally, BLE beacons and a compensated gyroscope sensor form a cooperative relationship in compensating each other.

2 Problem Statements

To apply gyroscope sensor for utilizing as PDR's data, several following problems must be dealt with. In this Section, we depict the problem of gyroscope sensor: an initial biased error δ_{IE} causes drift error [5], trembling errors δ_{TE} caused by a user's movement [4].

Ideally, the current device's rotation angle is calculated by the integral of angular velocity w_t (1). The device angle a_t denotes how much a device has rotated theoretically from the beginning to the time t (1). The device's angular velocity w_t implies the errors: $\delta_{IE,t}$, $\delta_{TE,t}$ (2). Each error differs during the time t.

$$Device\,Angle\,a_t = \int_0^t w_t\,dt \tag{1}$$

$$w_t = \left(\omega_t + \delta_{IE,t} + \delta_{TE,t}\right) \tag{2}$$

The gyroscope errors δ_{IE}, δ_{TE} results in the drift of heading estimation in the user's application (see Fig. 5.'s blue line). The initial biased error δ_{IE} varies depending on the user's device-embedded sensor and it is noisy during the time t. The gyroscope sensor is vulnerable to the trembling of a user's motion. The more a user tends to move, the more motion error δ_{TE} will be accumulated rapidly (see Fig. 1). As a result, the linear and continuous error accumulate as in Fig. 1 [5]. It means it is required to compensate those errors to estimate a user's heading orientation precisely. In this paper, we propose

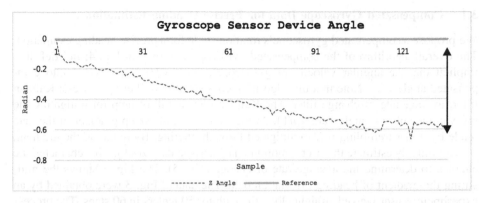

Fig. 1. Gyroscope's cumulative error angle when a user is in motion. The user walked through a hallway of 50 m for 60 steps in a minute.

a temporarily reliable compensated gyroscope data for heading orientation. The specific algorithm for it is also discussed in the Sect. 3-Proposed Algorithm.

3 Proposed Algorithm

In the below overall structure, The overall algorithm starts from the process of judging RSSI. The device constantly calculates RSSI inputs to find out whether a user is in BLE-built zone or not.

The Sect. 3.1 deals with the algorithm of none BLE-built zone using compensated gyroscope algorithm. The Sect. 3.2 will deals with the BLE-built zone algorithm.

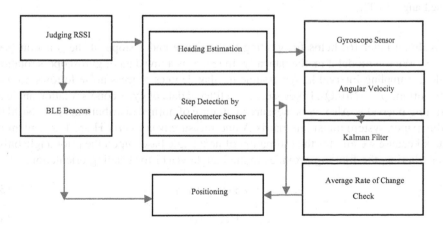

Fig. 2. The proposed algorithm architecture.

3.1   Compensated Gyroscope Data for Precise Heading Estimation

We propose a compensated gyroscope's rotation angle for a precise heading estimation. The overall algorithm of the compensated gyroscope is shown in Fig. 4; In brief, it is implicit that the angular velocity of gyroscope sensor $w_i(t)$ is converted into compensated angle $d_i(t)$. Note that this algorithm consists of two filtering process: Kalman filter, average rate of change filter. The Kalman filter is full of help for mitigating the big variance of raw angle data $\alpha(nT)$. At the same time, the steep gradient of the data made by user's trembling is also mitigated through filtering. By reducing the gradient, we can roughly estimate the user's motion characteristic data $\hat{\alpha}(nT)$. This characteristic enables to determine the average rate of change δ in (5). The Fig. 3 shows the mitigating the gradient of heading orientation data. The data of Fig. 3 were obtained by an experiment; a user walked straight along the path for 50 meters in 60 steps. The process of Fig. 3 is the record of angles during the experiment.

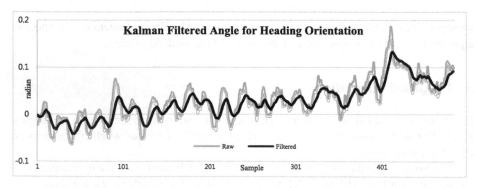

Fig. 3. The Kalman filter mitigates the slope of gyroscope sensor's angular data $\alpha(nT)$ into refined angle $\hat{\alpha}(nT)$.

Kalman Filter [6] helps for grading the variance and a slope of the function (see Fig. 3). We can model it mathematically. In (3), α_k is a raw data of gyroscope's rotation angle at sampling interval k. We assume ω_k that denotes process noise follows normal distribution, $\omega \sim N(0,Q)$. Likewise, $\hat{\alpha}_k$ is a filtered data of gyroscope's rotation angle at sampling interval k. Also, v_k is measurement noise of normal distribution, $v_k \sim N(0,U)$. In this paper, system transition matrix A and measurement matrix H are 1×1 identity matrix because we aim for the estimation of next angle based upon the prior angle only. After filtering, the filtering provides refined angle $\hat{\alpha}(nT)$ for heading orientation.

$$\alpha_k = A\alpha_{k-1} + \omega_{k-1} \tag{3}$$

$$\hat{\alpha}_k = H\alpha_k + v_{k-1} \tag{4}$$

The average rate of change δ can be calculated through followings (5): n denotes sampling period of supervised time over experiments, T is the sampling interval.

We continuously compare our Kalman filtered gyroscope sensor data $\hat{\alpha}_n$ with the criterion δ during the Average Rate of Change Filter.

$$Average\ Rate\ of\ Change\ \delta = \frac{\sum_n (\hat{\alpha}_n - \hat{\alpha}_{n-1}) * T}{n} \tag{5}$$

After average rate of change filter, only of the filtered angle $\hat{\alpha}(nT)$ that is bigger than δ will be selected for current heading orientation. These filtered angles are regarded as actual user's motion. Otherwise, it will be regarded as noise caused by user's trembling motion. Finally, we get $d(nT)$ which is current heading orientation. $d(nT)$ is determined by the summation of the previous heading orientation $d((n-1)T)$ and the current Average Rate of Change Filter value $\hat{\alpha}(nT)$.

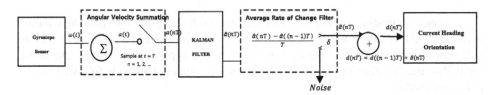

Fig. 4. The algorithm structure the compensated gyroscope sensor

Now we can move our compensated gyroscope sensor onto actual experiment. The Fig. 5 shows a sample trajectory of each condition: a compensated gyroscope, raw data of gyroscope, reference data. In this experiment, the user walked along the path of 100 meters from the west wing of the building. Each PDR steps are marked as small dots; the user's actual trajectory is lined black. The continual fluctuating characteristics of raw gyroscope angle data results in a distorted trajectory, 21.2 m error. In contrast to it, the compensated gyroscope provides a better performance, 1.12 m error. The experiment shows that it is able to clear away noises caused by the user's trembling motion.

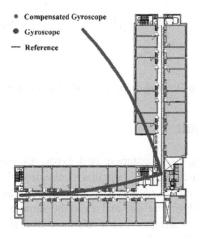

Fig. 5. The trajectory comparison between compensated gyroscope and a normal gyroscope in the building (Color figure online)

3.2 Localization Using BLE Beacons

RSS-Based Localization

In wireless networks, the signal transmitting through the medium is not intact at all from the point of receiver. When transmitting signal from BLE beacon, the received signal strength (RSS) gradually reduces. To make use of this fact, a user's current location can be estimated by calculating RSS from multiple beacons (see Fig. 6). Mostly, this attenuation is an exponential function of the distance. Each parameter of (6) denotes followings: P is the RSS measurement in decibels,P_0 is the RSS measurement at the distance d from an access point (node) is d_0, μ is the path loss constant.

$$P = P_0 + \mu dog_{10} \frac{d}{d_0} \qquad (6)$$

μ varies depending on the beacons and the built location. The path loss model parameters can be automatically calculated by the user's device. The node in proximity will send the stronger signal because of path loss. To apply multiple node calculation for RSS-based localization, maximum likelihood estimation (MLE) is followed by the path loss model.

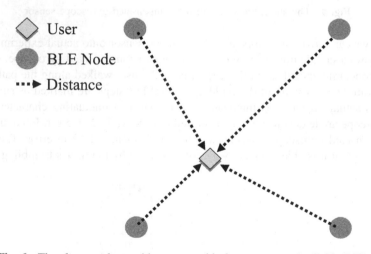

Fig. 6. The closer node provides a user with the stronger and reliable RSSs

4 Performance Result

In Fig. 7 shows the comparison compensated gyroscope with collaboration of two skills: a compensated gyroscope, BLE beacons localization. In both of figure of Fig. 7, a user walked the reference path of 200 meters along the black line. The user departs from the west wing of the building and arrives at the same point after 200 meters. The arrow mark of black line means the user's reference direction. Each PDR steps were marked as a red dot. In the right figure, green dots were added. They mean the

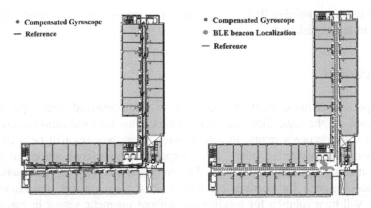

Fig. 7. The left figure used only compensated gyroscope while the right one used both of compensated gyroscope and BLE beacons. (Color figure online)

localization points using BLE beacons. Each green dot is marked as soon as they are calculated from the user's device.

In the left figure, the compensated gyroscope provides 3.26 m of positioning error during 300 steps. The compensated gyroscope prevents trembling noise by the average rate of change filter. So, it provides a stable trajectory in the hallway while the errors occur at the corner of the building. When a user turns at the corner, there is a possibility that gyroscope sensor may not calculate precise rotation angle [7]. As a result, it is required to compensated these errors occurring at the turning point.

In the right figure, we built five BLE beacons at the corner to compensate the heading orientation error of the compensated gyroscope. When a user approaches at the BLE-built zone, PDR stops and BLE beacons localization starts. At this point, the device calculates the time when a user arrives at the edge of BLE-built zone. When the user arrives at the end of zone, the magnetic sensor provides the initial orientation. After getting the initial orientation, the heading estimation starts again by the compensated gyroscope. The total average positioning error was 1.45 m. As the compensated gyroscope provides temporarily stable heading orientation, its heading orientation also needs to be compensated by BLE beacons (See Fig. 7's left).

In Table 1, the average error rate of each component for this system is shown. We confirm that BLE localization itself provides a reliable error rate enough to utilize as a landmark for compensated gyroscope sensor. The compensated gyroscope sensor obviously copes with the drift problem of gyroscope sensor itself in the more extended

Table 1. The average error of indoor positioning in the building depending on each condition

	BLE Localization (In 6 × 3 m space)	Gyroscope (100 m path)	Compensated Gyroscope (200 m path)	Compensated Gyroscope + BLE (200 m path)
Average Error	1.12 m	21.2 m	3.26 m	1.45 m

path. As a result, combining BLE and compensated gyroscope sensor results in 1.45 m error rate in the experiments.

5 Conclusion

In this paper, we demonstrate the performance of a compensated gyroscope algorithm to clear noise. At the same time, we used BLE beacons for localization to correct the compensated gyroscope's errors. In actual usage, these beacons will be used for initializing a user's positioning point. The compensated gyroscope algorithm provides robust positioning value in the hallway. The current technology uses magnetic sensor and gyroscope sensor for heading estimation [2, 4, 7]. However, the compensated gyroscope will be a solution for localization without magnetic sensor in magnetically unstable areas.

The current compensated gyroscope sensor copes with the drift problem of gyroscope sensor temporarily. Another algorithm should be added to it for a more robust system. The future work will be applying particle filter in addition to the compensated gyroscope sensor. The idea of particle filter based on 'Map-matching' can be applied to this problem. If the positioning particle gets out of the map, we can regard it as low-weighted particle and get more reliable particles for correcting a user's position.

Acknowledgement. This research was supported by Basic Science Research Program through the National Research Foundation of Korea (NRF) funded by the Ministry of Science, ICT & Future Planning (NRF-2016R1A2B1015032)

References

1. Li, X., Wang, J., Liu, C.: A Bluetooth/PDR integration algorithm for an indoor positioning system. Sensors (Basel) **15**, 24862–24885 (2015)
2. Sabatini, A.M.: Quaternion-based extended kalman filter for determining orientation by inertial and magnetic sensing. IEEE Trans. Biomed. Eng. **53**, 1346–1356 (2006). In IEEE International Conference Systems (2011)
3. Kriz, P., Maly, F., Kozel, T.: Improving Indoor Localization Using Bluetooth. Hindawi Publishing Corporation Mobile Information Systems (2016)
4. Kang, W., Nam, S., Han, Y., Lee, S.: Improved heading estimation for smartphone-based indoor positioning systems. In: IEEE 23rd International Symposium on Personal, Indoor and Mobile Radio Communications (2012)
5. Barthold, C., Subbu, K.P., Dantu, R.: Evaluation of gyroscope-embedded mobile phones. In: IEEE International Conference Systems (2011)
6. Simon, D.: Optimal State Estimation: Kalman, H Infinity, and Nonlinear, 2nd edn. Wiley, Hoboken (2006)
7. Kang, W., Han, Y.: SmartPDR: smartphone-based pedestrian dead. IEEE Sensors J. (2012)

Stretchable sEMG Electrodes Conformally Laminated on Skin for Continuous Electrophysiological Monitoring

Wentao Dong[1,2], Chen Zhu[1,2], Youhua Wang[1,2], Lin Xiao[1,2],
Dong Ye[1,2], and YongAn Huang[1,2(✉)]

[1] State Key Laboratory of Digital Manufacturing Equipment and Technology,
Huazhong University of Science and Technology, Wuhan 430074, China
yahuang@hust.edu.cn
[2] Flexible Electronics Research Center,
Huazhong University of Science and Technology, Wuhan 430074, China

Abstract. Current electrophysiological monitoring is based on invasive electrodes or surface electrodes. Here, a surface electromyography (sEMG) electrode with self-similar serpentine configuration is designed to monitor biological signal. Such electrode can bear rather large deformation (such as >30%) under an appropriate areal coverage. And the electrode conformally attached on the skin surface via van del Waal interaction could furthest reduce the motion artifacts from the motion of skin. The capacitive electrodes that isolates the electrodes from the body also provide an effective way to minimize the leakage current. The sEMG electrodes have been used to record physiological signals from different parts of the body with sharp curvature, such as index finger, back neck and face, and they exhibit great potential in application of human-machine interface in the fields of robots and healthcare. Integrating wireless data transmission capabilities into the wearable sEMG electrodes would be studied in future for intelligent could healthcare platform.

Keywords: Stretchable capacitive electrode · Stretchability · Conformability · Biological signal monitoring

1 Introduction

Surface electromyography (sEMG) is a non-invasive technique for measuring electrical signals generated by the contraction of skeletal muscles [1, 2]. sEMG electrode that involves sensors placed on the skin, is exploited for many clinical and research purposes, ranging from diagnosing neuromuscular disorders, studying muscle pain [3, 4], and controlling prosthetic or orthotic devices [5–9]. The sEMG electrodes laminated onto the skin surface should satisfy large deformations and conformally contact with the skin surface for continuous electrophysiological (EP) monitoring.

W. Dong and C. Zhu contributed equally to this work.

Y. Huang et al. (Eds.): ICIRA 2017, Part III, LNAI 10464, pp. 77–86, 2017.
DOI: 10.1007/978-3-319-65298-6_8

Conventional capacitive sensors with rigid electrodes affixed on the skin with adhesive tapes for biological signal monitoring which limits mounting locations to large, relatively flat regions of the body, such as wrist and chest [10]. It also causes noise in the biopotential measurement as the motion artifacts due to the relative slippage at electrode-skin interface [1, 11]. Recently, epidermal electronic system with ultrathin, low-modulus, lightweight, stretchable substrate that conformably laminates onto the skin surface has been designed to overcome such problems [11–14]. And tremendous affords are developed to achieving stretchability for biological monitoring via designing proper materials and structures [12, 15]. Multifunctional epidermal electronics system conformally contact with the skin surface has also been proposed to record physiological signals, such as, temperature, pulse beating, strain, and sEMG, electrocardiograms (ECG) [16]. It is considered that an optimal parameters of the filamentary serpentine structure can significantly improve the performance of the sEMG electrode [12, 17]. sEMG electrode with self-similar serpentine interconnect is designed to satisfy larger deformation for the motion of skin as an EP sensor [18, 19]. However, beyond the deformation, the conformability of the capacitive electrode should be taken into consideration for the biological signals monitoring in the parts of the body with sharp curvature.

The capacitive electrode with self-similar serpentine structure is designed to monitor the biological signals, which could satisfy large deformability (>30%) and high areal coverage. The design of sEMG includes three parts: stretchability, conformability, and capacitive epidermal sensing system. The stretchability is analyzed by experiments and finite elements modelling (FEM), and the conformability is measured by the interfacial peeling experiment. The capacitive sensing electrodes conformally contact with the skin surface via van der Waals interaction alone. The capacitive electrodes are used to record physiological signals from different parts of human body with sharp curvature, such as index finger, back neck and face. It is also used to record the ECG signal from the chest.

2   Design and Fabrication Process

2.1   Principle of Capacitive Sensing Electrodes

Figure 1(a) shows the optical image of the capacitive electrode conformally laminated onto the skin surface. The cross-section of sEMG electrode is shown in Fig. 1(b) with multi-layers. The sEMG electrode with the PDMS/PI/Au/PI multilayer structure guarantees the Au layer at the neutral plane of the sEMG for improving the stretchability of the sEMG electrodes. Figure 1(c) depicts the equal electrical model of capacitive electrode with a voltage follower. Higher signal gains are realized with higher input impedance and low-input capacitance to improve the signal quality of the sEMG electrodes. In the skin-electrode interface with PI dielectric layer, the capacitance is represented as $C_{in} = \varepsilon S_{effective}/h_{PI} = \varepsilon \alpha S_{sEMG}/h_{PI}$. S_{sEMG} is the area of the electrode, α is the areal coverage of the electrode, h_{PI} is the thickness of the dielectric layer PI. R_B is a bias resistance between the ground and the input of the operational amplifier (op amp). R_{in} and C_{in} are the resistance and capacitance of the voltage

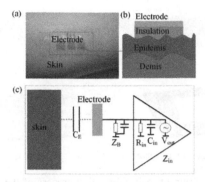

Fig. 1. (a) Photograph of the capacitive electrode; (b) Schematic graph of the electrode-skin interface; (c) The equal electrical model of capacitive electrode with a voltage follower.

follower. The resistance and capacitance circuit is used to depict the skin-electrode interface property. Both the stratum corneum (SC) and insulation layers are equivalent to a RC parallel component and therefore the transducing function are calculated by the equivalent skin/electrode model.

2.2 Fabrication

Figure 2 shows the main fabrication process of the sEMG electrode with mutli-layer structure. Figure 2(a) depicts the sEMG electrode is fabricated by microelectrome-chanical system (MEMS) techniques on a carrier wafer with polymethyl methacrylate (PMMA) sacrificial layer. The PDMS substrate on the glass substrate is used to pick up the sEMG electrode after the PMMA sacrificial layer is dissolved in acetone bathing shown in Fig. 2(b). The whole structure (PI/Au/PI) is patterned in stretchable form, released from the carrier Si wafer and transferred to the PDMS substrate. Figure 2(c) depicts the sEMG electrode is delaminated from the glass substrate with help of the polyvinyl alcohol (PVA) film which reduces the interfacial energy at the electrode/glass interface. Figure 2(d) shows the sEMG electrode is directly printed onto the surface of the skin. The electrode adheres onto the surface conformally after dissolving the PVA film in the water for sEMG signal recordings.

The electrical performance of the sEMG electrode mainly depends on material selection and geometry layout. Second order serpentine structure is adopted to improve the areal coverage and stretchability compared to the first order serpentine structure. The sEMG electrode with self-similar configuration is designed to monitor the EP signals for satisfying larger stretchability (>30%). The sEMG electrode with ultrathin construction and neutral mechanical plane configuration improves the stretchability and contact behavior. Figure 3(a) depicts optical graph of the sEMG designed to quantify biological signal measurement with reference, ground, and measurement electrodes with self-similar serpentine structure. The node connection of the unit cells in the network sEMG electrode forms triangular lattices. The line width of electrode is 50 μm. Figure 3(b) shows the local enlarged graph of the network sEMG electrode

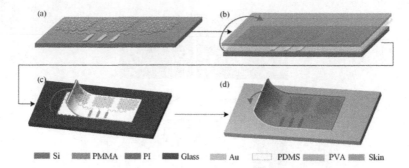

Fig. 2. Schematic graph of main fabrication process of the sEMG electrode

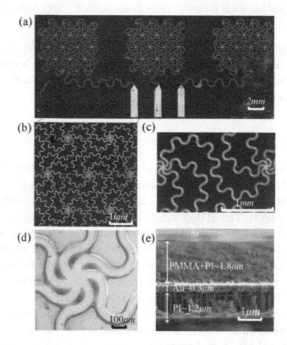

Fig. 3. (a) Optical image of the capacitive EMG electrode; (b) Local graph of the electrode with one regular hexagon formed by self-similar second order serpentine structure; (c) Optical image of the self-similar serpentine structure; (d) Enlarged image of electrode with one node; (e) SEM image of cross section of the electrode with multi-layers structure onto the carrier wafer.

with hexagon structure with self-similar interconnect. Figure 3(c) depicts optical image of the self-similar serpentine structure. Figure 3(d) depicts the local enlarged graph of the electrode with one node. The SEM image of cross section of the sEMG electrode with PI/Au/PI layers is shown in Fig. 3(e) where PMMA/PI ($\sim$1.8 µm), Au ($\sim$0.3 µm), and PI ($\sim$1.2 µm). Capacitive sEMG electrode is designed to measure the EP signals between the electrode-skin interfaces with PI polymer dielectric layer.

3 Mechanical and Electrical Performances

3.1 Conformability

Conformal contact at the electrode-skin interface can improve the accuracy of EP signal measurement. It reduces the motion artifact due to the movements of skin. The sEMG electrode with self-similar structure adheres onto the surface of the skin driven only by van der Waals interactions. The sEMG electrode mounts in this way exhibit excellent compliance and ability to follow skin motions without constraint or delamination, which satisfies large deformation of the skin. The total energy of the epidermal device adhered onto skin consists of bending energy and membrane energy of sEMG electrode, elastic energy of the skin and the device-skin interfacial adhesion energy. Figure 4(a) and (b) show the device adhered onto the surface of the skin in the compressed and stretched form respectively. Figure 4(c) depicts the sEMG electrode is delaminated from the surface of the skin for recycling and reusing the electrodes. The sEMG electrode is in the twisted form shown in Fig. 4(d). It shows that the sEMG electrode with low bending stiffness (thin, soft electrode/backing layer), smooth and soft skin, and strong adhesion all promote conformal contact.

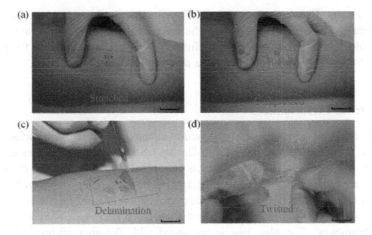

Fig. 4. Contact behavior at the sEMG electrode/skin interface in the (a) stretched; (b) compressed; (c) delaminated; and (d) twisted form. The scale bar is 1 cm.

3.2 Stretchability Analysis

The sEMG electrode with self-similar configuration satisfies large deformation of the skin surface. The stretchability of the sEMG electrode is evaluated quantitatively by FEM using the commercial package Abaqus 6.10. Figure 5(a) shows the 40% stretched state of the sEMG electrode by FEM simulation. The optical image of the electrode in the stretched format are shown in Fig. 5(b). Comparing the node parts Fig. 5(a1) and (b1), the deformation format of the optical image of the electrode is similar the FEM

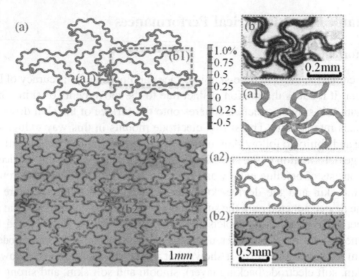

Fig. 5. Stretchability analysis of the capacitive electrode with triangular lattice up to 40% by (a) FEM and (b) experiments.

simulation results. Figure 5(a2) and (b2) show the distribution of maximum principal strain appear at the largest curvature of the microstructure. It is seen that the experimental results agree well with the FEM simulation results, which shows the patterned electrode structure satisfies 40% stretched deformation.

3.3 Electrical Performance

The sEMG electrode with self-similar structure laminated onto the skin surface forms a capacitance with PI dielectric layer [20]. The stretchable electrode reduces the contact impedance as it contact with skin surface conformally to increase the contact area. Capacitive electrodes with PI insulating layers exhibit higher stable gain over the frequency range 20–10000 Hz, due primarily to their relatively high CE. Figure 6(a) depicts the skin-electrode interface contact impedance decreases with the increasing scanning frequency. The blue line is measured with the area 40 mm^2. Figure 6(b) depicts the leakage current from the capacitive electrodes is rather small compared to the conventional medical biological electrodes. It shows that the currents below 0.2 mA when 10 mA passes through the electrodes. It is helpful to protect the soft biological tissues during biological signal monitoring.

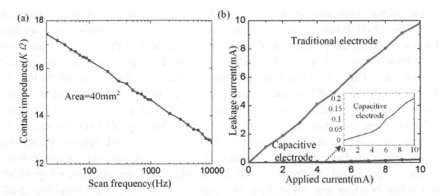

Fig. 6. Electrical parameters analysis of the capacitive EMG electrode. (a) Contact impedance; (b) The leakage current. (Color figure online)

4 Biological Signal Recordings

The electromyography signals are measured by the capacitive sEMG electrode adhered onto the skin surface. To minimize the motion artifact during movement, the electrodes should maintain conformal contact with the skin in case that external load applied to the epidermal biological electrode. A customized PCB is used to collect the sEMG data from human body. It yields analog data, converted to digital signals for recording using Lab-View software (National Instruments, USA) and analysis using Matlab (The Mathworks, Inc., Natick, MA). Photograph of the electrode on the wrist appears with different actions in Fig. 7(a) clenched; (b) half clenched; (c) loosened. Enlarged photograph of the

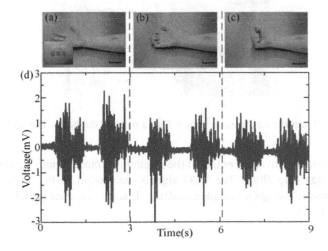

Fig. 7. The sEMG signals with the capacitive electrodes with different actions. The fist is in (a) loosened; (b) half clenched; (c) clenched state; and (d) The corresponding sEMG signal. The scale bar is 2 cm.

electrode on the wrist appears in Fig. 7(a). The corresponding biological signals are shown in Fig. 7(d). It has potential in clinics for diagnosis of neuromuscular disorders.

Figure 8 shows the physiological signals measurement of the capacitive electrode from different parts of human body, such as wrist, index finger and face with different motion of body. Capacitive electrode is even wrapped onto the finger, where bending creates high-quality sEMG signals, in spite of the relatively small area of the corresponding muscle Fig. 8(a) illustrates the stretchable capacitive electrodes for measuring sEMG signals on the forehead. The corresponding sEMG signals from the stretchable capacitive electrode are shown in Fig. 8(b) with frown and blink actions. Figure 8(c) shows the electrode laminated onto the index finger, as an example of a location where conventional gel-based electrode is not acceptable. And the corresponding sEMG signals when bending and unbending the finger in Fig. 8(d). The capacitive sEMG electrode laminated onto the surface of face shown in Fig. 8(e). Figure 8(f) depicts clenching the jaw, smiling, and moving the mouth create different types of sEMG signals, each of which is clearly distinguishable from the baseline noise. The biological signals from the parts are clearly distinguishable from the baseline noise, which demonstrated in these experiments suggests many possibilities in human-machine interface (HMI).

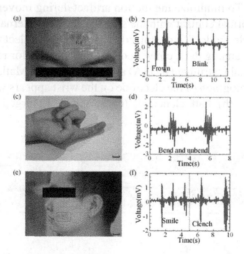

Fig. 8. Biological signal measurement from different parts of the body. The scale bar is 2 cm.

The measurements involve contact of the electrode laminated onto the chest for ECG signals recordings (Fig. 9(a)). The ECG signals are plotted in Fig. 9(b). Figure 9(c) shows the P wave, the QRS complex and the T wave are clearly defined from the capacitive electrode.

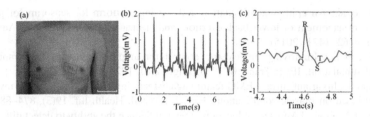

Fig. 9. ECG recordings from the electrodes. The scale bar is 5 cm.

5 Conclusion

The work reported here illustrates advantages in EP measurements that follow from the stretchable capacitive electrode with self-similar serpentine structure, which could satisfy larger deformation (>30%) validated by FEM simulation and experiments. The resulting electrodes offer enhanced levels of wearability, expanded options in electrodes sterilization and reuse, and minimized artifacts from body motions compared to previously reported technologies. The sEMG electrode with thin, soft substrate is conformal contact with the surface of the skin via van del Waal interaction alone, and it reduced the motion artifacts caused by the motion of the skin. The EP measurements from different parts of body were gotten by the capacitive sEMG electrodes. Exploring these options biological signal monitoring and human machine interface capabilities into the capacitive sEMG electrode represents promising directions for future research in intelligent cloud healthcare.

Acknowledgement. The authors acknowledge supports from the National Natural Science Foundation of China (51322507, 51635007).

References

1. Sun, Y., Yu, X.B.: Capacitive biopotential measurement for electrophysiological signal acquisition: a review. IEEE Sens. J. **16**(9), 2832–2853 (2016)
2. Lopez, A., Richardson, P.C.: Capacitive electrocardiographic and bioelectric electrodes. IEEE Trans. Biomed. Eng. **16**(1), 99 (1969)
3. Liao, L.D., et al.: Design, fabrication and experimental validation of a novel dry-contact sensor for measuring electroencephalography signals without skin preparation. Sensors (Basel) **11**(6), 5819–5834 (2011)
4. Al-Ajam, Y., et al.: The use of a bone-anchored device as a hard-wired conduit for transmitting EMG signals from implanted muscle electrodes. IEEE Trans. Biomed. Eng. **60** (6), 1654–1659 (2013)
5. Jeong, J.W., et al.: Materials and optimized designs for human-machine interfaces via epidermal electronics. Adv. Mater. **25**(47), 6839–6846 (2013)
6. Lim, S., et al.: Transparent and stretchable interactive human machine interface based on patterned graphene heterostructures. Adv. Funct. Mater. **25**(3), 375–383 (2015)

7. Xu, B., et al.: An epidermal stimulation and sensing platform for sensorimotor prosthetic control, management of lower back exertion, and electrical muscle activation. Adv. Mater. **28**(22), 4462–4471 (2015)

8. Pan, L., et al.: Improving myoelectric control for amputees through transcranial direct current stimulation. IEEE Trans. Biomed. Eng. **62**(8), 1927–1936 (2015)

9. He, J., et al.: Invariant surface EMG feature against varying contraction level for myoelectric control based on muscle coordination. IEEE J. Biomed. Health Inf. **19**(3), 874–882 (2015)

10. Mercer, J.A., et al.: EMG sensor location: does it influence the ability to detect differences in muscle contraction conditions? J. Electromyogr. Kinesiol. **16**(2), 198–204 (2006)

11. Jeong, J.W., et al.: Capacitive epidermal electronics for electrically safe, long-term electrophysiological measurements. Adv. Healthc. Mater. **3**(5), 642–648 (2014)

12. Kim, D.H., et al.: Epidermal electronics. Science **333**(6044), 838–843 (2011)

13. Kim, J., et al.: Next-generation flexible neural and cardiac electrode arrays. Biomed. Eng. Lett. **4**(2), 95–108 (2014)

14. Huang, X., et al.: Epidermal impedance sensing sheets for precision hydration assessment and spatial mapping. IEEE Trans. Biomed. Eng. **60**(10), 2848–2857 (2013)

15. Hammock, M.L., et al.: 25th anniversary article: the evolution of electronic skin (e-skin): a brief history, design considerations, and recent progress. Adv. Mater. **25**(42), 5997–6038 (2013)

16. Yeo, W.H., et al.: Multifunctional epidermal electronics printed directly onto the skin. Adv. Mater. **25**(20), 2773–2778 (2013)

17. Viventi, J., et al.: A conformal, bio-interfaced class of silicon electronics for mapping cardiac electrophysiology. Sci. Trans. Med. **2**(24), 24ra22 (2010)

18. Jang, K.I., et al.: Soft network composite materials with deterministic and bio-inspired designs. Nat. Commun. **6**, 6566 (2015)

19. Fan, J.A., et al.: Fractal design concepts for stretchable electronics. Nat. Commun. **5**(2), 3266 (2014)

20. Wang, L.-F., et al.: PDMS-based low cost flexible dry electrode for long-term EEG measurement. IEEE Sens. J. **12**(9), 2898–2904 (2012)

Isotropy Analysis of a Stiffness Decoupling 8/4-4 Parallel Force Sensing Mechanism

Jiantao Yao[1,2(✉)], Danlin Wang[1], Xueyan Lin[1], Hong Zhang[1],
Yundou Xu[1,2], and Yongsheng Zhao[1,2]

[1] Hebei Provincial Key Laboratory of Parallel Robot and Mechatronic System,
Yanshan University, Qinhuangdao 066004, China
jtyao@ysu.edu.cn
[2] Key Laboratory of Advanced Forging and Stamping Technology
and Science, (Yanshan University), Ministry of Education of China,
Qinhuangdao 066004, China

Abstract. A stiffness decoupling 8/4-4 parallel force sensing mechanism (PFSM) is presented. Its mathematic model is established with screw theory. The force mapping relation is studied and the stiffness matrix is found to be a diagonal matrix, which proves the stiffness decoupling characteristics of the mechanism. According to the concept of fully isotropy, the isotropy conditions are analyzed, the parameters which meet fully isotropy are given. The 8/4-4 PFSM's configuration under isotropy parameters is analyzed. Based on this configuration, an 8/4-4 mechanism cluster which meets the fully isotropy is presented. The cluster's configuration is classified and induced into four main configurations according to the different parameter conditions.

Keywords: Stiffness decoupling · Parallel force sensing mechanism · Isotropy analysis · 8/4-4 mechanism cluster

1 Introduction

Parallel mechanism possesses the distinguishing advantages including high rigidity, high accuracy and easy decoupling. Due to these advantages, parallel mechanism has been widely used in many fields of science and engineering, such as micro-manipulators [1, 2], machine tools [3], wind tunnel experiments [4, 5] and so on.

The force sensing mechanism based on generalized parallel structure has been widely studied by many scholars. Gaillet and Reboulet [6] firstly proposed the application of the Stewart platform to six-axis parallel force sensing mechanism (PFSM). Kerr [7] considered the axial stiffness of the branch in the studied of Stewart PFSM and enumerated some design criteria for the sensor structure. Dwarakanath and Venkatesh [8] presented a six-axis parallel mechanism based force/torque sensor with no mechanical joint. Yao et al. [9, 10] presented the theoretical analysis and experiment research of a novel statically indeterminate six-axis PFSM. Zhen Gao and Dan Zhang [11] designed a multidimensional acceleration sensor based on fully decoupled compliant parallel mechanism. Dwarakanath and Bhutani [12] proposed a beam type PFSM based on "joint less" connector configuration for the transmission of axial forces.

© Springer International Publishing AG 2017
Y. Huang et al. (Eds.): ICIRA 2017, Part III, LNAI 10464, pp. 87–97, 2017.
DOI: 10.1007/978-3-319-65298-6_9

Yao et al. [13] presented a novel six-component force sensor based on parallel and flexible mechanisms and manufactured the sensor prototype with 3-D printing technology.

As can be seen from the literature survey, isotropy is a very significant principle in the design of six-axis PFSM. Xiong [14] presented the concept "isotropy" for robot's force sensors on the basis of Fisher's information matrix. A Fattah and AMH Ghasemi [15] presented the concept of kinematic isotropy and used as a criterion in the design of various parallel manipulators with ideal kinematic and dynamic performance. JIN Zhenlin [16] defined the sensitivity isotropy evaluation criteria of the force sensing mechanism, investigated the relationships between the criteria and the parameters of all the transducers based on the Stewart platform within the geometric model of the solution space. Gogu G [17, 18] firstly defined fully isotropy and concluded the characteristics under the isotropy configuration. Yao et al. [19] defined the modified isotropy indices which considered stiffness coupling effect. [20] presented four kinds of redundant six-axis PFSM, analyzed the four mechanisms' isotropy performance based on modified isotropy indices.

It is known that stiffness decoupling is the precondition of isotropy, but the current stiffness decoupling idea is to get the numerical solution when coupled elements all equal to zero, its decoupled characteristics can only be achieved under certain parameter conditions, the coupled elements are not eliminated fundamentally. In this paper, a stiffness decoupling 8/4-4 PFMS without coupled element is proposed, fundamentally eliminates the coupled characteristics. Then its isotropy performance and derivative structure are analyzed. Finally, an 8/4-4 mechanism cluster based on 8/4-4 PFMS is presented.

2 Structure Model and Mathematic Model

2.1 The Structure Characteristics of the Generalized Six-Axis PFSM

Generalized six-axis PFSM is shown in Fig. 1, which is composed of a measuring platform, a fixed platform, and n flexible measuring branches connecting two platforms with spherical joints. The measuring coordinate system is in the center of the measuring platform. In order to ensure the integrity of the measurement model, the number of branches should be no less than 6. When external force applied on the measuring platform, ignoring the branches' weight and spherical joints' friction, the branches bear only the force acting along its axis.

Based on the screw theory, the force and moment applied on the measuring platform are distributed on all branches. For the equilibrium of the measuring platform, the Eq. (1) can be obtained:

$$F + \in M = \sum_{i=1}^{n} f_i \$_i \quad (n \geq 6) \tag{1}$$

where F and M respectively represent the force vector and moment vector acted on the measuring platform, f_i represents the reacting force produced on the i-th flexible

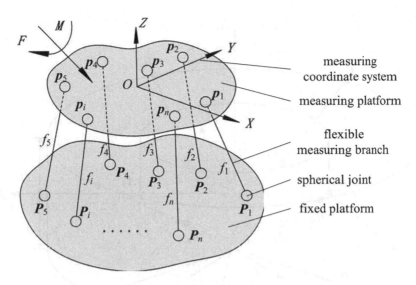

Fig. 1. Generalized six-axis PFSM

measuring branch and $\$_i$ represents the unit line vector along the axis of the i-th flexible measuring branch.

Then Eq. (1) can be expressed as

$$F_w = Gf \qquad (2)$$

where $F_w = [F\ M] = [F_x\ F_y\ F_z\ M_x\ M_y\ M_z]^T$ is the wrench acted on the measuring platform, $f = [f_1\ f_2\ f_3\ ...f_n]^T$ $(n \geq 6)$ is the vector composed of the axial force of each branch, G is the force Jacobian matrix which will directly influence the performance of the force sensing mechanism. G can be expressed as:

$$G = \begin{bmatrix} S_1 & S_2 & S_3 & \cdots & S_n \\ S_{01} & S_{02} & S_{03} & \cdots & S_{0n} \end{bmatrix} (n \geq 6) \qquad (3)$$

where (S_i, S_{0i}) represents the unit line vector along the axis of the i-th branch.

2.2   Structure and Mathematic Model of 8/4-4 PFSM

The structure diagram of the 8/4-4 PFSM is shown in Fig. 2. The mechanism's model is composed of three parts, outer platform for fixing, inner platform for measuring and 8 flexible measuring branches connecting between the two platforms. The inner and outer platforms are coaxial arrangement. O is the geometric center of mechanism, A_i are the intersections of i-th branches and inner platform, B_i are the intersections of i-th branches and outer platform. A_i are distributed on the same circumference of the inner platform. B_i are alternately distributed in two circumferences of the outer platform, the distance between two circumferences is $2H$. The angle between OA_i and OB_i is θ,

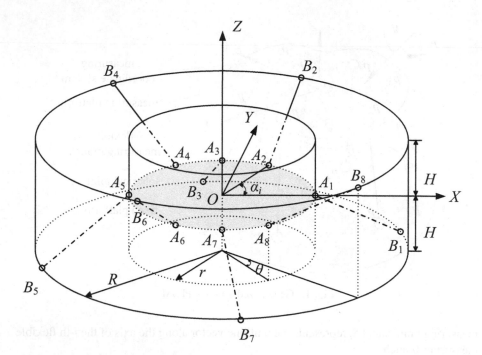

Fig. 2. Structure diagram of the 8/4-4 g of the 8/4-4 PFSM

8 branches are radial and tilted to one side, as shown in Fig. 3. According to the right-hand rule, the references coordinate system in Fig. 2. is established with O as the origin, OA_1 as the positive direction of axis X, and the coordinates of A_i and B_i can be expressed as

$$
\begin{aligned}
A_i &= \begin{bmatrix} r \cos \alpha_i & r \sin \alpha_i & 0 \end{bmatrix} \\
B_i &= \begin{bmatrix} R \cos(\alpha_i + \theta) & R \sin(\alpha_i + \theta) & (-1)^i H \end{bmatrix} \\
\alpha_i &= \frac{\pi}{4}(i - 1)
\end{aligned}
\tag{4}
$$

where r and R represent the inner and outer platform radius respectively, α_i is the angle between A_i and X axis, $\alpha + \theta$ is the angle between B_i and X axis.

Based on the screw theory, the force Jacobian matrix of the mechanism is obtained:

$$
G = \begin{bmatrix} \dfrac{A_1 - B_1}{|A_1 - B_1|} & \dfrac{A_2 - B_2}{|A_2 - B_2|} & \dfrac{A_3 - B_3}{|A_3 - B_3|} & \cdots & \dfrac{A_8 - B_8}{|A_8 - B_8|} \\[2mm] \dfrac{B_1 \times A_1}{|A_1 - B_1|} & \dfrac{B_2 \times A_2}{|A_2 - B_2|} & \dfrac{B_3 \times A_3}{|A_3 - B_3|} & \cdots & \dfrac{B_8 \times A_8}{|A_8 - B_8|} \end{bmatrix}
\tag{5}
$$

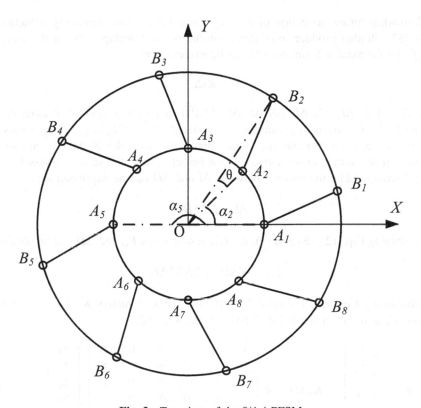

Fig. 3. Top view of the 8/4-4 PFSM

where

$$A_i - B_i = [\,r \cos \alpha_i - R \cos(\alpha_i + \theta) \quad r \sin \alpha_i - R \sin(\alpha_i + \theta) \quad (-1)^{i-1} H\,]$$
$$|A_j - B_j| = \sqrt{R^2 + r^2 - 2Rr \cos \theta + H^2} \tag{6}$$
$$B_i \times A_i = [\,(-1)^{i-1} Hr \sin \alpha_i \quad (-1)^i Hr \cos \alpha_i \quad -Rr \sin \theta\,]$$

3 Decoupling Analysis of 8/4-4 PFSM

Under the action of generalized external force F_w, the PFSM will have the corresponding deformation ΔD, which can be expressed as:

$$\Delta D = [\Delta d^T, \Delta \theta^T] = [\Delta x, \Delta y, \Delta z, \Delta \alpha, \Delta \beta, \Delta \gamma]^T \tag{7}$$

where, Δd is the resulting linear deformation, $\Delta \theta$ is the resulting rotational deformation.

Meanwhile under the action of reacting force, the flexible measuring branches of the PFSM will also produce axial deformation, the relationship between the reacting force f and the axial deformation Δl can be expressed as:

$$f = K_l \Delta l \tag{8}$$

where $\Delta l = [\Delta l_1 \ \Delta l_2 \ \Delta l_3 \ \Delta l_4 \ \Delta l_5 \ \Delta l_6 \ \Delta l_7 \ \Delta l_8]^T$ is a vector composed of axial deformation of 8 flexible measuring branches. $K_l = \mathrm{diag}(k_1 \ k_2 \ k_3 \ k_4 \ k_5 \ k_6 \ k_7 \ k_8)$ represents the stiffness matrix of 8 flexible measuring branches, since the 8 flexible measuring branches are the same, so we suppose the 8 branches have the same stiffness k_l.

By literature [21], the relation between Δl and ΔD can be expressed as:

$$\Delta l = G^T \Delta D \tag{9}$$

Combining Eqs. (2), (8) and (9), the relation between F_w and ΔD can be obtained:

$$F_w = K \Delta D = G K G^T \Delta D \tag{10}$$

Substituting Eqs. (5), (6) into Eq. (10), the stiffness matrix K_s and the relation between F_w and ΔD of the 8/4-4 PFSM can be obtained:

$$F_w = \begin{bmatrix} F_x \\ F_y \\ F_z \\ M_x \\ M_y \\ M_z \end{bmatrix} = K_s \Delta D = k_l \begin{bmatrix} \Lambda_1 & 0 & 0 & 0 & 0 & 0 \\ 0 & \Lambda_1 & 0 & 0 & 0 & 0 \\ 0 & 0 & \Lambda_2 & 0 & 0 & 0 \\ 0 & 0 & 0 & \Lambda_3 & 0 & 0 \\ 0 & 0 & 0 & 0 & \Lambda_3 & 0 \\ 0 & 0 & 0 & 0 & 0 & \Lambda_4 \end{bmatrix} \begin{bmatrix} \Delta x \\ \Delta y \\ \Delta z \\ \Delta \alpha \\ \Delta \beta \\ \Delta \gamma \end{bmatrix} \tag{11}$$

where

$$\begin{aligned} \Lambda_1 &= 4(R^2 + r^2 - 2Rr\cos\theta) \\ \Lambda_2 &= 8H^2 \\ \Lambda_3 &= 4H^2 r^2 \\ \Lambda_4 &= 8R^2 r^2 \sin^2\theta \end{aligned} \tag{12}$$

Equation (11) is the analytical solution of 8/4-4 PFSM. It can be seen that the stiffness matrix is a diagonal matrix, which proves the decoupled characteristic of the 8/4-4 PFSM. When a certain direction force/moment is applied on the origin of reference coordinate system, the deformation will also be on this certain direction but will not interfere with other directions. No matter what the value of the parameters are, as long as the geometric constraints in the Sect. 2.2 are satisfied, the decoupled characteristics of 8/4-4 PFSM will not be affected.

4   Isotropy Analysis of 8/4-4 PFSM

Based on the analysis of decoupled performance of 8/4-4 PFSM, it is found that when the decoupled performance is satisfied, the deformation of each direction is independent and easy to compare. Therefore, the decoupled characteristics and its constraints are considered as the premise and basis of isotropy. In order to obtain better isotropy properties, the decoupled performance must be considered at the beginning. However, stiffness decoupling does not mean that the isotropy is satisfied, a parametric model with isotropy constraints is still needed for the isotropy configuration.

When the force sensing mechanism is fully isotropy, the sensitivity of the force/moment in the three-dimensional direction is consistent. From this, we can conclude that the Eq. (13) must be satisfied when 8/4-4 PFSM is fully isotropy.

$$\begin{cases} \Lambda_1 = \Lambda_2 \\ \Lambda_3 = \Lambda_4 \end{cases} \tag{13}$$

Substituting Eq. (12) into Eq. (13), the parametric equation can be obtained:

$$\begin{cases} 4(R^2 + r^2 - 2Rr\cos\theta) = 8H^2 \\ 4H^2 r^2 = 8R^2 r^2 \sin^2\theta \end{cases} \tag{14}$$

Simplifying Eq. (14), we acquire:

$$R^2 + r^2 - 2Rr\cos\theta = 4R^2 \sin^2\theta \tag{15}$$

Dividing the Eq. (15) with the R^2:

$$1 + \left(\frac{r}{R}\right)^2 - 2\left(\frac{r}{R}\right)\cos\theta = 4\sin^2\theta \tag{16}$$

Suppose $t = r/R$, then Eq. (16) can be simplified to the Eq. (17) which takes θ as the independent variable and t as the dependent variable:

$$t^2 - 2t\cos\theta + 4\cos^2\theta - 3 = 0 \tag{17}$$

Solving Eq. (17) for t, we get:

$$t = \cos\theta \pm \sqrt{3}\sin\theta = 2\sin\left(\frac{\pi}{6} \pm \theta\right) \tag{18}$$

Substituting $t = r/R$ into Eq. (18), the parameters' solution for fully isotropy configuration is obtained:

$$\begin{cases} r = 2R\,\sin\left(\frac{\pi}{6} \pm \theta\right) \\ H = \pm\sqrt{2}R\,\sin\theta \end{cases} \tag{19}$$

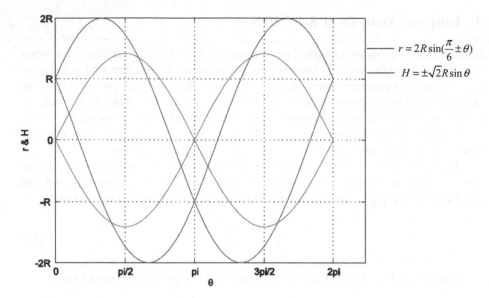

Fig. 4. The fully isotropy solution

According to Eq. (19), the isotropy of 8/4-4 PFSM is determined by r, R, H and θ, and the fully isotropy solution is a set of analytical solutions. In order to express the parameters' constraint relation more intuitively, the fully isotropy solution is drawn in Fig. 4. It can be seen that when R is determined, each θ (except at the intersection point) corresponds to the two groups' solution of R and H, that is, each θ (except the intersection point) corresponds to the four fully isotropy configurations.

Although all the solutions satisfying the isotropy can be expressed in Fig. 4, there also exists some repeated solutions due to the strong symmetry and repeatability of the Fig. 4, and the central symmetry of the 8/4-4 PFSM. It is necessary to simplify the configuration solutions obtained in Fig. 4. First, parameter H is simplified, because of the central symmetry of the 8/4-4 PFSM, $H = \sqrt{2}R \sin \theta$ or $H = -\sqrt{2}R \sin \theta$ does not have any real impact on the 8/4-4 PFSM, so $H = \pm\sqrt{2}R \sin \theta$ can be simplified as $H = \sqrt{2}R \sin \theta$. Then, parameter r is simplified, according to Fig. 4, we found that for $r = 2R\sin (\pi/6 + \theta)$ and $r = 2R\sin (\pi/6 - \theta)$ two cases, the change of r is a completely opposite process. The change of $r = 2R\sin (\pi/6 + \theta)$ from 0 to 2π is actually the same process of $r = 2R\sin (\pi/6 - \theta)$ from 2π to 0, so $r = 2R\sin (\pi/6 \pm \theta)$ can be simplified as $r = 2R\sin (\pi/6 + \theta)$. The parameters' solution for fully isotropy configuration is simplified as:

$$\begin{cases} r = 2R \sin(\frac{\pi}{6} + \theta) \\ H = \sqrt{2}R \sin \theta \end{cases} \tag{20}$$

Finally, the definition domain of θ is simplified, by observing the curves of Eq. (20) in Fig. 4, it is found that if $\theta = \theta_1$, $H = H_1$, $r = r_1$ is the solutions of Eq. (20), $\theta = \theta_1 + \pi$, $H = -H_1$, $r = -r_1$ is also the solutions of Eq. (20). For the central symmetry structure, $\theta = \theta_1$ and $\theta = \theta_1 + \pi$ have the same configuration. So the definition domain of θ can be simplified as $(0, \pi)$.Combined with Eq. (20), the parameters' solution for fully isotropy configuration after simplification can be expressed as:

$$\begin{cases} r = 2R \sin(\frac{\pi}{6} + \theta) \\ H = \sqrt{2}R \sin \theta \end{cases} \quad \theta \in (0, \pi) \tag{21}$$

Then, according to Eq. (21), the fully isotropy 8/4-4 mechanism cluster is obtained.

The 8/4-4 mechanism cluster is classified according to the parameters and configuration. First, according to the relationship between R and r, the mechanism cluster is divided into $r < R$ and $r > R$ two categories. Then subdivide again according to whether the branches intersect with the smaller platform on the top view. Finally, the configuration of four groups which can represent the 8/4-4 cluster structure is obtained, as shown in Table 1. The 8/4-4 PFSM proposed in Sect. 2 is included in the first class. Compared with the previous 8/4-4 PFSM, the fixed platform of the 8/4-4 mechanism cluster is the upper and lower platforms, the measuring platform is the middle platform, the flexible measuring branches interlaced distribute among three platforms. It is worth mentioning that the 8/4-4 mechanism cluster is a derivative of the previous 8/4-4 PFSM. So the cluster also has the stiffness decoupling characteristics.

Table 1. Schematic diagram of 8/4-4 mechanism cluster.

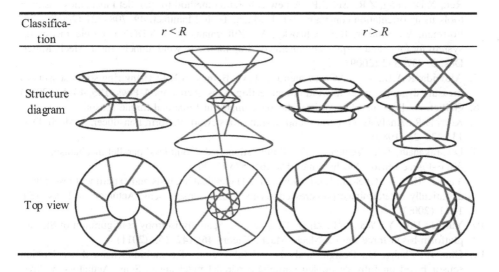

Classification	$r < R$		$r > R$	
Structure diagram				
Top view				

The proposal of 8/4-4 PFSM and 8/4-4 mechanism cluster, fundamentally eliminates the interference of coupling elements to isotropy, simplifies constraints and provides a simple isotropic mechanism cluster for the isotropy design of sensor's elastic body.

5 Conclusion

This paper presents an 8/4-4 stiffness decoupling PFSM. Its structure model and mathematic model are analyzed. The physical meaning and constraint conditions of decoupling are given. The 8/4-4 PFSM's decoupled characteristics is verified based on its mathematic model. The fully isotropic constraint conditions of 8/4-4 PFSM are deduced and analyzed. The parameters' solution for fully isotropy configuration is obtained. According to the solution, the fully isotropy 8/4-4 mechanism cluster is proposed, which enriches and simplifies the isotropy design of sensor's elastic body.

Acknowledgements. This research is sponsored by the financial support of National Natural Science Foundation of China (No. 5167052346) and Heibei Provincial Natural Science Foundation (No. E2015203165).

References

1. Dong, Y., Gao, F., Yue, Y.: Modeling and experimental study of a novel 3-RPR parallel micro-manipulator. Robot. Comput. Integr. Manufact. **37**, 115–124 (2016)
2. Liang, Q., Zhang, D., Chi, Z., Song, Q., Ge, Y., Ge, Y.: Six-DOF micro-manipulator based on compliant parallel mechanism with integrated force sensor. Robot. Comput. Integr. Manufact. **27**, 124–134 (2011)
3. Ren, X.D., Feng, Z.R., Su, C.P.: A new calibration method for parallel kinematics machine tools using orientation constraint. Int. J. Mach. Tools Manufact. **49**, 708–721 (2009)
4. Portman, V., Sandler, B.Z., Chapsky, V., Zilberman, I.: A 6-DOF isotropic measuring system for force and torque components of drag for use in wind tunnels. Int. J. Mech. Mater. Des. **5**, 337–352 (2009)
5. Almeida, R.A.B., Vaz, D.C., Urgueira, A.P.V., Borges, A.R.J.: Using ring strain sensors to measure dynamic forces in wind-tunnel testing. Sens. Actuators A Phys. **185**, 44–52 (2012)
6. Gaillet, A., Reboulet, C.: An isostatic six component force and torque sensor
7. Kerr, D.R.: Analysis, properties, and design of a stewart-platform transducer. J. Mech. Des. **111**, 25–28 (1989)
8. Dwarakanath, T.A., Venkatesh, D.: Simply supported, 'Joint less' parallel mechanism based force–torque sensor. Mechatronics **16**, 565–575 (2006)
9. Yao, J., Hou, Y., Chen, J., Lu, L., Zhao, Y.: Theoretical analysis and experiment research of a statically indeterminate pre-stressed six-axis force sensor. Sens. Actuators A Phys. **150**, 1–11 (2009)
10. Yao, J., Hou, Y., Wang, H., Zhou, T., Zhao, Y.: Spatially isotropy configuration of Stewart platform-based force sensor. Mech. Mach. Theory **46**, 142–155 (2011)
11. Gao, Z., Zhang, D.: Design, analysis and fabrication of a multidimensional acceleration sensor based on fully decoupled compliant parallel mechanism. Sens. Actuators A Phys. **163**, 418–427 (2010)
12. Dwarakanath, T.A., Bhutani, G.: Beam type hexapod structure based six component force–torque sensor. Mechatronics **21**, 1279–1287 (2011)
13. Yao, J., Zhang, H., Xiang, X., Bai, H., Zhao, Y.: A 3-D printed redundant six-component force sensor with eight parallel limbs. Sens. Actuators A Phys. **247**, 90–97 (2016)
14. Xiong, Y.: On isotropy of robot's force sensors (1996)

15. Fattah, A., Ghasemi, A.M.H.: Isotropic design of spatial parallel manipulators. Int. J. Robot. Res. **21**, 811–826 (2002)
16. Jin, Z.L., Gao, F.: Optimal design of a 6-axis force transducer based on Stewart platform related to sensitivity isotropy. Chin. J. Mech. Eng. **16**, 146–148 (2003)
17. Gogu, G.: Structural synthesis of fully-isotropic translational parallel robots via theory of linear transformations. Eur. J. Mech. A/Solids **23**, 1021–1039 (2004)
18. Gogu, G.: Structural synthesis of fully-isotropic parallel robots with Schönflies motions via theory of linear transformations and evolutionary morphology. Eur. J. Mech. A/Solids **26**(2), 242–269 (2007)
19. Yao, J., Hou, Y., Wang, H., Zhao, Y.: Isotropic Design of Stewart Platform-Based Force Sensor. DBLP (2008)
20. Yao, J., Zhang, H., Zhu, J., Xu, Y., Zhao, Y.: Isotropy analysis of redundant parallel six-axis force sensor. Mech. Mach. Theory **91**, 135–150 (2015)
21. Merlet, J.P.: Parallel Robots (Solid Mechanics and Its Applications). Springer, New York (2006)

Preliminary Results of EMG-Based Hand Gestures for Long Term Use

Peter Boyd, Yinfeng Fang, and Honghai Liu[✉]

School of Computing, University of Portsmouth, Portsmouth, UK
{peter.boyd,yinfeng.fang,honghai.liu}@port.ac.uk

Abstract. The application of pattern recognition techniques to Electromyography (EMG) signals has shown great potential for robust, natural, prostheses control. Despite promising development in EMG pattern recognition techniques, the non-stationary properties of these signals may render these techniques ineffective after a period of time, subsequently demanding frequent recalibration during long term use. Potentially one method to reduce the impact of non-stationary traits of EMG signals is through attempting to construct a training dataset that represents this gradual change in the signal. In this paper, we investigate the potential impact of data selection schemes for inter-day motion recognition, across a period of five days of high density data recording with an LDA classifier, and present our preliminary findings. This paper proves that training a classifier with data from several spaced points of a single day can improve its inter-day performance which subsequently supports the long term use of prosthesis. Therefore the work presented here may aid in furthering our understanding of the physiological changes in EMG signals and how they may be exploited to further improve the robustness of pattern recognition methods for long term use.

Keywords: Surface electromyography (sEMG) · Hand gesture recognition · Dataset optimisation · Prosthesis · Robustness · Pattern recognition

1 Introduction

Across a given period of time, classification error can be expected to increase. Unfortunately, the exact reason for the increase in classification error is not properly understood, pushing researchers to attempt to quantify the shift in EMG data and classification error across time [1]. Kaufmann [11] compared how classification accuracy degraded when training a classifier with early data sets, recent datasets, and gradually updated datasets. It was suggested that there exist some variable components within EMG signals, based on the degrading accuracy over time. If it can be assumed that there exists variable components of the EMG signal, there exists potential to construct a dataset that can reasonably predict this change, whilst remaining computationally efficient.

All of these traits are natural during daily use of a prosthesis, and therefore likely to cause classification failures during long periods of usage. Through recalibration, the effective usage of a prosthesis can be extended. There main methods

© Springer International Publishing AG 2017
Y. Huang et al. (Eds.): ICIRA 2017, Part III, LNAI 10464, pp. 98–108, 2017.
DOI: 10.1007/978-3-319-65298-6_10

for recalibration is through supervised means, and unsupervised means. Supervised recalibration requires the prosthesis user to manually decide if the prosthesis is not performing optimally, and to collect a new training dataset for the device. The new set of training data will either be appended to the existing data or shall replace the previous training data, this retraining has often been done through screen guided training [16]. A prosthesis guided training method was proposed by Chicone et al. [4], upon recalibration, the prosthesis would perform a series of gestures for the user to mimic. It was found that the prosthesis guided approach both provided better training samples and reduced the need for any additional devices to the prosthesis. Unsupervised methods, however, require no conscious input on the prosthesis users behalf. Adaptive algorithms are used to regularly update the existing training data based on different adaptive strategies. There has been successful research papers on improving classification accuracy with adaptive techniques [13,14,20]. An example would be Sensinger et al. [15], noted that unsupervised methods that rely on high confidence of classification could provide implementable degrees of accuracy, yet could suffer over-training over long periods of time. It was further found that adaptive methods based on low confidence of classification could decrease over training but were not reliable when unsupervised.

Although Adaptive methods are very capable of improving the overall time between conscious calibration, and may subsequently provide higher degrees of long term accuracy, the data selection schemes are not fully dependable for clinical use. Two general problems must be addressed with selecting a training scheme. Firstly, in schemes that directly modify the training set for retraining, the overall size of the training dataset should be minimal as to avoid over-training and to reduce the computation time during training. Secondly, that concept drift within EMG signals is not fully understand and therefore difficult to adequately model the change over time, or to capture a training data set that is representative of the range of gestures. Generally the selection of original data could be more important for initial use, largely due to the variable nature of EMG signals throughout a day.

Electromyography (EMG) is the method of analysing the electrical impulses from various muscles within the human body. Through pattern recognition, a machine may be trained to both understand and predict motions from monitored muscles. Although, in a laboratory setting, results are promising, clinical success of pattern recognition based prosthesis is still lacking. In academia, there exists increasingly promising methods for recognition of motion intent, with accuracy rates commonly reaching above 90%. However, a large quantity of academic research focuses on training with offline datasets, whilst online classification results are not as satisfactory. There are several potential causes for decreased accuracy during online classification such as: skin-electrode impedance [17], electrode displacement and shift [19], and fatigue [12].

In this study, an investigation into methods of data selection to improve the long term inter-day stability of a trained dataset shall be conducted. The following sections shall cover data collection, data selection, and processing, before a discussion on the findings of this research, ending in a conclusion and suggestions towards future work.

2   Methodology

2.1   Experimental Setup

Data Collection Device. To collect EMG data, a 16 channel device was used with 12 bits ADC resolution and 1Khz sampling frequency. The range for collected EMG data was 10 Hz to 500 Hz through a band pass filter in the hardware, 50 Hz powerline noise was removed with a notch filter in the hardware and a comb filter in the software. The EMG data was transferred from the device via USB, and processed on a Windows based PC. More details may be found in the previous works [7,8].

Device Wearing. The devices structure is that of a sleeve with 18 embedded electrodes, with two bias electrodes. The exact configuration of electrodes within the sleeve are arranged as to improve the signal repeatability [5]. A second sleeve was worn over the electrode sleeve to ensure a tight fit on all electrodes. The sleeve was worn on the subjects dominant arm, or preferred arm in cases of Cross-dominance. To ensure each day saw the same fitting, a marker was used to denote the location of both bias electrodes on the subjects arm, in-line with the top and bottom of the sleeve, as shown in Fig. 1a. The degree of shift experienced on the sleeve was updated by new markings before each instance of data collection. The benefit of the dual sleeve arrangement in this device is that it maintains both pressure and contact upon a larger section of the arm. Through the increased surface contact, the device is subsequently less likely to shift through and dramatic degree during wearing. During the course of this study it was found that once the sleeve had begun to imprint upon the skin, that any shift would gradually re-align with the original rotation on the arm. The flexible design of the elasticated fabric further meant that relative positions of the electrodes to the arm are very stable.

Issues such as sweating may be a factor in warm climates, or if the wearer is performing much physical activity, however, neither of these issues were encountered during the data collection. In a situation where the device was exposed to cold climates the wearer would allow the device to return to the ambient room temperature before data collection. The potential reasoning for this is that colder climates lead to a drier skin surface and subsequently high skin-electrode impedance.

Data Collection. Data collection began at 10:00 for each day and ended at 18:00, a 30 min window was granted between data collection providing a total of 16 sessions per day. The 30 min window is designed to keep constant records of EMG data without placing unnatural fatigue on the subject. EMG capture was performed by the subject sitting in front of a computer monitor, with elbows in a fixed position. Thirteen separate hand gestures, as shown in Fig. 1b, were shown to the subject with 5 s to transfer between gestures and 5 s of holding a gesture. The series of gestures were presented to the subject in a pseudo-random

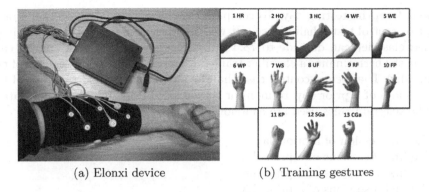

(a) Elonxi device   (b) Training gestures

Fig. 1. Elonxi device and training gestures

order, as to prevent the subject learning the order of gestures and transferring to a new gesture too early. The transition period both allows the participant to move between gestures and to have a steady gesture held for the next recording period.

Data Processing and Classification. Signal collection was performed through custom software, and processed in Matlab. As stated in the data collection section, the custom software would strip any remaining powerline noise and output what would hopefully be the EMG signal with minimal interference. Once the raw signal had been stripped of any interference and prepared for processing, the outputted EMG signal data was stripped of the transient signal between gestures. This is to say, the 5 second window of transition between each gesture was removed as to ensure each set of data is fully representative of representative class.

The feature extraction was performed in a 250 ms window, with a 50ms sliding window, the window size was chosen as it. The chosen features for this study were a combination of Auto-Regressive (AR) to the fourth order and root Mean Square (RMS), as the combination of these features had previously been shown to be both strong features and robust for long term classification [3,9]. For classification, Linear Discriminant Analysis (LDA) was used with a bayes-optimal classifier. Although Support Vector Machines (SVM) and Multi-Layer Perceptrons (MLP) are good algorithms for classification, LDA has previously been shown to be more robust during long term classification, although at a cost of native classification accuracy whilst MLP and SVM decayed rapidly [11].

2.2   Experiments

Data Selection. To compare the relative performance of each method of selecting data, several data selection strategies were evaluated using a single day as training data and alternative days as testing days, these strategies were firstly applied without any testing day data and secondly with one of the first recorded sessions from the testing day. To compare the efficiency of the selection strategies, two control tests shall be used. It was found by Jain et al. [10] that a single trained classifier can degrade to 78.4 ± 2.33%, therefore the first control case shall be the performance a classifier trained from one of the first 3 sessions of the testing day. The second control case shall be a classifier trained with one of the first sessions of a prior day, to monitor the decay of using minimal prior data.

The selection rules used are as follows:

1. The first session of a prior day
2. The first 2 sessions of a prior day
3. The first 3 sessions of a prior day
4. The first 4 sessions of a prior day
5. All 16 sessions of a prior day
6. 3 sessions, randomly selected from 3 points of a prior day
7. 4 sessions, randomly selected from 4 space points of a prior day
8. A single session of a prior day

The first 3 selection cases are to observe the impact of adding sequential sessions into the training dataset. The second and third case also provide a direct comparison of sequential data selection, as opposed to representative data selection. The usage of an entire day (16 sessions) is to observe the effect of using a larger dataset to represent an entire day of EMG signals.

The final non control cases are both within the proposed method of representative data selection. In this method, a given day is separated into descriptive periods such as: morning, middle day, and evening. The intention of this method is to provide a good representation of the shifts in the EMG signal during any given day whilst using a minimal quantity of recording points in aforementioned day. An example of selection of random data points for training data is demonstrated in Fig. 2.

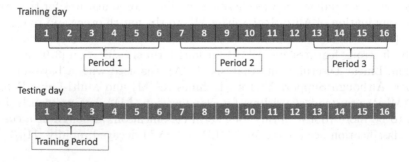

Fig. 2. Proposed Selection Rule using 3 prior day sessions and one testing day session

Data Analysis. As mentioned in the Data selection segment, several strategies of data selection are to be tested. As this study focuses on the general performance of the data selection strategies, the temporal change in accuracy along consecutive days has not been recorded. Instead, each day in the dataset is considered as a unique training day against all other days which shall be considered as testing days, without considering for their appearance in sequence. Each training day will therefore be tested against every other day using the different selection strategies, each test shall begin with the first session of the testing day and then continue until the final session, recording the accuracy of each internal session. This process shall be repeated for every other day until the accuracy of every training method on every possible training day has been collected. The final resulting interday accuracy shall then be calculated as the outputted mean when computing firstly the mean of each testing days internal performance, and then acquiring the mean of each testing days final performance. Ideally the output of acquiring the overall mean performance will grant insight into the generalisation of these methods.

3 Results

Figure 3 shows the averaged interday performance of the four data selection methods, all are shown with and without a testing day session whereas Table 1 shows the complete list of accuracy and the standard deviation of each method. As demonstrated by other authors [18], the interday accuracy of a single training set is very poor at 61.9%, although this level of accuracy is very similar to that of the intraday accuracy, as an early morning session alone is unable to fully represent transient changes within the EMG signal.

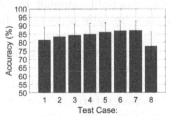 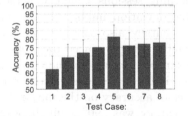

(a) With Testing Day Data   (b) Without Testing Day Data

Fig. 3. Results of all data selection rules listed. Case 1: First session only Case 2: First 2 sessions, Case 3: First 3 sessions Case 4: First 4 sessions Case 5: Entire Day (16 Sessions), Case 6: 3 Spread out sessions, Case 7: 4 Spread out sessions, Case 8: Single Session from testing day only

Table 1. Results of using sessions from prior day and first session of testing day as training data against the entirety of the testing day

Classification results	Without testing day data	With testing day data
Training day data	Accuracy (%)	Accuracy (%)
Testing day data only	77.8 ± 8.6	77.8 ± 8.6
Session 1	61.9 ± 8.0	81.2 ± 7.6
Session 1–2	68.3 ± 7.9	83.3 ± 7.3
Sessions 1–3	71.7 ± 7.8	84.2 ±6.7
Sessions 1–4	75.0 ± 7.8	84.8 ± 6.5
Entire day of 16 sessions	81.2 ± 7.0	86.0 ± 5.7
3 Spaced sessions	76.0 ± 7.6	86.9 ± 5.7
4 Spaced sessions	76.9 ± 7.4	87.0 ± 5.5

The other selection methods managed to achieve constantly acceptable accuracy rates of 85% ± 2% accuracy on interday testing sets. Showing that a single days worth of data is capable of providing a good representation of a persons EMG signals.

The highest result was found with using the proposed methods of the selecting defining periods of each training day with a single early session of the testing day with an accuracy of 86.9% ± 5.7% 87.0% ± 5.5%, for both the 3 and 4 sessions. When the training set has prior knowledge of the testing day, the result reduces to 76.0% ± 7.6% and 76.9% ± 7.4% which is very similar to using the first session alone of any given day (77.8%). The results of the entire day could be considered to be the effect of gathering a large enough dataset to accurately represent the gradual change of the signal during an average day. Using 4 morning sessions managed achieve a reasonable amount of accuracy, although the deviation implies that it may not be fully acceptable for daily usage, the accuracy also saw a large decrease when not using a testing day dataset. Finally, using 3 spaced datasets and 1 testing day set as training data resulted in similar accuracy to the entire day training set.

As shown in the confusions matrix in Fig. 4a, the proposed method achieved a high degree of accuracy on the majority of simple gestures, where the average inter-day accuracy was 90%, which poses itself as a very strong method for reliable long term EMG prosthesis control. Generally, the majority of simple classes were easily classified. The classes which routinely scored most poorly were the two complex gestures numbered 12 and 13 in Fig. 1b, where accuracy was rarely above 70% in any test case.

Actual Value / Predicted Value

Motion Class	1	2	3	4	5	6	7	8	9	10	11	12	13
1	91	0	0	0	0	5	5	0	0	0	0	0	0
2	1	82	0	4	0	1	0	1	1	0	0	15	2
3	0	1	93	0	0	0	0	0	0	0	0	0	0
4	0	0	1	84	0	0	0	0	0	0	0	0	0
5	0	0	0	0	98	1	0	0	2	0	0	1	2
6	1	0	0	1	0	86	3	0	1	0	1	0	0
7	2	0	1	0	0	1	86	0	0	0	0	0	0
8	0	5	0	9	0	0	1	98	0	0	0	0	0
9	0	0	0	0	0	1	0	0	91	0	0	0	0
10	2	0	2	0	1	3	2	0	4	96	3	0	2
11	0	0	2	0	0	1	0	0	0	2	94	0	0
12	0	10	1	1	0	0	1	0	0	0	1	65	22
13	0	1	1	0	0	0	0	0	0	1	1	18	69

(a) Confusion Matrix for Proposed Method of 3 Spaced Data Points

Actual Value / Predicted Value

Motion Class	1	2	3	4	5	6	7	8	9	10	11	12	13
1	93	0	0	0	0	6	5	0	0	0	0	0	0
2	1	82	0	5	0	1	0	1	1	0	0	16	1
3	0	1	86	0	0	0	0	0	0	0	0	0	0
4	0	0	1	79	0	1	0	0	0	0	0	0	0
5	0	0	0	0	95	1	0	0	4	0	0	0	2
6	1	0	0	2	0	85	2	0	1	0	1	0	0
7	1	0	0	0	0	2	88	0	0	0	0	0	0
8	0	2	0	11	0	0	1	97	0	0	0	0	0
9	0	0	0	0	1	1	0	0	88	1	0	0	0
10	1	1	2	0	3	3	2	0	6	96	6	0	1
11	0	0	4	0	0	1	0	0	0	1	90	0	0
12	0	12	6	1	0	0	1	0	0	0	1	60	17
13	0	2	0	0	0	0	0	0	0	1	1	23	76

(b) Confusion Matrix for Entire Day of Data

Actual Value / Predicted Value

Motion Class	1	2	3	4	5	6	7	8	9	10	11	12	13
1	71	0	0	0	0	3	5	0	0	0	0	0	0
2	2	63	0	6	0	4	1	2	1	0	0	5	1
3	0	1	83	1	0	0	0	0	0	0	2	0	0
4	0	0	2	76	0	1	0	0	0	0	0	0	0
5	0	0	0	1	98	1	0	0	11	4	1	1	16
6	2	0	0	1	0	83	4	0	2	0	1	0	0
7	6	0	0	0	0	0	75	0	0	0	0	0	0
8	0	13	0	10	0	0	2	97	0	2	5	2	5
9	0	0	0	0	0	1	1	0	76	0	1	0	0
10	17	0	3	0	1	6	11	0	9	91	5	0	4
11	0	0	1	0	0	0	0	0	0	1	80	0	0
12	0	22	2	4	0	0	0	1	1	0	3	73	27
13	0	2	8	0	1	0	1	0	0	1	1	18	44

(c) Confusion Matrix for Single Testing Day Session

Fig. 4. Confusion matrices of selected approaches with testing day data

4   Discussion

In this study, the effect of different training dataset selection strategies was investigated. The results demonstrated a stronger classifier, for inter-day EMG gesture recognition, can be built by using data that is representative of the EMG signal changes within a day.

The single sessions accuracy and reduction in accuracy when excluding a testing day dataset demonstrates the previous conclusion by Amsuss et al. [2] that classification accuracy decreases as a function of time. The accuracy of using 4 morning sessions is directly inline with those displayed by Kaufmann [11] when using LDA, where the error rate they had found was 21.27%, whilst this study found an error rate of 25% ± 7.8% when using their first 5 sessions. Unfortunately it was Kaufmann had not noted which periods of each day saw data collection, only that 5 datasets represented an entire day of data, whereas this study focused on the impact of when the data is collected within a day. However, when given any set of 4 consecutive hours provided similar performance on inter-day trials, suggesting that any 4 consecutive sessions will provide a seemingly reliable description of the EMG signal. It can further be agreed that using larger quantities of data, such as an entire day of data, would provide higher classification accuracy, however, the increase in accuracy is not statistically significant enough to justify the increase in computation and burden on the user.

Although the proposed method, with no prior knowledge of the testing day, achieves a similar degree of accuracy to training the algorithm in the morning of the testing day, the removal of the requirement to train the device immediately and achieve a reasonable degree of accuracy is very promising. Through the addition of the testing day data, the proposed method provided a very good inter-day accuracy without any modification to the training dataset. The extent of the inter-day accuracy is one where it could be expected to achieve high accuracy on future datasets, regardless of time difference between the original training day and present day.

As shown in the confusion matrices in Fig. 4, the simple gestures routinely were easy to classify from each test case, or at least represented the final accuracy best. The two complex gestures routinely scored very low, often being misclassified as one another. Due to this drop in accuracy being so localized, it could be suggested that using this method without those two classes could be more beneficial for the prosthesis user. The researcher is, however, of the opinion that it is important to promote a prosthesis control scheme which is both natural and intuitive for the amputee. Regarding the actual outputted results, differing methods would either provide higher accuracy on one complex gesture at the cost of reduced accuracy on the other gesture. Although the performance of the proposed algorithm for complex gestures is still not fully adequate for control, it displays a promising removal of the bias for either gesture. The usage of improved feedback methods during data collection may assist a prosthesis user in providing more and clear separable data for the complex gestures [6].

It should be noted that although the performance of this method is good for the majority of presented data, it has not been tested against fatigued data or data where the presence of shift exists. Subsequently, the robustness of this method has not yet been shown to be resilient to transient changes other than those which naturally throughout a day or across a period of time.

5   Conclusion

In this paper, the relationship between the periods of a day and the EMG signal was investigated. From the results of this preliminary study, there lies two main conclusions. Firstly, that selection of a data that can represent the change in the EMG signal throughout a standard day is capable of achieving an acceptable degree of accuracy in the long term. Further to this point, it can be assumed that such data must be fully representative of the day itself, at the risk of losing stability.

Secondly, the selection strategies in this paper suggest that usage of an entire days data provides a statistically insignificant increase in classification accuracy when compared to more simplistic and computationally efficient methods.

The intent of this research was to investigate the changes in the EMG signal during a day, its impact on classification accuracy over a series of days, and to find a method that was suited to promoting a robust training set for inter-day classification throughout each day.

It is suggested that future directions to expand upon this research is to look at ways of optimising or otherwise processing the original training dataset to better describe the shift in the EMG signal, either through promoting expansion or promoting clustering. Further cases of study may also investigate the relative change in performance with days that experience high degrees of fatigue or other transient changes during a single day, where external changes may impact the quality of the EMG signal.

References

1. Amsüss, S., Paredes, L.P., Rudigkeit, N., Graimann, B., Herrmann, M.J., Farina, D.: Long term stability of surface EMG pattern classification for prosthetic control. In: 2013 35th Annual International Conference of the IEEE Engineering in Medicine and Biology Society (EMBC), pp. 3622–3625, July 2013
2. Amsüss, S., Paredes, L.P., Rudigkeit, N., Graimann, B., Herrmann, M.J., Farina, D., Subjects, A.: Long term stability of surface EMG pattern classification for prosthetic control, pp. 3622–3625 (2013)
3. Chen, X., Zhang, D., Zhu, X.: Application of a self-enhancing classification method to electromyography pattern recognition for multifunctional prosthesis control. J. NeuroEng. Rehabil. 1–13 (2013)
4. Chicoine, C.L., Simon, A.M., Hargrove, L.J.: Prosthesis-guided training of pattern recognition-controlled myoelectric prosthesis. In: 2012 Annual International Conference of the IEEE Engineering in Medicine and Biology Society, pp. 1876–1879, August 2012
5. Fang, Y., Liu, H.: Robust sEMG electrodes configuration for pattern recognition based prosthesis control. In: 2014 IEEE International Conference on Systems, Man, and Cybernetics (SMC), pp. 2210–2215, October 2014

6. Fang, Y., Zhou, D., Li, K., Liu, H.: Interface prostheses with classifier-feedback based user training. IEEE Trans. Biomed. Eng. **PP**(99), 1 (2017)
7. Fang, Y., Liu, H., Li, G., Zhu, X.: A multichannel surface EMG system for hand motion recognition. Int. J. Humanoid Robot. **12**(02), 1550011 (2015). http://www.worldscientific.com/doi/abs/10.1142/S0219843615500115
8. Fang, Y., Zhu, X., Liu, H.: Development of a surface EMG acquisition system with novel electrodes configuration and signal representation. In: Lee, J., Lee, M.C., Liu, H., Ryu, J.-H. (eds.) ICIRA 2013. LNCS, vol. 8102, pp. 405–414. Springer, Heidelberg (2013). doi:10.1007/978-3-642-40852-6_41
9. Huang, Y., Englehart, K.B., Hudgins, B., Chan, A.D.C.: A Gaussian mixture model based classification scheme for myoelectric control of powered upper limb prostheses. IEEE Trans. Biomed. Eng. **52**(11), 1801–1811 (2005)
10. Jain, S., Singhal, G., Smith, R.J., Kaliki, R., Thakor, N.: Improving long term myoelectric decoding, using an adaptive classifier with label correction. In: Proceedings of the IEEE RAS and EMBS International Conference on Biomedical Robotics and Biomechatronics, pp. 532–537 (2012)
11. Kaufmann, P., Englehart, K., Platzner, M.: Fluctuating EMG signals: investigating long-term effects of pattern matching algorithms, pp. 6357–6360 (2010)
12. Lalitharatne, T.D., Hayashi, Y., Teramoto, K., Kiguchi, K.: Compensation of the effects of muscle fatigue on EMG-based control using fuzzy rules based scheme. In: 2013 35th Annual International Conference of the IEEE Engineering in Medicine and Biology Society (EMBC), pp. 6949–6952, July 2013
13. Orabona, F., Castellini, C., Caputo, B., Fiorilla, A.E., Sandini, G.: Model adaptation with least-squares SVM for adaptive hand prosthetics. Idiap-RR Idiap-RR-05-2009. Idiap, March 2009. Accepted in ICRA09
14. Pilarski, P.M., Dawson, M.R., Degris, T., Carey, J.P., Chan, K.M., Hebert, J.S., Sutton, R.S.: Adaptive artificial limbs: a real-time approach to prediction and anticipation. IEEE Robot. Autom. Mag. **20**(1), 53–64 (2013)
15. Sensinger, J.W., Lock, B.A., Kuiken, T.A.: Adaptive pattern recognition of myoelectric signals: exploration of conceptual framework and practical algorithms. IEEE Trans. Neural Syst. Rehabil. Eng. **17**(3), 270–278 (2009)
16. Simon, A.M., Hargrove, L.J., Lock, B.A., Kuiken, T.A.: A decision-based velocity ramp for minimizing the effect of misclassifications during real-time pattern recognition control. IEEE Trans. Biomed. Eng. **58**(8), 2360–2368 (2011)
17. Vidovic, M.M.C., Hwang, H.J., Amsüss, S., Hahne, J.M., Farina, D., Müller, K.R.: Improving the robustness of myoelectric pattern recognition for upper limb prostheses by covariate shift adaptation. IEEE Trans. Neural Syst. Rehabil. Eng. **24**(9), 961–970 (2016)
18. Vidovic, M.M., Hwang, H.J., Amsüss, S., Hahne, J.M., Farina, D., Müller, K.R.: Improving the robustness of myoelectric pattern recognition for upper limb prostheses by covariate shift. Adaptation **24**(9), 961–970 (2016)
19. Young, A.J., Hargrove, L.J., Kuiken, T.A.: The effects of electrode size and orientation on the sensitivity of myoelectric pattern recognition systems to electrode shift. IEEE Trans. Biomed. Eng. **58**(9), 2537–2544 (2011)
20. Zhang, Y., Wang, Z., Zhang, Z., Fang, Y., Liu, H.: Comparison of online adaptive learning algorithms for myoelectric hand control. In: 2016 9th International Conference on Human System Interactions (HSI), pp. 69–75, July 2016

Research on Variable Stiffness and Damping Magnetorheological Actuator for Robot Joint

Xiaomin Dong[✉], Weiqi Liu, Xuhong Wang, Jianqiang Yu, and Pinggen Chen

State Key Laboratory of Mechanical Transmission, Chongqing University, Chongqing, China
{xmdong, 20160713100, 2016070213t, yjq2012, pgchen}@cqu.edu.cn

Abstract. Aiming at the limitation of the traditional flexible robot's single adjustment stiffness or damping, a magnetorheological (MR) actuator of which stiffness and damping can be adjusted simultaneously and independently is proposed for the robot joint. The principle of equivalent variable stiffness and damping is analyzed theoretically, and the adjustment range of stiffness and damping is deduced. As the first step, the performance of variable damping is evaluated with experiment by using a MR damper. The preliminary results show that the magnetorheological actuator is capable of changing the damping by controlling the current applied to the damper.

Keywords: Variable stiffness and damping · Flexible robot joints · Magneto-rheological fluid

1 Introduction

Rigid robots, which have been widely used in traditional industries, are scarce of flexibility. In some cases, the rigidity will cause the harmfulness for human beings [1]. Therefore, flexible robot joint is becoming more and more concerned especially to avoid injure of human being in the uncertain environment [2, 3].Early studies of flexible joints, such as the flexible joint manipulator of the German aerospace, the harmonic reducer and the torque sensor provide the factor of flexibility. However, the stiffness of the flexible joint is still very large, and the flexibility of the joint needs to be improved by a specific spring element. A solution for this problem is called *Series Elastic Actuator* (SEA) proposed by MIT leg laboratory [4]. The main feature of this method is the introduction of elastic actuators between the motor and the load. Grioli et al. [5] presented a *Variable Stiffness Actuator* (VSA). Compared with the fixed passive compliance units, the variable stiffness implementations possess the advantage of regulating stiffness and position independently and the wide range of stiffness and energy storage capabilities. However, the introduction of the variable stiffness component increases the order of the system and reduces the stability and the bandwidth [6–9]. A novel force control actuator system called *Series Damper Actuator* (SDA) [10] was proposed. It has been demonstrated that the use of physical damping in the transmission system can effectively improve its closed-loop performance [11–14].

© Springer International Publishing AG 2017
Y. Huang et al. (Eds.): ICIRA 2017, Part III, LNAI 10464, pp. 109–119, 2017.
DOI: 10.1007/978-3-319-65298-6_11

The researches above show that the method of variable stiffness and damping is inconvenient and needs to be further improved. In the last decades, a smart damping technology, magnetorheological (MR) technology, has been received much attention due to fast response, wide damping range, reversible, low energy requirement etc. [15–18]. In our previous study [19], we proposed a linear buffering device with adaptive variable damping and stiffness based on magnetorheological, and the results had shown that the design of the buffer device and control strategy is effective. Based on the MR technology, the design of an actuator for joint of robots to realize the simultaneous adjustment of stiffness and damping was proposed in this study. This study's organization is as follows. Section 2 analyses the principle of variable stiffness and variable damping. Section 3 focuses on description of the MR damper structure design while an experiment setup is described and performance of variable stiffness are presented in Sect. 4. The conclusions and future work are addressed in Sect. 5.

2 Analysis of Variable Stiffness and Variable Damping

In this paper, the variable stiffness and damping actuator, which consists of two magnetorheological damper c_1, c_2 and two torsional springs k_{s1}, k_{s2}, can achieve real-time damping and stiffness changes. As can be seen in Fig. 1(a), the magnetorheological damper c_2 and the torsion spring k_{s2} are connected in parallel to form a unit which is connected in series with the torsion spring k_{s1} and then in parallel with the magnetorheological damper c_1. Figure 1(b) shows the equivalent model of the joint, which has the equivalent damping coefficient c_s and the equivalent stiffness k_s.

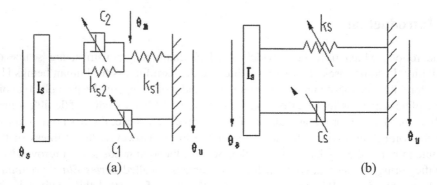

Fig. 1. Joint model of variable stiffness and damping and its equivalent model

When the rotary axis moves, the magnetorheological fluid generates damping at the gap between the shaft and the outer cylinder. The rheological properties of the MR fluid will change with the change of the applied magnetic field. After theoretical deduction and pre-analysis of the working characteristics of magnetorheological damper, the relationship between the damping force F_d, the relative speed of the shaft v and the applied current I can be expressed as [20]:

$$F_d = c_0 v + F_{MR}\text{sgn } v \tag{1}$$

$$F_{MR} = aI^2 + bI + c \tag{2}$$

Where c_0 is viscous damping coefficient, F_{MR} is coulomb damping force; a, b, c are constants obtained by fitting the test data.

Equation (1) shows that the damping force of the magnetorheological damper is composed of viscous damping force and Coulomb damping force. When the geometrical dimension of the damper is determined, the viscous damping force is only a function of the velocity of the piston. The Coulomb damping force is a function of the applied current, so it is possible to adjust the damping force by adjusting the magnitude of the applied current of the damper to realize the change of the damping coefficient.

According to Newton's second law, the dynamic equation of Fig. 1(a) is obtained:

$$I_s \ddot{\theta}_s - c_1 \left(\dot{\theta}_s - \dot{\theta}_u \right) - k_{s2}(\theta_s - \theta_m) - c_2 \left(\dot{\theta}_s - \dot{\theta}_m \right) = 0 \tag{3}$$

$$I_u \ddot{\theta}_u + c_1 \left(\dot{\theta}_u - \dot{\theta}_s \right) + k_{s1}(\theta_u - \theta_m) = 0 \tag{4}$$

$$k_{s2}(\theta_s - \theta_m) + c_2 \left(\dot{\theta}_s - \dot{\theta}_m \right) - k_{s1}(\theta_m - \theta_u) = 0 \tag{5}$$

The dynamic equation of Fig. 1(b) is obtained:

$$I_s \ddot{\theta}_s - c_s \left(\dot{\theta}_s - \dot{\theta}_u \right) - k_s(\theta_s - \theta_u) = 0 \tag{6}$$

$$I_u \ddot{\theta}_u + c_s \left(\dot{\theta}_s - \dot{\theta}_u \right) + k_s(\theta_s - \theta_u) = 0 \tag{7}$$

where I_s is the moment of inertia of the output shaft, θ_s is the turning angles of the output shaft, c_1 and c_2 are the damping coefficients of the two dampers, respectively, θ_m is the turning angles of the middle shaft, θ_u is the excitation of the joint.

Using the Laplace transform, we can get the transfer function between the output shaft angle and the input corner excitation of Fig. 1(a):

$$\frac{\theta_s(s)}{\theta_u(s)} = \frac{k_{s1} - k_{s1}^2(k_{s1}+k_{s2})/\left[(k_{s1}+k_{s2})^2 - c_2^2 s^2\right] + \left\{c_1 + k_{s1} - k^2/\left[(k_{s1}+k_{s2})^2 - c_2^2 w^2\right]\right\}s}{I_s s^2 + k_{s1} - k_{s1}^2(k_{s1}+k_{s2})/\left[(k_{s1}+k_{s2})^2 - c_2^2 s^2\right] + \left\{c_1 + k_{s1} - k^2/\left[(k_{s1}+k_{s2})^2 - c_2^2 s^2\right]\right\}s} \tag{8}$$

From Fig. 1(b) we can get the equivalent transfer function as:

$$\frac{\theta_s(s)}{\theta_u(s)} = \frac{k_s + c_s s}{I_s s^2 + k_s + c_s s} \tag{9}$$

The $s = jw$, w is the excitation frequency, into the Formulas (8) and (9).Then we can get:

$$\frac{\theta_s(jw)}{\theta_u(jw)} = \frac{k_{s1} - k_{s1}^2(k_{s1}+k_{s2})/\left[(k_{s1}+k_{s2})^2+c_2^2w^2\right] + \left\{c_1 + k_{s1}^2c_2/\left[(k_{s1}+k_{s2})^2+c_2^2w^2\right]\right\}jw}{-I_sw^2 + k_{s1} - k_{s1}^2(k_{s1}+k_{s2})/\left[(k_{s1}+k_{s2})^2+c_2^2w^2\right] + \left\{c_1 + k_{s1}^2c_2/\left[(k_{s1}+k_{s2})^2+c_2^2w^2\right]\right\}jw}$$

(10)

$$\frac{\theta_s(jw)}{\theta_u(jw)} = \frac{k_s + c_s jw}{-I_s w^2 + k_s + c_s jw}$$

(11)

Contrast Formulas (10) and (11) can be introduced:

$$c_s = c_1 + \frac{k_{s1}^2 c_2}{(k_{s1}+k_{s2})^2 + c_2^2 w^2} = c_1 + \frac{1}{\frac{(1+\eta)^2}{c_2} + c_2\left(\frac{w}{k_{s1}}\right)^2}$$

(12)

$$k_s = k_{s1} - \frac{k_{s1}^2(k_{s1}+k_{s2})}{(k_{s1}+k_{s2})^2 + c_2^2 w^2} = k_{s1}\left[1 - \frac{1+\eta}{(1+\eta)^2 + \left(\frac{c_2 w}{k_{s1}}\right)^2}\right]$$

(13)

Where η is the spring stiffness ratio,$\eta = \frac{k_{s2}}{k_{s1}}$,$w$ is the excitation frequency. And $\xi_1 = \frac{c_1}{2\sqrt{I_s k_{s1}}}$,$\xi_2 = \frac{c_2}{2\sqrt{I_s k_{s2}}}$,$\xi_1, \xi_2$ are the damping ratios of two dampers, respectively,.

When the excitation frequency w is constant, the Eqs. (12) and (13) are discussed as follows:

- The equivalent stiffness k_s of Fig. 1(b) is independent of the damping coefficient c_1; the equivalent stiffness k_s and the equivalent damping coefficient c_s have nonlinear function relationship with the damping coefficient c_2.
- Formula (12) shows that when c_2 is equal to zero or infinity, $c_s = c_1$; When c_2 increases from zero to infinity, the c_s will firstly increase and then decrease on, c_1 but it is always greater than c_1.
- Formula (13) shows that, when $c_2 = 0$, $k_s = \frac{k_{s1} \cdot k_{s2}}{k_{s1}+k_{s2}}$, equivalent to two springs in series. When $c_2 = \infty$, then $k_s = k_{s1}$.

From the above discussion, when the damper damping coefficient is adjusted, the equivalent damping and stiffness of the system will change at the same time. As a result, the equivalently variable stiffness range is $\left[\frac{k_{s1} \cdot k_{s2}}{k_{s1}+k_{s2}}, k_{s1}\right]$.

Assuming $I_s = 3.75 \times 10^{-5} kg.m^2$, $k_{s1} = 0.2 N.m/rad$, $\xi_1 = 0.01$, when giving ξ_2 different values, the frequency response characteristics of the system as shown in Fig. 2.

As can be seen from Fig. 2, with the damping ratio ξ_2 increasing, the resonance peak will shift to high frequency domain, which means that the equivalent stiffness will be increased.

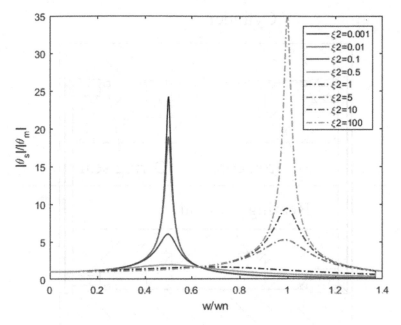

Fig. 2. The frequency response characteristics of the system

3 Structural Design of Magnetorheological Damper

To implement the variable stiffness and damping, the ability of variable damping is investigated firstly. A rotary MR damper is designed in Fig. 3. The damper consists of the outer cylinder, the iron core and the coil. In order to improve the magnetic field strength of the working area, the coil is designed as a double coil structure. The two-stage excitation coil is wound around the core through the center hole of the left end cap, and the size of the magnetic field can be changed by changing the current. The core and outer cylinder are positioned by bearings and fitted with O-rings to prevent leakage of magneto-rheological fluids. When the rotary axis moves, the magnetorheological fluid generates damping at the gap between the iron core and the outer cylinder.

Magnetic field analysis is an important part in the design of MR damper. It is necessary to simplify the structure of the damper when the finite element method is used to analyze the magnetic field. The structure and calculation model of the magnetic circuit can be simplified effectively by eliminating the components of the damper such as bearings and O-ring seals. In order to improve the computational efficiency of ANSYS calculation, the calculated model chooses the simplified 2-D axisymmetric model. The finite element model of the MR damper is shown in Fig. 4. The different areas are given the magnetic properties of the respective corresponding materials The current density applied to the magnetic field strength generated by the coil can be approximated by Eq. (14).

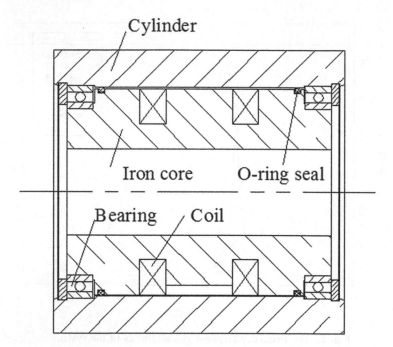

Fig. 3. The structure of MR damper

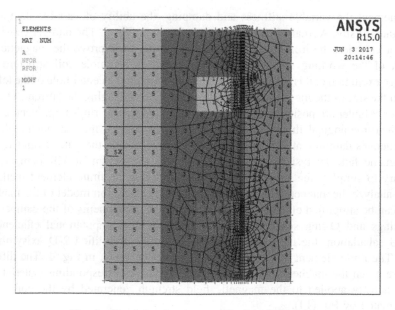

Fig. 4. The finite element model of the MR damper

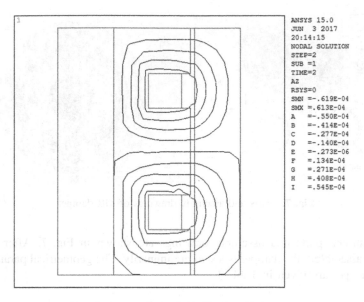

Fig. 5. Magnetic field distribution of damper

$$J = NI/A \tag{14}$$

Where N is the number of turns of the coil, I is the current size, and A is the cross-sectional area of the coil groove.

Through the previous steps, ANSYS calculated magnetic field distribution of damper shown in Fig. 5, the magnetic flux density of the distribution shown in Fig. 6.

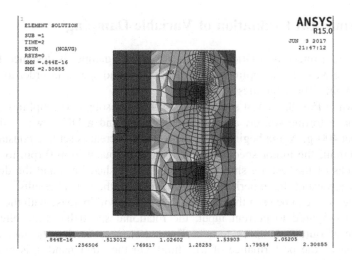

Fig. 6. The magnetic flux density of the distribution

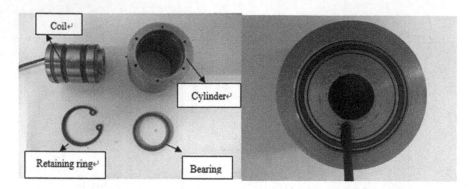

Fig. 7. Parts and assembly drawings of MR damper

The finished parts and assembly drawings are shown in Fig. 7. After the MR damper is assembled, the damper is tested preliminarily. The geometrical parameters of the MR damper are given in Table 1.

Table 1. Parameters of rotary MR damper

Parameter	Value
Gap	2 mm
Total length	44 mm
Working diameter	31 mm
Maximum diameter	42.7 mm
The number of turns of the coil	85
The resistance of the coil	2.5 Ω
The volume of MR fluid	3 mL

4 Experimental Evaluation of Variable Damping

In this section, a one-way rotation test of a rotary MR damper is designed to test the relationship between the damping torque, the current and speed, in order to verify the variable damping characteristics of actuator.

As shown in Fig. 8, the test system consists of a console, two couplings, a variable speed motor, a torque sensor, a rotary MR damper and a DC power. MR damper weighs about 400 g. At the beginning of the test, the current is set to a constant. Under the same current, the motor speed will change continuously from 0 rpm to 180 rpm.

The results of the test are shown in Fig. 9. The dashed lines and the dotted lines respectively represent the experimental results and the theoretical results.

From Fig. 9, it can be seen that the damping torque will increase with the increment of current. Compared to current input, the rotational speed has little effect on the performance of output damping torque. The reason may be that the increased temperature reduces the performance of MR fluid. On the other hand, this provides the good controllability. According to the test result, the obtained maximum damping

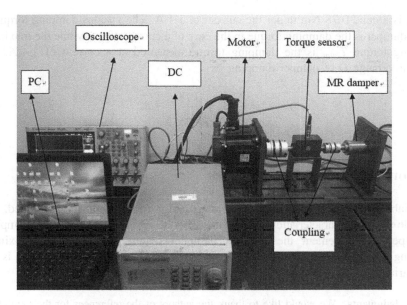

Fig. 8. The test system rotary MR damper

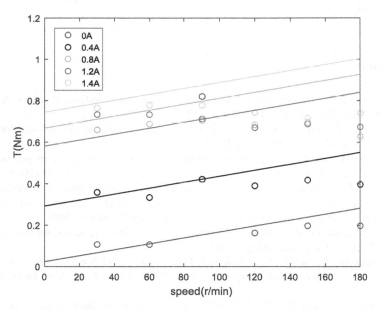

Fig. 9. Comparison of experimental and theoretical

torque is around 0.78 Nm under the current of 1.4 A. The viscous damping torque T_η of the damper is the damping torque in the case of a current of 0 A while the maximum damping torque T_{total} is the damping torque measured at a current of 1.4 A. The adjustable range D of damper damping is:

$$D = \frac{T_{total}}{T_\eta} = \frac{0.78}{0.1} = 7.8 \tag{15}$$

5   Conclusions and Future Work

A variable stiffness and damping MR actuator for robot joint is investigated. The variable stiffness characteristics are analyzed. As the first step, a rotary MR damper is developed to implement the variable damping feature. The obtained maximum damping torque reaches to 0.78 Nm. In addition, the damping dynamic range is 7.8. The variable stiffness feature will be investigated in the near future.

Acknowledgments. We would like to thank the authors of the references for their enlightenment. This research is also supported financially by the National Natural Science Foundation of People's Republic of China (Project No. 51275539 and 51675063), the Chongqing University Postgraduates' Innovation Project (No. CYB15017). These supports are gratefully acknowledged.

References

1. Vanderborght, B., Albu-Schaeffer, A., Bicchi, A.: Variable impedance actuators: A review. Robot. Auton. Syst. **61**(12), 1601–1614 (2013)
2. Visser, L.C., Carloni, R., Stramigioli, S.: Energy-efficient variable stiffness actuators. IEEE Trans. Robot. **27**, 865–875 (2011). IEEE Press
3. Bigge, B., Harvey, I.R.: Programmable springs: developing actuators with programmable compliance for autonomous robots. Robot. J. Auton. Syst. **55**(9), 728–734 (2007)
4. Robinson, D.W., Pratt, J.E., Paluska, D.J., et al.: Series elastic actuator development for a biomimetic walking robot. In: IEEE/ASME International Conference on Advanced Intelligent Mechatronics, Proceedings, pp. 561–568. IEEE Xplore (1999)
5. Grioli, G., Garabini, M., Catalano, M., et al.: Variable stiffness actuators: the user's point of view. IEEE Robot. Autom. Mag. (under review)
6. Laffranchi, M., Tsagarakis, N.G., Caldwell, D.G.: Analysis and development of a semiactive damper for compliant actuation systems. IEEE/ASME Trans. Mechatron. **18**(2), 744–753 (2013)
7. Laffranchi, M., Tsagarakis, N., Caldwell, D.G.: A compact compliant actuator (CompAct™) with variable physical damping. In: EEE International Conference on Robotics and Automation, pp. 4644–4650. IEEE Xplore(2011)
8. Laffranchi, M., Tsagarakis, N.G., Caldwell, D.G.: A variable physical damping actuator (VPDA) for compliant robotic joints. In: IEEE International Conference on Robotics and Automation, pp. 1668–1674. IEEE Xplore(2010)

9. Hurst, J.W., Rizzi, A.A., Hobbelen, D.: Series elastic actuation: potential and pitfalls. In: International Conference on Climbing and Walking Robots (2004)
10. Chew, C.M., Hong, G.S., Zhou, W.: Series damper actuator: a novel force/torque control actuator. In: IEEE/RAS International Conference on Humanoid Robots, pp. 533–546. IEEE Xplore(2004)
11. Garcia, E., Arevalo, J.C., Muñoz, G., et al.: Combining series elastic actuation and magneto-rheological damping for the control of agile locomotion. Robot. Auton. Syst. **59** (10), 827–839 (2011)
12. Radulescu, A., Howard, M., Braun, D. J., Vijayakumar, S.: Exploiting variable physical damping in rapid movement tasks. In: 2012 IEEE/ASME International Conference on Advanced Intelligent Mechatronics (AIM), pp. 141–148. IEEE, Taiwan (2012)
13. Enoch, A., Sutas, A., Nakaoka, S.I., Vijayakumar, S.: Blue: A bipedal robot with variable stiffness and damping. In: 2012 12th IEEE-RAS International Conference on Humanoid Robots (Humanoids), pp. 487–494. IEEE (2012)
14. Roy, N., Newman, P., Srinivasa, S.: CompAct™ Arm: a Compliant Manipulator with Intrinsic Variable Physical Damping. MIT Press (2012)
15. Imaduddin, F., Mazlan, S.A., Zamzuri, H.: A design and modelling review of rotary magnetorheological damper. Mater. Des. **51**(5), 575–591 (2013)
16. Jiménezfabián, R., Verlinden, O.: Review of control algorithms for robotic ankle systems in lower-limb orthoses, prostheses, and exoskeletons. **34**(4), 397–408 (2011)
17. Yu, J., Dong, X., Wang, W.: Prototype and test of a novel rotary magnetorheological damper based on helical flow. Smart Mater. Struct. **25**(2), 025006 (2016)
18. Chen, J., Liao, W.H.: Design and control of a Magnetorheological actuator for leg exoskeleton. In: International Conference on Robotics and Biomimetics, pp. 1388–1393. IEEE (2007)
19. Dong, X., Yu, M.: Absorbing control of magneto-rheological variable stiffness and damping system under impact load. Trans. Chin. Soc. Agric. Mach. (2010)
20. Yu, M., Liao, C., Chen, W., Huang, S.: Research on control method for MR damper. Chin. J. Chem. Phys. (2001) (in Chinese)

Physical Field-Enhanced Intelligent Space with Temperature-Based Human Motion Detection for Visually Impaired Users

Jiaoying Jiang[1], Kok-Meng Lee[1,2(✉)], and Jingjing Ji[1(✉)]

[1] State Key Laboratory of Digital Manufacturing Equipment and Technology, Huazhong University of Science and Technology, Wuhan 430074, Hubei, China
kokmeng.lee@me.gatech.edu, jijingjing@hust.edu.cn
[2] Woodruff School of Mechanical Engineering, Georgia Institute of Technology, Atlanta, GA 30332-0405, USA

Abstract. This paper presents a method utilizing temperature field to improve the indoor mobility of the blind/visually-impaired-people (Blind/VIP) in a physical field-enhanced intelligent space (iSpace) that takes advantages of the rapidly developing cloud-computing and personal mobile devices (generally built with sound, image, video and vibration alert capabilities) to share way-finding information among users. A method, which uses temperature fields and their gradients to detect face-orientation and analyze leg-postures for predicting the motion states of other humans in a traveling path, is introduced. The concept feasibility of the temperature-based human motion detection has been experimentally validated in a simulated school environment for the Blind/VIP where users are generally familiar with stationary objects but less confident in daily walking in a crowd where human motion is unpredictable. The experimental findings presented in this paper establish a basis for developing temperature filed-enhanced iSpace.

Keywords: Temperature filed · Motion detection · Blind/VIP

1 Introduction

Rapid advances of cloud-computing, along with the wide availability of personal mobile devices (capable of computing, communication and control, digital imaging, and global positioning) at affordable cost, offer potentials to enhance the global perception of the user's surrounds, particularly for people who lost the optical ability to interpret their surroundings. This paper explores the concept of a physical-field-enhanced intelligent space (iSpace) to help overcome some daily problems encountered in a typical school for the Blind/Visually-impaired-people (Blind/VIP). As the use of geomagnetic field effects on magnetic waypoints for way-finding has been discussed and experimentally validated in [1, 2], this paper focuses on developing an avoidance technique based on temperature fields to help the blind/VIP acquire the motion states of other humans along a traveling path around congested areas.

Methods on obstacle avoidance can be classified into human-generated (active) and nature-generated (passive) signals [3] categories for navigation. Active-signal methods

© Springer International Publishing AG 2017
Y. Huang et al. (Eds.): ICIRA 2017, Part III, LNAI 10464, pp. 120–129, 2017.
DOI: 10.1007/978-3-319-65298-6_12

requires an emitter/receiver pair to generate/detect signals (like ultrasonic and sonar) to determine the existence of the obstacles and their locations [4]; they are less sensitive to environmental influences but have limited range of operation. Passive-signal methods (for example, the uses of thermal infrared or optical fields [5, 6] naturally emitted/reflected from the obstacles to estimate the obstacle information), in general, have a potential to offer a low-cost solution. High-resolution imaging systems with advanced algorithms are now widely available in personal digital assistant (PDA) at relatively low-cost. Blind/VIP can share images/videos captured by a wearable camera through internet permitting a remotely guided navigation [7, 8]. More recently, rapid developing social networks (such as Twitter, Facebook and RSS) further enhance the global perception of the surrounding environments [9]. However, these technologies are generally ineffective for close range way-finding by a VIP.

In environments such as a typical school for the Blind/VIP, users are generally familiar with stationary objects but less confident in daily walking in a crowd where human motion is unpredictable. The human body, as a natural heat source, is significant in the infrared thermogram. For this reason, we explore the uses of thermal infrared to develop a temperature-based human motion detection. Temperature field is also a widely available physical field in nature and can be measured using infrared thermal imager; Applications include nondestructive testing [10, 11] (for detecting cracks, delamination and other defects), temperature map for supersonic wind tunnel testing in field of aerospace [12], temperature field reconstruction for monitoring tool wears in cutting [13] and diagnosing diseases [14, 15] in medicine.

The remainder of this paper offers the following:

- The concept of physical field enhanced intelligent space for the Blind/VIP is presented, where physical information based on temperature and magnetic fields is embedded in a navigation map.
- A method for detecting human motion state by using the temperature fields and their gradients is proposed. The infrared band (instead of the visible light band) with physically intuitive heat transfer model is exploited to simplify the detection of the face orientation, which provides a basis to determine the motion states of other humans in a traveling path.
- The concepts of the temperature-based face-orientation detection and leg-posture analysis have been experimentally illustrated and validated in a test environment simulating a Wuhan blind school for the Blind. As will be shown, the method effectively detect four different orientations of the face from the measured IR temperature fields, and differentiate the motion states from the leg posture analyzed using image processing routines.

2   Filed Enhanced iSpace for Blind/VIP

Figure 1 illustrates the design concept of intelligent space (iSpace) where geomagnetic and thermal fields are used to enhance building floor maps and complement optical fields commonly used with digital cameras that are now widely available in smart phones. An immediate application of an iSpace is an enhanced environment that helps the blind and visually impaired people (blind/VIP) users overcome problems encountered in daily way-finding. As an illustrative example and experimental setup, Fig. 1(a) shows the plan view of a teaching building in Wuhan Blind School and the path from the first floor to a classroom on the second floor through a corridor, where the trashcan①, water heater②, staircase handrail③ and iron fence④ along the path are existing ferromagnetic landmarks (or stationary magnetic waypoints) for identifying locations [1, 2]. The blind/VIP is generally familiar with common stationary objects (such as tables, chairs, etc.) in the room, but often find challenging to walk in a crowd; for examples, a person who moves in the same direction is unlikely an "obstacle", but a person who stands still or moves in opposite or a different direction may likely interfere with his/her walking. An effective method to identify "real barriers" in a crowd, rather than treating everyone as an obstacle, is a common problem to the blind/VIP.

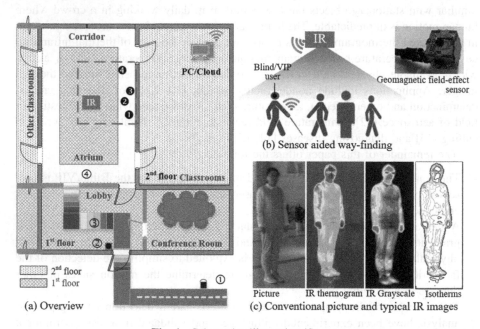

Fig. 1. Schematics illustrating iSpace

Figure 1(b, c) illustrates a real-time temperature-based method to improve the blind/VIP's perception of the surrounding, where the thermal images were captured using an infrared sensor to identify human bodies and their motion states around the blind/VIP. Unlike a conventional image which includes many other details in the scene making human detection a difficult image-processing task, the thermal image (Fig. 1c) significantly reduces the difficulty of human detection by taking advantages of the fact that human faces' temperature is generally much higher than the ambient as well as other objects in the surrounding, and can be used to determine his/her walking direction.

The iSpace being built for the blind and VIP users consists of a PC Cloud server (that includes floor maps enhanced with magnetic waypoints and navigation algorithms) and several infrared (IR) thermal imagers mounted at strategic locations. Once entering the iSpace, the user (with a standard VIP-adapted smart phone and geomagnetic sensor shown in Fig. 1b) communicates with the Cloud server that processes IR images (Fig. 1c) and shares information of other human around the user (through voice or vibration in existing VIP devices) judgment on the surrounding environment. In this paper, a corridor of a Wuhan blind school has been selected for concept-feasibility studies using temperature fields to enhance an iSpace, where the detection of human directions is typical in a relatively clouded space environment. The IR Imager mounted perpendicular to the corridor provides information complement to image data (viewed in the direction of the corridor), which are now widely available in personal mobile device such as a smart phone.

The use of geomagnetic field effects on magnetic waypoints for way-finding has been discussed in [1, 2], this paper focuses on developing an avoidance technique for daily way-finding, which helps the blind/VIP acquire the motion states of other humans along a traveling path particularly around congested areas.

2.1 Infrared Head Temperature Imaging for Face Orientation

IR radiation that corresponds to a wavelength range between visible light and microwave (0.75 ~ 1000 μm) has a significant thermal effect. All object in the natural temperature above the absolute zero (−273 °C) is constantly emitting infrared rays. The human body is essentially a natural bio-heat source producing heat to maintain the body temperature T at about 37 °C with a small range of fluctuations. The normal dry skin of the human body is close to the ideal blackbody surface ($\sigma = 5.67 \times 10^{-8}$ W/m$^2\cdot$K^4) with an emissivity ε of 0.98 to 0.99, by Stephen-Boltzmann's Law, emitting infrared heat energy $E = \varepsilon\sigma T^4$ which can be received by a detector and converted into digital images representing the thermal characteristics of the human body. Figure 1(c) compares four representations of a human image captured by an IR imager; conventional grayscale "optical picture" that includes many surrounding details in the image background, and IR temperature field presented in pseudo-colors, IR field in grayscale and isotherms. Unlike the grayscale picture (1st from left in Fig. 1c) that requires often tedious and time-consuming filtering and segmentation to identify the human body, the presence of the human body in the IR thermos-gram (2nd from left in Fig. 1c) can be easily identified for subsequent image processing (last from left in Fig. 1c).

The heat transfer mechanism of biological tissue is extremely complex. Given in Eq. (1), the Pennes' heat conduction model 16 that bases on the following assumptions has been most widely used in the study of biological heat transfers: The tissue is a continuous medium; the blood flow is equal to the heat transfer and heat deposition (or heat source) of the tissue; and the local structure and blood flow velocity speed are neglected. Please note that the first paragraph of a section or subsection is not indented.

$$\rho_t c_t \frac{\partial T}{\partial \tau} = k\nabla^2 T + \rho_b w_b c_b (T_b - T) + q_m + q_r \tag{1}$$

In Eq. (1) which can be solved with appropriately defined initial conditions and boundary conditions, ∇^2 is Laplacian operator; (ρ_t, c_t and k) are the (density, specific heat and thermal conductivity) of the tissue; w_b is the blood perfusion rate; T_b is the arterial Blood temperature; ρ_b and c_b are the density and specific heat of the blood, respectively; q_m is the heat production rate of tissue metabolism; and q_r is the external heat source heat. The biothermal model (1) reveals that the surface temperature of the human body is influenced by subcutaneous blood circulation, local tissue metabolism, skin thermal conductivity, skin and environmental temperature and humidity exchange constraints [17, 18]. Thus, the metabolism of various parts of the human body and heat dissipation are different; for examples, the parts with more blood vessels distribution (such as mouth, forehead, sulci nasolabialis, eyes) has a higher temperature than the parts covered by hair [19]. Combined with the infrared thermogram, this feature can be used to identify and determine the face and hence the motion states of the human.

2.2 Motion State from Gradient-Based Face Orientation

The detection of a human consists of two parts; namely, human face, and shape/motion characteristics of the human posture. Because human face can be characterized by a specific temperature range, this can be used to distinguish a human face form other thermogenic objects in the IR image. Once the human face is identified, other body parts can be derived from the overall shapes and motion state of the human in the field of view.

Figure 2 illustrates the theoretical basis orientation using heat transfer models for detecting the face (and hence the motion state of the human). As shown in Fig. 2(a), the human head (particularly the forehead) has a significantly higher temperature than the other parts of the body, and thus can be readily located by means of image

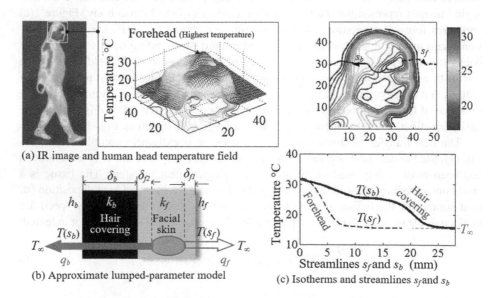

(a) IR image and human head temperature field

(b) Approximate lumped-parameter model

(c) Isotherms and streamlines s_f and s_b

Fig. 2. Human head temperature model for detecting face orientation

segmentation with an appropriately selected temperature threshold. In modeling the human head temperature, the radiative heat transfer is neglected as compared to the conduction and convection heat transfer.

For finding the face orientation from the IR image, the steady-state heat transfer of the human head is approximated by a lumped-parameter model of two homogeneous (forehead facial and hair covering) regions as shown in Fig. 2(b), where the two heat fluxes (qf and qb) flow along two opposite paths (sf and sb) in the (same and opposite) directions of walking as shown in Fig. 2(c). The highest temperature region is treated as a bio-heat source from which (qf and qb) flows to the ambient environment at temperature T_∞ through the two paths:

$$T\left(s_f\right) - T_\infty = q_f \left[k_f^{-1}\left(\delta_{f1} - s_f\right) + h_f^{-1} \right] \text{ where } s_f \leq \delta_{f1}. \tag{2a}$$

$$T(s_b) - T_\infty = q_b \begin{cases} k_f^{-1}\left(\delta_{f2} - s_b\right) + k_b^{-1}\delta_b + h_b^{-1} & \text{where } s_b \leq \delta_{f2} \\ k_b^{-1}\left(\delta_b + \delta_{f2} - s_b\right) + h_b^{-1} & \text{where } \delta_{f2} \leq s_b \leq \delta_b + \delta_{f2} \end{cases} \tag{2b}$$

In (2a, b), s_f and s_b are the two paths starting from the highest temperature T_m contour; (k_f and k_b,) and (h_f and h_b,) are the (thermal conductivities and convective heat transfer coefficients) of the two regions; and δ_{f1}, δ_{f2} and δ_b are the lengths of the paths defined in Fig. 2(c), through which the heat fluxes flow. Equations (2a, b) offer some intuitive insights into the solutions to the face orientation detection from IR images that measure surface temperatures. The arguments are best illustrated numerically with Fig. 2(c) where the temperatures along the two streamlines (perpendicular to the equal-temperature contours) of the side-facing human are compared:

- Since the parametric values for $(q_f, k_f, h_f, \delta_f)$ and for $(q_b, k_b, h_b, \delta_b)$ are different, the temperature gradients along the two streamlines are expected to be different.
- For the same temperature drop $(T_m - T_\infty)$, the temperature $T(s_f)$ along the streamline s_f through the forehead has a much steeper gradient than that $T(s_b)$ along the streamline s_b.
- Unlike $T(s_f)$ that linearly drops from T_m to T_∞, $T(s_b)$ experiences more than one gradients for the same temperature drop.
- Similar reasoning can be made to differentiate between the front-facing and back-facing of the same human, where $T(s_f)$ and $T(s_b)$ are symmetric and have similar gradients, but the forehead will show a higher maximum temperature T_m than the hair temperature.

Once the face orientation is known, the motion state can be determined from the leg postures of the human using image processing routines. For the front- and back-facing orientation, an additional view in the orthogonal direction can be used to simplify the identification. The procedure to determine the motion state from the leg posture for a side facing orientation consists of three steps: (1) Obtain the binary image of the human from the IR grayscale image with a thresholding operation. (2) Using morphological operations, reduce the image blob representing the human to a skeleton. (3) Compare the angle θ between the pair of legs with a threshold θ_{th} to determine the motion state:

$$State = \begin{cases} \text{In motion} & \theta \geq \theta_{\text{th}} \\ \text{Stationary} & \theta < \theta_{\text{th}} \end{cases} \tag{3}$$

3   Experimental Results and Discussion

Experiments were conducted to illustrate and evaluate the method to determine the motion state from the face orientation based on temperature gradients of the IR image and the leg-posture (using morphological skeleton processing of a binary image). The experimental results are given in Figs. 3, 4 and 5.

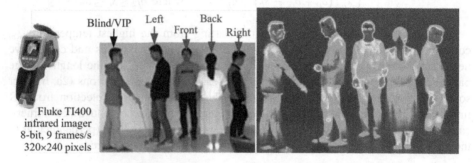

Fig. 3. Experimental setup (Image taken at a distance of 5.5 m)

Figure 3 shows the experiment setup simulating an environment commonly encountered in a daily walking in a crowd, where an 8-bit FLUKE Ti400 infrared thermal camera (320×240 pixels, 9 frames/s) was used to capture the scene at a distance of 5.5 m, where four peoples faced different directions in front of the blind/VIP to illustrate the face-orientation detection experimentally. The results demonstrating the face orientation detection based on temperature gradients are summarized in Fig. 4; each sub-figure shows the isotherms and temperatures along the two streamlines (on the left s1 and right s2) from the highest-temperature isotherm, and the temperature field.

The following are some observations drawn from Figs. 3, 4 and 5:

- For all the four different peoples (including Fig. 2c), the foreheads have the highest temperature (approximately 35 °C) where the starting points of the two streamlines (s_1 and s_2) can be clearly defined. On the other hand, the back of the head (primarily covered by hairs) exhibits a lower (but more uniformly distributed) temperature than the forehead; thus, the starting points of the two streamlines (s_1 and s_2) for the back of the head are arbitrarily defined.

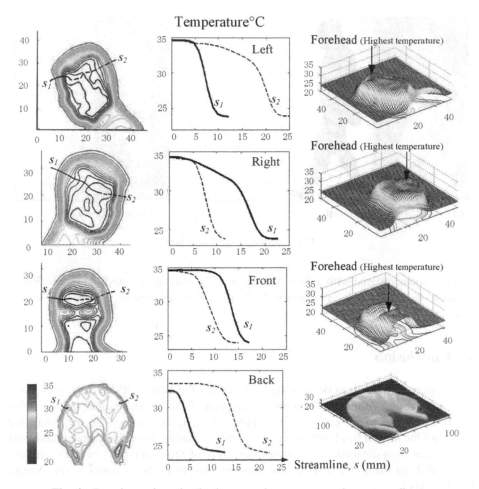

Fig. 4. Experimental results (isotherms and temperatures along streamlines)

– Unlike the front/back facing, the streamline-temperature for the side-facing is non-symmetric. The forehead temperature drops monotonically from 35 °C to the ambient temperature (24 °C) within a short distance of approximately 5 mm. On the other hand, two distinct temperature gradients (over a distance of 20 mm) can be observed along the path of conduction heat transfer on the hair-covering back. As compared in Fig. 4 (between the top two rows), the temperature field of the right-oriented face is opposite to that of the left-oriented face.

Figure 5 shows the results of the motion-states determined from the leg-postures for a side-facing orientation. As illustrated in Fig. 5(a), the procedure begins with a binary image obtained using a thresholding operation from the IR image; the image blob representing the human is then reduced to a skeleton by means of a morphological operation; next, straight-line segments are found by Canny edge detection followed by Hough transform; and finally the angle θ between the two legs is computed using inner

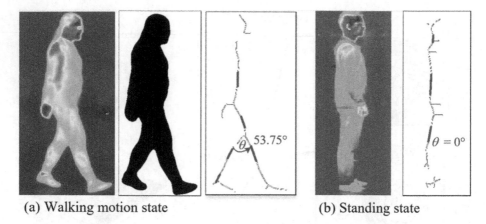

(a) Walking motion state   (b) Standing state

Fig. 5. Motion state detection

product and compared with a threshold (θ_{th} = 45° is used in this study). Figures 5(a, b) compare two different leg-postures with the same face orientation, θ = 53.75° walking and $\theta \approx 0°$ for standing.

4   Conclusion

The concept of multi-physics intelligent space for the Blind/VIP has been presented, where physical information from the geomagnetic field-effects and temperature fields are used to effectively enhance the traditional navigation map. The procedure for detecting the human head, face orientation and leg posture from the temperature-based images has been detailed. The concept feasibility of the method has been experimentally validated using a test setup that simulates a typical environment in the Wuhan Blind School.

The experimental findings presented in this paper have been encouraging, which will help establish a basis for developing temperature filed-enhanced iSpace in the future. Research is being directed towards integrating other physical fields in the experimental iSpace testbed and its practical implementation.

References

1. Lee, K.-M., Li, M.: Magnetic tensor sensor for gradient-based localization of ferrous object in geomagnetic field. IEEE Trans. Magn. **52**(8) (2016)
2. Lee, K.-M., Li, M., Lin, C.-Y.: Magnetic tensor sensor and way-finding method based on geomagnetic field effects with applications for visually impaired users. IEEE/ASME Trans. Mechatron. **21**(6), 2694–2704 (2016)
3. Dakopoulos, D., Bourbakis, N.N.G.: Wearable obstacle avoidance electronic travel aids for blind: a survey. IEEE Trans. Syst. Man Cybern. Part C: Appl. Rev. **40**(1), 25–235 (2010)

4. Ando, B., Baglio, S., Marletta, V., Valastro, A.: A haptic solution to assist visually impaired in mobility tasks. IEEE Trans. Hum.-Mach. Syst. **45**(5), 635–640 (2015)
5. Ando, B.: A smart multisensor approach to assist blind people in specific urban navigation tasks. IEEE Trans. Neural Syst. Rehabilit. Eng. **16**(6), 592–594 (2008)
6. Bourbakis, N., Makrogiannis, S.K., Dakopoulos, D.: A system-prototype representing 3D space via alternative-sensing for visually impaired navigation. IEEE Sens. J. **13**(7), 2535–2547 (2013)
7. Garaj, V., Hunaiti, Z., Balachandran, W.: Using Remote Vision: The Effects of Video Image Frame Rate on Visual Object Recognition Performance. IEEE Trans. Syst. Man Cybern. Part A: Syst. Hum. **40**(4), 698–707 (2010)
8. Hunaiti, Z., Garaj, V., Balachandran, W.: An assessment of a mobile communication link for a system to navigate visually impaired people. IEEE Trans. Instrum. Meas. **58**(9), 3263–3268 (2009)
9. Xiao, J., Joseph, S.L., Zhang, X., Li, B., Li, X., Zhang, J.: An assistive framework for the visually impaired. IEEE Trans. Hum. Mach. Syst. **45**(5), 635–640 (2015)
10. Jin, G.F., Zhang, W., Song, Y.J., Yang, Z.W., Wang, D.D.: Numerical simulation of ultrasonic infrared thermal wave detection with curvature structure crack. Sci. Technol. Eng. **13**(3), 776–779 (2013)
11. Serra, C., Tadeu, A., Prata, J., Simões, N.: Application of 3D heat diffusion to detect embedded 3D empty cracks. Appl. Therm. Eng. **61**(2), 596–605 (2013)
12. Cardone, G., Ianiro, A., Ioio, G.D., Passaro, A.: Temperature maps measurements on 3D surfaces with infrared thermography. Exp. Fluids **52**(2), 375–385 (2012)
13. Ji, J.J., Lee, K.M., Huang, Y., Li, C.Y.: An investigation on temperature measurements for machining of titanium alloy using IR imager with physics-based reconstruction. In: IEEE/ASME AIM2015, Pusan, Korea (2015)
14. Zhang, Y., Mahemuty, D., Jiang, S.C., Zhang, X.X.: Numerical simulation and heat transfer in A and therapy in Uyghur medicine. J. Zhongnan Univ. (Med. Sci.) (2010)
15. Li, K.Y., Dong, Y.G., Chen, C., Zhang, S.P.: The noninvasive reconstruction of 3D temperature field in a biological body with monte carlo method. Neurocomputing **72**(1–3), 128–133 (2008)
16. Pennes, H.H.: Analysis of tissue and arterial blood temperatures in the resting human forearm. J. Appl. Physiol. **85**(1), 5–34 (1998)
17. Kundu, B.: Exact analysis for propagation of heat in a biological tissue subject to different surface conditions for therapeutic applications. Appl. Math. Comput. **285**, 204–216 (2016)
18. Jones, B.F.: A reappraisal of the use of infrared thermal image analysis in medicine. IEEE Trans. Med. Imaging **17**(6), 1019–1027 (1999)
19. Zhang, D.: Surface temperature distribution and infrared image analysis of normal human body. Laser Infrared **17**(3), 52–56 (1994)

Optimal Design and Experiments of a Wearable Silicone Strain Sensor

Tao Mei[1], Yong Ge[1], Zhanfeng Zhao[2], Mingyu Li[1], and Jianwen Zhao[1(✉)]

[1] Department of Mechanical Engineering, Harbin Institute of Technology, Weihai 264209, China
zhaojianwen@hit.edu.cn
[2] Department of Electrical Engineering, Harbin Institute of Technology, Weihai 264209, China

Abstract. Motion capture of human body potentially holds great significance for exoskeleton robots, human-computer interaction, sports and rehabilitation research. Dielectric Elastomer Sensors (DESs) are excellent candidates for wearable human motion capture system because of their intrinsic characteristics of softness, lightweight and compliance. Fabrication process of the DES was developed, but a very few of optimal design is mentioned. To get greater measurement precision, in this paper, some optimization criteria was put forward and validated by some experiments. As a practical example, the sensor was mounted on the wrist to measure joint rotation. The experiment results indicated that there is a roughly linear relationship between the output voltage and the joint angle. Therefore, the DES can be applied to motion capture of human body.

Keywords: Dielectric elastomer sensor · Optimal design · Joint angle measurement

1 Introduction

In recent years, the demands of human motion capture have been increasing in the fields of exoskeleton robots, human-computer interaction, sports, and many others. Some technologies for motion capture are already applied in engineering, but they suffer from serious limitations. For example, optical system is limited to the room where the cameras are installed [1]; Inertial Measurement Units (IMUs) suffer from integration drift [2, 3]; goniometers are bulky and uncomfortable because of the hard elements [4].

Dielectric Elastomer Sensors (DESs) are soft, lightweight and compliant, can be worn directly on the human body, and overcomes the above disadvantages. It is an excellent candidate for wearable human motion capture system. The DES can be used as a pressure sensor or a stretchable sensor. The state of the art in DES pressure sensors is a growing field. In one study, Kasahara et al. [5] fabricated a flexible capacitive tactile sensor and could detect the direction and distribution of force. In another, Kim et al. [6] used plastic structures and dielectric elastomer capacitors to fabricate a six-axis force/torque sensor. In a third study, Böse et al. [7, 8] designed an innovative dielectric elastomer compression sensor which could convert the compression load to a tensile load by the various wave profile structures and had a high sensitivity. The research involving DES stretchable sensor is also being widely reaserched. In one example, Anderson et al.

© Springer International Publishing AG 2017
Y. Huang et al. (Eds.): ICIRA 2017, Part III, LNAI 10464, pp. 130–137, 2017.
DOI: 10.1007/978-3-319-65298-6_13

[9] developed a stretchable sensor to measure strain up to 80% with good accuracy. In addition, Considerable work has been done on self-sensing transducer namely on how to extract strain information during actuation [10–13].

The optimal design is a necessity work for DES to improve measurement precision. However, there are a few of researches about it. Therefore, this paper is focused on the optimal design of DES.

2   Working Principle of the DES

The typical DES consists of three layers of DE membranes and two layers of soft electrodes, as shown in Fig. 1(a). The middle layer is the dielectric, while the outer layers are used for protection. This structure creates an electrically flexible capacitor. The simplified equivalent electrical model is a variable capacitor C in series with two variable resistors R, as show in Fig. 1(b).

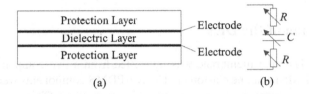

Protection Layer
Dielectric Layer
Protection Layer

(a)   (b)

Fig. 1. (a) Typical structure of the DES (b) Equivalent circuit of the DES

Both the capacitance and the resistance will change with deformation of the DES, and these changes can be used to infer the deformations. Compared to the series resistance, measuring changes in capacitance with deformation is more straightforward and reliable because the resistance is extremely dependent on the electrode uniformity and environmental temperature [14]. According to Eq. (1), the capacitance C is simply related to the overlapping electrode area A, the thickness of the dielectric layer d, the relative permittivity of the dielectric layer material ε_r, and the permittivity of free space ε_0.

$$C = \frac{\varepsilon_0 \varepsilon_r A}{d} \tag{1}$$

The schematic diagram of detection circuit is shown in Fig. 2. Its working principle is that the capacitance of the sensor is converted into the circuit output voltage by integrating the charging current of the capacitance. By setting appropriate circuit parameters, the relation between DES capacitance and circuit output voltage can be expressed as:

$$V_{OUT} = kC \tag{2}$$

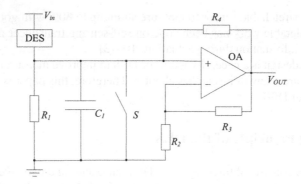

Fig. 2. The schematic diagram of detection circuit

Where k is a constant, decided by the circuit parameters, V_{OUT} is the circuit output voltage, C is the capacitance of DES.

3   Fabrication of the DES

The dielectric elastomer membrane was fabricated from a two-component polydime-thylsiloxane (PDMS) silicone elastomer. The two PDMS components were mixed, with the addition of a solvent (isooctane) using a planetary mixer (Thinky ARE-310) that also degassed the PDMS/solvent solution. The isooctane lowered the viscosity of the mixture, making it easier to cast. The mixture was cast on a substrate (polyethylene terephthalate, PET) using an automatic film coater (ZEHNTNER, ZAA2300) with a thickness determined by a universal applicator (ZEHNTNER, ZUA2000) [15]. The PDMS was then allowed to cure in the oven at approximately 80 °C. The soft electrode consisted of carbon particles (EC300 J) dispersed in a soft silicone matrix (MED4901, Nusil) with a 1:10 mass ratio [16]. The electrode was then applied to both sides of the dielectric layer and subsequently cured in an oven at approximately 80 °C. Finally, a short exposure to oxygen plasma was used to bond the electrode and protection layers together [17].

4   Optimal Design of the DES

To precisely measure human joint movement, high sensitivity is needed. For example, an output voltage noise has a smaller influence on measurement precision of the sensor with a higher sensitivity than a lower one for DESs. Besides, the great sensitivity can also improve the resolution.

4.1   Theoretical Analysis and Optimization Criteria

The undeformed shape of the electrode area of the DES was a rectangle. So the stretch ratio of sensor in three directions can be defined as:

$$\lambda_1 = \frac{l}{l_0} = \frac{l_0 + \Delta l}{l_0}$$

$$\lambda_2 = \frac{b}{b_0} = \frac{b_0 + \Delta b}{b_0} \tag{3}$$

$$\lambda_3 = \frac{d}{d_0} = \frac{d_0 + \Delta d}{d_0}$$

Where, l_0 is original length of electrode area of the sensor, b_0 is original width of electrode area of the sensor and d_0 is original thickness of the dielectric layer. Δl is variation of the length, Δb and Δd have similar definition.

So, the Formula (1) can be rewritten as:

$$C = \frac{\varepsilon_0 \varepsilon_r \lambda_2 b_0 l}{\lambda_3 d_0} \tag{4}$$

Because the volume of the sensor cannot be compressed and the sensor is uniaxial stretched, namely,

$$\lambda_2 = \lambda_3 \tag{5}$$

The Formula (4) can be rewritten as:

$$C = \frac{\varepsilon_0 \varepsilon_r b_0 l}{d_0} \tag{6}$$

The sensitivity of DES can be defined as:

$$S = \frac{dV_{OUT}}{dl} = \frac{k_1 \varepsilon_0 \varepsilon_r b_0}{d_0} \tag{7}$$

Based on Formula (7), sensitivity of the DES can be improved by increasing the width of electrode area, decreasing thickness of the dielectric layer or selecting the material with higher relative permittivity as the dielectric layer.

4.2   Experiments and Hysteresis Characteristic of the DES

To verify above optimization criteria, we fabricated some types of DESs with the same protection layers made of Silbione LSR 4305 and same length of electrode area. The DES was mounted on a one-dimensional motorized platform and could be stretched at a specified speed, starting position, and ending position. The corresponding output voltages were recorded by interval of displacement 500 μm.

Some DESs were fabricated by using same materials and different width of electrode area. Relationship between deformation and output voltage of the DES is shown in Fig. 3(a). According to the Fig. 3(b), it can be concluded that the sensor sensitivity will be higher with boarder electrode width, and is also proportional to the electrode width.

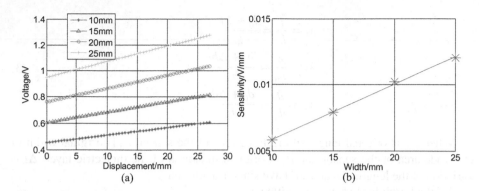

Fig. 3. (a) Relationship between deformation and output voltage of the DES (b) Sensitivity of the DES with different electrode width

Figure 4(a) shows similar characteristic of the DES with different thickness of dielectric layer. The thickness refers to the height set by the universal applicator, not the real thickness of DE membrane, but the two parameters are proportional [15]. According to the Fig. 4(b), it can be concluded, the thicker the dielectric layer, the higher the sensitivity will be. Sensitivity of the DES is roughly inversely proportional to the thickness of the dielectric layer.

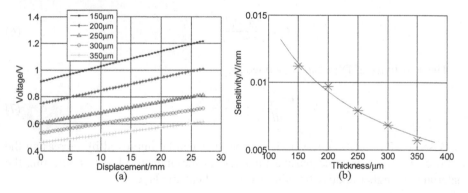

Fig. 4. (a) Relationship between deformation and output voltage of the DES (b) Sensitivity of the DES with different thickness of dielectric layer.

Figure 5 shows similar characteristic of the DES with different dielectric layer materials. Sylgard 186 (Dow Corning), Silbione LSR 4305 (Bluestar Silicones), and MED-4901 (NuSil) were chosen to fabricate the dielectric layer, and their sensitivities are 0.0079 V/mm, 0.0094 V/mm and 0.01 V/mm respectively.

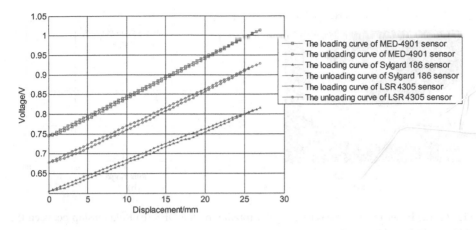

Fig. 5. Hysteresis loops of DESs with different materials of dielectric layer

The DESs have obvious hysteresis because of the inherit viscoelasticity of the dielectric material. The hysteresis loops of Sylgard 186, Silbione LSR 4305 and MED-4901 are shown in Fig. 5, and their maximum hysteresis are 1.34%, 0.86% and 0.79% respectively. Besides, if all points on the hysteresis loop were considered, the characteristic of MED-4901 is much better than Sylgard 186 and Silbione LSR 4305. Hysteresis will reduce precision of the DES directly, especially for dynamic measurement; therefore, MED-4901 is a suitable material to fabricate the dielectric layer among the three popular silicone materials.

According to the experiment results, sensitivity of the DES can be improved by increasing width of the electrode area, by decreasing thickness of the dielectric layer or by selecting suitable material with low hysteresis.

However, wider electrode area will make the DES and skin more difficult to fit well. Therefore, the better choice is to apply a thinner dielectric layer to fabricate the DES, and the thinner layer can make people more comfortable because of the lower rigidity. On the other hand, the material with lower hysteresis should be chosen to fabricate the dielectric layer, and the protection layer should have lower Young's modulus and higher strength to make the DES softer and more reliable.

5   Measuring of Joint Angle by the DES

To validate the DES in measurement of joint rotation, the DES was mounted on the wrist by adhesive tape, as shown ass Fig. 6(a). When the wrist was bending, the sensor would be also stretching. Meanwhile, the joint angle could be obtained by photographing and then picture processing. The experiment results are shown in Fig. 6(b). The fitting straight line is $y = 0.0017x + 0.5967$, and R-Squared is 0.9682. The result indicates that the output voltage of the DES is positively related to the joint angle, therefore, the DES can be applied to measure the joint rotation.

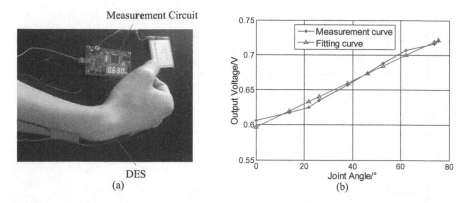

Fig. 6. (a) Experiment to measure the joint rotation by the DES (b) Relationship between the output voltage and the joint angle

6 Conclusion

To DES, a higher sensitivity also means greater discrimination that is expected to measure the joint rotation. Therefore, some optimization criteria were concluded to improve sensitivity and precision of the DES. Firstly, the wider the electrode area, the higher the sensitivity, but too wide DES will limit the fitting of skin and the DES, so, a suitable wider electrode area should be determined. Secondly, thinner dielectric layer can make higher sensitivity. Finally, the material with lower hysteresis should be chosen because the soft material has more obvious hysteresis.

In general, the dielectric layer should be fabricated by the material with lower hysteresis and should be as thinner as possible; to make the DES softer and keep the qualified strength, the protection layer should be fabricated by the material with lower Young's modulus and enough strength. Other technologies also can be applied to improve precision of the DES, such as using high specification electrical components, filtering and shielding. When the DES was used to measure joint angle, the experiment results indicated that the output voltage of the DES is positively related to the joint angle, therefore, the DES is qualified to be applied to capture human joint rotation.

Acknowledgments. This work was supported by the National Natural Science Foundation of China (Grant No. 91648106), the Natural Science Foundation of Shandong Province (Grant No. ZR2016EEM16) and the Foundation of State Key Laboratory of Mechanical Strength and Vibration (Grant No. SV2016-KF-13).

References

1. Goebl, W., Palmer, C.F.: Temporal control and hand movement efficiency in skilled music performance. PLoS ONE **8**(1), 96–98 (2013)
2. Zhou, H., Stone, T., Hu, H., et al.: Use of multiple wearable inertial sensors in upper limb motion tracking. Med. Eng. Phys. **30**(1), 123–133 (2008)

3. Zhou, H., Hu, H.F.: Inertial motion tracking of human arm movements in stroke rehabilitation. In: 2005 IEEE International Conference Mechatronics and Automation, vol. 3, pp. 1306–1311. IEEE (2005)
4. Mcgorry, R.W., Chang, C.C., Dempsey, P.G.F.: A technique for estimation of wrist angular displacement in radial/ulnar deviation and flexion/extension. Int. J. Ind. Ergon. **34**(1), 21–29 (2004)
5. Kasahara, T., Mizushima, M., Shinohara, H., et al.: Simple and low-cost fabrication of flexible capacitive tactile sensors. Jpn. J. Appl. Phys. **50**(1), 317–326 (2011)
6. Kim, D., Lee, C.H., Kim, B.C., et al.: Six-axis capacitive force/torque sensor based on dielectric elastomer. In: Proceedings of SPIE - The International Society for Optical Engineering, vol. 8687(3), 86872J (2013)
7. Böse, H., Fuß, E.F.: Novel dielectric elastomer sensors for compression load detection. In: Proceedings of SPIE - The International Society for Optical Engineering, vol. 9056, 905614 (2014)
8. Böse, H., Fuß, E., Lux, P.F.: Influence of design and material properties on the performance of dielectric elastomer compression sensors. In: Proceedings of SPIE - The International Society for Optical Engineering, vol. 9430, 943029 (2015)
9. Stretchsense Homepage. http://stretchsense.com. Accessed 31 May 2017
10. Jung, K., Kim, K.J., Choi, H.R.F.: A self-sensing dielectric elastomer actuator. Sens. Actuators A Phys. **143**(2), 343–351 (2008)
11. Gisby, T.A., O'Brien, B.M., Anderson, I.A.F.: Self-sensing feedback for dielectric elastomer actuators. Appl. Phys. Lett. **102**(19), 193703–193704 (2013)
12. Goulbourne, N.C., Son, S., Fox, J.W.F.: Self-sensing McKibben actuators using dielectric elastomer sensors. In: Proceedings of SPIE - The International Society for Optical Engineering, vol. 6524, 652414 (2007)
13. Matysek, M., Haus, H., Moessinger, H., et al.: Combined driving and sensing circuitry for dielectric elastomer actuators in mobile applications. In: Proceedings of SPIE - The International Society for Optical Engineering, vol. 7976(6), pp. 541–558 (2011)
14. O'Brien, B., Anderson, I., Calius, E.F.: Integrated extension sensor based on resistance and voltage measurement for a dielectric elastomer. In: Proceedings of SPIE - The International Society for Optical Engineering, vol. 6524(15), pp. 1–11 (2007)
15. Samuel, R., Araromi, O.A., Samuel, S., et al.: Fabrication process of silicone-based dielectric elastomer actuators. J. Visualized Exp. **108**, e53423 (2016)
16. Araromi, O., Poulin, A., Rosset, S., et al.: Thin-film dielectric elastomer sensors to measure the contraction force of smooth muscle cells. In: Spie Smart Structures and Materials + Nondestructive Evaluation and Health Monitoring. International Society for Optics and Photonics, vol. 9430, 94300Z (2015)
17. Araromi, O.A., Rosset, S., Shea, H.R.F.: High-resolution, large-area fabrication of compliant electrodes via laser ablation for robust, stretchable dielectric elastomer actuators and sensors. ACS Appl. Mater. Interfaces **7**(32), 18046–18053 (2015)

Mobile Robotics and Path Planning

Research and Implementation of Person Tracking Method Based on Multi-feature Fusion

Fang Fang[✉], Kun Qian, Bo Zhou, and Xudong Ma

Key Laboratory of Measurement and Control of CSE, School of Automation,
Southeast University, Nanjing 210096, Jiangsu, People's Republic of China
ffang@seu.edu.cn

Abstract. Aiming at the problem of person tracking for mobile robot in complex and dynamic environment, a multi-feature tracking strategy is proposed in this paper, by which the target can be determined based on the joint similarity. The joint similarity consists of motion model similarity, color histogram similarity and human HOG feature similarity. The tracking of target is realized by the method of joint likelihood data association. The above strategy can solve the problems such as similar color interference, target loss, and target occlusion. In addition, considering the lost target, a fast search strategy is proposed to search the target. Finally, the method is tested with the mobile robot. The experimental results show that the proposed method is robust and effective when the target is moving rapidly, and it can satisfy the real-time requirement of the system.

Keywords: Tracking · Multi-feature · Joint similarity · Joint likelihood data association

1 Introduction

Person tracking is one of the focuses in service robots field [1, 2]. In recent years, the RGB-D camera is successfully applied in mobile robots platform for the advantage of easy access to depth information as well as the good stability and cost-effective feature [3]. The person detecting and tracking based on Kinect camera are researched by Armando Pesenti Gritti, et al. realizing the person tracking by detecting the legs for the height limit of camera [4]. The researches of faces detection and upper body detection based on RGB-D camera are studied by Duc My Vo, realizing the tracking by the Kalman filter algorithm which can deal with the pose change and target occlusion problem [5]. Christian Dondrup used laser and RGB-D camera information as inputs and fused different types of sensor data, which finally realizes human tracking with the Kalman filter algorithm [6]. The face detection and recognition based on RGB-D camera are studied by Wolfgang Rosenstiel and finally the person tracking is achieved [7]. However, the above researches have the common precondition that the motion of target is smooth, which means it is easy to lose the target when the motion is not smooth. Therefore, to improve the performance of tracking with the challenges such as rapid motion and target occlusion, the tracking strategy based on multi-feature fusion is

© Springer International Publishing AG 2017
Y. Huang et al. (Eds.): ICIRA 2017, Part III, LNAI 10464, pp. 141–153, 2017.
DOI: 10.1007/978-3-319-65298-6_14

proposed in the paper, which contains the motion model similarity, color histogram similarity and human HOG feature similarity, realizing the person tracking by the method of joint likelihood data association. In addition, to deal with the target loss problem in tracking process, the quick search strategy is proposed which can deal with the problems such as similar color interference, target loss and occlusion.

2   Overall Design of Tracking Method

The person tracking method based on multi-feature fusion involves three aspects: (1) motion trend estimation of target person; (2) the feature of color distribution statistics for target area; (3) morphological feature of target. The three channels of information are fused and the tracking is realized by the joint likelihood data association, which can improve the stability of tracking with the challenge of dynamic interference. The overall design of tracking method is shown as Fig. 1.

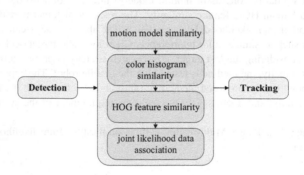

Fig. 1.   Framework of target person tracking method

As shown in Fig. 1, The calculated confidence of detected body is the input of the tracking module. Firstly, the motion model is built with the target position forecast by the particle filter algorithm, after which the similarity is calculated based on the predicted position and actual position. Besides, the color histogram similarity is calculated based on color histogram of target body and detected body, based on which the HOG feature similarity is calculated later. Finally, the tracking is realized by the method of joint likelihood data association which synthetically considers the motion model similarity, color histogram similarity and HOG feature similarity. The joint similarity is used to judge whether the detected body is the target person, which is stable and reliable in the dynamic environment.

3 Similarity Calculation

3.1 Motion Model Similarity

Motion Model Establishment

The motion model similarity can be calculated by the Mahalanobis distance between the predicted position and actual detected position. In dynamic and complex environment, the constant velocity model is established considering the interferential objects and occlusion, which can predicate the person position conveniently based on the particle filter algorithm [8]. Taking x and y to describe the position in environment. In motion model, the particle filter need to be updated based on the position of each frame. The state vector and observation vector are as Eqs. (1) and (2).

$$x_k = (x \quad y \quad \dot{x} \quad \dot{y})^T \tag{1}$$

$$z_k = (x \quad y)^T \tag{2}$$

The mean value is zero and the variance value is σ_a^2 with the assumption that the directions of x and y are satisfied with the normal distribution. The motion equation is shown as Eq. (3).

$$x_k = F \cdot x_{k-1} + G \cdot a_k \tag{3}$$

$$F = \begin{pmatrix} 1 & 0 & \Delta t & 0 \\ 0 & 1 & 0 & \Delta t \\ 0 & 0 & 1 & 0 \\ 0 & 0 & 0 & 1 \end{pmatrix}, G = \begin{pmatrix} \frac{\Delta t^2}{2} & 0 \\ 0 & \frac{\Delta t^2}{2} \\ \Delta t & 0 \\ 0 & \Delta t \end{pmatrix}, a_k = \begin{pmatrix} \ddot{x} \\ \ddot{y} \end{pmatrix} \tag{4}$$

The dynamic system model of filter is shown in Eq. (5):

$$\begin{cases} x_k = F \cdot x_{k-1} + w_k \\ z_k = H \cdot x_k + v_k \end{cases} \tag{5}$$

In Eq. (5), the process noise w_k stratifies the facts that mean value is zero and it is multi-parameter normal distribution. The variance is $Q(w_k \sim N(0, Q))$ where:

$$Q = G^T G \cdot \sum_a = \begin{pmatrix} \frac{\Delta t^2}{2} & 0 \\ 0 & \frac{\Delta t^2}{2} \\ \Delta t & 0 \\ 0 & \Delta t \end{pmatrix} \cdot \begin{pmatrix} \frac{\Delta t^2}{2} & 0 \\ 0 & \frac{\Delta t^2}{2} \\ \Delta t & 0 \\ 0 & \Delta t \end{pmatrix}^T \cdot \begin{pmatrix} \sigma_{ax}^2 \\ \sigma_{ay}^2 \end{pmatrix} \tag{6}$$

The parameters of observation model are as followings:

$$v_k \sim N(0, R_k)$$

$$H = \begin{pmatrix} 1 & 0 & 0 & 0 \\ 0 & 1 & 0 & 0 \end{pmatrix}$$

$$R_k = \begin{pmatrix} 1 & 0 \\ 0 & 1 \end{pmatrix} \cdot (\sigma_v^2 + \sigma_d^2)$$ (7)

The covariance matrix of noise is the sum of σ_v^2 with σ_d^2, which respectively represent the quantization error of voxel grid filter and the measuring error of RGB-D camera.

Target Position Prediction

The target position prediction is accomplished by the particle filter algorithm. The posterior probability density is estimated based on the priori probability density and current observation value. Firstly, the point cloud data is obtained from the RGB-D camera and the points are initialized as particles, after which each particle is assigned a weight and the weights normalization is finished. The target position can be replaced by the weighted average value of all particles [9].

① State probability distribution

The state probability density can be approximated by the empirical probability distribution. $z_{1:k}$ is the measurement set at time of k, $x_k^{(i)}$ represents the particle i, $\delta(\cdot)$ is the Dirac function, and $P(x_{0:k}|z_{1:k})$ is the probability density of x.

$$P(x_{0:k}|z_{1:k}) = \frac{1}{N} \sum_{i=1}^{N} \delta x_k^{(i)}(dx_{0:k})$$ (8)

② Particles generation

The particles $X_0^i \sim P(x_{0:k}|z_{1:k}), i = 1, \ldots, M$ can be sampled from $P(x_{0:k}|z_{1:k})$.

③ Weight calculation

To solve the difficulty in sampling from $P(x_{0:k}|z_{1:k})$, the known probability distribution function $q(x_{0:k}|z_{1:k})$ is introduced and the weight can be calculated as Eq. (9).

$$w_k^i = w_{k-1}^i \frac{P(z_k|x_k^i)P(x_k^i|x_{k-1}^i)}{q(x_k^i|x_{0:k-1}^i, z_{0:k})}$$ (9)

④ Weight normalization

Weights normalization can be calculated as Eq. (10).

$$w_k^{(i)} = \frac{w_k^i}{\sum\limits_{i=1}^{M} w_k^i} \tag{10}$$

⑤ Target position prediction

The target position can be replaced by the weighted average value of all particles. At the time of k, the weight of i-th particle is $w_k^{(i)}$, the target position can be predicted based on the following equation.

$$x_k = \sum_{i=1}^{N} x_k^{(i)} w_k^{(i)} \tag{11}$$

Motion Model Similarity Calculation
The motion model similarity can be calculated by the Mahalanobis distance between the target predicted position and the actual detected position. The Mahalanobis distance between target person i and detected person j can be calculated as Eq. (12), where $i = 1, j = 1, 2, 3, 4$.

$$D_M = \tilde{z}_k^T(i,j) \cdot S_k^{-1}(i) \cdot \tilde{z}_k(i,j) \tag{12}$$

$S_k(i)$ is the covariance matrix of tracking i, $\tilde{z}_k(i,j)$ is the residual vector between measurement vector $z_k(i,j)$ and the forecast vector $\hat{z}_{k|k-1}(i)$ as Eq. (13).

$$\tilde{z}_k(i,j) = z_k(i,j) - \hat{z}_{k|k-1}(i) \tag{13}$$

Where $z_k(i,j)$ consists of position information of measurement j and velocity information of tracking i. When $i = 1$, D_M can be represented as:

$$D_M = [D_{1,1}, D_{1,2} \ldots, D_{1,j}]^T \tag{14}$$

The similarity between predicted target position and actual detected position can be represented as Eq. (15), where $D_{\max} = \max\{D_{1,1}, D_{1,2}, \cdots, D_{1,j}\}$. The similarity is lower if the Mahalanobis distance is larger.

$$\rho_M = \begin{bmatrix} 1 - D_{1,1}/D_{\max} \\ 1 - D_{1,2}/D_{\max} \\ \cdots \\ 1 - D_{1,j}/D_{\max} \end{bmatrix} \tag{15}$$

3.2   Color Histogram Similarity

In the process of target tracking, it is necessary to compare the color feature information in the current view. The similarity between the color histogram feature of the regional image and the target is calculated to judge whether the detected object is the target person. Commonly used similarity calculation methods are Euclidean distance, Pasteur distance. The Pasteur coefficient is used in this paper to represent the similarity of the target area. The Bhattacharyya coefficient is a discrete probability density function. Take $\vec{p}_i$ and $\vec{q}_i$ respectively as the color histogram eigenvector of the target template and the region to be observed. The similarity coefficient is defined as Eq. (16).

$$\rho_{color} = \rho(\vec{p}, \vec{q}) = \sum_{i=1}^{m} \sqrt{\vec{p}_i, \vec{q}_i} \tag{16}$$

It can be seen from Eq. (16) that the Bhattacharyya coefficient is proportional to the similarity of the color histogram, which means that the larger the Bhattacharyya coefficient is, the higher the similarity. The distance between the target template and the observation area is calculated as in Eq. (17), where the smaller the Pasteur distance is, the greater the similarity.

$$d(\vec{p}, \vec{q}) = \sqrt{1 - \rho(\vec{p}, \vec{q})} \tag{17}$$

The color histogram feature of the target person is extracted, and whether the detected body is the target person is determined by calculating the similarity between the color histogram feature of the target person and each detected body.

3.3   Human HOG Feature Similarity

Human body detection based in HOG and SVM can only judge whether an object is a body, which may be the target person, the interferential person or human-like object. Therefore, it is necessary to calculate the human feature similarity to exclude the interferences. The HOG feature similarity can be obtained from the HOG feature of the human body. After a large number of experiments, the results can be summarized that the minimum value of confidence threshold min_*confidence* is −1.5 and the maximum confidence max_*confidence* is 2.0.

$$\rho_{HOG} = \frac{person\_confidence - \min\_confidence}{\max\_confidece - \min\_confidence} \tag{18}$$

The HOG feature confidence test is performed on the standing postures of different people, different sexes and similar human body. The similarity of HOG feature is calculated, and the test results are shown in Table 1. From the table it can be seen that the similarity is 0.35 when the human-like body is taken for the real person by mistaken, while the real body similarity is 0.70 or so, which means the similarity of human-like body is lower compared to the real human body.

Table 1. Similarity of human features based on HOG

Situation	HOG feature similarity
Stand upright	0.71
Stand side	0.73
Stand upright (Male)	0.69
Stand upright (Female)	0.71
Human-like body	0.35

4 Joint Likelihood Data Association

4.1 Correlation Probability Calculation

Data association is the key to tracking implementation. There are many methods of data association, such as Nearest Neighbor Data Association (NNDA), Probabilistic Data Association (PDA), Joint Probability Data Association (JPDA, often referred to as Joint Likelihood Data Association). The joint likelihood data association is a good compromise in the performance and computational loss when the human body is not very dense. Therefore the joint likelihood probability is used in this paper to deal with data. The correlation probability between each detection object j and the tracking target i is calculated as Eq. (19), which involves three likelihood probabilities which respectively based on motion model, color information and HOG feature.

$$L_{TOT}^{i,j} = L_{motion}^{i,j} \cdot L_{color}^{i,j} \cdot L_{HOG}^{i,j} \tag{19}$$

If the three likelihood probabilities are large, the combined likelihood probability obtained by multiplication is greater than a certain threshold, which means the detected body can be considered as the tracking target. The greater the value is, the greater the likelihood that the human body will be the target, that is the likelihood of detecting the human body as a tracking target is proportional to the joint likelihood. In order to simplify the calculation, Eq. (19) is changed to Eq. (20). After the conversion, the monotonic relationship has changed. The smaller $l_{TOT}^{i,j}$ is, the possibility is larger.

$$l_{TOT}^{i,j} = -\log(L_{TOT}^{i,j}) = \gamma \cdot D_M^{i,j} + \alpha \cdot c_{color}^{i,j} + \beta \cdot c_{HOG}^{j} \tag{20}$$

The $D_M^{i,j}$ is the Mahalanobis distance between the tracking i and detection j. The $c_{color}^{i,j}$ is the color histogram confidence between tracking i and detection j. The c_{HOG}^{j} is the HOG feature confidence of detection j. Coefficients γ, α and β are determined by the experience, which are respectively 0.9, 0.9 and 0.7, so the minimum value of joint confidence is 0.567.

4.2 Tracking Process Realization

Only the target person is in the robot view at the beginning. The detected body is initialized as the tracking target.

① Weight update

When the human body appears occlusion, the visibility of the human body color feature decreases, and the weight of the color feature need to be reduced. When the target is lost and then returned to the view of robot, the similarity of the motion model is unreliable, and the weight of the motion model based on the motion model needs to be reduced.

② Target distinction

If the interferential body has the same color with target person, it cannot be distinguished by the feature information based on the color histogram. In this situation, the target person position can be predicted by the particle filter algorithm. Since the position of the tracking target person is constant in a short time. Therefore, using the particle filter algorithm to predict the location of the human body can deal with the interference of same color feature, and the robot can distinguish the target to achieve tracking.

③ Tracking recovery

When the target is lost, the robot need to search the target (the search strategy will be introduced in next part). In the search process, firstly the color histogram similarity is calculated if a possible target is found and then the HOG feature similarity is calculated to judge whether the searched body is the human body.

5   Search Strategy for Target Loss

It was found that when the target person quickly disappeared form the robot vision, the robot would not detect the target person. In this situation, the target is lost and the robot need to retrieve the target again. There are many ways to search the target, such as blob detection, autonomous rotation detection. These methods are global search strategy, and the search takes a long time [11]. In order to improve the reliability, the quick search strategy need to be designed.

Assume that before the target is lost, the target center is defined as C, the angle of the movement as $\phi(\phi \in [-180°, +180°])$, and the outer rectangle of the target is determined by the height and width. According to the principle of motion continuity, the most likely position is the position near the center and at the range of ϕ.

The N is the search layer, the d is the search radius. On each layer, there are $8 * N$ search directions. If search starts at the direction n ($n = 0, 1, \ldots, 8 * N$), the following directions are: n, $(n+1) \bmod (8 * N)$, $(n-1) \bmod (8 * N)$, $(n+2) \bmod (8 * N) \cdots$. The search radius d is defined as $d \in ((N+1) * width, (N+1) * height)$, the n can be obtained by $n = floor(\frac{angle}{360/(N*8)}, n \subset \{0, 1, \ldots, 7\})$, where $angle = \begin{cases} \varphi, & if(\varphi \geq 0) \\ -\varphi, & if(\varphi < 0) \end{cases}$.

The $floor(x)$ means to get the maximum value that is not larger than x. For example, if $N = 1$, $\phi = 45°$. The search sequence is $(1, 0, 2, 7, 3, 6, 4, 5)$. Actually, if the first layer search is finished and the target is still not be found, the robot will stop searching next layer, alert and wait the target to back to view, avoiding useless search.

6   Experimental Results and Analysis

Considering the fact that the camera is moving, the joint likelihood tracking strategy involving the motion model, color features and HOG feature is proposed in this paper. In order to verify the effectiveness of the tracking strategy, the proposed method is validated by the mobile service robot experiment platform (Fig. 2). The tracking program is developed on the ROS platform. The tracking algorithm is implemented with OpenNI and point cloud library PCL. The tracking algorithm is evaluated by a series of experiments. For the particle filter algorithm, the number of particles is initialized to 100.

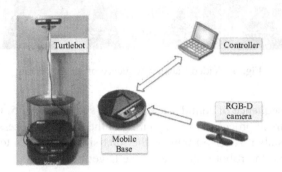

Fig. 2. The hardware platform

6.1   Target Tracking Experiment

The initial distance between the target person and the robot is 1.0 m, and a constant distance of 0.8 m is maintained between the robot and the target during the tracking process. In the case of single target without interference, the robot can reliably track the target person. During the tracking process, the robot and the target are moving. The speed of target person is about 0.3 m/s. The routes are shown in Fig. 3.

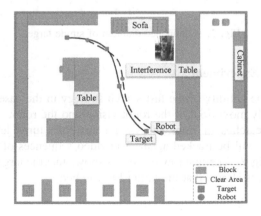

Fig. 3. Trajectory of tracking in case of single person

The video sequences of target tracking for single-person without interference are shown in Fig. 4, which are respectively the first frame, the 67th frame, the 189th frame, and the 265th frame. The first frame is at the beginning time. In the 67th frame, the target is moving forward. In the 189th frame, the target is turning left and the 265th frame shows that the target is going to stop. In this case, the similarity of the motion model is about 0.92, the similarity of the color feature is about 0.91, the similarity of the HOG feature is about 0.7, and the joint similarity is about 0.6. The experiment shows that the tracking is stable and effective.

Fig. 4. Video tracking sequence of single person

The tracking sequence of confidence diagram is shown in Fig. 5. It can be seen that there is an adjustment process at the beginning of the tracking process and the tracking similarity is not stable, After 60 frames, the joint similarity tends to be stable (about 0.6), at which time the robot stably tracks the target person.

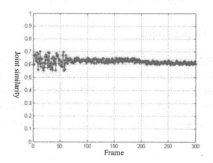

Fig. 5. Confidence diagram of single target

6.2   Target Loss Experiment

In order to verify the validity of the fast search strategy in the case of target loss, the target person quickly moves out of the robot vision and the robot will lose the target. The robot firstly searches in front, turns right and then turns left. If the target is searched, the target will be tracked again. The video sequences of tracking for target loss are shown in Fig. 6. The first row of images shows that the target quickly lost. The second row of images shows that the target be searched.

Fig. 6. Video tracking sequence of target loss

The tracking sequence of confidence diagram is shown in Fig. 7. It can be seen that when the target is tracked, the joint similarity is high at the beginning, when the target is quickly lost and the target is not searched in the right direction, the joint similarity is zero. When the target is lost, the motion feature will be unreliable and its coefficient will be zero, and the coefficients based on the color information and the HOG feature will be correspondingly larger.

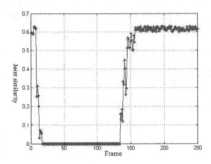

Fig. 7. Confidence diagram of tracking sequence for target loss

6.3 Target Occlusion Experiment

The video sequences of tracking for target occlusion are shown in Fig. 8. The third image shows that the target is occluded by the interferential body and the robot will search forward until the block disappears. The fourth image shows that the block disappears and the robot continues tracking. When the target is blocked, the coefficient based on the color feature will be unreliable and its coefficient will be zero, and the coefficients based on the motion model information and the HOG feature will be adjusted to one-half, respectively.

Fig. 8. Video tracking sequence in case of target occlusion

The results of the experiment are shown in Table 2 by using the joint likelihood data association algorithm, the particle filter tracking algorithm and the Cam-Shift tracking algorithm respectively. The experimental results show that joint likelihood data association algorithm has better tracking performance in scenes of lighting change, target occlusion, target loss and interference, compared with the other two algorithm.

Table 2. Comparison of experimental results of three algorithms

Method	Method in this paper	Particle filter	Cam-Shift
Normal	30	25	22
Lighting change	30	11	9
Target occlusion	28	20	5
Interference	30	21	23
Target loss	26	19	2

7 Conclusion

Aiming at the target tracking of mobile service robot in complex and dynamic environment, a multi-feature target tracking strategy is proposed in this paper. The key of the tracking strategy is the joint likelihood data association method which includes three similarities respectively based on the motion model, the color histogram feature and the HOG feature. The strategy can solve the problems such as same color interference, target loss and recovery, target occlusion, etc. In addition, considering the target loss, a quick search strategy is designed. Finally, the strategy is verified by experiments.

Acknowledgment. The authors would like to acknowledge the valuable support of Natural Sciences Foundation (NNSF) of China (No. 61573100, No. 61573101).

References

1. Takashi, Y., Nishiyama, M., Sonoura, T., et al.: Development of a person following robot with vision based target detection. In: Proceedings of the 2006 IEEE/RSJ International Conference on Intelligent Robots and Systems, Beijing, China, pp. 5286–5291. IEEE (2006)
2. Mekonnen, A., Lerasle, F., Herbulot, A.: Cooperative passersby tracking with a mobile robot and external cameras. Comput. Vis. Image Underst. 117(10), 1229–1244 (2013)
3. Huazhu, F., Dong, X., Stephen, L.: Object-based RGBD image co-segmentation with mutex constraint, pp. 4428–4436. School of Computer Engineering, Nanyang Technological University (2015)
4. Gritti, A., Tarabini, O., Guzzi, J., et al.: Kinect-based people detection and tracking from small-footprint ground robots. In: IEEE/RSJ International Conference on Intelligent Robots and Systems (IROS 2014), pp. 4096–4103. IEEE (2014)
5. Vo, D., Jiang, L., Zell, A.: Real time person detection and tracking by mobile robots using RGB-D images. In: IEEE International Conference on Robotics and Biomimetics (ROBIO), pp. 689–694. IEEE (2014)
6. Tasaki, T., Ozaki, F., Matsuhira, N., et al.: People detection based on spatial mapping of friendliness and floor boundary points for a mobile navigation robot. J. Robot. 1, 1–10 (2011)
7. Choi, W., Pantofaru, C., Savarese, S.: Detecting and tracking people using an RGB-D camera via multiple detector fusion. In: Computer Vision Workshops (ICCV Workshops), pp. 1076–1083 (2011)
8. Munaro, M., Menegatti, E.: Fast RGB-D people tracking for service robots. Auton. Robots 37(3), 227–242 (2014)
9. Bergen, J.R., Anandan, P., Hanna, K.J., Hingorani, R.: Hierarchical model-based motion estimation. In: Sandini, G. (ed.) ECCV 1992. LNCS, vol. 588, pp. 237–252. Springer, Heidelberg (1992). doi:10.1007/3-540-55426-2_27
10. Nummiaro, K., Koller-Meier, E., Van Gool, L.: An adaptive color-based particle filter. Image Vis. Comput. 21(1), 99–110 (2003)
11. Kailath, T.: The divergence and Bhattacharyya distance measures in signal selection. IEEE Trans. Commun. Technol. 15(1), 52–60 (1967)

Method and Experiment of the NAO Humanoid Robot Walking on a Slope Based on CoM Motion Estimation and Control

Qingdan Yuan[✉], Zhigang Xi, Qinghua Lu, and Zhihao Lin

School of Mechatronics Engineering,
Foshan University, Foshan 528000, Guangdong, China
yuanqingdan@126.com

Abstract. In this paper, a method of biped walking on a slope is studied, taking the NAO humanoid robot as research object. Firstly, we adopt the 3D linear inverted pendulum mode (3D-LIPM) to generate walking pattern, and obtain the reference trajectory of the center of mass (CoM). Then the Denavit-Hatenberg (D-H) parameters of the leg chain are determined based on the analysis of the NAO leg configuration. Finally in the process of walking, an extended Kalman filter (EKF) via fusing sensor data is used to estimate the robot's CoM motion, in addition, an inverse kinematics (IK) controller is implemented which regulates the CoM position state in real time based on position tracking errors. On a slope with given angle, walking uphill and downhill experiments are conducted. The experimental results show that for the NAO humanoid robot, the deviation of walking direction can be controlled within 2 cm, so that it can keep walking stability for a long distance.

Keywords: NAO humanoid robot · Biped walking · CoM motion estimation

1 Introduction

As an important branch of mobile robots, from the robot morphology, humanoid robots are the universal mobile and operating platforms which are the most suitable for working together with human. In all imitation of human behaviors, one of the outstanding features that robots should have is the walking function. It is the most important task to achieve stable walking in humanoid robot research.

The study of humanoid robot walking is mainly focused on two aspects: one is to generate the best walking pattern to achieve the optimal gait planning; the second is the perception of robot walking information, and using the appropriate control strategy to ensure its stability. For the previous problem, a number of ways to generate the walking pattern are already formed [1–3], which are offline methods based on the ZMP stability criterion. The second problem, in fact, should be solved is how to generate walking pattern in real time and adjust the gait based on the pose of robot. In this case, specific trajectories, precision, and repeatability are not important factors and the motion can be different in each step, planning and control are unified in this approach. The online walking pattern generation method can be applied to uneven roads, but many online

© Springer International Publishing AG 2017
Y. Huang et al. (Eds.): ICIRA 2017, Part III, LNAI 10464, pp. 154–165, 2017.
DOI: 10.1007/978-3-319-65298-6_15

controllers are required to compensate for the modeling inaccuracy by using various sensory feedback data, and thus involving experimental hand tuning.

The problem of robot walking on the inclined floor is concerned by many scholars. Zheng and Shen make a simple modification of the robot's walking pattern on the horizontal surface, and apply it to the slope [4]. Chew et al. presented an approach to compute the hip height for dynamically stable walking on the inclined plane based on geometric consideration [5]. Kim proposed a walking control algorithm of biped humanoid robot on uneven and inclined floor [6]. Besides, there were some soft computing-based approaches to generate walking motion on the inclined plane [7, 8]. Most of the above studies are carried out for the prototype in the laboratory, so some methods are not universal.

The NAO is a small size and popular robot now, which is widely used in education, competition, service and other occasions. Due to the popularity of the NAO robot, more and more scholars take the NAO as a humanoid robot platform, and make a good progress in robot biped walking [9, 10]. Hornung studied the problem of increasing the localization accuracy while climbing stairs [11]. Wu designed a novel sensory mapping to identify the state when the NAO walks on a slope [12]. The two papers mentioned above mainly focus on the navigation of indoor environment and the perception of terrain.

The rest of the paper is organized as follows. The generation method of robot's walking pattern on a slope is mainly introduced in Sect. 2. The characteristics of the NAO leg configuration are analyzed to determine the D-H parameters of the leg chain in Sect. 3. The estimated method and the controller of the robot's CoM are mainly described in Sect. 4. Section 5 is the walking experiment. Finally, the conclusion is presented in Sect. 6.

2 Walking Pattern Generation

The schematic diagram of the NAO robot walking on a slope is shown in Fig. 1. The inclined angle of slope is α. The red arrow indicates the walking direction of the robot, the angle between the walking direction and the gradient vector is expressed by β. The green arrow represents the orientation of the robot's torso, which can be represented by a quaternion: $q = [q_0 q_1 q_2 q_3]^T$. The robot performs a longitudinal walk when $\beta = 0°$ or $\beta = 180°$, or an omnidirectional walk when β is any other angle.

Fig. 1. Schematic diagram of robot walking on a slope.

The humanoid robot is much more difficult than multi-legged robot to maintain dynamic stability while walking, especially when the floor is uneven. There are many researchers who have investigated the dynamic balance motion criterion and walking patterns such as zero moment point (ZMP), virtual supporting point (VSP), dual length linear inverted pendulum method (DLLIPM) [13–15]. In the above partial methods, there are many steps in the motion planning, and there is a big gap between the actual effect and the theoretical calculation. On account of the configuration of the NAO's leg freedom, this paper focuses on longitudinal walking on the inclined plane. The walking pattern generation method is the 3D linear inverted pendulum mode (3D-LIPM) [2]. In this method, the CoM trajectory is limited to a plane that is parallel to the slope plane. The motion equations for CoM can be expressed as formula (1):

$$
\begin{aligned}
\ddot{x} &= \frac{gcos\alpha}{z_c}x - gsin\alpha \\
\ddot{y} &= \frac{gcos\alpha}{z_c}y
\end{aligned}
\tag{1}
$$

z_c is the height of the CoM from the slop surface in the formula. The ZMP function can be obtained as the formula (2):

$$
\begin{bmatrix} \ddot{x} - \frac{\ddot{z}+gcos\alpha}{z}x + gsin\alpha \\ \ddot{y} - \frac{\ddot{z}+gcos\alpha}{z}y \end{bmatrix} = -\frac{\ddot{z}+gcos\alpha}{z}\begin{bmatrix} x_{zmp} \\ y_{zmp} \end{bmatrix}
\tag{2}
$$

The solutions of the linear differential formula (1), which give the CoM motion trajectory in 3D-LIPM on an inclined plane, can be expressed as formula (3):

$$
\begin{aligned}
x(t) &= (x(0) - z_c\tan\alpha)\cosh\left(\frac{t}{T_c}\right) + T_c\dot{x}(0)\sinh\left(\frac{t}{T_c}\right) \\
y(t) &= y(0)\cosh\left(\frac{t}{T_c}\right) + T_c\dot{y}(0)\sinh\left(\frac{t}{T_c}\right)
\end{aligned}
\tag{3}
$$

In the formula, $T_c = \sqrt{z_c/(g \cdot cos\ \alpha)}$. $(x(0), \dot{x}(0))$ and $(y(0), \dot{y}(0))$ are the initial position and velocity of CoM in the x direction and y direction respectively. The CoM trajectory corresponding to the pre-planned ZMP trajectory can be generated, as shown in Fig. 2. According to the CoM trajectory and the landing position, combined with the inverse kinematics of the leg chain, the complete gait data can be obtained, and the walking pattern on the slope is generated offline.

3 NAO's Leg Configuration and D-H Parameters of Leg Chain

The NAO is a 58 cm tall and about 5 kg weight humanoid robot manufactured by Aldebaran Robotics in France. It has 25 degrees of freedom, as shown in Fig. 3. Each leg has 5 degrees of freedom, the foot size is about 15 cm × 8 cm.

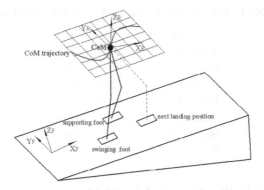

Fig. 2. Dynamic walking model.

The order of the joints in the NAO leg chain from bottom to top successively is AnkleRoll, AnklePitch, KneePitch, HipPitch and HipRoll. It should be noted that different from some humanoid robots with 6 independent degrees of freedom in each leg, the LHipYawPitch and RHipYawPitch of NAO are coupled on one serve and cannot move independently. NAO has a great difficulty in implementing omnidirectional walking because the legs movement must constitute an asymmetrical inverted pendulum motions. The HipYawPitch, HipRoll and HipPitch have intersecting axes at the hip. For the analysis of inverse kinematics, the joint angle of the HipYawPitch also needs to be calculated theoretically, because the HipYawPitch may affect the position and orientation of the foot. As shown in Fig. 3, in the leg chain, the thigh length and the tibia length are 100 mm and 102.9 mm respectively, which are two important parameters.

The Denavit-Hatenberg (D-H) model was used to calculate the inverse kinematics of the NAO. The predefined parameters are listed in Table 1, the origin is under the foot and the endpoint is the torso, in order to simplify the calculation. Because the HipYawPitch axes are not orthogonal to the other axes, both angles of the LHipYawPitch and RHipYawPitch are set to zero. As a result, the coupling effect of the HipYawPitch axes can be ignored when calculating the inverse kinematics.

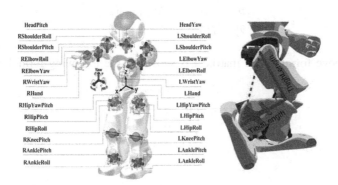

Fig. 3. NAO kinematic chains, joints and base frame

Table 1. D-H parameters from foot to torso on the NAO.

	$\theta(°)$	d(mm)	a(mm)	$\alpha(°)$
Joint1	0	0	45.11	0
Joint2	$\theta1$	0	0	−90
Joint3	$\theta2$	0	102.9	0
Joint4	$\theta3$	0	100	0
Joint5	$\theta4$	0	0	0
Hip	$\theta5$	0	0	90
Torso	0	0	85	−90

4   CoM Motion Estimation and Control

With regard to the offline walking pattern generated in Sect. 2, the prerequisite for the application is that the actual trajectory of the CoM must be basically consistent as the reference trajectory. However, when the robot is walking on the slope for a long distance, due to the asymmetry of the structure of the legs, the bad flatness of the slope surface or the slight slip of the foot and so on, it will lead to a certain degree of deviation from the CoM trajectory, and the walking direction will change also. When the deviation reaches a certain extent, the robot will lose stability and even fall. To avoid this situation, it is necessary to estimate the pose of the robot's torso, for the sake of simplicity, that is to accurately estimate the CoM motion and to adjust motor angle and angular velocity of each leg joint in real time, so as to achieve stable walking on a slope.

The CoM motion estimation of the robot includes the CoM position and orientation. Once the estimation is available, we may clearly reconstruct the corresponding information for any other part of the robot leg through direct kinematics. The NAO robot itself integrates a variety of sensors, an IMU located in the NAO robot's chest, the magnetic rotary encoders mounted on each joint, and the force sensitive resistors (FSRs) placed under each foot. In this paper, the sensor data are fused and the extended Kalman filter (EKF) is used to estimate the CoM motion.

The Kalman filter estimation model of the CoM position is represented as the formula (4):

$$\hat{p}_k = A\hat{p}_{k-1} + Bu_{k-1} + w^t_{k-1}$$
$$\tilde{z}^t_k = H\hat{p}_k + v^t_k \tag{4}$$

In the above formula, the matrix A is shown as the following formula:

$$A = \begin{bmatrix} 1 & T & T^2/2 \\ 0 & 1 & T \\ 0 & 0 & 1 \end{bmatrix} \tag{5}$$

In the formula, T is the sampling period. Since there is no control superimposed on the robot upper body, the control input $u_k = 0$. H matrix is equal to the formula (6):

$$H = \begin{bmatrix} 1 & 0 & 0 \\ 0 & 0 & 1 \end{bmatrix} \tag{6}$$

In the formula (4), w and v are vectors representing the standard process and measurement noise variables respectively, $\hat{p} = [\hat{p}, \dot{\hat{p}}, \ddot{\hat{p}}]^T$ is the estimated translation state of the NAO robot's CoM, $\tilde{z}^t = [\tilde{p}, \ddot{p}]$ is the measurement state , $\tilde{p}$ can be calculated by the forward kinematics formula, according to the landing position of supporting leg and the matrix of joint angle in the leg chain, as formula (7):

$$\tilde{p}(j) = \begin{bmatrix} \tilde{x}(j) \\ \tilde{y}(j) \end{bmatrix} = FK\left(\tilde{\Theta}(j), G(j-1)\right) \tag{7}$$

In the formula, $\tilde{\Theta}$ is an array composed of leg joint angles, G is landing position, j represents the j-th step. The landing position is changing constantly when the robot takes every step, $\tilde{p}$ is recalculated at the same time. The corresponding moment is triggered by a switching signal obtained by processing the FSRs output.

The estimated model of CoM rotational state can be represented by the following formula:

$$\begin{aligned} \hat{o}_k &= f\left(\hat{o}_{k-1}, 0, w^r_{k-1}\right) \\ \tilde{z}^r_k &= h\left(\hat{o}_k, v^r_k\right) \end{aligned} \tag{8}$$

The estimated value $\hat{o} = (\hat{q}, \hat{\omega})$ is composed of the torso orientation represented by quaternion $\hat{q} = [q_0, q_1, q_2, q_3]^T$ and spatial angular velocity $\hat{\omega}$. The measurement state $\tilde{z}^r - (\tilde{R}, \tilde{P}, \tilde{\omega})$ is composed of the torso orientation represented in roll, pitch and spatial angular velocity $\tilde{\omega} = [\omega_x, \omega_y, \omega_z]$. Both f and h are non-linear functions, where f is the relation between the previous state and current state under the assumption of constant angular speed motion, and h projects the orientation represented in orientation form onto the measurement torso orientation represented in roll, pitch and spatial angular velocity [16].

After obtaining the estimated value $\hat{q}$ of the torso orientation, the measurement of the torso linear acceleration $\tilde{a}$ in the inertial frame could be compensated by gravity, the measured acceleration of the CoM $\ddot{\tilde{p}}$ can be calculated, as shown in formula (9).

$$\ddot{\tilde{p}} = W(\hat{q}, \tilde{a}) \tag{9}$$

The entire process of the estimation of the CoM motion is as shown in Fig. 4.

There are a lot of controllers that control the robot's body pose, these controllers are mainly for some complex terrain, the controller algorithms are relatively complicated, and the robot walking speed is quite slow. On account of the terrain in this paper is a slope with given angle, an inverse kinematics controller is used to control the CoM

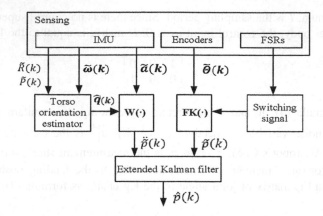

Fig. 4. Structure of the CoM motion estimator.

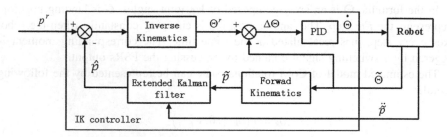

Fig. 5. The model of controller.

real-time trajectory of the NAO robot, which integrates the feedback loop in the joint space and is able to follow the desired joint motion as closely as possible. The model of controller is as shown in Fig. 5, in which the reference position of the CoM is represented as $p^r = [x^r, y^r]^T$.

5   Experiment

The experiment in this paper is carried out on the slope of a rectangular board, and the board size is 1.2 m × 0.9 m. In order to prevent the robot foot slipper while walking, a rubber pad of 2 mm thick is attached under the foot of the robot. The walking parameters on a fixed angle slope are shown in Table 2.

Table 2. Walking parameters.

Parameters	Uphill	Downhill
α	6°	6°
β	0°	180°
Step length(mm)	81	75
Step height(mm)	40	40
Step period(s)	4	4
Step width(mm)	110	110
Double support ratio (%)	25%	27%
$z_c(mm)$	305	320

Figure 6 shows the actual gait within one walking cycle when the robot climbs and goes down the 6° slope, using the walking pattern generated offline by the method in Sect. 2.

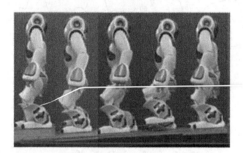

Fig. 6. The actual gait when the NAO walks on the slope.

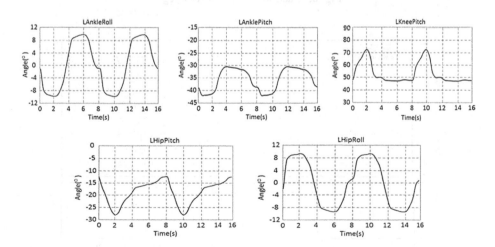

Fig. 7. Each joint angle of left leg.

Figures 7 and 8 show respectively the curves of each joint angle of left and right leg within two walking cycle when the robot climbs the slope.It can be seen that the movement of the left and right leg is not completely symmetrical, which is because the legs structure is asymmetrical, or bad flatness of the slope surface, or the slight slip of the foot.

There are four FSRs in each foot of NAO. In this paper, the pressure data collected are synthesized and filtered. Figure 9 shows the foot pressure changes within 4 steps from the starting point and the corresponding switch signal. The green dashed line represents the threshold value of 2.5 kg, and when the pressure of one foot rises above the threshold, the corresponding switching signal is generated. At this time the foot can be thought of as a supporting foot for kinematic calculation.

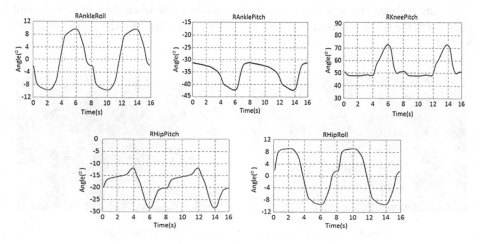

Fig. 8. Each joint angle of right leg.

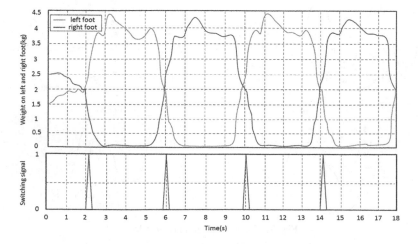

Fig. 9. Output signal of the FSRs and switching signal for the supporting foot.

(a) Uphill under open-loop control

(b) Uphill under closed-loop control

(c) Downhill under open-loop control

(d) Downhill under closed-loop control

Fig. 10. The top view of the NAO walking on the slope.

The top view of the NAO walking on the slope is shown in Fig. 10. Sub plot (a) and (c) are respectively the actual situation of the robot walking uphill and downhill with open-loop control. Sub plot (b) and (d) are respectively the actual situation of the robot walking uphill and downhill with closed-loop control. Our kinematic EKF runs at 60 Hz. The red line on the slope represents the reference line of the gradient vector.

The experiment demonstrates that under open-loop control, with walking distance increased, the walking direction of the robot is obviously right deviation, the oscillation of the upper body becomes violent, and walking is becoming increasingly unstable. While under closed-loop control, the walking direction of the robot is basically consistent with the gradient vector.

The four kinds of walking experiment described in Fig. 10 are respectively performed 10 times. We take the mean value of the Y direction from the middle point of adjacent step landing position, and connect all the mean points to obtain the line chart of the robot's walking direction, as shown in Fig. 11. It can be seen that when the number of walking steps increases, the deviation of the walking direction continues to increase with open-loop control. As the walking step reaches 12 steps, the maximum deviation in the y direction is about 8 cm, so the gait generated offline cannot guarantee stable walking. On the contrary, under closed-loop control, deviation of walking direction is controlled within 2 cm.

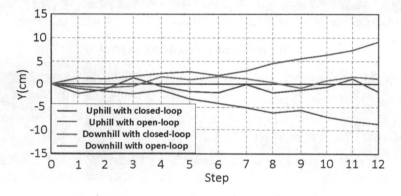

Fig. 11. Line chart of the robot's walking direction.

6 Conclusion

The problem of the robot walking stably on uneven road is always a hot topic for scholars. Although the method of generating online walking pattern can make the robot adapt to the changing terrain, there are still many technical difficulties need to be solved from current effect. The method proposed in this paper is very effective for a fixed angle slope. Offline walking pattern is generated by 3D-LIPM and we can obtain the reference trajectory of the robot's CoM. Then after fusing sensor data, extended Kalman filter is used to predict the robot's CoM motion, and the closed-loop control for walking process is realized. Experimental results demonstrate that the deviation of the robot walking direction is small, which can ensure the stable walking for a long distance. There are two aspects of the future work: firstly, an odometric localization will be formed integrated with the NAO robot's visual information; secondly, the robot should automatically recognize the angle of a slope, select the optimal walking pattern, and have the ability to accelerate the walking speed.

Acknowledgment. This work is supported by Foshan Science and Technology Project (FSTP) of China (2015AG10019) and it is also supported by Characteristic Innovation Project of Guangdong College (2015KTSCX149).

References

1. Kagami, S., Kitagawa, T., Nishiwaki, K., Sugihara, T., Inaba, M.: A fast dynamically equilibrated walking trajectory generation method of humanoid robot. Auton. Robots **12**(1), 71–82 (2002)
2. Kajita, S., Kanehiro, F., Kaneko, K., Yokoi, K.: The 3-D linear inverted pendulum mode: a simple modeling for a biped walking pattern generation. IEEE/RSJ Int. Conf. Intell. Robots Syst. **1**(1), 239–246 (2001)
3. Nagasaka, K., Inoue, H., Inaba, M.: Dynamic walking pattern generation for a humanoid robot based on optimal gradient method. IEEE Int. Conf. Syst. **6**(6), 908–913 (1999)
4. Zheng, Y.F., Shen, J.: Gait synthesis for the SD-2 biped robot to climb sloping surface. IEEE Trans. Robot. Autom. **6**(1), 86–96 (1990)
5. Chew, C.M., Pratt, J., Pratt, G.: Blind walking of a planar bipedal robot on sloped terrain. In: IEEE International Conference on Robotics and Automation, pp. 381–386. IEEE press, New York (1999)
6. Kim, J.Y., Park, I.W., Oh, J.H.: Walking control algorithm of biped humanoid robot on uneven and inclined floor. J. Intell. Robot. Syst. **48**(4), 457–484 (2007)
7. Yang, L., Chew, C.M., Zielinska, T., Poo, A.N.: A uniform biped gait generator with offline optimization and online adjustable parameters. Robotica **25**(5), 549–565 (2007)
8. Vundavilli, P.R., Pratihar, D.K.: Soft computing-based gait planners for a dynamically balanced biped robot negotiating sloping surfaces. Appl. Soft Comput. **9**(1), 191–208 (2009)
9. Liu, J., Urbann, O.: Bipedal walking with dynamic balance that involves three-dimensional upper body motion. Robot. Auton. Syst. **77**(C), 39–54 (2016)
10. Shahabazi, H., Jamshidi, K., Hasan, A.: Sensor-based programming of central pattern generators in humanoid robots. Int. J. Adv. Robot. Syst. **10**(3), 1 (2013)
11. Hornung, A., Oβwald, S., Maier, D., Bennewitz, M.: Monte carlo localization for humanoid robot navigation in complex indoor environments. Int. J. Humanoid Robot. **11**(2), 793–806 (2014)
12. Wu, C.M., Huang, C.P., Hsieh, C.H., Song, K.T.: A novel sensory mapping design for bipedal walking on a sloped surface. Int. J. Adv. Robot. Syst. **9**(4), 1 (2012)
13. Vukobratovic, M., Borovac, B.: Zero-moment point – thirty five years of its life. Int. J. Humanoid Robot. **1**(1), 157–173 (2004)
14. Tsuji, T., Ohnishi, K.: A control of biped robot which applies inverted pendulum mode with virtual supporting point. In: International Workshop on Advanced Motion Control, pp. 478–483 (2002)
15. Ali, F., Shukor, A.Z.H., Miskon, M.F., Nor, M.K.M., Salim, S.I.M.: 3-D biped robot walking along slope with dual length linear inverted pendulum method. Int. J. Adv. Robot. Syst. **10**(377), 1–12 (2013)
16. Lin, P.C., Komsuoglu, H., Koditschek, D.E.: A leg configuration measurement system for full-body pose estimates in a hexapod robot. IEEE Trans. Robot. **21**(3), 411–422 (2005)

TVSLAM: An Efficient Topological-Vector Based SLAM Algorithm for Home Cleaning Robots

Yongfu Chen[⊠], Chunlei Qu, Qifu Wang, Zhiyong Jin, Mengzhu Shen, and Jiaqi Shen

Huazhong University of Science and Technology, Wuhan 430074, China
chenyf@hust.edu.cn

Abstract. The Simultaneous Localization and Mapping problem limits the promotion of home cleaning robots in practical domestic environments. In this paper, a novel topological-vector based simultaneous localization and mapping (TVSLAM) algorithm is proposed to solve the problem. The algorithm involves four aspects. First, the ultra-wideband localization and dead reckoning localization are selected to develop a new combined localization algorithm which can improve the localization accuracy. In addition, a data acquisition algorithm which simplifies the process of data collection and demands much smaller memory size is proposed. Furthermore, a partitioning algorithm is developed to adapt to the various change rates of different rooms. Finally, an autonomous learning algorithm based on the regular and repetitive cleaning task is put forward. It makes the constructed map approach to the real environment with the increase of cleaning times. Overall, a novel topological-vector map is generated according to the above process of the algorithm. Simulation results show that the TVSLAM is an efficient and robust localization and mapping algorithm.

Keywords: SLAM · Topological-vector · Combined localization · Autonomous learning · Home cleaning robots

1 Introduction

Although home cleaning robots (HCRs) are the most popular service robots used in domestic, the coverage rate, overlap rate and efficiency often hamper their user acceptance [1]. To solve this problem, the HCR should be able to build the map autonomously [2] and locate itself within the map simultaneously [3]. These two tasks are called the Simultaneous Localization and Mapping (SLAM). In this paper, we focus on proposing an efficient SLAM algorithm.

The researchers have developed out a lot of creative solutions about localization after years of researches [4–8]. Biswas [4] proposed a WiFi localization algorithm that used Monte Carlo localization with Bayesian filtering. The localization error at a mean of 1.2 m was too large for HCRs. Gutmann [5] discovered that all of the wireless signals exist multipath and attenuation, due to the domestic environment contains multiple walls and barriers consisting of different materials. Segura [6] proposed an ultra-wideband (UWB) based localization system that solved the multipath identification problem owing

© Springer International Publishing AG 2017
Y. Huang et al. (Eds.): ICIRA 2017, Part III, LNAI 10464, pp. 166–178, 2017.
DOI: 10.1007/978-3-319-65298-6_16

to its high bandwidth. Cho [7] presented a dead reckoning localization algorithm with the rotary encoders and the inertial navigation system to reduce the accumulation errors. However, the existing localization methods all have limitations in practical domestic environments [8], such as low localization accuracy, high cost and so on.

Meanwhile, there are many possible ways have been proposed to represent the environment map [9]. The map representations can be categorized into two classifications. One is grid map [10–12] and the other is feature map [13–15]. The constructed map was modeled with a coarse-grain occupancy grid [10, 11]. Each square cell was of the size of sweeping tool and the partially covered cells were marked as real barriers. So this method was not accurate enough to describe the real environment. A few years later, Lee [12] adopted a high-resolution grid map representation with small cells, leading to a higher degree of freedom. There were also some grid maps that used the grids to store the environment image information [16] or the geomagnetic filed information [17, 18]. In spite of the information can be used to improve the localization accuracy, it doesn't get rid of the problem of demanding large storage size. The feature map was put forward as an alternative method of grid map at first, becoming research hotspot recently. Pfister [13] first provided a weighted line fitting algorithm to fit a line segment, which was adapted to most domestic environments. An [14, 15] proposed a line segment based map which was a compact map representation even in a cluttered environment. But the raw data, which was obtained from laser range finder, required a series of steps to get line segment [19]. The vector map is an advanced form of feature map, due to vector is a directed line segment [2] and the vector map has the same representation capability as grid map and topological map [20]. Compared to the feature map, its main advantage is that it simplifies the discrimination of barrier kinds [21].

The combined map [2, 9, 22] has been put forward nearly twenty years. Thrun [22] proposed the metric-topological map which was implemented by partitioning the grid map into coherent regions and building topological maps on the top of the regions. Due to the irregular shape of regions, it isn't possible to realize intelligent cleaning. Sohn [2] used a topological map as the robot trajectory to import the loop closing method to the mapping algorithm. In this paper, although some figures were a combination of topological map and vector map, there was no reference and application of combined map from start to the end. In the survey of Jelinek [9], the topological-metric map was considered to be a good representation in the nearest future.

It must be stressed that only in few papers that the SLAM problem was addressed with limited sensors. Abrate [23] proposed an EKF-based SLAM to enhance the autonomy of the educational robots with eight infrared range sensors only. Choi [19] presented a line feature based SLAM applied to a HCR to realize real-time control with low computing load. Although the sensor data of the two approaches was sparse and noisy, the amount of data was reduced significantly making it possible to use complicated algorithm. But the map accuracy was not enough for intelligent cleaning due to low data precision. Thus, partitioning algorithm [1, 2, 24] was used in the SLAM to improve the localization accuracy for low cost HCRs. Lee [1] proposed the incremental sector creation method which was proofed that the smaller sector could improve localization accuracy. Beak [25] used the sweep-invariant decomposition (SID). But the adjective cells should be merged together considering the efficiency. Dugarjav [24] modified the SID with oriented rectilinear decomposition which decomposed the map according to the exploring procedure and the longer boundary of the cell. Myung [26] decomposed the domestic environment into sub-regions based on the feature of map. And it used the photographs of rooms taken by CMOS camera. Although the above algorithms are based on the grid maps, we can refer the theory and then apply it to the new map representation.

The rest of this paper is organized as follows. Section 2 introduces the combined localization system and Sect. 3 explains the proposed direction control system and data acquisition algorithm. After that, Sect. 4 presents the topological-vector map building approach. Then, Sect. 5 presents the setup and results of simulations for verifying the proposed algorithm. Finally, we state the conclusion and future work in Sect. 6.

2 Combined Localization System

To generate a compact data structure and improve the localization accuracy, a multi-layer system is proposed. It's composed of three layers form low to high, the data collection layer, the map building layer and the data association layer. It's shown as Fig. 1.

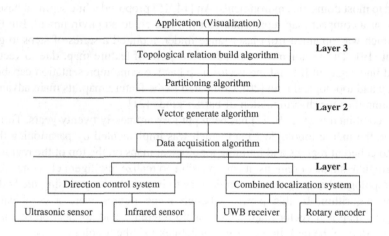

Fig. 1. The architecture of multi-layer system.

The true SLAM is difficult to realize without the aid of precise localization. So a combined localization system which is composed of dead reckoning localization and the UWB localization is proposed. The model of local localization is defined as follow:

$$
P_{i+1} = P_i + \Delta P = \begin{bmatrix} x_i \\ y_i \\ \theta_i \end{bmatrix} + \begin{bmatrix} \Delta x \\ \Delta y \\ \Delta \theta \end{bmatrix}
$$

$$
= \begin{bmatrix} x_i \\ y_i \\ \theta_i \end{bmatrix} + \begin{bmatrix} \int_{t_i}^{t_{i+1}} \frac{v_r+v_l}{2} dt \, \cos(\theta_{i-1} + \int_{t_i}^{t_{i+1}} \frac{2(v_r-v_l)}{W} dt) \\ \int_{t_i}^{t_{i+1}} \frac{v_r+v_l}{2} dt \, \sin(\theta_{i-1} + \int_{t_i}^{t_{i+1}} \frac{2(v_r-v_l)}{W} dt) \\ \int_{t_i}^{t_{i+1}} \frac{2(v_r-v_l)}{W} dt \end{bmatrix} \tag{1}
$$

where t_i is the time arrive at the previous point, t_{i+1} is the time arrive at the present point, v_r and v_l are the speeds of the right wheel and left wheel respectively, and W is the width between two wheels. The speed is calculated according to the sensor data of rotary encoder as follow:

$$\begin{cases} v_r = 2\pi \cdot R \cdot \frac{M_r}{N} \\ v_l = 2\pi \cdot R \cdot \frac{M_l}{N} \end{cases} \qquad (2)$$

where R is the radius of the active wheel, N is the pulse generated while the wheel rotates 360 degrees, M_r and M_l are the measured pulse number of left encoder and right encoder per second respectively.

The standard of IEEE P8021.5 presented a path loss model for indoor UWB signals of nominal center frequency of 5 GHz. The signal can penetrate one wall to maximum of 4 walls and the maximum effective distance is 20 m, enough for most families. But the localization accuracy will be decreased as the number of penetrating walls increasing. The accuracy after penetrating two walls still reaches the requirement. Thus, the transmitters are fixed at the center of ceiling of each room, the farthest 4 corners of the home and one receiver is installed on the HCR. So the required number of transmitters to ensure accuracy can be as few as possible. Furthermore, the absolute positions of transmitters should be measured before stimulation. The signals that meet the required strength are chosen to calculate the position in the global coordinate:

$$d = \frac{\alpha}{n} \sum_{i=1}^{n} \frac{1}{10^{\frac{P_i - PL_1 + S}{10\gamma}}} \qquad (3)$$

where P_i is the signal strength which the receiver receives from the ith transmitter. PL_1 is a constant that represents the loss of received signal strength at 1 m to the transmitter. The parameter γ has different values of different regions and obeys a normal distribution $N[\mu_\gamma, \sigma_\gamma]$.

The global position $G = (x_g, y_g, \theta_g)$ is got by solving the above equation every time when the robot arrives at a new turning point. Then, it is compared with the local position $P_i = (x_i, y_i, \theta_i)$ that is obtained by relative localization algorithm. If the position deviation Δd is bigger than the threshold σ or the orientation deviation $\Delta \alpha$ is larger than the threshold δ, the global position will be used to replace the local position.

3 Direction Control System and Data Acquisition Algorithm

3.1 Direction Control System

HCR uses the sensor data to control the behavior of the robot. According to the data generated by the sensor system composed of eight infrared range sensors and one ultrasonic sensor, the basic control system is produced. There are a total of 256 kinds of states, shown as Table 1.

Table 1. The different states of eight infrared range sensors.

	I_1	I_2	I_3	I_4	I_5	I_6	I_7	I_8
S_1	0	0	0	0	0	0	0	0
S_2	0	0	0	0	0	0	0	1
S_3	0	0	0	0	0	0	1	0
S_4	0	0	0	0	0	0	1	1
S_5	0	0	0	0	0	1	0	0
...	...	...	...	...	...	...	...	...
S_{256}	1	1	1	1	1	1	1	1

The state is calculated as follow:

$$S_i = \sum_{j=1}^{8} 2^{I_{ij}} \qquad (4)$$

where I_{ij} is the jth infrared range sensor signal at state i. When the signal level is high, it indicates that there is a barrier in this direction. According to the state S_i, the HCR decides which direction should go. The direction control system will be performed from the start to the end of the map building process. In addition, according to combined localization system which can help the robot to confirm its position, the specific control flow is shown as Fig. 2.

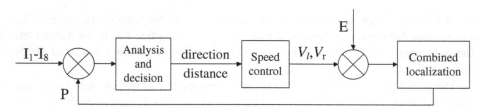

Fig. 2. The control flow of the direction control system.

3.2   Data Acquisition Algorithm

The outline data acquisition method only records points that the direction of cleaning robot changes. It is based on the right hand rule and wall following method. According to whether there is a barrier in front, the data points are sorted into two types, convex points and concave points. The storage form of point is defined as follow:

$$p_i = [(x_i, y_i), \theta_i, o_i, c_i] \quad i = 1, 2, \ldots \ldots, n. \qquad (5)$$

(x_i, y_i) is the current position of robot, θ_i is the current direction of robot, o_i is the sequence number of points and c_i represents the type of points.

The inside data acquisition method collects the data of inside barriers when the outline data acquisition is completed. The ideal domestic environment is that the barriers inside are composed of simple geometric shapes. If a new barrier is found, it will be appended to the end of inside barrier list. When multiple barriers encountered at the same time, the barriers are recorded from right to the left. According to the Euclidean distance, the robot computes the distance to each not searched barrier in the list to find the nearest barrier to search. When a data collection of a barrier is over, the barrier will be marked as searched. Then the end of the inside data acquisition is all of the barriers have been found. The search of different barriers is shown as Fig. 3:

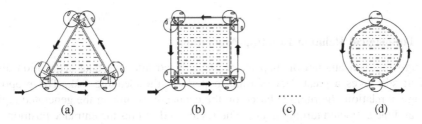

(a)   (b)   (c)   (d)

Fig. 3. The search of the inside barriers: (a) Triangle barrier, (b) Rectangle barrier, (c) Other polygons between (b) and (d), (d) Circle barrier.

4 Topological-Vector Map Building Approach

4.1 Vector Generating Method

According to the proposed specialized data acquisition algorithm, the vector is generated by connecting the collected data according to the time stamp t_i. The vector, $v_i^t \in M^t$, at time t, is defined as follow:

$$v_i^t = [(x_i, y_i), l_i, \alpha_i, \sigma_i] \tag{6}$$

where (x_i, y_i) are the start points, l_i is the length, α_i is the direction of the vector and σ_i is the standard deviation.

4.2 Curve Fitting Method

Most of the objects of domestic environments, like furniture and walls, can be compassed with lines in two-dimensional map. And the curves can also be spliced with short line segments. So the vector is used as the minimal element to build the environment map. When the simple line segment fitting method is used according to the fitting coefficient, a more satisfactory result is obtained. Thus, a curve fitting method that based on short lines is selected.

4.3   Partitioning Algorithm

The proposed partitioning algorithm is mainly based on the geometric relation. Here are three basic steps:

1. Extract all the concave points from the outline points set P_1. When it is an isolated point, the neighboring convex points are added to the list guaranteeing that there are at least two points.
2. Generate the linear equation with neighboring points belong to the same cluster.
3. Divide into sub-regions according to the geometric relation of the extracted vectors. The geometric relation just includes parallel and vertical.

4.4   Topological Relation Building Algorithm

Although the data amount of map is greatly reduced after the data acquisition algorithm, it still takes a great deal of time to perform the cleaning task. The proposed topological relation algorithm is based on the connectivity among the generated region nodes and the extracted turning nodes. The region node stores the entrance position and the information of including vector of the corresponding region. The turning node merely stores the turning points and the topological relations of connected nodes.

4.5   Map Updating Algorithm

In order to have a real-time map for exact path planning, the map is rebuilt every time before cleaning. But the majority regions remain the same, the rebuilding of total map is a waste of time. Therefore, a quick map updating method that only redraws the regions which have a relatively significant change is proposed. Whether the local map needs to rebuild, the referred formula of changing rate is as follow:

$$Rl_j = k\frac{m}{n} + (1-k)\sum_{i=1}^{m}\frac{o_i}{O} \quad j = 1, 2, \ldots\ldots.h \tag{7}$$

where n/O is the total number/area of barriers and m/o_i is the number/area of barriers that position has changed. k and $(1-k)$ are the weights of the number and area of barrier respectively. If Rl_j is bigger than the threshold, the local map of region j will be rebuilt. According to the Formula (7), the global updating formula is as follow:

$$R_g = \sum_{j=1}^{h}\frac{Rl_j}{h} \tag{8}$$

where h is the number of sub-regions.

4.6 Autonomous Learning Algorithm

The real map is difficult to build by exploring the environment only one time. So we propose an autonomous leaning system based on the two regular behaviors. First, the HCR usually does the cleaning task at least once a day. Second, the complexity of different regions is different. So the regions are treated respectively to improve the efficiency of map building. The formula is as follow:

$$t_i = k_1 \frac{l_i}{S_i} + k_2 \frac{sl_i}{S_i} + K \quad i = 1, 2, \ldots \ldots .h \tag{9}$$

t_i is the times of the region i that the HCR should explore. l_i and sl_i are the numbers of long and short line segment respectively. S_i is the area of the region i. k_1, k_2 and K are constants obtained by experiment.

5 Simulations

5.1 Simulation Setup

Although the simulation is different from real experiment fundamentally, we do our best to make it as close to the real environment as possible. A differential-drive cleaning robot is used as the basement of the simulation. The arrangement of sensors is shown as Fig. 4.

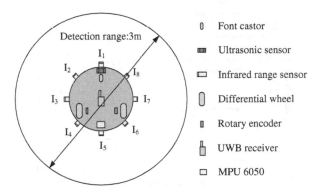

Fig. 4. The sensor configuration of home cleaning robot experiment platform.

5.2 Simulation Results

The evaluation of combined localization. Reliability: In this test, the HCR stops at 20 randomly chosen positions on the map. Figure 5(a) shows a plot of the localization error e at different points. There are two relatively higher points. But all the errors are within 0.08 m. Then the test is performed 10 times to compute the standard deviation σ.

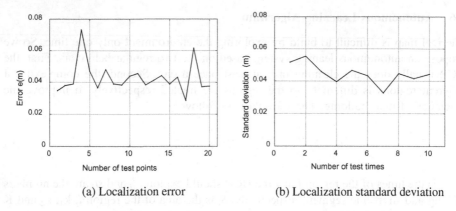

(a) Localization error   (b) Localization standard deviation

Fig. 5. The evaluation of combined localization.

The result is shown in Fig. 5(b), with a slight fluctuation around 0.05 m. Both of the results prove the reliability of the proposed localization method.

Accuracy: In this test, the robot again stops at 20 randomly chosen positions on the map. The mean estimated position is used to compare with the real position to eliminate the slight oscillation. The combined localization method shows the best accuracy with a mean error of 5 cm, shown as Fig. 6.

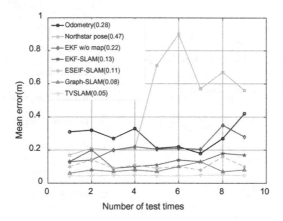

Fig. 6. Mean position error over all nine runs with average in parentheses.

The evaluation of topological-vector map. The vector map that changes from the grid map of SDR site B of University of Southern California was used for simulation, shown as Fig. 7(a) (http://radish.sourceforge.net/). Figure 7(b) is the converted original vector map composed of 458 vectors. It mainly consists of 6 sections according to the continuity of barrier.

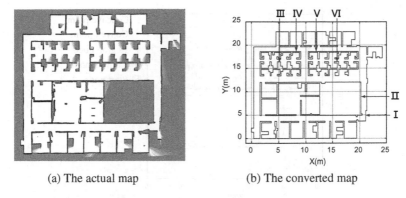

(a) The actual map   (b) The converted map

Fig. 7. The map used for simulation.

Because of the complexity of the environment, some regions of the constructed map are unexplored, shown as Fig. 8. However, the unexplored regions are almost found based on the proposed autonomous learning method with the increase of cleaning times. The ultimate constructed map agrees well with the real environment. In Fig. 9, a topological-vector map built with the proposed TVSLAM is shown. The topological map and vector map are linked through the turning nodes which belong to both of them.

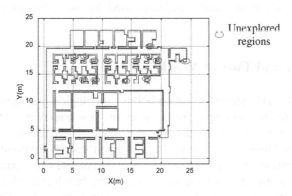

Fig. 8. A comparison between the constructed map and the real environment.

The proposed map representation has a lot of excellent performances compared with grid map and vector map, shown as Table 2. The memory size reduces to 13.96% of the vector map and 6.75% of the grid map significantly. And the consumption time also reduces by 56.93% of the grid map and 13.53% of the vector map.

As a result, the constructed topological-vector map is accurate enough for the HCRs to perform cleaning task. And the amount of data is fewer and computation burden is lighter compared to other map representations. Meanwhile, the data of the

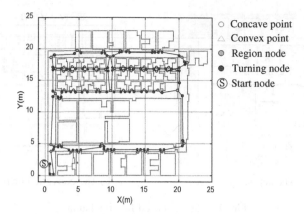

Fig. 9. The constructed topological-vector map.

Table 2. A comparison of memory and consumption time among different map representations.

	Grid map	Vector map	Topological-vector map
Memory (KB)	418	202	28.2
Consumption time (s)	7033.6	3503.4	3029.5

map has a high degree of data association used by the robot to realize accurate localization.

6   Conclusion and Future Work

In this paper, a TVSLAM algorithm which builds a topological-vector map for HCRs is presented. Compared to grid map or vector map, the new map representation based on the proposed data acquisition algorithm requires relatively smaller amount of memory and shorter map building time. Meanwhile, the map stores the environment data based on a multi-layer system which promotes the realization of real-time control. In addition, a new combined localization method which shows the best localization accuracy with an average error of 5 cm in simulation environments is presented. Furthermore, the proposed map updating algorithm avoids rebuilding the total map every time and increases the speed of map building. Finally, the application of autonomous learning algorithm makes the constructed map closer to the real environment gradually. Simulation results show that the TVSLAM algorithm is an efficient and robust online localization and mapping algorithm which is suitable for majority of the domestic environment. However, the environment of the simulation is relative simple compared to the real environment. As future work, we would like to improve the TVSLAM algorithm and then apply it on a HCR in the real world.

References

1. Lee, T.-K., Baek, S., Oh, S.-Y.: Sector-based maximal online coverage of unknown environments for cleaning robots with limited sensing. Robot. Autonom. Syst. **59**(10), 698–710 (2011)
2. Sohn, H.J., Kim, B.K.: VecSLAM: an efficient vector-based SLAM algorithm for indoor environments. J. Intell. Rob. Syst. **56**(3), 301–318 (2009)
3. Kang, J.-G., Kim, S., An, S.-Y., Oh, S.-Y.: A new approach to simultaneous localization and map building with implicit model learning using neuro evolutionary optimization. Appl. Intell. **36**(1), 242–269 (2012)
4. Biswas, J., Veloso, M.: WiFi localization and navigation for autonomous indoor mobile robots. In: IEEE International Conference on Robotics and Automation, pp. 4379–4384 (2010)
5. Gutmann, J.S., Eade, E., Fong, P., Munich, M.E.: Vector field SLAM—Localization by learning the spatial variation of continuous signals. IEEE Trans. Rob. **28**(3), 650–667 (2012)
6. Segura, M., Mut, V., Sisterna, C.: Ultra wideband indoor navigation system. Iet Radar Sonar Navig. **6**(5), 402–411 (2012)
7. Cho, B.-S., Moon, W.-S., Seo, W.-J., Baek, K.-R.: A dead reckoning localization system for mobile robots using inertial sensors and wheel revolution encoding. J. Mech. Sci. Technol. **25**(11), 2907–2917 (2011)
8. Mautz, R.: Indoor Positioning Technologies. Südwestdeutscher Verlag für Hochschulschriften (2012)
9. Jelinek, A.: Vector maps in mobile robotics. Acta Polytech. CTU Proc. **2**(2), 22–28 (2015)
10. Gonzalez, E., Alvarez, O., Diaz, Y., Parra, C., Bustacara, C.: BSA: a complete coverage algorithm. In: Proceedings of the 2005 IEEE International Conference on Robotics and Automation (ICRA), pp. 2040–2044 (2005)
11. Choi, Y.H., Lee, T.K., Baek, S.H., Oh, S.Y.: Online complete coverage path planning for mobile robots based on linked spiral paths using constrained inverse distance transform. In: The 2009 IEEE/RSJ International Conference on Intelligent Robots and Systems, pp. 5788–5793 (2009)
12. Lee, T.-K., Baek, S.-H., Choi, Y.-H., Oh, S.-Y.: Smooth coverage path planning and control of mobile robots based on high-resolution grid map representation. Robot. Autonom. Syst. **59**(10), 801–812 (2011)
13. Pfister, S.T., Roumeliotis, S.I., Burdick, J.W.: Weighted line fitting algorithms for mobile robot map building and efficient data representation. In: IEEE International Conference on Robotics and Automation (ICRA), pp. 1304–1311 (2003)
14. An, S.-Y., Kang, J.-G., Lee, L.-K., Oh, S.-Y.: SLAM with salient line feature extraction in indoor environments. In: International Conference on Control Automation Robotics and Vision, pp. 410–416 (2010)
15. An, S.-Y., Kang, J.-G., Lee, L.-K., Oh, S.-Y.: Line segment-based indoor mapping with salient line feature extraction. Adv. Robot. **26**(5–6), 437–460 (2012)
16. Lee, S., Lee, S., Baek, S.: Vision-based kidnap recovery with SLAM for home cleaning robots. J. Intell. Rob. Syst. **67**(1), 7–24 (2011)
17. Jung, J., Lee, S.M., Myung, H.: Indoor mobile robot localization and mapping based on ambient magnetic fields and aiding radio sources. IEEE Trans. Instrum. Meas. **64**(7), 1922–1934 (2015)
18. Lee, S.M., Jung, J., Kim, S., Kim, I.J., Myung, H.: DV-SLAM (Dual-sensor-based Vector-field SLAM) and observability analysis. IEEE Trans. Industr. Electron. **62**(2), 1101–1112 (2015)

19. Choi, Y.-H., Lee, T.-K., Oh, S.-Y.: A line feature based SLAM with low grade range sensors using geometric constraints and active exploration for mobile robot. Autonom. Robots **24**(1), 13–27 (2008)

20. Baizid, K., Lozenguez, G., Fabresse, L., Bouraqadi, N.: Vector maps: a lightweight and accurate map format for multi-robot systems. In: Kubota, N., Kiguchi, K., Liu, H., Obo, T. (eds.) ICIRA 2016. LNCS, vol. 9834, pp. 418–429. Springer, Cham (2016). doi:10.1007/978-3-319-43506-0_37

21. Sohn, H.J., Kim, B.K.: An efficient localization algorithm based on vector matching for mobile robots using laser range finders. J. Intell. Rob. Syst. **51**(4), 461–488 (2008)

22. Thrun, S.: Learning metric-topological maps for indoor mobile robot navigation. Artif. Intell. **99**(1), 21–71 (1998)

23. Abrate, F., Bona, B., Indri, M.: Experimental EKF-based SLAM for Mini-rovers with IR Sensors Only. In: EMCR (2007)

24. Dugarjav, B., Lee, S.-G., Kim, D., Kim, J.H., Chong, N.Y.: Scan matching online cell decomposition for coverage path planning in an unknown environment. Int. J. Precis. Eng. Manufact. **14**(9), 1551–1558 (2013)

25. Baek, S., Lee, T.-K., Se-Young, O.H., Ju, K.: Integrated on-line localization, mapping and coverage algorithm of unknown environments for robotic vacuum cleaners based on minimal sensing. Adv. Robot. **25**(13–14), 1651–1673 (2012)

26. Myung, H., Jeon, H.-M., Jeong, W.-Y.: Virtual door algorithm for coverage path planning of mobile robot. In: IEEE International Symposium on Industrial Electronics (ISIE), pp. 658–663 (2009)

Development of Wall-Climbing Robot Using Vortex Suction Unit and Its Evaluation on Walls with Various Surface Conditions

Jianghong Zhao and Xin Li[(✉)]

State Key Lab of Fluid Power Transmission and Control, Zhejiang University,
38 Zheda-Road, Hangzhou 310027, Zhejiang, People's Republic of China
{zhaojianghong, vortexdoctor}@zju.edu.cn

Abstract. This paper presents a wall-climbing robot called Vortexbot, which has a suction unit that uses vortex flow to generate a suction force. Unlike the traditional unit based on contact-type adsorption, the suction unit does not touch the wall surface, which greatly reduces the frictional resistance between the robot and wall and improves the passing ability of the robot. It first introduces the principle of the vortex suction unit. Then, the authors design the mechanical structure of Vortexbot. Furthermore, they survey the suction properties of the suction unit on a smooth wall surface. In addition, they study the effect of the roughness and shape (a raised obstacle and groove) of the wall surface on the suction performance of the suction unit. Finally, they experimentally verify the climbing performance of Vortexbot on several kinds of walls with different surface conditions.

Keywords: Climbing robot · Vortex suction unit · Non-contact gripping · Rough surface · Raised obstacle · Groove

1 Introduction

Today, robots play a major role in boosting industrial production and improving the quality of human life. Robots can replace humans in carrying out many dangerous and difficult tasks. A kind of specialised robot, the wall-climbing robot, has been widely used. Zhang H. et al. developed a series of climbing robots called Sky Cleaner I, II, III, and IV, which can clean the glass walls of high-rise buildings [1, 2]. NINJA I and II and ROMA I and II can carry out facade inspection and maintenance of high-rise buildings and bridges [3–6]. Roboclimber, a 3-ton spider robot, can autonomously execute slope consolidation tasks [7]. Alicia 1 and 2 can inspect equipment in petrochemical plants [8]. ROBICEN III and Robug II can carry out facade inspection and maintenance in nuclear plants [9, 10]. REST can carry out inspection, cleaning, and welding tasks in a ship hull [11]. Undoubtedly, with the development of technology, climbing robots will be able to accomplish more tasks for human beings.

During the research and design of climbing robots, one of the issues that researchers mainly consider is the type of suction method to be used [12, 13]. Vacuum adsorption is the most commonly used method [1, 2, 14–16]. It mainly generates negative pressure in the suction cup to provide suction force for climbing robots. This method has the

© Springer International Publishing AG 2017
Y. Huang et al. (Eds.): ICIRA 2017, Part III, LNAI 10464, pp. 179–192, 2017.
DOI: 10.1007/978-3-319-65298-6_17

following advantages. It can be used on walls of different materials. Its structure is relatively simple and can be manufactured easily at low cost. Existing vacuum technology can produce negative pressure close to full vacuum (i.e. 0 kPa (abs)); hence, theoretically, the suction cup can provide a large suction force, which can enable the robot to have a very strong load capacity. However, it is necessary to ensure very good sealing between the suction cup and wall surface to maintain the suction force. Otherwise, leakage might occur, which might lead to the failure of adsorption and might cause the robot to drop off the wall. This limitation restricts the use of the robot to walls with smooth and flat surfaces. Many researchers have adopted different approaches to overcome this problem (Fig. 1). Yanzheng Zhao et al. developed a wall-climbing robot with a single suction cup and adopted an air spring–regulating spring combination as the sealing mechanism (Fig. 1a) [17]. This mechanism enables the robot to work well on both glass and ceramic-tile surfaces. Longo D. et al. developed the second generation of the robot Alicia (Alicia 2) by using a larger cup, a much more powerful aspirator, and a sealing structure made by sandwiching Teflon and bristles (Fig. 1b) [8]. Therefore, Alicia 2 can work in much harsher environments such as a rough metal surface or concrete wall. Cromsci, developed by Schmidt D. et al., has a seven-chamber adhesion system and an adaptable inflatable rubber sealing (Fig. 1c); its surface is made of synthetic fibres, which make it suitable for large vertical concrete buildings [18]. Mo Koo et al. proposed a wall-climbing robot called LARVA, which contains an impeller-type adhesion mechanism [19]. To maintain the suction force, the robot uses a double-layered sealing mechanism consisting of a flexible bending layer and single straight layer (Fig. 1d). This mechanism allows it to climb rough concrete walls.

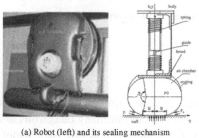

(a) Robot (left) and its sealing mechanism
(right), developed by Zhao et al. [17]

(b) Alicia 2 And its sealing mechanism [8]

(c) Cromsci and its seven-chamber adhesion system
and an adaptable inflatable rubber sealing[18]

(d) LARVA and its double-layered sealing mechanism [19]

Fig. 1. Wall-climbing robots with sealing mechanisms, which enable robots to climb walls with relatively rough surfaces

The abovementioned studies based on existing technology indicate that the most widely used method used for suppressing the leak between the suction cup and wall surface is to design and strengthen the sealing mechanism. However, while climbing a wall, the robot is influenced by the frictional resistance between the wall surface and the various kinds of complex sealing mechanisms. To ensure good sealing, the robot has to withstand a large frictional resistance because the frictional resistance is always in conflict with the sealing effect [20, 21]. In addition, the frictional resistance can cause the abrasion of the sealing material and simultaneously limit the walking speed of the robot because of the deformation of the sealing parts. Furthermore, because the actual nature of a wall surface can be very complex, the sealing mechanism cannot always ensure that it is close to the wall surface, which can lead to fluctuations in the suction force and hence hamper the smooth running of the robot [19]. We believe that it is very necessary and important to devise a new adsorption method that can avoid the series of problems encountered by vacuum suction methods.

2 The Suction Unit Based on Vortex Flow

Vortex levitation was proposed and studied by Li et al. in 2008 [22], and it has been used to develop a new kind of non-contact gripping mechanism [22–24]. The vortex gripper (Fig. 2) is mainly composed of a cylindrical vortex chamber and two tangential nozzles, which are created on the circular wall of the chamber. The compressed air blows into the vortex chamber through the tangential nozzles and forms a high-speed vortex flow along the internal face of the chamber. As in the case of a tornado, the centrifugal force produced by the vortex flow pushes the air in the center towards the peripheral region of the chamber, and thus, a negative pressure zone with rarefied air is created at the center. As a result, a considerable pressure difference is created between the top and bottom surfaces of the workpiece, which is placed under the vortex gripper. That is, the workpiece experiences an upward suction force. When the negative pressure is sufficiently large, the suction force becomes larger than the gravity of the workpiece. At this point, the workpiece is lifted up. In addition, because compressed air is constantly supplied through the nozzles into the vortex chamber (blue arrows in Fig. 2), air is continuously vented through the gap between the gripper and workpiece. This exhaust flow ensures that the gripper and workpiece do not contact each other.

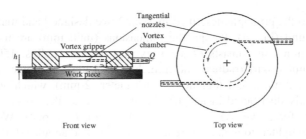

Fig. 2. Schematic of vortex gripper

3   The Proposed Wall-Climbing Robot, Vortexbot

We designed and manufactured a wall-climbing robot called Vortexbot (Fig. 3). It mainly consists of a car body and suction mechanism. It adopts the all-wheel-drive mode as the driving mode, which enables the robot to take full advantage of the suction force provided by the suction unit while climbing a vertical wall. Three wheels are installed on the based board of the car body and are driven by three geared DC motors. The suction unit is installed under the based board by screw connections. In addition, we ensure that the screw connections are located as close to the three wheels as possible. This design reduces the effect of force couples on the based board (i.e. the force couples formed by the counter force provided by the wall surface to each wheel and the suction force provided by the suction unit that spreads on each screw connection). Consequently, the design also reduces the strength requirement for the based board. Therefore, the based board is designed to be hollow in many sections, which makes the robot lightweight. A servomotor is used to change the direction of the front wheel and hence realise the steering motion of the robot. The drive circuit is installed on the based board and is used to control the movement of the robot. The robot is manufactured mainly using carbon fibre plate because it has both high strength and light weight.

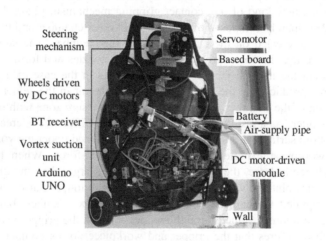

Fig. 3. Photograph of Vortexbot

According to the principle described in Sect. 2, we designed and manufactured the vortex suction unit (Fig. 4). Its specific dimensions (unit: mm) are indicated in the figure. The suction unit consists of a vortex ring (internal radius: 150 mm, height: 18 mm), an annular skirt (radius: 180 mm, height: 3 mm), an upper cover (radius: 160 mm, height: 2 mm), and four nozzles (diameter: 2 mm), which are tangentially and symmetrically distributed inside the chamber. Because the upper cover needs to bear the suction force, we used a high-strength carbon fibre plate. We used acrylic plates, which are lighter, to manufacture the vortex ring and annular skirt because these parts do not need to have high strength.

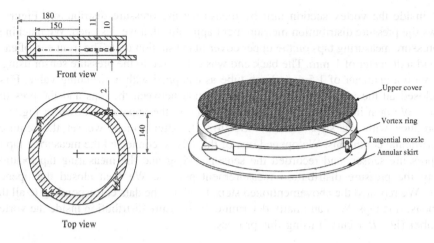

Fig. 4. Schematic structure and explosive view of vortex suction unit used in this study

4 Experimental Methods

4.1 Experimental Setup for Suction Force Measurement

We set up a measuring apparatus (Fig. 5) to study the change in the suction performance of the suction unit when the surface condition of the wall varies. The apparatus mainly consisted of two parts: a fixed trestle and a vertical mobile platform. The fixed trestle was used to fix the suction unit and force sensor, which was used to measure the suction force. The vertical mobile platform was set below the suction unit. We can install different simulated walls on the platform and change them. We can also adjust the distance between the platform and the suction unit by adjusting the feeding bolt. The following steps were performed during the experiment: (1) the vortex suction unit was fixed on the fixed trestle, and the simulated wall was installed on the vertical mobile platform; (2) the six leveling bolts were adjusted to ensure that the upper surface of the simulated wall just horizontally touched the lower surface of the suction unit; (3) the feeding bolt was adjusted to vertically lower the simulated wall and the distance between the simulated wall and the suction unit; (4) the gas switch was opened to provide compressed air at a certain flow rate to the suction unit; (5) a data acquisition card was used to record the change in the reading of the force sensor equal to the suction force and simultaneously obtain the distance value (i.e., h) from the screw micrometer on the lifting platform; and (6) steps 3–5 were repeated by adjusting h. We can obtain the F–h curve (i.e., the curve of the suction force against the distance) for the suction unit on the simulated wall with a given surface condition using the above steps.

4.2 Experimental Setup for Pressure Distribution Measurement

The air movement in the vortex suction unit controls the pressure distribution. In other words, the pressure distribution reflects the flow state. Therefore, we studied the flow state inside the vortex suction unit by measuring the pressure distribution. Figure 6 shows the pressure distribution measurement apparatus that we designed. We machined 37 pressure measuring taps on the upper cover of the suction unit. The front end of each tap had a diameter of 1 mm. The back end was connected to the pressure sensor using a tube with a diameter of 1.5 mm. Each tube as equipped with a switching valve. First, we closed all the valves to cut off the connection between the taps and the pressure sensor. We then installed the vortex suction unit on the apparatus shown in Fig. 5 in accordance with the steps listed in Section B. After which, we set the required experimental conditions. We opened one of the valves, connected the measuring tap to the pressure sensor, and recorded the sensor reading and the measuring tap location during the pressure distribution measurement process. We then closed the opened valve. We repeated the abovementioned steps to obtain the data corresponding to all the 37 measuring taps. We can finally determine the pressure distribution inside the vortex chamber (i.e., P–r curve) using this process.

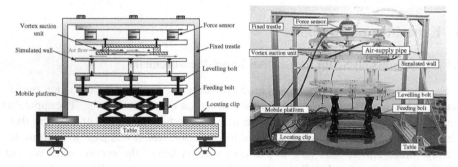

Fig. 5. Schematic and photograph of experimental setup used for measuring suction force

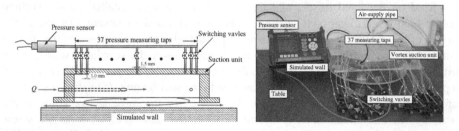

Fig. 6. Schematic and photograph of experimental setup used for measuring pressure distribution

5 Results and Discussion

5.1 Suction Performance of the Suction Unit on the Smooth Surface

We first obtained the F–h curve for the suction unit on a smooth surface. We obtained the F–h curves at different supply flow rates (Q = 200, 250, 300, 350, 400 L/min (ANR)) (Fig. 7). The suction force continuously increased as the flow rate increased. In addition, the changing trend of the F–h curves at different flow rates was almost the same. We rearranged the data in Fig. 7, as shown in Fig. 8, to more clearly express the tendency of change in the suction force when Q varies (i.e., F–Q curves). The variation tendency of the F–Q curves was almost the same when h changes. In addition, the F–Q curves can be well fitted by some quadratic curves. The correlation coefficient of each fitting curve was larger than 0.9953.

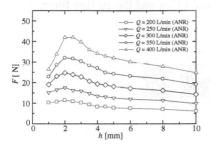

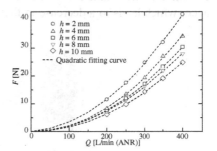

Fig. 7. F–h curves for vortex suction unit at different supply flow rates

Fig. 8. F–Q curves for vortex suction unit at different distances

This relationship is strong evidence for the adjustment of the suction force by changing the supply flow rate. For example, consider a case, where the suction unit is fixed to the robot, such that the distance between the wall and the suction unit is constant. Suppose that the original supply flow rate is Q_0. The suction force F 0 is kQ_0^2, where k is the proportional coefficient. The percentage change in the suction force can be expressed as follows if we change the flow rate to Q_1, then the percentage change in the suction force can be expressed as

$$\frac{\Delta F}{F_0} = \frac{kQ_1^2 - kQ_0^2}{kQ_0^2} \times 100\% = \left[\left(\frac{Q_1}{Q_0}\right)^2 - 1\right] \times 100\% \qquad (1)$$

A simple functional relationship exists between the percentage change in the suction force and the supply flow rate. That is, if we want to increase/decrease the suction force to a certain percentage ($a\%$), we can use this relationship to obtain the value of the required flow rate (i.e. $Q_1(= Q_0\sqrt{(1 \pm a\%)})$). This relationship is very important for adjusting and controlling the suction force of the climbing robot.

5.2   Suction Performance of the Suction Unit on the Rough Surfaces

Vortexbot will encounter wall surfaces with different roughness during real climbing. We used several pieces of sandpaper to study the effect of different surface roughness on the suction force of the vortex suction unit and simulate wall surfaces with different roughness. Figure 9 shows a smooth surface (i.e., acrylic plate) and three kinds of rough surfaces (i.e., sandpaper with mesh numbers of 400, 120, and 60). The wall surface becomes rougher as the mesh number decreases. Figure 10 shows the suction force variation on wall surfaces with different roughness for a supply flow rate of 350 L/min (ANR) and h of 4 mm. The suction force became smaller as the surface became rougher. The amplitude of the suction force decreased by 28.2% compared to that for the smooth surface when the roughness was P 60. It shows that the rotation flow inside the vortex chamber was weakened by the rough surface because the rough surface produces a resistance to the rotation flow. Moreover, the rougher the surface, the larger the resistance, which leads to a decrease in the rotating speed of air and the suction force.

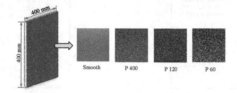

Fig. 9. Wall surfaces with different roughnesses

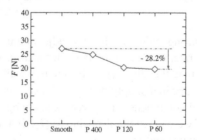

Fig. 10. Suction forces on wall surfaces with different roughnesses

5.3   Stepping Over Raised Obstacles

The process of the suction unit passing over a raised obstacle when Vortexbot walks on a wall surface with raised obstacles was a dynamic process. The climbing speed of the robot was almost negligible compared to the flow velocity of the rotating flow inside the vortex chamber. Hence, we can study such a dynamic process by studying the static pressure distributions and the suction forces when the raised obstacle was at different positions under the suction unit. Figure 11 shows the seven steps when the suction unit passes over a raised obstacle. Steps 1 and 7 represent the situations before and after the suction unit passes over the obstacle, respectively. Therefore, the wall surface under the

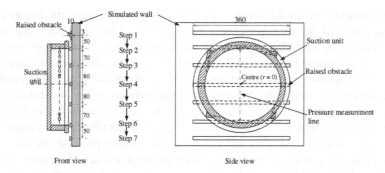

Fig. 11. Relative positions of obstacle and suction unit in different steps

suction unit at these two moments was smooth. Steps 2–6 represent different positions when the suction unit passes over the obstacle. The raised obstacle was 3 mm in height, 10 mm in width, and 360 mm in length.

Figures 12 and 13 show the variation in the suction force and pressure distribution, respectively, for every step. The suction performance of the suction unit was obviously affected when the suction unit passed over the obstacle. In step 1, the obstacle was outside the suction unit. At this moment, the suction force and the pressure distribution were the same as those on a smooth wall surface. In step 2, the obstacle was at the edge of the vortex chamber. Contrary to expectations, the suction force increased by 16%. The pressure distribution in step 2 indicated that the obstacle caused the negative pressure in the center of the vortex chamber to become lower than that in step 1. That means, the air rotation inside the suction unit became more severe at this moment, which might be because the obstacle can protect the rotation flow inside the vortex chamber when it was at the edge of the chamber.

Li indicated that air can easily discharge from the vortex chamber when the distance is large, which makes the air rotation inadequate and weakens the negative pressure distribution. We consider that the obstacle can prevent the air from being discharged from the chamber and can force the air to rotate in the chamber more adequately, which helps in the formation of a lower negative pressure distribution. The suction force considerably declines by 23% compared to the suction force in step 1

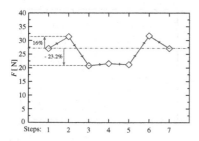

Fig. 12. Fluctuation in suction forces in different steps

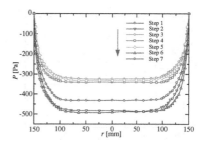

Fig. 13. P–r curves for different steps

when the obstacle reaches the position in step 3. Correspondingly, the negative pressure at the center of the chamber clearly increases. The obstacle hinders air rotation, which weakens the centrifugal effect of the rotation flow. The experimental data indicated that compared to step 3, the suction force and the pressure distribution were not affected by the movement of the obstacle when the obstacle as at the position in step 4. Accordingly, we observed nearly the same variation when the obstacle shifted out of the vortex chamber (steps 5, 6, and 7). In other words, the processes involved in the obstacle moving into and shifting out of the vortex chamber were symmetric.

The abovementioned analysis indicated that the relative positions of the obstacle and the suction unit can cause the suction force to fluctuate when Vortexbot steps over a raised obstacle on a wall surface. For the suction force to substantially decrease when the obstacle is inside the vortex chamber is dangerous. We should take some steps to counter such a situation (e.g., increase the supply flow rate to compensate for the decrease in the suction force or prevent the obstacle from entering the vortex chamber).

5.4   Stepping Over Grooves

As in the case of Fig. 11, we only replaced the obstacle by a groove. The groove was 3 mm in depth, 10 mm in width, and 360 mm in length. Figures 14 and 15 show the variation in the suction force and pressure distribution, respectively, in different steps. Compared to the raised obstacle (previous subsection), the groove had a much milder effect on the suction force and pressure distribution. The experimental results showed that the suction force slightly decreased when the groove entered the suction unit. The decrease in the suction force amplitude was within 10% compared to the suction force in step 1. In addition, the suction force almost remained constant when the suction unit passed over the groove, despite the change in the relative positions of the suction unit and the groove. We considered that this finding was obtained because the groove cannot hinder the rotation flow like the raised obstacle. Therefore, we only needed to slightly increase the supply flow rate when the robot steps a groove to ensure that the robot steadily passes over the groove.

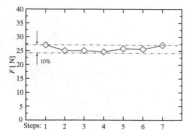

Fig. 14. Fluctuation in suction forces in different steps

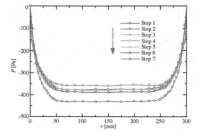

Fig. 15. $P–r$ curves for different steps

6   Practical Climbing Experiments

6.1   Climbing on Walls with Different Obstacles

Figure 16 shows the scene, where Vortexbot stepped over some raised obstacles and the specific dimensions (unit: mm). The results confirmed the major contribution of the new suction unit in improving the ability of the climbing robot to pass over obstacles. We can find that Vortexbot can pass over obstacles with heights ranging from 5 mm to 15 mm, which shows the powerful passing ability of Vortexbot.

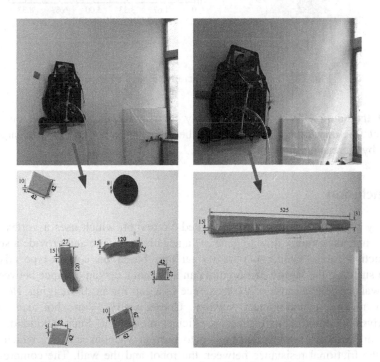

Fig. 16. Experiment on a smooth wall with raised obstacles

6.2   Payload Testing Experiment

We used dumbbells as the load to test the load capacity of Vortexbot (Fig. 17). The mass of Vortexbot itself (i.e., G_0) was 2388 g. The distance between the vortex suction unit and the wall surface was set as 2 mm. The suction unit only needed a supply flow rate (i.e., Q_0) of 330 L/min to make Vortexbot stably climb on the wall surface when the robot did not carry any payloads. We then gradually increased the payload (i.e., $G_t - G_0$) and the supply flow rate (i.e., Q). We obtained the $G_t - Q$ curve (Fig. 17(b)). We can see from the figure that: (1) Vortexbot can carry a payload with a mass of 2.5 kg when the supply flow rate was up to 484 L/min (the total mass of the robot at the moment was 4.888 kg ($G_t = 2.05 G_0$)); and (2) the load capacity was in direct proportion to the square of the supply flow rate, which was consistent with the quadratic

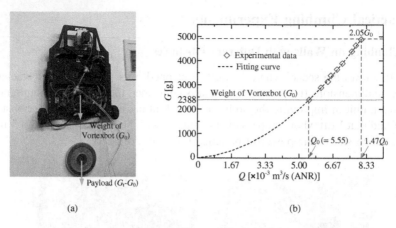

(a)   (b)

Fig. 17. Photograph and data of the payload experiment

curve of the suction force and the supply flow rate (Fig. 8), suppose the friction coefficient was unvarying. The result implied that the robot can have a larger load capacity by increasing the supply flow rate.

7   Conclusion

This study presents a climbing robot, called Vortexbot, which uses a vortex suction unit. This unit uses vortex flow to generate a negative pressure and provide a sufficient stable suction force. Unlike the traditional unit based on contact-type adsorption (vacuum suction), the vortex suction unit can maintain a certain distance between itself and the wall surface because of the existence of a gap for air discharging. Hence, this unit can realize a non-contact suction. Consequently, Vortexbot can adopt a wheel-driven locomotion to realize an efficient movement. Such a kind of suction mechanism needs no sealing mechanism to prevent vacuum leakage, which greatly reduces the frictional resistance between the robot and the wall. The compressed air vents from the gap between the suction unit and the wall surface after rotating in the vortex chamber. Hence, such kind of flow direction can avoid the effect of the dust and dropped items on the wall surface. As a result, Vortexbot can easily climb walls with various surface conditions, including surfaces that are rough, with raised obstacles, and with grooves. In this study, we obtained the effect of the supply flow rate, surface roughness, inclined condition, obstacles, and grooves on the suction unit performance based on the experimental results. We also tested Vortexbot on real wall surfaces. Moreover, we tested the payload capacity of Vortex.

The abovementioned results provide substantial evidence to justify a further study of the climbing robot. In the future, we intend to study the flow phenomenon inside the vortex suction unit by means of a flow field visualization and a numerical simulation. We hope to explain the reasons for the effect of the surface condition of the wall on the suction force from the viewpoint of hydromechanics. Moreover, we intend to continue optimizing the suction mechanism design and improve the obstacle-crossing ability and stability of the robot.

References

1. Zhang, H., Zhang, J., Wang, W., Liu, R., Zong, G.: A series of pneumatic glass-wall cleaning robots for high-rise buildings. Ind. Rob. Int. J. **34**, 150–160 (2007)
2. Zhang, H., Wang, W., Zhang, J.W.: High stiffness pneumatic actuating scheme and improved position control strategy realization of a pneumatic climbing robot. In: IEEE International Conference on Robotics and Biomimetics, pp. 1806–1811. IEEE Computer Society (2009)
3. Hirose, S., Arikawa, K.: Coupled and decoupled actuation of robotic mechanisms. In: IEEE International Conference on Robotics & Automation, vol. 1, pp. 33–39. IEEE (2001)
4. Hirose, S., Nagakubo, A., Toyama, R.: Machine that can walk and climb on floors, walls and ceilings. In: Proceedings of the International Conference on Advances in Robotics, pp. 753–758 (1991)
5. Balaguer, C., Gimenez, A., Abderrahim, M.: ROMA robots for inspection of steel based infrastructures. Ind. Rob. Int. J. **29**(3), 246–251 (2002)
6. Resino, J.C., Jardon, A., Gimenez, A., Balaguer, C.: Analysis of the direct and inverse kinematics of ROMA II Robot. In: Proceedings of the 9th International Conference on Climbing and Walking Robots, pp. 107–114 (2006)
7. Molfino, R., Armada, M., Cepolina, F.: Roboclimber the 3-ton spider. Ind. Rob. **32**(2), 163–170 (2005)
8. Longo, D., Muscato, G.: The Alicia(3) climbing robot. IEEE Rob. Autom. Mag. **13**(1), 42–50 (2006)
9. Savall, J., Avello, A., Briones, L.: Two compact robots for remote inspection of hazardous areas in nuclearpower plants. In: International Conference on Robotics and Automation, ICRA, Detroit, Michigan, USA, May, pp. 1993–1998 (1999)
10. Luk, B., Liu, K., Collie, A., Cooke, D., Chen, S.: Tele-operated climbing and mobile service robots for re-mote inspection and maintenance in nuclear industry. Ind. Rob. Int. J. **33**(3), 194–204 (2006)
11. Armada, M., Prieto, M., Akinfiev, T., Fernandez, R., Gonzalez, P., Garcia, E., Montes, H., Nabulsi, S., Pon-ticelli, R., Sarria, J., Estremera, J., Ros, S., Grieco, J., Fernandez, G.: On the design and development of climbing and walking robots for the maritime industries. Marit. Res. **2**(1), 9–32 (2005)
12. Chu, B., Jung, K., Han, C.S., Hong, D.: A survey of climbing robots: locomotion and adhesion. Int. J. Precis. Eng. Manufact. **11**(4), 633–647 (2010)
13. Schmidt, D., Berns, K.: Climbing robots for maintenance and inspections of vertical structures, a survey of design aspects and technologies. Rob. Autonom. Syst. **61**(12), 1288–1305 (2013)
14. Tummala, R.L., Mukherjee, R., Aslam, D., Dulimarta, H., Minor, M., Dangi, G.: Climbing the walls. IEEE Robot. Autom. Mag. **9**(4), 10–19 (2002)
15. Zhu, H., Guan, Y., Wu, W.: Autonomous pose detection and alignment of suction modules of a bipedwall-climbing robot. IEEE-ASME Trans. Mechatron. **20**(2), 653–662 (2015)
16. Guan, Y., Zhu, H., Wu, W., Zhou, X., Jiang, L., Cai, C., Zhang, L., Zhang, H.: A modular biped wall-climbing robot with high mobility and manipulating function. ASME Trans. Mechatron. **18**(6), 1787–1798 (2013). IEEE

17. Zhao, Y., Fu, Z., Cao, Q., Wang, Y.: Development and applications of wall-climbing robots with a single suction cup. Robotica **22**, 643–648 (2004)
18. Schmidt, D., Hillenbrand, C., Berns, K.: Omnidirectional locomotion and traction control of the wheel-driven, wall-climbing robot, CROMSCI. Rob. J **29**(7), 991–1003 (2011)
19. Koo, I.M., Trong, T.D., Lee, Y.H., Moon, H., Koo, J., Park, S.K., Choi, H.R.: Development of wall climbing robot system by using impeller type adhesion mechanism. J. Intell. Rob. Syst. **72**(1), 57–72 (2013)
20. Qian, Z., Zhao, Y., Fu, Z., Wang, Y.: Fluid model of sliding suction cup of wall-climbing robots. Int. J. Adv. Rob. Syst. **3**(3), 275–284 (2006)
21. Zhou, Q., Li, X.: Experimental comparison of Drag-wiper and Roller-wiper glass-cleaning robots. Ind. Rob. Int. J. **43**(4), 409–420 (2016)
22. Li, X., Kawashima, K., Kagawa, T.: Analysis of vortex levitation. Exp. Therm. Fluid Sci. **32**(8), 1448–1454 (2008)
23. Li, X., Kagawa, T.: Development of a new noncontact gripper using swirl vanes. Rob. Comput. Integr. Manufact. **29**, 63–70 (2013)
24. Zhao, J., Li, X.: Effect of supply flow rate on performance of pneumatic non-contact gripper using vortex flow. Exp. Therm. Fluid Sci. **79**, 91–100 (2016)

Motion Planning and Simulation of Multiple Welding Robots Based on Genetic Algorithm

Yongsheng Chao[✉] and Wenlei Sun

College of Mechanical Engineering, Xinjiang University, Urumqi 830047, China
cys21st@163.com

Abstract. To allocate welding tasks to multiple robots and find collision free paths for them, an approach of multi-robot motion planning and simulation based on genetic algorithm is proposed. Priorities of welding points are defined based on the sequence constraints of welding points. Welding points are allocated to multiply robots and the objective is to minimize the welding time of station. Adapted genetic algorithm is proposed to seek the optimized solution. The three dimension model of welding assembly line is built in eM-power software. The welding robots move along the allocated welding points in the virtual environment, which can find and settle collisions between two robots or between robot and parts, and free collision path is founded finally. Welding path simulation can sharply shorten the planning time for the task allocation of multiply robots.

Keywords: Welding robot · Motion planning · Genetic algorithm · Welding sequence constraints

1 Introduction

Welding robots have been widely applied to car body assembly lines. It can reduce the labor intensity and welding costs, and shorten the processing time, improve the welding quality and productivity [1]. A car body is usually composed of $300 \sim 500$ sheet stamping parts with complex shape and assembled through $4000 \sim 6000$ welding points [2]. In order to improve the production efficiency, there is a need to coordinate multiple robots to weld a car body. Therefore, How to allocate the welding points to multi-robot evenly and ensure them to complete task in the manner of free collision are important things for process planners [3, 4]. The traditional method of welding robot path planning is dependent on the experience and knowledge of process planners in robot welding industry. Process planners formulate a robot path according to the locations, number and welding sequence constraints of welding points [5]. Different process planners have different technical backgrounds, so they formulate different robot paths. This method of path planning lacks strict mathematical theory basis and path planning scheme is random to some extent, which often causes that unreasonable robot welding sequence, the long cycle time and the collision between robots or between robot and parts. The collision leads to the damage to equipments and robots. Therefore, if the collision takes place the process planner must replan new paths for robots from scratch until he finds collision avoidance paths.

© Springer International Publishing AG 2017
Y. Huang et al. (Eds.): ICIRA 2017, Part III, LNAI 10464, pp. 193–202, 2017.
DOI: 10.1007/978-3-319-65298-6_18

In recent years, with the development of the artificial technology, fuzz control, neural network and genetic algorithm have also been widely used in robot path planning. Mendes [6] proposes a adaptive fuzz control method to control the motion and force of industrial robot. Pashkevich [7] proposes a collision avoidance path planning approach for robot based on the topologically ordered neural network model. Panda [8] provides an approach to the dynamic path planning problems of robots in uncertain dynamic environments based on genetic algorithm. Kim [9] addresses welding task sequencing for robot arc welding process planning using genetic algorithm. Castillo [10] employs genetic algorithm for the problem of offline point-to-point autonomous mobile robot path planning. These artificial technologies provide the theoretical base for welding robot path planning in effective and fast way. Welding robot path are affected by many factors in the actual welding operation. So the path gained from optimization algorithm must be verified in a production environment. However, it is a time-consuming job and costs a lot of money to verify. The simulation in virtual environment is proved to be a powerful tool. It can shorten the time and save cost greatly. Based on the characteristics of robot welding production line, a mathematical model of the welding path is established. An adaptive genetic algorithm is proposed to solve the model and an optimal task sequencing is obtained. A robot path can be formed through passing every task point in sequence. Finally, virtual welding production line is established in eM-power software platform. Robot welding procedures are simulated in this platform to validate the feasibility of the path.

2   The Analysis of Robot Welding Constraints

2.1   Constraints Analysis

In order to obtain a higher degree of automation and make full use of production line, welding workstations must be divided reasonably according to the product structure. The number of clamping should be reduced as far as possible. To improve the dimensional accuracy and the production efficiency, welding procedure must meet the following constraints [11]: (1) Constraints of welding sequence. Different welding sequences lead to different welding deformations. To ensure the welding qualities, a reasonable welding sequence should be adopted. (2) Constraints of welding workspace. In generally, when multi-robot participates in the same welding task, the center distance of welding point should be far less than the size of the welding gun, or two adjacent welding points can't be welded using two welding guns at the same time. In addition, the collision between welding guns and parts should be considered in planning welding path. (3) Constraints of welding time. The tasks should be allocated to multiple robots evenly, which can make the task completion time of each robot as close as possible.

2.2   Criteria for Solving Constraints Problem

When two welding robots collide, workspaces of two robots are analyzed. Motion ranges of the end-effector are determined respectively. Non-overlapping workspace of two robot poses is selected. Changing the robot pose avoids collision between them. If there are too much overlapping workspace, priorities of welding points are assigned to make the only one robot work at the overlapping workspace at the same time. In addition, the top priority should be assigned to the robot which has the long welding path. The robot with the low priority should select the welding point in the non-overlapping workspace or wait until the robot with the top priority completes the task.

3 Mathematical Model of Sequencing Optimization for Car Body

Welding point allocation of multi-robot is a problem of multiple constraints. In the actual robot welding planning, several constraints must be considered in order to guarantee the welding quality of car body and to avoid the collision in the welding operation. Welding point allocation can be expressed as follows: given a robot set $P = \{p_1, p_2, \cdots, p_{n-1}, p_n\}$, which includes n welding robots. Given a welding point set $G = \{g_1, g_2, \cdots, g_{m-1}, g_m\}$, which includes m welding points. Given operation time set of welding points $T = \{t_1, t_2, \cdots, t_{m-1}, t_m\}$. Given welding sequence constraint set $E = \{e_{ij} \mid i \leq m, \ j \leq m\}$, e_{ij} indicates whether the welding point i is welded before welding point j.

$$e_{ij} = \begin{cases} 1 & \text{can't be welded before welding point } i \\ 0 & \text{can be welded before welding point } j \end{cases}$$

The mathematical model of welding task can be expressed as:

$$T_k = \sum_{i=1}^{m} \sum_{j=1}^{m} x_{kij} t_j + \sum_{i=1}^{m} \sum_{j=1}^{m} e_{ij} w_k \tag{1}$$

where x_{kij} is a decision variable, $x_{kij} = \begin{cases} 1 & \text{welding robot } k \text{ passes path}(i,j) \\ 0 & \text{doesn't pass the welding path}(i,j) \end{cases}$, T_k is the total welding time of robot k, w_k is the waiting time of robot k.

There are several robots performing the welding tasks in a workstation. Therefore, the workstation time is determined by the robot which spends most time. It can be expressed as follows:

$$T_{wt} = \max\{T_1, T_2, \cdots, T_{n-1}, T_n\} \tag{2}$$

4 Adaptive Genetic Algorithm for Task Allocation

4.1 The Gene Encoding for Task Allocation

Consider that m robots weld n welding points. A numeric string contains n digits. The locus of the string specifies the index of welding points sorting by ascending order.

The digit represents the index of robot to which the welding point is allocated. Each string corresponds to a task allocation solution, i.e. a chromosome. A task allocation solution is encoded in Fig. 1. The string which contains 10 digits represents 10 welding points to be allocated. Welding point 1, 5, and 6 are allocated to robot 2. Welding point 2, 3,9, and 10 are allocated to robot 1. Welding point 4,7, and 8 are allocated to robot 3 (Table 1).

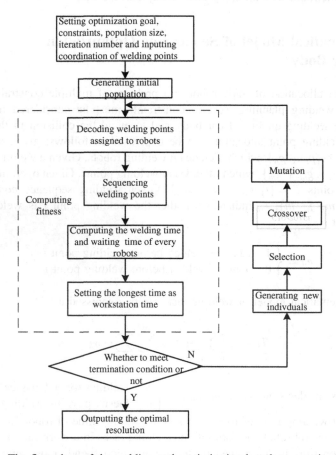

Fig. 1. The flow chart of the welding path optimization based on genetic algorithm

4.2   Selection Operation

A. Fitness function

In genetic algorithm, the value of fitness function should be positive. The optimized objective is the shortest welding time. The fitness function f is the reciprocal of time:

$$f = 1/T_{min} \qquad (3)$$

Table 1. Gene string encoding

Welding points	1	2	3	4	5	6	7	8	9	10
Robot index	2	1	1	3	2	2	3	3	1	1

B. Selection Operator

The selection operator determines how to select the chromosomes for creating the next generation. Therefore, it directly affects the quality of next genome. The selection operator should ensure selecting the good chromosomes. There are several kinds of selection operator in genetic algorithm. One of famous methods is the roulette selection. In this method, the probability of the individual selected is proportional to its fitness. The greater the individual fitness is, the higher probability of the individual selected is. The probability of selection is as follows:

$$p_i = f_i / \sum_{j=1}^{m} f_i \tag{4}$$

where m is the population size, f_i is the fitness of the i th individual.

C. Crossover Operator

The goal of crossover operator is to produce new individuals. Two individuals are selected randomly from the genome. The conventional two-point crossover is employed and two individuals are crossed over to form two new offspring with probability p_c.

D. Mutation Operator

Mutation is a unary operator. It can supplement crossover operators, improve exploration, and prohibit local optima. Mutation operator randomly selects one individual from genome. Two different integers ranging from 1 to the number of welding points are selected randomly with probability p_m. Genes between different integers are sorted randomly.

4.3 Adaptive Genetic Algorithm

The crossover and mutation probabilities in the traditional genetic algorithm are fixed. The search result is easy to fall into the local optimum in the early search stage and isn't easy to converge the optimal solution in the later stage. Therefore, the adaptive genetic algorithm is adopted to avoid the this problems. It has the following advantages: (1) It has strong global search ability and a weak local search ability to prevent premature convergence in the early search stage; (2) Its global search ability declines and local search ability increases in the late search stage, which contributes to find the optimal individual. The crossover and mutation probabilities of adaptive genetic algorithm are as follows:

$$p_c = \begin{cases} p_{c\_\max} - \left(\frac{(p_{c\_\max} - p_{c\_\min})}{it_{\max}}\right) * iter & , f_c > f_{avg} \\ p_{c\_\max} & , f_c \le f_{avg} \end{cases} \tag{5}$$

$$p_m = \begin{cases} p_{m\_\min} + \left(\frac{p_{m\_\max} - p_{m-\min}}{it_{\max}}\right) * iter & , f_m > f_{avg} \\ p_{m\_\min} & , f_m \le f_{avg} \end{cases} \tag{6}$$

where $p_{c\_\max}$ and $p_{c\_\min}$ are the maximum and minimum crossover probability respectively, $it_{\max}$ is the maximum iteration number, $iter$ is the current iteration number, f_{avg} is the average fitness of the genome, f_c is the larger of two fitness, f_m is the fitness of the individual which need to mutate.

5 Development of Optimization Program and Simulation

5.1 Generation of Welding Path

The optimization program for the welding path is developed by Visual C++. The optimization procedure based on genetic algorithm is showed in Fig. 1. Coordinates of welding points and constraints matrix are inputted through the document file in the optimization program. The number of robots allocated, the number of iteration, the number of populations, the maximum and minimum probability of crossover, the maximum and minimum probability of mutation are inputted in the interface. The shortest path and distance are outputted after the end of the search.

5.2 Simulation of Welding Path

eM-power platform is a digital planning system and developed in accordance with the principle of virtual manufacturing system. It is a powerful tool for enterprises to implement virtual manufacturing, which provides the digital design, process planning, manufacturing, simulation and analysis of assembly. The specific process of simulation for welding path in eM-power platform is indicated in Fig. 2.

The task allocation result is simulated in eM-power platform. The basis of simulation is the modeling of welding production line for robots. To facilitate the modeling of welding production line layout, we propose a new layout scheme in which the layout view of production line is modeled under the 2D environment and is browsed under the 3D environment. In this way, the interferences between various manufacturing resources are easy to be found and can be modified timely.

The layout of welding production line is shown in Fig. 3. First, the resource trees are established on eM-power platform. Resource trees include various kinds of resources and parts which are used in welding operation. Each node of resource trees corresponds to its 2D model and 3D model. The 2D model is connected with its 3D model by means of an intermediate module. In this way, when the position of the parts in a 2D view environment changes, the position in the 3D view environment will automatically changes. In addition, when the z-coordinates of parts in part trees or

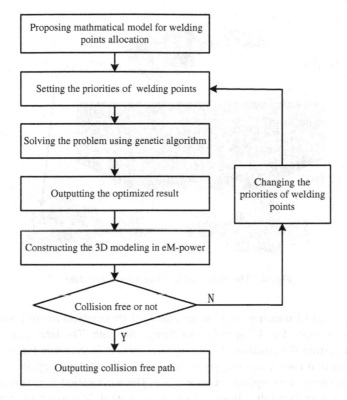

Fig. 2. Simulation procedure of welding path

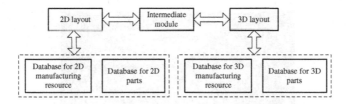

Fig. 3. Layout scheme of the welding production line

manufacturing resources in the resource tree are edited, their positions in the 3D view environment will change accordingly. Welding production line is modeled in eM-power platform. 3D and 2D process layout for welding production are shown in Fig. 4.

The optimized path gained from the genetic algorithm is simulated in the 3D process layout to ensure that there is no interference between robots and parts. Welding points assigned to each robot are inputted in the interface. Robot will move according to the order of welding points. If the interference happens process planner can find it. In Fig. 5, a collision is found between the welding torch and parts. In this situation we can

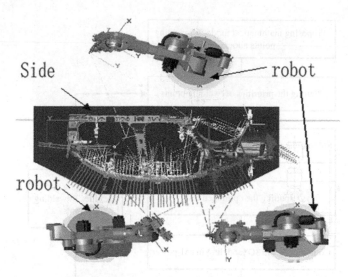

Fig. 4. The model of welding production line

adjust the position of welding torch or the clamp. If this method doesn't work we can change the sequence of welding points to change the path. The interference is eliminated after adjusting the position of welding torch, which is shown in Fig. 6. But if the above methods don't work welding points need to be allocated once again. The paths of each robot also need to be optimized once again. The workstation time is reduced from 120 s to 100 s after using the above optimization method. Moreover, the reliability of the welding operation also increases after simulation.

Fig. 5. The interference between the welding torch and the part

Fig. 6. Eliminating the interference after adjusting the position of the welding torch

6   Conclusions

This paper proposes an efficient and feasible approach of multi-robot welding path optimization. An adaptive genetic algorithm is proposed to optimize the welding path with the consideration of welding sequencing constraints, the task balancing of robots and the length of the welding path. The welding production line is modeled in eM-power software platform. The Optimized path is simulated in this platform. If the interference or collision occurs in the simulation, robot path is modified until a collision-free path is found. This method not only shortens the time of path planning, but also improves the efficiency of welding greatly. By means of the path simulation, the damage to robots and equipments can be avoided.

Although the proposed method solves the problem of the task allocation and path optimization to some extent, there are still many problems that need to be solved. There are many factors which need to be considered, such as equipment layout. In the welding procedure each robot base is fixed and only their arms and welding torch moves. Therefore, if there is a little change in the positions of parts or clamp, there may be a collision between the robot and the equipments. The welding path needs to be optimized once again. It is easy to find these problems in the simulation. Therefore, how to avoid repeated optimizations allowing for above problems in the optimization procedure is worth the further research.

Acknowledgement. This work was supported by the National Science Foundation of China [51565058].

References

1. Cederberg, P., Bolmsjö, G., Olsson, M.: Robotic arc welding - trends and developments for higher autonomy. Ind. Robot **29**, 98–104 (2002)
2. Yongsheng, C., Haijiang, L.: Feature model and case retrieval for body-in-white part. Int. J. Adv. Manufact. Technol. **54**, 231–237 (2011)
3. Gerkey, B.P., Mataric, M.J.: Sold!: auction methods for multirobot coordination. IEEE Trans. Robot. Autom. **18**, 758–768 (2002)
4. Matari, J.M., Sukhatme, G.S., Stergaard, E.H.: Multi-robot task allocation in uncertain environments. Autonom. Robots **14**, 255–263 (2003)
5. Quan, W.: Clamp Design and Welding Assembly Technology for Automobile. Press of Beijing University of Technology, Beijing (1996)
6. Mendes, N., Neto, P.: Indirect adaptive fuzzy control for industrial robots: a solution for contact applications. Expert Syst. Appl. **42**, 8929–8935 (2015)
7. Pashkevich, A., Kazheunikau, M., Ruano, A.E.: Neural network approach to collision free path-planning for robotic manipulators. Int. J. Syst. Sci. **37**, 555–564 (2006)
8. Panda, R.K., Choudhury, B.B.: An effective path planning of mobile robot using genetic algorithm. In: 2015 IEEE International Conference on Computational Intelligence and Communication Technology (CICT), pp. 287–291 (2015)
9. Kim, K.Y., Kim, D.W., Nnaji, B.O.: Robot arc welding task sequencing using genetic algorithms. IIE Trans. **34**, 865–880 (2002)
10. Castillo, O., Trujillo, L., Melin, P.: Multiple objective genetic algorithms for path-planning optimization in autonomous mobile robots. Soft. Comput. **11**, 269–279 (2007)
11. Schwarzer, F., Saha, M., Latombe, J.C.: Adaptive dynamic collision checking for single and multiple articulated robots in complex environments. IEEE Trans. Rob. **21**, 338–353 (2005)

Leader-Follower Formation Control Based on Artificial Potential Field and Sliding Mode Control

Xu Wang, Hong-an Yang[✉], Haojie Chen, Jinguo Wang,
Luoyu Bai, and Wenpei Zan

School of Mechanical Engineering,
Northwestern Polytechnical University, Xi'an 710072, China
yhongan@nwpu.edu.cn

Abstract. This paper presents a leader-follower formation integrated control method based on artificial potential field (APF) and sliding mode control (SMC) in an unknown environment with obstacles. Firstly, the online path planning in formation control is executed via APF to find a collision-free path for leader from the initial position to the goal position. Then, the trajectory tracking controller is designed via SMC method to adjust the linear velocity and angular velocity of the followers to form and maintain the predefined formation. Finally, the effectiveness of the proposed formation integrated control method has been verified by simulation.

Keywords: Leader-follower formation · Online path planning · Trajectory tracking · Artificial potential field (APF) · Sliding mode control (SMC)

1 Introduction

Multi-mobile robots formation control has been one of the interesting research topics in the control community in the past decade. It is a technology that under certain environmental constraints, the control of multi-mobile robots maintains a certain geometric relationship, quickly forms a stable formation and maintains formation moving, and through cooperation, completes assigned tasks. Multi-mobile robots formation has been widely used in many areas, such as exploration, surveillance, search & rescue, and transportation of large objects.

Two major problems in leader-follower formation control includes: (1) How to plan the path online to reach the goal position; (2) How to form and maintain the formation. Aiming at the first problem, many methods have been proposed, such as artificial potential field (APF) method [1–4], fuzzy logic method [5], neural network method [6], genetic algorithm [7]. Among these methods, APF stands out in real time computations and handling the dynamics of the robot because of its simple and effective motion planners for practical purpose.

As for the second problem, the trajectory tracking controller used to form and maintain the formation is presented. It makes robots move toward and maintain the predefined formation while tracking the desired trajectory as a group [8–10].

© Springer International Publishing AG 2017
Y. Huang et al. (Eds.): ICIRA 2017, Part III, LNAI 10464, pp. 203–214, 2017.
DOI: 10.1007/978-3-319-65298-6_19

The main controller design methods include: sliding mode control (SMC) [11, 12], backstepping control [13], the adaptive tracking control [14] and so on. The advantages of SMC include fast response, good transient performance and robustness regarding to parameter variations. This paper incorporates SMC method into the design of motion controller to achieve a fast convergence rate for the whole formation control framework.

In an unknown environment with obstacle, path planning and maintaining the stability of the formation is an integrated problem that needs to be addressed. Therefore, an integrated control system of APF based path planning and SMC based trajectory tracking controller is presented. In this method, APF is used to navigate and plan the formation path autonomously for leader and the SMC is applied in forming and keeping the formation.

This paper is organized as follows: In Sect. 2, leader-follower formation problem is described. In Sect. 3, the kinematic model of multi-mobile robots formation is presented. In Sect. 4, leader-follower formation integrated control system based on APF and SMC will be put forward. In Sect. 5, simulation results are presented. Finally, conclusions and future research of this work are given in Sect. 6.

2   Problem Description

In this paper, we focus on the path planning and leader-follower formation control. As shown in Fig. 1, in an unknown environment with static obstacles, the leader carries on the online path planning to search a near-optimal, collision-free and safety path, while the followers tracking the leader to form and keep the formation, finally multi-mobile robots reach the predefined goal.

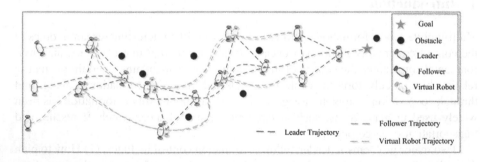

Fig. 1. Multi-mobile robots operating environment

In order to solve the problem of path planning and formation control, we take the two wheeled mobile robot as the research object, an integrated control method for path planning and trajectory tracking of multi-mobile robots is proposed based on leader-follower scheme. Under certain environmental constraints, the leader through online path planning, generates the main trajectory of the formation and the desired trajectory of the followers. The followers trace their desired trajectory using the trajectory tracking method. Finally, multi-mobile robots realize the formation, keep the formation, avoid obstacles and reach goal position.

3 Kinematic Model of Leader-Follower Formation

In this paper we consider the standard kinematic model of mobile robot in Fig. 2. The mobile robot's wheels are subjected to the Pfaffian constraint.

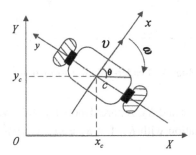

Fig. 2. Two-wheeled mobile robot

Where robot's position and velocity vectors are represented by $(x_c, y_c)^T$ and $(v, \omega)^T$, v and ω represents the line velocity and angular velocity respectively. θ represents the angle between the X axis and the axis of the body. So the kinematic model can be given by

$$\dot{x}_c \sin\theta - \dot{y}_c \cos\theta = [\sin\theta \quad -\cos\theta \quad 0] \begin{bmatrix} \dot{x}_c \\ \dot{y}_c \\ \dot{\theta} \end{bmatrix} = 0 \tag{1}$$

It is expressed as a symmetric affine system in the form:

$$\begin{bmatrix} \dot{x}_c \\ \dot{y}_c \\ \dot{\theta} \end{bmatrix} = \begin{bmatrix} \cos\theta & 0 \\ \sin\theta & 0 \\ 0 & 1 \end{bmatrix} \begin{bmatrix} v \\ \dot{\theta} \end{bmatrix} \tag{2}$$

In this paper, a typical triangle formation structure is selected as an example, as shown in Fig. 3. The main formation trajectory is determined by leader, the followers form and maintain formation via following its desired posture.

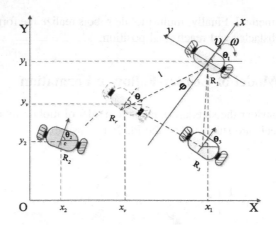

Fig. 3. Leader-Follower formation

In Fig. 3, R_1 is the leader and R_2, R_3 are denoted as followers. As the virtual robot of R_2, R_r is in $(l, \varphi)^T$-formation with R_1. Let $R_1 = (x_1, y_1, \theta_1)^T$, so the posture of R_r is the following:

$$\begin{cases} x_r = -l \cos \varphi \cos \theta_1 - l \sin \varphi \sin \theta_1 + x_1 \\ y_r = -l \cos \varphi \sin \theta_1 + l \sin \varphi \cos \theta_1 + y_1 \\ \theta_r = \theta_1 \end{cases} \tag{3}$$

R_r is the ideal posture of R_2 and can be obtained via formula (3). The same method can be used to R_3. As long as R_2 and R_3 are guaranteed to reach their desired posture in time, multi-mobile robots will be able to keep formation moving.

The actual posture of R_2 is $(x_2, y_2, \theta_2)^T$. The actual liner velocity and angular velocity of R_2 and R_r are $(v_2, \omega_2)^T$ and $(v_r, \omega_r)^T$, respectively.

In the local coordinate system $\ddot{x} - \ddot{y}$, the actual tracking error of R_2 is $(x_e, y_e, \theta_e)^T = (x_2 - x_r, y_2 - y_r, \theta_2 - \theta_r)^T$. Through the coordinate transformation, the tracking error of R_2 can be expressed as the global coordinate system:

$$\begin{bmatrix} x_e \\ y_e \\ \theta_e \end{bmatrix} = \begin{bmatrix} \cos \theta_1 & -\sin \theta_1 & 0 \\ \sin \theta_1 & \cos \theta_1 & 0 \\ 0 & 0 & 1 \end{bmatrix} \begin{bmatrix} x_2 - x_r \\ y_2 - y_r \\ \theta_2 - \theta_r \end{bmatrix} \tag{4}$$

Then, the error rate equation can be obtained:

$$\begin{bmatrix} \dot{x}_e \\ \dot{y}_e \\ \dot{\theta}_e \end{bmatrix} = \begin{bmatrix} y_e \omega_2 - v_2 + v_r \cos \theta_e \\ -x_e \omega_2 + v_r \sin \theta_e \\ \omega_r - \omega_2 \end{bmatrix} \tag{5}$$

4 Formation Integrated Control System Based on APF and SMC

In this paper, combined with the method of leader-follower formation trajectory tracking, we propose the combination of APF and SMC to construct the integrated control system of leader-follower formation, as shown in Fig. 4. The integrated control system of leader-follower formation is divided into three parts. The first part is multi-mobile robots system initialization, the second and third parts will be discussed in Sects. 4.1 and 4.2 respectively.

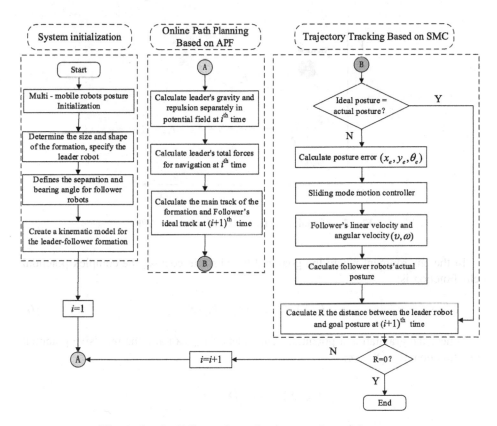

Fig. 4. Leader-Follower formation integrated control system

4.1 Online Path Planning Based on APF

In order to simplify the analysis, this paper puts forward the following assumptions:

(1) All robots will not fail and can detect the distance and angle in real time.
(2) The multi-mobile robots form a regular triangular shape with a length of l, with the velocity v. The radius of the robot is R. The leader robot position is

$X_1 = (x_1, y_1)$. The position of followers are $X_2 = (x_2, y_2)$ and $X_3 = (x_3, y_3)$ respectively.

(3) In the multi-mobile robots operating environment, the obstruction is a circle with radius r_{oi}. The obstacle positions are $X_{oi} = (x_{oi}, y_{oi})$.

Based on the above assumptions, $R_{obi} = R + r_{oi} + l \cdot \sqrt{3}/3$ is the radius of the obstacle.

Under the action of the repulsive force and the gravitational force affected by the obstacle and the goal, the multi-mobile robots carry on the online path planning. The force that the leader robot received in advance is shown in Fig. 5.

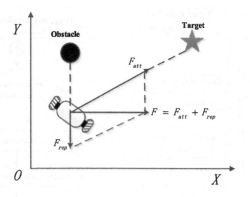

Fig. 5. Virtual attractive force of robots in APF

In the environment where the goal and the obstacle coexist, the complex potential field function is:

$$U(X) = U_{att}(X) + U_{rep}(X) \tag{6}$$

Where the gravitational potential field function $U_{att}(X)$ and the repulsive potential field function $U_{rep}(X)$ are:

$$U_{att}(X) = \frac{1}{2} k_{att}(X - X_g)^2 \tag{7}$$

$$U_{rep} = \begin{cases} \frac{1}{2} k_{rep} \left(\frac{1}{\rho(X, X_0)} - \frac{1}{\rho_0} \right)(X - X_g)^n, & \rho(X, X_0) \leq \rho_0 \\ 0, & \rho(X, X_0) > \rho_0 \end{cases} \tag{8}$$

In the complex potential field, the mobile robot is subjected to the force:

$$F(X) = -\nabla[U(X)] = F_{att}(X) + F_{rep}(X) \tag{9}$$

Where the gravitational function $F_{att}(X)$ and the repulsive function $F_{rep}(X)$ are:

$$F_{att}(X) = -\nabla U_{att}(X) = -k_{att} |X - X_g| \tag{10}$$

$$F_{rep} = -\nabla U_{rep}(X) = \begin{cases} F_{rep1} + F_{rep2}, X - X_0 \le \rho_0 \\ 0, X - X_0 > \rho_0 \end{cases} \tag{11}$$

In formula (11), the repulsive force F_{rep1} and F_{rep2} are:

$$F_{rep1} = k_{rep} \left(\frac{1}{\rho(X, X_0)} - \frac{1}{\rho_0} \right) \frac{1}{\rho^2(X, X_0)} (X - X_g)^n \frac{\partial \rho(X, X_0)}{\partial X} \tag{12}$$

$$F_{rep2} = \frac{n}{2} k_{rep} \left(\frac{1}{X - X_0} - \frac{1}{\rho_0} \right)^2 (X - X_g)^{|n-1|} \frac{\partial \rho(X - X_g)}{\partial X} \tag{13}$$

In the formulas (6)–(13), X is the position of the robot. X_g is the position of the goal. K_{rep} is the repulsive coefficient. K_{att} is the gravitational factor. $\rho(X, X_0)$ is the distance between the obstacle and the robot. ρ_0 is the influence range of obstacles, n is any positive real number [2]. And in the formulas (12) and (13), $(X - X_g)^n$ as the impact factor, makes target point is always the smallest point in the market potential.

4.2 Trajectory Tracking Based on SMC

Leader generates the main trajectory of the formation and followers' ideal trajectory. Using the motion controller, the followers adjust the linear velocity and angular velocity, tracking the ideal trajectory, to achieve leader-follower formation. This paper combines the backstepping method and SMC to design the corresponding motion controller. The specific design method is as follows:

Firstly, we use the backstepping method to get the sliding mode switching function. When $x_e = 0$, select the Lyapunov function:

$$V_y = 0.5 y_e^2 \tag{14}$$

Suppose $\theta_e = -\arctan(v_r y_e)$, we can get $\dot{V}_y = y_e \dot{y}_e$.

Based on the leader-follower formation kinematic model, the follower posture error differential Eq. (5), we can get:

$$\dot{V}_y = y_e \dot{y}_e = -v_r y_e \sin(\arctan(v_r y_e)) \tag{15}$$

By formula (15), we can see that whatever the value of $v_r y_e$, we can get $\dot{V}_y \le 0$. According to the Lyapunov stability definition, if x_e converges to 0, θ_e converges to

— arctan$(v_r y_e)$. Then y_e of the followers in the multi-mobile robots system converges to 0. Therefore, the follower robot's switching function can be designed as:

$$s = \begin{bmatrix} s_1 \\ s_2 \end{bmatrix} = \begin{bmatrix} x_e \\ \theta_e + \arctan(v_r y_e) \end{bmatrix} \tag{16}$$

Combined with the design method of the SMC, the motion controller is designed to make the error movement point in a limited time smoothly converge to the sliding surface and make $s_1 \to 0, s_2 \to 0$.

In order to improve the dynamic quality of the error movement point, double exponential approach is used to control the error point, reducing the chattering and smoothing the transition to the sliding surface.

$$\dot{s} = \begin{bmatrix} \dot{s}_1 \\ \dot{s}_2 \end{bmatrix} = \begin{bmatrix} \dot{x}_e \\ \dot{\theta}_e + \frac{\dot{v}_r y_e + v_r \dot{y}_e}{1 + (v_r y_e)^2} \end{bmatrix} = \begin{bmatrix} y_e \omega - v + v_r \cos \theta_e \\ \omega_r - \omega + \frac{v_r(-x_e \omega + v_r \sin \theta_e + \dot{v}_r y_e)}{1 + (v_r y_e)^2} \end{bmatrix}$$
$$= \begin{bmatrix} -k_{11}|s_1|^{a_1} \frac{s_1}{|s_1| + \delta_1} - k_{12}|s_1|^{a_2} \frac{s_1}{|s_1| + \delta_1} \\ -k_{21}|s_2|^{a_1} \frac{s_2}{|s_2| + \delta_2} - k_{22}|s_2|^{a_2} \frac{s_2}{|s_2| + \delta_2} \end{bmatrix} \tag{17}$$

Through the formula (17), we can obtain follower robot's sliding mode motion control variables $[v, \omega]^T$, as shown in Eq. (18):

$$\begin{bmatrix} v \\ \omega \end{bmatrix} = \begin{bmatrix} y_e \omega + v_r \cos \theta_e + k_{11}|s_1|^{a_1} \frac{s_1}{|s_1| + \delta_1} + k_{12}|s_1|^{a_2} \frac{s_1}{|s_1| + \delta_1} \\ A(\omega_r + \frac{v_r^2 \sin \theta_e + \dot{v}_r y_e}{1 + (v_r y_e)^2} + (k_{21}|s_2|^{a_1} + k_{22}|s_2|^{a_2}) \frac{s_2}{|s_2| + \delta_2}) \end{bmatrix} \tag{18}$$

Where $A = \frac{1 + (v_r y_e)^2}{(v_r y_e)^2 + v_r x_e + 1}$.

Equation (18) is follower robot's sliding mode motion controller. For leader-follower formation trajectory tracking, all follower robots have the same sliding mode variable structure controller.

5 Integrated System Simulation Verification of Leader-Follower Formation

To prove the effectiveness of the proposed strategy and algorithms, the simulations have been implemented in MATLAB.

(1) Three robots form a triangular formation and track the static target without obstacle.

The parameters of the simulations are chosen as $\rho_0 = 20$, $K_{att} = 12$, $a = 0.5$, $K_{11} = 0.6$, $K_{12} = 1.4$, $\delta_1 = 0.02$, $\delta_2 = 0.02$, $a_1 = 0.2$, $b_1 = 1.2$, $a_2 = 0.2$, $b_2 = 1.2$. And assume that the desired formation of leader and follower robots is defined by $(l, \varphi)^T$, of which the desired distance $l = 10$ m and the desired bearing angle $\varphi = \pi/6$.

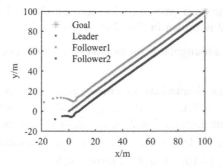

Fig. 6. Leader-Follower formation target tracking

The initial states $(x, y, \theta)^T$ for the leader robot are $(0, \ 0, \ \pi/6)^T$, and for follower robots are given as $(-18, 9, \pi/6)^T$, $(-10, -8, \pi/6)^T$, respectively.

Simulation results depicted in Fig. 6, Multi-mobile robots under the action of APF and sliding mode motion controller, achieve online path planning, after 56 s, from the discrete state to form leader-follower triangular formation, and maintain the formation to reach the goal.

The velocity and tracking error of Follower1 and Follower2 during the whole process are shown in Fig. 7. In Fig. 7(a), under the action of the SMC, Follower1 adjusts its angular velocity and linear velocity to track the desired posture in the formation. As shown in Fig. 7(b), Follower1 can quickly track the ideal pose from the

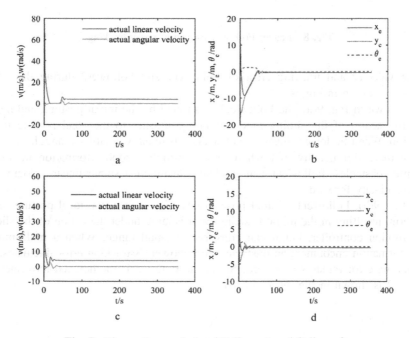

Fig. 7. The motion analysis of Follower1 and Follower2

initial state and keep the pose error zero forward. Similar to Follower1, the detailed movement of Follower2 is shown in Fig. 7(c) and (d).

(2) Three robots form a triangular formation and track the static target with stationary obstacles.

The parameters of the simulations are chosen as $\rho_0 = 10$, $K_{att} = 10$, $K_{rep} = 1600$, $a = 0.5$, $K_{11} = 0.6$, $K_{12} = 1.4$, $\delta_1 = 0.02$, $\delta_2 = 0.02$, $a_1 = 0.2$, $b_1 = 1.2$, $a_2 = 0.2$, $b_2 = 1.2$ and the desired distance $l = 2$ m, the desired bearing angle $\varphi = \pi/6$. The initial states $(x, y, \theta)^T$ for the leader are $(-5, -7, \pi/6)^T$, and followers' initial states are given as $(-8, 5, \pi/6)^T$, $(-10, -8, \pi/6)^T$, respectively.

As shown in Fig. 8, in the unknown environment with obstacle, the multi-mobile robots form a regular triangular formation from the discrete initial posture under the action of APF and sliding mode motion controller.

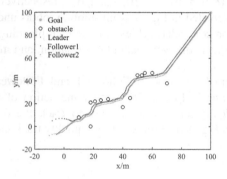

Fig. 8. Leader-Follower formation target tracking

The velocity and tracking error of Follower1 and Follower2 during the whole process are shown in Fig. 9.

As shown in Fig. 9(a), the Follower1 is affected by the initial posture, and moves under the action of the sliding mode motion controller to the ideal posture in the formation. When Follower1 meets the obstacles, its linear velocity and angular velocity will fluctuate significantly, but when the multi-mobile robots formation no longer encounter obstacles, Follower1 can gradually maintain a stable linear velocity and angular velocity forward.

In Fig. 9(b), Follower1's initial pose distance is far from the ideal posture in the formation, resulting in the initial tracking error is large under the action of the sliding mode motion controller, but it can converge to small range. When a multi-mobile robots formation encounters an obstacle, the Follower1's tracking error fluctuates, but the tracking error remains near zero. Similar to Follower1, the detailed movement of Follower2 is shown in Fig. 9(c) and (d).

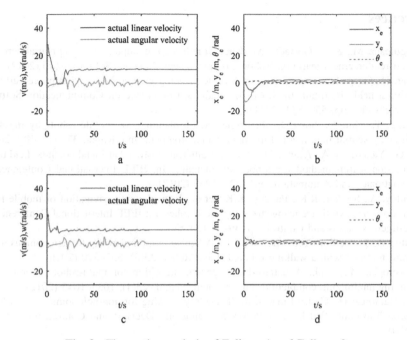

Fig. 9. The motion analysis of Follower1 and Follower2

6 Conclusion

Based on the simulation results, it can be seen that the leader navigates itself by artificial potential field, and generates the main track of the formation and followers' ideal trajectory. Followers tracks its ideal trajectory under the action of SMC to form and maintain the formation tracking goal effectively.

Thus, in this paper, effectiveness of the combination method of APF and SMC is investigated and get a better insight of how the formation can be maintained in the obstacle circumstance. Under the action of APF and SMC, multi-mobile robots can effectively form formation from discrete states, perform on-line path planning, and keep formation track target. In future work, we are planning to make multi-mobile robots keep the formation, avoid the obstacles and track the target in the dynamic environment where the target and the barriers are moving.

Acknowledgments. This work was sponsored by the Seed Foundation of Innovation and Graduate Students in Northwestern Polytechnical University (Z2017018).

References

1. Duguleana, M., et al.: Obstacle avoidance of redundant manipulators using neural networks based reinforcement learning. Robot. Comput. Integr. Manuf. **28**(2), 132–146 (2012)
2. Bing, H., Gang, L., Jiang, G., et al.: A route planning method based on improved artificial potential field algorithm. In: IEEE International Conference on Communication Software and Networks, pp. 550–554 (2011)
3. Zhang, T., Zhu, Y., Song, J.: Real-time motion planning for mobile robots by means of artificial potential field method in unknown environment. Ind. Robot. **37**, 384–400 (2010)
4. Li, G., Yamashita, A., Asama, H., et al.: An efficient improved artificial potential field based regression search method for robot path planning. In: IEEE International Conference on Mechatronics and Automation, pp. 1227–1232 (2012)
5. Pandey, A., Sonkar, R.K., Pandey, K.K., et al.: Path planning navigation of mobile robot with obstacles avoidance using fuzzy logic controller. In: IEEE International Conference on Intelligent Systems and Control, pp. 39–41 (2015)
6. Belter, D., Łabęcki, P., Skrzypczyński, P.: Adaptive motion planning for autonomous rough terrain traversal with a walking robot. J. Field Robot. **33**(3), 337–370 (2016)
7. Kapanoglu, M., et al.: A pattern-based genetic algorithm for multi-robot coverage path planning minimizing completion time. J. Intell. Manuf. **23**(4), 1035–1045 (2012)
8. Sira-Ramírez, H., Castro-Linares, R.: Trajectory tracking for non-holonomic cars: A linear approach to controlled leader-follower formation. In: Decision and Control, pp. 546–551 (2010)
9. Ailon, A., Zohar, I.: Controllers for trajectory tracking and string-like formation in Wheeled Mobile Robots with bounded inputs. In: Melecon 2010 - 2010 IEEE Mediterranean Electrotechnical Conference, pp. 1563–1568 (2010)
10. Kowalczyk, W., Kozłowski, K.R., Tar, J.K.: Trajectory Tracking for Formation of Mobile Robots. Springer, London (2009)
11. Wang, Z.P., Yang, W.R., Ding, G.X.: Sliding mode control for trajectory tracking of nonholonomic wheeled mobile robots based on neural dynamic model. In: Second WRI Global Congress on Intelligent Systems, pp. 270–273 (2010)
12. Solea, R., Filipescu, A., Nunes, U.: Sliding-mode control for trajectory-tracking of a wheeled mobile robot in presence of uncertainties. In: Asian Control Conference, ASCC 2009, pp. 1701–1706 (2009)
13. Sadowska, A., Huijberts, H.: Formation control design for car-like nonholonomic robots using the backstepping approach. In: Control Conference, pp. 1274–1279 (2013)
14. Park, B.S., Yoo, S.J.: Adaptive leader-follower formation control of mobile robots with unknown skidding and slipping effects. Int. J. Control Autom. Syst. **13**, 587–594 (2015)

Trajectory Tracking by Terminal Sliding Mode Control for a Three-Wheeled Mobile Robot

Jia-Xin Shao[1], Yu-Dong Zhao[1], Dong-Eon Kim[1],
and Jang-Myung Lee[2(✉)]

[1] Department of Electrical and Computer Engineering,
Pusan National University, Busan 609-735, Korea
{jiaxin1696,yudong1696,dongeon1696}@pusan.ac.kr
[2] Department of Electronic Engineering,
Pusan National University, Busan 609-735, Korea
jmlee@pusan.ac.kr

Abstract. In this paper, the issue of trajectory tracking for a non-holonomic three-wheeled mobile robot is researched and a controller with two layers' structure which separately deals with the kinematic and dynamic characteristics of the mobile robot system is constructed. For the kinematics, a conventional and common kinematic controller is chosen to transform the position tracking errors into command velocity which will be taken as the reference input into the proposed dynamic controller. For the dynamics, a terminal sliding mode control strategy is designed to track the command velocity generated by the kinematic controller. The stability of the proposed control scheme has been proved by Lyapunov theory. Practical experiments are carried out to verify the effectiveness and accuracy of the proposed control algorithm.

Keywords: Wheeled mobile robot · Trajectory tracking · Terminal sliding mode control

1 Introduction

Autonomous mobile robots have been widely used in the fields of industry, military and civilian application. It thus results in amounts of research attentions to the control of mobile robot. Within all the issues, high-precision trajectory tracking control is the basis for Autonomous mobile robot control. Wheeled mobile robots, however, are characterized as non-holonomic systems, that is, it does not satisfy the Brockett smooth and stable condition. So, it is necessary to do research on trajectory tracking for autonomous wheeled mobile robot.

Sliding mode control (SMC) has been applied to a lot of system due to its fast response, good transient performance, and robustness under parameter variations and disturbances. For high precision control, though the linear sliding mode control can achieve the asymptotic convergence, it cannot realize a fast convergence in finite time. Thus, in [1] a Terminal Sliding Mode Control (TSMC) is proposed. Rigid robotic manipulators, as a multi-input and multi-output system, has been researched in [2] with

© Springer International Publishing AG 2017
Y. Huang et al. (Eds.): ICIRA 2017, Part III, LNAI 10464, pp. 215–225, 2017.
DOI: 10.1007/978-3-319-65298-6_20

TSMC while researches on the application to a second-order nonlinear uncertain systems is carried out in [3]. The conventional TSMC, however, has three disadvantages, such as singularity problem, slower convergence to the equilibrium system state is far away from the equilibrium. Besides, chattering problem still exists due to discontinuous control law and frequent switching action near sliding surface. In order to overcome singularity problem, non-singular TSMC(NTSMC) scheme is proposed and researched in [4, 5]. And NTSMC techniques have got successful application in a lot of systems, such as nonlinear magnetic bearing system [6]. Besides, when the system state is far away from the designed sliding surface the conventional TSMC has a slower reaching speed than the conventional SMC. So, a fast TSMC(FTSMC) strategy is proposed in [7]. Research on FTSMC with exponential decay rate is carried out in [8]. And in [9] application of FTSMC submarine control is realized. Besides, disturbance observer is designed and researched to overcome chattering problem of conventional TSMC in [10].

In this paper, a TSMC scheme based on kinematic control algorithm is proposed to do trajectory tracking control for a three wheeled non-holonomic mobile robot.

The rest part of the paper is organized as follows. Section 2 is the derivation of the mathematical model of the three-wheeled mobile robot; Sect. 3 shows the controller design. Section 4 introduces the experiments which is carried out to verify the effectiveness and feasibility of the proposed algorithm. Finally, the conclusion is drawn in Sect. 5.

2 Mathematic Model

Figure 1 shows the schematic structure of the wheeled mobile robot. Neglecting the surface friction, the kinematics equations for the non-holonomic mobile robot is as follows,

$$\dot{q}_v = \begin{bmatrix} \dot{x} \\ \dot{y} \\ \dot{\varphi} \end{bmatrix} = \begin{bmatrix} \cos\varphi & 0 \\ \sin\varphi & 0 \\ 0 & 1 \end{bmatrix} \begin{bmatrix} \dot{R} \\ \dot{\varphi} \end{bmatrix} = J_v(q_v)v. \tag{1}$$

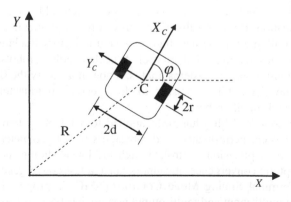

Fig. 1. Wheeled mobile robot.

Where $q_v = [x \quad y \quad \varphi]^T$ is the actual position and orientation of the mobile plat-form. $\dot{R}$ denotes the linear velocity of the COM of WMR, $\dot{\varphi}$ denotes the angular velocity of the COM of WMR, and $J_v(q_v)$ is the so called Jacobian matrix.

The non-holonomic velocity constraint declares that driving wheels purely roll and do not slip. And it can be represented as follows,

$$\dot{y}\cos\varphi - \dot{x}\sin\varphi = 0. \tag{2}$$

For simplification, the above constraints equation can be denoted as,

$$A^T(q_v)\dot{q}_v = 0. \tag{3}$$

where $A^T(q_v) = [-\sin\varphi \quad \cos\varphi \quad 0]^T$.

The generalized dynamic equation can be denoted as follows,

$$M_v(q_v)\ddot{q}_v + C_v(q_v,\dot{q}_v)\dot{q}_v + F_v(\dot{q}_v) + G_v(q_v) + \tau_{dv} = B_v\tau_v - A^T(q_v)\lambda \tag{4}$$

where $M_v(q_v)$ is a symmetric and positive definite inertia matrix, $C_v(q_v,\dot{q}_v)$ is a vector of velocity-dependent centripetal and Coriolis forces, $F_v(\dot{q}_v)$ is a surface friction, $G_v(\dot{q}_v)$ is a gravitational vector, τ_{dv} is a disturbance, B_v is input transformation matrix, τ_v is the control input, and λ is a Lagrange multiplier associated with the constraints.

By differentiating Eq. (1), we can get

$$\ddot{q}_v = J_v(q_v)\dot{v} + \dot{J}_v(q_v)v. \tag{5}$$

Neglecting the surface fraction between the wheel and the contact surface, Eq. (4) can be written as

$$M_vJ_v(q_v)\dot{v} + (M_v\dot{J}_v(q_v) + C_vJ_v(q_v))v + F_{vm} + \tau_{dv} = B_v\tau_v - A^T(q_v)\lambda \tag{6}$$

Because of $J_v^T(q_v)A^T(q_v) = 0$, by multiplying $J_v^T(q_v)$ on two sides of Eq. (6), then the dynamic equation of mobile platform can be derived as follows,

$$M_v'\dot{v} + C_v'v + F_v' = B_v'\tau_v. \tag{7}$$

where

$$M_v' = J_v^T(q_v)M_vJ_v(q_v),$$
$$C_v' = J_v^T(q_v)(M_v\dot{J}_v(q_v) + C_vJ_v(q_v)),$$
$$F_v' = J_v'(q_v)(F_{vm} + \tau_{dv}), \text{ and } B_v' = J_v^T(q_v)B_v.$$

3   Controller Design

3.1   Kinematic Controller for Mobile Platform

As derived previously, the steering system of WMR is given as

$$\dot{q}_v = \begin{bmatrix} \dot{x} \\ \dot{y} \\ \dot{\varphi} \end{bmatrix} = \begin{bmatrix} \cos\varphi & 0 \\ \sin\varphi & 0 \\ 0 & 1 \end{bmatrix} \begin{bmatrix} \dot{R} \\ \dot{\varphi} \end{bmatrix}. \tag{8}$$

Figure 2 illustrates trajectory tracking problem. Within this figure, we define a virtual reference mobile robot that generates a trajectory for the actual one to follow. (x, y, φ) is the position of the real mobile robot, (x_r, y_r, φ_r) is the position of the virtual mobile robot, and (e_1, e_2, e_3) is the relative posture tracking error. The reference steering system is defined as follows,

$$\dot{q}_{vr} = \begin{bmatrix} \dot{x}_r \\ \dot{y}_r \\ \dot{\varphi}_r \end{bmatrix} = \begin{bmatrix} \cos\varphi_r & 0 \\ \sin\varphi_r & 0 \\ 0 & 1 \end{bmatrix} \begin{bmatrix} \dot{R}_r \\ \dot{\varphi}_r \end{bmatrix}. \tag{9}$$

where $q_{vr} = [x_r \quad y_r \quad \varphi_r]^T$ denotes the desired position and orientation. $\dot{R}_r$ and $\dot{\varphi}_r$ are the reference linear and rotation velocity, respectively.

It is necessary to find the velocity control law $V_c = [v_c \quad \omega_c]$, such that $q_v \to q_{vr}$ as $t \to \infty$.

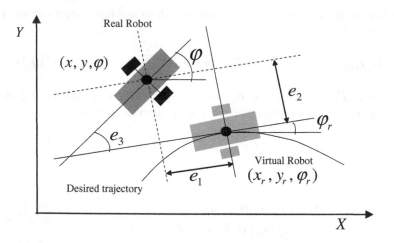

Fig. 2. Posture tracking for the mobile platform.

The absolute posture tracking error between virtual and actual posture is defined as $\tilde{q}_v = q_{vr} - q_v$, and also relative posture tracking error e_{q_v} is defined to transfer the reference posture q_{vr} to a frame fixed on the actual mobile robot. e_{q_v} is denoted as follows,

$$e_{q_v} = \begin{bmatrix} e_1 \\ e_2 \\ e_3 \end{bmatrix} = \begin{bmatrix} \cos\varphi & \sin\varphi & 0 \\ -\sin\varphi & \cos\varphi & 0 \\ 0 & 0 & 1 \end{bmatrix} \begin{bmatrix} x_r - x \\ y_r - y \\ \varphi_r - \varphi \end{bmatrix}. \tag{10}$$

And the error rate can be written as,

$$\begin{bmatrix} \dot{e}_1 \\ \dot{e}_2 \\ \dot{e}_3 \end{bmatrix} = \dot{R} \begin{bmatrix} -1 \\ 0 \\ 0 \end{bmatrix} + \dot{\varphi} \begin{bmatrix} e_2 \\ -e_1 \\ -1 \end{bmatrix} + \begin{bmatrix} \dot{R}_r \cos e_3 \\ \dot{R}_r \sin e_3 \\ \dot{\varphi}_r \end{bmatrix}. \tag{11}$$

Based on Lyapunov stability theory, the command velocity is derived as follows,

$$V_c = \begin{bmatrix} v_c \\ \omega_c \end{bmatrix} = \begin{bmatrix} \dot{R}_r \cos e_3 + k_1 e_1 \\ \dot{\varphi}_r + k_2 \dot{R}_r e_2 + k_3 \dot{R}_r \sin e_3 \end{bmatrix}. \tag{12}$$

where k_1, k_2 and k_3 are positive constants.

3.2 Dynamic Controller for Mobile Platform

For mobile platform, a terminal sliding mode controller is designed. The dynamic equation in velocity form can be expressed as follows,

$$M_v' \dot{v} + C_v' v + F_v' = B_v' \tau_v. \tag{13}$$

The auxiliary tracking error and new states are defined as follows,

$$e_v = v - V_c. \tag{14}$$

$$\dot{r}_v = V_c - k_v \left| \int_0^t e_v dt \right|^\gamma. \tag{15}$$

where $V_c = [v_c \quad \omega_c]^T$ is the command velocity calculated by kinematic controller, and $e_v = [e_v \quad e_\omega]^T$. The filtered second-order error sliding mode surface is chosen as

$$s_v = e_v + k_v \left| \int_0^t e_v dt \right|^\gamma = v - \dot{r} \tag{16}$$

where $0 < \gamma < 1$ is a constant.

We select the control law as follows,

$$\tau_v = \tau_{veq} + \tau_{vfb}. \tag{17}$$

where

$$\tau_{veq} = K_{mv}\ddot{r}_v + K_{cv}\|v\|\dot{r}_v,$$

$$\tau_{vfb} = -c_{v1}s_v - c_{v2} \cdot |s_v|^{\gamma} \cdot \frac{s_v}{\|s_v\| + \kappa_v}.$$

τ_{veq} is feed forward control law. It is used to compensate the dynamic property. τ_{vfb} is the feedback control law. It is used to make the system reach at the previously described sliding mode in finite time. $c_{v1} = diag(c_{v1v}, c_{v1\omega})$ and $c_{v2} = diag(c_{v2v}, c_{v2\omega})$ are positive constant matrix and $\kappa_v \geq 0$. And $0 < \gamma < 1$ is a constant.

3.3   Stability Analysis by Lyapunov Theory

It is assumed that $\|F'_v\|$. Then dynamic equation can be written as

$$M'_v\dot{s}_v = -C'_v s_v - M'_v\ddot{r}_v - C'_v\dot{r}_v - F'_v + B'_v\tau_v. \tag{18}$$

Define the following Lyapunov function

$$V_v = \frac{1}{2}s_v^T M'_v s_v + \frac{1}{2}(e_1^2 + e_2^2) + \frac{1 - \cos e_3}{k_2}. \tag{19}$$

And its derivation is

$$\dot{V}_v = -s_v^T(C'_v - \frac{1}{2}M'_v)s_v + s_v^T(-M'_v\ddot{r}_v - C'_v r_v + F'_v + B'_v\tau_v)$$

$$+ e_1\dot{e}_1 + e_2\dot{e}_2 + \frac{\dot{e}_3 \sin e_3}{k_2} \tag{20}$$

$$= s_v^T(-M'_v\ddot{r}_v - C'_v r_v + F'_v + B'_v\tau_v) - k_1e_1^2 - \frac{k_3\dot{R}_r \sin^2 e_3}{k_2}.$$

Assumption. There are positive real numbers, Δ_{mv}, Δ_{cv} and Δ_{uv} that satisfy the following conditions:

$$\begin{cases} \|K_{mv} - M'_v\| \leq \Delta_{mv} \\ \|K_{cv} - C'_v\| \leq \Delta_{cv} \\ \|F'_v\| \leq \Delta_{uv} \end{cases}. \tag{21}$$

where $K_{mv} \in R^{2\times2}$ and $K_{cv} \in R^{2\times2}$ are positive finite diagonal matrices which are determined through trial and error method.

Substituting the Eq. (17) into Eq. (18), then we obtain,

$$
\begin{aligned}
\dot{V}_v &= s_v^T \left[(K_{mv} - M_v')\ddot{r}_v + (K_{cv}\|v\| - C_v')\dot{r}_v \right] - c_{v1}\|s_v\|^2 \\
&\quad + s_v^T F_v' - c_{v2}|s_v|(\|s_v\| + \kappa_v)^{-1} - k_1 e_1^2 - \frac{k_3 \dot{R}_r \sin^2 e_3}{k_2} \\
&\leq \|s_v\| \left[\left\| K_{mv} - M_v' \right\| \|\ddot{r}_v\| + \left\| K_{cv}\|v\| - C_v' \right\| \|\dot{r}_v\| \right] - c_{v1min}\|s_v\|^2 \\
&\quad + \|s_v\| \Delta_{mv} - c_{v2}|s_v|(\|s_v\| + \kappa_v)^{-1} \\
&\leq \|s_v\| (\|\Delta_{mv}\| \|\ddot{r}_v\| + \Delta_{cv}\|\dot{r}_v\| + \Delta_{cv} - c_{v2min}(\|s_v\| + \kappa_v)^{-1}).
\end{aligned}
\tag{22}
$$

where $c_{v1min} = \lambda_{min}(c_{v1})$. If $c_{v2min} = \lambda_{min}(c_{v2})$ is selected such that $(\Delta_{mv}|\ddot{r}_v| + \Delta_{cv}|\dot{r}_v| + \Delta_{vu})(\|s_v\| + \kappa_v) \leq c_{v2min}$ is guaranteed, we then obtain

$$
\dot{V}_v \leq 0.
\tag{23}
$$

Therefore, according to the Lyapunov stability theory $s_v \to 0$ as $t \to \infty$. This leads to the fact that $e_v \to 0$ and $e_{q_v} \to 0$ as $t \to \infty$.

4 Experiment

Figure 3 shows the real structure of the wheeled mobile robot. The real-time experiments are carried out by a three-wheeled mobile robot. It has two coaxial differential driving wheels and one auxiliary universal wheel. Besides, STM32f407 is chosen to be the control unit, which generates two PWM signals for each wheel. And the motor driver is NT-DC20A. Zigbee communication is utilized to transfer experimental data from mobile robot to computer.

Fig. 3. Structure of mobile robot

The physical parameters of the wheeled mobile robot are listed in Table 1 as follows,

Table 1. Parameters of the wheeled mobile robot

Symbol	Parameter	Value
m_p, m_w	Mass of body and wheel	5 kg, 0.58 kg
d	The distance between the center point and wheel	0.145 m
r	Radius of wheel	0.075 m

For mobile platform, a reference trajectory which contains linear and rotation movement is chosen as bellow,

$$\begin{cases} v_r = 0.1\,(m/s) & 17.85*n(s) < t < = 17.85*n+10(s) \\ v_r = 0.2(m/s) & t > 17.85*n+10(s) \\ \omega_r = 0(rad/s) & 17.85*n(s) < t < = 17.85*n+10(s) \\ \omega_r = 0.2(rad/s) & t > 17.85*n+10(s) \end{cases} \tag{24}$$

where, v_r and ω_r are the reference linear velocity and rotation velocity, respectively. $n = 0, 1, 2, 3$.

The initial configuration are defined as $q_{vr} = \begin{bmatrix} x_r & y_r & \varphi_r \end{bmatrix}^T = \begin{bmatrix} 0.1 & 0.2 & 0.785 \end{bmatrix}^T$ and $q_v = \begin{bmatrix} x & y & \varphi \end{bmatrix}^T = \begin{bmatrix} 0 & 0 & 0 \end{bmatrix}^T$ for reference and real trajectory, respectively (Fig. 4).

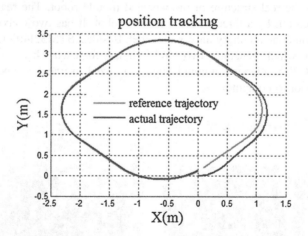

Fig. 4. Position tracking result of mobile robot

The real experimental results by proposed terminal sliding mode control based on kinematic controller are as follows:

According to Figs. 5, 6 and 7, we can see that the tracking errors of x, y and rotation angle all converge to 0 in a finite time. In Figs. 8 and 9, the short fluctuations result from the changing of linear velocity from 0.1 m/s to 0.2 m/s and rotation velocity from 0 m/s to 0.2 m/s, or vice versa.

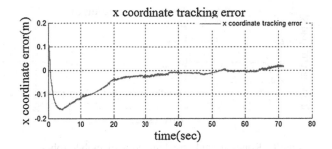

Fig. 5. x coordinate tracking error of mobile robot

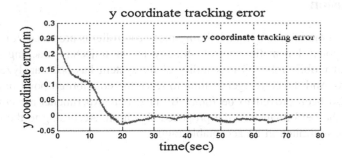

Fig. 6. y coordinate tracking error of mobile robot

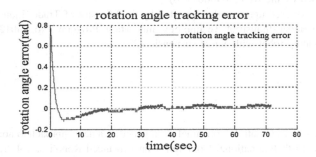

Fig. 7. Rotation angle tracking error of mobile robot

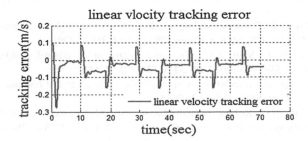

Fig. 8. Rotation angle tracking error of mobile robot

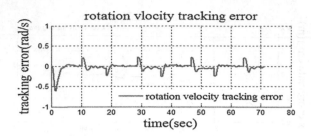

Fig. 9. Rotation angle tracking error of mobile robot

5 Conclusion

In this paper, a model free terminal sliding mode controller is proposed. It is combined with a traditional kinematic control strategy to do trajectory tracking control of three-wheeled mobile robot. The stability of the system has been analyzed by Lyapunov theory. And the convergence time to reach the equilibrium point in the sliding mode has also been proved to be finite. Besides, real experiments are conducted to demonstrate the performance of the proposed algorithm. According to the results, the tracking error is limited in an acceptable range.

Acknowledgment. "This research was supported by the MOTIE (Ministry of Trade, Industry & Energy), Korea, under the Industry Convergence Liaison Robotics Creative Graduates Education Program supervised by the KIAT (N0001126)."

"This material is based upon work supported by the Ministry of Trade, Industry & Energy (MOTIE, Korea) under Industrial Technology Innovation Program. No. G01201605010048 '40 km/h of balancing robot with active suspension'."

References

1. Venkataraman, S.T., Gulati, S.: Terminal sliding modes: a new approach to nonlinear control synthesis. In: Fifth International Conference on Advanced Robotics, vol. 1, pp. 443–448 (1991)
2. Man, Z.H., Paplinski, A.P., Wu, H.R.: A robust MIMO terminal sliding mode control scheme for rigid robotic manipulators. IEEE Trans. Autom. Control **39**(12), 2464–2469 (1994)
3. Park, K.B., Tsuji, T.: Terminal sliding mode control of second-order nonlinear uncertain systems. Int. J. Robust Nonlinear Control **9**(11), 769–780 (1999)
4. Wu, Y.Q., Yu, X.H., Man, Z.H.: Terminal sliding mode control design for uncertain dynamic systems. Syst. Control Lett. **34**(5), 281–287 (1998)
5. Zuo, Z.Y.: Non-singular fixed-time terminal sliding mode control of non-linear systems. IET Control Theory Appl. **9**(4), 545–552 (2015)
6. Chen, S.Y., Lin, F.J.: Robust nonsingular terminal sliding-mode control for nonlinear magnetic bearing system. IEEE Trans. Control Syst. Technol. **19**(3), 636–643 (2011)

7. Ghasemi, H., Rezaie, B., Rahmani, Z.: The design of fast terminal sliding mode kinematic control for wheeled mobile robots. Int. J. Mechatron. Electr. Comput. Technol. **4**, 660–692 (2014)

8. Mobayen, S.: Fast terminal sliding mode tracking of non-holonomic systems with exponential decay rate. IET Control Theory Appl. **9**(8), 1294–1301 (2014)

9. An, X.Y., Ren, Z., Lu, W.: Terminal sliding mode control of attitude synchronization for autonomous docking to a tumbling satellite. In: International Conference on Mechatronic Sciences, Electric Engineering and Computer, Shenyang, China (2013)

10. Xu, K., Chen, M.: Terminal sliding mode control with disturbance observer for autonomous mobile robots. In: Proceedings of the 34th Chinese Control Conference, Hangzhou, China (2015)

Research and Development of Ball-Picking Robot Technology

Hengbin Yu, Shoujun Wang, Haibo Zhou[✉], Lu Yang, and Xu Zhou

Tianjin Key Laboratory for Advanced Mechatronic System Design and Intelligent Control, National Demonstration Center for Experimental Mechanical and Electrical Engineering Education (Tianjin University of Technology), Tianjin 300384, China
haibo_zhou@163.com

Abstract. Apply ball-picking robot in the golf, table tennis, tennis and other small-ball training court can enhance the efficiency of picking up, reduce the operating costs of the stadium, and improve the level of technical service. This paper explores the research of ball-picking robot. Analyzes the research results and application of key technologies in this field, such as navigation and location, path planning and wireless communication. Discusses the trend of intelligent ball-picking robot technology under the background of rapid development of intelligent core technology such as Internet, multi-robot cooperation, machine visual and multi-sensor fusion. The results show that, in addition to golf-picking equipment, the current ball-picking robot technology is still in the primary test stage. The degree of intelligence is still relatively low. There are few achievements can be popularized and applied. The industrial robot in recognition and localization, autonomous navigation, artificial intelligence have a rapid development. They will promote the research on ball-picking intelligent service robot technology. And promote the development of entertainment fitness industry.

Keywords: Ball-picking robot · Navigation and location · Path planning · Wireless communication · Internet

1 Introduction

With the development of the world economy and improvement of people's living standards, small ball games such as golf, table tennis and tennis, which are suitable for people of all ages, are increasingly popular [1]. At the same time, the demands for the service of the site and infrastructure facilities have also become increasingly prominent. For example, in the golf court, the work of picking up golf had to be done when the golf court is closed, or even the practice has to be stopped for the picking work because of the high-speed of the golf. Therefore, the golf picking work is inefficient, dangerous as well as costly [2, 3]. For some small balls are not only high-speed, but also hit high frequency, picking work is more important. Intelligent robot can perceive the external environment, automatically identify the target object, automatically make the path planning, thus it can replace people to do the dangerous work. The application of ball-picking robot in training court can not only improve the efficiency of ball recycling, reduce the

© Springer International Publishing AG 2017
Y. Huang et al. (Eds.): ICIRA 2017, Part III, LNAI 10464, pp. 226–236, 2017.
DOI: 10.1007/978-3-319-65298-6_21

operating costs of the stadium, but also improve the overall technical ball-picking service of training court. Therefore, the study of ball-picking robot is very important, and has a very broad application prospects. This paper explores the research on ball-picking robot, analyzes and summarizes the key results such as navigation and location technology, path planning and wireless communication technology, and also discusses the development trend and challenges of intelligent ball-picking robot in the future.

2   Status of the Ball-Picking Robot

Facing the social demand, the ball-picking robot, as a major category of service robot, has become a hot issue in the field of robot research in recent years. The current studies on golf, table tennis, tennis and other training venues, the kinds ball-picking robot are including mainly to blade gripping, spiral cleaning, air suction, brush roller and other special equipment.

2.1   Golf-Picking Robot and Equipment

Considering the heavy work of picking up golf, Pacheco L., Ribeiro F., and De Looze [1, 2, 4] have started to study the conceptual physical since 1995. By 2012, Nino Pereira of the University of Minho in Portugal and Daniel Whitney of MIT in the United States jointly designed a gripping autonomous golf recycling robot (Fig. 1) [5]. The application of multi-sensor (non-machine vision) information fusion technology such as GPS, gyroscope and sonar, dual RRT algorithm for path planning, and wireless sensor technology for human-computer interaction [6] makes it possible to achieve the goal of independent navigation in the training ground, obstacle avoidance, picking and other functions. It

Fig. 1. Blade gripping autonomous golf recycling robot

Fig. 2. Blade gripping golf equipment

can pick up 10,000 balls and meeting the needs of a court with 25,000 m², which is efficient and capacious.

The hand-push single link, power machine-driven single link, double and triple link ball-picking equipment produced by Shenzhen City Green Rui Golf Technology Co., Ltd and Guangzhou Hufa Golf Products Co., Ltd etc. (Fig. 2) have a picking width of up to 3.18 m, capacity of about 2000PCS, and the working speed of 10 km/h, suitable for large and medium-sized complex ground of the driving range, with the characteristics of a large picking width, high efficiency, good maneuvering property and so on.

2.2 Tennis-Picking Robot

In 2011, the twelfth "Challenge Cup" national college students' extracurricular academic and technological works contest also develop a tennis-picking robot research. The works include embedded intelligent tennis-picking robot, tennis-picking intelligent robot prototype based on smart car and camera recognition technology, and smart tennis-picking robot [7–9] (Fig. 3).

Fig. 3. The twelfth "Challenge Cup" tennis-picking robots

In 2012, Southeast University, Beijing University of Posts and Telecommunications, and Alvin Tan Wei, University of Bradford, UK [10, 11], develop kinds of drum-type, roller-type and spring-type tennis-picking robot (Fig. 4) which mainly relies on machine vision, remote control or GPS planning path to complete the process of picking up. In 2013, Lanzhou University of Technology designs the intelligent cleaning-type tennis recycling robot based on the visual recognition and infrared sensor obstacle avoidance [12]. It is composed of image acquisition system, obstacle avoidance system, power system, pickup counting system and information processing system. The data display function is realized between the wireless communication network and the host computer.

In 2011, Michigan College of Shanghai Jiao Tong University developed a kind of manipulator picking robot (Fig. 5a) [13]. In 2013, Shanghai Dianji University developed a kind of intelligent manipulator picking tennis machine (Fig. 5b) [14], mainly including central processing unit, infrared sensor, ultrasonic sensor, AD conversion circuit, manipulator and drive system. Complete the ball recognition, ball gripping using 2-DOF manipulator to. In 2016, Jiangnan University developed a vision-based tennis-picking robot system (Fig. 5c) [15].

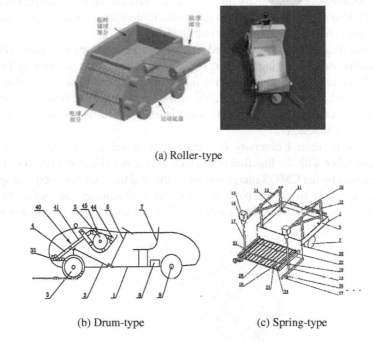

(a) Roller-type

(b) Drum-type   (c) Spring-type

Fig. 4. Tennis-pick robot

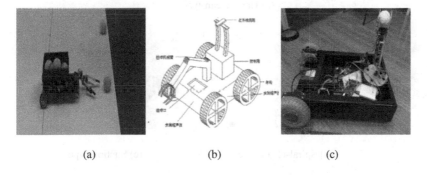

(a)   (b)   (c)

Fig. 5. Manipulator-type tennis-picking robot

2.3 Table Tennis-Picking Robot

On the table tennis picking research, Yang Yi [16], from Beijing Institute of Technology proposes an autonomous cleaning robot with indoor global camera. The robot can find the ball, pick up the ball, pour away the ball and charge automatically with the control system. With the path planning method of "Minimum Exercise Cost Short Distance Ball Priority", the robot has two modes of operation, namely single-machine operation and multi-machine collaboration. It applies to different sizes of balls, especially for tennis and table tennis.

In 2008, Northeastern University invented an autonomous table tennis-picking robot (Fig. 6a) [17]. Use vision technology to locate the table tennis, to select the final path and to pick up the ball through the front holding hand.

In 2014, Southeast University invented a cleaning table tennis-picking robot (Fig. 6b) [18]. It is composed of walking mechanism, picking mechanism, control system and power supply system. With machine vision and infrared sensor, it can complete the identification, tracking and picking up table tennis. Clean the table tennis ball into the channel by the picking gripper in the front, with automatic obstacle avoidance, automatic tracking and other functions.

In 2014, Guangzhou University of Technology developed a suction-type table tennis-picking robot with the function of obstacle avoidance (Fig. 6c) [19]. It can locate of the table tennis by the CMOS image sensor on the real time, so that robot can quickly reach the table tennis to do the picking work. Infrared sensors can make the robot effectively to avoid obstacles and enhance the ability of adapting to the environment.

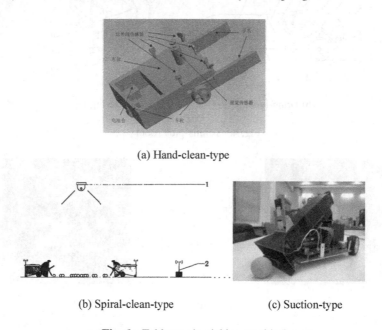

(a) Hand-clean-type

(b) Spiral-clean-type   (c) Suction-type

Fig. 6. Table tennis-picking machine

2.4 Badminton-Picking Robot

A badminton-picking robot is also searched through the network (Fig. 7) [20]. It mainly includes monocular camera, the brush roller, synchronous belt lifting mechanism, the ball barrel and the ARM control system. It has the function of image recognition and visual navigation. It picks badminton with brush roller, counting, sorting and other functions.

(a) Roller-type   (b) Synchronous belt schedules

Fig. 7. Roller badminton-picking robot

3 Key Technology

3.1 Navigation and Location

Ball-picking robot's working environment is usually unstructured. It requires that the ball-picking robot can accurately measure their position and the distribution of obstacles in the environment, thus achieving the goal of independent movement and obstacle avoidance. So the ball-picking robot must be removable, easy to operate, with high level of intelligence and adaptability and so on. To achieve this requirement, navigation and location technology is a core key technology [21]. The common navigation methods of mobile robot are magnetic navigation, inertial navigation, laser navigation, visual navigation, GPS navigation, gyroscope navigation, magnetic compass navigation, photoelectric encoder navigation [22, 23].

Magnetic navigation determines the movement path for the robot, through the underground magnetic line buried in advance. Although this navigation technology is more mature, it is not fit for the ball-picking robot working in the non-structural environment. Laser navigation, ultrasonic navigation or infrared navigation [14, 18–21]emits the wave signal automatically, then judges the obstacles according to the reflected wave signal, and finally makes obstacle avoidance navigation. If they are added to the ball-picking robot, its performance of obstacle avoidance can be improved [24].

GPS navigation and location technology has been applied widely in military or civilian equipment [25]. It can be achieved to apply GPS location to navigation for ball-picking robot. However, the position accuracy of the mobile GPS receiver is affected by the satellite signal condition and the road environment, as well as by the clock error, spread error, receiver noise and many other factors. Therefore, the accuracy is relatively low if only GPS navigation technology is used, and the reliability is not enough [26]. In the robot navigation applications field, magnetic compass, photoelectric encoder are usually also used to navigate [27]. So the GPS navigation technology can greatly improve the location and picking up accuracy of ball-picking robot, if a variety of other navigation supplementation technology is used.

Visual navigation obtains the robot's surrounding environment visual information through the camera, completes the obstacles and signs recognition through the image processing, and finally gets navigation parameters and implements navigation and location [28]. The camera can be installed on a robot, and gets the navigation parameters as the robot moves [17, 21]. It can also be installed around the active area [16] to locate the robot. The former is difficult to process large number of image data, so it is suitable for outdoor picking robot in unstructured environments, oppositely, the latter is only suitable for small area or indoor robot location [29]. Visual navigation can get more complete environmental information, has the ability to adapt to the environment. But visual navigation need to handle large amount of information and the accuracy of long-range navigation is low [30], the visual algorithm needs to be improved or combined with other navigation methods.

Each navigation and location technology has its advantages and disadvantages. In the actual application process, ball-picking robot needs usually choose several methods to integrate advantages, thus to improve the accuracy of navigation and location.

3.2 Path Planning

Ball-picking robot is a mobile robot. Mobile robot needs to solve the problem of path planning, that is to find the optimal path of avoiding obstacle from the initial state to the target state in the movement space according to some optimization criteria (such as the minimum cost of work, the shortest walking route, the shortest time [31, 32]. A good path planning method requires rationality, completeness, optimality, real-time, environment-adaptability, and satisfies all constraints [33].

The path planning method is divided into global path planning and local path planning [34]. The global path planning has large computation and poor real-time, so it is difficult to adapt to the dynamic non-deterministic environment. The main methods are: visibility graph, free space method [6] and grid method, etc. [35–37]. Local path planning is more real-time and practical, and more adaptable to dynamic environment. The disadvantage is that sometimes local extreme points or oscillation resulted from only relying on local information can't guarantee that the robot reach the target point successfully. The main methods of local path planning are: artificial potential field method, genetic algorithm, fuzzy logic algorithm and neural network method, etc. [38].

Ball-picking robot usually works in unstructured environment, which means that the environment information can't be got accurately in advance. In other words, the ball-picking robot will rely entirely on some sensors to obtain environmental information. So the local path planning is more practical for ball-picking robot.

3.3 Wireless Communication

Multi-robot collaborative operations, human-computer interaction and human-machine collaboration of ball-picking robot require a lot of online operations. A fixed host computer and a moving slave computer are usually used for information processing and command delivery, and the support of robot communication technology is also necessary, there are wire communication and wireless communication. For the continuous movement of ball-picking robot, the wireless communication [12] is the only choice to achieve the exchange of data and control command information because of the uncertain location and the environment.

At present, the main wireless communication technology in the robot field includes WiFi, Bluetooth, ZigBee and so on.

WiFi, as a wireless LAN technology, has a wide range of applications due to its conventional networking and strong ductility since its appearance [39]. Wireless access and high-speed transmission are the main technical advantages of WiFi [40, 41]. WiFi network, as a data transmission platform, takes high-speed computers, MCU or embedded systems as data processing center, ball-picking robot can be controlled in a long distance. And its functions can be extended, its transmission affection much better in the 10 m range. But WiFi communication has so large power consumption that it can't be used for a long-distance.

ZigBee wireless communication technology has the advantages of low power consumption, low cost, short time, large network capacity, reliability, safety and so on [42]. If the obstacle is less outdoors, it can reach to 100 m transmission distance. ZigBee technology itself is designed for a low-speed communications, and its maximum speed of only 259 kb/s. For some large amount of data communication occasions, it is not appropriate.

Bluetooth technology is a short-range wireless communication technology, it is better generally in the transmission range of about 1 m. It uses radio frequency identification devices (RFID) technology to achieve the wireless interconnection between devices. For its stronger penetrating ability, it can be transmitted in all directions with a very reliable data transmission capability. Low power and price make its application more and more widely.

4 Ball-Picking Robot Outlook

4.1 Deep Integration with Vision Technology

Visual information is huge, and the information obtained is comprehensive and the utilization rate is high [43]. The ball-picking robot, working in the non-structural environment, can make huge information integration with vision technology, which can not only identify the picking environment effectively, but also make the ball recognition and localization more accurate, as well as improve its ability to navigate and avoid obstacles. It can be seen that the development of the intelligent ball-picking robot depends on vision technology.

4.2 Multi-robot Collaborative Operations and AI

When ball-picking robot works in the multi-ball environment, it still needs to improve picking efficiency and collaborative operations. Therefore, a ball-picking robot can be designed in multi-finger and multi-arm structure. At the same time, multi-robot collaborative operations and artificial intelligence (AI) [4] are also the effective ways to improve the efficiency.

4.3 Human-Machine Collaboration

Human-machine collaboration is the trend of the development of the robot, so the ball-picking robot must be safe and friendly to work with people. It needs to show a human-computer interaction, human-machine cooperation, human-machine integration and other obvious intelligent service technology features [4]. It needs to join the man-computer interface device, and integrate with human-computer interaction technology, to achieve the combination of manual control and automatic control.

4.4 Internet Technology

Ball-picking robot, combining with the Internet technology, can achieve a better remote operation and real-time control. It is necessary that use the Internet cloud platform to solve the robot problems of three-dimensional environment perception, path planning, navigation and human-like smart operation. To achieve the goal of data exchange, management and multi-robot control in "user- cloud platform-robot", it requires more Internet technology support.

5 Conclusion

The wide demand of ball-picking in public service areas, characteristics of unstructured environment in training ground and other complex multi-barrier areas, the ball-picking robot has important scientific significance and application prospect. Based on navigation and location technology, path planning technology, wireless communication technology, ball-picking robot has basically achieved the picking ability. However, in the complex multi-barrier non-structural environment conditions, the development of intelligent ball-picking robot, which can achieve automatic identification and picking, autonomous navigation and obstacle avoidance, efficient collaborative operation, is still slower relatively. With the further integration of machine vision, multi-robot collaborative operations and AI, human-machine collaboration and Internet communications, the ball-picking robot can be more intelligent, and achieve independent environmental identification, independent picking ball, man-machine collaboration and other key technologies.

Acknowledgments. This paper contains the results of researches on Tianjin Municipal Natural Science Foundation Project (17JCZDJC30400 and 16JCQNJC04100) supported by the Tianjin Municipal Science and Technology Commission, and the National College Students' Innovative Entrepreneurial Training program (201610060050) of the People's Republic of China.

References

1. de Golfe, F.P.: Estatísticas do golfe (2011), http://portal.fpg.pt/web/guest
2. Pacheco, L., de Oliveira André, J.B., Ribeiro A.F.: Mobile robot for autonomous golf balls picking. In: Proceedings of the Portuguese Conference on Automatic Control Vila Real, pp. 814–818 (2008)

3. Ribeiro, F., Moutinho, I., Silva, P., et al.: Mobile robot construction for edutainment application. Revista RobÓTica **69**, 12–16 (2007)
4. Gao, F., Guo, W.: Thinking of the development strategy of robots in China. J. Mech. Eng. **52**(7), 1–5 (2016)
5. Pereira, N., Ribeiro, F., Lopes, G., et al.: Autonomous golf ball picking robot design and development. Ind. Robot. **39**(6), 541–550 (2012)
6. Wang, Z.Y., Liu, Y.L.: Research of the indoor service robot based on zigBee positioning. Appl. Mech. Mater. **734**, 179–182 (2015)
7. The Twelfth "Challenge Cup" National College Students Extracurricular Academic and Technical Works Contest (2011), http://www.tiaozhanbei.net/project/18174/
8. The Twelfth "Challenge Cup" National College Students Extracurricular Academic and Technical Works Contest (2011), http://www.tiaozhanbei.net/project/12921/
9. The Twelfth "Challenge Cup" National College Students Extracurricular Academic and Technical Works Contest (2011), http://www.tiaozhanbei.net/project/8437/
10. Zesheng, X.: Design of tennis ball-picking robot. Mech. Manuf. Autom. **41**(2), 140–141 (2012)
11. Automatic Tennis-Picking Robot (2014), http://v.ku6.com/show/U01F3LT6DQPQ7My80slVVg...html
12. Jie, C., Fei, M., Libo, N., et al.: A method of finishing based on visual recognition and multi - sensor data fusion. CN201310294142.9. Lanzhou University of Technology (2013)
13. Automatic Tennis-picking Robot, Michigan College Design Exhibition (2014), http://v.youku.com/v_show/id_XNjg3NjM5MjQ4.html
14. Tengfei, L., Li, L., Binglin, H., et al.: Development of intelligent tennis ball picking machine. Electron. World **12**, 68–69 (2013)
15. Shijun, S.: Vision-based tennis self-picking method reality. Light Ind. Mach. **34**(6), 62–65 (2016)
16. Yi, Y., Yacheng, L., Xin, Y., et al.: Self-Picking Robot. CN200810183233.4. Beijing Institute of Technology (2012)
17. Danyang, A.: Configuration and Research of Key Technologies of New Autonomous Seizure Table Tennis Robot. Northeastern University, Shenyang (2008)
18. Su, S.: A kind of ping pong ball picking robot. CN201410179165.X. Southeast University (2014)
19. Xu, D.W., Liu, J., Lin, G.: Design and realization of table tennis picking robot. Mach. Tool Hydraulics **42**(3), 17–19, 50 (2014)
20. Badminton Pickup Robot Based on ARM (2016), http://v.ku6.com/show/t5Hy8iwFSAu8bZ3zjZ_GPg...html
21. Meng, Q., Lee, M.H.: Design issues for the elderly. Adv. Eng. Inform. **20**, 171–186 (2006)
22. Wen, W.Z., Ge, G.: Present situation and future development of mobile robot navigation technology. Robot **25**(5), 470–474 (2003)
23. Zheng, X., Xiong, R.: The technologies of mobile robot navigation and location. Electr. Mech. Eng. **20**(5), 35–37 (2003)
24. Barawid Jr., O.C., Mizushima, A., Ishii, K., et al.: Development of an autonomous navigation system using a two-dimensional laser scanner in an orchard application. BioSyst. Eng. **96**(2), 139–149 (2007)
25. Kaplan, E.D.: Understanding GPS Principles and Applications, 2nd edn. HÄFTAD, Engelska (2006)
26. Aboul-Enein, Y.H.: The precision revolution: GPS and the future of aerial warfare. Technol. Cult. **45**(4), 883–884 (2003)

27. Liu, J., Hao, J.: Research on autonomous navigation robot navigation and location technology. Technol. Appl. **1**, 23–26 (2005)
28. Chugo, D., Matsushima, S., Yokota, S., et al.: Camera-based indoor navigation for service robots. In: SICE Annual Conference, pp. 1008–1013 (2010)
29. Wu, S., Cheng, M., Hsu, W.: Design and implementation of a prototype vision-guided golf-ball collecting mobile robot. In: IEEE International Conference on Mechatronics, pp. 611–615 (2005)
30. Budiharto, W., Santoso, A., Purwanto, D., et al.: A navigation system for service robot using stereo vision and kalman filtering. In: Control, Automation and Systems, pp. 1771–1776 (2011)
31. Kim, W.S.: Virtual reality calibration for telerobotic servicing. In: Proceedings of the IEEE International Conference on Robotics and Automation, pp. 2769–2775 (1994)
32. Song, Y.: Research on Coordinated Motion Accuracy and Off-Line Programming Calibration Technology of Automatic Welding Robot System. Harbin Institute of Technology (2002)
33. Liu, H., Yang, J., Lu, J., et al.: A survey of mobile robot motion planning. Chin. Eng. Sci. **8**(1), 85–94 (2006)
34. Latomabe, J.C.: Robot motion planning. Kluwer, Norwell (1991)
35. Pere: Automatic planning of manipulator movements. IEEE Trans. Syst. Man Cyb. **11**(11), 681–698 (1981)
36. Dijkstra, E.W.: A note on two problems in connection with graphs. Numcr. Math. **1**, 269–271 (1959)
37. Alexopoulos, C., Griffin, P.M.: Path Planning for a Mobile Robot. IEEE Trans. Syst. Man Cybern. **22**(2), 318–322 (1992)
38. Yong, K., Wang, H., Abuja, N.: A potential field approach to path planning. IEEE Trans. Robot. Autom. **8**(1), 23–32 (1992)
39. Li, Y.: The principle and application of Wifi technology. Technol. Inf. **4**(2), 59–61 (2010)
40. Shuqi, W., Ge, S.: Design of Wifi terminal based on S3C2410 in mine tunnel. Microcomput. Inf. **23**(4), 186–188 (2007)
41. Jiang, F., Zhang, L., He, C.: Application of Wifi technology in mine remote monitoring system. Coal Mine Saf. **42**(5), 62–65 (2010)
42. Wang, D., Zhang, J.R., Wei, Y., et al.: Building Wireless Sensor Networks (Wsns) by ZigBee Technology. J. Chongqing Univ. **29**(8), 95–100 (2006)
43. Zhao, P.: Research and Development of Machine Vision. Science Press, Beijing (2012)

Mobile Indoor Localization Mitigating Unstable RSS Variations and Multiple NLOS Interferences

Kyuchang Kwon[1]($\boxtimes$) and Younggoo Kwon[2]($\boxtimes$)

[1] Konkuk University, Seoul 05029, Republic of Korea
kcks0501@naver.com
[2] Konkuk University, No. 120 Neungdong-ro, Gwangjin-gu, Room-C320,
Engineering Building No. 21, Seoul 05029, Republic of Korea
ygkwon@konkuk.ac.kr
http://iotlab.konkuk.ac.kr/

Abstract. Many researches have studied indoor localization techniques in past decades. Depending on a wireless sensor network, current distance-based localization techniques exploit different measurements of Received Signal Strength (RSS) values between RF devices. It is simple to implement and costly efficient, while the estimation accuracy is significantly reduced in indoor environments. In this paper, we focus on localization methods in real indoor environments using IEEE 802.15.4 standard Zigbee network. The objective is to mitigate the instability and divergence of signal strength by using successional RSS evaluations with Kalman filtering and avoid multiple NLOS interferences by using LOS node set identification procedure.

Keywords: Indoor localization · Received signal strength · Unstable RSS variation · Kalman filter · Least square estimation · Multiple nlos problem

1 Introduction

During recent years, indoor localization technologies using wireless sensor networks (WSNs) have become a hot topic for many researches. The development of WSNs gives the possibilities of modeling localization systems. These localization systems can be divided into range-free and range-based techniques. The most popular method is to utilizing the Received Signal Strength (RSS) values as localization parameter considering the simplicity and low hardware costs. In fingerprinting method, the empirical or theoretical model of the signal propagation is translated into the position or distance estimates [1,2]. While it is easy to implement, fingerprint takes a lot of effort for premeasurement procedure. The range based method is another approach for RSS-based indoor localization technique [4,5]. This method depends on the measurement of distances between

© Springer International Publishing AG 2017
Y. Huang et al. (Eds.): ICIRA 2017, Part III, LNAI 10464, pp. 237–245, 2017.
DOI: 10.1007/978-3-319-65298-6_22

transceivers by radio propagation models [6]. Cost efficiency, real-time and accurate localization properties are the big advantages for the range based method.

In RF network channels of indoor environments, it is hard to find true locations because of interferences. These interferences are occurred by various causes such as multipath problems, mobile target, furniture, obstacles and etc. Some interferences make scattering and unstable localization results under unstable RF channel conditions [6,7]. In this problem, the measured RSS has more unstable and stochastic characteristics than the conventional environments. Another problem is Non Line of Sight (NLOS) interference [8]. In indoor environments, the Line of Sight (LOS) path can be blocked and the communication is conducted through reflections and diffractions. This interference leads to attenuation and variation of signal level and finally produce the estimation errors in localization results.

In this paper, we focus on the indoor localization system combined with Kalman filter and NLOS avoidance method. In unstable channel conditions, the Kalman filter estimates RSS state with successional evaluating to reduce RSS variations [9,10]. From Kalman filter, we use this filtered RSS values for location estimation by Least Square Estimation (LSE) method [11,12]. If the target node is in NLOS channel condition, it tries to find LOS reference node set iteratively and mitigates NLOS error distance from the estimated location. This algorithm does not require the radio map collecting processes and provides the simplicity of implementation in wireless sensor network.

The remainder of this paper is structured as follows. In Sect. 2, the unstable channel and NLOS channel problems are presented. The proposed algorithm is explained in Sect. 3, and the experimental results and the performance analyses are presented in Sect. 4. We conclude the paper in Sect. 5.

2   Problem Statement

The accurate estimations of signal strength versus distance affect to system performances in most RSS-based indoor localization approaches. The major challenge for accurate RSS measurement is mitigating RSS variations which come from the dynamic and unpredictable interferences. In this paper, we deal with two types of channel conditions: the unstable RSS variation channel problem and Non-Line of Sight (NLOS) channel problem.

The Fig. 1(a)–(c) show the RSS-based indoor localization under the stable LOS, unstable and NLOS channel conditions. Figure 1(a) shows a LOS channel condition. In this interference-free environment, the estimated locations are close to the target node. This ideal localization condition is not common in real indoor RF environments. Figure 1(b) shows an unstable channel problem with interferences. The divergence and instability characteristic of RSS measurements decay the localization accuracy, and the estimated locations are scattering over the target node. These undesirable interferences are come from multipath propagations, movement of the target node and signal interferences from other wireless networks. Figure 1(c) shows an example of NLOS channel problem. In this figure, the estimated locations are far from target node because of the obstacle which

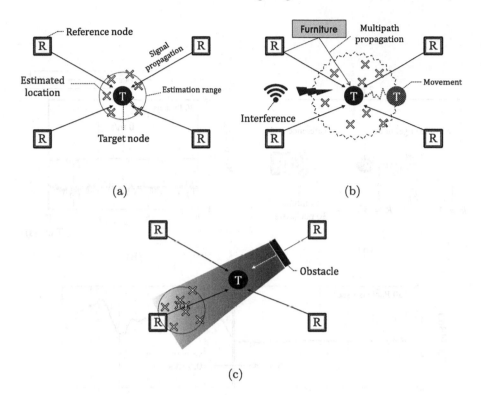

Fig. 1. The channel interference in indoor localization. (a) Stable LOS channel, (b) Unstable channel, (c) NLOS channel

blocks the way of signal propagation. As the measured RSS values have big attenuations with weak signal strength level, the estimated locations are located far from target node. Unlike unstable channel condition problem, the errors in NLOS condition are determined by the position of obstacles and the signal attenuations. We separate the unstable and the NLOS channel problems to improve the localization system implementation performance.

3   Algorithm Implementation

We propose a new RSS-based indoor localization algorithm which has Least Square Estimation (LSE) with successional RSS evaluation using Kalman filtering and the method of mitigating multiple NLOS interferences by LOS subset identification procedure in wireless sensor networks.

3.1   Successional RSS Evaluation Using Kalman Filtering

The received RSSs from the same point at different time show high degree of fluctuations. The real time RSS measurements are not credible for the

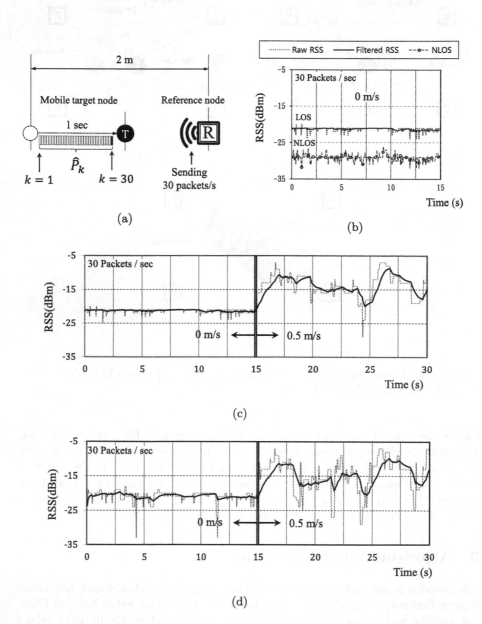

Fig. 2. The successional RSS evaluation using Kalman filtering. (a) A illustration of successional evaluating, (b) Kalman filtering in stable LOS and NLOS channel, (c) Mobile target node in stable LOS channel, (d) Mobile target node in unstable channel

accurate localizations. To decrease the effect of signal variations, we use the successional RSS evaluation using Kalman filtering before estimating the blind target node location. In Kalman filtering procedure, the system state and measurement model can be described as follows [9,13]:

$$P_k = AP_{k-1} + w_{k-1} \tag{1}$$

$$\tilde{P}_k = HP_k + v_{k-1} \tag{2}$$

P_k and $\tilde{P}_k$ represent RSS state and measurement RSS at the time step k. The w and v represent process noise and measurement noise with Gaussian random variable $w_k \sim N(0, Q)$ and $v_k \sim N(0, U)$. The system transition matrix A and measurement matrix H are 1×1 identity matrices. The availability of the measurement $\tilde{P}_k$ changes the estimate of a constant P_k as follows:

$$C_k = AC_{k-1}A^T + Q \tag{3}$$

$$K_k = C_k H^T (HC_k H^T + U)^{-1} \tag{4}$$

$$\hat{P}_k = \hat{P}_k + K_k(\tilde{P}_k - H\hat{P}_{k-1}) \tag{5}$$

$$C_k = (1 - K_k H)C_{k-1} \tag{6}$$

where the $\hat{P}_{k-1}$ and C_{k-1} are the estimated RSS state and its covariance before the measurement $\tilde{P}_k$ is processed, the $\hat{P}_k$ and C_k are the estimated RSS state and its covariance after the measurement $\tilde{P}_k$ is processed, and K_k is the Kalman gain. Using this filtering procedure, we can obtain the distance $\hat{d}$ as follows [6]:

$$\hat{P}_k = \hat{P}_0 - 10\mu log_{10}(\hat{d}/d_0) \tag{7}$$

where $\hat{P}_k$ denotes the recently filtered RSS value as shown in Eq. (5), μ is the path loss constant, d_0 is a reference distance, and $\hat{P}_o$ is the filtered RSS value at $d = d_0$.

From the filtered RSS values and N reference nodes, the Euclidean coordinate of the target node, $\hat{x} = (x_0, y_0)$, is obtained as follows [11,12].

$$\hat{x} = argmin_{x \in L} \sum_{i=1}^{N} |\, \|x - r_i\| - \hat{d}_i| \tag{8}$$

where the $r_i = (x_i, y_i)$ is the location of ith reference node, the L is target space where the unknown target node location exist, $\hat{d}_i$ is estimated distance using the Eq. 7.

Figure 2(a) shows the illustration of successional evaluating RSS values. Each target node receives 30 packets from one reference node and total 180 packets from six reference nodes during estimation period t_{est} using MTM-CM3300 MSP devices. The mobile target node moves with 0.5 m/sec speed after 15 s to the reference node in Fig. 2(a). The t_{est} is set as 1 s, and the mobile target node evaluates 30 packets/1 s successionally. The performance of Kalman filtering is shown in Fig. 2(c) and (d). In Fig. 2(c), the RSS variance is decreased from 0.23 to 0.09 at 0 m/s and 0.94 to 0.29 at 0.5 m/s. In Fig. 2(d), the RSS variance is decreased from 0.34 to 0.12 at 0 m/s and 2.59 to 0.65 at 0.5 m/s. Kalman filtering has the efficient noise reducing capability with unstable radio channel conditions for the mobile target. For LSE procedure, the localization system uses the latest RSS values, which are successionally evaluated within $k_1 \sim k_{30}$.

3.2   LOS Node Set Identification

The successional RSS evaluation has several limitations when considering NLOS channel conditions in real indoor environment. As shown in Fig. 2(b), the Kalman filter cannot mitigate the NLOS problem fully because of the signal attenuation level. To avoid the NLOS problem, we propose the LOS node set identification procedure. First, the localization system preliminarily estimates a location vector of the target node $\hat{x}$ from N reference nodes using Eq. (8). The $\binom{n}{n-1}$ number of reference subsets of $N-1$ nodes estimate location vectors $l_1 \sim l_N$ and its function values $f_1 \sim f_N$ using LSE. The NLOS condition is detected by observing how far these location vectors of subsets are scattered to each other. This scattering is computed by NLOS error distance D which comes from $l_1 \sim l_N$ as follows:

$$D = \max\{\, \text{dist}(\, l_a, l_b \,) \mid 1 < a < n,\, a < b < n) \} \tag{9}$$

where l_i is the location vector of the ith subset and $\text{dist}(l_i, l_j)$ is the Euclidean distance between l_i and l_j. For the NLOS detection, the NLOS error distance is compared with a threshold value D_{th} which is selected as 1.0 m in our case. If the largest Euclidean error distance is smaller than the threshold value D_{th}, the system assumed that the current channel is in LOS state and the estimated LSE location is chosen as the target node's location. If $D > D_{th}$, an NLOS reference node is identified and iteration is continued to mitigate the interferences. In this case, the system selects the new estimated location vector, which has the smallest square result of f_i, as the new reference node set. The minimum value of f_i means that the reference node subset S_i has the smallest NLOS interference among all reference node subset during the current iteration stage. The iteration of location determination procedure is repeated until a LOS subset is found or $N < 4$. After the target node location determination, the system terminates the current iteration cycle and restarts the next localization phase.

4   Experiment Results

We performed the experiment at the 9m×12m lab-scale testing area which is shown in Fig. 3(a). Ten MSP MTM-CM3300 RF modules are placed as the reference nodes which support MSP430 MCU and CC2420 radio chip [14,15]. Two obstacles are placed at the left-bottom side of the corner to create the NLOS channel region, and WiFi AP device is placed at the right side as interference. The mobile target node moves along the path as shown in Fig. 3(a). Total 30 locations are estimated during the movement. Six closest reference nodes with

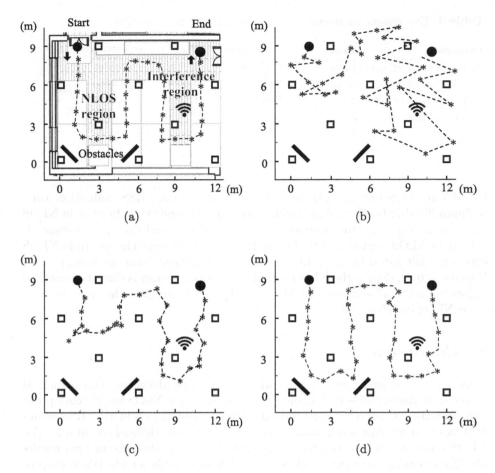

Fig. 3. The location estimation. (a)Network configuration (b) LSE only, (b) LSE with Kalman filtering, (c) Proposed algorithm.

the highest RSS values in order are used for location estimation. The experiment parameters and the average estimation error of the Fig. 3(b)–(d) are shown in Tables 1 and 2.

The parameters in Table 1 are obtained from the experimental results of Figs. 1 and 2. The evaluated path loss exponent is 3.0 ∼ 3.6 and the RSS is −7 dBm at the reference distance of 1 m. The Table 2 compares the average localization error distances in three cases: LSE only, LSE with successional RSS evaluation with Kalman filtering, and proposed algorithm. In Fig. 3(b), only the LSE is used and the average error distance is 4.11 m in NLOS region and 3.98 m in interference region. The estimated locations in NLOS region are estimated far from the real locations, and the estimated locations in interference region are widely scattered over the region. In Fig. 3(c), the successional RSS evaluation is used for LSE. The average error is decreased by 3.03 m in interference region and

Table 1. Experiment parameters

Parameters	Value
d_0 (Reference distance)	1.0 m
P_0 (RSS at d_0)	-7 dBm
μ (Path loss constant)	$3.0 \sim 3.6$
σ (STD of RSS)	0.8
D_{Th}	1.0 m

Table 2. Performance results

Used algorithm	NLOS region	Interference region
LSE only	4.11 m	3.98 m
LSE-KF	3.56 m	0.95 m
Proposed	1.11 m	0.91 m

by 0.55 in NLOS region. This shows that the successional RSS evaluation using Kalman filtering is effective in interference channel conditions, but not in NLOS channel. In Fig. 3(d), the proposed algorithm is used and the error distance is 1.11 m in NLOS region and 0.91 m in interference region. The multiple NLOS errors are mitigated by using LOS node set identification and the average error is reduced by 2.45 m in the NLOS region. This outcome means that the proposed algorithm is useful to detect the NLOS condition and mitigate the error distance from NLOS interferences.

5   Conclusion

This paper presented a new mobile indoor localization algorithm. Our main goal is to reduce the unstable RSS variation and mitigate the NLOS interference problem from the estimated target locations. The successional evaluating RSS values with Kalman filtering is adapted for various target node channel conditions. The NLOS condition is detected by checking the Euclidean distance of LSE results obtained from the reference node subsets. The LOS node set identification procedure by using the reference node subsets eventually finds the reference nodes which are in LOS conditions only for localization procedure to overcome multiple NLOS interferences. The performed experiments show that the proposed approach improves the localization performances in various interference channel. With Kalman filtering, the unstable channel effects from WiFi/Bluetooth devices, reflected obstacles and moving objects, are reduced much. Iterative location determination procedure by using the reference node subsets removes the multiple NLOS interferences, and improves the indoor localization performance significantly.

Acknowledgement. This research was supported by Basic Science Research Program through the National Research Foundation of Korea (NRF) funded by the Ministry of Science, ICT & Future Planning (NRF-2016R1A2B1015032).

References

1. Faragher, R., Harle, R.: Location fingerprinting with bluetooth low energy beacons. IEEE J. Sel. Areas Commun. **33**(11), 2418–2428 (2015). doi:10.1109/JSAC.2015. 2430281
2. Yang, Z., Wu, C., Liu, Y.: Locating in fingerprint space: wireless indoor localization with little human intervention. In: Proceedings of the 18th Annual International Conference on Mobile Computing and Networking. ACM (2012). ISBN: 978-1-4503-1159-5
3. Lin, K., et al.: Enhanced fingerprinting and trajectory prediction for IoT localization in smart buildings. IEEE Trans. Autom. Sci. Eng. **13**(3), 1294–1307 (2016). doi:10.1109/TASE.2016.2543242
4. Luo, X., OBrien, W.J., Julien, C.L.: Comparative evaluation of Received Signal-Strength Index (RSSI) based indoor localization techniques for construction job-sites. Adv. Eng. Inf. **25**(2), 355–363 (2011). http://doi.org/10.1016/j.aei.2010.09.003
5. Rapinski, J., Cellmer, S.: Analysis of range based indoor positioning techniques for personal communication networks. Mob. Netw. Appl. **21**(3), 539–549 (2016). doi:10.1007/s11036-015-0646-8
6. Sklar, B.: Digital Communications, vol. 2. Prentice Hall, Upper Saddle River (2001)
7. Wang, J., et al.: Robust device-free wireless localization based on differential RSS measurements. IEEE Trans. Ind. Electron. **60**(12), 5943–5952 (2013). doi:10.1109/TIE.2012.2228145
8. Seidel, S.Y., et al.: Path loss, scattering and multipath delay statistics in four European cities for digital cellular and microcellular radiotelephone. IEEE Trans. Veh. Technol. **40**(4), 721–730 (1991). doi:10.1109/25.108383
9. Faragher, R.: Understanding the basis of the Kalman filter via a simple and intuitive derivation [lecture notes]. IEEE Sign. Process. Mag. **29**(5), 128–132 (2012). doi:10.1109/MSP.2012.2203621
10. Bekkali, A., Sanson, H., Matsumoto, M.: RFID indoor positioning based on probabilistic RFID map and Kalman filtering. In: Third IEEE International Conference on Wireless and Mobile Computing, Networking and Communications, WiMOB 2007. IEEE (2007). doi:10.1109/WIMOB.2007.4390815
11. Savvides, A., Han, C.-C., Strivastava, M.B.: Dynamic fine-grained localization in ad-hoc networks of sensors. In: Proceedings of the 7th Annual International Conference on Mobile Computing and Networking. ACM (2001). doi:10.1145/381677. 381693
12. Ficco, M., Esposito, C., Napolitano, A.: Calibrating indoor positioning systems with low efforts. IEEE Trans. Mob. Comput. **13**(4), 737–751 (2014). doi:10.1109/TMC.2013.29
13. Simon, D.: Optimal State Estimation Kalman, H Infinity, and Nonlinear Approaches. John Wiley and Sons, New York (2006)
14. Maxfor, MTM-CM3300MSP User Manual, Maxfor technology Inc. (2006). http://tip.maxfor.co.kr/data/MTM-CM3300MSP20Manual.pdf
15. Texas Inst., 2.4 GHz IEEE 802.15. 4 ZigBee-ready RF Transceiver Datasheet, CC2420, Chipcon products from Texas Instruments (2006)

The Integrated Indoor Positioning
by Considering Spatial Characteristics

Dongjun Yang and Younggoo Kwon[(⊠)]

Konkuk University, No. 120 Neungdong-ro, Gwangjin-gu,
Room-C320, Engineering Building No. 21, Seoul 05029, Republic of Korea
ydj0336@naver.com, ygkwon@konkuk.ac.kr
http://iotlab.konkuk.ac.kr

Abstract. The indoor localization relies on the IMU sensor on the smartphone because there is no signal like GPS that compensates for location in real time like the outdoor. However, if IMU sensors are not periodically reset, the localization error will increase. Although various methods are suggested, it is difficult to apply it to any indoor location because there are few methods considering the spatial characteristics of the indoor. In this paper, we divided the indoor characteristics into two parts and suggested an integrated system considering the spatial characteristics.

Keywords: Mobile sensor · Indoor positioning · Landmark

1   Introduction

Recently, there are many methods recognizing the position in the room by utilizing the inertial sensors (usually acceleration, gyro, compass) in the smartphone. These techniques usually use an acceleration sensor to count the user's step count and a dead-reckoning method utilizing a gyro or magnetic field sensor to estimate the heading estimation called Pedestrian Dead Reckoning (PDR) [1,2]. However, PDR requires a way to initialize errors accumulated by sensors over time. PDR initially has an exact position but as time passes it differs from the real position due to the noise of mobile sensor. In an outdoor, periodic GPS signal helps to correct the position of user. However, this method can not be applied in indoor because there is no signal that can be periodically received like GPS. Other studies in the past have typically estimated their location in indoor by using wireless signals. Most of them are fingerprint method based on wifi or bluetooth, or using triangulation method. However, since wireless signals are irregular and are affected by materials and obstacles inside the building, the reliability of locations obtained from RF signals is reduced [3]. The reliability of the location is degraded in the NLOS(Non-line-of-sight) state where the wireless environment is unstable due to the influence of the building. Therefore, it is necessary to perform the location recognition by distinguishing the LOS(Line-of-sight) having good wireless environments and the NLOS. In addition, since

© Springer International Publishing AG 2017
Y. Huang et al. (Eds.): ICIRA 2017, Part III, LNAI 10464, pp. 246–253, 2017.
DOI: 10.1007/978-3-319-65298-6_23

the accuracy of location recognition varies depending on the number of APs, it is necessary to apply the range-based method, which is a localization method even when the number of AP(Access Point)s is insufficient. We propose a way to apply the localization differently according to spatial characteristics. Using the landmark to initialize the sensor's error in the smartphone effectively increases the accuracy of the location in the interior space, such as the corridor. However, the above method is not effective in a lobby-like space where there is no limit to the direction in which the user moves. Therefore, we used the range based method predicting the user 's position using the AP. We used BLE which measure more precise position than wifi when wireless environment is not good. In this paper, we suggest integrated localization that using landmark method in hallway and using RF-based method with multiple BLEs in open space.

2 Problem Statement

Various types of indoor localization have been introduced, but experiments considering spatial characteristic are insufficient. The physical constraints of the building space, such as a corridor, and the open space with no constrains, have different characteristics. So we need to use different localization to distinguish the two. We will prove this by experimenting and analyzing the characteristics of the two spaces and suggest ways to apply different solutions based on each problem.

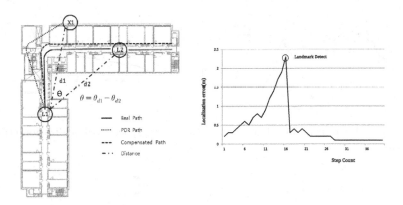

Fig. 1. (a) Compensating heading estimation with multiple landmarks (b) Localization error in hallway

Problems that occur to spaces with limited direction are due to accumulated errors over time of gyro sensors and magnetic field sensors. Table 1 shows the initial heading estimation problem due to the magnetic and gyro sensors. The user's initial direction is usually estimated by using a magnetic field sensor. However, the sensor is influenced by strong magnetic field around it so it is hard

Table 1. Initial heading estimation problem

Degree(°)	0–30	30–60	60–90	90–120
Number(%)	9(45)	5(25)	4(20)	2(10)

to get precise initial direction in [4]. In addition, due to the pedestrian step length predicted by the acceleration sensor and the inaccurate step detection, the difference between the actual location and the predicted location increases with time [1]. There are many ways to correct the gyro and the magnetic field sensor value, but it is not effective against the place where the influence of the magnetic field is high [5]. To solve this problem, we will use the smartphone sensors to detect landmarks with specific patterns and modify the user's position and direction by utilizing multiple landmarks.

In an open space, such as lobby, since the pedestrian's walking patterns is complex, the localization error increases more than the hallway. Figure 2 shows that compensation using multiple landmarks is not corrected effectively in open space for the long distance between two detected landmarks and a complexity of pedestrian's walking pattern. So Least square Method(LSM) using BLE is

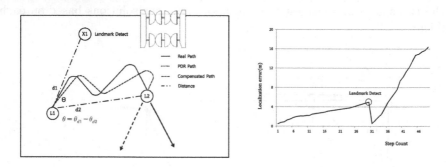

Fig. 2. (a) Problem using multiple landmarks in open space (b) Localization error in open space

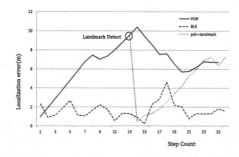

Fig. 3. Experimental result of three methods in open space

used to estimate current location. Unlike the method using multiple landmarks in PDR compensation, the LSM is independent of the previous position. So it improve localization accuracy comparing with other two methods shown in Fig. 3. The coordinates of the target node, (x, y), is obtained as follows (1 and 2). N is the number of the reference nodes and M is the set of locations separated by 1 m in open space. The (x_i, y_i) is the coordinates of ith reference node, d_i is the calculated distance to the ith reference node and r_i denotes the distance from designated point(x, y) to reference node.

$$r_i = d_i - \sqrt{(x_i - x_0)^2 + (y_i - y_0)^2} \tag{1}$$

$$(\mathbf{x}, \mathbf{y}) = argmin_{(x,y)\in M} \sum_{\forall i \in N} r_i^2 \tag{2}$$

However, since the strength of the radio signal is irregular, it is necessary to compensate the strength by applying Kalman filter (3) where Z_k is measured rss value, K_k kalman gain and P_{k-1} is previous estimated rss value.

$$P_k = K_k * Z_k + (1 - K_k) * \hat{P}_{k-1} \tag{3}$$

Some obstacles and interferences affect localization error so we need to determine wireless condition, At first, NLOS(Non-line-of-sight) or LOS condition are determined and the nodes not satisfying the condition are removed. Second, after ignoring the distorted nodes, iterate the process until LOS condition is determined. Last, estimate the location using LSM with nodes satisfying the condition.

3   Algorithm Implementation

In this section, we divide three parts to explain the integrated indoor localization considering spatial characteristics. First is how to integrate different localization methods in the hallway and open space. Second, compensation for heading estimation using multiple landmarks. Third, range based localization using BLE.

3.1   How to Integrate Localization Between Hallway and Open Space

The open space and the localization method in the hallway have different methods, but the two methods have to work organically according to the situation.

$$n \geq N - 2 \tag{4}$$

$$(\mathbf{x}, \mathbf{y}) \in M \tag{5}$$

The number of BLEs used in the experiment is assumed to be N, and the number of BLEs having the intensity suitable for the condition is set to n. Using the localization method of hallway based on pdr, open space localization method is used when the above conditions (4) and (5) are satisfied. The first condition determining open space is the number of BLEs(n) satisfying a specific condition. Second condition is whether the estimated location$(\mathbf{x}, \mathbf{y})$ is within the open space(M). If both conditions are satisfied, apply LSM using multiple landmarks.

3.2 Compensated Heading Estimation Using Multiple Landmarks

Magnetic landmark shows the unique pattern of the magnetic field sensor value generated by the server room. Using these patterns, you can apply the minimum and maximum values appropriately to recognize the differences between them and compare the rate of change to assign the pattern as a landmark [5].

$$S = \frac{1}{|A|} \sum_{\forall a \in A} \frac{min(f_1(a), f_2(a))}{max(f_1(a), f_2(a))} \tag{6}$$

The wifi signal specifies the landmark at which the similarity(S) difference in (6) between the previously scanned location (l_1) and the current scanned location (l_2) changes abruptly [5]. Let $A = A_1 \cup A_2$ and $f_i(a)$ is the signal strength in location (l_i).

$$\tilde{\theta} = \theta + \hat{\theta} \tag{7}$$

$$S_L = S_l * \frac{d_1}{d_2} \tag{8}$$

Using these two detected landmarks, the angle(θ) between the landmarks is used to correct the current heading estimation($\hat{\theta}$) to get compensated heading estimation ($\tilde{\theta}$) shown in (7). Let (d_1) and (d_2) denote that the distance between PDR and previous landmark and distance between current landmark and previous landmark. (8) shows the compensated steplength(S_L) using current steplength(S_l) and weighted value($\frac{d_1}{d_2}$). After calibration, the localization accuracy increases using the compensated heading estimation($\tilde{\theta}$) and steplength(S_L) like Fig. 1.

3.3 Range Based RSS Localization

In a place where the building is complicated, applying the method shown in Fig. 1(a) does not improve the localization error efficiently as shown in Fig. 2(b). The distance between the previously detected landmark and the currently detected landmark is too far to compensate for any heading estimation errors. So we should not use PDR method applied to the hallway but the method that provides independent location in real time. Applying a LSM using a bluetooth low-energy(BLE) module that receives radio signals in open space allows independent results of the previous location. Figure 3 shows LSM using BLE is effective to use in open space than other methods that dependent to previous location. However, because of the lack of reliability of the raw signals, the LSM must be used with the filtered signal(P_k) using the Kalman Filter (3). Based on the filtered RSS (Received Signal Strength) of the BLE modules, the LOS(Line-of-sight) and NLOS are determined by the distance(d) calculated by applying path-loss-model (9).

$$P_k = P_0 - 10\mu log_{10}(d/d_0) \tag{9}$$

If the distance(d) is greater than the threshold value we set, NLOS state is determined.

$$d \geq D_{th} \tag{10}$$

In case of LOS state, all BLE nodes are used to estimate the location. In the case of NLOS state, a recursive method are used to find and remove the node causing NLOS state and apply LS method to the remaining nodes satisfying LOS condition.

4   Experiment Result

We used smartphone to estimate location and the BLE[RECO (2016)] modules are used to broadcast wireless signal. Android api which uses various sensors to measure the heading estimation are utilized to set the initial heading estimation. Pressure sensor data is sent to server and server checks the landmark information in DB corresponding to the sensor value. After finished, the user can get the location of BLE and other landmark locations.

4.1   Experimental Areas

We experimented on two spaces: the hallway, the lobby. Figure 4(a) shows the hallway has two sections with a length of 3 m and a height of 50 m. The initial point was set in front of the elevator. Figure 4(b) shows the lobby with a length of 12 m and a height of 12 m in the lobby. There are 5 BLEs in the open space to estimate user's location.

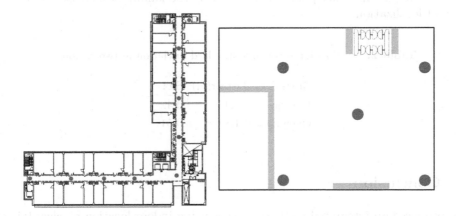

Fig. 4. (a) Experiment in hallway (b) Experiment in open space

4.2   Hallway and Open Space Performance

In hallways, we compare the localization error with predetermined trace and estimated trace. Figure 5(a) shows that the compensated heading estimation using multiple landmarks is effective in hallway. The location error in hallway is closed to the real path and the error is within 1.5 m. Figure 5(b) shows the

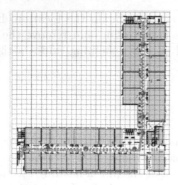 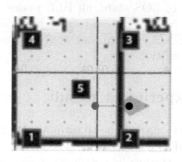

Fig. 5. (a) Experiment result in hallway (b) Experiment result in open space (Color figure online)

location error between real location (red) and estimated location (black). So, calculate the error every second using the euclidean distance between real and estimated location.

Table 2 shows that the localization error and standard deviation (STD). The localization error and STD is higher in open space than in hallway because the wireless signal is distorted and obstacles makes the signal weak. Multiple landmarks makes heading estimation compensated so localization error is reduced and starting at the exact point when detecting the landmark increase the accuracy of localization.

Table 2. Localization error and standard deviation in two regions

Areas	Error(m)	STD
Hallway	1.43	0.42
Open space	1.83	0.51

5 Conclusion

We proposed an integrated method of recognizing indoor location by classifying spatial characteristics. In the following paper, we will apply the unsupervised method of creating the landmark rather than the recorded. And the LSM will be applied for the diverse open space like lobby and seminar room and auditorium. We will also show simultaneous localization with open space and hallway.

Acknowledgments. This research was supported by Basic Science Research Program through the National Research Foundation of Korea (NRF) funded by the Ministry of Science, ICT & Future Planning (NRF-2016R1A2B1015032)

References

1. Kang, W., Han, Y.: SmartPDR: smartphone-based pedestrian dead reckoning for indoor localization. IEEE Sens. J. **15**(5), 2906–2916 (2015)
2. Harle, R.: A survey of indoor inertial positioning systems for pedestrians. IEEE Commun. Surv. Tutorials **15**(3), 1281–1293 (2013)
3. Bahl, P., Padmanabhan, V.N.: RADAR: an in-building RF-based user location and tracking system. In: Nineteenth Annual Joint Conference of the IEEE Computer and Communications Societies, Proceedings, INFOCOM 2000. IEEE, vol. 2 (2000)
4. Kang, W., et al.: Improved heading estimation for smartphone-based indoor positioning systems. In: 2012 IEEE 23rd International Symposium on Personal Indoor and Mobile Radio Communications (PIMRC). IEEE (2012)
5. Wang, H., et al.: No need to war-drive: unsupervised indoor localization. In: Proceedings of the 10th International Conference on Mobile Systems, Applications, and Services. ACM (2012)

Characterization of the Sick LMS511-20100Pro Laser Range Finder for Simultaneous Localization and Mapping

Wenpeng Zong[✉], Guangyun Li, Minglei Li, Li Wang,
and Yanglin Zhou

Zhengzhou Institute of Surveying and Mapping, Longhai Middle Road 66,
Zhengzhou 450052, Henan, People's Republic of China
zong.wen.peng@stu.xjtu.edu.cn

Abstract. This paper presents a characterization of the Sick LMS511-20100Pro laser range finder. With high accuracy and good robustness, this range sensor is suitable for various mobile robotic applications both indoors and outdoors. However, very few studies concerning the performance characterization of the LMS511-20100Pro can be consulted for better understanding and utilizing this sensor in practice. Therefore, some factors that could influence the sensor performance, such as drift effect, target distance, angular resolution, target material and mixed pixel, were tested and further analyzed. The effect of target distance and material on intensity information was also investigated. All the performed experiments demonstrate that the performance of the LMS511-20100Pro exceeds the specified values and can meet the requirements of simultaneous localization and mapping, although its practical performance might be significantly affected by some certain factors.

Keywords: LRF characterization · Drift effect · Mixed pixel · Intensity · SLAM

1 Introduction

Robots have become smarter and smarter, more and more powerful thanks to the astonishing progress of advanced sensor technology over the past few decades. Range sensing is one of the key technologies for most mobile robots to realize autonomous navigation and 3D mapping within complex environments, enabling the accomplishment of a variety of tasks. Though there are various range sensors available for robot researchers, including 2D and 3D laser range finder (LRF, also known as laser scanner or LiDAR, i.e. Light Detection and Ranging), stereo vision system, ultrasonic sensor, structured light sensor and depth camera, their cost and performance vary from one to another [1]. Due to its high precision, efficiency and especially affordable cost, the 2D LRF has been widely used in all kinds of applications, such as power line inspection [2], pipeline deformation detection [3], autonomous driving [4], precision agriculture [5], etc.

In general, the performance of 2D LRFs is proportional to their prices. According to ranging accuracy and applied domain, the 2D LRFs can be divided into two categories:

© Springer International Publishing AG 2017
Y. Huang et al. (Eds.): ICIRA 2017, Part III, LNAI 10464, pp. 254–266, 2017.
DOI: 10.1007/978-3-319-65298-6_24

survey-grade (with millimeter accuracy for ranging, e.g. Riegl VMX-450) and navigation-grade (usually with centimeter accuracy, e.g. Sick and Hokuyo series LRFs). The former is mainly used in surveying and mapping field while the latter was originally developed for safety protection and afterwards began to be extensively utilized in robotics community since 1990s. In the following, only the navigation-grade LRFs are involved.

The practical performance of the LRF is affected by numbers of factors, while insufficient attention has been paid to related research. It is essential to perform a detailed characterization of the LRF before use, which will be beneficial to sensor selection and feasibility analysis specific to the application. Ye et al. [6] might be the first one to present a comprehensive characterization of the 2D Sick LMS200 LRF, involving a series of experiments for analyzing the effect of the data transfer rate, drift, target properties and the incidence angle of the laser beam, where a linear motion table was applied to provide the ground truth and distance changes. Most following literatures leveraged the same or similar experimental setup and procedure. Cai et al. [7] tested and analyzed the ranging accuracy of the Sick LMS291, while Lee et al. [8] compared the LMS200 and the Hokuyo URG-04LX in terms of measurement drift over time, the effect of material and colour, and the ability to map different surface patterns. Kneip et al. [9] and Okubo et al. [10] provided insight into the same LRF URG-04LX from different perspectives. The characterization of the Hokuyo UBG-04LX-F01 was presented by Park et al. [11]. The Hokuyo UTM-30LX was assessed respectively by Pouliot et al. [2] and Demski et al. [12] for specific applications. Pomerleau et al. [13] proposed uncertainty models based on empirical results for the URG-04LX, the UTM-30LX and the Sick LMS151. Olivka et al. [14] deeply discussed and researched potential error sources resulted from manufacturing inaccuracies and introduced a complex calibration method suitable for the URG 04LX. In addition to general performance tests, Zhang et al. [15] investigated the influence of smoke on the accuracy of the Galaxy Electronic GLDT-1-B2BN LRF. Almost all of the ranging accuracy experiments involved in these literatures except [12] were restricted to relatively small scales (1–5 m) due to the limited length of the linear track applied.

A mid-range Sick LMS511-20100Pro LRF will be utilized for our simultaneous localization and mapping (SLAM) application, however, there is little available information regarding its practical performance to refer to. Therefore, a series of characterization experiments were performed based on previously published studies. This paper is structured as follows: In Sect. 2, an overview of the LMS511 LRF is given. The basic experiment setup is then introduced in Sect. 3, followed by presentation of the measurement results and analysis of various experimental conditions in Sect. 4. Finally, the conclusion is detailed in Sect. 5.

2 The LMS511-20100Pro LRF

As shown in Fig. 1, the Sick LMS511-20100Pro LRF (LMS511, for short) is a new and high-performance laser measurement sensor for ranges of up to 80 m. The LRF can be used both indoors and outdoors with outstanding performance even in adverse environmental conditions due to multi-echo technology.

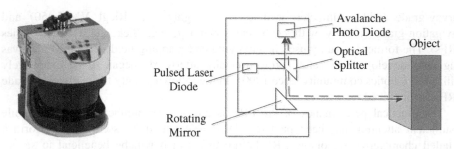

Fig. 1. The Sick LMS511-20100Pro LRF and its operating principle

Differing from the phase shift-based ranging principle of the URG-04LX, the LMS511 applies the time-of-flight (TOF) principle to measure the distance to the object. As depicted in Fig. 1, an infrared laser beam is emitted by the pulsed laser diode and then is split into two sub-beams by the optical splitter. One sub-beam directly enters the avalanche photo diode (APD, as the receiver) and simultaneously triggers the timer, while the other one is reflected onto the object by the mirror. When the return beam arrives at the APD, the timer stops and the distance from the sensor optical center to the object is calculated using the time lag and the velocity of light. The continuously rotating mirror deflects the laser beam in a fixed step angle such that the planar scanning is realized.

Although the LMS511 provides at most five echo measurements, only the first echo measurement is utilized in the present paper. Meanwhile, the received power for each echo can be recorded and output as the so-called received signal strength indicator (RSSI), which is a crucial value for certain applications. Furthermore, the LRF can be conveniently connected to a computer via RS232/RS422/Ethernet/USB interface through the SOPAS software provided by Sick AG. The angular resolution and scan frequency are both selectable, but a high resolution corresponds to a relatively low scan frequency. With a broad field of view of 190°(−5° to 185°), the nominal accuracy of the LMS511 is ±12 mm for the object with 100% remission at 6 m distance and the statistical errors for the 10% remission object are ±7 mm (1 to 10 m) and ±9 mm (10 to 20 m), respectively. The LRF can also receive encoder data through "I/O" interface, which is an important reason for its being selected. The key technical specifications for the LMS511 are listed in Table 1.

Table 1. Key technical specifications for the LMS511

Item	Value	Unit
Operating range	0.7–80	m
Field of view	190	deg
Angular resolution	0.167/0.25/0.5/0.667/1	deg
Scan frequency	25/35/50/75/100	Hz
Measurement accuracy	±12	mm
Divergence angle	4.7	mrad
Laser wavelength	905	nm
Weight	3.7	kg

3 Experimental Setup

To provide maximum flexibility, parameter setting and data acquisition of the LMS511 were implemented through a specialized software developed with C++ language in Visual Studio 2012 environment instead of applying SOPAS, while the data processing was completed by Matlab 2012. Since a linear motion table or track was not available, a novel experimental setup based on a Trimble S8 total station and retro-reflective targets was designed to characterize the mid-range LRF. The accuracy of total station is superior to 1 mm and thus its measurement can be used as the ground truth. As depicted in Fig. 2, the total station, the LRF and the rotatable target are mounted on three tripods that stand in a line. The distance between the total station and the LRF is 8 m while the distance from the LRF to the target is variable and depends on the specific test item. The default target is a piece of white paper stuck on a block of tile. Four retro-reflective targets with a cross-hair are utilized for the total station to determine the accurate horizontal distance between the LRF and the target.

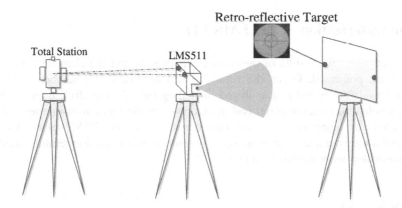

Fig. 2. The experimental setup

The LMS511 is connected to a computer via the Ethernet interface, which is the only way to transmit full scan data in real time. By default, the angular resolution is set to $0125°$ and the scan frequency 25 Hz. Accordingly, a full scan contains 1141 range measurements and the 571th one corresponds to the so-called center beam (i.e. 90° scanning direction). However, it is quite difficult to exactly align the target with the LRF. In practice, the target was adjusted to be approximately perpendicular to the center beam and we recorded the minimum distance value obtained around the center beam as the measured horizontal distance between the LRF and the target.

Before performing the characterization, it is essential to investigate the distribution of the measured range. The target was placed at a fixed distance and we collected 10000 range samples from the center beam. The statistical result showed that the measured range data follows Gaussian distribution. In the following experiments, the root mean squared error (RMSE) and the standard deviation (SD) are exploited to evaluate the ranging accuracy and repeated accuracy, respectively. The corresponding calculating formulas are as follows:

$$\sigma = \pm\sqrt{\frac{\sum\limits_{i=1}^{n}(d_i - d_t)^2}{n}} \tag{1}$$

$$s = \pm\sqrt{\frac{\sum\limits_{i=1}^{n}(d_i - \bar{d})^2}{n-1}} \tag{2}$$

where σ denotes RMSE and s denotes SD; while d_i is the range value measured by the LRF and d_t is the range value measured by the total station; n is the number of the LRF measurements and $\bar{d}$ is the mean value of n measurements. In the following, the standard number of measurements for each experiment equals to 5000.

4   Characterization of the LMS 511

In this section, all the results of performed experiments and the characteristics of the LMS511 are presented. Given the future SLAM application, a number of relevant characteristics were tested and analyzed, including the effect of drift, target distance, angular resolution, target material and mixed pixel on the range measurement, as well as the effect of target distance and target material on the RSSI value. All these experiments were conducted in an indoor environment where the temperature, humidity and illumination were relatively stable.

4.1   Drift Effect

Time drift is a common phenomenon for most sensors, which reflects the varying sensor characteristic over time and is primarily caused by the change of external condition and sensor itself. In this experiment, the target was placed at about 3.04 m in front of the LRF and 150,000 full scans in total were acquired after 100 min' continuous measuring. Only all the 571th range measurements were considered and we calculated the average value and standard deviation of each 1500 values (within every minute).

As is apparently shown in Fig. 3(a), the average value of the range measurements stabilizes after about 35 min. However, even during the first 35 min the average range decreases only by 2.5 mm, which is insignificant and tolerable for most mobile robot applications. Figure 3(b) shows that the standard deviation fluctuates around 3.5 mm within a quite small range of 0.5 mm. Therefore, compared to other LRFs in its class, the LMS511 is generally a delicate sensor. Additional experiments confirmed the same result. In order to exclude the drift effect, all the following experiments were conducted after the LRF had been operating for at least one hour.

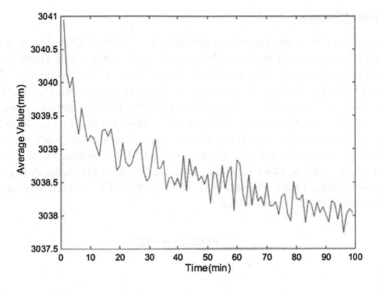

(a) Drift of the average value over time

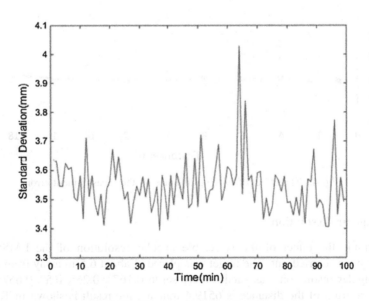

(b) Drift of the standard deviation over time

Fig. 3. Drift effect during 100 min at a distance of about 3.04 m

4.2 Target Distance

In this experiment, the same target was placed in front of the LRF at distances varying from 1 m to 20 m with a step of 2 m. Meanwhile, the corresponding ground truth was obtained through the total station with the help of retro-reflective targets. Then the RMSE σ and standard deviation s of measured ranges at each distance were calculated. The result is visualized in Fig. 4. As was expected, the RMSE roughly increases with distance, but not meeting a simple linear model applied by [6, 10, 11, 15] or a cubic polynomial model used by [9]. On the whole, the measurement accuracy accords with the nominal value. In particular, the accuracy of the LMS511 within a distance of 10 m surpasses 10 mm. On the other hand, the standard deviation maintains steady within a scale of 3.6–4.8 mm. That is, the repeated accuracy of the LRF is much higher than the specified value from the manufacturer.

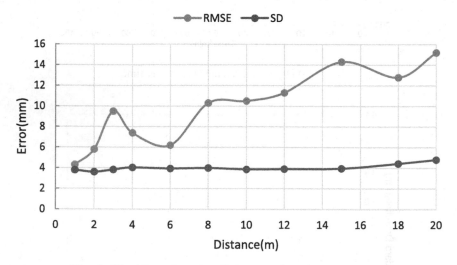

Fig. 4. The effect of varying distance on the measurement error

4.3 Angular Resolution

To determine the effect of the selectable angular resolution of the LMS511, this experiment was carried out. The target was placed at about 6.5 m away from the LRF, whose angular resolution was successively set to 0.167°, 0.25°, 0.5°, 0.667° and 1°. The ground truth of the distance is 6519.4 mm and the result is shown in Table 2.

Table 2. Result of different scanning angular resolution

Resolution	0.1667°	0.25°	0.5°	0.667°	1°
Mean (mm)	6514.1	6513.9	6514.5	6514.7	6515.2
RMSE (mm)	6.2	5.9	5.6	5.4	5.1
SD (mm)	3.6	3.6	3.6	3.6	3.6

Obviously, the angular resolution has a certain influence (max: 1 mm) on the measurement accuracy of the LMS511, while the repeated accuracy is completely unaffected.

4.4   Target Material

Different materials have different surface reflection properties to the laser, which influences the ranging performance of the LRF. To study the correlation between target material and ranging accuracy and repeated accuracy, several kinds of targets easily available from our lab were tested at the same distance of 6 m to the LMS511. Considering the result shown in Fig. 5, target material has a noteworthy effect on the ranging accuracy. For instance, the RMSE of the yellow hardboard is twice of the black anodized aluminium plate'. By contrast, the effect of target material on the repeated accuracy is slight and even could be neglected.

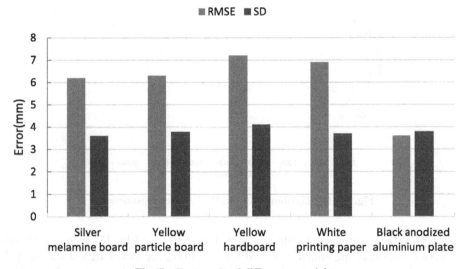

Fig. 5. Test result of different materials

4.5   Mixed Pixel

Despite good directivity, the laser beam emitted by the LRF can be considered as Gaussian beam and always has a definite divergence angle. According to the specification, the corresponding value of the LMS511 is 4.7 mrad. Accordingly, the light spot size at the front screen is 13.6 mm, but it increases to 99 mm at 18 m measuring range. Consequently, such a large spot might finally be located at the edges of two or more objects. In this situation, LRFs will output an intermediate range between the distances to the foreground and background objects, which is usually called mixed pixel phenomenon [16].

To verify whether this problem occurs in terms of the LMS511, we performed an experiment. The foreground object was a cuboid carton placed at about 2200 mm in front of the LRF, while the background was the white wall 7070 mm far from the LRF. A total of 3000 full scans were sampled and respective mean values in each beam direction were calculated. A mixed pixel was found at the 589th range measurement and an average value between 2200 mm and 7070 mm, i.e. 6465.6 mm, was acquired. Whereas, one should note that it is the result of averaging values of multiple scans, rather than a measured range in a single scan recorded by the LMS511, exactly on the contrary to the findings of [6, 7, 10]. As depicted in Fig. 6, the number of range measurements around 7070 mm is much larger than that around 2200 mm, resulting in the average value closer to the background object.

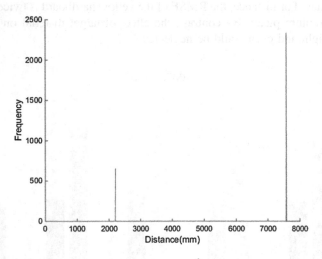

Fig. 6. The distribution of the 589th range measurement

To provide a comparison with the Sick LMS291 tested by [7] in terms of the sensitive angle occurring mixed pixel (MPSA, for short), the following test was carried out by utilizing the experimental setup similar to that in [7]. As is shown in Fig. 7, the LMS511 was clockwise actuated by the precision rotating platform and the 571th range measurement was observed 1000 times per 0.005°. This test discovered the MPSA for the LMS511 is 0.02°, which is far smaller than 0.19° for the LMS291 [7]. The laser used in the LMS511 has a much smaller divergence, which might account for this difference.

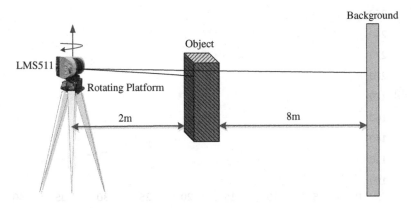

Fig. 7. The experimental setup for testing MPSA

4.6   Effect of Target Distance and Target Material on RSSI

Besides geometrical information, RSSI provided by the LMS511, varying from 1 to 254 for a valid range measurement, is another type of useful information, which can be applied for object recognition and classification. With an arbitrary unit, this value is generated through a logarithmic model and depends on the target remission. To make better use of the RSSI values in future applications, we conducted an experiment to inquire into the variation of the RSSI value with regard to target material and distance. The above mentioned five materials and a circular retro-reflective target were tested at a fixed distance of 6 m and the result is shown in Table 3. According to the result, the RSSI value of the retro-reflective target is much higher than that of other tested materials except the only metal, i.e. the black anodized aluminium plate. Taking into account the shape and size, the point cloud of circular retro-reflective targets can be extracted from that of the whole environment with the aid of RSSI values.

Table 3. The mean and SD of RSSI for different materials

Material	Mean of RSSI	SD of RSSI
Silver melamine board	225.4	1.2
Yellow particle board	220.0	1.2
Yellow hardboard	201.0	1.3
White printing paper	213.0	1.1
Black anodized aluminium plate	254.0	1.2
retro-reflective target	25.39	0.6

The other experiment took the silver melamine board as the test target, of which the RSSI values at different distances were tested. As depicted in Fig. 8, the RSSI value dramatically drops with the increasing distance. Hence, a strict calibration model is needed to better utilize the RSSI value.

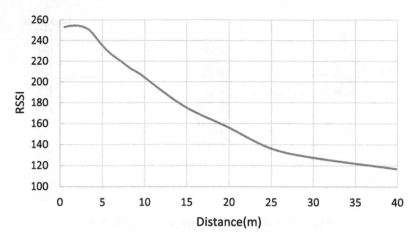

Fig. 8. Variation of the RSSI value with regard to target distance

5 Conclusion

This paper presents a comprehensive characterization of the Sick LMS511-20100Pro and analyses various potential issues affecting the performance and usage of the LRF, involving drift effect, target distance, angular resolution, target material, mixed pixel and the effect of target material and distance on RSSI, which determines the usability of the sensor in our future SLAM application.

The experiments suggest that, in comparison to the LRFs reported in previous works, the LMS511 suffers neither a long warm-up time nor a significant variation of the range measurement. Further, apart from some special applications demanding higher accuracy, there is no need to warm up the LMS511 before use, because the variation of the range measurement induced by the drift effect is at most merely 2.5 mm during about 35 min' stabilization time. Through the experiment at different distances (1–20 m), the measurement accuracy (10 mm within 10 m) of the LMS511 slightly exceeds the nominal accuracy while the repeated accuracy (3.6–4.8 mm within 20 m) that remains relatively stable with the increasing distance is much better than specifications. The angular resolution influences the measurement accuracy of the LMS511 to a certain degree (max: 1 mm), but the repeated accuracy is completely unaffected. The target material has a remarkable effect on the measurement accuracy, however, inappreciable on the repeated accuracy. It is demonstrated that mixed pixel problem doesn't exist in a single scan for the LMS511. Along with a quite small MPSA (0.02°), this problem doesn't need to be taken into account for general applications.

As the measurement of power received by the LRF, the RSSI value is principally affected by the target distance and material. The experiment concerning the target material indicates the possibility to utilize RSSI values for object recognition and classification, as well as the possibility to use retro-reflective targets for point cloud registration and evaluation of positioning precision in 3D navigation and mapping. Nevertheless, an appropriate calibration model regarding RSSI is needed in order to eliminate the influence of the target distance and other factors.

The measurement error of the LMS511 originates from numerous aspects of the target, the sensor itself and the environment such that a linear or polynomial model is not universally applicable. In addition to the issues involved in this paper, the performance of the LRF is also associated with the incidence angle of laser, lifetime of the sensor, target shape, working environment, etc. Thus, it is an intractable issue to calibrate the LRF using a universal formula in terms of each error source, given which the present paper doesn't provide any calibration model. Notwithstanding, a comprehensive characterization of the LMS511 is of great value and will benefit the designers and developers of related systems. For future works, other factors such as the effect of the incidence angle and the outdoor adverse environment will be addressed to complement the present characterization.

References

1. Wong, U., et al.: Comparative evaluation of range sensing technologies for underground void modelling. In: 2011 IEEE/RSJ International Conference on Intelligent Robots and Systems, San Francisco, CA, pp. 3816–3823 (2011)
2. Pouliot, N., Richard, P., Montambault, S.: LineScout power line robot: characterization of a UTM-30LX LIDAR system for obstacle detection. In: IEEE/RSJ International Conference on Intelligent Robots and Systems, Algarve, Portugal, pp. 4327–4334 (2012)
3. Qiu, J., Song, Z., Zhang, J.: A new method for detecting pipeline deformation by an inspection robot with a moving 2D laser range finder. In: IEEE International Conference on Robotics and Biomimetics, Phuket, Thailand, pp. 987–992 (2011)
4. Ahn, K., Lee, T., Kang, Y.: A method of digital map construction with road feature extraction using 2-dimensional laser range finder. In: 15th IEEE International Conference on Control, Automation and Systems, pp. 509–512 (2015)
5. Wang, Y., et al.: A UGV-based laser scanner system for measuring tree geometric characteristics. In: Proceedings of SPIE, International Symposium on Photoelectronic Detection and Imaging 2013: Laser Sensing and Imaging and Applications, 890532, Beijing, China, vol. 8905 (2013)
6. Ye, C., Borenstein, J.: Characterization of a 2D laser scanner for mobile robot obstacle negotiation. In: IEEE International Conference on Robotics and Automation, Washington, USA, vol. 3, pp. 2512–2518 (2002)
7. Cai, Z., et al.: Design of the 3D perceptive system based on laser scanner for a mobile robot. Int. J. Comput. Sci. Netw. Secur. **6**(3), 20–29 (2006)
8. Lee, K., Ehsani, R.: Comparison of two 2D laser scanners for sensing object distances, shapes, and surface patterns. Comput. Electron. Agric. **60**(2), 250–262 (2008)
9. Kneip, L., et al.: Characterization of the compact Hokuyo URG-04LX 2D laser range scanner. In: IEEE International Conference on Robotics and Automation, Kobe, Japan, pp. 1447–1454 (2009)
10. Okubo, Y., Ye, C., Borenstein, J.: Characterization of the Hokuyo URG-04LX laser rangefinder for mobile robot obstacle negotiation. In: Proceedings of SPIE, Conference on Unmanned, Robotic, and Layered Systems, Orlando, USA, vol. 7332 (2009)
11. Park, C., et al.: Characterization of the Hokuyo UBG-04LX-F01 2D laser rangefinder. RoMan IEEE **58**, 385–390 (2010). Viareggio, Italy

12. Demski, P., Mikulski, M., Koteras, R.: Characterization of Hokuyo UTM-30LX laser range finder for an autonomous mobile robot. In: Nawrat, A., Simek, K., Świerniak, A. (eds.) Advanced Technologies for Intelligent Systems of National Border Security. Studies in Computational Intelligence, vol 440, pp. 143–153, Springer, Heidelberg (2013). doi:10. 1007/978-3-642-31665-4_12
13. Pomerleau, F., et al.: Noise characterization of depth sensors for surface inspections. In: IEEE 2nd International Conference on Applied Robotics for the Power Industry, Zurich, Switzerland, pp. 16–21 (2012)
14. Olivka, P., et al.: Calibration of short range 2D laser range finder for 3D SLAM usage. J. Sens. **2016**(5), 1–13 (2016)
15. Zhang, Y., Xiao, J., Zhang, H.: Characterization of the GLDT laser range finder. In: IEEE International Conference on Information and Automation, Lijiang, China, pp. 2340–2344 (2015)
16. Adams, M.D., Probert, P.J.: The interpretation of phase and intensity data from AMCW light detection sensors for reliable ranging. Int. J. Robot. Res. **15**(5), 441–458 (1996)

Performance Metrics for Coverage of Cleaning Robots with MoCap System

Kuisong Zheng, Guangda Chen, Guowei Cui, Yingfeng Chen, Feng Wu[✉],
and Xiaoping Chen

Multi-Agent Systems Lab, School of Computer Science and Technology,
University of Science and Technology of China, Hefei 230027, Anhui, China
{kszheng,cgdsss,cuigw,chyf}@mail.ustc.edu.cn,
{wufeng02,xpchen}@ustc.edu.cn

Abstract. Nowadays there are a lot of kinds of cleaning robots which produced by different manufacturers come into people's lives. But it is still a problem that how to evaluate each robot's performance to check whether the quality is acceptable. In this paper, we make the first trial to evaluate the complete coverage path planning algorithm which is the core algorithm of a cleaning robot with Mocap system, and three simple metrics were proposed to evaluate overall performance of the algorithm. Lastly, the comparisons between different kinds of robots are presented.

Keywords: Cleaning robot · Mocap system · Coverage algorithm

1 Introduction

A robotic vacuum cleaner (or robot cleaner or cleaning robot) is a service robot [7,15] which designed to help people clean their houses efficiently, one of the popular robot cleaners is Roomba [13]. A typical robot cleaner is composed of a mobile base, cleaning units, collision sensors and attachments. Nowadays many robot cleaners are equipped with lasers, they will build a map to help them generate a more efficiently cleaning path. Robot cleaners from different manufacturers vary in price, size, motors, sensors, efficiency and so on. There are already many evaluation indexes of a robot cleaner such as price, noise, suction. However, these evaluation indexes are often performed by unofficial third party evaluation organizations, and the process of evaluation is often carried out manually and the data are not recorded for post review. Therefore, a general framework of performance evaluation of a robot cleaner is needed to address this problem.

We will focus on evaluation the performance of the coverage algorithm of a robot cleaner in this paper, because the algorithm has greatest contribution to the intelligence of the robot. It is urgent to have a general framework to evaluate the performance of a robot cleaner for the following reasons: (1) A customer wants to know which type of robot cleaner is more efficient or more suitable for their needs. (2) Currently evaluation methods rely on experience, for example,

© Springer International Publishing AG 2017
Y. Huang et al. (Eds.): ICIRA 2017, Part III, LNAI 10464, pp. 267–274, 2017.
DOI: 10.1007/978-3-319-65298-6_25

A tester sprinkles scraps on the ground and see how many scraps are left with eye after the cleaning robot finish cleaning, this kind of method is not scientific. (3) Different kinds of robot cleaners vary in hardware and software, therefore we can only rely on external measurement to evaluate their performance.

In this paper, we proposed a general platform for evaluating the performance of coverage of a cleaning robot. Then we discussed the three criterias for measuring the performance. Lastly, we conduct the experiments to evaluate the performance of complete coverage algorithms of three robots which manufactured by different companies.

2   Related Work

Complete coverage algorithms have applications in floor cleaning [14], and lawn mowing [1]. Obviously, the goal of the algorithm is to generate the shortest path that cover the entire area as much as possible. A large body of algorithms are developed [10–12,26]. Different kinds of coverage algorithms were compared in [11], *randomized coverage algorithms* which don't need a laser sensor are very simple but inefficient, while *sensor-based coverage algorithms* use sensors information to help the robot cleaners make a plan.

Cleaning robots in the market can be divided into two types according to whether they are equipped with laser sensors. The strategies of randomized robots are mostly pre-defined heuristic behaviors, for example, perform spiral walking in free space or following a wall detailed in [2] and so on. While laser-based robots need to build a map of the home environment, the map can be an occupancy grid map or geometry map or others which detailed in [21]. They also need to use algorithms such as those proposed in [20] to localize where they are in the map. With the sensors and computational resources, they are also possible to make high-level decision-making or long-term planning [4,25] in the future.

As a product, a cleaning robot has a lot of indexes of performance evaluation criteria as detailed in [18]. However we focus on evaluation of the performance of the coverage algorithm of different kinds of cleaning robots. Sylvia used two simple methods (percentage of coverage and distance travelled) for measuring the performance of complete coverage algorithms [22], the position of the robot is calculated by computer vision, this method has the following disadvantages: (1) The camera needs to be calibrated in advance. (2) The method cannot be carried out in real home environment. (3) The accuracy of the position of the robot can't be guaranteed. (4) The process of evaluation needs a lot of manual intervention. Therefore we use Motion Capture System (MCS) as our evaluation tool in virtual of it can provide realtime, accurate movement data of measured objects. Actually MCS has already been used in robotics in many tasks [8,9,19,27,28].

Our goal in this paper is to develop benchmarks problems for comparing the performance of different cleaning robots. Many benchmark problems have been proposed to evaluate robot systems, e.g., robot soccer [3,5], search and rescue tasks [16,17,23,24], etc. However, to the best of our knowledge, none has been proposed for the cleaning robots in the literature.

3   Performance Metrics for Coverage

Coverage task is difficult for a cleaning robot because the environment is dynamic and complex. The coverage algorithm itself is not covered in this paper, we want to evaluate the performance of the cleaning robots which can be buied off the shell. The following three performance metrics are used.

Compute the coverage rate. To compute the coverage rate of a cleaning robot, we build a map with SLAM [6] to compute the free space A_{free} and get the area A_{robot} robot covers with Algorithm 1.

$$R_{cover} = \frac{A_{robot}}{A_{free}} \tag{1}$$

3.1   Computing the Repeat Coverage Area Ratio

We record positions of the cleanning robots with MoCap while they run freely in the simulated home environment, and we de-sample the positions so that each position is at least 2 *cm* or 2 *deg* away from its neighbors. Then we compute the repeat coverage area ratio by feeding the positions to Algorithm 1.

3.2   Computing the Distance Travelled

This metric is very simple, we can compute the distance D_{robot} that the robot travelled by summing up the distance of each two adjacent positions. D_{min} is the least length which a robot travels just covers the whole filed. The efficiency of the coverage E can be calculated by:

$$E = \frac{D_{robot}}{D_{min}} \tag{2}$$

Algorithm 1. Computing the repeat coverage area ratio

1: Each grid of map M has two filed, $\langle covered, repetition\_time \rangle$
2: Let C indicate whether a grid of map is covered
3: Let R indicate times that a grid of map is covered
4: **Input:** Robot positions P
5: **procedure** COMPUTEREPETITION(P)
6:     $I' \leftarrow \emptyset$
7:     **for** $p \in P$ **do**
8:         Compute indexes I that the robot covers in position p
9:         **for** $i \in I - I'$ **do**
10:             $C[i] \leftarrow 1$
11:             $R[i] \leftarrow R[i] + 1$
12:         $I' \leftarrow I$

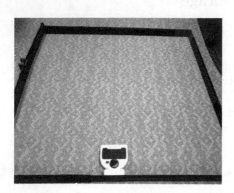

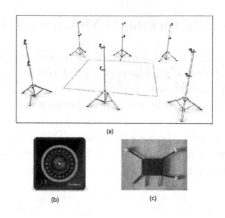

Fig. 1. Experiment site

Fig. 2. (a) MoCap system (b) Optical camera (c) Markset

4   Experimental Setup

We tested three popular cleaning robots among which two cleaning robots are laser-based and one cleaning robot is collision-based. Experiment environment is shown in Fig. 1, the area of test platform is about 9.5 m^2. MoCap system shown in Fig. 2 can track markerset which fixed on the robot in realtime. The cleaning robot will run with no interference once started. Every robot performances the cleaning task several times and the trajectories are recorded.

The collected trajectories and coverage areas are shown in Fig. 3. The first row of images are the coverage areas of the three cleaning robots, the second row of pictures are the trajectories of the three cleaning robots. Obviously, robot A travelled the farthest distance to finish the task, robot C is better than robot A because it's trajectory is more ordered, and the trajectory of robot B is as ordered as robot C except that distance between sweeping lines is larger than robot C.

4.1   Result and Comparisons

We compared these three cleaning robot in travelled distance, covered area, sweeping time as shown in Fig. 4. Obviously, all robots successfully covered the whole field, robot A can't plan according to the environment in advance because it is not equipped with a laser sensor, so it travelled according it's designed behavior which is almost randomly, therefore, it waste a lot of time and it's efficiency is worst among the three robots, and robot B finishes the cleaning task with the least time among the three robots. The repetition times of the three cleaning robots is shown in Fig. 5, we can see robot B has least repeation ratio.

Fig. 3. Areas are in the upper row and trajectories are in the lower row, the color of the area which the cleaning robot covers many times is brighter.

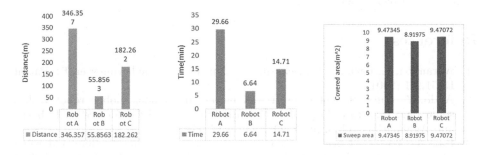

Fig. 4. Comparisons of three robots

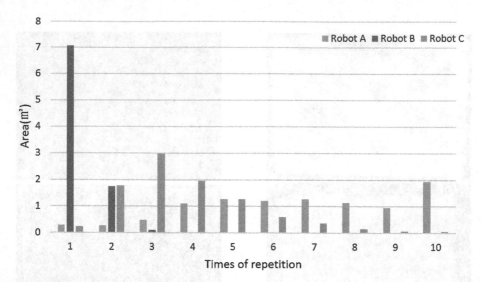

Fig. 5. Repetition times of the robots

5   Conclusions

In this paper, we proposed three performance evaluation indices of the complete coverage algorithm of a cleanning robot, they are coverage area percentage, repeat coverage area ratio, distance travelled respectively, these metrics show the performance of coverage algorithm of cleaning robots from the whole level. We use Mocap system as an external measurement because it is accurate, realtime and easy to use. Then, we test the three robots buied in market to see how well they perform actually, we can conclude that laser-based cleaning robots are better than randomized cleaning robots from the result.

Acknowledgments. This work was supported in part by National Natural Science Foundation of China under grant No. 61603368, the Youth Innovation Promotion Association of CAS (No. 2015373), and Natural Science Foundation of Anhui Province under grant No. 1608085QF134.

References

1. Abramson, S., Levin, S., Tikochinsky, Y., Zur, E.: Robotic lawnmower. US Patent D451,931 (2001)
2. Ando, Y., Yuta, S.: Following a wall by an autonomous mobile robot with a sonar-ring. In: Proceedings of 1995 IEEE International Conference on Robotics and Automation, vol. 3, pp. 2599–2606. IEEE (1995)
3. Bai, A., Wu, F., Chen, X.: Towards a principled solution to simulated robot soccer. In: Chen, X., Stone, P., Sucar, L.E., Zant, T. (eds.) RoboCup 2012. LNCS (LNAI), vol. 7500, pp. 141–153. Springer, Heidelberg (2013). doi:10.1007/978-3-642-39250-4_14

4. Bai, A., Wu, F., Chen, X.: Bayesian mixture modelling and inference based thompson sampling in monte-carlo tree search. In: Proceedings of the Advances in Neural Information Processing Systems (NIPS), pp. 1646–1654, Lake Tahoe, United States (2013)
5. Bai, A., Wu, F., Chen, X.: Online planning for large markov decision processes with hierarchical decomposition. ACM Trans. Intell. Syst. Technol. (ACM TIST) **6**(4), 45 (2015)
6. Bailey, T., Durrant-Whyte, H.: Simultaneous localization and mapping (slam): Part ii. IEEE Robot. Autom. Mag. **13**(3), 108–117 (2006)
7. Chen, Y., Wu, F., Shuai, W., Wang, N., Chen, R., Chen, X.: KeJia robot–an attractive shopping mall guider. In: Tapus, A., André, E., Martin, J.C., Ferland, F., Ammi, M. (eds.) Social Robotics. LNCS, vol. 9388, pp. 145–154. Springer, Cham (2015). doi:10.1007/978-3-319-25554-5_15
8. Chen, Y., Wu, F., Wang, N., Tang, K., Cheng, M., Chen, X.: *KeJia-LC*: a low-cost mobile robot platform — champion of demo challenge on benchmarking service robots at RoboCup 2015. In: Almeida, L., Ji, J., Steinbauer, G., Luke, S. (eds.) RoboCup 2015. LNCS, vol. 9513, pp. 60–71. Springer, Cham (2015). doi:10.1007/978-3-319-29339-4_5
9. Cheng, M., Chen, X., Tang, K., Wu, F., Kupcsik, A., Iocchi, L., Chen, Y., Hsu, D.: Synthetical benchmarking of service robots: a first effort on domestic mobile platforms. In: Almeida, L., Ji, J., Steinbauer, G., Luke, S. (eds.) RoboCup 2015. LNCS (LNAI), vol. 9513, pp. 377–388. Springer, Cham (2015). doi:10.1007/978-3-319-29339-4_32
10. Choi, Y.H., Lee, T.K., Baek, S.H., Oh, S.Y.: Online complete coverage path planning for mobile robots based on linked spiral paths using constrained inverse distance transform. In: 2009 IEEE/RSJ International Conference on Intelligent Robots and Systems, pp. 5788–5793 (2009) doi:10.1109/IROS.2009.5354100
11. Choset, H., Pignon, P.: Coverage path planning: the boustrophedon cellular decomposition. In: Zelinsky, A. (eds) Field and Service Robotics, pp. 203–209. Springer, London (1998). doi:10.1007/978-1-4471-1273-0_32
12. Huang, W.H.: Optimal line-sweep-based decompositions for coverage algorithms. In: Proceedings of 2001 ICRA IEEE International Conference on Robotics and Automation (Cat. No.01CH37164), vol. 1, pp. 27–32 (2001). doi:10.1109/ROBOT.2001.932525
13. Jones, J.L.: Robots at the tipping point: the road to irobot roomba. IEEE Robot. Autom. Mag. **13**(1), 76–78 (2006)
14. Jones, J.L., Mack, N.E., Nugent, D.M., Sandin, P.E.: Autonomous floor-cleaning robot. US Patent 6,883,201 (2005)
15. Lu, D., Zhou, Y., Wu, F., Zhang, Z., Chen, X.: Integrating answer set programming with semantic dictionaries for robot task planning. In: Proceedings of the 26th International Joint Conference on Artificial Intelligence (2017)
16. Ramchurn, S.D., Huynh, T.D., Wu, F., Ikuno, Y., Flann, J., Moreau, L., Fischer, J.E., Jiang, W., Rodden, T., Simpson, E., Reece, S., Roberts, S., Jennings, N.R.: A disaster response system based on human-agent collectives. J. Artif. Intell. Res. (JAIR) **57**, 661–708 (2016)
17. Ramchurn, S.D., Wu, F., Fischer, J.E., Reece, S., Jiang, W., Roberts, S., Rodden, T., Greenhalgh, C., Jennings, N.R.: Human-agent collaboration for disaster response. J. Auton. Agents Multi-Agent Syst. (JAAMAS), pp. 1–30 (2015)
18. Rhim, S., Ryu, J.C., Park, K.H., Lee, S.G.: Performance evaluation criteria for autonomous cleaning robots. In: International Symposium on Computational Intelligence in Robotics and Automation, CIRA 2007, pp. 167–172. IEEE (2007)

19. Röwekämper, J., Sprunk, C., Tipaldi, G.D., Stachniss, C., Pfaff, P., Burgard, W.: On the position accuracy of mobile robot localization based on particle filters combined with scan matching. In: 2012 IEEE/RSJ International Conference on Intelligent Robots and Systems (IROS), pp. 3158–3164. IEEE (2012)

20. Rusinkiewicz, S., Levoy, M.: Efficient variants of the ICP algorithm. In: Proceedings of the Third International Conference on 3-D Digital Imaging and Modeling, pp. 145–152. IEEE (2001)

21. Thrun, S., et al.: Robotic mapping: a survey. Exploring Artif. Intell. New Millennium 1, 1–35 (2002)

22. Wong, S.C., Middleton, L., MacDonald, B.A., Auckland, N.: Performance metrics for robot coverage tasks. In: Proceedings of Australasian Conference on Robotics and Automation, vol. 27, p. 29 (2002)

23. Wu, F., Jennings, N.R.: Regret-based multi-agent coordination with uncertain task rewards. In: Proceedings of the 28th AAAI Conference on Artificial Intelligence (AAAI), Quebec City, Canada, pp. 1492–1499 (2014)

24. Wu, F., Ramchurn, S.D., Jiang, W., Fischer, J.E., Rodden, T., Jennings, N.R.: Agile planning for real-world disaster response. In: Proceedings of the 24th International Joint Conference on Artificial Intelligence (IJCAI), Buenos Aires, Argentina, pp. 132–138 (2015)

25. Wu, F., Zilberstein, S., Chen, X.: Trial-based dynamic programming for multi-agent planning. In: Proceedings of the 24th AAAI Conference on Artificial Intelligence (AAAI), Atlanta, United States, pp. 908–914 (2010)

26. Yang, S.X., Luo, C.: A neural network approach to complete coverage path planning. IEEE Trans. Syst. Man, Cybern. Part B (Cybern.) 34(1), 718–724 (2004)

27. Zhang, H., Cao, R., Zilberstein, S., Wu, F., Chen, X.: Toward effective soft robot control via reinforcement learning. In: Proceedings of the 10th International Conference on Intelligent Robotics Applications (2017)

28. Zheng, K., Chen, Y., Wu, F., Chen, X.: A general batch-calibration framework of service robots. In: Proceedings of the 10th International Conference on Intelligent Robotics Applications (2017)

A General Batch-Calibration Framework of Service Robots

Kuisong Zheng, Yingfeng Chen, Feng Wu$^{(\boxtimes)}$, and Xiaoping Chen

Multi-Agent Systems Lab, School of Computer Science and Technology,
University of Science and Technology of China, Hefei 230027, Anhui, China
{kszheng,chyf}@mail.ustc.edu.cn, {wufeng02,xpchen}@ustc.edu.cn

Abstract. Calibration is important to service robot, but the process of calibration is time consuming and laborious. With the popularity of service robot, an automatic and universal calibration system is urgent to be developed, therefore we propose a general batch-calibration framework, Motion Capture System is adopt as an external measurement device in virtual of it can provide realtime, accurate movement data of measured objects. We will show that the system is effective and promising with a case study of odometry calibration.

Keywords: Robot calibration · MoCap system

1 Introduction

Service robots have attracted increasing attention from the commercial companies as well as the research groups in recent years [11]. A service robot is often defined as a robot which autonomously performs daily services for humans, aiming to improve their life quality [17]. Essentially, it is also an artificial electro-mechanical machine as same as the traditional industrial robot, but expected to be more intelligent, human-interactive, humanoid and safe. Therefore, the tangible service robots have to facing the calibration problem inevitably, which is a common issue with all kinds of robots [2,10,12,28], and it's even more pressing for service robots [13].

The process of tuning the parameters of the kinematic and the dynamic models of a robot is called calibration, this operation is so important that nearly all robots need to be calibrated to perform better after manufacture. Unfortunately, most calibration work is completed by a human operator manually, the process of calibration is time-consuming and laborious. The effects of calibrations usually depend on the experience and knowledge of the operator and devices or tools used. What is worse, to calibrate batch robots which are not coincide with the original design drawing, the same operation is repeated over and over.

As to service robots, an automatic and universal calibration system is much more urgent to be developed for these reasons: (1) A large quantity of service robots will be made to satisfy the demands of families in the future, this market is more huge than industrial robots in long term, it's impossible for manufacturer to

© Springer International Publishing AG 2017
Y. Huang et al. (Eds.): ICIRA 2017, Part III, LNAI 10464, pp. 275–286, 2017.
DOI: 10.1007/978-3-319-65298-6_26

calibrate every robot manually, a pipeline of automatic calibration is a necessary choice. (2) The cost of the service robots will decrease to a certain amount that is acceptable to most consumer, it means the components of robots may not strictly the same due to their relative low cost, so the calibration is required to mask the difference of hardware. (3) The users of the service robots are ordinary consumers, who are not robotic experts, the calibration should be done before selling to them.

In this paper, we raise the problem of batch calibration of service robots, which needs to be addressed with the coming population explosion of service robots. To solve this problem, we summarize and analyse previous calibration methods, and then propose a general framework of batch calibration, which is especially applicable for service robots. Afterwards, a proof-of-concept of the proposed framework is implemented with an optical Motion Capture System (Mocap) as global measuring tool. Lastly, we conduct the calibration experiments of estimating the odometry parameters of a mobile robot platform as a case study under this framework.

2   Related Work

Robot calibration has a long research history, as Roy et al. said in [23]: "the need for calibration is as old as the field of robotics itself". Many AI planning and learning algorithms [3,4] that can be run on robotic system usually require that the robots have been well calibrated. The mechanical structure of robot systems often slightly change or drift due to wear of parts, reassembling of components, and loosening of joints. The task of calibration is to correct these alterations and keep an accurate model, which describes the relationship between the input control values and actual outputs.

In overviews [15,22], they classified the robot calibration into three levels: The level 1 is simply to ensure the reading from a joint sensor yields the correct joint positions; The level 2 is to extend one joint (in level 1) to multiple joints – i.e., a complex kinematic model; The level 3 considers the dynamic models, deflection of robot links, gear backlash and so on, beyond the kinematics. Each level include four steps: modeling, measurement, identification, and correction.

Large majority of the kinematic calibration studies in literature [9,20,27] are related to the industrial robots, on account of that they are engaged in manu-facturing processes which require precise positioning and force control. Follow the same pattern, they employed the Denavit-Hartenberg model or its variants to establish the transformation matrix based on forward kinematics, then deter-mined the unknown parameters with corresponding measurements.

Robot calibration usually involves the collecting of actual measurements, the measurement data are then used to compute the parameters of undetermined models. In general, effective and accurate measurements could greatly improve the precision of the results. The measurements mainly come from two sources: (1) Internal sensors mounted on robots – this method exploits the constrains of measurements from sensors to estimate the model parameters. (2) External mea-suring equipments – The equipments provide global measurements, usage of these data are straight-forward since the model equations are often included in the models.

A typical case using internal sensors is hand-eye calibration. Here, the cameras are fixed on an end effector or a pan-tilt unit, and vision techniques are applied to acquire the absolute or relative positions of tailor-made signs (e.g., checkerboards) which are easy-recognized. The work [26] solved common formulations of calibration problems using nonlinear optimization with a eye-in-hand systems. Maier et al. [18] also cast a similar hand-eye calibration of Nao's whole-body as a least-squares optimization problem, which is settled with g^2o graph optimization library.

Lots of work focus on the problem how to select optimal measurement configurations for accurate robot calibration, minimizing the variance of the parameters to be estimated. The observability index ([6,16]) was proposed to evaluate the utility of different configuration sets and several criteria were studied out form theoretical points. The recent work [7,18] also designed specific algorithms to get better practical effects. However, our work has not touch on this topic, it's still worth mentioning these work.

Overall, the calibration methods have been extensive researched and have achieved good results in application, but still lack of a general calibration platform which are expected to be automatic and efficient. Our work tries to make a contribution on the exploration and designing of such an calibration prototype under the background – population explosion of service robots. The conception of general calibration platform will be described in next section, in Sect. 3 we introduce our preliminary implement integrating with Mocap system. Lastly, a case study of odometry calibration is presented to show that the system is effective and promising.

3 Conception of the Calibration Platform

3.1 Motivation and Objectives

As mentioned in Sect. 1, the expectable advance of service robots will bring about mass production, but how to raise the efficiency of the indispensable calibration procedure and ensure the quality is still not formally put forward. Currently in practice, this process not only depends on human experience, but also is inefficient and time-consuming (shown in Fig. 1). Our motivation is to establish a platform that could simplify the process, imagine that a new-made robot enter in the platform, just executing a set of benchmark, then individual parameter is set to make every robot in optimal configuration. As shown in Fig. 2, the calibration and quality control is automatically accomplished by the platform without manual intervention.

The design objectives of the platform are:

1. *Automatical*: Our primary aim is to substitute manual operations to increase working efficiency, so the desired platform must be highly automatic, minimizing the need for human intervention to an extreme. Generally, the manual

work almost all concentrate on the measuring, therefore, choosing a automatic measurement tools is crucial for creating such a platform.

2. *Batch-oriented*: The platform is expected to perform calibration in bulk to maintain coordination with quantity production. Robots go through the platform with different process in order just like common product on assembly line.

3. *General*: The platform could apply to different kind of robots with the same way, methods and procedures are roughly changeless, and this principle is reflected in the architecture design of the platform which will describe in the following section.

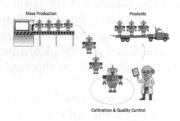

Fig. 1. Traditional robots calibration procedure

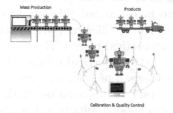

Fig. 2. Calibration in our proposed platform

3.2   Architecture of Calibration Platform

The architecture of our proposed calibration platform is shown in the Fig. 3. The function of this platform is to provide a general way to calibrate all kinds of parameters of robots, and calibration requires measurement data both from robots' internal sensors and external equipments. The internal measurements could be collected from the robot itself by executing certain motion commands. The external measurements are captured by the *Automatic Measuring System* (AMS). Beyond measurement data, a model is necessary to define of a particular calibration, which show clearly what are the unknown parameters and the relationship between observation. Obviously, the models are highly related to specific robots kinematics and calibration cases. In order to improve the generality of this platform, we propose a universal model description language extended from *Unified Robot Description Format* (URDF) [19], which is an *XML* format for representing a robot model. Given the measurements and models, the *General Calibration Solver* is responsible for figuring out the corresponding parameters mostly based on data fitting techniques. Since our purpose is to test the performance of the robots, once the parameters of the models are identified, the calibrated robots will take *Standard Test* and the results are compared with *Performance Criterion*.

Our proposed platform highly generalizes the calibration procedure of service robots, the users could just keep attentions on modeling calibration problem

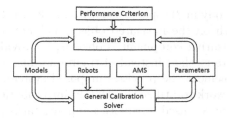

Fig. 3. The concept map of our proposed platform

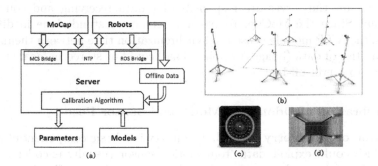

Fig. 4. (a) Modules in our implementation (b) Diagram of Mocap system (c) Optical camera (d) markset

with unified description language and their criterion of performance. Our aim is also to develop a calibration platform, which could perform batch calibration automatically for kinds of robots, meanwhile, it's friendly for users and even a black box.

4   Preliminary Implementation and Case Study

In this section, we firstly present an implementation of the calibration platform proposed above, which employs the *Optical MoCap* as *Automatic Measuring System*. Then, the calibration problems of both odometry and sensor pose are introduced systematically, and the solution is presented by mean of the calibration system.

4.1   Implementation of Calibration System with MoCap

Mocap system is originally used in computer animation for television, cinema, and video games as the technology matured. As shown in Fig. 3, our MoCap system consists of 12 cameras equipped with infrared LED around the camera lens. The reflective markers are attached on the measured objects, the centers of the marker images are matched from the various camera views using triangulation to compute their frame-to-frame positions in 3D space. In order to tract the 6D pose of rigid body in 3D space, usually a markset (at least with 3 marks) is assembled to attach on the measured object. The advantage of introducing such an external measuring equipment is obvious, it provide an automatic and high-accuracy measure method instead of traditional manual measuring, which makes it easy to implement the previous proposed system. As far as we know, this may be the first work to utilize the MoCap into robotic calibration domain though it has already been used in robotics in many tasks [29].

Our system is shown in Fig. 4, the measurements of MoCap and robots are sending to the server realtime through network, the *MCS Bridge* module and *ROS Bridge* on the server are responsible for data receiving and conversing respectively. Since the MoCap, robots and server are stand-alone in different machines, the *NTP* module is used to synchronization time between them. Some large quantities of data (images or point cloud) could be stored in local and then transferred to server off-line.

4.2   Calibration of Odometry Model and Sensor Pose

Calibration of Odometry Model. Odometry is a basic component of mobile robot, which could exploit data from motion sensor (usually encoder or camera) to estimate pose change over time. For many robot application, such as localization and mapping, odometry plays an indispensable role as the input of prior knowledge, hence accuracy odometry could simplify the subsequent process. However, odometry often suffers from kinds of systematic and nonsystematic errors, resulting in a significant decrease in performance [5]. The purpose of odometry calibration is to identify the effective parameters of motion model, which often have small difference between nominal values and therefore cause most of the systematic errors.

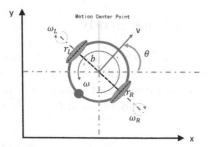
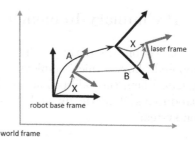

Fig. 5. The structure of differential-driven wheels

Fig. 6. Illustration of sensor calibration

The odometry motion model in our case is a typical differential wheel structure (shown in Fig. 5). Ignoring the nonsystematic errors in odometry (such as wheel-slippage, uneven floor and etc.), the kinematic model could be represented compactly as following:

$$\begin{bmatrix} v \\ \omega \end{bmatrix} = C * \begin{bmatrix} \omega_R \\ \omega_L \end{bmatrix} \qquad C = \begin{bmatrix} \frac{r_R}{2} & \frac{r_L}{2} \\ \frac{r_R}{b} & \frac{-r_L}{b} \end{bmatrix} \tag{1}$$

In Eq. 1, the v and ω indicate the translational and angular velocity of robot respectively. The $\omega_L(\omega_R)$ and the $r_L(r_R)$ are the rotate speed and radii of left (right) wheel, the b is the distance between two wheels. Once parameter matrix C is determined, the odometry calibration is done.

To collect data under the calibration system, robot are driven to perform movements. Meanwhile, the pose of robot and the encoder data are recorded, thus the relevant parameters would be figured out. Via proper formula manipulation, we could exploit linearity of the parameter matrix C and convert the calibration to a least-squares estimation problem [1] as following:

$$\begin{bmatrix} \theta_{N_1,1} - \theta_{0,1} \\ \vdots \\ \theta_{N_p,p} - \theta_{0,p} \end{bmatrix} = \begin{bmatrix} \Phi_{\theta,1} \\ \vdots \\ \Phi_{\theta,p} \end{bmatrix} * \begin{bmatrix} C_{2,1} \\ C_{2,2} \end{bmatrix} \tag{2}$$

$$\Phi_{\theta,p} = T \left[\sum_{i=0}^{N_p-1} w_{R,i} \ \sum_{i=0}^{N_p-1} w_{L,i} \right]$$

In Eq. 2, the $\theta_{0,p}$ and $\theta_{N_p,p}$ are the robot's directions at 0 and N_p moments in the pth trajectory sample, the pth trajectory contains the encoder data of $0, 1, \cdots, N_p$ moments (totally $N_p + 1$ moments in this trajectory, namely, N_p equal intervals with time T), and $\Phi_{\theta,p}$ is the overall angular change of the pth trajectory sample, which is the sum of minor changes in all N_p intervals $\sum_{i=0}^{N_p-1} \Delta\theta_i$. Thus, we can establish a deterministic regressor for this problem by sampling p trajectories. In the same way, we can get the rest of the parameters ($C_{1,1}$ and $C_{1,2}$) in C, which could be further investigated in [1] and not detail here on account of the page limitation.

Thus far, we have elicited the calibration procedure from Eqs. 1 and 2 mathematically, but in practice, the problem is not fully solved since the pose of the motion center on robot could hardly be measured directly. Actually, the motion center point is a virtual point and unobservable. The solution to this issue will be presented in Sect 4.2.

Calibration of Sensor Pose. In our case, our aim is to calibration the plane transformation between the frame of installed laser and the frame of robot base (originated at the motion center point). As shown in Fig. 6, the unknown transformation X in plane has three degree of freedom, thus can be denoted as $X = (l_x, l_y, l_\theta)$. The $A = (A_x, A_y, A_\theta)$ is the related transformation of robot's motion center between two different poses, and the $B = (B_x, B_y, B_\theta)$ is the related transformation of two different laser poses. Hence, we could get an equation by the operation of the homogeneous coordinates transformation [21]:

$$AX = XB \tag{3}$$

In fact, this is the common form of hand-eye calibration problem and has been extensive researched through several approaches ([14, 24, 25]). Although our problem is a simple case in which the axis of two related transformation is parallel in 2D, previous methods are not fully applicable since they are designed for general cases in 3D. To deduce a solution from Eq. 3 is straightforward:

$$\Psi * \begin{bmatrix} l_x & l_y & sinl_\theta & cosl_\theta \end{bmatrix}^T = \begin{bmatrix} -A_x \\ -A_y \end{bmatrix}$$

$$\Psi = \begin{bmatrix} cosA_\theta - 1 & -sinA_\theta & B_y & -B_x \\ sinA_\theta & cosA_\theta - 1 & -B_x & -B_y \end{bmatrix} \tag{4}$$

Ideally, a consistent system of two solvable homogeneous transform equations of the form $A_1 X = X B_1$ and $A_2 X = X B_2$ has a unique solution. Considering the existence of noise in measurements, we collect N sets of data and convert the constrain linear problem to an optimization problems:

$$min \quad \varphi^T (\sum_i^N \Psi_i^T * \Psi_i) \varphi$$

$$s.t. \quad \varphi_3^2 + \varphi_4^2 = 1 \tag{5}$$

Find the optimum $\varphi^* = [l_x^*, l_y^*, sinl_\theta^*, cosl_\theta^*]^T$ would solve this problem successfully.

From the view of practical operation, the measurements of B could be acquired by pairwise scan-matching, while the A is actually the movements of robot center point as same as the odometry calibration in previous section. Therefore, how to determine the motion center point and recovery its pose from the MoCap system is an important step.

Determine the Motion Center Point of Robot. Our MoCap system could track the pose of the 3D pose of the rigid body attached on marksets. However, we couldn't put on the markset on the robot's motion center point in practice since it's intangible. So we need a method to acquire the transformation between the base frame and mark-set frame, this is similar to the problem of calibration of sensor pose. Fortunately, a easy method is found to this issue and avoid the interdependence of the two problems. As shown in Fig. 7, the pose of markset in world frame is captured by the MoCap system at any time, based on these information, we could figure out the $X = (h_x, h_y, h_\theta)$ by certain specific movements. Firstly, we command the robot spin on the spot, assuming that the robot's center point is fixed during the operation, thus we could get the radius R and the circle center P_1 of the trajectory. Then we drive the robot forward along the direction of its x axis, meanwhile, the pose of the start and end moment are record as P_2 and P_3.

$$\theta_{aux} = acos(\overrightarrow{P_1 P_2} \cdot \overrightarrow{P_2 P_3} / \|\overrightarrow{P_1 P_2}\| * \|\overrightarrow{P_2 P_3}\|)$$

$$h_x = R * cos\theta_{aux}$$

$$h_y = R * sin\theta_{aux}$$

$$h_\theta = Angle(\overrightarrow{P_1 P_2}) - Angle(P_2) \tag{6}$$

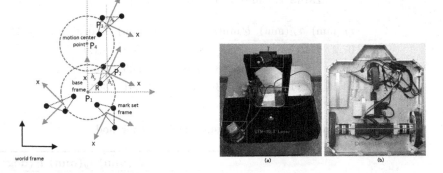

Fig. 7. Illustration of the method to determine the robot's motion center

5   Experiments

In this section, we present our experiments under the proposed calibration system, our aim is to identify the odometry and sensor parameters of the *KeJia* mobile robot.

5.1   Equipment and Environment

The robot used in this experiment is the *KeJia* robot, which has participated in consecutive RoboCup@Home Competitions and once won the world champion. In our experiment, we focus on the basis of the robot since only the wheels and the laser are considered in calibration.

5.2   Configuration and Data Set

We use the following three methods to perform the calibration: (1) Direct measurement of the odometric parameters; (2) The technique described in [8]; (3) The method proposed in this paper.

Method (1) is very straightforward and do not need to collect trajectories. For our robot, the raddi R_L and R_R to are both estimated to be about 96 mm, and the wheelbase is about 420 mm long, we are not able to measure the exact laser pose manually because the center of laser can not be determined manually. For Method (2) and Method (3), We performed three different configurations for the laser pose on the same robot, and collect different trajectories for each configuration, such as straight, circle, S-shape, rotating in place and on only one wheel. For Method (2), we combined the trajectories together and feed them to the method. For our method, we first get the odometry model parameters. For calibration of odometry model parameters trajectories with open path and constant-sign curvature are preferable as detailed in [1], so we cut the closed circles into circle segments to avoid compensation of curvatures. We label the three configurations A, B, and C.

Table 1. Calibration results using [8]

	r_L(mm)	r_R(mm)	b(mm)	l_x(mm)	l_y(mm)	l_θ(deg)
A	97.2	97.3	427.7	81.7	−1.1	−1.9
B	96.6	96.8	422.68	90.06	174.39	−57.3
C	94.56	94.9	414.63	52.15	177.91	38.9

$\alpha = 0.005, N = 8.$

Table 2. Calibration results using our method

	r_L(mm)	r_R(mm)	b(mm)
mean	99.2	97.0	0.4299
std	0.0013	0.0013	0.0014

		l_x(mm)	l_y(mm)	l_θ(deg)
A	mean	81.4	2.3	-2.4
	std	0.028	0.017	0.009
B	mean	89.8	172.6	-55.7
	std	0.025	0.014	0.012
C	mean	53.1	178.3	37.2
	std	0.013	0.007	0.021

5.3   Result and Comparisons

For each of configurations A, B, C, we collect multiple trajectories and divided them into 10 subsets in order to calculate mean and standard deviation.

Obviously, manual measurement cannot reach the precision of millimeters and tenths of degrees. As shown in Tables 1 and 2, the results of method 2 and our method are nearly the same, and it's hard to say which is superior, but our method can separate the calibration problems of odometry and laser pose since we know the ground truth of robot pose from MoCap and the odometry model will be not influenced by laser noise model.

6   Conclusions

In this paper, we claim that a general calibration system is urgent for service robots, to address this problem, the proposed platform highly generalizes the calibration procedure of service robots, and we use odometry calibration as a case study to show that our system is effective and promising.

Acknowledgments. This work was supported in part by National Natural Science Foundation of China under grant No. 61603368, the Youth Innovation Promotion Association of CAS (No. 2015373), and Natural Science Foundation of Anhui Province under grant No. 1608085QF134.

References

1. Antonelli, G., Chiaverini, S., Fusco, G.: A calibration method for odometry of mobile robots based on the least-squares technique: theory and experimental validation. IEEE Trans. Rob. **21**(5), 994–1004 (2005)

2. Bai, A., Wu, F., Chen, X.: Towards a principled solution to simulated robot soccer. In: Chen, X., Stone, P., Sucar, L.E., Zant, T. (eds.) RoboCup 2012. LNCS (LNAI), vol. 7500, pp. 141–153. Springer, Heidelberg (2013). doi:10.1007/978-3-642-39250-4_14

3. Bai, A., Wu, F., Chen, X.: Bayesian mixture modelling and inference based thompson sampling in monte-carlo tree search. In: Proceedings of the Advances in Neural Information Processing Systems (NIPS), pp. 1646–1654. Lake Tahoe, United States (2013)

4. Bai, A., Wu, F., Chen, X.: Online planning for large markov decision processes with hierarchical decomposition. ACM Trans. Intell. Syst. Technol. (ACM TIST) **6**(4), 45 (2015)

5. Borenstein, J., Feng, L.: Measurement and correction of systematic odometry errors in mobile robots. IEEE Trans. Rob. Autom. **12**(6), 869–880 (1996)

6. Borm, J.H., Meng, C.H.: Determination of optimal measurement configurations for robot calibration based on observability measure. Int. J. Rob. Res. **10**(1), 51–63 (1991)

7. Carrillo, H., Birbach, O., Taubig, H., Bauml, B., Frese, U., Castellanos, J., et al.: On task-oriented criteria for configurations selection in robot calibration. In: 2013 IEEE International Conference on Robotics and Automation (ICRA), pp. 3653–3659. IEEE (2013)

8. Censi, A., Franchi, A., Marchionni, L., Oriolo, G.: Simultaneous calibration of odometry and sensor parameters for mobile robots. IEEE Trans. Rob. **29**(2), 475–492 (2013)

9. Chen, H., Fuhlbrigge, T., Choi, S., Wang, J., Li, X.: Practical industrial robot zero offset calibration. In: 2008 IEEE International Conference on Automation Science and Engineering, CASE 2008, pp. 516–521. IEEE (2008)

10. Chen, S., Wu, F., Shen, L., Chen, J., Ramchurn, S.D.: Decentralized patrolling under constraints in dynamic environments. IEEE Trans. Cybern. 1–13 (2015)

11. Chen, Y., Wu, F., Shuai, W., Wang, N., Chen, R., Chen, X.: Kejia robot - an attractive shopping mall guider. In: Proceedings of the 7th International Conference on Social Robotics, pp. 145–154 (2015)

12. Chen, Y., Wu, F., Wang, N., Tang, K., Cheng, M., Chen, X.: *KeJia-LC*: a low-cost mobile robot platform — champion of demo challenge on benchmarking service robots at RoboCup 2015. In: Almeida, L., Ji, J., Steinbauer, G., Luke, S. (eds.) RoboCup 2015. LNCS, vol. 9513, pp. 60–71. Springer, Cham (2015). doi:10.1007/978-3-319-29339-4_5

13. Cheng, M., Chen, X., Tang, K., Wu, F., Kupcsik, A., Iocchi, L., Chen, Y., Hsu, D.: Synthetical benchmarking of service robots: a first effort on domestic mobile platforms. In: Almeida, L., Ji, J., Steinbauer, G., Luke, S. (eds.) RoboCup 2015. LNCS (LNAI), vol. 9513, pp. 377–388. Springer, Cham (2015). doi:10.1007/978-3-319-29339-4_32

14. Daniilidis, K.: Hand-eye calibration using dual quaternions. Int. J. Rob. Res. **18**(3), 286–298 (1999)

15. Elatta, A., Gen, L.P., Zhi, F.L., Daoyuan, Y., Fei, L.: An overview of robot calibration. Inf. Technol. J. **3**(1), 74–78 (2004)

16. Hollerbach, J.M., Wampler, C.W.: The calibration index and taxonomy for robot kinematic calibration methods. Int. J. Rob. Res. **15**(6), 573–591 (1996)
17. Lu, D., Zhou, Y., Wu, F., Zhang, Z., Chen, X.: Integrating answer set programming with semantic dictionaries for robot task planning. In: Proceedings of the 26th International Joint Conference on Artificial Intelligence (2017)
18. Maier, D., Wrobel, S., Bennewitz, M.: Whole-body self-calibration via graph-optimization and automatic configuration selection. In: 2015 IEEE International Conference on Robotics and Automation (ICRA), pp. 5662–5668. IEEE (2015)
19. Meeussen, W., Hsu, J., Diankov, R.: Urdf-unified robot description format (2012)
20. Omodei, A., Legnani, G., Adamini, R.: Three methodologies for the calibration of industrial manipulators: experimental results on a scara robot. J. Rob. Syst. **17**(6), 291–307 (2000)
21. Paul, R.P.: Robot Manipulators: Mathematics, Programming, and Control: The Computer Control of Robot Manipulators (1981)
22. Roth, Z., Mooring, B., Ravani, B.: An overview of robot calibration. IEEE J. Rob. Autom. **5**(3), 377–385 (1987)
23. Roy, N., Thrun, S.: Online self-calibration for mobile robots. In: 1999 Proceedings of IEEE International Conference on Robotics and Automation, vol. 3, pp. 2292–2297. IEEE (1999)
24. Shah, M., Eastman, R.D., Hong, T.: An overview of robot-sensor calibration methods for evaluation of perception systems. In: Proceedings of the Workshop on Performance Metrics for Intelligent Systems, pp. 15–20. ACM (2012)
25. Shiu, Y.C., Ahmad, S.: Calibration of wrist-mounted robotic sensors by solving homogeneous transform equations of the form ax = xb. IEEE Trans. Rob. Autom. **5**(1), 16–29 (1989)
26. Strobl, K.H., Hirzinger, G.: Optimal hand-eye calibration. In: 2006 IEEE/RSJ International Conference on Intelligent Robots and Systems, pp. 4647–4653. IEEE (2006)
27. Whitney, D., Lozinski, C., Rourke, J.M.: Industrial robot forward calibration method and results. J. Dyn. Syst. Meas. Contr. **108**(1), 1–8 (1986)
28. Wu, F., Ramchurn, S., Chen, X.: Coordinating human-UAV teams in disaster response. In: Proceedings of the 25th International Joint Conference on Artificial Intelligence (IJCAI), pp. 524–530 (2016)
29. Zhang, H., Cao, R., Zilberstein, S., Wu, F., Chen, X.: Toward effective soft robot control via reinforcement learning. In: Proceedings of the 10th International Conference on Intelligent Robotics Applications (2017)

Autonomous Navigation Control
for Quadrotors in Trajectories Tracking

Wilbert G. Aguilar[1,2,3(✉)], Cecilio Angulo[3],
and Ramón Costa-Castello[4]

[1] Dep. DECEM, Universidad de las Fuerzas Armadas ESPE,
Sangolquí, Ecuador
wgaguilar@espe.edu.ec
[2] CICTE Research Center, Universidad de las Fuerzas Armadas ESPE,
Sangolquí, Ecuador
[3] GREC Research Group, Universitat Politècnica de Catalunya,
Barcelona, Spain
[4] IOC Research Institute, Universitat Politècnica de Catalunya, Barcelona, Spain

Abstract. In this paper, we describes a novel proposal for the autonomous navigation control of quadrotor micro aerial vehicles for trajectories tracking in the XY plane. The quadrotor vehicle is an AR.Drone 1.0 from the company Parrot with a nonlinear behavior. The proposal includes system modeling, controller design, planning and simulation of the results. In our approach, we separate the model into two primary models: A linearity for the steady state and a nonlinearity for the dynamic transition.

Keywords: Automatic control · Systems identification and modeling · Motion planning · Quadrotor

1 Introduction

A Quadrotor is an aerial vehicle with four coplanar rotors which movements depend on the difference between the generated torque of each one of them [1–6]. There are several research line for these platforms, commonly classified in the literature as: Control [7–9], Navigation [10–19] and Guidance [19–30].

This article is focused on control, specifically in the XY plan, where the internal control of motors is considered as a black box [1, 7–11, 31, 32]. Pitch and roll angles are the inputs with values between 0 and 1, and the position and velocity in X and Y are the outputs in the model.

The model that relates angular inputs of pitch and roll with the velocities and positions in the XY plane (Fig. 1) can be estimated based on data given by the CEA [5]. Z axis value will be fixed on a specific altitude. The proposed model is a combination of a static nonlinearity and a dynamic linearity.

The paper has been organized in the following sections: The second chapter describes the estimated model of the quadrotor, including static nonlinearity and dynamic linearity. The third chapter refers to the proposal design of the controller and the planner. Finally, the fourth and fifth chapters respectively present the results and the conclusions.

© Springer International Publishing AG 2017
Y. Huang et al. (Eds.): ICIRA 2017, Part III, LNAI 10464, pp. 287–297, 2017.
DOI: 10.1007/978-3-319-65298-6_27

Fig. 1. Inputs and outputs of the AR.Drone 1.0.

2 Model of the Quadrotor

The estimation of the dynamic movement model of the AR.Drone 1.0, which relates the angle inputs with the velocities and positions, is based on data provided by the Automation Spanish Committee (CEA). The strong relation between the pitch angle with the velocity in X, and between the roll angle and the velocity in Y, is shown in Fig. 2.

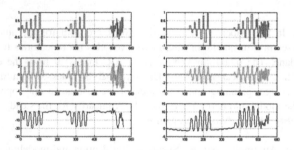

Fig. 2. 1st Right: Pitch. 2nd Right: velocity X. 3rd Right: Position X. 1st Left: Roll. 2nd Left: velocity Y. 3rd Left: Position Y.

Based on data, some hypothesis can be considered for solving the estimation problem:

1. Models for each movement axis can be decoupled and defined by the following relation: Pitch angle with velocity and position in X, Roll angle with velocity and position in Y.
2. The relation between angles and velocities is described by a static nonlinear model combined with a dynamic linear model.
3. The sources of external noise cause the difference between the estimated velocity, based on integrated position data, and the velocity data.

Based on these hypothesis, the model is obtained in 3 steps:

- Estimation of the static nonlinear model.
- Estimation of the dynamic linear model.
- Estimation of the noise sources.

2.1   Estimation of the Static Model

The nonlinearity of the model is estimated as a polynomial that relates the angular inputs of pitch and roll with the stationary values of velocity in X and Y axis, respectively. In this way, there is necessary to define an approximated stationary values of velocities for each angle value as an input. Part of the proposed solution for the model is that the nonlinearity between the values of angle and velocity would be a polynomial. We computed a sixth grade polynomial as a nonlinear estimation:

$$P(q) = aq^6 + bq^5 + cq^4 + dq^3 + eq^2 + fq + g \tag{1}$$

The sixth grade polynomial respect to the real values is shown in Fig. 3.

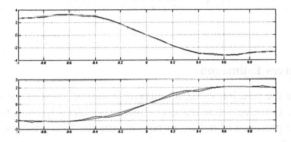

Fig. 3. Polynomial and real data.

There is important to remember that the estimated polynomial represents the relation between angles and the stationary value of velocity. Consequently, the stationary values of velocity are obtained as an output of the polynomial. This data set will be defined as the stationary velocity.

2.2   Dynamic Model Estimation

A dynamic linear model, that relates the polynomial output with the velocity data, can be identified once the nonlinearity is estimated. Figure 4 shows overlapping graphics of stationary velocity and real velocity.

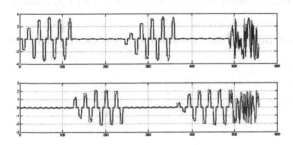

Fig. 4. Stationary velocity and real velocity.

This is achieved using the model identification, and considering:

- The input variable corresponds to the stationary velocity.
- The output variable corresponds to the real velocity.
- The system is modeled as a linear process.
- The transference function is of second order, with two real poles.

Over 90% precision can be achieved using these criteria. The transfer function for Vx as for Vy is:

$$H_x(s) = \frac{4.9286}{s^2 + 4.968s + 5.0553} \tag{2}$$

$$H_y(s) = \frac{3.2228}{s^2 + 3.6765s + 3.3766} \tag{1}$$

2.3   Noise Sources Estimation

In order to obtain the position value, an integrator was added after the transfer function of each velocity. However obtained data differs considerably from the position data. This is due to the presence of noise sources. We use the data provided by the CEA for their identification.

As shown in the third criteria for the model estimation, the difference function (blue function from the first graphic in Figs. 5 and 6) between the estimated velocity, from the integrated velocity data, and the position data, is due to external noise sources.

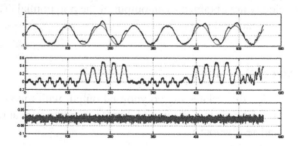

Fig. 5. Noise sources in the X axis. (Color figure online)

This difference function has a wave form that can be estimated as a sinusoidal model:

$$Ruido_1(t) = A \sin(wt) \tag{4}$$

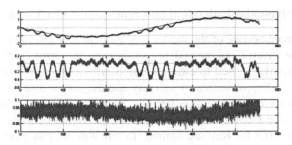

Fig. 6. Noise sources in the Y axis. (Color figure online)

Where A is estimated from the average difference of the function amplitude, and is the sinusoidal angular frequency. The mathematical expressions of first noise source for X and Y are:

$$RuidoX_1(t) = 0.9\sin(0.0800t) \tag{5}$$

$$RuidoY_1(t) = -1.205\sin(0.0101t) \tag{6}$$

There are 3 additional noise sources:

- Noise from velocity in the same axis, estimated from the velocity amplitudes:

$$RuidoX_2(V_X) = -\frac{1}{35}V_X \tag{7}$$

$$RuidoY_2(V_Y) = -\frac{1}{40}V_Y \tag{8}$$

- Noise from the position in the other axis, estimated from the values of position

$$RuidoX_2(X) = \frac{1}{25}X \tag{9}$$

$$RuidoY_2(Y) = \frac{1}{50}Y \tag{10}$$

- Random noise with variance 0.0005 (Figs. 5 and 6).

3 Controller Design

The second chapter presents a decoupled model with high performance according to the criteria of the simulator provided by the CEA (This results are presented in chapter 4). The controller design is reduced to perform three compensations:

- Inverse polynomial.
- Sinusoidal noise compensation.
- Linear model controller.

3.1   Inverse Polynomial Estimation and Sinusoidal Noise Compensation

In Sect. 2.1 the nonlinearity of the model was estimated as a sixth grade polynomial. We calculate the inverse polynomial to compensate the error in the feedback, the adjustment polynomial is used to estimate this polynomial, in this case the input data are the estimated values of stationary velocity and the output data the Pitch reference.

This polynomial is noninvertible because there are different values of stationary velocity for the same Pitch value. For this reason, the values between −0.6 and 0.6 for Pitch and −0.7 and 0.7 for Roll are limited. In this way we can work in the bijective area (see Fig. 7). These values correspond to the maximum and minimum value of stationary velocity achievable for X and Y respectively.

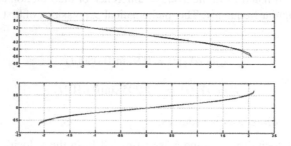

Fig. 7. Inverse polynomial.

The adjustment is obtained with a fifth grade polynomial:

$$P_i(q) = a_i q^5 + b_i q^4 + c_i q^3 + d_i q^2 + e_i q + f_i \tag{11}$$

3.2   Linear Model Controller

Once the inverse polynomial and the sinusoidal noise are compensated, the controller design is based on the feedback model that minimizes the error between the current output and the set point.

The velocity and position must be compensated at the same time locating the poles in the left semi-plane without destabilizing the system (Figs. 8 and 9).

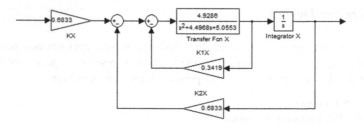

Fig. 8. X Controller.

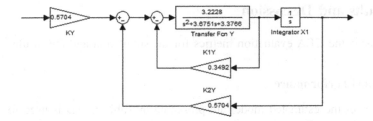

Fig. 9. Y Controller.

Analyzing the feedbacked system we can estimate the values of K1, K2 and K. These second order systems are presented as:

$$H(s) = \frac{A}{s^2 + Bs + C} \tag{12}$$

By feedbacking both compensations in parallel, of K1 and K2 for velocity and position respectively, and multiplying by the gain K, we get:

$$\frac{AK}{s^2 + Bs + (C + K1A)s + K2A} \tag{13}$$

The feedback system form is:

$$\frac{AK}{(s + s1)(s + s2)(s + s3)} \tag{14}$$

$$\frac{AK}{s^3 + (s1 + s2 + s3)s^2 + (s1s2 + s1s3 + s2s3)s + s1s2s3} \tag{15}$$

Solving the equation system:

$$s3 = B - s1 - s2 \tag{16}$$

$$K1 = (s1 * s2 + s2 * s3 + s1 * s3 - C)/A \tag{17}$$

$$K2 = s1 * s2 * s3/A \tag{18}$$

In order to locate roots that are the farthest away from cero in the left semiplane, we define s1 = s2 = s3B/3. To minimize error between the current position and the set point, the gain K = K2.

The gains obtained in X and Y are:

$$K1x = 0.3419; K2x = 0.6833; Kx = 0.6833;$$
$$K1y = 0.3492; K2y = 0.5704; Ky = 0.5704;$$

4   Results and Discussion

We are using the CEA evaluation metrics for the simulation and the results.

4.1   Model Performance

The results of the estimated model are presented in Table 1. AD is used for Highly-Desired, and D for Desired.

Table 1. Model performance

Variable	GPP: 0.90108		
	Value (m)	Grade	Percentage
Dist. X Media	0.06272	AD	31.3598%
Dist. Y Media	0.016007	AD	8.0035%
Dist. X Máx.	0.99527	D	38.0977%
Dist. Y Máx.	0.21274	AD	42.5478%

4.2   Controllers Performance

The controller performance respect to the straight trajectory distance and time is presented in Table 2.

Table 2. Model performance

Variable	GPP: 28.272		
	Value	Grade	Percentage
Medium Dist.	0.07347 m	D	46.9403%
Maximum Dist.	0.20151 m	T	51.5123%
Travel time	45.96 seg	D	84.4636%

An interpolator that generates intermediate points as local commands, where the vehicle must go, is used to reduce error distance from the straight trajectory (see Figs. 10 and 11). There is important to remember that if higher precision is required during the trajectory tracking, the time increases inevitably. For this reason, there is important to look for a balance between time and precision.

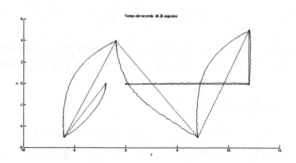

Fig. 10. Route without interpolator.

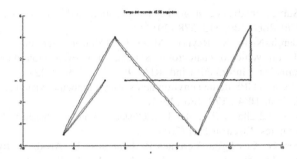

Fig. 11. Route with interpolator.

5   Conclusions and Future Work

After estimating the nonlinear model that relates the input angles Pitch and Roll with velocities and positions in the XY plane, and designing the trajectory tracking controller in the same plane, the conclusions are: The model that relates angles from the quadrotor with translation and velocity in each axis is decoupled, i.e., each axis depends only on one angle. The relations between each angle and the corresponding axis can be estimated as a combination of a static nonlinearity and a dynamic linearity. The nonlinearity is fitted with a sixth grade polynomial with high reliability. The linearity corresponds to a second order system of real poles for the velocity and one additional integrator for the position. There are 4 noise sources between the estimated position from the integrated velocity and the measured position: a sinusoid, a source dependent on the velocity in the same axis, a source dependent on the position on the other axis, and random noise. The system can be controlled by an inverse fifth grade polynomial that removes the nonlinearity, the noise source compensation, and a parallel compensator control.

Acknowledgement. This work is part of the projects VisualNavDrone 2016-PIC-024 and MultiNavCar 2016-PIC-025, from the Universidad de las Fuerzas Armadas ESPE, directed by Dr. Wilbert G. Aguilar. A special thanks to the Automation Spanish Committee CEA for providing original data from the AR.Drone 1.0.

References

1. Aguilar, W.G., Angulo, C.: Real-time model-based video stabilization for microaerial vehicles. Neural Process. Lett. **43**(2), 459–477 (2016)
2. Aguilar, W.G., Angulo, C.: Real-time video stabilization without phantom movements for micro aerial vehicles. EURASIP J. Image Video Process. **2014**(1), 46 (2014)
3. Aguilar, W.G., Angulo, C.: Robust video stabilization based on motion intention for low-cost micro aerial vehicles. In: 2014 11th International Multi-Conference on Systems, Signals Devices (SSD), pp. 1–6 (2014)

4. Kendoul, F.: Survey of advances in guidance, navigation, and control of unmanned rotorcraft systems. J. Field Robot. **29**, 315–378 (2012)
5. Blasco, X., García-Nieto, S., Reynoso-Meza, G.: Control autónomo del seguimiento de trayectorias de un vehículo cuatrirrotor Simulación y evaluación de propuestas. Rev. Iberoam. Automática e Informática Ind. RIAI **9**(2), 194–199 (2012)
6. Autonomous flight in GPS-denied environments using monocular vision and inertial sensors. J. Aerosp. Inf. Syst. **10**(4), 172–186 (2013)
7. Aguilar, W.G., Casaliglla, V.P., Pólit, J.L.: Obstacle avoidance based-visual navigation for micro aerial vehicles. Electron **6**(1) (2017)
8. Aguilar, Wilbert G., Casaliglla, Verónica P., Pólit, José L., Abad, V., Ruiz, H.: Obstacle avoidance for flight safety on unmanned aerial vehicles. In: Rojas, I., Joya, G., Catala, A. (eds.) IWANN 2017. LNCS, vol. 10306, pp. 575–584. Springer, Cham (2017). doi:10.1007/978-3-319-59147-6_49
9. Aguilar, W.G., Casaliglla, V.P., Polit, J.L.: Obstacle avoidance for low-cost UAVs. In: Proceedings of the IEEE 11th International Conference on Semantic Computing, ICSC 2017 (2017)
10. Engel, J., Sturm, J., Cremers, D.: Scale-aware navigation of a low-cost quadrocopter with a monocular camera. Rob. Auton. Syst **62**(11), 1646–1656 (2014)
11. Engel, J., Sturm, J., Cremers, D.: Camera-based navigation of a low-cost quadrocopter. In: 2012 IEEE/RSJ International Conference on Intelligent Robots and Systems, pp. 2815–2821 (2012)
12. Pierre-Jean, B.: The navigation and control technology inside the AR.Drone Micro UAV, pp. 1477–1484 (2011)
13. François, P.B., David, C., Jemmapes, D.: The navigation and control technology inside the AR. Drone micro UAV. In: Proceedings of the 18th IFAC World Congress, pp. 1477–1484 (2011)
14. Fernandez, A., Diez, J., de Castro, D., Silva, P.F., Colomina, I., Dovis, F., Friess, P., Wis, M., Lindenberger, J., Fernandez, I.: ATENEA: advanced techniques for deeply integrated GNSS/INS/LiDAR navigation. In: 2010 5th ESA Workshop on Satellite Navigation Technologies and European Workshop on GNSS Signals and Signal Processing (NAVITEC), pp. 1–8 (2010)
15. Aguilar, W.G., Morales, S.G.: 3D environment mapping using the Kinect V2 and path planning based on RRT algorithms. Electron **5**(4) (2016)
16. Aguilar, W.G., Morales, S., Ruiz, H., Abad, V.: RRT* GL based optimal path planning for real-time navigation of UAVs. In: Rojas, I., Joya, G., Catala, A. (eds.) IWANN 2017. LNCS, vol. 10306, pp. 585–595. Springer, Cham (2017). doi:10.1007/978-3-319-59147-6_50
17. Cabras, P., Rosell, J., Pérez, A., Aguilar, W.G., Rosell, A.: Haptic-based navigation for the virtual bronchoscopy. In: IFAC Proceedings Volumes (IFAC-PapersOnline), vol. 18, no. PART 1 (2011)
18. Dong, Z., Zhang, G., Bao, H: Robust monocular SLAM in dynamic environments. In: 2013 IEEE International Symposium on Mixed and Augmented Reality (ISMAR), pp. 209–218 (2013)
19. Geiger, A., Lenz, P., Urtasun, R.: Are we ready for autonomous driving? the KITTI vision benchmark suite. In: Proceedings of the IEEE Computer Society Conference on Computer Vision and Pattern Recognition, pp. 3354–3361 (2012)
20. Liu, S., Yuan, L., Tan, P., Sun, J.: SteadyFlow: spatially smooth optical flow for video stabilization. In: 2014 IEEE Conference on Computer Vision and Pattern Recognition, pp. 4209–4216 (2014)
21. Liu, F., Niu, Y., Jin, H.: Joint subspace stabilization for stereoscopic video. In: 2013 IEEE International Conference on Computer Vision, pp. 73–80 (2013)

22. Grundmann, M.: Computational Video: Post-Processing Methods for Stabilization, Retargeting and Segmentation. Georgia Institute of Technology (2013)
23. Cho, S., Wang, J., Lee, S.: Video deblurring for hand-held cameras using patch-based synthesis. ACM Trans. Graph **31**(4), 1–9 (2012)
24. Benenson, R., Omran, M., Hosang, J., Schiele, B.: Ten years of pedestrian detection, what have we learned? In: Agapito, L., Bronstein, Michael M., Rother, C. (eds.) ECCV 2014. LNCS, vol. 8926, pp. 613–627. Springer, Cham (2015). doi:10.1007/978-3-319-16181-5_47
25. Miksik, O., Mikolajczyk, K.: Evaluation of local detectors and descriptors for fast feature matching. In: 2012 21st International Conference on Pattern Recognition (ICPR), pp. 2681–2684 (2012)
26. Dollár, P., Wojek, C., Schiele, B., Perona, P.: Pedestrian detection: an evaluation of the state of the art. IEEE Trans. Pattern Anal. Mach. Intell **34**(4), 743–761 (2012)
27. Leutenegger, S., Chli, M., Siegwar, R.Y.: BRISK: binary robust invariant scalable keypoints. In: Proceedings of the IEEE International Conference on Computer Vision, pp. 2548–2555 (2011)
28. Rublee, E., Rabaud, V., Konolige, K., Bradski, G: ORB: an efficient alternative to SIFT or SURF. In: Proceedings of the IEEE International Conference on Computer Vision, pp. 2564–2571 (2011)
29. Aguilar, W.G., Luna, M.A., Moya, J.F., Abad, V., Ruiz, H., Parra, H., Angulo, C.: Pedestrian detection for UAVs using cascade classifiers and saliency maps. In: Rojas, I., Joya, G., Catala, A. (eds.) IWANN 2017. LNCS, vol. 10306, pp. 563–574. Springer, Cham (2017). doi:10.1007/978-3-319-59147-6_48
30. Aguilar, W.G., Luna, M.A., Moya, J.F., Abad, V., Parra, H., Ruiz, H.: Pedestrian detection for UAVs using cascade classifiers with meanshift. In: Proceedings of the IEEE 11th International Conference on Semantic Computing, ICSC 2017 (2017)
31. Mourikis, A.I., Roumeliotis, S.I.: A Multi-state constraint kalman filter for vision-aided inertial navigation. In: Proceedings 2007 IEEE International Conference on Robotics and Automation, pp. 3565–3572 (2007)
32. Engel, J., Cremers, D.: Accurate figure flying with a quadrocopter using onboard visual and inertial sensing. In: IMU (2012)

On-Board Visual SLAM on a UGV Using a RGB-D Camera

Wilbert G. Aguilar[1,3,4(✉)], Guillermo A. Rodríguez[2,3], Leandro Álvarez[2,3],
Sebastián Sandoval[2,3], Fernando Quisaguano[2,3], and Alex Limaico[2,3]

[1] Dep. DECEM, Universidad de las Fuerzas Armadas ESPE, Sangolquí, Ecuador
wgaguilar@espe.edu.ec
[2] Dep. DEEE, Universidad de las Fuerzas Armadas ESPE, Sangolquí, Ecuador
[3] CICTE Research Center, Universidad de las Fuerzas Armadas ESPE, Sangolquí, Ecuador
[4] GREC Research Group, Universitat Politècnica de Catalunya, Barcelona, Spain

Abstract. We present a approach to real-time localization and mapping using a RGB-D camera, such as Microsoft Kinect, and a small and powerful computer Intel Stick Core M3 Processor. Our system can run the computation and sensing required for SLAM on-board the UGV, removing the dependence on unreliable wireless communication. We make use of visual odometry, loop closure and graph optimization to achieve this purpose. Our approach is able to perform accurate and efficient on-board SLAM, and we evaluate its performance thoroughly with varying environments and illumination conditions. The experiments demonstrate that our system can robustly deal with difficult data in indoor and outdoor scenarios.

Keywords: SLAM · RGB-D · Loop closure detection · Graph optimization · Visual odometry · RANSAC · UGVs

1 Introduction

To create a robot that could be placed in an unknown environment and obtain the location without any previous knowledge of relative position is one of the most desired goals of mobile robotics [1–3]. A truly autonomous robot would have applications in many fields, such as medicine, military, entertainment and many others. SLAM (Simultaneous Localization and Mapping) is a versatile solution to this problem. Many approaches have been developed over the last decade, with methods such as bundle adjustment [4–6] or filtering-based estimation [7, 8], providing accurate approximation, but giving up important information in the process for obtaining relevant reference points that compromises the accuracy of the entire system when operating in rough conditions.

Interest in RGB-D cameras like Microsoft Kinect has grown in recent years in the robotic community because of their ability to provide real-time color images and depth maps. New approaches for SLAM combine the scale information of 3D depth sensing with the visual information of the cameras to create accurate 3D environment maps, in which a robot could navigate autonomously. Some methods use dense visual odometry [9–11], showing accurate results on the main problematic, but making use of an external computer to process the visual data. This represents a considerable complication when

© Springer International Publishing AG 2017
Y. Huang et al. (Eds.): ICIRA 2017, Part III, LNAI 10464, pp. 298–308, 2017.
DOI: 10.1007/978-3-319-65298-6_28

implementing the methods on UGVs, which cannot support a large amount of weight. This inconvenience can be solved using onboard micro-computers, such as the Intel Stick core M3, which is enough to run all the algorithms needed for SLAM as we demonstrate in Sect. 4.

Our approach consists in an on-board implementation of SLAM for an UGV, using the dense color and depth images obtained from RGB-D cameras [12]. We based our work on contributions made in loop closure detection [13] and graph-based SLAM [14], using the Rtab_map library from ROS as a starting point. Our system is able to perform on-board SLAM on an UGV using an Intel Stick M3 computer, optimizing communication resources and simultaneously processing the algorithms in real-time.

2 Related Works

The SLAM problem has been an important topic in the history of computer vision [15–17] and mobile robotics [18–22]. Visual SLAM systems [23, 24] are used to extract interest points from the camera images [25–27], simplifying data association. For online loop closure detection, bag-of-words [28] approach has been used [29, 30]. The bag-of-words approach consists in representing each image by visual words taken from a vocabulary. Graph pose optimization approaches [31, 32] can be employed to reduce odometry errors using poses and link transformations inside each map and between the maps, optimizing the final results of the system.

Different type of sensors are used to achieve SLAM, including 2D scanners [33, 34], monocular cameras [35, 36], and recently RGB-D sensors such as the Microsft Kinect [37]. SLAM approaches for mobile ground systems include 3D motion estimation to create a map of the environment [38], but are unable to process the information on-board. In [39–42] the autonomous mapping and localization system in outdoor environments performs the hard processing work off-board in a ground station, taking away the real time capabilities from the system.

3 Our Approach

In this article, the basic structure of the map is a graph with nodes and links. These nodes store important information, like the odometry poses for every location taken in the map, the visual information of the RGB, depth images of the Kinect, and the visual words used for loop closure detection. As for the links, they save the rigid geometrical transformations within nodes. Close links are added among the current and the previous nodes with their corresponding odometry transformation. Loop closure links are added when a loop closure is detected within the current node and one from the same or previous map. We use the combined data of visual odometry, loop closure detection [9] and optimization to run on-board real time SLAM in an Intel Stick M3 approach and an UGV. The Fig. 1 shows the implementation of our on the UGV.

Fig. 1. Implementation of the system on the UGV.

3.1 Visual Odometry

We use visual odometry to estimate the trajectory of the RGB-D sensor from a specific region within the image. This method is useful for the pose estimation problem which will be linked directly to the measurements given by the RGB-D sensor via a non-linear model. This model is responsible for the 3D geometric configuration of the current environment. The Fig. 1 shows the visual odometry performed by the system. For a robust visual odometry model, the approach from [43] is used. This method defines a RGB-D sensor with a color brightness function $I(p, t)$ and a depth function $D(p, t)$, where $p = (u, v)$ are pixel coordinates within the image acquired at time t. After defining a series of motion models and mathematical transformations, a non-linear least square cost function

$$C(x) = \sum_{P^* \in \mathcal{R}^*} \left(\mathcal{I}\left(w\left(P^*; T(x)\hat{T}\right)\right) - \mathcal{I}^*(P^*) \right)^2 \tag{1}$$

is obtained, where $P^* = \{p, D\} \in \mathbb{R}^{nx3}$ and are the 3D points associated with the depth image and the image points p, the current image $\mathcal{I} = \{I, D\}$ which is the set containing both brightness and depth, w the motion model that defines the 3D geometric deformation of a structured light RGB-D camera, $T(x)$ the incremental pose to be estimated, $\hat{T}$ the estimated pose of the current image and $\mathcal{I}^*$ the reference image. By minimizing the cost function (1), the pose and trajectory of the camera can be estimated. The minimization algorithm from [43] estimates these parameters (Fig. 2).

Fig. 2. Visual odometry performed by the system

3.2 Loop Closure Detection

Our approach uses a Bayesian filter to evaluate loop closure hypotheses over all previous images, based on the method described in [13]. The loop closure detector uses a bag-of-words (visual words, which are SURF features quantized to an incremental visual dictionary) approach to determinate the likelihood that a new image comes from a previous or new location. When a loop closure hypothesis is accepted, a new constraint is added to the map graph, following by a graph optimizer that minimizes the errors in the map. A loop closure is detected when a pre-defined threshold H is reached by the loop closure hypothesis. We use the map memory management approach of [14] to limit the number of locations used for loop closure detection and graph optimization algorithms, thus respecting the real-time limitations on large-scale environments.

The visual words are extracted from the RGB image. This image is registered with the depth image, knowing that a 3D position can be computed using the calibration matrix and the depth information provided by the depth image for each point in the RGB image. The RANSAC algorithm [44] uses the 3D visual word matches to compute the rigid transformation between the corresponding images when a loop closure is detected. The loop closure is accepted and a link with this transformation between the loop closure hypothesis node and the current node is added to the graph, only in the case of a minimum of I inliers are found (Fig. 3).

Fig. 3. Loop closure detection performed by the system

3.3 Graph Optimization

We use the tree based parametrization [19] to describe an efficient configuration of the nodes in the graph. It can construct a graph with the given trajectory of the UAV. The pose and link transformations are used as the limitations. Errors produced by the visual odometry estimation can be propagated to all the links when a loop closure is found, and correcting the map at the same time. In order to decrease the computational cost of the algorithms, we are not using the approach proposed in [13]. Instead, we will use a more straightforward method, using the tree based parametrization algorithm [19] to create a tree from the map graph with only one map. By this procedure the tree of the algorithm will only have one root, removing the requirement of a robust memory management for the system.

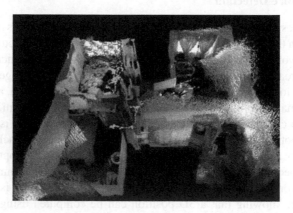

Fig. 4. Results of the graph optimization performed by the system.

4 Results and Discussion

The metric of evaluation used to obtain the error on the distance estimation is the percent error of measurements taken at different instances, defined as follows:

$$Error\% = \frac{|Experimental\ value - Reference\ value|}{Reference\ value} * 100\% \qquad (2)$$

For obtaining the error on the pose estimation, the absolute error was used and defined as follows:

$$Error\% = |Experimental\ value - Reference\ value| \qquad (3)$$

We test the difference in the pose referred to an initial reference point taking two different measurements, one before (Initial) and another after (Final) performing the mapping trajectory. We tested the system in two different trajectories, a linear and a close-loop trajectory. The poses were measured in meters.

4.1   Linear Trajectory

The first trajectory is a 4,50 m straight line, as shown in Fig. 5. This location was used to test the mapping capabilities of the system in a linear trajectory. As mentioned earlier, we used a reference point to obtain the values of the poses before and after the mapping process.

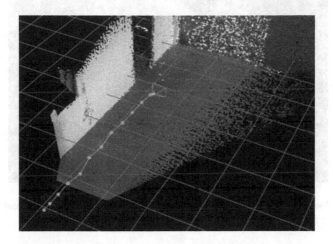

Fig. 5. Linear trajectory performed by the system. The teal lines represent the trajectory estimated by the system.

To obtain the values of the poses we took the average value of five measurements acquired by our system. The results of these measurements and the error calculated is shown in the following table.

Table 1. Error in the distance estimation of a linear trajectory

Real (m)	Estimated (m)	Error (%)
4,50	4,571	1.577

In Table 1, we can appreciate that the error in the measurements of the pose is less than 2%. The error in the measurements is low, and the mapping was performed with success, showing the robustness of our system for linear trajectories.

4.2   Close-Loop Trajectory

The next trajectory performed is a 4,50 m × 2,30 m square, as shown in Fig. 6. Similar to the last scenery, this location was used to test the mapping capabilities of the system in a close loop trajectory.

Fig. 6. Close loop trajectory performed by the system. The teal lines represent the trajectory estimated by the system.

Same as the last trajectory, we took the average value of five measurements of pose acquired by our system. The following table shows the error in the distance estimation in the furthest corner of the trajectory.

In Table 2, we can appreciate that the error between the estimated distance and the real distance is less than 3%. This shows the robustness of our system in closed trajectories. The table that follows shows the error in the pose after performing the trajectory.

Table 2. Error in the distance estimation of a closed loop trajectory

Axis	Real (m)	Estimated (m)	Error (%)
X	4,50	4,567	1,48
Y	2,30	2,349	2,13

Table 3. Error in the pose estimation of a closed loop trajectory

Pose	Before mapping	After mapping	Difference in measurements
X	0.0505983	−0.253923	0,3045213
Y	0.0353084	−0.141439	0,1767474
Z	−0.0508969	−0.461427	0,4105301
Roll	−0.027007	−0.103477	0,07647
Pitch	−0.0331638	0.0484087	−0,0815725
Yaw	−0.141548	−0.0844838	−0,0570642

In Table 3, we can observe that the error in the measurements of the pose is less than 5 cm, showing that the mapping capabilities of the system are strong.

4.3   Distance Estimation Error

To estimate the error in the distance measured by the device, we take a similar approach to the previous tests. Based on a known distance to a reference point, we calculate the error at different instances of the mapping process. The results are shown in Table 4. The measured error is 2.506% after mapping. This shows that our system is accurate enough to perform SLAM on board.

Table 4.   Distance taken at different times and the error in the measurements

Real distance (m)	Distance before mapping (m)	% Error before mapping	Distance after mapping (m)	% Error after mapping
2,139	2,153	0,650	2,194	2,506

4.4   CPU Usage and Power Consumption

Using the tools provided by the operating system, we measure the use of each CPU from the computer, which is shown in Table 5.

Table 5.   CPU usage

CPU core number	No processing (%)	Performing SLAM (%)	Increment (%)
1	9,4	38,4	29
2	6,0	43,8	37,8
3	9,9	35,0	25,1
4	12,0	44,2	32,2

We carried out the same procedure to obtain the power consumption of the system before and after mapping, shown in Table 6. Even though there is notable increase in the CPU usage and power consumption, the computer works better than expected and is able to run real-time SLAM without lagging.

Table 6.   Power consumption

No processing (W)	Performing SLAM (W)	Increment (W)
5,48	8,20	2,72

5   Conclusions

Our proposal for on-board SLAM can estimate the UGV position with high accuracy and run on an Intel Stick core M3 computer in real time, removing the dependence on unreliable wireless communication. As shown in Fig. 6, our system can perform mapping in real time with a small error in the distance, as shown in Table 1. The system does not behave in the same way while doing the mapping with a weaker light source. In order to obtain the best results from the system, the mapping needs to be performed in the day.

Our approach uses the loop closure detection algorithm to relate reference points in different instances in time. We have shown that the system is capable of updating the map in order to adapt to the environment, which is suitable for UGV applications.

Acknowledgement. This work is part of the projects 2016-PIC-024 and 2016-PIC-025, from the Universidad de las Fuerzas Armadas ESPE, directed by Dr. Wilbert G. Aguilar.

References

1. Aguilar, W.G., Verónica, C., José, P.: Obstacle avoidance based-visual navigation for micro aerial vehicles. Electronics **6**(1), 10 (2017)
2. Aguilar, W.G., Casaliglla, V.P., Pólit, J.L., Abad, V., Ruiz, H.: Obstacle avoidance for flight safety on unmanned aerial vehicles. In: Rojas, I., Joya, G., Catala, A. (eds.) IWANN 2017, Part II. LNCS, vol. 10306, pp. 575–584. Springer, Cham (2017). doi:10.1007/978-3-319-59147-6_49
3. Aguilar, W.G., Verónica, C., José, P.: Obstacle avoidance for low-cost UAVs. In: IEEE 11th International Conference on Semantic Computing (ICSC), San Diego (2017)
4. Huang, A.S., Bachrach, A., Henry, P., Krainin, M., Maturana, D., Fox, D., Roy, N.: Visual odometry and mapping for autonomous flight using an RGB-D camera. In: International Symposium on Robotics Research (ISRR) (2011)
5. Mur-Artal, R., Montiel, J., Tardo, J.: ORB-SLAM: a versatile and accurate monocular SLAM system. IEEE Trans. Robot. **31**(5), 1147–1163 (2015)
6. Triggs, B., McLauchlan, P.F., Hartley, R.I., Fitzgibbon, A.W.: Bundle adjustment — a modern synthesis. In: Triggs, B., Zisserman, A., Szeliski, R. (eds.) IWVA 1999. LNCS, vol. 1883, pp. 298–372. Springer, Heidelberg (2000). doi:10.1007/3-540-44480-7_21
7. Engel, J., Schöps, T., Cremers, D.: LSD-SLAM: large-scale direct monocular SLAM. In: Fleet, D., Pajdla, T., Schiele, B., Tuytelaars, T. (eds.) ECCV 2014, Part II. LNCS, vol. 8690, pp. 834–849. Springer, Cham (2014). doi:10.1007/978-3-319-10605-2_54
8. Kohlbrecher, S., von Stryk, O., Meyer, J., Klingauf, U.: A flexible and scalable SLAM system with full 3D motion estimation. In: IEEE International Symposium on Safety, Security, and Rescue Robotics (2011)
9. Whelan, T., Kaess, M., Leonard, J., McDonald, J.: Deformation based loop closure for large scale dense RGB-D SLAM. In: IEEE/RSJ International Conference on Intelligent Robots and Systems (2013)
10. Nister, D., Naroditsky, O., Bergen, J.: Visual odometry. In: Computer Vision and Pattern Recognition, pp. 652–659 (2004)
11. Konolige, K., Agrawal, M., Sola, J.: Large-scale visual odometry for rough terrain. In: International Symposium Robotics Research (2007)
12. Kerl, C., Sturm, J., Cremers, D.: Dense visual slam for RGB-D cameras. In: Proceedings of the International Conference on Intelligent Robot Systems (IROS) (2013)
13. Michaud, M., Labbe, F.: Appearance-based loop closure detection for online large-scale and long-term operation. IEEE Trans. Robot. **29**(3), 734–745 (2013)
14. Michaud, M., Labbe, F.: Online global loop closure detection for large-scale multi-session graph-based SLAM. In: Proceedings of the IEEE/RSJ International Conference on Intelligent Robots and Systems, pp. 2661–2666 (2014)

15. Aguilar, W.G., Luna, M.A., Moya, J.F., Abad, V., Ruiz, H., Parra, H., Angulo, C.: Pedestrian detection for UAVs using cascade classifiers and saliency maps. In: Rojas, I., Joya, G., Catala, A. (eds.) IWANN 2017, Part II. LNCS, vol. 10306, pp. 563–574. Springer, Cham (2017). doi: 10.1007/978-3-319-59147-6_48

16. Aguilar, W.G., Luna, M., Moya, J., Abad, V., Parra, H., Ruiz, H.: Pedestrian detection for UAVs using cascade classifiers with meanshift. In: IEEE 11th International Conference on Semantic Computing (ICSC), San Diego (2017)

17. Stühmer, J., Gumhold, S., Cremers, D.: Real-time dense geometry from a handheld camera. In: Goesele, M., Roth, S., Kuijper, A., Schiele, B., Schindler, K. (eds.) DAGM 2010. LNCS, vol. 6376, pp. 11–20. Springer, Heidelberg (2010). doi:10.1007/978-3-642-15986-2_2

18. Thrun, S.: Robotic mapping: a survey. In: Exploring Artificial Intelligence in the New Millennium (2003)

19. Grisetti, G., Grzonka, S., Stachniss, C., Pfaff, P., Burgard, W.: Efficient estimation of accurate maximum likelihood maps in 3D. In: Proceedings of the International Conference on Intelligent Robots and Systems (IROS) (2007)

20. Dellaert, F.: Square root SAM. In: Proceedings of the Robotics: Science and Systems (RSS), pp. 177–184 (2005)

21. Kaess, M., Ranganathan, A., Dellaert, F.: iSAM: incremental smoothing and mapping. IEEE Trans. Robot. **24**(6), 1365–1378 (2008)

22. Jin, H., Favaro, P., Soatto, S.: Real-time 3-D motion and structure of point features: front-end system for vision-based control and interaction. In: IEEE Conference on Computer Vision and Pattern Recognition (CVPR) (2000)

23. Murray, G., Klein, D.: Parallel tracking and mapping for small AR workspaces. In: Proceedings of the IEEE and ACM International Symposium on Mixed and Augmented Reality (ISMAR) (2007)

24. Strasdat, H., Montiel, J.M., Davison, A.: Scale drift-aware large scale monocular SLAM. In: Proceedings of Robotics: Science and Systems (2010)

25. Aguilar, W.G., Angulo, C.: Real-time model-based video stabilization for microaerial vehicles. Neural Process. Lett. **43**(2), 459–477 (2016)

26. Aguilar, W.G., Angulo, C.: Real-time video stabilization without phantom movements for micro aerial vehicles. EURASIP J. Image Video Process. **1**, 1–13 (2014)

27. Aguilar, W.G., Angulo, C.: Robust video stabilization based on motion intention for low-cost micro aerial vehicles. In: 11th International Multi-Conference on Systems, Signals & Devices (SSD), Barcelona, Spain (2014)

28. Zisserman, J., Sivic, A.: Video google: a text retrieval approach to object matching in videos. In: Proceedings of the 9th International Conference on Computer Vision, pp. 1470–1478 (2003)

29. Botterill, T., Mills, S., Green, R.: Bag-of-words-driven, single-camera simultaneous localization and mapping. J. Field Robot. **28**(2), 204–226 (2011)

30. Konolige, K., Bowman, J., Chen, J., Mihelich, P., Calonder, M., Lepetit, V., Fua, P.: View-based maps. Int. J. Robot. Res. **29**(8), 941–957 (2010)

31. Christensen, H.I., Folkesson, J.: Closing the loop with graphical SLAM. IEEE Trans. Robot. **23**(4), 731–741 (2007)

32. Johannsson, H., Kaess, M., Fallon, M., Leonard, J.J.: Temporally scalable visual SLAM using a reduced pose graph. In: RSS Workshop on Long-term Operation of Autonomous Robotic Systems in Changing Environments (2012)

33. Montemerlo, M., Thrun, S., Koller, D., Wegbreit, B.: FastSLAM: a factored solution to the simultaneous localization and mapping problem. In: Proceedings of the National Conference on Artificial Intelligence (AAAI) (2012)

34. Grisetti, G., Stachniss, C., Burgard, W.: Improved techniques for grid mapping with rao-blackwellized particle filters. IEEE Trans. Robot. (T-RO) **23**, 34–46 (2007)
35. Weiss, S., Scaramuzza, D., Siegwart, R.: Monocular-SLAM-based navigation for autonomous micro helicopters in GPS-denied environments. J. Field Robot. **28**(6), 854–874 (2011)
36. Tardos, R., Mur-Artal, J.D.: Visual-inertial monocular SLAM with map reuse. IEEE Robot. Autom. Lett. **2**, 796–803 (2016)
37. Engelhard, N., Endres, F., Hess, J., Sturm, J., Burgard, W.: Realtime 3D visual SLAM with a hand-held RGB-D camera. In: RGB-D Workshop on 3D Perception in Robotics at the European Robotics Forum (2011)
38. Kohlbrecher, S., Meyer, J., Von Stryk, O., Klingauf, U.: A flexible and scalable SLAM system with full 3D motion estimation. In: Proceedings of the IEEE International Symposium on Safety, Security, and Rescue Robotics, Kyoto (2011)
39. Fraundorfer, F., Heng, L., Honegger, D., Lee, G., Meier, L., Tanskanen, P., Pollefeys, M.: Vision-based autonomous mapping and exploration using a quadrotor MAV. In: Intelligent Robots and Systems (IROS) (2012)
40. Aguilar, W.G., Morales, S.: 3D environment mapping using the kinect V2 and path planning based on RRT algorithms. Electronics **5**(4), 70 (2016)
41. Aguilar, W.G., Morales, S., Ruiz, H., Abad, V.: RRT* GL based optimal path planning for real-time navigation of UAVs. In: Rojas, I., Joya, G., Catala, A. (eds.) IWANN 2017, Part II. LNCS, vol. 10306, pp. 585–595. Springer, Cham (2017). doi:10.1007/978-3-319-59147-6_50
42. Cabras, P., Rosell, J., Pérez, A., Aguilar, W.G., Rosell, A.: Haptic-based navigation for the virtual bronchoscopy. In: 18th IFAC World Congress, Milano, Italy
43. Audras, C., Comport, A., Meilland, M., Rives, P.: Real-time dense appearance-based SLAM for RGB-D sensors. In: Australasian Conference on Robotics and Automation (2011)
44. Bolles, M.A., Fischler, R.C.: Random sample consensus: a paradigm for model fitting with Apphcatlons to image analysis and automated cartography. Commun. ACM **24**, 381–395 (1981)

Projective Homography Based Uncalibrated Visual Servoing with Path Planning

Zeyu Gong, Bo Tao[⊠], Hua Yang, Zhouping Yin, and Han Ding

State Key Laboratory of Digital Manufacturing Equipment and Technology,
Huazhong University of Science and Technology,
Wuhan, Hubei, People's Republic of China
taobo@hust.edu.cn

Abstract. Path planning is a beneficial strategy in improving the robustness of uncalibrated visual servoing. In this paper, a novel controller based on projective homography along with a new expression of optimal camera path are proposed to accomplish visual servoing tasks under totally uncalibrated situations. The interpolated poses of optimal path is express in the current planned frame, and the expression of optimal path is completely free of camera parameters and homography decomposition. The tracking of planned path is realized in projective homography space directly, based on the novel uncalibrated controller denoted as PHUVS. The simulation results prove the effectiveness of the proposed path optimization method and PHUVS controller.

Keywords: Uncalibrated visual servoing · Path planning · Projective homography

1 Introduction

Visual servoing has been extensively studied in recent years due to the flexibility and robustness it can offer to robot control. From the view of task function, it can be classified into 3 categories: image based visual servoing (IBVS) [1], position based visual servoing (PBVS) [2] and the hybrid one which combines both image information as well as 3D information(HBVS) [3].

Traditional visual servoing often requires accurate model information such as camera intrinsic matrix and hand-eye relationships, which depends on tedious and costly calibration process with certain level of expertise [4]. Hence, tremendous interests are boosted in the so-called uncalibrated visual servoing. Given that IBVS is free of 3D reconstruction, it is more suitable for uncalibrated visual servoing [5, 6]. The main strategy of uncalibrated IBVS is to employ different method to estimate the so-called image Jacobian [5, 7–14].

IBVS suffers from global convergence problems thus is tend to fail when the initial error is large [15]. Moreover, there will be undesirable motion in Euclidian space [16], especially in presence of large rotation around optical axis [17]. A possible solution is using path planning for visual servoing to interpolate images between initial and desired image. Classical path planning for IBVS often requires camera parameters and target model [18–20], thus is not suitable for uncalibrated visual servoing.

© Springer International Publishing AG 2017
Y. Huang et al. (Eds.): ICIRA 2017, Part III, LNAI 10464, pp. 309–319, 2017.
DOI: 10.1007/978-3-319-65298-6_29

Some methods were proposed to relieve the reliance on models of target and camera. A representative strategy is using the scaled Euclidian reconstruction approach to obtain the rotation and the scaled translation motion to interpolate camera path. In [21], optimal path in collineation is obtained by separating translation and rotation displacement in homography. Similarly, robust reconstruction and decomposition of homography or essential matrix is adopt to distinguish rotation and translation in [22–24]. Note that in reconstruction process, the camera parameters are required more or less, and decomposition algorithms usually need additional information (image, e.g.) to eliminate ambiguity. Another strategy is employing additional image(s) to separate rotation and translation component. For eye-in-hand configuration, one additional image is required [25, 26], while two additional images are required for fixed camera configuration [27]. Neither camera parameter nor target model are needed.

The main contribution of this paper lies in two aspects. First, we proposed a new expression of optimal path of camera in projective homography space. The second innovation is a novel tracking controller based on projective homography, denoted as PHUVS.

2 Geometric Modeling and Notations

2.1 Camera Model and Notations

Suppose the target is observed by a camera at two poses. Define one of poses is the desired pose while the other is the initial one. Denote the frame of camera at desired pose as $\mathcal{F}_1$, while the frame at initial pose as $\mathcal{F}_0$. For point $\mathcal{P}_i$ in 3D space, the coordinate with respect to $\mathcal{F}_0$ is $\boldsymbol{\chi}_i^0 = \left[X_i^0\, Y_i^0\, Z_i^0 \right]^T$. Similarly, the coordinate with respect to $\mathcal{F}_1$ is denoted as $\boldsymbol{\chi}_i^1 = \left[X_i^1\, Y_i^1\, Z_i^1 \right]^T$.

$$\boldsymbol{\chi}_i^0 = \mathbf{R}_{01} \boldsymbol{\chi}_i^1 + \mathbf{b}_{01} \tag{1}$$

Where $\mathbf{R}_{01} \in \mathbb{SO}(3)$ is the rotation matrix describing the orientation of $\mathcal{F}_1$ with respect to $\mathcal{F}_0$; $\mathbf{b}_{01}$ is the translation between $\mathcal{F}_1$ and $\mathcal{F}_0$ with respect to $\mathcal{F}_0$. $\mathcal{P}_i$ is projected to $\mathbf{p}_i^1 = \left[u_i^1\, v_i^1\, 1 \right]^T$ in the desired image and $\mathbf{p}_i^0 = \left[u_i^0\, v_i^0\, 1 \right]^T$ in the initial image. The perspective projection can be described by:

$$Z_i^0 \mathbf{p}_i = \mathbf{K} \boldsymbol{\chi}_i^0, \; Z_i^1 \mathbf{p}_i^1 = \mathbf{K} \boldsymbol{\chi}_i^1 \tag{2}$$

where $\mathbf{K}$ is the intrinsic matrix of the camera. For uncalibrated visual servoing tasks, $\mathbf{K}$ is unknown or inaccurate. In this paper, by "uncalibrated" we mean it is totally unknown and is not used in the whole control process.

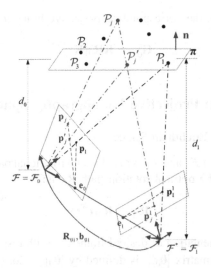

Fig. 1. Geometry of two views

2.2 Projective Homography

Suppose there is a plane π in Euclidean space, as shown in Fig. 1, and for all points on the plane, it is well known there is a Euclidean homography relates the coordinates of the same point in initial and desired frames, which is denoted $\mathbf{H}_{01}$ as below:

$$\chi_i^0 = \underbrace{\left(\mathbf{R}_{01} + \frac{\mathbf{b}_{01}}{d_1} \mathbf{n}_1^T \right)}_{\mathbf{H}_{01}} \chi_i^1 \tag{3}$$

where d_1 is the distance between the plane π and origin of $\mathcal{F}_1$; $\mathbf{n}_1$ is the normal vector of plane π expressed in $\mathcal{F}_1$. Obviously, when the plane is at infinity, i.e. $d_1 \to \infty$, the Euclidean homography at infinity, denoted as $\mathbf{H}_{01}^\infty$, is:

$$\mathbf{H}_{01}^\infty = \mathbf{R}_{01} \tag{4}$$

That means $\mathbf{H}_{01}^\infty$ is only related to the orientation difference between two frames.

Based on (2) and (3), the projective homography matrix related to the projected images is defined by the following equation:

$$\mathbf{p}_i^0 = \underbrace{\frac{Z_i^1}{Z_i^0}}_{\alpha_i} \underbrace{\left(\mathbf{K} \mathbf{H}_{01} \mathbf{K}^{-1} \right)}_{\mathbf{G}_{01}} \mathbf{p}_i^1 \tag{5}$$

For plane at infinity, the corresponding projective homography at infinity is:

$$G_{01}^{\infty} = KR_{01}K^{-1} \tag{6}$$

3 Optimal Path in Projective Homography Space

3.1 Optimal Path in Euclidean Space

The optimal path between $\mathcal{F}_1$ and $\mathcal{F}_0$ is divided into rotational part and translational part. One expression of the optimal rotation path R_{t1} is:

$$R_{t1} = R_{01}e^{[\theta_0]_{\times}t} \tag{7}$$

R_{t1} describes the orientation of desired frame $\mathcal{F}_1$ with respect to the current frame $\mathcal{F}_t$ at time t. The skew matrix $[\theta_0]_{\times}$ is defined by $[\theta_0]_{\times} = \log(R_{01}^T)$. t is a normalized time variable. Another expression of the optimal rotation path is:

$$R_{01} = U\Delta_{01}U^{-1} \tag{8}$$

$$\Delta_{01} = \begin{bmatrix} \cos\theta_0 + i\sin\theta_0 & 0 & 0 \\ 0 & \cos\theta_0 - i\sin\theta_0 & 0 \\ 0 & 0 & 1 \end{bmatrix} \tag{9}$$
$$U = \begin{bmatrix} u_1 & u_2 & u_3 \end{bmatrix}$$

where diagonal element of Δ_{01} is only related to the rotation angle correspond to R_{01}, while U groups the eigenvectors of R_{01} and is only depending on the unit rotation axis u_3 of R_{01} which corresponds to the eigen value 1. When revolve around a fixed axis, the rotation angle changes while axis remain constant, thus the optimal path of R_{t1} is:

$$R_{t1} = U\Delta_{t1}U^{-1} \tag{10}$$

$$\Delta_{t1} = \begin{bmatrix} \cos\theta(t) + i\sin\theta(t) & 0 & 0 \\ 0 & \cos\theta(t) - i\sin\theta(t) & 0 \\ 0 & 0 & 1 \end{bmatrix} \tag{11}$$

$$\theta(t) = (1-t)\theta_0, \quad \begin{cases} \theta(0) = \theta_0 \\ \theta(1) = 0 \end{cases} \tag{12}$$

As for translational part, the optimal path is a straight line segment denoted by b_{01} with respect to $\mathcal{F}_0$ connecting $\mathcal{F}_0$ and $\mathcal{F}_1$. When describing this desired path with respect to $\mathcal{F}_t$, the translation b_{t1} between $\mathcal{F}_t$ and $\mathcal{F}_1$ should be:

$$b_{t1} = R_{t0}((1-t)b_{01}) \tag{13}$$

where $t \in [0, 1]$ is normalized variable of time. $\mathbf{R}_{t0}$ is the rotation matrix between $\mathcal{F}_t$ and $\mathcal{F}_0$. This is slightly different from the results in [21].

3.2 Optimal Path of Projective Homography

Proposition 1. The optimal path of the projective homography between current image and desired image, denoted as $\mathbf{G}_{t1}^p$ is:

$$\mathbf{G}_{t1}^p = \mathbf{G}_{t1}^\infty + (1 - t)\mathbf{G}_{t1}^\infty \left(\mathbf{G}_{01}^\infty\right)^{-1} \left(\mathbf{G}_{01} - \mathbf{G}_{01}^\infty\right) \tag{14}$$

where $\mathbf{G}_{01}^\infty$ is the projective homography at infinity between initial and desired image and $\mathbf{G}_{t1}^\infty$ is projective homography at infinity between current and desired image.

The eigen-decomposition of $\mathbf{G}_{01}^\infty$ is:

$$\mathbf{G}_{01}^\infty = \mathbf{V}\boldsymbol{\Lambda}_{01}\mathbf{V}^{-1},$$

$$\boldsymbol{\Lambda}_{01} = \begin{bmatrix} \cos\theta_0 + i\sin\theta_0 & 0 & 0 \\ 0 & \cos\theta_0 - i\sin\theta_0 & 0 \\ 0 & 0 & 1 \end{bmatrix} \tag{15}$$

where $\mathbf{V}$ and $\boldsymbol{\Lambda}_{01}$ groups eigenvectors and eigenvalues of $\mathbf{G}_{01}^\infty$ respectively. And $\mathbf{G}_{t1}^\infty$ can be obtained by:

$$\mathbf{G}_{t1}^\infty = \mathbf{V}\boldsymbol{\Lambda}_{t1}\mathbf{V}^{-1},$$

$$\boldsymbol{\Lambda}_{t1} = \begin{bmatrix} \cos\theta(t) + i\sin\theta(t) & 0 & 0 \\ 0 & \cos\theta(t) - i\sin\theta(t) & 0 \\ 0 & 0 & 1 \end{bmatrix}, \tag{16}$$

$$\theta(t) = (1 - t)\theta_0$$

Proof. Based on the discussion above, the optimal path should be:

$$\mathbf{G}_{t1}^p = \mathbf{K}\mathbf{R}_{t1}\mathbf{K}^{-1} + \mathbf{K}\mathbf{b}_{t1}\mathbf{n}_1^T\mathbf{K}^{-1}/d_1 \tag{17}$$

$$\mathbf{G}_{01}^\infty = \mathbf{K}\mathbf{U}\boldsymbol{\Delta}_{01}(\mathbf{K}\mathbf{U})^{-1} \tag{18}$$

$$\mathbf{G}_{01}^\infty = \mathbf{V}\boldsymbol{\Lambda}_{01}\mathbf{V}^{-1} \tag{19}$$

Obviously $\boldsymbol{\Lambda}_{01} = \boldsymbol{\Delta}_{01}$ and $\mathbf{V} = \mathbf{K}\mathbf{U}$.

Futhermore, we have

$$\mathbf{G}_{t1}^{\infty} = \mathbf{V}\mathbf{\Lambda}_{t1}\mathbf{V}^{-1} \tag{20}$$

$$\mathbf{K}\mathbf{b}_{t1}\mathbf{n}_1^T\mathbf{K}^{-1}/d_1 = (1-t)(\mathbf{K}\mathbf{R}_{0t}\mathbf{K}^{-1})^{-1}\mathbf{K}\mathbf{b}_{01}\mathbf{n}_1^T\mathbf{K}^{-1}/d_1 \tag{21}$$

Since $\mathbf{R}_{01} = \mathbf{R}_{0t}\mathbf{R}_{t1}$, we can obtain:

$$\mathbf{G}_{01}^{\infty} = \mathbf{G}_{0t}^{\infty}\mathbf{G}_{t1}^{\infty} \tag{22}$$

$$\mathbf{K}\mathbf{b}_{t1}\mathbf{n}_1^T\mathbf{K}^{-1}/d_1 = (1-t)\mathbf{G}_{t1}^{\infty}(\mathbf{G}_{01}^{\infty})^{-1}(\mathbf{G}_{01} - \mathbf{G}_{01}^{\infty}) \tag{23}$$

4 Projective Homography Estimation

From Proposition 1, apparently how to estimate $\mathbf{G}_{01}$ and $\mathbf{G}_{01}^{\infty}$ is the foundation of realizing path optimization in projective homography space.

The estimation of projective homography $\mathbf{G}_{01}$ is mature according to references. For non-planar object, a virtual parallax method could be used if at least 8 points is available [28] (3 points are used to define the plane as shown in Fig. 1). Without additional information, the projective homography can only be estimated up to scale, that is:

$$\mathbf{G}_{01} = \beta\hat{\mathbf{G}}_{01} \tag{24}$$

where β is a scaling factor. And for points on the plane:

$$\mathbf{p}_i^0 = (Z_i^1/Z_i^0)\beta\hat{\mathbf{G}}_{01}\mathbf{p}_i^1 \tag{25}$$

As for $\mathbf{G}_{01}^{\infty}$, given that it is only dependent on rotation matrix. In this paper, the estimation method with an additional image proposed in [26] is adopted. The additional image is easy to acquire. Moreover the depth ratios under $\mathcal{F}_1$ and $\mathcal{F}_0$ can be determined

$$\hat{Z}_i^0 = \zeta Z_i^0, \hat{Z}_i^1 = \zeta Z_i^1, i \in \{1, 2, \dots m\} \tag{26}$$

where $\hat{Z}_i^0$ and $\hat{Z}_i^1$ are the estimated depth of $\mathcal{P}_i$ with respect to $\mathcal{F}_0$ and $\mathcal{F}_1$ respectively. Therefore:

$$\beta = (\mathbf{p}_i^0)_1\hat{Z}_i^0 / (\hat{\mathbf{G}}_{01}\mathbf{p}_i^1)_1\hat{Z}_i^1 \tag{27}$$

Where $(\mathbf{v})_j$ denotes the jth elements of vector $\mathbf{v}$. Then one can obtain $\mathbf{G}_{01}$ with by (24).

5   Uncalibrated Controller Design Based on Projective Homography

In this section, a novel method based on projective homography for uncalibrated visual servoing, denoted as PHUVS, is proposed to realize path tracking.

Define a matrix function to depict the error between current and planned frame:

$$\mathbf{E}(t) = \mathbf{G}_{t1}^{p} - \mathbf{G}_{t1}^{c} \tag{28}$$

The task function can be achieved by concatenating the rows of $\mathbf{E}$.

$$\mathbf{e} = [\,\mathbf{E}_1 \quad \mathbf{E}_2 \quad \mathbf{E}_3\,]^{T} \tag{29}$$

where $\mathbf{E}_i$ is the rows of $\mathbf{E}$.

The derivative of rotation matrix $\mathbf{R}_{t1}^{c}$ is:

$$\dot{\mathbf{R}}_{t1}^{c} = -[\boldsymbol{\omega}]_{\times}\mathbf{R}_{t1}^{c} \tag{30}$$

where $[\boldsymbol{\omega}]_{\times}$ denote skew-symmetric form of angular velocity $\boldsymbol{\omega}$.

The derivative of translation vector b_c is:

$$\dot{\mathbf{b}}_{t1}^{c} = -\,\mathbf{v} - [\boldsymbol{\omega}]_{\times}\mathbf{b}_{t1}^{c} \tag{31}$$

where $\mathbf{v} \in \mathbb{R}^3$ denotes the translational velocity of the camera with respect to $\mathcal{F}_t$

$$-\dot{\mathbf{G}}_{t1}^{c} = \left(\mathbf{K}[\boldsymbol{\omega}]_{\times}\mathbf{K}^{-1}\mathbf{G}_{t1}^{c} + \left(\mathbf{K}\mathbf{v}\mathbf{n}_1^{T}\mathbf{K}^{-1}\right)/d_1\right) \tag{32}$$

Define $\mathbf{g}_c$ by concatenating rows of $-\mathbf{G}_{t1}^{c}$:

$$\mathbf{g}_c = -\left[\,(\mathbf{G}_{t1}^{c})_1 \quad (\mathbf{G}_{t1}^{c})_2 \quad (\mathbf{G}_{t1}^{c})_3\,\right]^{T} \tag{33}$$

where $(\mathbf{G}_{t1}^{c}) = \left[\,(\mathbf{G}_{t1}^{c})_1 \quad (\mathbf{G}_{t1}^{c})_2 \quad (\mathbf{G}_{t1}^{c})_3\,\right]^{T}$.

Then equation relating the derivative of $\mathbf{g}_c$ and the camera velocity is developed:

$$\dot{\mathbf{g}}_c = \mathbf{L}_{gc} \cdot \mathbf{u}_c \tag{34}$$

where $\mathbf{L}_{gc} \in \mathbb{R}^{9 \times 6}$ is the Jacobian matrix and $\mathbf{u}_c = \left[\,v_c^{T} \quad \omega_c^{T}\,\right]^{T} \in \mathbb{R}^{6 \times 1}$ is the velocity of camera.

Furthermore, the hand-eye Jacobian is defined as:

$$\mathbf{u}_c = \mathbf{L}_{cr} \cdot \mathbf{u}_r \tag{35}$$

where $\mathbf{L}_{cr} \in \mathbb{R}^{6\times6}$, $\mathbf{u_r} = \begin{bmatrix} \boldsymbol{v}_r^T & \boldsymbol{\omega}_r^T \end{bmatrix}^T \in \mathbb{R}^{6\times1}$ is the velocity of robot end effector. Thus

$$\dot{\mathbf{g}}_c = \mathbf{L} \cdot \mathbf{u_r} \tag{36}$$

where composite Jacobian $\mathbf{L} = \mathbf{L}_{gc} \cdot \mathbf{L}_{cr} \in \mathbb{R}^{9\times6}$.

The derivative of $\mathbf{G}_{t1}^p$ is independent of the camera's motion:

$$\dot{\mathbf{G}}_{t1}^p = \dot{\mathbf{G}}_{t1}^\infty + \left((1-t)\dot{\mathbf{G}}_{t1}^\infty - \mathbf{G}_{t1}^\infty\right)\left(\mathbf{G}_{01}^\infty\right)^{-1}\left(\mathbf{G}_{01} - \mathbf{G}_{01}^\infty\right) \tag{37}$$

Denote $\mathbf{g}_p$ by concatenating rows of $\mathbf{G}_{t1}^p$:

$$\mathbf{g}_p = \begin{bmatrix} \left(\mathbf{G}_{t1}^p\right)_1 & \left(\mathbf{G}_{t1}^p\right)_2 & \left(\mathbf{G}_{t1}^p\right)_3 \end{bmatrix}^T \tag{38}$$

Obviously, the discrete form of the derivative of task function is:

$$\Delta\mathbf{e}(k) = \mathbf{L}_k\mathbf{u_r}(k)\Delta t + \partial\mathbf{g}_p/\partial t\big|_k\Delta t = \tilde{\mathbf{L}}_k\tilde{\mathbf{u}}_r(k) \tag{39}$$

where k is integer that indicate the kth control period.

Similar to [5], the dynamic tracking controller is designed as:

$$\mathbf{u}_r(k) = -\lambda\left(\hat{\mathbf{L}}_k^T \cdot \hat{\mathbf{L}}_k\right)^{-1} \cdot \hat{\mathbf{L}}_k^T \cdot (\mathbf{e}(k) + \mathbf{e}_t(k)) \tag{40}$$

where λ is a positive gain factor, $\hat{\mathbf{L}}$ is the estimated Jacobian matrix of composite Jacobian, and the term $\mathbf{e}_t(k)$ is the prediction of task function variation.

In this paper, the Broyden-Gauss-Newton (BGN) method in state space [5] is chosen to realize the Jacobian estimation.

6 Simulation Verifications

6.1 System Description

A PUMA 560 manipulator equipped with an in-hand camera is simulated using a pervasively applied robotics toolbox in MATLAB by Corke [29] to verify the proposed mothed. The in-hand camera is a simulated acA1300-60gc camera. The imaging procedure is simulated using Epipolar Geometry Toolbox by Mariottini [30].

To simulate a real uncalibrated application environment, the camera frame do not coincide with the frame on the end effector of PUMA 560 manipulator, as shown in Fig. 2(a). In controlling process, the camera-manipulator configuration is totally unknown. A 3D target with at least 8 feature points is adopted in this paper. Three of them are chosen to define a plane, which is depicted as a triangle in Fig. 2(b).

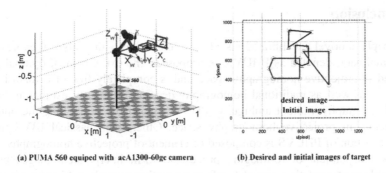

(a) PUMA 560 equiped with acA1300-60gc camera   (b) Desired and initial images of target

Fig. 2. System description

6.2 Simulation of Large Initial Error

To better illustrate the performance of the proposed path optimization method as well as the tracking controller, a challenging case with large rotation angle around optical axis as shown in Table 1, is implemented. It is widely recognized to be problematic for visual servo tasks. The results of PHUVS-P are quite remarkable, as shown in Fig. 3. Besides the satisfactory path in Euclidean space, the trajectory of in joint space is quite smooth as shown in bottom middle of Fig. 3.

Table 1. Large initial error of camera pose with respect to desired frame

Translation/mm	$\Delta t_x = -400$	$\Delta t_y = -550$	$\Delta t_z = -300$
Rotation/degree	$rot_x = 20$	$rot_y = 20$	$rot_z = 150$

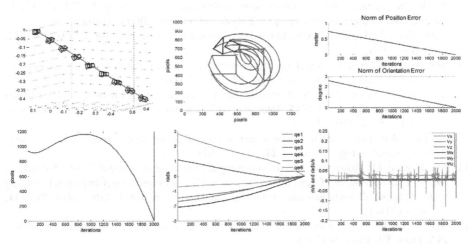

Fig. 3. Results of large initial error case. (top left) camera trajectory. (top middle) image paths of feature points. (top right) the evolution of camera pose error norm. (bottom left) the evolution of image error's norm. (bottom middle) evolution of joint errors. (bottom right) evolution of camera velocities

7 Conclusion

In this paper, a novel uncalibrated visual servo controller with analytical expression of the optimal path, denoted as PHUVS, is proposed. The expression of optimal path is described in projective homography space and is completely free of camera intrinsic parameters as well as traditional homography decomposition. By estimating the projective homography without ambiguity, PHUVS is capable of tracking the optimized path directly in projective homography space. Other than classical IBVS method, the task function of PHUVS is composed of element of projective homography, thus is not only independent of re-imaging process to form interpolated images but also controllable in computation cost of Jacobian estimation procedure. Simulations of large initial error are conducted under eye-in-hand camera-robot configuration, and the results reveal the effectiveness of the proposed method.

References

1. Espiau, B., Chaumette, F., Rives, P.: A new approach to visual servoing in robotics. IEEE Trans. Robot. Autom. **8**, 313–326 (1992)
2. Wilson, W.J., Hulls, C.C.W., Bell, G.S.: Relative end-effector control using cartesian position based visual servoing. IEEE Trans. Robot. Autom. **12**, 684–696 (1996)
3. Malis, E., Chaumette, F., Boudet, S.: 2-1/2-D visual servoing. IEEE Trans. Robot. Autom. **15**, 238–250 (1999)
4. Hua, C.C., Wang, Y.Q., Guan, X.P.: Visual tracking control for an uncalibrated robot system with unknown camera parameters. Robot. Comput. Integr. Manufact. **30**, 19–24 (2014)
5. Hao, M., Sun, Z.Q.: A universal state-space approach to uncalibrated model-free visual servoing. IEEE-ASME Trans. Mechatron. **17**, 833–846 (2012)
6. Shademan, A., Jagersand, M.: Three-view uncalibrated visual servoing. In: IEEE/RSJ 2010 International Conference on Intelligent Robots and Systems (IROS 2010) (2010)
7. Piepmeier, J.A., Lipkin, H.: Uncalibrated eye-in-hand visual servoing. Int. J. Robot. Res. **22**, 805–819 (2003)
8. Sebastian, J.M., Pari, L., Angel, L., Traslosheros, A.: Uncalibrated visual servoing using the fundamental matrix. Robot. Auton. Syst. **57**, 1–10 (2009)
9. Shademan, A., Farahmand, A., Jägersand, M.: Robust Jacobian estimation for uncalibrated visual servoing. In: 2010 IEEE International Conference on Robotics and Automation (ICRA), pp. 5564–5569 (2010)
10. Kosmopoulos, D.I.: Robust Jacobian matrix estimation for image-based visual servoing. Robot. Comput. Integr. Manufact. **27**, 82–87 (2011)
11. Ma, J., Zhao, Q.: Robot visual servo with fuzzy particle filter. J. Comput. **7**, 842–845 (2012)
12. Music, J., Bonkovic, M., Cecic, M.: Comparison of uncalibrated model-free visual servoing methods for small-amplitude movements: a simulation study. Int. J. Adv. Robot. Syst. **11**, 187–194 (2014)
13. Ma, Z., Su, J.: Robust uncalibrated visual servoing control based on disturbance observer. ISA Trans. **59**, 193–204 (2015)
14. Pari, L., Sebastian, J.M., Angel, L.: A comparative study between analytic and estimated image jacobian by using a stereoscopic system of cameras. In: IEEE/RSJ 2010 International Conference on Intelligent Robots and Systems (IROS 2010), pp. 6208–6215 (2010)

15. Chaumette, F.: Potential problems of stability and convergence in image-based and position-based visual servoing. In: The Confluence of Vision and Control, pp. 66–78. Springer (1998)
16. Kazemi, M., Gupta, K., Mehrandezh, M.: Path-planning for visual servoing: a review and issues. In: Visual Servoing via Advanced Numerical Methods, pp. 189–207. Springer, Heidelberg (2010)
17. Corke, P.I., Hutchinson, S.A.: A new partitioned approach to image-based visual servo control. IEEE Trans. Robot. Autom. **17**, 507–515 (2001)
18. Kyrki, V., Kragic, D., Christensen, H.: New shortest-path approaches to visual servoing. In: 2004 IEEE/RSJ International Conference on Intelligent Robots and Systems, (IROS 2004), Proceedings, pp. 349–354 (2004)
19. Hayet, J.B., Esteves, C., Murrieta-Cid, R.: A motion planner for maintaining landmark visibility with a differential drive robot. Algorithmic Found. Robot. Viii **57**, 333–347 (2010)
20. Allotta, B., Fioravanti, D.: 3D motion planning for image-based visual servoing tasks. In: 2005 IEEE International Conference on Robotics and Automation (ICRA), vol. 1–4, pp. 2173–2178 (2005)
21. Mezouar, Y., Chaumette, F.: Optimal camera trajectory with image-based control. Int. J. Robot. Res. **22**, 781–803 (2003)
22. Chesi, G., Hung, Y.S.: Global path-planning for constrained and optimal visual servoing. IEEE Trans. Robot. **23**, 1050–1060 (2007)
23. Chesi, G., Prattichizzo, D., Vicino, A.: Straight line path-planning in visual servoing. J. Dyn. Syst. Meas. Control Trans. ASME **129**, 541–543 (2007)
24. Chesi, G.: Visual servoing path planning via homogeneous forms and LMI optimizations. IEEE Trans. Robot. **25**, 281–291 (2009)
25. Schramm, F., Micaelli, A., Morel, G.: Calibration free path planning for visual servoing yielding straight line behaviour both in image and work space. In: 2005 IEEE/RSJ International Conference on Intelligent Robots and Systems, vol. 1–4, pp. 2688–2693 (2005)
26. Schramm, F., Morel, G.: Ensuring visibility in calibration-free path planning for image-based visual servoing. IEEE Trans. Robot. **22**, 848–854 (2006)
27. Liang, X.-W., Huang, X.-H., Min, W.: Uncalibrated path planning in the image space for the fixed camera configuration. Acta Automatica Sinica **39**, 759–769 (2013)
28. Malis, E., Chaumette, F.: 2 1/2 D visual servoing with respect to unknown objects through a new estimation scheme of camera displacement. Int. J. Comput. Vis. **37**, 79–97 (2000)
29. Corke, P.: MATLAB toolboxes: robotics and vision for students and teachers. IEEE Robot. Autom. Mag. **14**, 16–17 (2007)
30. Mariottini, G.L., Prattichizzo, D.: EGT for multiple view geometry and visual servoing: robotics vision with pinhole and panoramic cameras. IEEE Robot. Autom. Mag. **12**, 26–39 (2005)

A Fully Cloud-Based Modular Home Service Robot

Yili Wang, Naichen Wang, Zhihao Chen, and Wenbo Chen[✉]

Shanghai Institute of Technology, No. 100, Haiquan Road,
Fengxian District, Shanghai, China
wyl_stloy@hotmail.com, wangroger0801@outlook.com,
remakejobs@163.com, chenwenbo860@126.com

Abstract. This paper proposed a Fully Cloud-based Modular (FCM) Robot technology that enables almost all of the business logic program of the home service robot running on the cloud and the robot execute the order locally on control aspects. Moreover, its modular feature can reduce the cost of the robot and, meanwhile, provide a flexibility of customization, upgrades, maintaining. The traditional home service robot often uses cloud computing algorithms and cloudDB, but the basic calculations have been solved locally, which leads to a condition that the robot have to carry many additional facilities. Some concepts of robotic intelligence system have been put forward in the paper including: integrating single-board computers and microcontrollers as the distributed computing unit on the robot, a cloud server structure for FCM robot, finite-state machines and network protocol for solving common problems caused by fully depending on the cloud service. Finally, we built a home service robot following the concept of FCM technology. Comparing with traditional service robots, several experiments are conducted to verify the performance of the FCM technology.

Keywords: Cloud robotics · Modular robot · Intelligent algorithm · Cloud computing

1 Introduction

Internet has become more and more popular in the world, Internet of things, cloud computing, artificial intelligence and other technologies are developing rapidly. Our inspiration is derived from the existing robot theories as follows. (1) Network Function Virtualization (NFV), has been proposed and become the main method of virtualization services and the key to the technologies of internet. This technology has many advantages, such as reducing hardware costs, reallocating network resources, flexible network deployment and energy saving. In addition, Service Function Chaining (SFC) technology makes different traffic work-flows traverse different network functions in some specific order to provide different levels of service in data centers. If we combine these two services with the two core algorithms which are called Multi-layer Worst-Fit (MWF) and Multilayer Best Fit (MBF), the experimental results show that MWF can reduce bandwidth consumption by 15% while only increase the number of used servers by 1% compared to the traditional Best-fit algorithm. (2) We not only

© Springer International Publishing AG 2017
Y. Huang et al. (Eds.): ICIRA 2017, Part III, LNAI 10464, pp. 320–334, 2017.
DOI: 10.1007/978-3-319-65298-6_30

apply cloud computing technology to home service robots for enhancing its additional computing power and more advanced features, but also put forward the concept of modular home service robot. In this way, all the function terminals are modularized so that they can be executed individually or work together to meet the needs of the users. The modular design can also be easy to maintain and upgrade in the future.

Although the existing cloud services are fully developed, the common service robots are using the local computer to transfer and access the cloud data, a lot of computational tasks are still handled on the local computer. On the other hand, modular home service robots are rarely to be seen in daily life. Although the related technology was quite developed, few manufacturers are willing to customize the robots.

In order to solve these tough problems, the technology which we put forward in this paper focuses on the Fully Cloud-based Modular (FCM) home service robot. Compared with traditional cloud service robots or modular robots, this idea emphasizes that all the data collected by the sensors will be uploaded to the cloud, the cloud computer will use the well-developed algorithms to process these data, then return the instructions to the robot so that it can assign the instructions to the components. In this way, the cost of the robot can be reduced and the cycle efficiency of robot self-learning can be increased. On the other hand, the method of modular can make the robot be customized in order to meet the needs of users. The basic structure of the home service robot described in this paper is similar to the existing robots, but its control system is based on the technology of cloud computing. Our home service robot can not only maintain the original function, but also reduce the computing pressure to the local hardware, extending the life time of the robot. This solution also simplifies robot maintenance and reduces upgrade costs. It can also solve some bottlenecks of the extensive use of home service robots i.e. robot pricing, computing pressure, upgrade etc.

This paper is divided into four chapters, the second chapter introduces the principles of robots, logic layer, state machine, and hardware. The third chapter uses this design idea, select the appropriate hardware to establish the entity and compare with the traditional home service robot, which shows the advantages of FCM home service robot. Finally, we analysis the experiment and the result, planning the future development direction of the idea.

2 Design Concept and Logic Structure

2.1 Design Concept

According to the International Federation of Robotics (IFR) for robotic classification, robots are roughly divided into three categories: industrial robot, personal/home service robots and special service robots. The development history of robots can be divided into four categories: mechanical robots, electromechanical robots, program control robots and intelligent robots. A complete robot should consist of four parts: mechanical systems, sensor systems, actuators and control systems. The mechanical system is the main structure just like the skeleton of human, allowing the home service robot to move freely and smoothly. The sensor system is used to obtain the information of the surroundings around the robot then understand the pose and location information of the home service robot itself. Actuators are used to make the corresponding action. The main actuators are as following: robot arms, steering gears, DC motors, etc. The most important part of the robot is the control system, like the brain of human. The information processing and decision-making ability is the main task of the control system, it can understand the order and control the actuators to solve various tasks autonomously. The common robotic control system is mostly integrated in the local hardware, which require the expensive hardware and powerful information processing capabilities. All the factors make the home service robot are not easy to be maintained and upgraded in the future.

This paper focuses on the FCM home service robot. In order to ensure that the robot have a swift response and follow the orders from the master correctly, all the data include the navigation data, object recognizing, the biological information would be collected from local sensor, then the raspberry pi will deliver the data to the cloud server which runs a variety of optimization algorithms. The FCM home service robot is also the direction of development in the future and the key to popularize home service robot.

2.2 Software Structure

The software structure of the service robot could be split into three layers: the conversation layer aims at the communication between human and robot and the data flow from components to communication model, forward to the cloud service; logic layer is designed as the model of finite state machine following the concept of automation; modular layer, which provided the independence and flexibility of components. According to this model, it could provide a robust and flexible structure between human, robot and the cloud server (Fig. 1).

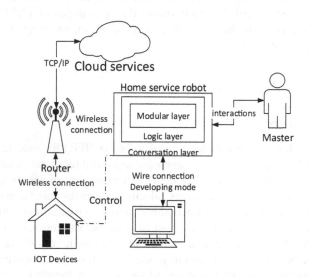

Fig. 1. Logic structure and the communication with real world.

The cloud servers could be private Cloud Computing Platform or any Cloud Computing Operators in the market i.e. Ali Cloud, Microsoft Azure, AWS services. The way of dominating the FCMs are various. For example, ROS (Robot Operating System), which provides libraries and tools to help software developers create robot applications, needs to be deployed on the cloud servers for deciding the action for the FCM robots. The format of the cloud servers can be visual machines or Hadoop Cluster. The interaction processing like face recognition, object identification could be completed via cognitive services e.g. Microsoft cognitive services, Face++, Iflytek. Additionally, a user interface for user to control the robot also can be an application on smart-phones or on websites. The connection through these cloud services can be linked by SFC technology (Fig. 2).

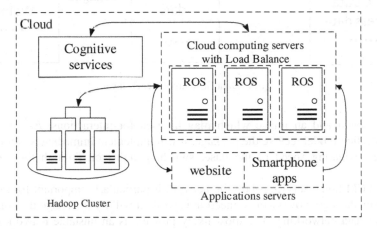

Fig. 2. The Cloud server structure designed for FCM robots.

Conversation Layer The human-machine interaction nowadays is not limited to key board or touching buttons. By the deep learning and artificial intelligence technology growth, talking and using body language is becoming a new way of contacting with machines. On the view of human, the start of the conversation is defined as a long signal which could be shaking hands or calling robot's name, or, it could be assigned by the cloud sever triggered from a specific timer or the message from users' online commands. On the view of machine, each time, the robot sends data packages to the cloud server then wait for a reply from the cloud service, meanwhile, it is the cloud that decided the start of the conversation. The core protocol the robot used to communicate with the service is TCP/IP communication. The communication flow on the view is illustrated as the follow figure (Fig. 3).

Each package sent to the cloud contains, at least, the identification code of the robot which should be per-registered in the cloud, the state of the components and the current data received from its sensors. If some kinds of the data need a span to collect, like a voice record, it will be sent as soon as the data is completely prepared by the sensor.

The received package from the service includes but not limited to the switching command to the sensors, the motion command to the moveable components, and media

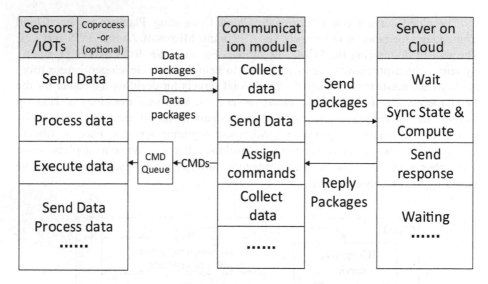

Fig. 3. Data flow among each layer.

data like sound and an image which may be used for interactions. As soon as the communication model receive the package, it will implement immediately, for example changing navigating speed of the base, switching lights on through the local area network.

As a FCM home service robot, the network is particularly important for executing tasks, when the signal of Wi-Fi connection is weak, it will try to change the connection to cellular data networking. The most likely problem is an unstable network connection. During these interval, the data package sent would not get any response, meanwhile, if the robot is navigating, serving, even pouring water, it wouldn't know what should do next. This could be the biggest challenge of the robust of the fully cloud based robot. The solution to this situation is that the cloud service has to send additional commands to the related components that what to do if there is some problem, and these commands will be saved locally in the CMD queue.

Logic Layer: Finite-State Machines Finite-state machines (FSM) is a mathematical model of computation which is maturely designed in software or engineering area. And it implemented widely in program like games, AI, computer system, ROS and intelligent machines like drones, cars and PLC controllers on robots. In this case, the logic layer of the FCM modular home service robot is modeled as an automation-like machine. On the cloud side, multiple approaches can be programmed depending on the business and tasks. In this paper, the whole FSM can be divided into six states and illustrated by the UML state machine in Fig. 4. Mention that an approach of optimizing the network flow consuming is put forward: the LFCS (low frequency communication state) and the HFCS (high frequency communication mode).

The FSM is described as the follows:

(1) Closing state means that all the components and system are closed and nothing can be operated through network. The moveable components like robot arm or wheels are not energized.
(2) During starting process, all the components will start self-checking and report the state. Thereafter, calibration will be executed if needed.
(3) Sleeping state is the main sensors and motion components are disabled while the network part is working. During this state, master can wake the robot not only by the button on the robot but through the network. However, there will probably be a delay depending on the communication frequency.
(4) Low frequency state is a way of communication that the robot maintaining short communication at each time interval with the cloud service. Each data package sent from robot includes the robot current inner state, recent camera frames, laser data and, if installed, the state of other sensors.
(5) High frequency connection state. As soon as the robot start making conversation with people or processing tasks, robot should change to high frequency connection state which is able to send and receive correct data package to/from the cloud service in a high speed. In this case we test our robot, its theoretical download speed is and upload speed is under a frequency of 30fps.

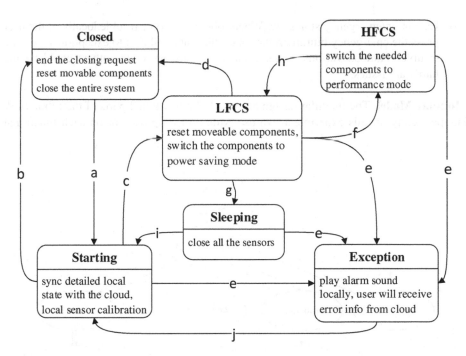

Fig. 4. State diagram of the logic layer. The detailed descriptions in the box represent for the actions the robot would done when entering the state.

Table 1. State diagram of the logic layer.

	Closed	Starting	LFCS	HFCS	Sleeping	Exception
Closed	0	a	0	0	0	0
Starting	b	0	c	0	0	E
LFCS	d	0	0	f	g	E
HFCS	0	0	h	0	0	E
Sleeping	0	i	0	0	0	E
Exception	0	j	0	0	0	0

(a) Closed - > Starting: Power button pushed (open the robot)
(b) Start - > Closed: Power Button Pushed (close the robot)
(c) Starting - > LFCS: initial process finished
(d) LFCS - > Close: instructions
(e) Exception/emergency stopped: when error occurred or emergency button pushed
(f) LFCS - > HFCS: instructions detected
(g) LFCS - > Sleeping: Sleep instructions received
(h) HFCS - > LFCS: Task finished
(i) Sleeping - > Starting: wake up instruction received
(j) Exception - > Starting: Problem solved

(6) Exception/emergency stop state. When robot meet some trouble locally or the user press the emergency button on the robot, the state will turn to exception state. The trouble could be various, some could be a network issues others could be the hardware issues.

Modular Model The modular design allows all the functional parts of the robot work independently. In this experiment, we use Arduino as a processor on each functional

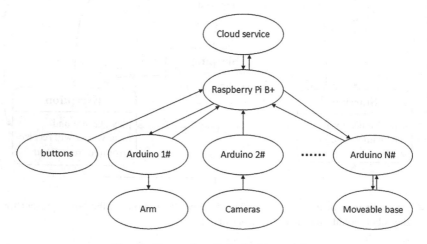

Fig. 5. Connections linked with modular.

terminal, the responsibility of it is giving orders to actuators and collecting data to the Raspberry Pi. Arduino and Raspberry Pi uses serial communication or Electric potential to connect with each other. Raspberry Pi and cloud server uses TCP/IP protocol, through the Wi-Fi to transport data. With the help of Arduino, the functional terminal can not only receive the Electric potential signals, analog signals, strings which sent by the sensors, but also to operating steering gear, motor and other actuators. The I/O pin on Raspberry Pi have been appointed in advanced, every functional terminal has its own pin. This is convenient for users to build up and install extra functional terminals (Fig. 5).

Hardware Structure Different kinds of functional terminal, such as arm, camera, voice sensor, laser sensor and move base make up the whole hardware part of the robot. An Arduino embedded devices has been installed on every functional terminal which makes it can be used independently. Furthermore, it is precisely because of this feature, we can add more functional terminal or remove some of it to meet the specific requirements from customers, reducing unnecessary hardware maintenance and costs as well.

Raspberry Pi B+ has been used as the core device, Raspberry Pi B+ equipped with Broadcom BCM2836 processor, there are four Cortex-A7 architecture cores, clocked at 900 MHz. Raspberry Pi B+ through the serial port or network port to connect with the underlying hardware, such as Arduino or any other IOT hardware. Its responsibility is collecting data from different kinds of installed functional terminals with the data of string and packing to the cloud server, which realizes the advantages of modular feature and cloud computing.

The minimum configuration of the home service robot includes the following components: two or three Maxon DC motors (with different encoder 500 cpr), two DC motor drives, two drive wheels, universal wheel, a 24v battery, power module, Raspberry Pi B+ , Arduino and Hokuyo URG-04LX laser. The approximate cost of the model based on the minimum configuration is about 750 dollars. The cost of the more expensive model could be about 1500 dollars.

3 Performance

Following the theory of the FCM modular home service Robot, we built a service robot to test the performance. Figure 6 shows the appearance of the robot. The moveable module of robot was a chassis with electric motor, three Omni-directional wheels, and speedometers. The Hokuyo laser sensor URG-04LX with a wide-range of 5600 mm 240 is in the front of the robot above the chassis. The trunk is a lift platform with an Arduino controller is inside the chassis. On the top of the robot a digital 720 P camera with a microphone. We chose a Raspberry Pi B+ running Linux for the control model. Microsoft Azure is a growing collection of integrated cloud services that contains: infrastructure as a service (IaaS), platform as a service (PaaS), and Software as a service (SaaS). In our case, we chose Azure Cloud Service IaaS: Virtual Machines running Linux Ubuntu. So far, we tested the data transfer frequency, accuracy benchmark, navigation test and interaction test.

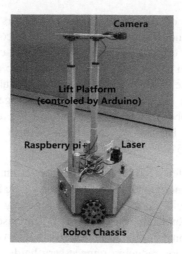

Fig. 6. The appearance of the robot.

3.1 Data Transfer Frequency Test

To evaluate the max data transforming rate in different data size between one services to one service robot, a test that using TCP protocol parallelly under Python had been made. By changing the size of the data package sent from the Raspberry Pi to the Azure Cloud service, we measured the average transfer rate in 1000 rounds of connections. Note that the max rate is matured under light inner computation and few executing statements. Figure 7 illustrates the transfer rate. The max data package size is 65535 bits and min size is 32bit.

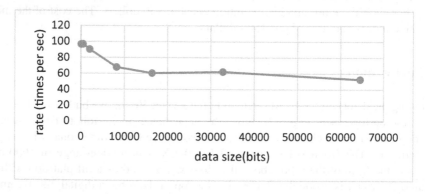

Fig. 7. Communication rate vs data size rate (times per sec).

The minimum rate is above 70 times per seconds which means this structure can satisfy the needs of real-time controlling communication.

3.2 Accuracy Benchmark

The accuracy test aims at testing the navigation accuracy by navigating to specific waypoints (the data is shown in Table 1). In the field, robot is required to reach each of these 8 waypoints in a specific order and the data we collected in the Tables 1 and 2 is the average of our multiple measurements (Figs. 8 and 9).

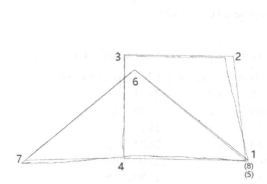

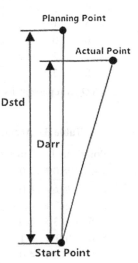

Fig. 8. Path recoded of two benchmark tests number 1 to 8 stand for target points sorted by order.

Fig. 9. Test value diagram.

The accuracy Acc is measured by the deviation is distance between the standard waypoints d_{std} and the arrival waypoints d_{arr1} and calculated by Eq. 1.

$$Acc = \frac{d_{std} - d_{arr}}{d_{std}} \times 100\%$$ (1)

In order to avoid errors created by the angle of robot in the measured values, we collect the vertical distance from the actual point to the planning point for measuring its accuracy.

The time consuming t is one of the benchmark evaluation. The accuracy to time ratio can evaluate the efficiency E of navigation function.

$$E = \frac{acc}{t}$$ (2)

The cloud server continued receiving the mileage from the robot and computed then assigned the next command. The local code based robot realized the aforementioned actions on a laptop. As the data shows in Tables 1 and 2, it illustrates the

Table 2. Accuracy results between two approaches of controlling robot.

Point	D_{arr1} (cm)	D_{arr2} (cm)	Points	D_{std} (cm)	$Acc1$	$Acc2$	Avg Acc1
1	0	0	/	/	/	/	
2	3	4	1–2	150	98.0%	97.3%	98.96%
3	2	2	2–3	150	98.6%	98.6%	
4	1	1.7	3–4	150	99.3%	98.8%	Avg Acc2
5	1	1	4–5	150	99.3%	99.3%	
6	0	0	5–6	170	100%	100%	98.94%
7	3	2	6–7	170	98.2%	98.8%	
8	2	1	7–8	250	99.2%	99.6%	/

(D_{arr1} represent local robot, D_{arr2} represent FCM robot.)

Table 3. Accuracy results between two approaches of controlling robot.

Point	$Acc1$	Time1(sec)	E 1	$Acc2$(sec)	Time2(sec)	E2	Avg Efficiency1
1–2	98.0%	3.9	25.12	97.3%	4.0	24.33	
2–3	98.6%	4.0	24.66	98.6%	4.2	23.49	22.58
3–4	99.3%	4.3	23.10	98.8%	4.1	24.11	
4–5	99.3%	4.0	24.83	99.3%	4.1	24.22	Avg
5–6	100%	4.1	24.39	100%	4.0	25.00	Efficiency2
6–7	98.2%	4.0	24.55	98.8%	4.0	24.70	22.53
7–8	99.2%	8.7	11.40	99.6%	8.4	11.85	

(E1 represent local robot, E2 represent FCM robot.)

average accuracy and efficiency between the FCM robot and local robot are almost the same (Table 3).

3.3 Navigation Test

Navigation is a basic function of service robot. To prove the feasibility of the FCM robot, a navigation test had been conducted. The test field was a closed area (5.4 meters in length and 3 meters in width). The field was arranged with serval random obstacles. The red arrows in Fig. 10 are indicated as the current direction of the robot and the combination of all the red arrows expresses the route plan. The robot needed to navigate from one side to the opposite side avoiding the obstacles. When the robot finishes its task, it will record the time of the whole test.

The algorithms we use for navigation is Dynamic Window Approach. On the cloud service, the Ubuntu Virtual machine was running the ROS platform as the controller of the FCM robot.

And the local code based robot had the same algorithms and platform as the FCM robot. Because of the characteristics of the ROS navigation system, even the robot in the same field, the time of the route planning will be little different according to the

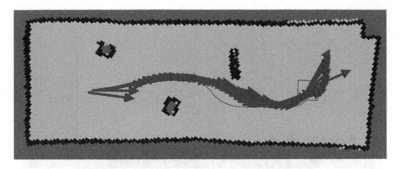

Fig. 10. Navigation path of the FCM robot.

Table 4. Navigation time executed by two kinds of robots.

Number of tests	FCM tech robot	Local code based robot
1	19.4 s	17.8 s
2	18.5 s	17.5 s
3	18.4 s	18.1 s
4	19.8 s	17.9 s
5	17.8 s	18.5 s
Average time	18.78 s	17.96 s

PC's computing power and Data transmission. So, we put two kinds of robots in the same field to do the same navigation test and the time has been recorded in Table 4.

Both of two robots can complete the test without crash. The test results by two robots is similar in path but the FCM robot took a little bit longer time in finishing the navigation. If using more powerful computer in the cloud, providing FCM services for hundreds of robots at the same moment, the robots based on the FCM technology will have more advantages and better efficiency in the future.

3.4   Interaction Test

The interaction test included the object detection test and speaking and listening tests. In daily life, vision and listening is the most common interaction among people, it is important that the robot can understand people then giving them a reply and can tell the category and name of an object correctly and efficiency. We used the computer vision model implemented under OpenCV and text to speech, speech to text model provided by Microsoft cognitive service testing the performance of robot interaction with the world and human under fully could-based service robot. The test sample include two sections, (1) weather report, (2) object recognition. The entrance of the task is that people ask the robot "what's the weather today?" and "what is it?" just like daily talking with others (Fig. 11).

Fig. 11. Object detection test – detected.

Because the tests are not very complicated, FCM tech robot has a bit advantage in response time (Table 5).

Table 5. Interaction test time executed by two kinds of robots.

Number of tests	FCM tech robot	Local code based robot
1	1.27 s	1.35 s
2	1.35 s	1.46 s
3	1.31 s	1.29 s
4	1.41 s	1.45 s
5	1.37 s	1.43 s
Average time	1.342 s	1.396 s

3.5 Experimental Summary

Based on the experiment we did all above, we make a summary table for comparison. The table is as follows (Table 6).

Table 6. Comparison of the test results.

	Core hardware resources	Accuracy	Efficiency	Navigation	Interaction
FCM robot	Raspberry Pi+, Arduino	98.94%	22.53	18.78 s	1.342 s
Local robot	Two Laptop, FPGA, MCU	98.96%	22.58	17.96 s	1.396 s

According to the data in the table, we can know that FCM robot in these tests can fully qualified for our released task. On the other hand, FCM technology robot use less core hardware resources, which leads to longer battery life and cheaper prices. More complex tasks and multiple robots control will reflect more advantages of the FCM technology in the future.

4   Conclusion and Future Work

We propose the technology of Fully Cloud-based Modular (FCM) home service robot in this paper, which is a forward-looking design of home service robot. The common home service robot usually uses local computer as a core to calculate and analyze the data and the local computer usually needs to be powerful and requires many unnecessary functional modules, such as high-capacity batteries, strong carrying capacity robot chassis, etc. Therefore, the Fully Cloud-based Modular (FCM) technology has been put forward. In order to prove the feasibility of this design, we have chosen the suitable hardware to do a lot of tests. It is found form the results of the tests that the FCM robot have completed all the tests. The performance is similar to the traditional robot. If the tasks are arranged more complex such as cooperation of multiple robots, difficult operations, our FCM robot will be much more better and, meanwhile, reducing more costs. Furthermore, modular customization can cater the requirements of the target user perfectly and minimize the robot as much as possible, it also can provide the possibility to the future upgrades. Not only home service robot can use the FCM technology but other kinds of moveable robot platform can do as well, we will focus on developing and promoting FCM technology and implement it on other robot platforms.

References

1. Han., B., Gopalakrishnan, V., Ji, L.S., Lee, S.J.: Network function virtualization: challenges and opportunities for innovations. IEEE Commun. Mag. **53**(2), 90–97 (2015)
2. Battula, L.R.: Network security function virtualization (NSFV) towards Cloud computing with NFV Over Openflow infrastructure: Challenges and novel approaches. In: International Conference on Advances in Computing, Communications and Informatics, pp. 1622–1628. IEEE (2014)
3. Hsieh, C.H., Chung, J.W., Chen, C., Lu, S.H.: Network-aware service function chaining placement in a data center. In: 2016 18th Asia-Pacific Network Operations and Management Symposium (APNOMS), Kanazawa, pp. 1–6 (2016)
4. Li, C.Y., Hsu, C.C., Wang, W.Y., Chien, Y.H., Li, I.H.: Cloud computing based localization for mobile robot systems. In: 2014 CACS International Automatic Control Conference (CACS 2014), Kaohsiung, pp. 238–242 (2014)
5. Chen, Y., Du, Z., García-Acosta, M.: Robot as a service in cloud computing. In: 2010 Fifth IEEE International Symposium on Service Oriented System Engineering, Nanjing, pp. 151–158 (2010)
6. Salmeron-Garcia, J., Inigo-Blasco, P., Diaz-del-Rio, F., Cagigas-Muniz, D.: Mobile robot motion planning based on Cloud Computing stereo vision processing. In: 41st International Symposium on Robotics, ISR/Robotik 2014, Munich, Germany, pp. 1–6 (2014)
7. Ahn, H.S., Beak, Y.M., Sa, I.-K., Kang, W.S., Na, J.H., Choi, J.Y.: Design of reconfigurable heterogeneous modular architecture for service robots. In: 2008 IEEE/RSJ International Conference on Intelligent Robots and Systems, Nice, pp. 1313–1318 (2008)
8. Zhang, Y.-C.: Discussion on robot and its classification from the angle of standardization. Home Appliances Technol. **6**, 30–33 (2016)
9. Wang, Y.-Q.: Application of humanoid mobile service robot. J. Inner Mongolia Univ. Nationalities (Nat. Sci. Ed.)

10. ROS wiki page, http://wiki.ros.org/ROS/, Accessed 10 Apr 2017
11. Risoldi, M., Amaral, V., Barroca, B., Bazargan, K., Buchs, D., Cretton, F., Falquet, G., Calvé, A., Malandain, S., Zoss, P.: A language and a methodology for prototyping user interfaces for control systems. In: Lalanne, D., Kohlas, J. (eds.) Human Machine Interaction. LNCS, vol. 5440, pp. 221–248. Springer, Heidelberg (2009). doi:10.1007/978-3-642-00437-7_9
12. Phillips, R., Sutherland, H.: Performance evaluation of alternative robot controller architectures using a finite-state machine emulator. In: Proceedings of IEEE International Conference on Robotics and Automation, pp. 638–643. IEEE (1985)
13. Run-time Verification of Regularly Expressed Behavioral Properties in Robotic Systems with Logic-Labeled Finite State Machines
14. Azure Homepage, http://azure.microsoft.com, Accessed 10 Apr 2017
15. Fox, D., Burgard, W., Thrun, S.: The dynamic window approach to collision avoidance. IEEE Robot. Autom. Mag. 4(1), 23–33 (1997)

People Tracking in Unknown Environment Based on Particle Filter and Social Force Model

Yang Wang$^{(\boxtimes)}$, Wanmi Chen, and Yifan Luo

School of Mechatronic Engineering and Automation,
Shanghai University, Shanghai 200072, China
wy7ang@163.com, wanmic@163.com

Abstract. In this paper, we introduce a novel scheme for tracking moving person based on particle filter and social force model. The tracking process contains two parts: the predict model and the decision model. We adopt the particle filter algorithm to predict the position and velocity of human. According to the result of prediction, we adapt a sophisticated motion model to calculate the value of social force. Finally, we can control the velocity of robot dynamically through the value of social force.

Keywords: Human tracking · Particle filter · Social force model

1 Introduction

Human localization and tracking has been extensively studied in plenty of field in the past years. Many researchers have developed different algorithms and models to study this problem. The tracking processing can be generally divided into three steps: (1) distinguish between human and non-human object. (2) verify the distance between human and robot, and the position of human. (3) compute the preferred velocity and update the information.

In [1], there is a method to predict the velocity and position of human by using more sophisticated motion models. While the literature [2] proposes a method for real-time tracking people by using multiple laser range-finders. In [3], it presents a novel system for tracking people in an open space by employing multiple single-row laser scanners and one video camera. In [4], it proposes mixture particle filter combined with KLD sampling method to track human dynamically. Similarly, particle filter is generally applying in the prediction of human position and state before tracking [5, 6].

Based on the above discussions, we attempt to combine particle filter algorithm with social force model in human tracking which can make this process more stable and faster. As the researches did in [7], to increase the robustness and accuracy in moving object tracking, particle filter algorithm can select necessary particle information between two adjacent frames through the histogram and other key factors. As pointed out in [8], particle filters may perform poorly when the posterior is multi-modal as the result of ambiguities or multiple targets, it proposes to use a cascaded Adaboost algorithm to guide particle filter.

© Springer International Publishing AG 2017
Y. Huang et al. (Eds.): ICIRA 2017, Part III, LNAI 10464, pp. 335–342, 2017.
DOI: 10.1007/978-3-319-65298-6_31

In this paper, we adopt particle filter to predict the distance between human and robot, use Social force module to compute the preferred force to control the robot for tracking.

2 Particle Filter and Action Model

2.1 Particle Filter

In standard particle filtering [7], we approximate the posterior $P(x_t|z_{0:t})$ with a Dirac measure using a set of N particles $\{x_t^i\}i = 1, ...N$. To accomplish this, we sample candidate particles from an appropriate proposal distribution $x_t^i \sim q(x_t|x_{0:t-1}, z_{0:t})$ [8], and weight these particles according to the following importance ratio:

$$\omega_k^i \approx \omega_{k-1}^i \frac{p\left(z_k|x_k^i\right)p(x_k^i|x_{k-1}^i)}{q(x_k|x_{k-1}^i, z_k)} \tag{1}$$

We resample the particles by using their importance weights to generate an unweight approximation of $P(X_t|Y_{0:t})$.

The particles are used to obtain the approximation [9] of the posterior distribution as following:

$$p(x_t|z_{1:t}) \approx \sum_i^N \omega^{(i)} \delta_{x^{(i)}}(x_t) \tag{2}$$

$\delta_{x^{(i)}}(x_t)$ denotes the vector($[x^{(i)}, y^{(i)}]$)of every particles, $\omega^{(i)}$ is the weight of every particles.

In [10], it introduces a strategy by adopting Kullback-Leibler divergence [11] (KLD) sampling which makes the mixture particle filter less intensive. Suppose that we have two distributions p and q, KLD is defined as:

$$KLD(p, q) = \sum_x p(x) \log\left(\frac{p(x)}{q(x)}\right) \tag{3}$$

KLD is always positive and zero if the distributions are identical. Equation (4) gives the number [12] of particles n that guarantees with probability $1 - \delta$ that KLD is less than e.

$$n = \frac{k-1}{2e}\left\{1 - \frac{2}{9(k-1)} + \sqrt{\frac{2}{9(k-1)}}z_{1-\delta}\right\} \tag{4}$$

Where $z_{1-\delta}$ is the upper $1 - \delta$ quantile of the standard normal distribution and k is the number of bins of the posterior density estimate with support.

2.2   Action Module

The action module means the system state transition, the transform process of particles. Particle transmission is a kind of random motion process, obey the first order ARP [13] equation:

$$X_t = AX_{t-1} + BW_{t-1} + C \tag{5}$$

X_t denotes the object status in the t moment, W_{t-1} is the normalized noise. A, B and C are the constants. State transition process is not relative to the current moment of observation in t moment.

The action status parameters of particle N_i are given by:

$$P^i_{x_t} = A_i P^i_{x_{t-1}} + B_1 W^i_{t-1} + C \tag{6}$$

$$P^i_{y_t} = A_i P^i_{y_{t-1}} + B_1 W^i_{t-1} + C \tag{7}$$

Where A_1, A_2, B_1, B_2 are constants, normally, A is 1, B is the radius of particle propagation, W is random value in $[-1, 1]$.

2.3   Social Force Model

The social forces model for pedestrian dynamics [14] is generally used for evacuation dynamic [15] and directing crowd in complex environment [16]. In [17], human body could be positioning accurately through the particle filter and the social force model. In this scheme, we consider that such force also exists between the robot and the tracked human body. In addition, this force F can be decomposed into two parts: a repulsive force component F^{rep} that prevents the agent from colliding with human body in the environment and an attractive force F^{att} which can control the agent tracking the human [16]. In [18], it is found that the agent can track human and keep a desired distance through the concept of social force model. The formula of social force is given by:

$$F = m\frac{dv}{dt} = F^{att} + F^{rep} \tag{8}$$

Where:

$$F^{att} = m\frac{v^{pre} - v(t)}{\tau} \tag{9}$$

$$F^{rep} = A exp\left[\frac{r_{ij} - d_{ij}}{B}\right] n_{ij} \tag{10}$$

Here, A and B are constants, m denotes the mass of the robot, $v(t)$ means the actual velocity and v^{pre} means the prediction velocity of human. τ represents the certain characteristic time, $r_{ij} = (r_i + r_j)$ denotes the safety distance between human and robot,

$d_{ij} = \|r_i - r_j\|$ denotes the actual distance between human and robot. If the distance d_{ij} is less than r_{ij}, the robot may bump into human body. $n_{ij} = \left(n_{ij}^1, n_{ij}^2\right) = (r_i - r_j)/d_{ij}$ is the normalized vector pointing from robot j to human body i. According to the value of F, we send the dynamic speed message to base controller and robot will alter the velocity dynamically.

3 Experiments

3.1 Tracking Step

Step 1: Manually extract target template in the initial frame of tracked sequence, using it as a first frame of the state vector. The object initial status parameter is $P^{init}\left(P_X^{init}, P_Y^{init}\right)$, the initial weight of particle is $\omega^i = \frac{1}{N}$. Then sample several vector from the priori model according to the probability, the new particle can be used to estimate the real distribution.

Step 2: This step includes the state transition (Eq. (5)) using ARP.

Step 3: Predict the status [18] of the real distribution using the adaptive particle filter. When the particle filter receives $P(Z_k|X_k^i)$, update the weight of every particles $\omega_i = \omega_{i-1}P(Z_k|X_k^i)$, after normalization: $\omega_k^i = \omega_k^i/\sum_{i=1}^N \omega_k^i$. Finally the particle filter can get the posterior distribution:

$$p(x_t|z_{1:t}) \approx \sum_i^N \omega^{(i)}\delta_{x^{(i)}}(x_t) \tag{11}$$

Step 4: When the social force model gets the status $S_k(x_k, y_k)$ from the particle filter, it will calculate the current speed as following equation in (12)–(15)

$$V_{diff} = \left[\frac{X_k - X_{k-1}}{\tau}, \frac{Y_k - Y_{k-1}}{\tau}\right]^T \tag{12}$$

The V_{diff} is the velocity vector differential between human and robot, and the expected velocity of robot is:

$$V_{pre} = V_{diff} + V_{k-1} \tag{13}$$

V_{k-1} is the actual velocity at prior moment $k - 1$, and the distance between person and robot:

$$d_{ij} = \sqrt{x_k^2 + y_k^2} \tag{14}$$

Hence, the final force is as follow:

$$F = m\frac{v^{pre} - v_k}{\tau} + Aexp\left[\frac{r_{ij} - \sqrt{x_k^2 + y_k^2}}{B}\right]n_{ij} \qquad (15)$$

The v_k is the current velocity of robot, it can be calculated from many sensors like odometer or kinect.

$$r_i = v_{pre}/|v_{pre}| \qquad (16)$$

$$r_j = v_k/|v_k| \qquad (17)$$

$$n_{ij} = \left(n_{ij}^1, n_{ij}^2\right) = (r_i - r_j)/d_{ij} \qquad (18)$$

n_{ij} is the normalized vector pointing from robot r_j to human body r_i.

3.2 The Experimental Results

We simulated a uniform movement with white noise, the result shows two different methods (Kalman filter, Particle filter with Social force model). Figure 1 illustrates the fluctuation of velocity. Figure 2 demonstrates the whole process in displacement

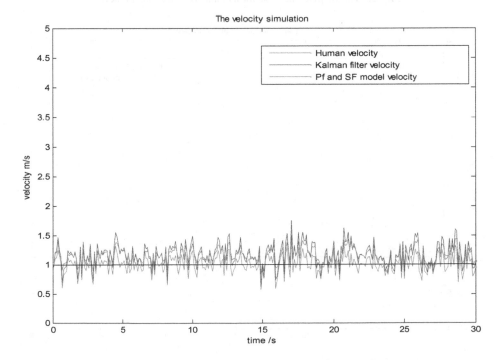

Fig. 1. The velocity of human and KF and our method

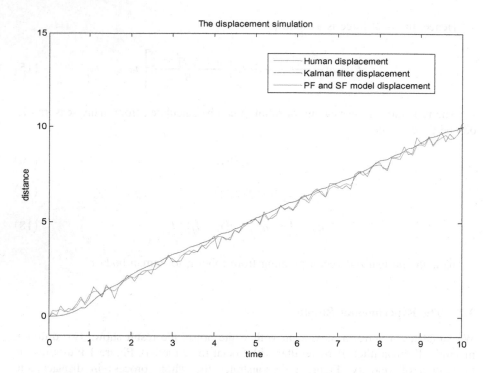

Fig. 2. The displacement of human and KF and our method

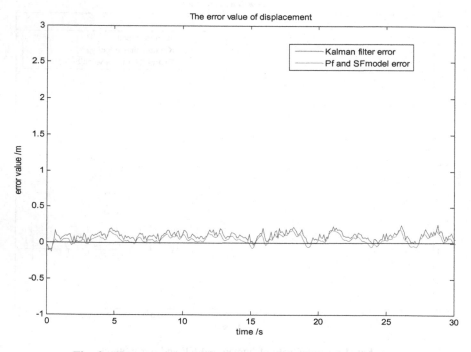

Fig. 3. The error value of displacement about KF and our method

Table 1. The error data from three Experiments.

Experiments times	Mean error of KF	Mean error of new method	Maximum error of KF	Maximum error of new method
I	0.182	0.144	0.252	0.240
II	0.207	0.173	0.277	0.262
III	0.156	0.127	0.242	0.237

during 30 s. Figure 3 shows the relative displacement error of robot and human using two different methods.

4   Conclusion

In this note, we have considered the human tracking problem in unknown environment using particle filter and social force module.

The first result of the experiment shows the velocity in human movement, kalman filter and our new algorithm. It can not be easy to distinguish the difference between kalman filter and new algorithm. So the second result shows the displacement between two methods. It reveals that the new algorithm is more inclined to the human displacement. The third result shows the relative displacement error of robot and human using kalman filter and the new method with particle filter and social force module. Table 1 demonstrates the data statistics from the three experiments. The results of Table 1 can definitely show our method has lower mean error and better performance in maximum error and minimum error.

Finally all the results above can easily prove the superiority of the particle filter and social force module.

Acknowledgment. This work is supported by Shanghai University, and we would like to appreciate the senior engineer, Dr. Wanmi Chen for the support of our paper. We also would like to appreciate the equipment and experiment experience supported by Shanghai Robotics Society.

References

1. Luber, M., Stork, J.A., Tipaldi, G.D., et al.: People tracking with human motion predictions from social forces. In: IEEE International Conference on Robotics and Automation, pp. 464–469. IEEE (2010)
2. Fod, A., Howard, A., Mataric, M.A.J.: Laser-based people tracking. In: 2002 Proceedings of IEEE International Conference on Robotics and Automation, ICRA, pp. 3024–3029. IEEE Xplore (2002)
3. Cui, J., Zha, H., Zhao, H., et al.: Tracking multiple people using laser and vision. In: IEEE/RSJ International Conference on Intelligent Robots and Systems, pp. 2116–2121. IEEE (2005)

4. Madapura, J., Li, B.: Multi-target tracking based on KLD mixture particle filter with radial basis function support. In: IEEE International Conference on Acoustics, Speech and Signal Processing, pp. 725–728 IEEE (2008)
5. Salmond, D.J., Birch, H.: A particle filter for track-before-detect. In: 2001 Proceedings of the American Control Conference, vol. 5, pp. 3755–3760. IEEE Xplore (2001)
6. Rollason, M., Salmond, D.: A particle filter for track-before-detect of a target with unknown amplitude. In: Target Tracking: Algorithms and Applications, vol. 1, pp. 14/1–14/4. IET (2001)
7. Iwahori, Y., Takai, T., Kawanaka, H., Itoh, H., Adachi, Y.: Particle filter based tracking of moving object from image sequence. In: Gabrys, B., Howlett, Robert J., Jain, Lakhmi C. (eds.) KES 2006. LNCS, vol. 4252, pp. 401–408. Springer, Heidelberg (2006). doi:10.1007/11893004_52
8. Okuma, K., Taleghani, A., Freitas, N., Little, J.J., Lowe, D.G.: A boosted particle filter: multitarget detection and tracking. In: Pajdla, T., Matas, J. (eds.) ECCV 2004. LNCS, vol. 3021, pp. 28–39. Springer, Heidelberg (2004). doi:10.1007/978-3-540-24670-1_3
9. Hue, C., Cadre, J.P.L., Perez, P.: A particle filter to track multiple objects. In: 2001 Proceedings of IEEE Workshop on Multi-Object Tracking, pp. 61–68. IEEE (2001)
10. Arulampalam, S., Maskell, S., Gordon, N.J., Clapp, T.: A tutorial on particle filters for on-line nonlinear/non-gaussian Bayesian tracking. IEEE Trans. Sig. Process. 50, 174–188 (2002)
11. Fox, D.: Adapting the sample size in particle filters through KLD-sampling. Int. J. Rob. Res. 22, 985–1003 (2003). doi:10.1177/0278364903022012001
12. Boers, Y.: On the number of samples to be drawn in particle filtering. In: IEE Colloquium on Target Tracking: Algorithms and Applications (1999)
13. Bollerslev, T.: Generalized autoregressive conditional heteroskedasticity. J. Econometrics 31(3), 307–327 (1986)
14. Helbing, D., Molnár, P.: Social force model for pedestrian dynamics. Phys. Rev. E: Stat. Phys. Plasmas Fluids 51(5), 4282 (1995)
15. Helbing, D., Farkas, I., Vicsek, T.: Simulating dynamical features of escape panic. Nature 407, 487–490 (2000). cond-mat/0009448
16. Patil, S., Jur, V.D.B., Curtis, S., et al.: Directing crowd simulations using navigation fields. IEEE Trans. Visual. Comput. Graphics 17(2), 244 (2011)
17. Feng, P., Wang, W., Dlay, S., et al.: Social force model based MCMC-OCSVM particle PHD filter for multiple human tracking. IEEE Trans. Multimed. 19(4), 725–739 (2017)
18. Wang, C., Li, Y., Ge, S.S., et al.: Adaptive control for robot navigation in human environments based on social force model. In: IEEE International Conference on Robotics and Automation, pp. 5690–5695. IEEE (2016)

Time-Jerk Optimal Trajectory Planning for a 7-DOF Redundant Robot Using the Sequential Quadratic Programming Method

Li Jiang, Shaotian Lu, Yikun Gu[✉], and Jingdong Zhao

State Key Laboratory of Robotics and System, Harbin Institute of Technology,
Harbin 150080, China
guyikun@hit.edu.cn

Abstract. In order to improve the efficiency and trajectory tracking accuracy of a robot and reduce its vibration, this paper uses the sequential quadratic programming (SQP) method to perform time-jerk (defined as the derivative of the acceleration) optimal trajectory planning on a 7-degrees-of-freedom (DOF) redundant robot. Kinematic constraints such as joint velocities, accelerations, jerks, and traveling time are considered. When utilizing the SQP method, the initial input is set as average time intervals, and the output is optimal time intervals. Trajectory planning simulations in joint space are performed with optimal time intervals, the results showed that the SQP method is effective and feasible for improving working efficiency and decreasing vibration.

Keywords: Trajectory planning · Sequential quadratic programming · Cubic splines · Jerk

1 Introduction

Recently, many researchers conducted research on the time-jerk optimal trajectory planning in order to improve the efficiency and smoothness of the robot's movement and reduce its vibration. Typical examples of such trajectory planning are shown in [1–5]. Gasparetto et al. proposed a method for optimal trajectory planning of robot manipulators. Their objective function is composed of two parts along the trajectory, which are the proportional to the total execution time and the proportional to the integral of the squared jerk (indicated the derivative of the acceleration). They took into account velocity, acceleration, jerk and trajectory intervals constraints, and the total execution time of the trajectory was not set in advance. The simulation results showed that this algorithm was effective in performing an optimal trajectory planning [1]. Zanotto et al. provided an experimental validation for the method proposed in Ref. [1]. This method did not require the dynamic model of the robot, and the experimental results demonstrated that the effectiveness of the algorithm by comparing it with some classical algorithms and it can be implemented in any industrial manipulator without upgrading its hardware [2]. Cao et al. solved the traveling time problem by using the SQP method so that the traveling time and jerk of the manipulator in the movement process achieved

© Springer International Publishing AG 2017
Y. Huang et al. (Eds.): ICIRA 2017, Part III, LNAI 10464, pp. 343–353, 2017.
DOI: 10.1007/978-3-319-65298-6_32

a comprehensive optimum to a certain degree, and this was followed by obtaining the minimum time-jerk optimal trajectories that satisfied nonlinear kinematic constraints. The experimental results indicated that the planned trajectories of shotcrete manipulators obtained by using this method were reasonable and effective [3]. Zhong et al. presented a new method based on chaotic particle swarm optimization in order to find the time-jerk optimal trajectory of the manipulator by using the cubic splines. The simulation results revealed the effectiveness of the presented method when it was used to solve the conflicting problem between high executive efficiency and low mechanical vibrations utilizing limited control energy [4]. Liu et al. proposed an improved adaptive genetic algorithm to realize the time-jerk optimal model search, and the joint velocity, acceleration, jerk constraints under the heavy load and the high-speed handling characteristics conditions for electrolytic copper transport robot were considered. The simulation results showed the optimization model was effective [5]. Cong et al. presented a new method based on fuzzy genetic algorithm in order to decide the time-jerk synthetic optimal trajectory of robot in joint space by using the cubic splines [6].

After summarizing the aforementioned typical time-jerk optimal trajectory planning research, we can find the main research steps can be defined as follows: First, an optimal objective function is utilized to describe the optimal problem. Second, corresponding optimal method is selected to solve the optimal objective function according to problem characteristics. Third, optimal results are analyzed and corresponding results are chosen based on requirements.

Currently, many researchers perform the trajectory planning research in joint space instead of the operating space mainly because of the following reasons: On one hand, the control system performs effect on the manipulator joints rather than on the end effector, so it is more easily to regulate the trajectory according to the demands in the joint space [1]. On the other hand, it is beneficial for avoiding kinematic singularities.

At present, the interpolating functions which are utilized to perform trajectory planning in joint space mainly involve cubic spline and fifth-order B-spline etc. And the major optimal methods consist of genetic algorithm, particle swarm optimization (PSO) and SQP algorithms etc. The current literature about time-jerk optimal trajectory planning demonstrates two problems. First, the research in this field especially on the redundant robots is still relatively less; especially the research on the redundant robots is less. So it is necessary to carry out further study. Second, many studied objects are simple, so it is needed to study time-jerk optimal trajectory planning on more complicated robot.

In this paper, we apply the SQP method to implement time-jerk optimal trajectory planning on a redundant 7-DOF robot. Because the initial input has a large effect on the result when the SQP method is used, we modify the set style of the initial input to make it more properly for the redundant 7-DOF robot.

The remainder of this paper is organized as follows: A 7-DOF redundant robot which is used in this paper is presented in Sect. 2. The time-jerk optimal trajectory planning problem is described in Sect. 3. Section 4 shows the simulation results of the SQP method. Section 5 is the conclusion.

2 The 7-DOF Redundant Robot and Its Inverse Kinematics

A 7-DOF configuration robot is shown in Fig. 1. The corresponding parameters of the robot are shown in Table 1. This robot has 7 modular joints. It resembles the configuration of the Space Station Remote Manipulator System (SSRMS). In order to make the transformation matrix can be computed more conveniently, the coordinate system $(x_0' \ y_0' \ z_0')$ is added to the base coordinate system $(x_0 \ y_0 \ z_0)$. Firstly, it is necessary to calculate the transformation matrix ${}_{0'}^{0}T$ from base coordinate system $(x_0 \ y_0 \ z_0)$ to the coordinate system $(x_0' \ y_0' \ z_0')$, then the following transformation matrix can be obtained based on the coordinate system $(x_0' \ y_0' \ z_0')$. The pose transformation matrix between adjacent coordinate systems is described as follows:

$$
{}_{i}^{i-1}T = \begin{bmatrix} c\theta_i & -s\theta_i & 0 & a_{i-1} \\ s\theta_i c\alpha_{i-1} & c\theta_i c\alpha_{i-1} & -s\alpha_{i-1} & -d_i s\alpha_{i-1} \\ s\theta_i s\alpha_{i-1} & c\theta_i s\alpha_{i-1} & c\alpha_{i-1} & d_i c\alpha_{i-1} \\ 0 & 0 & 0 & 1 \end{bmatrix} \tag{1}
$$

where $c = cos, s = sin$.

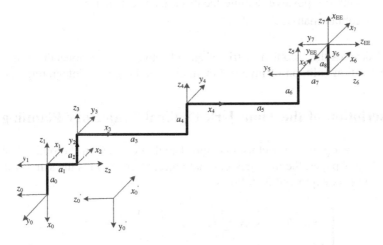

Fig. 1. The coordinate frame of the 7-DOF redundant robot

Table 1. Parameters of the 7-DOF redundant robot

Number	a_{i-1}	α_{i-1}	d_i	θ_i
1	0	90°	a_0	θ_1 (0°)
2	0	90°	a_1	θ_2 (0°)
3	0	−90°	a_2	θ_3 (−90°)
4	a_3	0°	a_4	θ_4 (0°)
5	a_5	0°	a_6	θ_5 (90°)
6	0	90°	a_7	θ_6 (0°)
7	0	−90°	a_8	θ_7 (0°)
8	0	90°	0	θ_8 (90°)

Before implementing the optimal trajectory planning of a robot, its inverse kinematics should be obtained. We adopt the configuration control method to solve the inverse kinematics of the 7-DOF redundant robot. This method is an effective method for solving the inverse kinematics of a redundant robot. It can ensure the inverse kinematics is unique and make a redundant robot achieves cyclic motion, which is fits for a robot that needs to realize repetitive motion [7, 8]. Besides, this method can be computed fast, so it is especially suitable for real-time control. The damped-least-squares (DLS) expression of the configuration control scheme is described as follows:

$$\dot{q} = W_v^{-1} J^T [J W_v^{-1} J^T + \lambda^2 W^{-1}]^{-1} (\dot{X}_d + K E_e) \tag{2}$$

where
W ——symmetric positive-definite weighting matrix, $W = \text{diag}[W_e, W_c]$, W_e and W_c are also symmetric positive-definite weighting matrices for the basic task and additional task, respectively;
W_v ——symmetric positive-definite weighting matrix;
λ ——positive scalar constant;
X_d ——the desired behavior of the robot;
E_e ——error, $E_e = X_d - X$;
K ——symmetric positive-definite feedback gain (constant);
J ——Jacobian matrix, $J = (J^{ee}, J^\psi)^T$.

where J^{ee} denotes the Jacobian matrix of end effector, and ψ denotes the arm angle [9, 10]. The position solution of inverse kinematics can be got by integrating (2).

3 Description of the Time-Jerk Optimal Trajectory Planning

The function of optimal trajectory planning largely depends on the optimal objective function. In this paper, the time-jerk optimal objective function is the same with that in Ref. [3], and it is expressed as follows:

$$\begin{cases} \min f(h) = K_T N \sum_{i=1}^{n-1} h_i + \alpha K_J \sum_{i=1}^{N} \sum_{i-1}^{n-1} \left[\frac{(\ddot{Q}_{j,i+1} - \ddot{Q}_{j,i})^2}{h_i} \right] \\ s.t. \\ \max\{|\dot{Q}_{j,i}(t_i)|, |\dot{Q}_{j,i}(t_i*)|, |\dot{Q}_{j,i}(t_{i+1})|\} - V_{jm} \le 0 \\ \max\{|\ddot{Q}_{j,1}(t_1)|, |\ddot{Q}_{j,2}(t_2)|, \cdots, |\ddot{Q}_{j,n}(t_n)|\} - A_{jm} \le 0 \\ \max \left| \frac{\ddot{Q}_{j,i}(t_{i+1}) - \ddot{Q}_{j,i}(t_i)}{h_i} \right| - J_{jm} \le 0 \\ \sum_{i=1}^{n-1} h_i - T_m \le 0 \end{cases} \tag{3}$$

where $j = 1, \ldots, N \, \forall \, i = 1, \ldots, n-1$.

In optimal problem (3), $K_T + K_J = 1$. Table 2 lists the meaning of the symbols in (3). The traveling time and integral of the squared jerk may hold large difference in quantity grade, so the elastic coefficient α is presented to balance their effects. In this way,

Table 2. Meaning of symbols

Name	Meaning
N	Number of robot joints
K_T	Time weighting coefficient
h_i	Time interval between two plan-points
A_{jm}	Acceleration limit for the jth joint (symmetrical)
T_m	Traveling time limit
n	Number of via-points
$\dot{Q}_j(t)$	Velocity of the jth joint
$\ddot{Q}_j(t)$	Acceleration of the jth joint
$\dddot{Q}_j(t)$	Jerk of the jth joint
V_{jm}	Velocity limit for the jth joint (symmetrical)
K_J	Jerk weighting coefficient
J_{jm}	Jerk limit for the jth joint (symmetrical)

they can be kept in the same quantity grade. In reality, the total traveling time and integral of the squared jerk could achieve optimum values through altering K_T, K_J and α according to different requirements. The optimal time intervals are obtained via solving the optimal problem (3), afterwards, the time-jerk optimal trajectory under constraints can be gained after using a method to plan its trajectory.

We adopt the cubic splines to implement trajectory planning in joint space because they hold many advantages: First, compared with higher order polynomials, they do not show excessive oscillations and overshoot between any pair of planning points. Second, the produced trajectories have continuous acceleration values. Third, the cubic splines are simple and can be easily implemented. So we apply the cubic splines presented in Refs. [1–3] to plan the trajectory in joint space.

4 Apply the SQP Method to Determine Optimal Time Intervals

The initial input has a large effect on the optimal result with the SQP method. So we modify the initial input to obtain a more reasonable result. Equation (4) is introduced to decide the initial input H_0.

$$H_0 = \Lambda H_I \tag{4}$$

where H_I denotes the average time interval between two plan-points. And Λ is decided as follows:

$$
\begin{cases}
\varLambda_1 = \max_{j=1,\dots,N}\left\{\max_{t\in[t_i,t],i=1,\dots,n-1}\left\{\dfrac{|\dot{Q}_{j,i}(t)|}{V_{jm}}\right\}\right\}\\[2ex]
\varLambda_2 = \max_{j=1,\dots,N}\left\{\max_{t\in[t_i,t],i=1,\dots,n-1}\left\{\dfrac{|\ddot{Q}_{j,i}(t)|}{A_{jm}}\right\}\right\}\\[2ex]
\varLambda_3 = \max_{j=1,\dots,N}\left\{\max_{t\in[t_i,t],i=1,\dots,n-1}\left\{\dfrac{|\dddot{Q}_{j,i}(t)|}{J_{jm}}\right\}\right\}
\end{cases}
\tag{5}
$$

where $\varLambda = \max\{1\varLambda_1\,\varLambda_2\,\varLambda_3\}$.

The kinematic constraints of the joints are given as follows: $V_{jm} = 3$ deg/s, $A_{jm} = 3$ deg/s^2, and $J_{jm} = 5$ deg/s^3 ($j = 1, \dots, N$). In addition, some other parameters are set as follows: the initial traveling time constraint of the robot $t = 150$ s, and the elastic coefficient $\alpha = 100$. After determining the initial input, the SQP method is utilized to calculate the optimal time intervals. And the function fmincon from Matlab R2014 is used to perform relevant programs.

5   Time-Jerk Optimal Trajectory Planning of the 7-DOF Redundant Robot

Only when the inverse kinematics of the 7-DOF redundant robot is obtained can the optimal trajectory planning be carried out. In this paper, the DLS method of the configuration control scheme presented in Sect. 2 is used to resolve the inverse kinematics of the 7-DOF robot. The geometric parameters of the 7-DOF redundant robot are given as follows: $a_0 = 716.1$ mm, $a_1 = 430$ mm, $a_2 = 430$ mm, $a_3 = 2080$ mm, $a_4 = 387$ mm, $a_5 = 2080$ mm, $a_6 = 430$ mm, $a_7 = 430$ mm, and $a_8 = 716.1$ mm.

Some other parameters are given as follows: $W_e = \mathrm{diag}[1, 1, 1, 1, 1, 1]$, $W_c = 1$, $\lambda = 1$, $W_v = \mathrm{diag}[1, 1, 1, 1, 1, 1, 1]$, $K = \mathrm{diag}[1, 1, 1, 1, 1, 1, 1]$, $\dot{X}_d = (0,0,0,0,0,0,0)^{\mathrm{T}}$. The robot will be in a singular position when the joint angles defined in Table 1 are used directly. In order to avoid this condition, the initial joint angles are set as follows: $\theta_1 = 0°$, $\theta_2 = 0°$, $\theta_3 = -90°$, $\theta_4 = 60°$, $\theta_5 = 90°$, $\theta_6 = 0°$ and $\theta_7 = 0°$.

In order to execute time-jerk optimal trajectory planning, we select a tested mission as tracking a circle. The initial position of the end effector of the 7-DOF robot corresponds to $(-2679.2, -2173.7, -3765.0)$ (mm) according to the base coordinate frames $(x_0y_0z_0)$. The radius R and the center of the tested circle are 1000 mm and $(-2679.2, -2173.7, -2765.0)$ (mm), respectively. The trajectory plan-points in the circle are acquired from the circular central angel. The planned circular central angle is divided into three sections along the counter-clockwise rotation, which are uniform acceleration section, uniform velocity section and uniform deceleration section. At the uniform acceleration section, the robot begins accelerating from initial position to the end with the angular acceleration of $120/(1/80 \times n)^2$ °/s^2 (n denotes the amount of planned via-points). The robot finishes the uniform acceleration section when it moves along the circle for $60°$, then it goes into the uniform velocity section and moves $240°$. And finally the robot goes into the uniform deceleration section with the angular deceleration of $120/(1/80 \times n)^2$ °/s^2. It decelerates until returning to the initial point, then it stops moving. Thus, the robot finishes the whole movement. Here, we set $n = 8$.

Table 3. Input data for trajectory planning

Joint	Plan-points (deg)								
	1	2	3	4	5	6	7	8	9
1	0	Virtual	−32.66	−37.74	−15.44	4.21	20.26	Virtual	−0.14
2	0		−0.00	−0.00	−0.00	0.00	0.00		0.00
3	−90.00		−79.19	−79.02	−77.37	−79.99	−95.51		−89.78
4	60.00		88.73	116.13	112.24	81.75	55.89		58.95
5	90.00		91.92	84.66	75.24	79.18	87.45		95.52
6	0		0.00	0.00	0.00	0.00	−0.00		−0.00
7	0		−8.79	−24.03	−34.67	−25.15	−8.08		−4.56

Table 3 lists the input data for trajectory planning. Point-1 is set as the initial point. When the robot finishes a cyclic movement, the point-1 coincides with the point-9. So there are 9 plan-points while the number of via-points is 8.

The weighting coefficients arc sct as follows: The K_T are set as 1, 0.8, 0.5, 0.2, and 0, respectively, and the corresponding values of K_J can be acquired through calculating. Table 4 demonstrates the typical time-jerk optimal trajectory planning simulation results after implementing the SQP method. Table 4 reveals that the traveling time gradually increases along with K_T gradually decreases from 1 to 0, and the time-jerk optimal objective function has the same varied tendency.

Figure 2 displays the joint trajectories prior to optimization. When K_T is given as 1, 0.5, and 0, respectively, the relevant simulation results are demonstrated in Figs. 3–5. When K_T is given as 1, the optimal objective function only includes the minimum-time function. As illustrated in Fig. 3, some joint velocities reach the velocity limit at some

Table 4. Results of trajectory optimization

Parameters	Numerical results ($\alpha = 100$)					
	Initial	$K_T = 1$ $K_J = 0$	$K_T = 0.8$ kJ = 0.2	$K_T = 0.5$ kJ = 0.5	$K_T = 0.2$ kJ = 0.8	$K_T = 0$ $K_J = 1$
h_1	18.75	0.0856	1.3243	2.3704	3.5994	9.6299
h_2	18.75	21.7232	17.7275	17.2235	19.9953	25.9146
h_3	18.75	10.6204	11.9293	11.9572	10.4663	17.0299
h_4	18.75	10.0535	10.3676	10.2987	11.5058	20.1355
h_5	18.75	11.6834	12.3444	12.3622	12.0545	18.5251
h_6	18.75	12.0821	11.5718	11.0709	13.0009	23.1325
h_7	18.75	9.1368	8.5757	11.5181	14.7662	26.0372
h_8	18.75	0.2585	2.7984	3.9257	4.8098	9.0483
$\sum h_i$	150	75.6434	76.6390	80.7266	90.1982	149.4530
$\sum_{j=1}^{N} \sum_{i=1}^{n-1} \left(\frac{(\alpha_{j,i+1}-\alpha_{j,i})^2}{h_i} \right)$	0.0205	7.9239	1.1114	0.5212	0.2255	0.0148
$\min f(h)$	/	529.5041	451.4067	308.6041	144.3187	1.4836

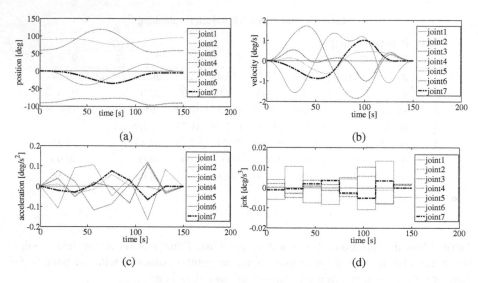

Fig. 2. The joint trajectories prior to optimization

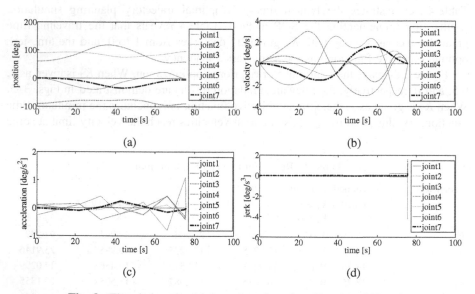

Fig. 3. Time-jerk optimal joint trajectories when $K_T = 1$ and $K_J = 0$

moments, and the velocities, accelerations, and jerks in general are larger than those in Fig. 2. Because the working efficiency of the robot is high, the traveling time is short. Therefore, this condition suits a robot which requires strict working efficiency. In Fig. 4, the traveling time is larger than that in Fig. 3. And the jerks are generally smaller than those in Fig. 3. Moreover, some joint velocities approach the velocity limit at some moments. Hence, this condition suits a robot which needs both relatively

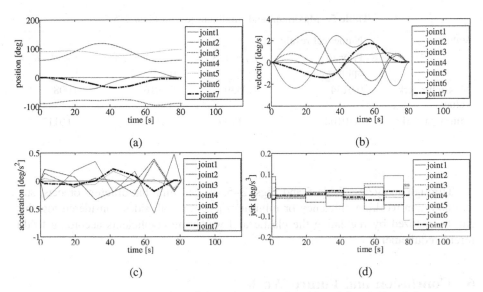

Fig. 4. Time-jerk optimal joint trajectories when $K_T = 0.5$ and $K_J = 0.5$

strict working efficiency and less vibration demands. In Fig. 5, the traveling time is almost equal to the initial constraint time t. Yet, the velocities, accelerations, and jerks on the whole are smaller than those in Fig. 2; therefore, the advantage of SQP method can be reflected at this condition. In order to clarify the extreme values conveniently, the maximum and minimum jerk values in Figs. 2–5 are shown in Table 5.

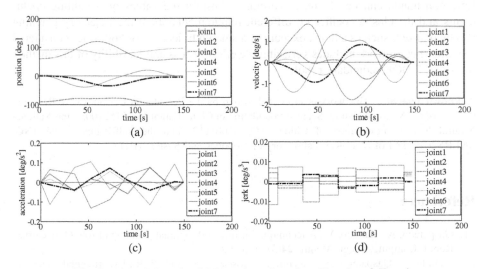

Fig. 5. Time-jerk optimal joint trajectories when $K_T = 0$ and $K_J = 1$

Table 5. Maximum and minimum jerk in Figs. 2–5

Value	Numerical results			
	Prior to optimization	$K_T = 1$, $K_J = 0$	$K_T = 0.5$, $K_J = 0.5$	$K_T = 0$, $K_J = 1$
Maximum jerk value	0.0134	1.5539	0.0916	0.008
Minimum jerk value	–0.0108	–4.1999	–0.1458	–0.0117

Through analyzing Table 5 and Figs. 2–5, we can find that different objects, such as maximum working efficiency or minimum jerks or both sides considered together, can be realized by regulating the elastic and weighting coefficients according to different requirements.

6 Conclusion and Future Work

In this paper, based on the inverse kinematics analysis using the DLS expression of the configuration control scheme, the SQP method with modified initial input is presented to solve the traveling time intervals for a 7-DOF redundant robot. With the average time intervals as the initial input, the optimal time intervals are obtained as the result. By using SQP method, the total traveling time and the proportional to the integral of the squared jerk of the robot during the movement process realize a comprehensive optimum. The trajectory planning in joint space is implemented. And the simulation results show the time-jerk optimal trajectories which meet kinematics and the travelling time constraints can be obtained. Through adjusting the values of weighting coefficients and the elastic coefficient, different requirements about the trajectory planning can be realized, such as a quick execution, a smooth trajectory, or both sides considered together. Future work includes verifying the effectiveness and feasibility of this method on a real 7-DOF redundant robot.

Acknowledgments. This work is supported in part by the Foundation for Innovative Research Groups of the National Natural Science Foundation of China (Grant No. 51521003), the National Natural Science Foundation of China (NO. 61603112) and the Self-Planned Task (NO. SKLRS201721A) of State Key Laboratory of Robotics and System (HIT).

References

1. Gasparetto, A., Zanotto, V.: A technique for time-jerk optimal planning of robot trajectories. Robot. Comput. -Integr. Manuf. **24**(3), 415–426 (2008)
2. Zanotto, V., Gasparetto, A., Lanzutti, A., Boscariol, P., Vidoni, R.: Experimental validation of minimum time-jerk algorithms for industrial robots. J. Intell. Robot. Syst. **64**, 197–219 (2011)

3. Cao, Z.Y., Wang, H., Wu, W.R., Xie, H.J.: Time-jerk optimal trajectory planning of shotcrete manipulators. J. Cent. South. U. **44**(1), 114–121 (2013)
4. Zhong, G.L., Kobayashi, Y., Emaru, T.: Minimum time-jerk trajectory generation for a mobile articulated manipulator. J. Chin. Soc. Mech. Eng. **35**(4), 287–296 (2014)
5. Liu, F., Lin, F.: Time-jerk optimal planning of industrial robot trajectories. Int. J. Robot. Autom. **31**(1), 1–7 (2016)
6. Cong, M., Xu, X.F., Xu, P.: Time-jerk synthetic optimal trajectory planning of robot based on fuzzy genetic algorithm. In: Proceedings of the 15th International Conference on Mechatronics and Machine Vision in Practice (M2VIP), Auckland, pp. 274–279 (2008)
7. Seraji, H.: Configuration control of redundant manipulators: theory and implementation. IEEE Trans. Robot. Autom. **5**(4), 472–490 (1989)
8. Glasst, K., Colbaught, R., Lim, D., Seraji, H.: On-line collision avoidance for redundant manipulators. In: Proceedings of the IEEE International Conference on Robotics and Automation, Atlanta, pp. 36–43 (1993)
9. Kenneth, K.D., Long, M., Seraji, H.: Kinematic analysis of 7-DOF manipulators. Int. J. Robot. Res. **11**(4), 469–481 (1992)
10. Xu, W.F., Zhang, J.T., Yan, L., Wang, Z.Y.: Parameterized inverse kinematics resolution method for a redundant space manipulator with link offset. J. Aeronaut. **36**(1), 33–39 (2015)

Nonlinear Control of Omnidirectional Mobile Platforms

Víctor H. Andaluz$^{(\boxtimes)}$, Oscar Arteaga, Christian P. Carvajal, and Víctor D. Zambrano

Universidad de las Fuerzas Armadas ESPE, Sangolquí, Ecuador
{vhandaluz1, obarteaga, vdzambrano}@espe.edu.ec,
chriss2592@hotmail.com

Abstract. This work presents kinematic modeling and a kinematic nonlinear controller of an omnidirectional mobile platform that generates saturated reference velocity commands for path following problem. The dynamic compensation controller is considered through of a platform-inner-loop system to independently track four velocity commands. Stability and robustness of the complete control system are proved through the Lyapunov method. Finally, simulation results are presented and discussed, which validate the proposed controller.

Keywords: Omnidirectional · Kinematic modeling · Nonlinear controller · Lyapunov · Non-linear systems

1 Introduction

Throughout the technological and industrial development a number of risks and accidents have been involved in human operators due to the manipulation and transfer of objects in tasks in dangerous environments for the operator [1]. The needless of supplant physically the human resource in these tasks has been a topic of interest in the last years. It has been sought to develop machines or robots that can transport in a certain way the operation capacity of the workers for places where their security is threatened [2].

Mobile Robotics is an active research area where researchers from all over the world find new technologies to improve mobile robots intelligence and areas of application. Among different mobile platforms, omnidirectional mobile robots are frequently used to perform different tasks due to their good mobility and simple configuration. This robot structure has been used in various applications like surveillance, floor cleaning, and industrial load transportation using autonomous guided vehicles [3].

A variety of control strategies have been developed to solve the problem of an autonomous vehicle to converge to a path and follow it, without any time specification, *i.e.*, problem of path following. The different proposed strategies consider that the vehicle moves with constant speed, however, this parameter is not always correct because in real tasks, non-structured environments, a vehicle can not always move at a constant speed [4]. In addition there are control proposals in which only a desired vector is considered $\mathcal{P}_d \in \Re$ that represents the desired path to been follow by the

© Springer International Publishing AG 2017
Y. Huang et al. (Eds.): ICIRA 2017, Part III, LNAI 10464, pp. 354–364, 2017.
DOI: 10.1007/978-3-319-65298-6_33

vehicle to any speed, for which different methods are proposed for obtaining $\mathcal{P}_d$ [5]. Let $\mathcal{P}_d(s)$ be a desired geometric path parameterized by the curvilinear abscissa $s \in \mathfrak{R}$. In the literature it is common to find different control algorithms for path following where $s(t)$ is considered as an additional control input. In [6–10], the rate of progression ($\dot{s}$) of a virtual vehicle has been controlled explicitly. On the other hand, in the literature one can find works based on vision and/or laser these methods are based on the processing of data in order to detect the way to be followed [11, 12].

The path following problem has been well studied and many solutions have been proposed and applied in a wide range of applications. Let $\mathcal{P}_d(s) \in \mathfrak{R}$ be a desired geometric path parameterized by the curvilinear abscissa $s \in \mathfrak{R}$. In the literature it is common to find different control algorithms for path following where $s(t)$ is considered as an additional control input.

As described in this paper is presented a non-linear control for the tracking path of an omnidirectional mobile robot with a four mecanun wheel configuration.

Additionally, we present the omnidirectional mobile plataform, the same one that allows to carry a load up to 100[kg] or place as working tools two robotic arms to their ends. The proposed control scheme is divided into two subsystems, each one being a controller itself: the first one is a kinematic controller with saturation of velocity commands, which is based on the omnirctional mobile robot's kinematics. The path following problem is addressed in this subsystem. It is worth noting that the proposed controller does not consider $s(t)$ as an additional control input as it is frequent in literature; and the second one is a dynamic compensation controller is considered through of a platform-inner-loop system to independently track four velocity commands. Additionally, both stability of the controller is proved through Lyapunov's method. To validate the proposed control algorithms, experimental results are included and discussed.

This project have 5 sections including the Introduction. The Sect. 2 describes the kinematics modeling of the mobile platform in which considerates two lineal speeds and one angular speed like a input. The design and stability analysis of the nolineal control algorithm based on the theory of Lyapunov is presented on the Sect. 3. The Sect. 4 shows the experimental results, discussion and analysis of control scheme proposed. Finally, the conclusions are presented in the Sect. 5.

2   Omnidrectional Mobile Platform Model

It is assumed that the omnidirectional mobile platform moves on a planar horizontal surface. Let $<\mathcal{R}, \mathcal{X}, \mathcal{Y}, \mathcal{Z}>$ be any fixed frame with $\mathcal{Z}$ vertical. The *location of the platform* is given by a vector $\xi(t) \in \mathfrak{R}^m$ of $m = 3$ coordinates which define the position and the orientation of the mobile platform in $<\mathcal{R}>$. They are called the *operational coordinates of the omnirectional mobile platform*. It is written as

$$\xi(t) = \begin{bmatrix} x_p & y_p & \phi_p \end{bmatrix}^T$$

where x_p and y_p represents the position in the axis $\mathcal{X}$ and $\mathcal{Y}$ of the system $<\mathcal{R}>$, respectively, of the point of interest - control point - of the mobile platform; while ϕ_m is the orientation of the platform with respect to the $\mathcal{X}$ axis of the reference system $<\mathcal{R}>$. The set of all the locations constitutes the *operational space of the platform*, denoted by $\mathcal{M}$

$$f : N \to \mathcal{M}$$
$$(\mathbf{q}_p) \mapsto \xi = f(\mathbf{q}_p)$$

where, N is the *configuration space* of the omnidirectional mobile platform.

2.1   Kinematic Model

The omnidirectional mobile platform is composed by a set of three velocities represented at the spatial frame $<\mathcal{P}>$. The displacement of the mobile platform is guided by the two linear velocities u_{pi} and u_{pj} defined in a rotating right-handed spatial frame $<\mathcal{P}>$, and the angular velocity ω_p, as shown in Fig. 1.

Fig. 1. Schematic of the omnidirectional mobile platform

Each linear velocity is directed as one of the axes of the frame $<\mathcal{P}>$ attached to the center of gravity of the mobile platform: u_{pi} points to the frontal direction and u_{pj} points to the left-lateral direction. The angular velocity ω_p rotates the referential system $<\mathcal{P}>$ counterclockwise, around the axis $\mathcal{Z}$ (considering the top view). In other words, the Cartesian motion of the mobile platform at the inertial frame $<\mathcal{R}>$ is defined as, $<\mathcal{P}>$

$$\begin{bmatrix} \dot{x}_p \\ \dot{y}_p \\ \dot{\psi}_p \end{bmatrix} = \begin{bmatrix} \cos\phi_p & -\sin\phi_p & -a\sin\phi_p \\ \sin\phi_p & \cos\phi_p & a\cos\phi_p \\ 0 & 0 & 1 \end{bmatrix} \begin{bmatrix} u_{pi} \\ u_{pj} \\ \omega_p \end{bmatrix}$$

$$\dot{\xi}(t) = \mathbf{J}(\phi_p)\boldsymbol{\eta}_p(t) \tag{1}$$

where $\dot{\xi}(t) \in \Re^m$ represents the vector of axis velocity of the system $<\mathcal{R}>$; $\mathbf{J}(\phi_p) \in \Re^{m \times n}$ is the Jacobian matrix that defines a linear mapping between the vector of the mobile platform velocities $\boldsymbol{\eta}_p(t)$ and the vector of the interest point velocity $\dot{\xi}(t)$; the control of maneuverability of the mobile platform is defined $\boldsymbol{\eta}_p(t) \in \Re^n$; and a is a distance.

The angular speeds of each wheel of the mobile platform in function of the speeds of maneuverability of the robot is represented as,

$$
\begin{bmatrix} \omega_{fl} \\ \omega_{rl} \\ \omega_{rr} \\ \omega_{fr} \end{bmatrix} = \begin{bmatrix} \frac{1}{R} & -\frac{1}{R} & -\frac{b+c}{R} \\ \frac{1}{R} & \frac{1}{R} & -\frac{b+c}{R} \\ \frac{1}{R} & -\frac{1}{R} & \frac{b+c}{R} \\ \frac{1}{R} & \frac{1}{R} & \frac{b+c}{R} \end{bmatrix} \begin{bmatrix} u_{pi} \\ u_{pj} \\ \omega_p \end{bmatrix}
$$

$$
\boldsymbol{\omega}_\omega(t) = \mathbf{J}_\omega \boldsymbol{\eta}_p(t) \tag{2}
$$

where R represents the radio of each wheel of the robot; b and c are distance; $\mathbf{J}_\omega$ is a constant matrix, see Fig. 2. The subscripts fl and fr represents the frontal right and left wheels, respectively; while rl and rr, respectively, are the back right anf left wheels.

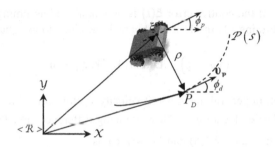

Fig. 2. Problem of control.

3 Controller Design and Stability Analysis

As represented in Fig. 2, the path to be followed is denoted as $\mathcal{P}(s)$; the actual desired location $P_D = (x_P(s_D), y_P(s_D))$ is defined as the closest point on $\mathcal{P}(s)$ to the interest point of the mobile platform, with s_D being the curvilinear abscissa defining the point P_D; $\tilde{\xi}_x = x_P(s_D) - x_p$ is the position error in the $\mathcal{X}$ direction and $\tilde{\xi}_y = y_C(s_D) - y_p$ is the position error in the $\mathcal{Y}$ direction; $\rho = \sqrt{\tilde{\xi}_x^2 + \tilde{\xi}_y^2}$ represents the distance between the interest point of the mobile platform $\xi(x,y)$ and the desired point P_D on inertial frame $<\mathcal{R}(\mathcal{X}, \mathcal{Y}, \mathcal{Z})>$.

The controller proposed to solve the path following problems of the omnidirectional mobile platform aims to calculate at every time $\tilde{\xi}_x(t), \tilde{\xi}_y(t)$ and $\tilde{\xi}_z(t)$ and use these measures to drive the mobile platform in a direction which decreases the control errors. The proposed kinematic controller is based on the kinematic model (1). Hence following control law is proposed,

$$\begin{bmatrix} u_{pi} \\ u_{pj} \\ \omega_p \end{bmatrix} = \mathbf{J}^{-1} \left(\begin{bmatrix} |\boldsymbol{v}_P| \cos(\beta) \\ |\boldsymbol{v}_P| \sin(\beta) \\ 0 \end{bmatrix} + \begin{bmatrix} l_x \tanh\left(\frac{k_x}{l_x}\tilde{\xi}_x\right) \\ l_y \tanh\left(\frac{k_y}{l_y}\tilde{\xi}_y\right) \\ l_\phi \tanh\left(\frac{k_\phi}{l_\phi}\tilde{\xi}_\phi\right) \end{bmatrix} \right) \tag{3}$$

Now (3) can be represented in a compact forme by

$$\boldsymbol{\eta}_p(t) = \mathbf{J}^{\#}\left(\boldsymbol{v}_P + \mathbf{L_K}\tanh\left(\mathbf{L_K^{-1}K}\tilde{\boldsymbol{\xi}}\right)\right) \tag{4}$$

where $\boldsymbol{v}_P$ is the reference velocity input of the aerial mobile manipulator for the controller; $\mathbf{J}^{\#}$ is the matrix of inverse kinematics for the mobile platform; while that $l_x > 0, k_x > 0, l_y > 0, k_y > 0, l_\phi > 0$ and $k_\phi > 0$ area gain constants of the controller that weigh the control error respect to the inertial frame $<\mathcal{R}>$; and the $\tanh(.)$ represents the function saturation of maniobrability velocities in the aerial mobile manipulator.

The behaviour of the control error $\tilde{\boldsymbol{\xi}}(t)$ is now analyzed assuming perfect velocity tracking. By substituting (4) in (1) it is obtained the close loop equation,

$$\left(\boldsymbol{v}_P - \dot{\tilde{\boldsymbol{\xi}}}\right) + \mathbf{L}\tanh\left(\mathbf{L_K^{-1}K}\tilde{\boldsymbol{\xi}}\right) = \mathbf{0} \tag{5}$$

Remember that, in general, the desired velocity vector $\boldsymbol{v}_P$ is different from the time derivative of the desired location $\dot{\boldsymbol{\xi}}_d$. Now, defining difference signal $\Upsilon = \dot{\boldsymbol{\xi}}_d - \boldsymbol{v}_P$ and remembering that $\dot{\tilde{\boldsymbol{\xi}}} = \dot{\boldsymbol{\xi}}_d - \dot{\boldsymbol{\xi}}$, (5) can be written as

$$\dot{\tilde{\boldsymbol{\xi}}} + \mathbf{L_K}\tanh\left(\mathbf{L_K^{-1}K}\tilde{\boldsymbol{\xi}}\right) = \Upsilon. \tag{6}$$

Remark 1: $\boldsymbol{v}_P$ is collinear to $\dot{\boldsymbol{\xi}}_d$ (tangent to the path), then Υ is also a collinear vector to $\boldsymbol{v}_P$ and $\dot{\boldsymbol{\xi}}_d$.

For the stability analysis the following Lyapunov candidate function is considered $V\left(\tilde{\boldsymbol{\xi}}\right) = \frac{1}{2}\tilde{\boldsymbol{\xi}}^T\tilde{\boldsymbol{\xi}}$. Its time derivative on the trajectories of the system is, $\dot{V}\left(\tilde{\boldsymbol{\xi}}\right) = \tilde{\boldsymbol{\xi}}^T\Upsilon - \tilde{\boldsymbol{\xi}}^T\mathbf{L_K}\tanh\left(\mathbf{L_K^{-1}K}\tilde{\boldsymbol{\xi}}\right)$. A sufficient condition for $\dot{V}\left(\tilde{\boldsymbol{\xi}}\right) < 0$ to be negative definite is,

$$\left|\tilde{\boldsymbol{\xi}}^T\mathbf{L_K}\tanh\left(\mathbf{L_K^{-1}K}\tilde{\boldsymbol{\xi}}\right)\right| > \left|\tilde{\boldsymbol{\xi}}^T\Upsilon\right| \tag{7}$$

For large values of $\tilde{\boldsymbol{\xi}}$, the condition in (7) can be reinforced as, $\left\|\tilde{\boldsymbol{\xi}}^T\mathbf{L_K'}\right\| > \|\tilde{\boldsymbol{\xi}}\|\|\Upsilon\|$ with $\mathbf{L_K'} = \mathbf{L_K}\tanh(k_{aux}\mathbf{i})$, where k_{aux} is a suitable positive constant and $\mathbf{i} \in \mathfrak{R}^m$ is the vector of unity components. Then, $\dot{V}$ will be negative definite only if

$$\|\mathbf{L}\| > \frac{\|\Upsilon\|}{\tanh(k_{aux})} \tag{8}$$

hence, (8) establishes a design condition to make the errors $\tilde{\xi}$ to decrease.

Now, for small values of $\tilde{\xi}$, condition (7) will be fulfilled if

$$\left\| \tilde{\xi}^T \mathbf{K} \frac{\tanh(k_{aux})}{k_{aux}} \tilde{\xi} \right\| > \|\tilde{\xi}\| \|\Upsilon\|$$

which means that a sufficient condition for $\dot{V}\left(\tilde{\xi}\right) < 0$ to be negative definite is,

$$\|\tilde{h}\| > \frac{\|\Upsilon\|}{\lambda_{\min}(\mathbf{K}) \tanh(k_{aux})}$$

thus implying that the error $\tilde{\xi}$ is ultimately bounded by,

$$\|\tilde{\xi}\| \le \frac{k_{aux} \|\Upsilon\|}{\varsigma \lambda_{\min}(\mathbf{K}) \tanh(k_{aux})}; \text{ with } 0 < \varsigma < 1 \tag{9}$$

It is important to indicate that the desired velocity is $\mathbf{v_{hd}} = \dot{\tilde{\xi}}_d - \Upsilon$. Once the control error is inside the bound (9), that is with small values of $\tilde{\xi}$, $\mathbf{L_K} \tanh\left(\mathbf{L_K^{-1} K \tilde{\xi}}\right) \approx \mathbf{K \tilde{\xi}}$. Now, we prove by contradiction that this control error tends to zero. The closed loop Eq. (7) can be written as, $\dot{\tilde{\xi}} + \mathbf{K \tilde{\xi}} = \Upsilon$ or after the transient, in Laplace transform,

$$\dot{\tilde{\xi}}(s) = \frac{1}{s\mathbf{I} + \mathbf{K}} \Upsilon(s). \tag{10}$$

According to (10) and recalling that $\mathbf{K}$ is diagonal positive definite, the control error vector $\tilde{\xi}$ and the velocity vector Υ can not be orthogonal. Nevertheless both vectors are orthogonal by definition (see Remark 1 and remember the minimum distance criteria for $\tilde{\xi}_d$ on P). Therefore the only solution for steady state is that $\tilde{\xi}(t) \to \mathbf{0}$ asymptotically.

Remark 2: The kinematic characteristics of the omnidirectional mobile platform with four rude mechanun allows that the maneuverability angular velocity ω_p to independently control the orientation of the mobile platform, Sect. 3. Hence $\tilde{\xi}_\phi(t) \to 0$ asymptotically independent of the control errors of $\tilde{\xi}_x$ and $\tilde{\xi}_y$.

4 Experimental Results

In order to illustrate the performance of the proposed controller, several experiments were carried out for path following control of an omnidirectional mobile platform. Some of the most representative results are presented in this Section. The experimental test was implemented on a SASHA robot, which admits two linear velocities and an angular velocities as input reference signals, see Fig. 3.

Fig. 3. SASHA omnidirectional mobile robot

The experiment corresponds to the control system shown in Fig. 4. Note that for the path following problem, the desired velocity of the mobile robot will depend on the task, the control error, the angular velocity, etc. For this experiment, it is consider that the reference velocity module depends on the control errors. Then, reference velocity in this experiment is expressed as $|\mathbf{v_{hd}}| = \upsilon_P/(1 + k\rho)$, where k is a positive constant that weigh the control error module. Also, the desired location is defined as the closest point on the path to the mobile robot.

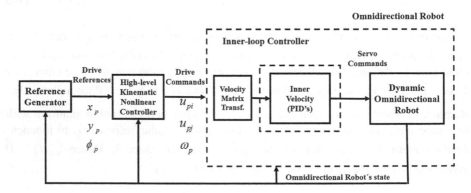

Fig. 4. Block diagram of the propoused controller.

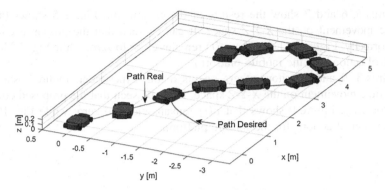

Fig. 5. Stroboscopic movement of the mobile platform in the path following problem.

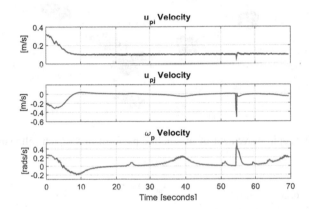

Fig. 6. Maniobrability velocity of the mobile platform

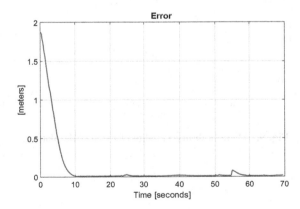

Fig. 7. Distance between the mobile platform position and the closest point on the path.

Figures 5, 6 and 7 show the results of the experiment. Figure 5 shows the stroboscopic movement on the *X-Y-Z* space. It can be seen that the proposed controller works correctly. Figure 6 shows that $\rho(t)$ remains close to zero; while Fig. 7 illustrates the control actions for the mobile platform.

Figures 8, 9 and 10 show others results of the experiment. Figure 8 shows the stroboscopic movement on the *X-Y-Z* space. It can be seen that the proposed controller works correctly. Figure 9 shows that $\rho(t)$ remains close to zero; while Fig. 10 illustrates the control actions for the mobile platform.

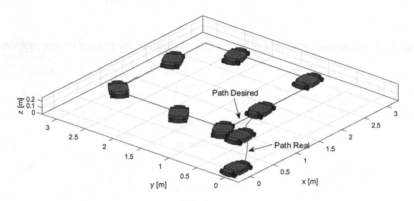

Fig. 8. Stroboscopic movement of the mobile platform in the path following problem.

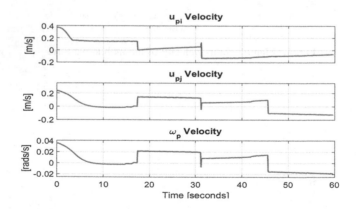

Fig. 9. Maniobrability velocity of the mobile platform

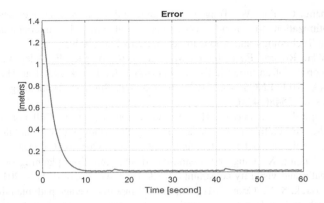

Fig. 10. Distance between the mobile platform position and the closest point on the path.

5 Conclusions

In this work, it is proposed kinematic nonlinear controller for solving the path following problem of an omnidirectional mobile platform. In addition it is presents the mechanical design of the platform, which has reference velocities as control signals to the robot. Also, in the kinematic modeling of the platform, the control point $\xi(t) \in \Re^m$ is considered at a distance a from the center of mass of the robot. Position in which the load is located, or work tool, *i.e.*, laser sensor, robotic arm. Finally, the stability is proved by considering the Lyapunov's method, and the performance of the proposed controller is shown through real experiments.

References

1. Mandal, S., Singh, K., Behera, R.K., Sahu, S.K., Raj, N., Maiti, J.: Human error identification and risk prioritization in overhead crane operations using HTA, SHERPA and fuzzy VIKOR method. J. Expert Syst. Appl. **42**, 7195–7206 (2015). Elsevier
2. Rubio, F., Llopis-Albert, C., Valero, F., Suñera, J.L.: Industrial robot efficient trajectory generation without collision through the evolution of the optimal trajectory. J. Rob. Autonom. Syst. **86**, 106–112 (2016). Elsevier
3. Wang, X., Fu, M., Ma, H., Yang, Y.: Lateral control of autonomous vehicles based on fuzzy logic. J. Control Eng. Pract. **34**, 1–17 (2015). Elsevier
4. Kanjanawanishku, K., Hofmeister, M., Zell, A.: Path following with an optimal forward velocity for a mobile robot. IFAC Proc. **43**, 19–24 (2010). Elsevier
5. Lee, J.-K., Park, J.-B., Choi, Y.-H.: Tracking control of nonholonomic wheeled mobile robot based on new sliding surface with approach angle. IFAC Proc. **46**, 38–43 (2013). Elsevier
6. Soeanto, D., Lapierre, L., Pascoal, A.: Adaptive non-singular path-following, control of dynamic wheeled robots. In: Proceedings of 42nd IEEE/CDC, Hawaii, USA, pp. 1765–1770 (2003)
7. Egerstedt, M., Hu, X., Stotsky, A.: Control of mobile platforms using a virtual vehicle approach. IEEE Trans. Rob. Autom. **46**(11), 1777–1782 (2001)

8. Xu, Y., Zhang, C., Bao, W., Tong, L.: Dynamic sliding mode controller based on particle swarm optimization for mobile robot's path following. In: International Forum on Information Technology and Applications, pp. 257–260 (2009)

9. Andaluz, V.H., Roberti, F., Toibero, J.M., Carelli, R., Wagner, B.: Adaptive dynamic path following control of an unicycle-like mobile robot. In: Jeschke, S., Liu, H., Schilberg, D. (eds.) ICIRA 2011. LNCS, vol. 7101, pp. 563–574. Springer, Heidelberg (2011). doi:10. 1007/978-3-642-25486-4_56

10. Wangmanaopituk, S., Voos, H., Kongprawechnon, W.: Collaborative nonlinear model-predictive collision avoidance and path following of mobile robots. In: ICROS-SICE International Joint Conference 2009, pp. 3205–3210 (2009)

11. Huang, Y., Zhang, X., Fang, Y.: Vision-based minimum-time planning of mobile robots with kinematic and visibility constraints. IFAC Proc. **47**, 11878–11883 (2014). Elsevier

12. Lee, T.-K., Baek, S.-H., Choi, Y.-H., Oh, S.-Y.: Smooth coverage path planning and control of mobile robots based on high-resolution grid map representation. Rob. Auton. Syst. **59**, 801–812 (2011). Elsevier

13. Mazzolani, F.M.: Design criteria for aluminium structures: technology, codification and applications. In: Mazzolani, F.M. (ed.) Aluminium Structural Design. ICMS, vol. 443, pp. 1–87. Springer, Vienna (2003). doi:10.1007/978-3-7091-2794-0_1

14. Fuji, H., Maeda, M., Nogi, K.: Tensile properties and fracture locations of friction-stir welded joints of 6061-T6 aluminum alloy. J. Mater. Sci. Lett. **22**, 1061–1063 (2003). Kluwer Academic Publishers

15. Menzemer, C.C., Fei, L., Srivatsan, T.S.: Design criteria for bolted connection elements in aluminum alloy 6061. ASME – J. Mechan. Des. **121**, 348–351 (1999)

16. Maddox, S.J.: Review of fatigue assessment procedures for welded aluminium structures. Int. J. Fatigue **25**, 1359–1361 (2003). Elsevier

17. Burgess, S.C.: A backwards design method for mechanical conceptual design. ASME – J. Mech. Des. **134** (2012)

18. Li, J., Gao, X., Huang, Q., Du, Q., Duan, X.: Mechanical design and dynamic modeling of a two-wheeled inverted pendulum mobile robot. In: IEEE - International Conference on Automation and Logistics, China (2007)

19. Kong, L., Sabbaghian, M.: Design of an engagement guiding mechanism for roller chain drives. ASME – J. Mech. Des. **118**, 538–543 (1996)

20. Troedsson, I., Vedmar, L.: A method to determine the static load distribution in a chain drive. ASME – J. Mech. Des. **121**, 402–408 (1999)

21. Rolovic, R., Tipton, S.M., Sorem Jr., J.R.: Multiaxial stress concentration in filleted shafts. ASME – J. Mech. Des. **123**, 300–303 (2001)

22. Bagci, C.: Computer-aided fatigue design of power transmission shafts using three-dimensional finite shaft element and updated mean stress diagram. ASME – J. Mech. Trans. Autom. Des. **109**, 533–540 (1987)

23. Efendi, J., Salih, M., Rizón, M., Yaacob, S., Adom, A.H., Mamat, M.R.: Designing omni-directional mobile robot with mecanum wheel. Am. J. Appl. Sci. **3**, 1831–1835 (2005). Science Publications

24. Udengaard, M., Lagnemma, K.: Analysis, design, and control of an omnidirectional mobile robot in rough terrain. ASME – J. Mech. Des. **131**, 121002 (2009)

25. Wang, T., Hopperstad, O.S., Larsen, P.K., Lademo, O.G.: Evaluation of a finite element modelling approach for welded aluminium structures. Comput. Struct. **84**, 2016–2032 (2006). Elsevier

Virtual Reality and Artificial Intelligence

Leaf Recognition for Plant Classification Based on Wavelet Entropy and Back Propagation Neural Network

Meng-Meng Yang[1], Preetha Phillips[2,3], Shuihua Wang[1,4], and Yudong Zhang[1,5(✉)]

[1] School of Computer Science and Technology, Nanjing Normal University, Nanjing 210023, Jiangsu, China
yudongzhang@ieee.org, zhangyudong@njnu.edu.cn
[2] School of Natural Sciences and Mathematics, Shepherd University, Shepherdstown, WV 25443, USA
[3] West Virginia School of Osteopathic Medicine, 400 N Lee St., Lewisburg, WV 24901, USA
[4] Department of Electrical Engineering, The City College of New York, CUNY, New York, NY 10031, USA
[5] Jiangsu Key Laboratory of Advanced Manufacturing Technology, Huaiyin 223003, Jiangsu, China

Abstract. In this paper, we proposed a method for plant classification, which aims to recognize the type of leaves from a set of image instances captured from same viewpoints. Firstly, for feature extraction, this paper adopted the 2-level wavelet transform and obtained in total 7 features. Secondly, the leaves were automatically recognized and classified by Back-Propagation neural network (BPNN). Meanwhile, we employed K-fold cross-validation to test the correctness of the algorithm. The accuracy of our method achieves 90.0%. Further, by comparing with other methods, our method arrives at the highest accuracy.

Keywords: Feature extraction · Classification · Back-Propagation · K-fold cross-validation · Pattern recognition

1 Introduction

With acceleration of the extinction rate of plant species, it is essential urgent to protect plants [1, 2]. A preliminary task is the classification of plant type, which is challenging, complex and time-consuming. As we all know, plants includes the flowering plants, conifers and other gymnosperms and so on. Most of them do not blooming and have fruit, but almost all of them contain leaves [3]. Therefore, in this paper we focus on feature extraction and classification of leaves.

Recent studies are analyzed below: Heymans, Onema and Kuti [4] proposed a neural network to distinguish different leaf-forms of the opuntia species. Wu, Bao, Xu, Wang, Chang and Xiang [5] employed probabilistic neural network (PNN) for leaf recognition system. Wang, Huang, Du, Xu and Heutte [6] classified plant leaf images with complicated background. Jeatrakul and Wong [7] introduced Back Propagation Neural Network (BPNN), Radial Basis Function Neural Network (RBFNN), Probabilistic Neural Network (PNN) and compared the performances of them. Dyrmann, Karstoft and Midtiby [8] used convolutional neural network (CNN) to classify plant species.

© Springer International Publishing AG 2017
Y. Huang et al. (Eds.): ICIRA 2017, Part III, LNAI 10464, pp. 367–376, 2017.
DOI: 10.1007/978-3-319-65298-6_34

Zhang, Lei, Zhang and Hu [9] employed semi-supervised orthogonal discriminant projection for plant leaf classification.

Although these methods have achieved good results, ANN has the higher accuracy and less time-consuming in classification than other approaches. Therefore, in this paper, we employ the BPNN algorithm on the leaves of the classification of automatic identification.

Our contribution in this paper includes: (i) We proposed a five-step preprocessing method, which can remove unrelated information of the leaf image. (ii) We developed a leaf recognition system.

2   Methodology

2.1   Pretreatment

We put a sheet of glass over the leaves, so as to unbend the curved leaves. We pictured all the leaves indoor (put the leaves on the white paper) using a digital camera with (Canon EOS 70D) by two cameraman with experience over five years. The pose of the camera is fixed on the tripod during the imaging. Two light-emitting diode (LED) lights are hanged 6 inches over the leaves. Those images out of focus are removed.

2.2   Image Preprocessing

During the imaging, the background information is captured, which will inference the detection of the leaves. Therefore, the necessary preprocessing is to remove the irrelevant background and color channels.

First, we suppose texture is related to leave category, and color information is of little help. Hence, we removed the background of the image. Second, we convert the RGB image into a grayscale image. Table 1 shows the steps of image preprocessing.

Table 1. Steps of image preprocessing

Step 1: Segment the leaves manually
Step 2: Crop the left and right margins, to form a rectangular image
Step 3: Resize to the size of [200, 200].
Step 4: Fill the background with black color (value = 0)
Step 5: Convert RGB to gray-level image

Figure 1(a) shows the original image. Figure 1(b) shows the image of background with black color. Figure 1(c) shows the grayscale image

(a) Original image   (b)background with black color   (c) Grayscale image

Fig. 1. Illustration of leaves

2.3   Feature Extraction

Fourier transform (FT) [10] is a method of signal analysis, which decomposes the continuous signal into harmonic waves with different frequency [11]. It can more easily deal with the original signal than traditional methods. However, it has limitations for non-stationary processes, it can only get a signal which generally contains the frequency, but the moment of each frequency is unknown. A simple and feasible handling method is to add a window. The entire time domain process is decomposed into a number of small process with equal length, and each process is approximately smooth, then employing the FT can help us to get when each frequency appears. We call it as Short-time Fourier Transform (STFT) [12, 13].

The drawback of STFT is that we do not know how to determine the size of windowing function. If the window size is small, the time resolution will be good and the frequency resolution will be poor. On the contrary, if the window size is large, the time resolution will be poor and the frequency resolution will be good.

Wavelet transform (WT) [14–16] has been hailed as a microscope of signal processing and analysis. It can analyze non-stationary signal and extract the local characteristics of signal. In addition, wavelet transform has the adaptability for signal in signal processing and analysis [17], therefore, it is a new method of information processing which is superior to the FT and STFT.

In this paper, the feature extraction [18, 19] of the original leaf images is carried out by 2-level wavelet transform, which it decompose the leaves images with the low-frequency and high-frequency coefficients. However, a massive of features not only increase the cost of computing but also consume much storage memory does little to classification [20]. We need to take some measures to select important feature. Entropy [21, 22] is used to measure the amount of information of whole system in the information theory and also could represent the texture of the image.

$$\begin{bmatrix} X \\ p(x) \end{bmatrix} = \begin{bmatrix} x_1 \ x_2 \ x_3 \ \cdots \ x_n \ x_{n+1} \\ p_1 \ p_2 \ p_3 \ \cdots \ p_n \ p_{n+1} \end{bmatrix}, \\ 0 \leq p_i \leq 1, \sum_{i=1}^{n+1} p_i = 1 \tag{1}$$

where X is a discrete and random variable, $p(x)$ is the probability mass function, the amount of information contained in a message signal x_i can be expressed as

$$I(x_i) = -\log p_i \tag{2}$$

where $I(x_i)$ is a random variable, it can't be used as information measure for the entire source [23], Shannon, the originator of modern information theory, defines the average information content of X as information entropy [24, 25]:

$$H(X) = \mathrm{E}[I(x_i)] = -\sum p_i \log p_i \tag{3}$$

where E is the excepted value operator, H represents entropy. Figure 2 shows the entropy of source.

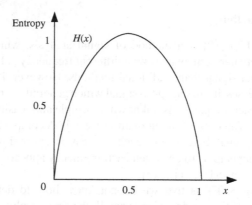

Fig. 2. The entropy of source

There are seven comprehensive indexes (as shown in Fig. 3) are obtained after adopting 2-level WT, with four in size of 50×50 and three 100×100. The entropy of these matrices is calculated as the input to the following BP neural network (BP).

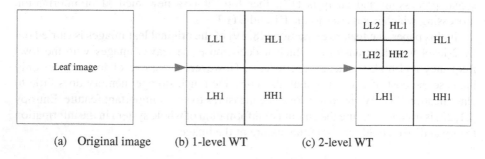

(a)   Original image   (b) 1-level WT   (c) 2-level WT

Fig. 3. 2-level WT

2.4   Back-Propagation Neural Network

Back Propagation algorithm is mainly used for regression and classification. It is one of the most widely used training neural networks model. Figure 4 represents the diagram of BP, the number of nodes in the input layer and output layer is determined but the number of nodes in the hidden layer is uncertain. There is an empirical formula can help to determine the number of hidden layer nodes, as follow

$$t = \sqrt{r + s} + c \tag{4}$$

where t, r, s represents the number of nodes in the hidden layer, input layer, and output layer, respectively. c is an adjustment constant between zero and ten.

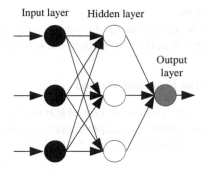

Input layer   Hidden layer

Output
layer

Fig. 4. The diagram of BP

In our experiment, the features of leaves were reduced to seven after feature selection, i.e. $r = 7$. Meanwhile, there is one output unit, which stands for the predictable result, so $s = 1$. Because c is constant, according to formula (4), we set $t = 15$.

BPNN algorithm is part of a supervised learning method. The following is the main ideas of the BP algorithm learning rule:

Known vectors: input learning samples $\{P^1, P^2, \ldots \ldots, P^q\}$, the corresponding output samples $\{T^1, T^2, \ldots \ldots, T^q\}$.

Learning objectives: The weights are modified with the error between the target vector $\{T^1, T^2, \ldots \ldots, T^q\}$ and the actual output of the network $\{A^1, A^2, \ldots \ldots, A^q\}$ in order to A^i ($i = 1, 2, \ldots, q$) is as close as possible to the excepted T^i, i.e., error sum of squares of the network output layer is minimized.

BPNN algorithm has two parts: the forward transfer of working signal and error of the reverse transfer. In the forward propagation process, the state of each layer of neurons only affects the state of the next layer of neurons.

$$x_j = \sum_i w_{ij} P^i \tag{5}$$

where x_j represents the input of hidden layer. w_{ij} is the weight between input layer and the hidden layer.

$$x_j' = f(x_j) = 1/(1 + e^{(-x_j)}) \tag{6}$$

Equation (6) indicates the function of hidden layer.

$$A^i = \sum_j w_{jk} x_j' \tag{7}$$

A^i is the actual output of output layer.

$$e = \frac{1}{2} \sum_i \left(A^i - T^i\right)^2 \tag{8}$$

where T^i is the desired output.

If the desired output T^i is not obtained at the output layer, we will use the Eq. (8) and the error value will be calculated, then the error of the reverse transfer is carried out.

2.5   Implementation of the Proposed Method

The aim of this study was to distinguish the type of leaves with high classification accuracy. The steps of proposed system are sample collection, pretreatment, preprocessing, feature extraction, classifier classification (Fig. 5). Table 2 shows the pseudocode of the proposed system.

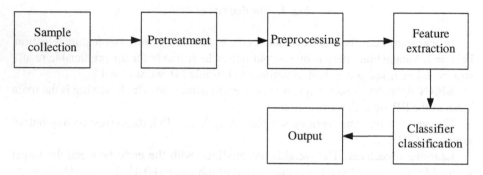

Fig. 5. Pipeline of our proposed method

Table 2. Pseudocode of the proposed system

Proposed Approach

Input: grayscale image.
Parameter: *NIM*, total image number. *NK*, number of k-fold cross validation.
Phase1. 2-level WT.
For *i*=1:NIM
      Read in the image fle.
      A matrix I[$200 \times 200 \times NIM$]is applied to store all grayscale images
      Employ the WT to extract 7 coefficients.
      A matrix out[$1 \times NIM$] is employed to store the classification of leaves
End
Phase2.Extract entropy from the grayscale image
For *j*=1:*NIM*
      Extract entropy from matrix $I[200 \times 200 \times NIM$].
      Put the entropy into a matric $M[7 \times NIM$].
End
Phase3.BP Classification using 5×5 cross validation.
Divide the input data $M[7 \times NIM$]and target data out[$1 \times NIM$] into five different groups randomly.
For *i*=1:*NK*
      Regard the *i*th group as test, and other 4 groups to train the BPNN.
        Output the classification results.
End
Calculate the average accuracy.

2.6   Five-Fold Cross Validation

The training criterion of the standard BP neural network is to require that the error sum of squares (or the fitting error) of the expected and output values of all samples is less than a given allowable error ε. In general, the smaller the value of ε is, the better the fitting accuracy. Nevertheless, for the actual application, the prediction error decreases with the decrease of the fitting error at the first. However, when the fitting error on training set decreases to a certain value, the prediction error on test set increases, which indicates that the generalization ability decreases. This is the "over-fitting" phenomenon which the BPNN modeling encountered. In this paper, cross-validation is used to prevent over-fitting.

The basic of the cross validation method is that the original dataset of the neural network were divided into two parts: the training set and the validation set. First of all, the classifier was trained by training set, and then test the training model using the validation set, through the above results to evaluate the performance of the classifier.

The steps of K-fold cross validation:

Step 1.   All training set S is divided into k disjoint subsets, suppose the number of training examples in S is m, then the training examples of each subset is m/k, the corresponding subset is called $\{s_1, s_2, \ldots., s_k\}$.

Step 2.   A subset is selected as test set from subset $\{s_1, s_2, \ldots., s_k\}$ each time, and the other $(k-1)$ as the training set.

Step 3.   Training model or hypothesis function is obtained through training.

Step 4.   Put the model into the test set and gain the classification rate.

Step 5.   Calculate the average classification rate of the k-times and regard average as the true classification rate of the model or hypothesis function.

3   Experiment and Discussions

The method of BP is implemented in MATLAB R2016a (The Mathworks, Natick, MA, USA). This experiments were accomplished on a computer with 3.30 GHz core and 4 GB RAM, running under the Window 8 and based on 64-bit processors.

3.1   Database

We used leaves as the object of study, which are all size of 200×200 and in jpg format. The input dataset contains 90 images, which 30 of ginkgo biloba, 30 of Phoenix tree leaf and 30 of Osmanthus leaves. As mentioned above, we use the 5-fold cross validation to prevent the case of over-fitting, Thence, we ran one trials with each 72 (25 ginkgo biloba and 26 Osmanthus and 21 Phoenix tree) are used for training and the left 18 (5 ginkgo biloba and 4 Osmanthus and 9 Phoenix tree) are used for test. Figure 6 show one trail of 5-fold cross-validation.

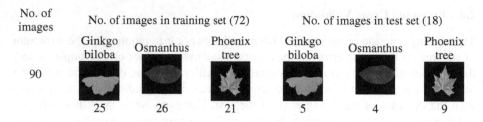

Fig. 6. One trail of five-fold cross-validation

3.2   Algorithm Comparison

The wavelet-entropy features were fed into different classifiers. We compared our method (BPNN) with Medium KNN [26], Coarse Gaussian SVM [27], Complex Tree [28], Cosine KNN [29]. The results of comparing were shown in Table 3.

Table 3. Comparison with different methods

Method	Accuracy
Medium KNN [26]	84.4%
Coarse Gaussian SVM [27]	66.7%
Complex Tree [28]	81.1%
Cosine KNN [29]	82.2%
BPNN (Our)	90.0%

(KNN = k-nearest neighbors, SVM = support vector machine, BPNN = back-propagation neural network)

Results in Table 3 are the central contribution of our experiment, we can see that the accuracy of our method achieves 90.0%, and it is better than other methods. Why BPNN performs the best among the fiver algorithms? The reason lies in the universal approximation theorem proven in reference [30].

4   Conclusion and Future Research

This paper introduced a novel automatic classify method for the leaves images. With the combination of BP neural network and wavelet-entropy, the accuracy of our method achieves 90.0%.

It costs 1.2064 s to finish feature extraction for 90 leaves images and the average time for each image is 0.0134 s.

Nevertheless, there are still several problems remaining unsolved: (1) we try to improve the accuracy of the algorithm; (2) we can employ other methods of feature extraction in order to decrease the time of extraction; (3) we may extend this method to other type of images, such as car image, tree image, etc.

Acknowledgment. This paper is financially supported by Natural Science Foundation of China (**61602250**), Natural Science Foundation of Jiangsu Province (**BK20150983**).

References

1. Carro, F., Soriguer, R.C.: Long-term patterns in Iberian hare population dynamics in a protected area (Donana National Park) in the southwestern Iberian Peninsula: effects of weather conditions and plant cover. Integr. Zool. **12**, 49–60 (2017)
2. Lim, S.H., et al.: Plant-based foods containing cell wall polysaccharides rich in specific active monosaccharides protect against myocardial injury in rat myocardial infarction models. Sci. Rep. **6**, 15 (2016). Article ID: 38728
3. Du, J.X., et al.: Computer-aided plant species identification (CAPSI) based on leaf shape matching technique. Trans. Inst. Meas. Control **28**, 275–284 (2006)
4. Heymans, B.C., et al.: A neural network for Opuntia leaf-form recognition. In: IEEE International Joint Conference on Neural Networks, pp. 2116–2121. IEEE (1991)
5. Wu, S.G., et al.: A leaf recognition algorithm for plant classification using Probabilistic Neural Network. In: International Symposium on Signal Processing and Information Technology, p. 120. IEEE (2007)
6. Wang, X.F., et al.: Classification of plant leaf images with complicated background. Appl. Math. Comput. **205**, 916–926 (2008)
7. Jeatrakul, P., Wong, K.W.: Comparing the performance of different neural networks for binary classification problems. In: Eighth International Symposium on Natural Language Processing, Proceedings, pp. 111–115. IEEE (2009)
8. Dyrmann, M., et al.: Plant species classification using deep convolutional neural network. Biosyst. Eng. **151**, 72–80 (2016)
9. Zhang, S.W., et al.: Semi-supervised orthogonal discriminant projection for plant leaf classification. Pattern Anal. Appl. **19**, 953–961 (2016)
10. Meier, D.C., et al.: Fourier transform infrared absorption spectroscopy for quantitative analysis of gas mixtures at low temperatures for homeland security applications. J. Testing Eval. **45**, 922–932 (2017)
11. Tiwari, S., et al.: Cloud point extraction and diffuse reflectance-Fourier transform infrared spectroscopic determination of chromium(VI): A probe to adulteration in food stuffs. Food Chem. **221**, 47–53 (2017)
12. Garrido, M.: The feedforward short-time fourier transform. IEEE Trans. Circ. Syst. II-Express Briefs **63**, 868–872 (2016)
13. Saneva, K.H.V., Atanasova, S.: Directional short-time Fourier transform of distributions. J. Inequal. Appl. **10**, Article ID: 124 (2016)
14. Huo, Y., Wu, L.: Feature extraction of brain MRI by stationary wavelet transform and its applications. J. Biol. Syst. **18**, 115–132 (2010)
15. Ji, G.L., Wang, S.H.: An improved reconstruction method for CS-MRI based on exponential wavelet transform and iterative shrinkage/thresholding algorithm. J. Electromag. Waves Appl. **28**, 2327–2338 (2014)
16. Yang, M.: Dual-tree complex wavelet transform and twin support vector machine for pathological brain detection. Appl. Sci. **6**, Article ID: 169 (2016)
17. Liu, A.: Magnetic resonance brain image classification via stationary wavelet transform and generalized eigenvalue proximal support vector machine. J. Med. Imaging Health Inform. **5**, 1395–1403 (2015)

18. Bezawada, S., et al.: Automatic facial feature extraction for predicting designers' comfort with engineering equipment during prototype creation. J. Mech. Des. **139**, 10 (2017). Article ID: 021102
19. Gerdes, M., et al.: Decision trees and the effects of feature extraction parameters for robust sensor network design. Eksploat. Niezawodn. **19**, 31–42 (2017)
20. Zhang, Y.: Binary PSO with mutation operator for feature selection using decision tree applied to spam detection. Knowl. Based Syst. **64**, 22–31 (2014)
21. Yang, J.: Preclinical diagnosis of magnetic resonance (MR) brain images via discrete wavelet packet transform with Tsallis entropy and generalized eigenvalue proximal support vector machine (GEPSVM). Entropy **17**, 1795–1813 (2015)
22. Phillips, P., et al.: Pathological brain detection in magnetic resonance imaging scanning by wavelet entropy and hybridization of biogeography-based optimization and particle swarm optimization. Prog. Electromag. Res. **152**, 41–58 (2015)
23. Sun, P.: Pathological brain detection based on wavelet entropy and Hu moment invariants. Bio-Med. Mater. Eng. **26**, 1283–1290 (2015)
24. Wei, L.: Fruit classification by wavelet-entropy and feedforward neural network trained by fitness-scaled chaotic ABC and biogeography-based optimization. Entropy **17**, 5711–5728 (2015)
25. Yang, J.: Identification of green, Oolong and black teas in China via wavelet packet entropy and fuzzy support vector machine. Entropy **17**, 6663–6682 (2015)
26. Zhou, X.-X.: Comparison of machine learning methods for stationary wavelet entropy-based multiple sclerosis detection: decision tree, k-nearest neighbors, and support vector machine. Simulation **92**, 861–871 (2016)
27. Sharma, B., et al.: Traffic accident prediction model using support vector machines with Gaussian kernel. In: Fifth International Conference on Soft Computing for Problem Solving, pp. 1–10. Springer, Berlin (2016)
28. Maleszka, M., Nguyen, N.T.: Using subtree agreement for complex tree integration tasks. In: Selamat, A., Nguyen, N.T., Haron, H. (eds.) ACIIDS 2013. LNCS, vol. 7803, pp. 148–157. Springer, Heidelberg (2013). doi:10.1007/978-3-642-36543-0_16
29. Anastasiu, D.C., Karypis, G.: Fast parallel cosine k-nearest neighbor graph construction. In: 6th Workshop on Irregular Applications: Architecture and Algorithms (IA3), pp. 50–53. IEEE (2016)
30. Nguyen, H.D., et al.: A universal approximation theorem for mixture-of-experts models. Neural Comput. **28**, 2585–2593 (2016)

A Registration Method for 3D Point Clouds with Convolutional Neural Network

Shangyou Ai, Lei Jia, Chungang Zhuang$^{(\boxtimes)}$, and Han Ding

School of Mechanical Engineering, Shanghai Jiao Tong University,
Shanghai 200240, People's Republic of China
cgzhuang@sjtu.edu.cn

Abstract. Viewpoint independent 3D object pose estimation is one of the most fundamental step of position based vision servo, autopilot, medical scans process, reverse engineering and many other fields. In this paper, we presents a new method to estimate 3D pose using the convolutional neural network (CNN), which can apply to the 3D point cloud arrays. An interest point detector was proposed and interest points were computed in both source and target point clouds by region growing cluster method during offline training of CNN. Rather than matching the correspondences by rejecting and filtering iteratively, a CNN classification model is designed to match a certain subset of correspondences. And a 3D shape representation of interest points was projected onto an input feature map which is amenable to CNN. After aligning point clouds according to the prediction made by CNN, iterative closest point (ICP) algorithm is used for fine alignment. Finally, experiments were conducted to show the proposed method was effective and robust to noise and point cloud partial missing.

Keywords: CNN · Point clouds · Point detector · Registration · Rigid transformation

1 Introduction

The scope of object pose estimation ranges from medical data process to automation in industry. For example, position based vision servo (PBVS) [1] is one of the two basic approaches in the field of visual servo control, and it necessitates the pose of the robot with respect to a specific coordinates prior to be known before subsequent execution. The variation of illumination conditions, background clutter, and occlusion makes conventional image-based techniques ineffective. Since 3D LIDAR scanner is far more accessible in recent years, one can obtain the 3D point cloud of an object much easier than before. It becomes very attractive to do the registration work for 3D point clouds as well as for images [2–5]. Among plenty of approaches, iterative closest points (ICP) [6] is a well-known method to solve the registration problem numerically. However, it always suffers from the local minima because of the non-convex characteristic and the iteration nature of the ICP approach. [4] Provides a globally optimal ICP solution based on branch-and-bound method, in exchange of time consumption. Here we focus on coarse registration method to provide initial transformation before using ICP.

© Springer International Publishing AG 2017
Y. Huang et al. (Eds.): ICIRA 2017, Part III, LNAI 10464, pp. 377–387, 2017.
DOI: 10.1007/978-3-319-65298-6_35

Some research focuses on intelligence method for point cloud processing, e.g. convolutional neural networks (CNN). In most of cases, CNN deals with the feature maps which have intuitive interpretation, like the image [7]. In order to deal with point clouds using CNN, an elaborate feature map for point clouds have to be generated. In [8], a Hough accumulator is designed associated with every points in 3D point cloud for normal estimation, and the image-like structure of the accumulator is amenable to CNN.

Since massive unstructured point clouds are difficult to find the point-to-point correspondences between target and source point clouds, point detectors are always designed for reduction of computation complexity [9]. Interest points are selected by detectors according to a specific criterion, which is invariant to rigid transformation. And a correspondence is identified by point descriptor if the similarity between two points greater than a threshold. Many research focuses on design distinctive point descriptor for 3D point clouds [3]. In [10], the geodesic graph model (GGM) was proposed, the method utilized the fact that geodesic-like distance is an invariant structure feature during non-rigid deformation.

Once the interest points are detected, a typical method for estimating the transformation is Random Sample Consensus (RANSAC) [11]. The RANSAC method estimates a transformation for a given set of correspondences iteratively, and yields to the best one that eliminates most of outliers. In this paper, instead of using RANSAC for transformation estimation, we treat the correspondence matching problem as a classification task using CNN. Owing to the effectiveness of our designed point detector, only a few points were efficient for transformation estimation. As mentioned before, a new feature map associated with interest points is also derived to be fed to CNN. After matching the correspondences predicted by CNN, singular value decomposition (SVD) is used for transformation estimation. And ICP is used as fine registration method.

2 Methodology

2.1 Registration Problem

Given two 3D point clouds, addressed as source point cloud S and target point cloud T respectively (source point cloud is available as reference, and target point loud is often acquired by a 3D scanner). We want to find a rigid transformation $\mu(R, p)$ which minimize the error E:

$$E(R, t) = \sum_{i=1}^{N} \|(Rs_i + p) - t_i\|^2 \tag{1}$$

where the set $\{(s_i, t_i) \text{ with } s_i \in S, t_i \in T, i \in 1 \cdots N\}$ forms the correspondences between source and target point clouds. In the cases that different number of points in two point clouds (e.g. partial missing in the target point cloud), only a part of matches are expected, and a rejection scheme is sometimes desirable that discards the points without counterparts [9]. In addition, accurate pair-wise matching for all the points is infeasible in practice due to the high cardinality of point clouds.

3D interest point detectors are always designed to reduce the complexity in correspondences matching [12–14]. The consistency of detected interest points should be guaranteed with the presence of noise and outliers during rigid transformation, i.e., the point detector have to be as discriminative as possible to keep the local shape information invariant to rigid transformation and robust to other disturbances. Interest points are detected in the source and target point cloud respectively, and we obtain interest points set $P^S = \{p_1^S, \ldots, p_{K_S}^S\} \subset S$, $P^T = \{p_1^T, \ldots, p_{K_T}^T\} \subset T$.

A correspondence is identified usually by using point descriptors which describe local neighborhood of each interest point [15]. A correspondence (a, b) hold if:

$$\|S(D(a)) - S(D(b))\| > \tau \tag{2}$$

where D is the descriptor function that mapping local neighborhood of a point to a set of scalar, S is a similarity measurement function, and τ is a predefined threshold. We do not require the correspondences identification by descriptor function in this paper, only interest points are required to match correspondences.

We proposed the interest point detector to denoise the information underlying in point cloud transformation. To this end, a region growing clustering is implemented to ensure the consistency of detected interest points.

2.2 Point Detector

A set of interest points are detected to represent the pose of point cloud. As a premise, sampling strategy is implemented to select the salient points in the model of the point cloud, thereby the original point cloud is represented by small number of points, a region growing clustering is carried out and the interest points are designed to be the center of the clusters with most amount of points. Details are described as follows.

First, we down-sample the source and target point clouds respectively by the strategy of choosing the salient points with a significance metric proposed by [12].

For every point in the point cloud, the covariance matrix is computed according to its K nearest neighbors:

$$COV(p_i) = \sum_{j=1}^{K} (p_j - p_i)(p_j - p_i)^T \tag{3}$$

the smallest eigenvalue of $COV(p_i)$ was chosen to be the significance assigned to each points, which measures the variance of its neighborhood. And the salient points were selected to be the top $\eta_s \times n$ points among all the points in terms of their significance. Here η_s is the sampling rate and n is the total number of points. Second, the salient points were gathered into clusters by a region growing method [16]. A seed point is randomly chosen which had not been clustered, and its neighbor points are gathered into a cluster. Intra-cluster distance threshold T_i was set to keep differences between points in clusters subtle, and inter-cluster distance threshold T_c was set to avoid difference between clusters too small, thus decrease the ambiguity for correspondences matching. Finally, the interest points were set to be centers of the L clusters with most number of points. Note that while increasing the number of interest points can increase

Fig. 1. Visualization of interest point detection. From left to right: Origin point cloud; salient points selected; clusters by region growing clustering (shown in colorful blobs) and detected 7 interest points (shown in red crosses). (Color figure online)

the robustness of pose representation to noise and occlusion, but also increase the cost of computation during registration period. Figure 1 shows the process of interest point detection.

2.3 Matching with CNN

Because the results of region growing clustering will deviate a bit based on the initial state of iteration and other disturbances, the final clusters result will not be identical for the same point cloud in every experiment. Then a deterministic algorithm which sort interest points into a canonical order for correspondence matching will not work. Therefore, we proposed the CNN classification model to achieve automatic correspondences matching. The representation of internal relationship of interest points is set to be the input feature map of CNN. Since the internal relationship between interest points are invariant to rigid transformation, the CNN helps to recover the complex mapping from the representation of interest points to correct correspondences.

As for source and target point clouds, interest points were computed in the source set $\{p_1^S \dots p_{K_S}^S\}$ and target set $\{p_1^T \dots p_{K_T}^T\}$ respectively. In the training step, K_T interest points in the source set were randomly selected, and for the chosen K_T points, the corresponding selection is assigned to a given category, which will be set as the training set target of the CNN. The categorical procedure is specified as follows.

Assume that every detected points in the target set can be matched to a detected point in the source set, there will be a total number of $C_{K_S}^{K_T}$ possible combinations. Then the category of one possible combinations is assigned to one of the $C_{K_S}^{K_T}$ selections, and the mapping from the selection to the category is trivial. Note that though the rapid growth of possible combinations against the number of interest points will increase the complexity of computation incredibly, relatively small number of interest points are chosen in practice (at least three for rigid transformation) make it feasible for point cloud registration.

Instead of input the raw point coordinates to the neural networks, the weighted adjacency matrix of interest points is computed for the input feature map of CNN. Regarding interest points as vertexes in a complete graph, the K_T interest points were then mapped to a weighted adjacency matrix M_t using the Euclidian distance between interest points as weight:

$$M_t = (m_{ij})_{K_T \times K_T}, \ m_{ij}^T = \sqrt{(p_i^T - p_j^T)^2} \qquad (4)$$

Figure 2 illustrates the matching procedure with CNN.

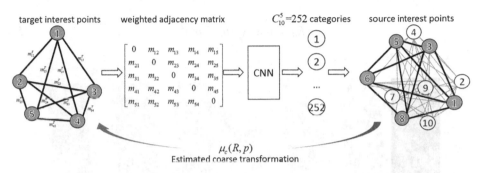

Fig. 2. Illustration of our proposed CNN matching process. 10 and 5 interest points are detected in the source point cloud and target point cloud respectively. Here m_{ij}^T indicates the weight in target cloud. The prediction made by CNN is a set of source interest points. We consider the graph as undirected graph and $m_{ij}^T = m_{ji}^T$.

The reason for this procedure is twofold. First, in order to match the correspondences between the source and target point cloud, the feature map should be invariant to rigid transformational, and robust to noise and outliers, and Euclidian distance between points meet the requirements. Second, the dataset can be transfer from raw arrays of coordinates into an organized feature map, which is amenable for CNN. And taking advantages of local conjunctions detection and shared weights of CNN, the point correspondences which woven in a tangle way originally can be found correctly by the information encoded in the weighted complete graph.

2.4 Pose Estimation

In the online registration step, the weighted adjacency matrix was computed from target set by the same pipeline. Applying the prediction of CNN, a set of points in the source set are assigned as correspondences. And the least-square error transformation is estimated by the SVD method [17], which is used to associate correspondences by all the possible permutation, and select the one with least error in (1).

After coarse registration computed by SVD, a fine registration is performed by implementing the ICP method.

3 Experiments

3.1 Region Growing Cluster

We choose the Stanford happy Buddha and a valve model for evaluation of our proposed method. Figure 3(b) and (c) show the results of region growing clustering. The sampling rate for choosing the salient point is set to be 20%, and the thresholds T_i, T_c for clustering are defined according to the range R_d of point cloud.

$$R_d = \max \sqrt{(p_i^S - p_j^S)^2} \tag{5}$$

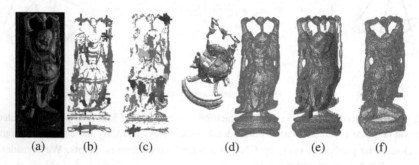

Fig. 3. Registration result of happy Buddha. (a) Model demonstration. (b) and (c) Interest points detection of source and target point cloud respectively. Colorful blobs show the result of top 15 clusters, the final interest points were shown in red crosses after rejection. Note that three correspondences can be found, since interests points can be found in the same region of source and target point cloud. (d) Initial state before registration, target and source point cloud were shown in red and green respectively. (e) Estimated coarse registration. (f) Estimated fine registration using ICP. (Color figure online)

Here we set $T_i = (1/30)R_d$, $T_c = (1/10)R_d$. Figures 3(b) and (c) show the results of clustering for happy Buddha, clusters are shown in colorful blobs. The target point cloud was scanned by laser scanner, and only the points in the front of the model were present. A rejection scheme is implemented to reject clusters according to two parameters for any cluster. First one $\phi_j = \sum_{i=1}^{L}(p_i - p_j)^2$ which measures the total distance from other clusters for cluster j, the second is $\psi_j = \max\sqrt{(p_m - p_n)^2}$ with $m, n \in j$, which indicates the diameter of cluster j. We compute the two parameters for all clusters, and reject the clusters that the ratio ψ_j/ϕ_j are larger than others. After the rejection, final detected interest points were supposed to be the most distinguishable points, and were shown in red crosses in Figs. 3(b) and (c).

3.2 CNN Architecture

The CNN classification model with architecture is shown in Fig. 4. The input to the network is $K_T \times K_T$ weighted adjacency matrix of interest points, we choose $K_T = 3$ here for preliminary experiment.

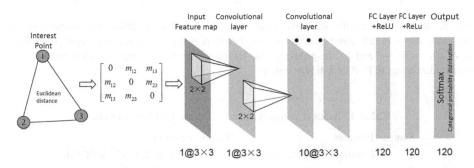

Fig. 4. Architecture of the CNN classification model. Layer's size may be changed according to the input size $K_T \times K_T$.

The first hidden layer convolves 10 filters of kernel size 2×2 with stride 1 and zero padding 1 with the input feature map, and then apply a rectified linear unit (abbreviated as ReLu).

The second layer is a convolutional layer of kernel size 2×2 with stride 1 and zero padding 1. Two fully-connected layer with 120 neurons is followed behind, and the final layer is a softmax layer.

The training data was generated from source point cloud by randomly choosing 10000 weighted adjacency matrices with corresponding categories mentioned in Sect. 2.3. A normalization is also implemented to reduce influence of the variations of scale of point clouds. We implement normalization by multiplying input feature map by a constant α inverse proportional to R_d so that $\alpha R_d = 200$.

3.3 Performance Analysis and Results

The proposed CNN model was trained from scratch, and after applied the proposed method to the test data, matching accuracy achieves 91%. Figures 3 and 5 shows the registration result of proposed method on both happy Buddha and the valve model. Since the number of interest points is relatively small, predicting correspondences from the CNN require less than 0.1 s on a 3.3 GHz Core i5 machine with 8 GB memory. We test the samples which have correct prediction by CNN, and reach the final RMS error 0.0082 (divided by R_d achieves relative error 4.1%) and angular-axis error 0.0837 in average without fine registration. Accuracy can be improved conceivably by increasing the number of interest points, in exchange for more time consumption and complexity of the CNN model. Tables 1 and 3 present an example of detected interest points in happy Buddha and valve model respectively, the prediction made by CNN is No.2,

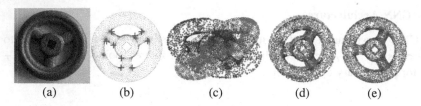

(a)   (b)   (c)   (d)   (e)

Fig. 5. Registration results of valve model. (a) Model demonstration. (b) Interest points detected, shown in red crosses. (c) Initial state before registration, source and target point cloud were shown in '+' and 'o' respectively. (d) Estimated coarse registration. (e) Estimated fine registration using ICP. (Color figure online)

Table 1. Interest points detected in happy Buddha example.

Source	Position	Target	Position
1	(−0.0190, 0.0245, −0.0188)	1	(0.0224, 0.1132, −0.0093)
2	(0.0210, 0.1166, −0.0098)	2	(0.0300, 0.2395, −0.0138)
3	(−0.0074, 0.1880, 0.0004)	3	(−0.0241, 0.0563, 0.0259)
4	(0.0282, 0.1722, −0.0232)		
5	(−0.0312, 0.1151, −0.0233)		
6	(0.0257, 0.2421, −0.0166)		
7	(−0.0112, 0.0885, −0.0179)		
8	(−0.0199, 0.0566, 0.0200)		
9	(−0.0029, 0.0580, 0.0215)		
10	(−0.0190, 0.2448, −0.0188)		

Table 2. Weighted adjacency matrices in happy Buddha example. (Predicted points are the No.2, No.6, and No.8 points in the source set of Table 1).

Source			Target		
0	0.126	0.078	0	0.127	0.081
0.126	0	0.194	0.127	0	0.195
0.078	0.194	0	0.081	0.195	0

No.6, and No.8 points in the source set for happy Buddha, No.1, No.6, No.8 points for valve model. Tables 2 and 4 present the corresponding matrices of target set and predicted points in source set. Comparing with the ground truth rigid transformation, the computed transformation using SVD is 0.0077 for RMS error, 0.0923 for angular-axis error.

Research points out that point detectors may have the drawback of being sensitive to noise [5]. Experiments have been conducted on the valve model. We randomly generate considerable number of noise in the bounding box of point cloud and Fig. 6 shows the linear growth of error against noise. The experiments indicate that with the help of CNN, correspondences matching using local interest points can be robust to noise.

Table 3. Interest points detected in valve model example.

Source	Position	Target	Position
1	(−18.39, 4.877, 0.1504)	1	(−15.55, 3.789, 11.92)
2	(−12.17, 1.412, 0.1009)	2	(1.141, −6.646, −16.45)
3	(−13.84, 13.76, −0.1177)	3	(20.81, −1.991, 2.022)
4	(19.02, 5.172, −0.0981)		
5	(9.177, 10.65, −0.0247)		
6	(13.69, 13.85, −0.0095)		
7	(−4.979, −10.49, −0.0814)		
8	(−5.225, −19.22, 0.0657)		
9	(5.080, −19.26, −0.0089)		

Table 4. Weighted adjacency matrices in valve model. (Predicted points are the No.1, No.6, and No.8 points in the source set of Table 3).

Source	Target
$\begin{bmatrix} 0 & 33.31 & 38.10 \\ 33.31 & 0 & 27.46 \\ 38.10 & 27.46 & 0 \end{bmatrix}$	$\begin{bmatrix} 0 & 34.53 & 38.12 \\ 34.53 & 0 & 27.38 \\ 38.12 & 27.38 & 0 \end{bmatrix}$

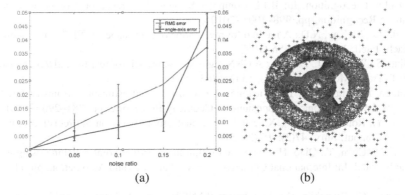

(a)   (b)

Fig. 6. Result of test on sensitivity to noise. (a) RMS error and angle-axis error against noise in average, bars indicate the range of error. (b) Registration result with 20% of noise.

4 Conclusion

We proposed a 3D point cloud registration method, with convolutional neural network for correspondences matching. In this method, only interest points are required to be detected and no requirement for correspondences identification by point descriptors. The feature map of the CNN is the weighted adjacency matrix of complete graph generated by detected interest points. Experimental results show the effectiveness of our proposed method. This method presents a new potential application of CNN in correspondences matching, where limitless ground truth data can be generate to be fed into CNN, and a set of interest points which are detected in target point cloud can be matched to the correct counterparts. Our future research includes utilizing other local descriptions, feature map representation, and some strategies focusing on rejecting interest points.

Acknowledgements. This work is partially supported by the National Natural Science Foundation of China (51375309).

References

1. Chaumette, F., Hutchinson, S.: Visual servo control. I. basic approaches. IEEE Robot. Autom. Magvol **13**(4), 83–90 (2006)
2. Jiang, J., Cheng, J., Chen, X.: Registration for 3-D point cloud using angular-invariant feature. Neurocomputing **72**, 3839–3844 (2009)
3. Rusu, R.B., Blodow, N., Beetz, M.: Fast point feature histograms (FPFH) for 3D Registration. In: IEEE International Conference on Robotics Automation, pp. 1848–1853 (2009)
4. Yang, J., Li, H., Jia, Y.: Go-ICP: solving 3D registration efficiently and globally optimally. IN: IEEE International Conference on Computer Vision, pp. 1457–1464 (2013)
5. Drost, B., Ulrich, M., Navab, N., Ilic, S.: Model globally, match locally: efficient and robust 3D object recognition. In: IEEE Computer Society Conference on Computer Vision and Pattern Recognition, pp. 998–1005 (2010)
6. Besl, P.J., McKay, N.D.: A method for registration of 3-D shapes. IEEE Trans. Pattern Anal. Mach. Intell. **14**(2), 239–256 (1992)
7. Miao, S., Wang, Z.J., Liao, R.: A CNN regression approach for real-time 2D/3D registration. IEEE Trans. Med. Image **35**(5), 1352–1363 (2016)
8. Boulch, A., Marlet, R.: Deep learning for robust normal estimation in unstructured point clouds. In: Eurographics Symposium on Geometry Processing, pp. 281–290 (2016)
9. Diez, Y., Roure, F., Llado, X., Salvi, J.: A qualitative review on 3D registration methods. ACM Comput. Surv. **47**(3), 45 (2015)
10. Qian, D., Chen, T., Qiao, H.: A new algorithm for non-rigid point matching using geodesic graph model. In: International Conference on Mechatronics and Automation, pp. 1174–1180 (2015)
11. Papazov, C., Burschka, D.: An efficient RANSAC for 3D object recognition in noisy and occluded scenes. In: Kimmel, R., Klette, R., Sugimoto, A. (eds.) ACCV 2010. LNCS, vol. 6492, pp. 135–148. Springer, Heidelberg (2011). doi:10.1007/978-3-642-19315-6_11
12. Y. Zhong.: Intrinsic shape signatures: A shape descriptor for 3D object recognition. International Conference on Computer Vision Workshop 3D representation Recognition, pp. 689–696 (2010)
13. Mian, A.S., Bennamoun, M., Owens, R.A.: On the repeatability and quality of keypoints for local feature-based 3D object retrieval from cluttered scenes. Int. J. Comput. Vision **89**(2–3), 348–361 (2008)
14. Chen, H., Bhanu, B.: 3D free form object recognition in range images using local surface patches. Pattern Recogn. Lett. **28**(10), 1252–1262 (2007)

15. Salti, S., Tombari, F., Stefano, L.D.: A performance evaluation of 3D keypoint detectors. In: IEEE International Conference on 3D Imaging, Modeling, Processing, Visualization, and Transmission, pp. 236–243 (2011)
16. Pratt, W.K.: Digital Image Processing, 4th edn., pp. 590–595. Wiley, LosAltos (2007)
17. Arun, K.S., Huang, T.S., Blostein, S.D.: Least-squares fitting of two 3-D point sets. IEEE Trans. Pattern Anal. Machine Intell. **9**, 698–700 (1987)

Tool Wear Condition Monitoring Based on Wavelet Packet Analysis and RBF Neural Network

Tao Li, Dinghua Zhang, Ming Luo[✉], and Baohai Wu

Key Laboratory of Contemporary Design and Integrated Manufacturing
Technology, Ministry of Education, Northwestern Polytechnical University,
Xi'an 710072, China
luoming@nwpu.edu.cn

Abstract. CNC milling is widely used in manufacturing complex parts of aerospace fields, and the development of the intelligent tool wear monitoring can improve the utilization of the tool during the milling process while ensuring the surface quality of the processed parts. In this paper, a novel method based on wavelet packet analysis and RBF neural network was proposed for monitoring the tool wear condition during milling. Firstly, cutting force signals were measured during milling, and filtered by filter function. Secondly, the cutting vibration signals caused by tool wear were separated by the wavelet packet decomposition from initial data, and the energy of the reconstructed signals was characterized for analyzing tool wear during the milling process. Then, the filtered cutting force and the cutting vibration features were trained by RBF neural network. Fifteen groups of features were trained by RBF neural network, and three groups of features were used to test RBF neural network. Finally, the results show that the method can accurately monitor the flank wear of milling cutter within a short time, which provides a theoretical basis and experimental scheme for further implementing the on-line tool wear monitoring.

Keywords: CNC milling · Tool wear · Cutting force · Wavelet packet analysis · RBF neural network

1 Introduction

As an important part of advanced manufacturing technology, online tool condition monitoring technology has become a research topic in recent years [1]. The tool, a direct implementation of the cutting process, inevitably exists wear, breakage and other conditions during the cutting process. Changes in the tool condition directly lead to the increase of cutting force and cutting temperature, the rise of workpiece surface roughness, workpiece size out of tolerance, cutting color change and cutting chatter [2]. Therefore, there has an urgent need to monitor the tool wear condition. Tool wear condition monitoring means that the computer acquires a variety of sensor signal changes during the product processing process, real-time predicts the tool wear or breakage through data fusion method, and uses the alarm devices to accurately prompt the tool change time. Research shows that CNC machine tools with tool monitoring

© Springer International Publishing AG 2017
Y. Huang et al. (Eds.): ICIRA 2017, Part III, LNAI 10464, pp. 388–400, 2017.
DOI: 10.1007/978-3-319-65298-6_36

system can reduce downtime by 75%, improve production efficiency by 10–60%, and increase machine tool utilization by 50% [3].

The intelligent monitoring of tool wear condition is generally divided into two steps: features extraction and features recognition. Referring to the feature extraction methods there are three types: time domain analysis, frequency domain analysis, time-frequency domain analysis. Time domain analysis is a method of signal processing directly in time domain. Haber et al. [4] extracted the mean and peak value of the vibration signal as the features, so as to predict the tool wear. However, the use of time domain signal as features extraction is often affected by the noise signal or the signal changes. As for frequency domain analysis, it transforms the signal from time domain to frequency domain by Fourier series or Fourier transform. Kopac and Sali [5] measured the sound pressure at 0.5 mm from the cutting area by a condenser microphone and analyzed in the frequency domain from 0 to 22 kHz. However, while transforming from time domain to frequency domain using Fourier transform, signals from all the time domain has to be used. The time-frequency domain analysis method can be used to describe the time domain and frequency domain of the signal, which can clearly describe the change of signal frequency with time. Zhu et al. [6] introduced the wavelet analysis method and applied it in tool condition monitoring. Li [7] proposed the wavelet packet analysis of the AE signal during the rotation of the tool. The above research has demonstrated that the time-frequency analysis method can accurately be used for signal time-frequency analysis, the signal in time domain and frequency domain can achieve high resolution.

In recent years, neural network has been widely used in data fusion and recognition. Wang [8] proposed a method for monitoring tool wear based on automatic combined neural network. Elanayar and Shin [9] proposed a dynamic model for tool wear monitoring. Pai et al. [10] proposed a method using a radial basis function network to predict the rake wear in end milling operation. It is found that the prediction results for tool wear using RBF neural network is more robust and high accuracy than resource allocation network. All of these methods are based on fuzzy algorithm and neural network algorithm, choosing the appropriate recognition algorithm plays an important role in improving the recognition convergence speed and improving the local minima during the fitting process.

The above reference only uses the wavelet packet analysis algorithm to divide the signal into time-frequency domain, so as to filter out the noise signal or the environmental signal which is not related to the main signal. In this paper, the wavelet packet analysis algorithm is used to extract the cutting force signals due to vibration which related to tool wear except the cutting force signal, which is the contribution of this paper. The wavelet packet is used to segment the frequency of the cutting force signal in the time-frequency domain. In addition to the cutting force signal at the low frequency, the reconstructed signal energy value of the intermediate frequency band is also related to the tool wear, and the signal is extracted as a vibration signal related to the tool wear. And then this paper adopts RBF neural network to make data fusion for input feature, in a short period of time to predict the tool wear accurately. In Sect. 2, the mathematical model of wavelet packet analysis and RBF neural network are discussed. In Sect. 3, the cutting force signal in the machining process is measured by the milling experiment, and the signal value of the cutting force signal due to vibration in the cutting force signal is separated by wavelet packet as the feature. These two types of characteristic signals are input into the RBF neural network, and the network output is the tool flank wear. In Sect. 3.2, from the analysis data, the tool wear monitoring method can accurately monitor the tool wear in time. Finally, Sect. 4 contains the conclusions.

2 Mathematical Model of Tool Wear Monitoring

In this paper, the theory of wavelet packet analysis and RBF neural network are used to monitor the tool wear. Firstly, the wavelet packet theory is used to analyze the relevant signals in the process, and then the signal energy related to tool wear is extracted as features and input of the RBF neural network model according to the prior knowledge, so as to effectively identify and predict the tool wear. In this section, the wavelet packet theory analysis model and RBF neural network model are discussed.

2.1 A Mathematical Model of Energy Feature Extraction Based on Wavelet Packet Analysis

Wavelet packet theory analysis. In the wavelet packet analysis, the scale function of a standard orthogonalization $\psi(x)$ is used, and with the help of two scale difference recursive equations, functions are generated as formula (1) called orthogonal wavelet packet of $\psi(x)$

$$\begin{cases} w_{2n}(x) = \sqrt{2}\sum_{k=Z} h_k w_n(2x - k) \\ w_{2n+1}(x) = \sqrt{2}\sum_{k=Z} g_k w_n(2x - k) \end{cases} \tag{1}$$

where $w_0 = \psi(x)$, h_k, g_k are respectively a pair of conjugate quadrature filter coefficients derived from $\psi(x)$. In theory, according to the observation signal $s(t) \in L^2(R)$, the discrete orthogonal wavelet packet transform is defined as the projection coefficient of $s(t)$ on the orthogonal wavelet packet base $\{w_{n,j,k}(t)\}_{n \in Z/Z^-, j \in Z, k \in Z}$ [11], namely:

$$P_s(n,j,k) = \langle s(t), w_{n,j,k}(t) \rangle = \int_{-\infty}^{+\infty} s(t)\left[2^{-\frac{j}{2}}\overline{w_n}(2^{-j}t - k)\right]dt \tag{2}$$

where $\{P_s(n,j,k)\}_{k \in Z}$ is the wavelet packet transform coefficient sequence of $s(t)$ on the orthogonal wavelet packet space U_j^n. In fact, according to a certain observation signal $s(t)$, set up a group of low and high conjugate quadrature filter coefficients of $\{h_k\}_{k \in Z}$ and $\{g_k\}_{k \in Z}$, then the wavelet packet transform coefficients can be expressed as the following recursive formula [11]:

$$\begin{cases} P_s(2n,j,k) = \sum_{k=Z} h_{l-2k} \cdot P_s(n,j-1,l) \\ P_s(2n+1,j,k) = \sum_{k=Z} g_{l-2k} \cdot P_s(n,j-1,l) \end{cases} \tag{3}$$

In order to overcome the shortcomings of traditional wavelet transform, wavelet packet transform is used to divide the signal at the time-frequency domain. The wavelet packet can be improved with the increase of resolution 2^j, and the broadened spectrum window has the fine quality of further segmentation. For a given signal, the signal can be divided into any frequency band by a group of low- and high-pass combinatorial quadrature filters H, G, which has high time resolution and frequency resolution in low frequency and high frequency [11]. Firstly, the three-layer wavelet packet decomposition is carried out to process the cutting force signal. The cutting force signal and the cutting force signal due to vibration which related to the wear are extracted from the reconstructed wavelet packet, details about three-layer wavelet packet decomposition process is shown as follows, j is the number of layers (Fig. 1).

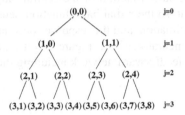

Fig. 1. Tree structure of wavelet packet decomposition

Extraction of the signal eigenvalues related to tool wear using wavelet packet analysis. In the wavelet packet decomposition process, each time the conjugate quadrature filter is used, the signal length is shortened by half. Suppose that the original signal $s(t)$ is decomposed by L-layer wavelet packet. If the original data length is 2^N, after the decomposition of the L layer, the length of each signal becomes 2^{N-L}. Using the wavelet packet, the signal can be decomposed into the corresponding frequency band according to the characteristics of any time-frequency resolution (which satisfies the Heisenberg uncertainty principle), and the signals in different frequency bands are decomposed into the corresponding frequency range, according to the effective frequency of prior knowledge extraction, the signal is reconstructed to restore the signal length to 2^N. At this point, the energy distribution $E(j, n)$ of the reconstructed signal in the time-frequency domain is defined as follows:

$$E(j, n) = \sum_{k=z} [P_s(n, j, k)]^2 \tag{4}$$

In the formula, the discrete numerical calculation of the wavelet packet transform coefficients $P_s(n, j, k)$ is obtained by using the recursive algorithm of formula (3). The key components of the orthogonal wavelet packet space are selected, the principle of which is to select the orthogonal wavelet packet space with relatively concentrated energy. This method can make full use of prior information, so that the main wavelet packet energy eigenvalues of the original signal can be enhanced, and the dimension of the eigenvalue can be reduced. Of course, if the prior information is uncertain or the feature dimension can reduce the classification and recognition ability of the model, it is necessary to retain the wavelet packet energy eigenvalue as the final feature vector. In this paper, the reconstruction signal energies of cutting force signals and the cutting force signals due to vibration which related to tool wear are extracted as features to identify subsequent neural network. The cutting force fluctuation signals are decomposed from the original cutting force by wavelet packet method.

2.2 A Mathematical Model for Identifying Tool Wear Based on RBF Neural Network

Artificial neural network (ANN) is a nonlinear dynamic network system which is based on the research of modern neurophysiology and psychology. As a new method of knowledge processing, it has been widely used in many fields. Among them, when the input signal near the radial basis function of the central range, the hidden layer node will produce a larger output, so the radial basis function neural network (RBFNN) has the capacity of local approximation, and the response speed and recognition accuracy is also better than that of BP neural network. Figure 2 is the structure of the RBF neural network, it is a three-layer feedforward network including the input layer, hidden layer and output layer [12].

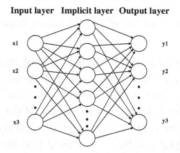

Fig. 2. RBF neural network structure

In the RBF neural network, RBF is used as the excitation function in the hidden layer and the output layer is a simple linear function [13]. The input layer to the hidden layer is nonlinear transmission, and the clustering learning algorithm is adopted. The output layer is trained by least squares algorithm [14–16]. The role of the hidden layer node is to analyze some local areas of the input signal. The important parameter of the hidden layer node is the center and the width of the radial basis function, which are denoted by c and σ, as shown in Eq. (5). When the input vector is closest to the center of a hidden layer node, the output of the hidden layer node is the largest, and the output of the hidden layer node will decrease with the distance between the vector and the radial center.

$$R_j(x) = e^{-\frac{\left\| x - c_j \right\|^2}{2\sigma_j^2}} \tag{5}$$

where $||x - c_j||^2 = \sum_{i=1}^{n} (x_i - c_{ji})^2$, x is the input vector, c_j is the clustering center of the jth Gaussian function in the n-dimensional vector, and can be randomly selected according to the input vector. σ_j^2 is the normalized scale parameter, which is equivalent to the variance. R_j represents the output of the hidden layer function. The output of the hidden layer node is passed to the output layer linearly by weight ω, and the corresponding function expression is shown in formula (6).

$$y = \sum_{i=1}^{n} w_i R_i \tag{6}$$

where w_i is the output weight, and y is the expected output of the RBF neural network. In this paper, the energy value of the reconstructed signal of the cutting force signal and the cutting force fluctuation signals due to vibration are extracted by wavelet packet analysis as the input feature vector. The output layer is the tool flank wear.

3 Experimental Validation and Discussion

3.1 Experimental Setup

The experiments were carried out for up milling, and a Kistler 9123C rotating dynamometer was used to measure the force during the milling process Tool wear was measured with a non-contact optical measurement equipment Alicona InfiniteFocus. And the cutting parameters of this experiment are:

Feed per tooth: $f_z = 0.05$ mm/t;   Spindle speed: $n = 1200$ r/min;
Radial cutting depth: $a_e = 1$ mm;   Axial cutting depth: $a_p = 2$ mm;

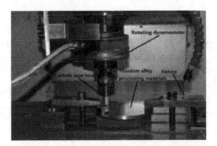

Fig. 3. The experimental setup

To avoid the influence of cutter runout, only one insert was installed, and the machining process was dry cutting. For each cutting parameters group, the same cutting experiments were carried out repeatedly until the tool wear reaches 0.3 mm (Fig. 3).

3.2　Results and Discussion

With the increase of cutting length (L), tool wear gradually increased. In this experiment, 18 appropriate groups are carried out for tool flank wear measurement. The experimental data are shown in Table 1.

Table 1. The tool flank wear of insert cutter

Number	4L	8L	12L	16L	19L	22L	25L	28L	31L
Tool wear/mm	0.042	0.075	0.098	0.128	0.119	0.121	0.126	0.135	0.181
Number	34L	37L	40L	43L	46L	59L	52L	55L	58L
Tool wear/mm	0.19	0.195	0.215	0.232	0.24	0.242	0.244	0.248	0.265

Feature extraction of tool wear based on cutting force. In the actual machining process, the signal of cutting force, vibration and acoustic emission will change significantly with the increase of tool wear. However, the acquisition of all relevant signals will increase the cost and complexity of the device. Therefore, this paper proposes to collect only the cutting force signal during the cutting process. By analyzing the original force signal using the wavelet packet analysis, the signal related to the tool vibration is decomposed. The eigenvalues related to tool wear are extracted from these two kinds of signals, which are used to identify the tool wear. The following are specific analysis of how to extract the tool wear eigenvalues.

The spectrum transform of cutting force based on fast Fourier transform. Firstly, the cutting force of three cycles is described by MATLAB. Secondly, the spectrum analysis is carried out by Fourier transform to transform the signal from the time domain to frequency domain. The fast Fourier transform is used to shift the frequency of unequal Fourier transform data, and the frequency range becomes: $-fs/2 \sim fs/2$. As shown in Figs. 4 and 5.

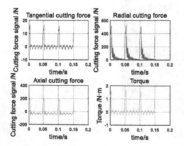
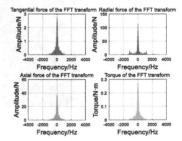

Fig. 4. Three-way cutting force and torque signal

Fig. 5. Spectrum after fast Fourier transform

Extraction of the cutting force fluctuation energy values related to tool wear. According to the analysis of the frequency spectrum, the cutting force signal amplitude is high at low frequency, and the signal amplitude decreases with the increase of frequency. When the frequency is greater than 1.2 kHz, the signal amplitude is small enough to be ignored. According to the experiment, when number N is 8, the time domain waveform becomes smooth, and the frequency characteristic is also good, so choosing the db8 as the wavelet packet basis is more appropriate. According to the results of the spectrum analysis, the $0 \sim 1.2$ kHz signal is decomposed into 8 frequency bands by using db8 wavelet packet decomposition, and the width of each frequency band is 150 Hz, and then the signal is reconstructed for each band. Figure 6 shows the reconstruction of the decomposition of the 8 bands.

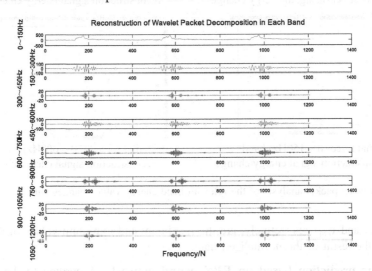

Fig. 6. Reconstruction diagram of three-layer wavelet packet decomposition

As shown in Fig. 6, the signal reconstruction of the $0 \sim 150$ Hz frequency band is regarded as cutting force signal. The energy values of these reconstructed signals by frequency band are analyzed. It is found that except that the energy of the reconstructed signal in the energy spectrum of $0 \sim 150$ Hz is related to the tool wear condition in the energy spectrum of the tangential force f_t, the energy value of the reconstructed signal of the second band $(150 \sim 300$ Hz$)$ is also related to the tool wear condition. The frequency band signal can be identified as the cutting force fluctuation signal energy value, denoted by V_t. In the energy spectrum of the radial force f_r, the energy value of the reconstructed signal in the fourth band $(450 \sim 600$ Hz$)$ is also related to the tool wear condition. The frequency band is identified as the energy value of the cutting force fluctuation signal in the radial force, denoted by V_r. Figure 7 shows the trend of changes of the signal energy in the frequency band with the increase of the tool wear.

In order to facilitate the training and testing of RBF neural network model, the above 18 groups of data need to be normalized.

$$x'_i = \frac{x - x_{min}}{x_{max} - x_{min}} \tag{7}$$

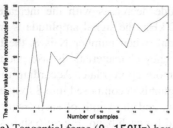
a) Tangential force (0~150Hz) band
reconstruction signal energy change

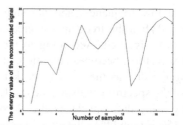
b) Tangential force (150~300Hz) band
reconstruction signal energy change

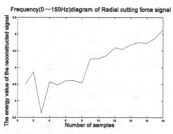
c) Radial force (0 ~ 150Hz) band
reconstruction signal energy change

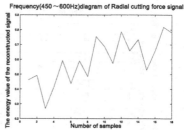
d) Radial force (450 ~ 600Hz) band
reconstruction signal energy change

Fig. 7. Wavelet packet analysis of the relevant band energy value changes with the tool wear

where x_{max} and x_{min}, respectively, are the maximum and minimum values of the data, and x_i' is the result of the normalized data.

Tool wear prediction based on RBF neural network. According to the general description about the mathematical model of RBF neural network in Sect. 2, the number of neuron nodes at the input of neural network is determined by the type of features. The number of hidden neuron nodes can be determined by the empirical formula method (8). The number of output neuron nodes is determined by the expected output value.

$$m = \sqrt{n+l} + \alpha \tag{8}$$

In this paper, the input node are the features associated with the tool wear, and there are 4 features, which are the tangential and radial cutting forces in the cutting process, as well as the cutting force fluctuation signals due to vibration peeling off from the tangential and radial cutting forces (the cutting force fluctuation signal is represented by the symbol V), and the output node is the tool flank wear (denoted by VB). The cutting tool is machined 58 times on the workpiece flank side, and the tool flank wear of the 18 groups were measured as the output values of the neural network (L represents the number of the same tool machined times). The statistics of input vectors and output data are shown in Table 2.

Table 2. Tool wear related features

	F_t/N	F_r/N	V_t/N	V_r/N	VB/mm
4L	95.0	3492.2	9.0	462.3	0.042
8L	131.0	3866.5	14.7	492.2	0.075
12L	90.8	2629.7	14.6	268.7	0.098
16L	122.9	3570.3	12.9	413.7	0.128
19L	115.7	3476.7	17.2	592.4	0.119
22L	120.8	3603.8	16.3	437.0	0.121
25L	119.2	3610.4	19.8	590.8	0.126
28L	125.4	3537.1	17.4	484.1	0.135
31L	137.0	4255.7	16.5	754.2	0.181
34L	141.2	4268.7	17.8	688.0	0.19
37L	146.4	4363.0	19.9	573.1	0.195
40L	131.4	4594.2	20.8	786.1	0.215
43L	125.5	4546.4	11.4	657.9	0.232
46L	139.9	4687.7	13.4	733.5	0.24
49L	134.8	4750.2	18.7	528.5	0.242
52L	139.5	4734.3	20.2	658.1	0.244
55L	141.8	4869.6	20.9	817.2	0.248
58L	145.9	5140.1	20.1	780.5	0.265

According to the data of the 18 groups in Table 2, the model is first normalized, and the model is trained and predicted by MATLAB RBF neural network toolbox. Set the input layer neural network element number 4. The number of output layer is 1. The number of hidden layer is automatically selected by the network, and the neural network diffusion coefficient is 0.3. 19L, 31L, 52L three groups are used to test, and the other tool wear data are for RBF neural network training, as shown in Figs. 8 and 9.

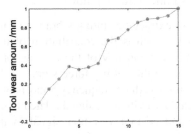

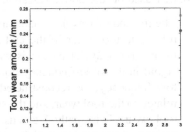

Fig. 8. RBF neural network training results   **Fig. 9.** RBF neural network test results

According to the analysis results of RBF neural network in Sect. 3, the prediction error of the flank wear is calculated by the formula (9). The error analysis results are shown in Table 3.

Table 3. Error analysis of tool flank wear prediction

Tool wear amount/mm	Test groups/group		
	First group	Second group	Third group
Real value (*VB*)	0.119	0.181	0.267
Predictive value (*VB$_m$*)	0.1135	0.1773	0.244
RBFNN predicted error	4.62%	2.04%	9.43%

$$\delta = |VB - VB_m|/VB \times 100 \qquad (9)$$

Among them, *VB* represents tool flank wear, *VB$_m$* on behalf of RBF neural network predictive value. The error of the tool wear prediction shows that the average error of the flank wear of the three groups is 5.36%, and the average time used to predict the three groups using the MATLAB RBF neural network toolbox is 0.13 s. The error can meet the tool wear monitoring accuracy requirements, and the average time can also achieve the effect of on-line monitoring of tool flank wear.

4   Conclusions

In this paper, the wavelet packet analysis and RBF neural network are used to monitor the flank wear of milling cutter. The wavelet packet analysis is used to analyze the signal features related to tool wear from the original cutting force signal, the cutting force signal and the cutting vibration signal which related to the tool wear are extracted, and the energy of the reconstructed signal is taken as the extracted eigenvalue. For the extracted features, the data fusion is carried out by RBF neural network, and then accurately identify the flank wear of milling cutter within a short period of time, which provides theoretical support for the real-time on-line monitoring of the subsequent tool wear. Based on the above research, the article is summarized as follows:

1. The method of wavelet-packet analysis is used to extract the cutting vibration signal related to tool wear in the cutting force signal, which is the contribution of this paper. The wavelet packet is used to segment the frequency of the cutting force signal in the time-frequency domain. In addition to the cutting force signal at the low frequency, the reconstructed signal of the intermediate frequency band is also related to the tool wear, and the signal is extracted as a vibration signal related to the tool wear. This method avoids the measurement of tool vibration signal in the process of machining, which reduces the cost and complexity of the experimental.
2. Three groups of data for RBF test are selected during the process of tool wear, and the collected cutting force signal can be transformed into the input features for RBF neural network entry through software, and then use the trained RBF neural network to output tool wear, so as to achieve on-line monitoring of tool wear.

3. In this paper, RBF neural network is used to identify two kinds of features of tool wear. The method can quickly and effectively identify the flank wear of the milling cutter. The average recognition time can reach 0.13 s, and the average recognition error is 5.36%, which provides a good theoretical basis and experimental scheme for further accurate tool wear monitoring.

Acknowledgements. This project is supported in part by National Basic Research Program of China (Grant No. 2013CB035802) and National Natural Science Foundation of China (Grant No. 51575453).

References

1. Ertunc, H.M., Loparo, K.A.: A decision fusion algorithm for tool wear condition monitoring in drilling. Int. J. Mach. Tools Manufact. **41**(9), 1347–1362 (2001)
2. Li, X.: A brief review: acoustic emission method for tool wear monitoring during turning. Int. J. Mach. Tools Manufact. **42**(2), 157–165 (2002)
3. Rehorn, A.G., Jiang, J., Orban, P.E.: State-of-the-art methods and results in tool condition monitoring: a review. Int. J. Adv. Manufact. Technol. **26**(7–8), 693–710 (2005)
4. Haber, R.E., Jiménez, J.E., Peres, C.R., et al.: An investigation of tool-wear monitoring in a high-speed machining process. Sens. Actuators A Phys. **116**(3), 539–545 (2004)
5. Kopač, J., Šali, S.: Tool wear monitoring during the turning process. J. Mater. Process. Technol. **113**(1), 312–316 (2001)
6. Zhu, K., San, W.Y., Hong, G.S.: Wavelet analysis of sensor signals for tool condition monitoring: A review and some new results. Int. J. Mach. Tools Manufact. **49**(7), 537–553 (2009)
7. Wang, G., Cui, Y.: On line tool wear monitoring based on auto associative neural network. J. Intell. Manufact. **24**(6), 1085–1094 (2013)
8. Elanagar, V.T.S., Shin, Y.C.: Design and implementation of tool wear monitoring with radial basis function neural networks. In: Proceedings of the 1995 American Control Conference, vol. 3, pp. 1722–1726. IEEE (1995)
9. Srinivasa, P., Nagabhushana, T.N., Rao, P.K.R.: Flank wear estimation in face milling based on radial basis function neural networks. Int. J. Adv. Manufact. Technol. **20**(4), 241–247 (2002)
10. Wu, Y., Du, R.: Feature extraction and assessment using wavelet packets for monitoring of machining processes. Mech. Syst. Sig. Process. **10**(1), 29–53 (1996)
11. Rothweiler, J.: Polyphase quadrature filters–a new subband coding technique. In: IEEE International Conference on Acoustics, Speech, and Signal Processing, vol. 8, pp. 1280–1283, ICASSP 1983. IEEE (1983)
12. Elanayar, S., Shin, Y.C.: Robust tool wear estimation with radial basis function neural networks. Trans. Am. Soc. Mech. Eng. J. Dyn. Syst. Measure. Control **117**, 459–467 (1995)
13. Dimla, D.E., Lister, P.M., Leighton, N.J.: Neural network solutions to the tool condition monitoring problem in metal cutting—a critical review of methods. Int. J. Mach. Tools Manufact. **37**(9), 1219–1241 (1997)
14. Kuo, R.J., Cohen, P.H.: Multi-sensor integration for on-line tool wear estimation through radial basis function networks and fuzzy neural network. Neural Netw. **12**(2), 355–370 (1999)

15. Panda, S.S., Chakraborty, D., Pal, S.K.: Flank wear prediction in drilling using back propagation neural network and radial basis function network. Appl. Soft Comput. **8**(2), 858–871 (2008)
16. Yao, Y., Li, X., Yuan, Z.: Tool wear detection with fuzzy classification and wavelet fuzzy neural network. Int. J. Mach. Tools and Manufact. **39**(10), 1525–1538 (1999)

Research on Modeling and Simulation of Distributed Supply Chain Based on HAS

Wang Jian[1,2(✉)], Huang Yang[1,2], and Wang ZiYang[1,2]

[1] School of Automation, Huazhong University of Science and Technology,
Wuhan, China
wj0826_can@mail.hust.edu.cn
[2] Key Laboratory for Image Information Processing and Intelligent
Controlling of Ministry of Education, Wuhan, China

Abstract. A general information model for the interaction communication between supply chain members is not proposed in the simulation modeling methods of distributed supply chain based on HLA (High Level Architecture) and SCOR (Supply Chain Operation Reference). A single simulation model is used for the modeling of internal structure of federates, by which a clear description of the internal structure and action of supply chain members can not be obtained. In order to solve these problems, a simulation modeling method of distributed supply chain based on HAS (HLA-Agent-SCOR) is put forward, according to characteristics of modeling of HLA, SCOR and Agent. Firstly, the supply chain structure modeling based on HLA that used to build structure model of supply chain is discussed. Secondly, modeling of Agent blocks integrating processes from SCOR and modeling of federates based on Agent are illustrated to create model of supply chain members. Thirdly the simulation framework of distributed supply chain system is designed. Lastly, a simulation example is given and the feasibility of simulation modeling method of distributed supply chain based on HAS is verified.

Keywords: Supply chain · Distributed · HLA · Agent · SCOR · Simulation modeling

1 Introduction

At present, HLA-based distributed supply chain simulation modeling methods are divided into two types: one type, according to the FOM/SOM specification, a supply chain federal structure model can be created based on the framework and rules of HLA entirely [1]; the other type, through the integration of HLA and SCOR (Supply Chain Operation Reference) model [2], using the business process in the SCOR model, a supply chain simulation process description can be given and the corresponding federal structure model can be created at the same time [3]. These two types of simulation modeling methods can satisfied the requirement of modeling of the macrostructure and behavior of the supply chain for distributed simulation. However, when modeling the internal structure and

© Springer International Publishing AG 2017
Y. Huang et al. (Eds.): ICIRA 2017, Part III, LNAI 10464, pp. 401–412, 2017.
DOI: 10.1007/978-3-319-65298-6_37

action of the members of the supply chain, a single simulation model is adopted, by which a clear description of the internal structure and action of supply chain members can not be obtained. Moreover, A general information model for the interaction communication between supply chain members is not provided by the existing method when a federated structural model is built. As a consequence, low efficiency of the simulation modeling development of the supply chain system and the low reusability of the simulation objects inside the federate occur.

In this paper, a new distributed supply chain modeling method based on HLA, Agent and SCOR (HAS) model is proposed, and then a distributed supply chain system simulation framework is designed. Firstly, according to the thought of federation and federates and the specification of FOM/SOM, the structural model of the whole supply chain system and the interactive information model among the members are created. Secondly, the business processes from the SCOR model are integrated into the Agent entities, then Agent entities and federates are integrated to create a node model. On the one hand, by integrating HLA, Agent and SCOR into this simulation modeling method, the supply chain system simulation model can be established quickly, and the reusability of the internal simulation object model is increased at the same time. On the other hand, according to the well-designed simulation framework, a corresponding distributed simulation system of supply chain can be developed easily and quickly.

2   Simulation Modeling Method of Distributed Supply Chain Based on HAS

HLA-based modeling is based on the idea of federation and federates, using HLA FOM/SOM rules to describe the modeling objects [3]. In Agent-based modeling, the components of a system are seen as Agent entities with autonomy, sociality, reactivity and intelligence [4,5]. SCOR-based modeling uses the business processes to model the supply chain system and the performance indicators to evaluate. HAS-based distributed supply chain simulation modeling integrates the three modeling ideas that are suitable for complex supply chain system simulation modeling. As is shown in Fig. 1, HAS-based distributed supply chain simulation modeling has three components. They are supply chain structure modeling based on HLA, Agent entities integrating business processes of SCOR model and federate model based on Agent entities. In other words, there are two steps: macromodeling and micromodeling. In the first step, the supply chain structure modeling based on HLA is used to model the whole structure of supply chain in macro level, namely building the structure model. In the second step, the Agent entities integrating business processes of SCOR model and federate model based on Agent entities are used to model the member of supply chain, namely building node model.

2.1   Supply Chain Structure Modeling

A typical four- echelon supply chain system consists of a number of decentralized suppliers, manufacturers, distributors and customers, as well as information flow

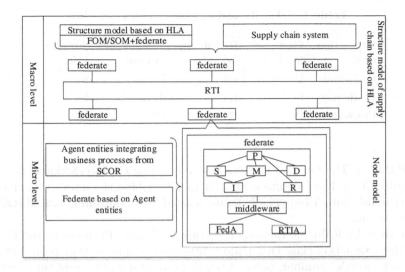

Fig. 1. HAS-based distributed supply chain simulation modeling

and material flow connected to them. According to the thoughts of federation and federates in HLA, the whole supply chain system can be seen as a federation and the supply chain members can be seen as federates. Therefore, four types of federate with the feature of corresponding supply chain members are designed. They are supplier federate, manufacturer federate, distributor federate and customer federate, which can perform the behaviors of supply chain members. Apart from that, the information flow and material flow among supply chain members can be seen as interaction information, according to HLA standard, it can be converted into interaction data. Figure 2 describes the supply chain structure model based on HLA. Besides corresponding federates and interaction data, Run Time Infrastructure and manage federate are need. According to HLA rules, all of the information exchanged in a federation is identified in the federation object model (FOM) [1]. In order to accelerate the development of simulation model, a general FOM suitable for the supply chain structure model based on HLA is proposed. Table 1 is the subscribing and publishing of interaction class. Here:

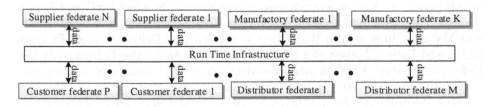

Fig. 2. Supply chain structure model based on HLA

Table 1. Subscribe and publish of Interaction class

Interaction class	Customer federate	Distributor federate	Manufactory federate	Supplier federate
OrderInfo	P	PS	PS	S
Prodelivery	S	PS	PS	P
ProReturn	P	PS	PS	S

P(Publish): The federate is capable of publishing the interaction class.

S(Subscribe): The federate is capable of subscribing the interaction class.

PS(Publish Subscribe): The federate is capable of publishing and subscribing the interaction.

The three basically interaction classes are OrderInfo, Prodelivery and ProReturn. The interaction class OrderInfo contains some information about demand order such as order number, product name, product quantity, sent time and due time. Both of the interaction class Prodelivery and ProReturn are demand meet. The former contains information about product delivery such as order number, delivery product name, quantity and receive time, the latter is about defective products return such as order number, product name, quantity, and return time. For each type of federate, the parameters of the basically interaction class are the same. It is easy and quickly to build a macro level structure model of supply chain system based on HLA with the provided four types of federate and general special FOM. After the structure model is built, the remaining work is to configure federates internal structures and their relationships, as well as the business process elements and functions of federates.

2.2 Supply Chain Member Modeling

2.2.1 Agent Entities Integrating Business Processes of SCOR Model

In the SCOR model, there are five core processes of Plan, Source, Make, Deliver and Return in the level-1 processes [6]. Also, these level-1 processes can be divided into sub processes called level-2 processes. The Plan is divided into Plan supply chain (sP1), Plan source (sP2), Plan make (sP3), Plan deliver (sP4) and Plan return (sP5). As make-to-order and make-to-stack categories are introduced in this paper, the Source is divide into Source Stocked Product (sS1) and Source make-to-order products (sS2); the Make is divided into Make-to-Stock (sM1) and Make-to-Order (sM2); the Deliver is divided into Deliver Stocked Product (sD1) and Deliver Make-to-Order Product (sD2). As for the Return, only the Defective Product is taken into consideration, so the Return is divided into Source Return Defective Product (sSR1) and Deliver Return Defective Product (sDR1). According to the specific responsibilities, as well as behaviors or actions and characteristics of the members in the supply chain system, the corresponding standard business processes of level-2 from SCOR model can be selected to describe its logistics characteristic.

In order to provide a way to build a processes model inside the supply chain members for distributed simulation, agent models for sP1, sP2, sP3, sP4, sP5, sS1, sS2, sM1, sM2, sD1, sD2, sSR1 and sDR1 are created as sP1-Agent, sP2-Agent, sP3-Agent, sP4-Agent, sP5-Agent, sS1-Agent, sS2-Agent, sM1-Agent, sM2-Agent, sD1-Agent, sD2-Agent, sSR1-Agent and sDR1-Agent. There is no doubt that inventory or stock must be existent, thus an agent function model of inventory agent (sI) is created as sI-Agent, The processes in level-2 can be divided into its own sub processes called level-3 processes. For example, the sS1 contains the sub processes elements of Schedule product deliveries (sS1.1), Receive product (sS1.2), Verify product (sS1.3), Transfer product (sS1.4) and Authorize supplier payment (sS1.5). Table 2 lists part of the agent function models integrating the standard business processes from the SCOR model. As is shown in Table 2, the five core processes are mapped as these agent function models integrating the standard business processes in level-3 from the SCOR model. Therefore, for the specific business functions internal of supply chain members and their structural model based on HLA, federates can be filled by corresponding Agent entities and the modeling of internal business processes of supply chain is accomplished.

Table 2. Subscribe and publish of Interaction class

Name	symbol	Process category	Sub processes or functions
sP1-Agent	sP1-A	Plan supply chain (sP1)	Identify, prioritise and aggregate supply chain requirements (sP1.1), Identify, prioritise and aggregate supply chain resources(sP1.2), Balance supply chain resources with supply chain requirements (sP1.3), and Establish and communicate supply chain plans (sP1.4)
sS1-Agent	sS1-A	Source Stocked Product (sS1)	Schedule product deliveries (sS1.1), Receive product (sS1.2), Verify product (sS1.3), Transfer product (sS1.4) and Authorise supplier payment (sS1.5)
sM1-Agent	sM1-A	Make-to-Stock (sM1)	Schedule production activities (sM1.1), Issue Material (sM1.2), Produce and test (sM1.3), Package (sM1.4), Stage finished product (sM1.5), Release finished product to deliver (sM1.6) and Waste disposal (sM1.7)
sD1-Agent	sD1-A	Deliver Stocked Product (sD1)	Process inquiry and quote (sD1.1) Receive, configure, enter and validate order (sD1.2), Reserve inventory and determine delivery date (sD1.3), Consolidate orders (sD1.4), Build loads (sD1.5), Route shipments (sD1.6), Select carriers and rate Shipments (sD1.7), Receive product from source or make (sD1.8), Pick product (sD1.9), Pack product (sD1.10), Load Vehicle and generate shipping docs (sD1.11), Ship product (sD1.12), Receive and verify product by customer (sD1.13), Install product (sD1.14) and Invoice (sD1.15)
sSR1-Agent	sSR1-A	Source Return Defective Product (sSR1)	Identify Defective Product Condition (sSR1.1), Disposition Defective Product (sSR1.2), Request Defective Product Return Authorization (sSR1.3), Schedule Defective Product Shipment (sSR1.4), Return Defective Product (sSR1.5)
sDR1-Agent	sDR1-A	Deliver Return Defective Product (sDR1)	Authorize Defective Product Return (sDR1.1), Schedule Defective Return Receipt (sDR1.2), Receive Defective Product (includes verify) (sDR1.3), Transfer Defective Product (sDR1.4)
sI-Agent	sI-A	Store Products	Manage product inventory (sI.1), Manage incoming product (sI.2), Manage import/export requirements (sI.3) and Manage finished goods inventory (sI.4)

2.2.2 Federate Model Based on Agent Entities

In order to model the internal structure of the members of the supply chain and the relationship among them well, federate model based on Agent entities is proposed. Due to the inconsistency between Agent entities and federates, and in order to maximally maintaining the advantages of them, a special strategy for integration is adopted: not to change anything of federates and Agent entities, and a middleware that bridges federate and Agent entity is added in the internal

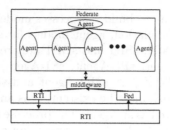

Fig. 3. Federate model based on Agent entities

of federate. Figure 3 is federate model based on Agent entities. In the internal of the federate, there are several ellipses which represent simulation agent, an adaptive middleware, an RTIAmbassador and a FedAmbassador. These agents are used to perform the simulation functions of supply chain and can communicate with each other as well as communicating with RTI through the middleware. In other words, these agents can communicate with external federates.

3   Simulation Framework of Distributed Supply Chain System

By using computer platform, distributed modeling and simulation technology based on HLA and modeling and simulation technology based on Agent, this simulation framework helps to build a simulation system of distributed supply chain based on pRTI and JADE platform. As is shown in Fig. 4, the simulation framework has four layers. The details of each layer are as follow.

(1) The first layer is the physical layer, which provides the most basic hardware support. Computer hardware, Ethernet and operating systems provide the most basic communication and computing power for the simulation system.

(2) The second layer is the JAVA layer, which provides a good programming environment. JAVA platform is distributed, multi-threaded, reliable and safe, and it is useable for distributed simulation system programming language. Whats more, platform independence and good portability of JAVA platform provide much flexibility for the supply chain simulation system.

(3) The third layer is HLA-pRTI layer, a key layer for the simulation system framework. HLA-pRTI provides development API based on JAVA programming language, and the distributed simulation services can run successfully on the JAVA platform. These distributed simulation services are divided into three types. They are simulation environment management services, RTI interface management services and simulation operation management services. By using simulation environment management services, the creation and management of federation and federates as well as the configuration of RTI can be realized. Receiving and sending of interaction class and interaction queue can be accomplished through using the RTI interface management services. The configuration

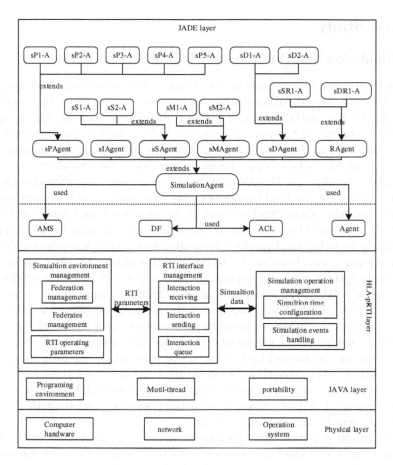

Fig. 4. Simulation framework of distributed supply chain system

of simulation time and disposing of simulation events can be completed via using the simulation operation management services.

(4) The fourth layer is JADE layer which provides DF (Directory Facilitator), AMS (Agent Management System) and ACL (Agent Communication Language) services that meet FIPA (The Foundation for intelligent Physical Agent) specification. In this layer, a SimulationAgent class is designed. This class is the super class of all other classes, by which DF, AMS and ACL services can be used. The corresponding sub classes are sPAgentsIAgentsMAgent sDAgent and sRAgent. They are used to simulate the logistics business processes of plan, source, make, deliver and return.

4   Case Study

4.1   Simulation Model of Supply Chain Based on HAS

In this case study, a four-echelon supply chain system consisting of one supplier, one manufactory, one distributor and one customer is assumed. The supplier is responsible for raw materials supply with an infinite inventory and delivering raw materials to the plant; the plant is responsible for sourcing raw materials from supplier and storing them in the inventory, then assembling or manufacturing them into final products, storing the final products in the inventory, and finally delivering the products to the distributor; the distributor is responsible for storing and transferring the final products to the customer, including sourcing final products from the plant, storing them in its inventory, and finally delivering them to the customer; the customer is responsible for consumption of final products, namely sourcing final product from distributor with a stochastic demand and consuming them. Besides the responsibilities above, the four members of supply chain system are both responsible for returning of defective product.

According to the proposed HAS-based distributed supply chain simulation modeling method, in the first step, the supply chain structure modeling based on HLA is used to model the whole structure of supply chain in macro level and build the structure model. As a consequence, the federation of this four-echelon supply chain system can be formed by picking up corresponding federates designed in supply chain structure modeling based on HLA. At the same time, the table of subscribing and publishing of interaction class is obtained by referring to the proposed general FOM. As a result, structure modeling based on HLA of this four-echelon supply chain system is built.

In the second step, the Agent entities integrating business processes of SCOR model and federate model based on Agent entities are used to model the member of supply chain and create node model. The internal business processes of this four-echelon supply chain members are as follow.

Supplier: sP4, sP5, sD1 and sDR1

Manufactory: sP1, sP2, sP3, sP4, sP5, sS1, sM1, sD1, sSR1 and sDR1 (and inventory management)

DistributorsP2, sP4, sP5, sS1, sD1, sSR1 and sDR1 (and inventory management)

Customer: sP5, sP2, sS1 and sSR1

Therefore, the corresponding Agent entities is picked up to fill in the federate and the relationship of these Agent entities is confirmed, and node model of each supply chain member is completed. Figure 5 is the simulation model of supply chain based on HAS. As is shown in Fig. 5, there are four simulation federates connected with each other through HLA-RTI. The internal structure and business of federates are deployed by Agent entities and middleware. The communication between federates and Agent entities, as well as the communication among Agent entities are implemented through the middleware.

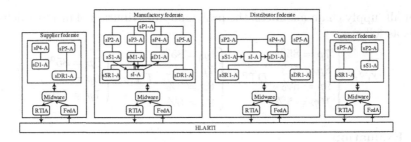

Fig. 5. The simulation model of supply chain based on HAS

4.2 Simulation Environment and Simulation Parameters Configuration

In this case, the simulation environment is composed of five computers with the configuration: Intel(R) Core(TM) i3 3.20 GHz CPU, 4G Memory, windows 10 enterprise. Software Prti 1.3 is installed on one of these five computers to be a server. Each of the rest four computers runs a federate and all federates belong to the same RTI platform. The whole platform is developed based on JAVA JDK 1.7.

In this simulation, the parameters configuration is made as follows. The supply chain simulation is driven by demand order. The demand that follows uniform distribution will be generated after each time interval by the customer [7]. Here, $D(i)$ represents the amount of the demand i, and $D(i) \sim U(20, 25)$. $T(i)$ represents the time interval between demand i and demand $i + 1$. Here,

$$T(i) = 1 + t(i), t(i) \sim p(1)$$

and $p(1)$ is a Poisson distribution [8].

The distributor has the sub business processes of source and deliver and has an inventory buffer. Firstly, the replenishment policy (s, S) is used, here $s = 180$ and $S = 250$, and the initial inventory amount is 250 units. Secondly, the delivery time of order i for distributor is DT_i^d. Thirdly, defective product may occur in each delivery order, the probability is 5 and the amount of defective product follows uniform distribution $U(1, 4)$.

As for the manufactory, firstly the replenishment policy (s, S) is used too, here $s = 300$ and $S = 450$ is for finished product. $s = 600$ and $S = 900$ is for raw material, and the initial inventory amount of finished product is 300 units and the initial inventory amount of raw material is 800 units. Secondly, the delivery time of order i for manufactory is DT_i^p. Thirdly, the make time of order i is $MT_i^p = 5$. Lastly, the probability of defective product occur is 5% and the amount of defective product follows uniform distribution $U(1, 9)$.

As for the supplier, an infinite inventory is assumed. The delivery time of order i for supplier is DT_i^s. The probability of defective product occur is 5% and the amount of defective product follows uniform distribution $U(1, 20)$.

For all supply chain members, the probability distribution of product delivery time is as follows:

$$DT_i^d = \begin{cases} 7p = 85\% \\ 8p = 5\% \\ 9p = 5\% \\ 10p = 5\% \end{cases} \quad DT_i^p = \begin{cases} 6p = 85\% \\ 7p = 5\% \\ 8p = 5\% \\ 9p = 5\% \end{cases} \quad DT_i^s = \begin{cases} 5p = 85\% \\ 6p = 7.5\% \\ 7p = 7.5\% \end{cases}$$

4.3 Evaluation

In this simulation example, the detailed information about sourcing, producing, delivery and inventory changes can be observed dynamic, and then corresponding simulation data can be collected for evaluation and analysis of supply chain performance. In order to evaluate and analyze the supply chain performance, three basic and important metrics like reliability, responsiveness and inventory changes from SCOR model are taken into consideration. For supply chain reliability, the perfect order fulfilment is used. The definition of a perfect order is that products are delivered to customer on right time and right quantity without defective product occurrence. Seen from the definition of a perfect order, the accuracy rate of deliver time and deliver quantity have tremendous influence on perfect order fulfilment. As is shown in Fig. 6, rate of perfect order fulfilment of manufactory is the highest while the supplier is the lowest. In addition, the perfect order fulfilment rate is decided by the accuracy rate of delivery time and delivery amount together. The delivery amount accuracy may be the highest while the delivery time accuracy is the lowest. As a consequence, the perfect order fulfilment rate is not the highest. For responsiveness, the order fulfillment cycle time is selected. The order fulfillment cycle time of distributor consists of source time and delivery time. The order fulfillment cycle time of Plant is made up of source time, make time and delivery time. Information about inventory changes is important for inventory management, Fig. 7 shows the inventory change of distributor and manufactory. For the manufactory, on the one hand, with the products manufactured in the Make-Agent, the amount of the product inventory increases and the amount of the material inventory decreases. On the other hand, with products delivered to the distributor and materials transported to the manufactory, the amount of the product inventory decreases and the amount of the material inventory increases. For the distributor, when the products are delivered to the customer, the amount of the product inventory decreases. When the products are delivered to the distributor from the manufactory, the amount of the product inventory increases.

Through the analysis above, it could be found that with any given supply chain system, the HAS-based supply chain simulation model can be built quickly through the proposed simulation modeling method of distributed supply chain based on HAS, and then a simulation experiment can be carried out. Furthermore, when the structure of supply chain changes, in order to build a new simulation model, there is no need to change the Agent entities. Simply changing the amount of federate and FOM of the old simulation model can fulfill the requirement. Therefore, the reusability of the simulation objects inside

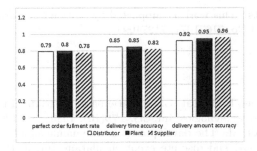

Fig. 6. Rate of perfect order fulfilment

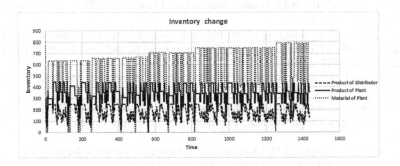

Fig. 7. Inventory change of distributor and manufactory

the federate is improved and the develop efficiency of modeling and simulation is accelerated to a certain extent.

5   Conclusion

In this paper, a new approach of distributed supply chain simulation modeling based on HAS is proposed. HAS-based distributed supply chain simulation modeling has three components. They are supply chain structure modeling based on HLA used to build structure model of the whole supply chain, Agent entities integrating business processes of SCOR model and federate model based on Agent entities used to build node model of supply chain member. Then simulation framework of distributed supply chain system is put forward. Finally, a case study of four-echelon supply chain system is presented, and in this case study the supply chain simulation model based on HAS is built through the proposed modeling method, also a simulation experiment is performed on the developed simulation system. The experiment shows that simulation model of distributed supply chain system can be established quickly through the proposed simulation modeling method. The performance of supply chain can be evaluated effectively and the reusability of the simulation objects inside the federate can be improved.

Acknowledgements. This work is supported by National Natural Science Foundation of China (Grand No. 71671071, 71271094).

References

1. Uygun, Ö., Öztemel, E., Kubat, C.: Scenario based distributed manufacturing simulation using HLA technologies. Inf. Sci. **179**(10), 1533–1541 (2009)
2. Persson, F., Araldi, M.: The development of a dynamic supply chain analysis tool Integration of SCOR and discrete event simulation. Int. J. Prod. Econ. **121**(2), 574–583 (2009)
3. Mustafee, N., Katsaliaki, K., Taylor, S.J.: A review of literature in distributed supply chain simulation. In: 2014 Winter Simulation Conference (WSC), pp. 2872–2883. IEEE, December 2014
4. Long, Q.: Distributed supply chain network modelling and simulation: integration of agent-based distributed simulation and improved SCOR model. Int. J. Prod. Res. **52**(23), 6899–6917 (2014)
5. Li, J., Chan, F.T.: An agent-based model of supply chains with dynamic structures. Appl. Math. Model. **37**(7), 5403–5413 (2013)
6. Supply Chain Council: Supply chain operations reference (SCOR) model version 10.0. The Supply Chain Council, Inc. SCOR. The Supply Chain Reference (binder) (2010)
7. Azevedo, E.M., Leshno, J.D.: A supply and demand framework for two-sided matching markets. J. Polit. Econ. **124**(5), 1235–1268 (2016)
8. Altay, N., Litteral, L.A., Rudisill, F.: Effects of correlation on intermittent demand forecasting and stock control. Int. J. Prod. Econ. **135**(1), 275–283 (2012)

Robust EMG Pattern Recognition with Electrode Donning/Doffing and Multiple Confounding Factors

Huajie Zhang, Dapeng Yang(✉), Chunyuan Shi, Li Jiang, and Hong Liu

State Key of Laboratory of Robotics and System, Harbin Institute of Technology,
Harbin 150080, People's Republic of China
yangdapeng@hit.edu.cn

Abstract. Traditional electromyography (EMG) pattern recognition did not take into account confounding factors such as electrode shifting, force variation, limb posture, etc., which lead to a great gap between academic research and clinical practice. In this paper, we investigated the robustness of EMG pattern recognition under conditions of electrode shifting, force varying, limb posture changing, and dominant/non-dominant hand switching. In feature extraction, we proposed a method for threshold optimization based on Particle Swarm Optimization (PSO). Compared with the traditional trail & error method, it can largely increase the classification accuracy (CA) by 10.2%. In addition, the hybrid features integrated with discrete Fourier transform (DFT), wavelet transform (WT), and wavelet packet transform (WPT) were proposed, which increased the CA by 30.5%, 25.4%, 22.9%, respectively. We introduced probabilistic neural network (PNN) as a new classifier for EMG pattern recognition, and reported the CA's obtained by a large variety of features and classifiers. The results showed that the combination of DFT_MAV2 (a novel feature based on DFT) and PNN reached the best CA (45.5%, 14 motions, validated on different hands without re-training).

Keywords: Myoelectric signal · Electrode shifting · Dynamic limb posture · Feature extraction · Pattern recognition

1 Introduction

The myoelectric signal (also termed as Electromyography, EMG) is a bioelectrical signal that generated alongside muscular contractions. Accompanying with different motion patterns, the human body can produce different EMG signals, both in the time domain (TD) and frequency domain (FD). Thus, by extracting kinematic information from EMG signals, it is possible to accomplish a motion/force control of external devices, such as prosthesis [1, 2], exoskeleton [3], wheelchair [4] and virtual keyboard [5]. Compared with traditional human-machine interfaces (HMI) such as handle, mouse and keyboard, the novel HMI constructed on EMG can free our hands to a certain extent. Especially for amputees, EMG-based HMI can be provided with customized prosthesis, mobile phone, game consoles and other communications or entertainment equipment. These devices are helpful to their activities of daily life (ADL's), as well as, their physical and mental rehabilitation.

© Springer International Publishing AG 2017
Y. Huang et al. (Eds.): ICIRA 2017, Part III, LNAI 10464, pp. 413–424, 2017.
DOI: 10.1007/978-3-319-65298-6_38

In the literature, the studies on EMG pattern recognition (EMG-PR) have achieved tremendous achievements. However, there are still many issues to be resolved in its practical applications. According to the statistics [2, 6], only 50% ~ 70% of amputees use their myoelectric prosthesis regularly, citing the limited functionality, low control precision and lack of sensory feedback as main reasons. It seems like that all HMI's constructed based on EMG-PR have similar problems. The great gap between academics and clinics may be attributed to the differences between ideal experimental conditions and real clinical conditions. In practice, the EMG-PR is largely influenced by confounding factors including, but not limited to, limb posture [7, 8], electrode shifting [9], muscle contraction strength [10], and so on.

A considerable number of studies have been put forward to mitigate these interference factors independently. Fougner et al. [7] used sensor fusion of EMG and accelerometers to resolve the effect of limb posture and suggested to train the classifier with the signals collected from multiple limb positions. Geng et al. [8] proposed a cascade classifier with accelerometers and EMG. However, these studies only took into account the effect of limb posture, but not multiple factors simultaneously. In addition, the EMG signals were collected under a fixed limb posture, thus the influence of dynamic limb movements in real applications was not considered. For reducing the effect of electrode shifting, Young et al. [9] improved the robustness of the EMG-PR by changing the configuration of electrodes. However, the shifting direction is unpredictable and the distance of the electrodes is inconsistent, which does not conform to Young's hypothesis. Against the strength variation of muscle contraction, Al-Timenmy et al. [10] proposed a novel set of features to improve the performance of EMG-PR. The feature (Time-Dependent Power Spectrum Descriptors, TD-PSD) achieved better results compared with the other widely-accepted feature sets such as reduced spectral moments, TD features (waveform length, zero crossings, slope sign changes, etc.), the root mean square and autoregressive (RMS + AR), and the wavelet features. However, they did not consider other confounding factors in their study.

Sometimes, we expect that the EMG-based HMI's have generalization ability in some similar but not identical neuromuscular systems. For example, the EMG interface can be easily switched between the left hand and the right hand without any re-training. Hence, it is meaningful to validate the robustness of various EMG-PR schemes under both dominant and non-dominant side. We assume that the EMG-PR method that considers only one confounding factor could not be robust enough when dealing with multiple factors. According to this hypothesis, we aim to improve the accuracy of EMG-PR under multiple factors such as electrode shifting, limb posture, and dominant/non-dominant hand switch. Our results could help improve the robustness of EMG control methods in clinical practice.

2   Materials and Methods

2.1   Subjects

Six healthy, able-bodied subjects (male, aged between 23 and 30 years; abbreviation as Sub1–Sub6) volunteered to participate in our experiment. They have no records of neuromuscular diseases. Informed consent form was signed before experiments. All experiments were approved by the Ethical Committee of Harbin Institute of Technology and conformed to the Declaration of Helsinki.

2.2 Hand Motions

We inspected 14 significant wrist and hand motions (shown in Fig. 1), since they largely contribute to hand's operational dexterity. In every trail of collection, the subjects were asked to complete all the 14 motions sequentially. All motions, with exception of HR, require maintaining a base muscular contraction (nearly 30% maximum voluntary contraction, MVC) for 7.5 s; meanwhile, several contraction fluctuations (30% ~ 50% MVC) were encouraged for introducing the factor of force variation. In each session of collection, there were four replicas of trails. Full rests were given between any trials for getting rid of muscle fatigue.

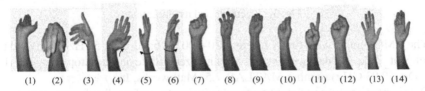

(1)   (2)   (3)   (4)   (5)   (6)   (7)   (8)   (9)   (10)   (11)   (12)   (13) (14)

Fig. 1. Hand motions: (1) wrist extension (WE); (2) wrist flexion (WF); (3) ulnar deviation (UD); (4) radial deviation (RD); (5) wrist pronation (WP); (6) wrist supination (WS); (7) lateral grasp (LG); (8) spherical grasp (SG); (9) cylinder grasp (CG); (10) tripod pinch(TP); (11) index point (IP); (12) hand close (HC); (13) hand open (HO); and (14) hand rest (HR)

2.3 Limb Posture

In the process of each hand motion, the subjects were asked to complete a dynamic limb posture:
(P1) → (P2) → (P1) → (P3) → (P1) → (P4) → (P1) → (P5) → (P1) → (P6) → (P1) →
(P7) → (P1), as shown in Fig. 2.

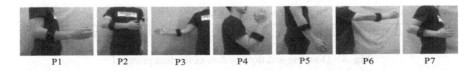

P1   P2   P3   P4   P5   P6   P7

Fig. 2. Dynamic limb postures: P1, Humerus hanging at side, forearm horizontal; P2, Forearm in front of the chest; P3, Forearm swing out, perpendicular to the body; P4, Humerus hanging at side, forearm 60° above horizontal; P5, Straight arm hanging at side; P6, Straight arm reaching forward horizontally; P7, Humerus back, about 90° with forearm.

2.4 EMG Data Acquisition

A wireless myoelectric signal acquisition system, TrignoTM Wireless system (Delsys INC, USA), was employed to record the EMG signals. The electrodes and a self-developed armband for easily fixing the electrodes are shown in (Fig. 3-a) and (Fig. 3-b). Eight electrodes were placed equidistantly around the circumference of the forearm (about 1/3 forearm length from elbow), as shown in (Fig. 3-c) and (Fig. 3-d).

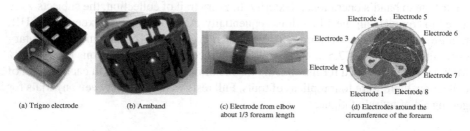

(a) Trigno electrode   (b) Armband   (c) Electrode from elbow   (d) Electrodes around the
                                           about 1/3 forearm length   circumference of the forearm

Fig. 3. Electrode and its placement

The EMG signals were firstly filtered (20–500 Hz band-pass, 50 Hz notch) and then sampled at 2 kHz. All data were stored and analyzed offline on a desktop computer (Intel Core Duo, 2.93 GHz) running MATLAB 7.5 (Mathworks, Inc.)

We propose a new experiment protocol to integrate confounding factors such as force variation, limb posture, electrode shifting and dominant/non-dominant hand switch, as much as possible. The EMG data collected under this protocol may be equivalent to the one in real practice. We termed this new data-collection protocol as donning & doffing experiment.

In this donning & doffing protocol, the subjects were firstly asked to wear (don) the electrodes on their *left* arm. After completing a session of data collection (four trials), they were instructed to take off (doff) the electrodes and have a rest of five minutes. Then, the subjects put the electrodes on their *right* forearm for another data-collection session. The subjects should repeat this process until 12 sessions of collection are completed (abbreviation as Session 1–Session 12). The whole process is shown in Fig. 4.

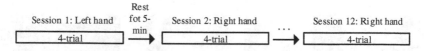

Fig. 4. Donning and doffing data-collection protocol

This donning & doffing protocol includes many confounding factors in EMG data collection. Obviously, donning and doffing the armband on the same arm could not guarantee exactly the same electrode location, which inevitably led to electrode movements. Hence, the effect of electrode shifting was well considered in the protocol. The subjects were asked to elicit dynamical muscle contractions, which introduced the factor of force variation. Moreover, the limb movements compassed necessary dynamic uncertainty in real arm/hand operations. In addition, the EMG signals collected from both limbs (dominant and non-dominant switch) could simulate the data acquired under similar autonomy structures.

2.5   Feature Extraction

The raw EMG signals were firstly segmented into a series of 256 ms windows with 50% overlap. The features were extracted within each window. A large variety of features were examined.

Common Single Feature

At present, the EMG features can be divided into time domain (TD), frequency domain (FD) and time-frequency domain (TFD) features. After a survey on these features, we selected 21 common single features among them (see Table 1).

Table 1.   Common single features

Features types	Common single features
TD features	MAV [11]; MAV1 [12]; MAV2 [12]; SSI [12]; DASDV [12]; VORDER [13]; MYOP [12]; RMS [14]; LOG [13]; WL [11]; VAR [13]; WAMP [13]; SSC [13]; ZC [11]
FD features	MDF [10]; SM3 [10]; MDA [10]
TFD features	WTM [15]; WTSVD [15]; WPTM [16]; WPTSVD [16]

Note: MAV = mean absolute value; MAV1 = modified mean absolute value type 1; MAV2 = modified mean absolute value type 2; SSI = simple square integral; DASDV = difference absolute standard deviation; VORDER = v-Order; MYOP = myopulse percentage; RMS = root mean square; LOG = log detector; WL = waveform length; VAR = variance of EMG; WAMP = Willison amplitude; SSC = slope sign change; ZC = zero crossing; MDF = median frequency; SM3 = the 3rd spectral moments; MDA = median amplitude spectrum; WTM = wavelet transform maximum; WTSVD = singular value decomposition of wavelet transform coefficient; WPTM = wavelet packet transform maximum; WPTSVD = singular value decomposition of wavelet packet transform coefficient.

Feature Threshold Optimization Based on Particle Swarm Optimization

According to [11–13], a threshold needs to be set beforehand to extract the features of ZC, MYOP, WAMP and SSC. However, there still lacks an efficient method to determine a suitable threshold value (usually, individual experience and trial & error), letting alone the threshold within each channel may be different. Conventional methods are time-consuming, also they cannot guarantee the threshold is well selected. To address this issue, we proposed an automatic scheme for optimizing the threshold based on particle swarm optimization (PSO) [17], which is an intelligence optimization algorithm widely used in parameter optimization. The classification accuracy of 10-fold cross-validation was adopted as the value of fitness function.

Feature Extraction Based on Discrete Fourier Transform

The discrete Fourier transform (DFT) has been utilized in the EMG PR to extract robust features against varying contraction level [18]. Here, we apply it with TD method to extract new feature sets to enhance the robustness of EMG PR. In this method, the segmented EMG signals were firstly transformed into its frequency domain by DFT. Then, the amplitude (such as MAV, RMS, etc.) of the data in different frequency bands

(21–100 Hz, 101–180 Hz, 181–260 Hz, 261–340 Hz, 341–420 Hz and 421–500 Hz) was calculated. At last, the 48-dimensional EMG features (1 feature/band × 6 bands/ channel × 8 channels) were normalized. The features based on this measure were termed as DFT_*, hereafter, such as DFT_MAV and DFT_RMS.

Feature Extraction Based on Wavelet Transform and Wavelet Packet Transform
The wavelets transform (WT) and wavelet packet transform (WPT) are common TFD analysis methods in PR based EMG control [15, 16]. Although the features from TFD transform may contain useful information, the conventional feature representations (coefficients, band energy, etc.) perform very limited in improving the classification accuracy. In this paper, we proposed a feature extraction method combining wavelet (packet) transform and TD feature representation together. For the wavelet (packet) transform, the "db2 (db3)" family was used and a 5-level (3-level) decomposition was performed. Then, on each level of decomposition, a TD feature (MAV, RMS, or WL) was extracted from the approximation and detail coefficients. At last, 48-dimensional WT features (1 feature/band × 6 bands/channel × 8 channels) and 64-dimensional WPT features (1 feature/band × 8 bands/channel × 8 channels) are extracted. The features based on these measures were named as WT_* and WPT_*, respectively.

2.6 Pattern Classification

In this paper, five widely-used classifiers, Artificial Neural Network (ANN) [18], Linear Discriminant Classifier (LDA) [19, 20], Support Vector Machine (SVM) [16, 21, 22], K Nearest Neighbor algorithm (KNN) [13] and Ensemble Learning [23], were inspected.

Though ANN (such as BP neural network, etc.) has the ability of self-learning, self-organization and self-adaptation, it has numerous parameters to be adjusted and, some-times, it is very time-consuming. In this paper, the probabilistic neural network (PNN) [24] was adopted as the classifier, which has a few of parameters and fast convergence speed. The PNN consists of four layers as input layer, model layer, summation layer and output layer.

We also attempt to integrate the LDA and AdaBoost [25] in Ensemble Learning, to construct a new classifier as AdaLDA. The AdaBoost algorithm improves the weak classifier (LDA) to a strong one by weighting them all.

3   Results and Discussions

This paper mainly investigated the robustness of EMG-PR under two conditions: (1) electrode donning/doffing on the same forearm, and (2) electrode donning/doffing on different forearms. To validate the effectiveness of different features, we chose SVM (RBF kernel $C = 32$, $\gamma = 0.01$) as a unified classifier. The results were analyzed using repeated measures ANOVA.

(1) *Electrode donning/doffing on the same forearm.* In this case, any two sessions collected from the same limb (left or right) were selected as the training set and test set, respectively. In total, there were 60 ($2 \times P(6,2)$) different combinations.

(2) *Electrode donning/doffing on different forearms.* In this case, the EMG data collected from one limb was used as the training set, while the data collected from the other limb was used as the test set. In total, there were 72 $(2 \times P(6,1) \times P(6,1))$ combinations.

3.1 Classification Result on the Same Forearm

How electrode shifting and dynamic limb movement affect the raw EMG signals is still unknown. Figure 8 illustrates an example of this effect by giving the raw EMG signals (HR) under two totally different conditions. Figure 5-(a) shows the EMG signals with a static posture (P1); while Fig. 5-(b) shows the EMG signals collected with dynamic limb movements after doffing and donning the electrodes.

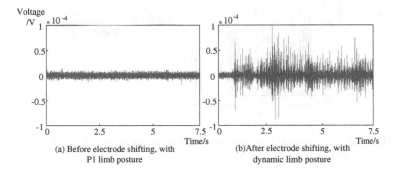

(a) Before electrode shifting, with
P1 limb posture

(b)After electrode shifting, with
dynamic limb posture

Fig. 5. The raw EMG signals of the HR in the P1 posture

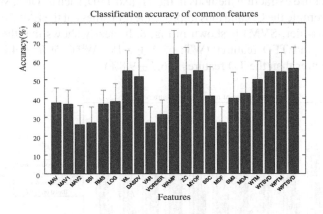

Fig. 6. The CA of common single features

Under the experiment condition (1), Fig. 6 shows the classification accuracy (CA) acquired by common single features, which are generally very low (26%–64%). The highest CA was acquired by WAMP, $63.3 \pm 12.6\%$, which is still not suitable for a real application.

To improve the CA, we proposed novel feature extraction schemes, including optimizing the feature's threshold based on PSO, feature extraction based on DFT, feature extraction based on WT, and feature extraction based on WPT.

Feature Threshold Optimization Based on PSO

After applying the PSO algorithm, the threshold used in every single feature can be optimized to a certain extent. Compared with the traditional methods (individual experience, or trial & error experiment), the resultant CA after threshold optimization can be significantly improved 10.2% on average, as shown in Fig. 7.

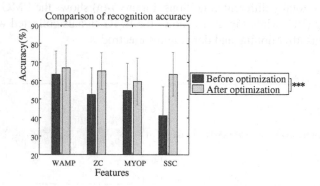

Fig. 7. The result of the method with non-optimization and optimization (*** denotes $p < 0.001$, ** denotes $p < 0.005$, * denotes $p < 0.01$).

New Features   Because the CA acquired by TD features was very low (Fig. 6), we tried several new feature extraction methods in the TD and TFD, such as DFT, WT and WPT, intending to promote the CA and thus improve the robustness of the EMG-PR. The result of the CA (classifier: SVM) is shown in Fig. 8. It clearly shows that the FD features ($67.3 \pm 0.3\%$) and TFD features (WT: $62.2 \pm 5.1\%$, WPT: $59.7 \pm 4.0\%$) performs significantly better than the TD features ($36.7 \pm 9.8\%$).

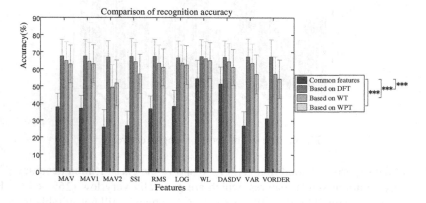

Fig. 8. The CA obtained by common single features and new features

Motion Classification Details

Although the new features can largely improve the CA (increased by 30.5%, 25.4%, 22.9% on average for DFT, WT and WPT, respectively), the overall CA is still not favorable for practice. We try to find a new motion set, within which each motion can get an acceptable accuracy. From this view, the DFT_MAV and SVM are selected to draw the confusion matrix, as shown in Fig. 9.

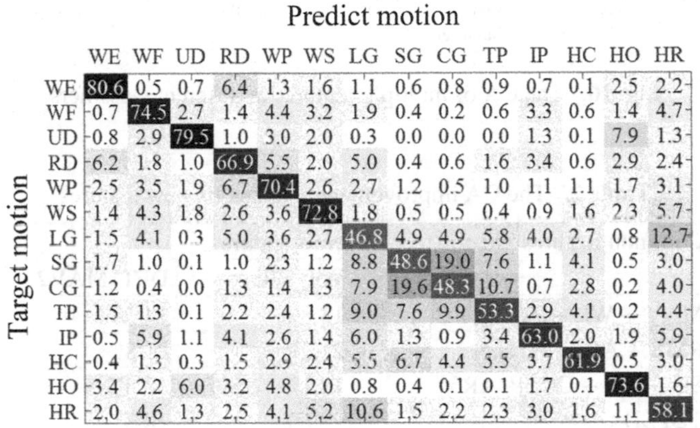

Predict motion

	WE	WF	UD	RD	WP	WS	LG	SG	CG	TP	IP	HC	HO	HR
WE	80.6	0.5	0.7	6.4	1.3	1.6	1.1	0.6	0.8	0.9	0.7	0.1	2.5	2.2
WF	0.7	74.5	2.7	1.4	4.4	3.2	1.9	0.4	0.2	0.6	3.3	0.6	1.4	4.7
UD	0.8	2.9	79.5	1.0	3.0	2.0	0.3	0.0	0.0	0.0	1.3	0.1	7.9	1.3
RD	6.2	1.8	1.0	66.9	5.5	2.0	5.0	0.4	0.6	1.6	3.4	0.6	2.9	2.4
WP	2.5	3.5	1.9	6.7	70.4	2.6	2.7	1.2	0.5	1.0	1.1	1.1	1.7	3.1
WS	1.4	4.3	1.8	2.6	3.6	72.8	1.8	0.5	0.5	0.4	0.9	1.6	2.3	5.7
LG	1.5	4.1	0.3	5.0	3.6	2.7	46.8	4.9	4.9	5.8	4.0	2.7	0.8	12.7
SG	1.7	1.0	0.1	1.0	2.3	1.2	8.8	48.6	19.0	7.6	1.1	4.1	0.5	3.0
CG	1.2	0.4	0.0	1.3	1.4	1.3	7.9	19.6	48.3	10.7	0.7	2.8	0.2	4.0
TP	1.5	1.3	0.1	2.2	2.4	1.2	9.0	7.6	9.9	53.3	2.9	4.1	0.2	4.4
IP	0.5	5.9	1.1	4.1	2.6	1.4	6.0	1.3	0.9	3.4	63.0	2.0	1.9	5.9
HC	0.4	1.3	0.3	1.5	2.9	2.4	5.5	6.7	4.4	5.5	3.7	61.9	0.5	3.0
HO	3.4	2.2	6.0	3.2	4.8	2.0	0.8	0.4	0.1	0.1	1.7	0.1	73.6	1.6
HR	2.0	4.6	1.3	2.5	4.1	5.2	10.6	1.5	2.2	2.3	3.0	1.6	1.1	58.1

(Target motion, row-label axis)

Fig. 9. Confusion matrix of DFT_MAV(%)

From Fig. 9, we find that the CA for LG, SG, CG, TP is relatively low. The average CA for these four motions is 49.3%; while the CA for the other 10 motions is 70.2%. It is very easy to confuse LG with HR, mainly because of the similarity of the EMG signals caused by varying limb postures. The SG and CG are also easily confused, mainly because the muscles for actuating these motions are very close that a large amount of signal crosstalk is introduced.

3.2 Classification Results on the Different Forearms

In this session, we discuss the CA when electrodes interchange between two forearms without retraining. Confounding factors including force variation and limb posture are also considered. Based on the result obtained in Sect. 3.1, WT_WL and DFT_MAV2 are selected in this session. Meanwhile, we select four other features, MAV + WL + ZC + SSC [11], RMS + AR5 [26], SE + WL + CC5 + AR5 [27] and TDPSD [10], as comparison.

The CA's obtained by different sets of features and classifiers are shown in Fig. 10. Compared with the other features, the WT_WL and DFT_MAV2 can achieve a relatively robust classification against many confounding factors (43.2%, 43.8% on average, respectively). The CA's improvement obtained by new features are shown in Table 2.

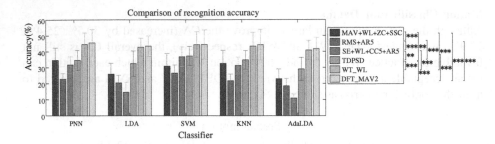

Fig. 10. The CA of multiple features in different classifiers

Table 2. The CA improvement obtained by new features

Common features	New features	
	WT_WL	DFT_MAV2
MAV + WL + ZC + SSC	13.8%	14.5%
RMS + AR5	21.2%	21.8%
SE + WL + CC5 + AR5	18.1%	18.8%
TDPSD	9.6%	10.3%

Different combinations of feature and classifier achieve significantly different performance. For example, the CA obtained by MAV + WL + ZC + SSC separately with PNN, LDA, SVM, KNN, AdaLDA is $34.6 \pm 7.8\%$, $25.9 \pm 6.9\%$, $30.8 \pm 8.0\%$, $32.5 \pm 8.2\%$, $22.9 \pm 6.1\%$, respectively. An attractive characteristic of the newly proposed features is that they are insensitive to classifiers, that is, they have a more stable CA in cooperation with different classifiers. For example, the CA obtained by WT_WL using different classifiers (PNN, LDA, SVM, KNN and AdaLDA) is $44.5 \pm 7.3\%$, $42.7 \pm 5.7\%$, $44.4 \pm 8.1\%$, $43.4 \pm 8.7\%$ and $40.8 \pm 6.1\%$, respectively. The CA obtained by DFT_MAV2 with these classifiers is $45.5 \pm 8.4\%$, $43.4 \pm 6.5\%$, $44.4 \pm 9.2\%$, $44.2 \pm 9.6\%$, $41.7 \pm 7.1\%$, respectively. The best combination of feature and classifier, in terms of CA, is DFT_MAV2 and PNN ($45.5 \pm 8.4\%$).

The results also show that there is no significant difference when using different classifiers. Compared with the features, the classifiers have very limited effect to improve the CA. It seems like that, for increasing the CA with confounding factors, the features play a much more important role than the classifiers. The feature extraction would become a focus to improve the robustness of EMG control in clinic practice.

4   Conclusion

We proposed a threshold optimization method based on PSO algorithm, which can increase the CA by 10.2% on average ($p < 0.001$) with comparison to traditional methods. Under confounding factors (electrodes shifting, limb posture, dominant/non-dominant hand switch), the proposed new features (the hybrid features integrated with DFT, WT, and WPT) have a strong classification robustness (the CA is increased by

30.5%, 25.4%, 22.9%, on average, respectively). The PNN has a better recognition than LDA. In addition, the influence of features is greater than classifier to the classification accuracy under confounding factors.

Acknowledgments. The authors would like to thank all the subjects participated in the experiments for their generous cooperation. The authors also appreciate the help of Qi Huang, Wei Yang, and Yuan Liu for their help in the experiments and paper drafting. This work is partially supported by the National Natural Science Foundation of China (No. 51675123, No. 61603112) and the Self-Planned Task of State Key Laboratory of Robotics and System (No. SKLRS201603B).

References

1. Ning, J., Dosen, S., Muller, K.R., Farina, D.: Myoelectric control of artificial limbs: is there a need to change focus? IEEE Signal Process. Mag. **29**, 148–152 (2012)
2. Castellini, C., Artemiadis, P., Wininger, M., Ajoudani, A., Alimusaj, M., Bicchi, A., Caputo, B., Craelius, W., Dosen, S., Englehart, K., Farina, D., Gijsberts, A., Godfrey, S.B., Hargrove, L., Ison, M., Kuiken, T., Markovic, M., Pilarski, P., Rupp, R., Scheme, E.: Proceedings of the first workshop on Peripheral Machine Interfaces: going beyond traditional surface electromyography. Front. Neurorob. **8**(22), 1–17 (2014)
3. Kinnaird, C.R., Ferris, D.P.: Medial gastrocnemius myoelectric control of a robotic ankle exoskeleton. IEEE Trans. Neural Syst. Rehabil. Eng. **17**(1), 31–37 (2009)
4. Bobtsov, A.A., Borgul, A.S.: Human-machine interface for mechatronic devices control. IFAC Proc. **46**(9), 614–618 (2013)
5. Jorgensen, C., Dusan, S.: Speech interfaces based upon surface electromyography. Speech Commun. **52**(4), 354–366 (2010)
6. Micera, S., Carpaneto, J., Raspopovic, S.: Control of hand prostheses using peripheral information. IEEE Rev. Biomed. Eng. **3**, 48–68 (2010)
7. Fougner, A., Scheme, E., Chan, A.D., Englehart, K., Stavdahl, O.: Resolving the limb position effect in myoelectric pattern recognition. IEEE Trans. Neural Syst. Rehabil. Eng. **19**(6), 644–651 (2011)
8. Geng, Y., Ping, Z., Li, G.: Toward attenuating the impact of arm positions on electromyography pattern-recognition based motion classification in transradial amputees. J. Neuroeng. Rehabil. **9**(1), 74 (2012)
9. Young, A.J., Hargrove, L.J., Kuiken, T.A.: Improving myoelectric pattern recognition robustness to electrode shift by changing interelectrode distance and electrode configuration. IEEE Trans. Biomed. Eng. **59**(3), 645–652 (2012)
10. Al-Timemy, A.H., Khushaba, R.N., Bugmann, G., Escudero, J.: Improving the performance against force variation of EMG controlled multifunctional upper-limb prostheses for transradial amputees. IEEE Trans. Neural Syst. Rehabil. Eng. **24**(6), 650–661 (2016)
11. Hudgins, B., Parker, P., Scott, R.N.: A new strategy for multifunction myoelectric control. IEEE Trans. Biomed. Eng. **40**, 82–94 (1993)
12. Phinyomark, A., Phukpattaranont, P., Limsakul, C.: Feature reduction and selection for EMG signal classification. Expert Syst. Appl. **39**(8), 7420–7431 (2012)
13. Zardoshti-Kermani, M., Wheeler, B.C., Badie, K., Hashemi, R.M.: EMG feature evaluation for movement control of upper extremity prostheses. IEEE Trans. Rehabil. Eng. **3**(4), 324–333 (1995)

14. Oskoei, M.A., Hu, H.: Support vector machine-based classification scheme for myoelectric control applied to upper limb. IEEE Trans. Biomed. Eng. **55**(8), 1956–1965 (2008)
15. Lucas, M.F., Gaufriau, A., Pascual, S., Doncarli, C., Farina, D.: Multi-channel surface EMG classification using support vector machines and signal-based wavelet optimization. Biomed. Signal Process. Control **3**(2), 169–174 (2008)
16. Xing, K., Yang, P., Huang, J., Wang, Y., Zhu, Q.: A real-time EMG pattern recognition method for virtual myoelectric hand control. Neurocomputing **136**, 345–355 (2014)
17. Kennedy, J., Eberhart, R.: Particle swarm optimization. In: 1995 Proceedings IEEE International Conference on Neural Networks, vol. 4, pp. 1942–1948 (1995)
18. He, J., Zhang, D., Sheng, X., Li, S., Zhu, X.: Invariant surface emg feature against varying contraction level for myoelectric control based on muscle coordination. IEEE J. Biomed. Health Inform. **19**(3), 874–882 (2015)
19. Purushothaman, G., Ray, K.K.: EMG based man–machine interaction—a pattern recognition research platform. Rob. Autonom. Syst, **62**(6), 864–870 (2014)
20. Liu, J., Sheng, X., Zhang, D., Jiang, N., Zhu, X.: Towards zero retraining for myoelectric control based on common model component analysis. IEEE Trans. Neural Syst. Rehabil. Eng. **24**(4), 444–454 (2016)
21. http://www.csie.ntu.edu.tw/~cjlin/libsvm 25 Mar 2017
22. Cortes, C., Vapnik, V.: Support vector network. Mach. Learn. **20**(3), 273–297 (1995)
23. Schapire, R.E.: The boosting approach to machine learning: an overview. In: Denison, D.D., Hansen, M.H., Holmes, C.C., Mallick, B., Yu, B. (eds.) Nonlinear Estimation and Classification. Lecture Notes in Statistics, vol 171, pp. 149–171. Springer, New York (2003). doi:10.1007/978-0-387-21579-2_9
24. Chasset, P.O.: PNN: probabilistic neural networks, pp. 109–118 (2013)
25. Freund, Y., Schapire, R.E.: A decision-throretic generalization of on-line learning and an application to boosting. J. Comput. Syst. Sci. **55**(1), 119–139 (1997)
26. Oskoei, M.A., Hu, H.: Myoelectric control systems—a survey. Biomed. Signal Process. Control **2**(4), 275–294 (2007)
27. Khushaba, R.N., Al-Timemy, A., Kodagoda, S., Nazarpour, K.: Combined influence of forearm orientation and muscular contraction on EMG pattern recognition. Expert Syst. Appl. **61**(11), 154–161 (2016)

A Robot Architecture of Hierarchical Finite State Machine for Autonomous Mobile Manipulator

Haotian Zhou[1], Huasong Min[1(✉)], Yunhan Lin[1], and Shengnan Zhang[2]

[1] Institute of Robotics and Intelligent Systems,
Wuhan University of Science and Technology, Wuhan, China
mhuasong@wust.edu.cn
[2] Wuchang Institute of Technology, Wuhan, China

Abstract. The intelligent robots have been participating in people's work increasingly, and challenging to accomplish work autonomously with fully understand human intention through voice interaction. We proposed a robot architecture of hierarchical finite state machine (HFSM) for autonomous mobile manipulator which run on robot operating system (ROS). The system has the abilities to analyze user's input information, communicate with user to obtain complete intention when the user's intention is incomplete and transfer the intention to mission plan for executing of tasks. In this paper, we described the operation procedure of system and each component, and designed the experiment scenario to verify the feasibility of the proposed architecture. The experiments result showed our autonomous mobile manipulator achieve the high performance of automation of tasks.

Keywords: Hierarchical Finite State Machine · Autonomous mobile manipulator · Robot architecture · Human-Robot Interaction

1 Introduction

Robots are developed towards intelligence and automation. Human beings hope that robot has flexible manipulator which can work in a non-structured environment, handle complex information in real time, interact with the user through Human-Robot Interaction (HRI), and understand the user's intention to complete specified task [1]. It requires robot not only have many sensors to perceive surroundings, but also accomplish complex functions such as autonomous moving, object recognition and motion control. Such a high intelligent and autonomous robot requires many components to collaborate to complete mission efficiently [2]. However, there are two important aspects of robot for considering to achieve intelligence and automation. One is that how to transform from voice to instruction; the other is how to combine many components effectively for automation.

© Springer International Publishing AG 2017
Y. Huang et al. (Eds.): ICIRA 2017, Part III, LNAI 10464, pp. 425–436, 2017.
DOI: 10.1007/978-3-319-65298-6_39

To solve the problems above, at first reasoning component and planning component are essential in the system for getting and parsing intention of user. Secondly, a sort of flexible system architecture must be designed. The system consists of many functional components, each component includes one or more nodes. If there is no appropriate architecture, the huge system makes data communication between programs complex, task switch complicated, and process disordered [3]. In addition, the later maintenance of system is difficult and it has a poor expansibility.

Finite State Machine (FSM) is a mathematical model that the number of states is limited and the state is transferred by events or conditions. FSM was originally used for video game and film fields, and then robot developers found that there are some similarities between the structure of character behavior program designed by FSM and robot program architecture. Afterwards it was applied in the field of robot [4]. The application of FSM in motion control of robot could improve the robustness of system, Li Q. et al. used FSM to manage the operation state transition of manipulator to grasp the unknown object, increased the efficiency of task [5]. Park H. et al. [6] and Da X. et al. [7] applied FSM on biped robot to control two legs for walk in different terrain; the experiments verified that it improved the stability of movement. Kurt A. et al. designed a mobile system which used FSM, it could choose path for obstacle avoidance by state transition [8]. In the aspect of robot system architecture, Jayawardena C. et al. presented a social assistance robot HealthBot, which was based on FSM; it guided users to enter relevant information to accomplish the function needed, satisfied the requirement of the old [9]. However, the achievement of function was based on complete intention, which will be helpless in front of incomplete intention. Nguyen H. et al. designed home robot behavior with ROSCO system upon FSM, the robot could execute complex task such as unlocking a door in new environment [4]. But the behavior construction was specific to expert users, unable to generate mission plan from user intention.

In this paper we presented a robot architecture of hierarchical finite state machine for autonomous mobile manipulator. The process of interaction to implementation was autonomous through conversion of state machine. HFSM just like a FSM embedded other FSM. It has a limitation of transfer, states from FSM embedded with no need for considering states from other FSM. There is isolation between irrelevant states which has a decrease of complexity of HFSM. We combined speech recognition, speech synthesis, reasoning and planning components with the first hierarchy FSM, which accomplished analysis of user input information, communicating with user to obtain complete intention when the users intention is incomplete and transferring the intention to task instruction. And the second hierarchy FSM integrated navigation, vision and manipulator operation components for improving robustness of task executing.

The paper is arranged as follows. In Sect. 2, we describe the hardware structure of system. In Sect. 3 we introduce the software architecture of system, including entire architecture, HFSM, components and pipeline of each part. The Sect. 4 is experiments and analysis, we designed the experimental scene to verify the feasibility of proposed architecture. In Sect. 5, conclusion and future work are presented.

2 Hardware Structure of System

As shown in Fig. 1 is the hardware structure of system, it consists of Clearpath Robotics Husky A200 mobile platform, Universal Robots UR5 manipulator and controller, Robotiq 2-finger gripper, Microsoft Kinect, Microstrain IMU, Garmin GPS, Mic array, sound and mini computers. Fig. 2 is the autonomous mobile manipulator. Considering the balance of system computational load, the devices were distributed on two computers equally. For reducing data transmission delay among each part, computers and UR controller are connected as local area network with Ethernet switch.

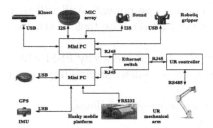

Fig. 1. Hardware structure of system   **Fig. 2.** Autonomous mobile manipulator

3 Software Architecture of System

3.1 HFSM

HFSM is the core part of system. In Fig. 3, Init, Idle, Interaction, Analysis, Planning, Implement and Clear state compose the first hierarchy, Decision, Navigation, Manipulator Operation and Vision state constitute the second hierarchy which are embedded in Implement State. The output signal of state control state transfer which relevant data are transmitted. In the first hierarchy FSM, it achieves voice interaction, intention acquisition, mission planning and delivering the task instruction to Implement State. In the second hierarchy FSM, it executes different tasks according to task instruction. So the converting from interaction to task instruction and task implementation are two separate parts. Robot will be incapable to deal with user information when executing task, while it never do any movement when talking with people. This hierarchical structure reduces interference.

- Init state: the first state when the robot is powered on. In this state, the system launches related nodes and initializes the relevant parameters. It will transform to idle state after completing initialization.
- Idle state: entering idle state indicates that robot is waiting for the input from users. When there is a voice input it will turn to interaction state for communication.

- Interaction state: the interaction state mainly achieves two functions: accepting the text from speech recognition component and sending the feedback text to the speech synthesis component for informing user. When receiving text of speech recognition component, it will go to analysis state. If analysis state and implement state produced feedback; they will turn to interaction state.
- Analysis state: it sends voice text to the reasoning component and receives result. The reasoning component will process the voice text to determine whether the user intention is complete. If not, Analysis state will receive feedback information. Otherwise it will transfer to planning state with complete user's intention.
- Planning state: it sends intention to the Planning component. And will turn to implement state after accepting the task instruction. The task instruction is a list includes many instructions.
- Implement state: implement state contains the second hierarchy FSM which executes tasks based on the list of task instructions. When entering implement state, it will transform to decision state at first.
- Decision state: the working process of decision state is as follows.
  (1) Getting the first task instruction from list.
  (2) Selecting a state from navigation, manipulator operation and vision state according to task instructions type.
  (3) Going to the state and send task instruction to corresponding component for execution. At the same time obtaining the result of performing.
  (4) Producing failure signal if task fails, go to interaction state for feedback.
  (5) Getting the next task instruction from list if task succeeds, skip to step (2).
  (6) Producing success signal if all tasks from instruction list succeed, then going to clear state.
- Navigation state: extracting target pose from static map according to target name in task instruction and sending to navigation component, which control robot moving to target.
- Vision state: sending instruction to vision component for object recognition and receiving object information recognized such as name, pose and volume. Then finding out target object in task instruction and taking related data into scene case.
- Manipulator Operation state: extracting related data of target from scene case according to task instruction and sending to manipulator operation state, which control manipulator to work.
- Clear state: clearing relevant information of mission and transforming to idle state for next mission.

Figure 4 shows the entire pipeline of the system. After robot initialization, it waits for voice input and obtains user intention by several conversations with user. Then autonomous navigation, visual recognition and robot manipulator operations tasks are executed on basis of the list order of task instruction.

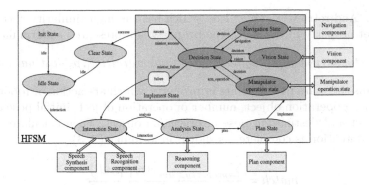

Fig. 3. Software architecture of system

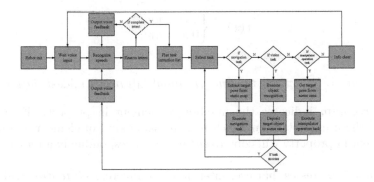

Fig. 4. System pipeline

3.2 Speech Recognition and Speech Synthesis Components

After receiving voice signal from Mic Array, speech recognition component converts it into speech text, in this paper we use Pocketsphinx speech recognition system. It is an open source voice recognition system with advantage of fast offline recognition speed, high precision. Which also support for Chinese and users could train model library by themselves [11]. The speech synthesis component converts the feedback text information into voice signal for speech, using the Ekho speech synthesis system. Ekho is an open source Chinese voice synthesis system, it has fast off-line synthesis speed and coherent pronunciation [12].

3.3 Reasoning Component

The reasoning component obtains user's intention through analysis of text, which used the CBR-BDI reasoning mechanism proposed by our research team [10]. When it receives text from the Analysis state, adopts the TF-IDF (Term Frequency-Inverse Document Frequency) of VSM (Vector Space Model) algorithm [13] to calculate text similarity between speech text and each case from case base. And retrieve the best case which has the high similarity with speech text to get property in the case. The property is represented as formula (1).

$$property_i = (p_1[action], p_2[name], p_3[num], p_4[initial], p_5[destination]) \quad (1)$$

p means there are five elements in property, which are action of robot operation, name of operation object, number of operation object, initial pose of operation object and destination pose of operation object. Then calculate whether the user's intention is complete, as shown in formulas (2, 3).

$$match = \frac{\sum_{property\_num}^{i} r(p_i \cap null)}{property\_sum} \quad (2)$$

$$r(x) = \begin{cases} 1 \ x \neq null \\ 0 \ x = null \end{cases} \quad (3)$$

$$intention = (t_1[action], t_2[name], t_3[num], t_4[initial], t_5[destination]) \quad (4)$$

property_num indicates the number of element in property. Formula (4) expresses the user's intention which has the same kind of element as *property*. The element in property will copy to *intention* correspondingly after calculating *match*.

match = 1 expresses that user intention is complete, so *intention* will be sent to analysis state. *match* < 1 means the intention is incomplete, so a solution must be found for feedback. As shown in formulas (5, 6, 7 and 8).

$$GC_i = (guidance_i, solution_i) \quad (5)$$

$$guidance_i = (g_1[action], g_2[name], g_3[num], g_4[initial], g_5[destination]) \quad (6)$$

$$sum = \sum_{i}^{property\_num} f(t_i, g_i) \quad (7)$$

$$f(x, y) = \begin{cases} 1 \ x = y \\ 0 \ x \neq y \end{cases} \quad (8)$$

The guidance case has many cases GC, each one includes *guidance* and *solution*. *guidance* has the same kind of element as *intention*, *solution* means the solution for feedback. Calculating the similarity between *intention* and each *guidance* in guidance case through formulas (7, 8). The solution with high similarity is the best feedback to user. When user hears the voice feedback, he or she

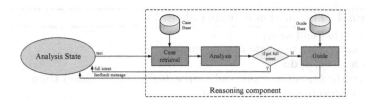

Fig. 5. Pipeline of reasoning component

will continue to communicate with robot until complete intention is obtained. Figure 5 is the pipeline of reasoning component.

An Example is presented below for understanding. In fact the dialogue between user and robot is Chinese.

[user]:**transport apple from area one to area two.**(through case retrieval it get intention $(transport, apple, null, areaone, areatwo)$.)

[robot]:**how many apples should be taken?**($match$ result < 1, find a solution in guidance case which ask the number of operation object.)

[user]:**two apples.**(get intention $(transport, apple, two, areaone, areatwo)$.)

[robot]: **transport two apples from area one to area two, right?**($match$ result $= 1$,get complete intention and query user for confirmation.)

[user]:**Yes.**

[robot]:**OK.**(after confirmation, send complete intention to analysis state.)

3.4 Planning Component

Planning component parses user intention from plan state to task instructions list. According to the robot's function, we designed six kinds of instructions.

(1) Shrink: manipulator operation instruction. It shrink manipulator for avoiding collision before movement of robot.
(2) Move $(target)$: navigation instruction. It controls robot moving to $target$ pose.
(3) Seek $(name)$: vision instruction. It uses Kinect for object recognition, recognize object $name$.
(4) Pickup $(name)$: manipulator operation instruction. It picks up object $name$.
(5) Place $(target)$: manipulator operation instruction. It places object on $target$.
(6) Back: navigation instruction. It control robot moving to original pose which the position is (0, 0, 0) and orientation is (0, 0, 0, 1) based on odometer coordinate frame.

Because the manipulator of robot operates only an object every time, $action$ and num in $intention$ determine the process in task instruction list. The procedure of mission planning algorithm is that extracting $action$ and num of $intention$ for matching a instruction template in template case. Then it extracts $name$, $initial$ and $destination$ adding to template. As shown in Algorithm 1.

Algorithm 1. The mission planning algorithm

1: extract *action* and *num*;
2: search instruction template in the template case
3: **for** each $i \in [1, n]$ **do**
4:     **if** ($instruction[i] = move$ and $instruction[i+1]! = place$) **then**
5:         $instruction[i] = moveinitial$
6:     **end if**
7:     **if** ($instruction[i] = move$ and $instruction[i+1] = place$) **then**
8:         $instruction[i] = movedestination$
9:         $instruction[i+1] = placedestination$
10:         $i++$
11:     **end if**
12:     **if** ($instruction[i] = seek$) **then**
13:         $instruction[i] = seekname$
14:         $instruction[i+1] = pickupname$
15:         $i++$
16:     **end if**
17: **end for**

Among Algorithm 1, n represents the amount of instruction in template, and *instruction*[i] means the *No.i* instruction in template. We give an example with the *intention* = (*transport, apple, two, areaone, areatwo*) in Sect. 3.3 above. So the template according to *action* and *num* is {shrink; move; seek; pickup; shrink; move; place; shrink; move; seek; pickup; shrink; move; place; back}. The task instruction list will be {shrink; move area one; seek apple; pickup apple; shrink; move area two; place area two; shrink; move one; seek apple; pickup apple; shrink; move area two; place area two; back} through Algorithm 1.

3.5   Navigation Component

Robot movement is controlled by navigation component. The structure is shown in Fig. 6. At first, it confirms the relative pose between robot and target based on odometer and target pose. Then producing command according to global path and local path. Global path is calculated by global map generated through map from SLAM and local path is calculated by local map generated through laser data. Afterwards it controls robot for moving through command and return result.

Because of the lack of laser sensor, we take the depth data from Kinect as laser data by ROS package named depthimage_to_laserscan [14] for SLAM of Gmapping [15], which is faster then SLAM of RGBD. The global map and local map are 2D costmap, we use Dijkstra algorithm calculates global path and DWA (Dynamic Window Approach) algorithm calculates local path. Odometer is calculated by GPS data and IMU data, using EKF (Extended Kalman Filter) algorithm for modifying cumulation error.

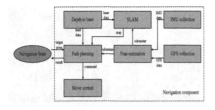
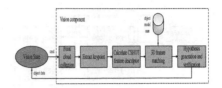

Fig. 6. Structure of navigation component   **Fig. 7.** Pipeline of vision component

3.6   Vision Component

Vision component recognizes object and calculates the pose of object through point cloud data. The pipeline is shown in Fig. 7. When receiving instruction from vision state, Kinect collects Point cloud data. Then it uses ISS (Intrinsic Shape Signatures) key points algorithm [16] to extract ISS key points in scene, and adopts CSHOT (Color Signature of Histograms of Orientations) feature descriptor algorithm [17] to calculate CSHOT feature vector from key points. The relative pose transformation matrix of object model to corresponding point in scene is generated through 3D feature matching with model in object model case. After that, Random Sample Consensus (RANSAC) algorithm is used for producing transformation hypothesis and iterative closest point algorithm is used for verifying hypothesis. Finally the object data is sent to vision state.

3.7   Manipulator Operation Component

Manipulator operation component controls the manipulator of robot. As shown in Fig. 8, RRT (Rapidly-exploring Random Tree) algorithm [18] is used for trajectory planning according to target pose and current pose of robot. Then it controls manipulator moving to target pose. And Gripper server calculates the stretch or close of gripper based on instruction, and sends command to gripper control module for operation.

4   Experiments and Analysis

The experimental scenario was designed to verify the feasibility of proposed architecture, as shown in Fig. 9. In order to simulate the working environment, four areas were arranged in the scene. There were many objects which could be grabbed by gripper in each area. Obstacles were placed among the four regions. In this paper we used Kinect for navigation which had weakness of narrow data acquisition range and low data accuracy. So the shape of obstacles was regular. Later we will use laser sensor instead of Kinect.

Ten different mission experiments were designed in this paper, in each mission experiment user let robot to transport one or more objects from one area to another. The voice information which user input was incomplete interaction

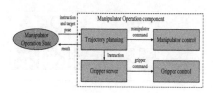

Fig. 8. Structure of manipulator operation component

Fig. 9. Experimental scenario

every time. So the robot realized voice interaction, mission planning, autonomous navigation, object recognition, manipulator pick up and manipulator place.

For example, in mission 2 the user communicated with robot several times to let robot to transport two apple from area three to area four. So the robot should execute autonomous navigation five times, object recognition two times, manipulator pick up two times and manipulator place two times according to the task instruction list. Mission 3 to mission 10 were the same experiment form. Through the mission experiments to verify whether the robot could accomplish the user specified tasks to achieve autonomous process of robot interaction, reasoning, planning and implement.

Table 1. Experimental results

Mission list	Getting complete intention	Autonomous navigation	Object recognition	Manipulator pick up	Manipulator place	Mission result
Mission 1	Yes	√√	√	√	√	Success
Mission 2	Yes	√√ × ⊘⊘	√⊘	√⊘	√⊘	Failure
Mission 3	Yes	√√√	√	√	√	Success
Mission 4	Yes	√√√	√	√	√	Success
Mission 5	Yes	√√√√	√√	√√	√√	Success
Mission 6	Yes	√√√√	√×	√×	√×	Failure
Mission 7	Yes	√√√√	√√	√√	√√	Success
Mission 8	Yes	√√√√√√	√√√	√√√	√√√	Success
Mission 9	Yes	√√√	√	√	√	Success
Mission 10	Yes	√√√√	√√	√√	√√	Success

Table 1 shows the experimental result. The second column indicates whether robot could get complete intention through communication with users. √ indicates the success of task, × indicates the failure of task and ⊘ indicates the no execution of task. The sum of √, × and ⊘ indicates the time robot needs to execute the task in a mission experiment. For example in mission 2, robot needs to execute autonomous navigation five times, object recognition two times, manipulator pick up two times and manipulator place two times according to the task instruction list after getting complete intention. While it failed autonomous navigation in the third time, so mission result is failure. There is no need to execute the second object recognition, manipulator pick up and manipulator place.

Each result of task in the mission affects the mission result. The automation of mission depends not only on the HFSM but also on success of every component. In the ten times mission experiments, mission 2 failed because of the failure of autonomous navigation and mission 6 failed because of the failure of object recognition. Based on the analysis, error data collected by Kinect led to the failure. Kinect is always affected by surrounding environment especially light. In the other eight experiments, robot has accomplished the specified mission, which proved that the proposed architecture is feasible and practical.

5   Conclusions and Future Work

In this paper, a robot architecture of hierarchical finite state machine for autonomous mobile manipulator was proposed. The experiment results expressed that the robot can achieve voice interaction with users for getting complete intention, plan task instruction according to intention and execute task based on task instruction. We could conclude that our proposed architecture accomplish the automation of task according to speech interaction.

In the future, we will improve our system at three aspects. Firstly a laser sensor and a better RGBD camera should be taken for improving the robustness of autonomous navigation and object recognition. In the second, we will take a research work on behavior tree and try to modify system architecture with behavior tree for adapting complex environment. The third point is that revising the reasoning mechanism in reasoning component for better interaction with people.

Acknowledgements. This work is supported by Natural Science Foundation of China. (Project No. 61673304) and (Project No. 61175094).

References

1. Rossi, S., Leone, E., Fiore, M., et al.: An extensible architecture for robust multimodal human-robot communication. In: 2013 26th IEEE/RSJ International Conference on Intelligent Robots and Systems, pp. 2208–2213. IEEE Press, New Jersey (2013)
2. Ma, Q., Zou, Y., Zhang, T.: Study of service robot architecture based on middleware and abstract environment. In: 2012 IEEE International Conference on Robotics and Biomimetics, pp. 1200–1205. IEEE Press, Washington DC (2012)
3. Foukarakis, M., Leonidis, A., Antona, M., et al.: Combining finite state machine and decision-making tools for adaptable robot behavior. In: 8th International Conference on Universal Access in Human-Computer Interaction, pp. 625–635. Springer Verlag, New York (2014)

4. Nguyen, H., Ciocarlie, M., Hsiao, K., et al.: ROS commander (ROSCO): behavior creation for home robots. In: 2013 IEEE International Conference on Robotics and Automation, pp. 467–474. IEEE Press, New Jersey (2013)

5. Li, Q., Meier, M., Haschke, R., et al.: Object dexterous manipulation in hand based on finite state machine. In: 2012 9th IEEE International Conference on Mechatronics and Automation, pp. 1185–1190. IEEE Computer Society Press, Washington DC (2012)

6. Park, H.W., Ramezani, A., Grizzle, J.W.: A finite-state machine for accommodating unexpected large ground-height variations in bipedal robot walking. IEEE Trans. Robot. **29**(2), 331–345 (2013)

7. Da, X., Hartley, R., Grizzle, J.W.: First steps toward supervised learning for underactuated bipedal robot locomotion, with outdoor experiments on the wave field. In: 2017 IEEE International Conference on Robotics and Automation. IEEE Press, New Jersey (2017)

8. Kurt, A., Özgüner, Ü.: Hierarchical finite state machines for autonomous mobile systems. Control Eng. Pract. **21**, 184–194 (2013)

9. Jayawardena, C., Kuo, I.H., Broadbent, E., et al.: Socially assistive robot healthbot: design, implementation, and field trials. IEEE Syst. J. **10**(3), 1056–1067 (2016)

10. Lin, Y., Min, H., Zhou, H., et al.: A RGB-D sensor and speech interaction based object manipulation system with improved CBR-BDI reasoning. In: 2017 IEEE International Conference on Control and Automation. IEEE Computer Society Press, Washington DC (2017)

11. Pocketsphinx Homepage. http://cmusphinx.sourceforge.net/wiki/. Accessed 16 June 2016

12. Ekho Chinese Text-to-Speech Software Homepage. http://www.eguidedog.net/cn/ekho_cn.php. Accessed 16 June 2016

13. Guo, Q.: The similarity computing of documents based on VSM. In: 32nd Annual IEEE International Computer Software and Applications Conference, pp. 585–586. IEEE Press, New Jersey (2008)

14. ROS Package Depthimage to Laserscan Homepage. http://wiki.ros.org/depthimage_to_laserscan. Accessed 16 June 2016

15. Omara, H.I.M.A., Sahari, K.S.M.: Indoor mapping using kinect and ROS. In: 1st International Symposium on Agents. Multi-Agent Systems and Robotics, pp. 110–116. IEEE press, New York (2015)

16. Tombari, F., Salti, S., Di, S.L.: A combined texture-shape descriptor for enhanced 3D feature matching. In: 2011 18th IEEE International Conference on Image Processing, pp. 809–812. IEEE Press, New Jersey (2011)

17. Zhong, Y.: Intrinsic shape signatures: a shape descriptor for 3D object recognition. In: 2009 IEEE 12th International Conference on Computer Vision Workshops, pp. 689–696. IEEE Press, New Jersey (2009)

18. Karaman, S., Walter, M.R., Perez, A., et al.: Anytime motion planning using the RRT. In: 2011 IEEE International Conference on Robotics and Automation, pp. 1478–1483. IEEE Press, New Jersey (2011)

A Diagnostic Knowledge Model of Wind Turbine Fault

Hongwei Wang[1,2] iD, Wei Liu[2(✉)] iD, and Zhanli Liu[1] iD

[1] Huadian Electric Power Research Institute, Hangzhou 310030, China
hongwei.wang@port.ac.uk
[2] University of Portsmouth, Portsmouth, PO1 3DJ, UK
Wei.Liu@port.ac.uk

Abstract. With the development of the wind power industry, wind power has become one of the main green generation energy. At the same time, with the wind power installed capacity increasing, the failure rate gradually growth. As wind turbine is a complex electromechanical equipment, the fault diagnosis for this kind of equipment is also a complicated process. Focused on the current shortage of fault diagnosis knowledge representation, this paper proposes a diagnostic knowledge model for wind turbine and also elaborates the model structure definition with a target to ensure the accuracy of fault diagnosis. Besides, this model can also offer assistance reference model for researchers in related fields to develop advanced methods for sharing and reuse of diagnostic knowledge.

Keywords: Wind turbine · Fault diagnosis · Knowledge model

1 Introduction

As the global economic development accelerates, economic models constantly tend to move towards sustainable development. Under this environment the utilisation of renewable energy has received extensive attention by most countries, especially wind energy in power generation has been widely used all over the world. According to the report of "The Wind in power 2016 European statistics" [1], 12.5 GW of gross additional wind capacity was installed in Europe only in 2016, with a total installed capacity of 153.7 GW, generating almost 300 TWh and covering 10.4% of the EU's electricity demand. With a stable growth, wind power has grown from the penultimate energy in 2005 to the second largest power form generation capacity in Europe in 2016. The rapid development of wind power has also brought great challenges to the wind equipment manufacturing industry and wind farms. Current studies of Wind Turbines (WT) are aimed at reducing the cost of energy to ensure that the wind power industry is more competitive compared with other sources of energy. Additionally, in order to increase power generation capacity, more and more large size WTs are designed and deployed.

However, the condition of transportation and installation of equipment have raised considerable challenges to the operation and maintenance of WTs. Moreover, WTs are often installed on extremely high towers over 20 m in height especially for the offshore wind farms, which greatly increases cost and difficulty of maintenance and repair. In this case, it is not unusual for the maintenance cost of WTs components to exceed their

© Springer International Publishing AG 2017
Y. Huang et al. (Eds.): ICIRA 2017, Part III, LNAI 10464, pp. 437–448, 2017.
DOI: 10.1007/978-3-319-65298-6_40

procurement cost [2, 3] and the Operation and Maintenance (O&M) occupies a significant share of the annual cost of WT downtime. For a turbine with a 20-year working life, the O&M costs of 750 kW turbines might account for about 25%–30% of the overall energy generation cost [4, 5]. Therefore reducing the O&M cost and improving the operation efficiency and reliability of WT are significant. In recent years, WT fault diagnosis system has been developed to monitor health condition and detected faults at an early time. However, WT is a complex electromechanical equipment with a variety of potential failures. And so far there are few detailed integrated diagnosis methods of WT in a complicated process. To address the disadvantages of current methods in developing an effective and complete fault diagnostic knowledge representation model, this paper will put forward such a model to improve accuracy of fault diagnosis and support effective capture and reuse of fault diagnosis knowledge.

2   Typical Categories of Faults

WT is a complex multi-component system which converts wind energy into mechanical energy and finally generates electricity feeding into the grid. Because most WTs are located in remote areas such as offshore wind farms, they are operating under very harsh weather and unstable load conditions. This leads to a lot of failures in different subsystems, which adversely affect the lifetime. The most common failures have root causes in subsystems including blades, shaft bearings, electrical control, yaw and pitch system, rotor brake, drive train, hydraulics, gearbox and generator [6]. Figure 1 illustrates the main components of a WT that are faced with all the above concerns [7]. Failures can be classified according to the failure modes and causes [8]. A failure cause is a root cause which can lead to the occurrence of a failure and has been defined as the circumstances during design, manufacture or use [9].

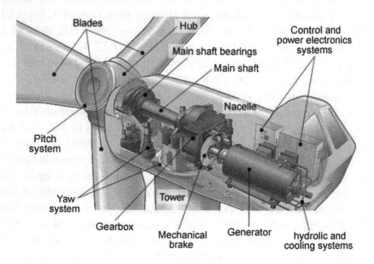

Fig. 1. The main components of WT [7]

2.1 Gearbox

As an indispensable part of a non-direct drive WT, the gearbox is one of the most frequently damaged subsystems in the turbine. A good amount of literature has indicated that gearbox is by far the most liable subsystem responsible for wind turbine downtime and maintenance cost [10]. The function of the gearbox is to transmit wind power into the generator through increasing the rotor speed from the low-speed shaft to the high-speed shaft. Because WTs operate in harsh environment conditions such as strong wind, lightning, air pressure and excessive loading due to dust wind, tribological conditions for drive trains of WTs such as wear, fatigue and corrosion, can very likely lead to increased component damages and machine malfunctions [10]. Unstable loading causes can also cause various failures such as bearing shaft misalignment, shaft imbalance, leaking oil, high oil temperature, bearing damage, etc. [11]. Other failures such as gear backlash, tooth breakage and cracks in the gear tooth and surface fractures may also occur during the operation of WTs.

2.2 Blades

Blades are the main rotors of WT, which transform wind power into up-lifting force through fluid dynamic physical design [6]. Because of poor rigidity of the blade, blade failures may occur when wind speed is too high or the vibration of the irregular operation during the course of rotation. Due to stress (both external and internal) and fatigue, crack and damage may gradually take place as time goes on. In the meanwhile, turbulent wind, lightning and production defects can also cause blade faults [12, 13], thus leading to performance deterioration of wind generation. These reasons bring the rotor imbalance, blades and hub corrosion, blade damages, crack, bend, and serious aeroelastic deflections [14]. Other failures caused by components inside a blade include cracks at the bonding resin, missing adhesive, delamination's within the glass fibre reinforced plastics (GFRP) or the sandwich crack on the web, excess of bonding resin, etc. [15].

2.3 Electronics Control

Electronics control occupies 11% of the failure rate in the whole cost of maintenance of a WT [16]. The common failures are damages in generator winding and transformer wirings, short-circuit and over voltage of electronics components [17]. The root causes include lightning, poor electrical installation, technical defects and resonances within Resistor-Capacitor (RC) circuits [18]. These sensors are coupled together with algorithms and architectures, which allow for efficient monitoring of the machine's condition [19].

2.4 Generator

The function of a generator is transforming mechanical energy into electricity energy. Due to the fact that generators operate in the electromagnetic environment with high failure rates. Failures in the wind turbine generator will cause several issues such as abnormal noises and excessive vibration in the generator [7]. Additionally, winding failure comes from defective insulation system, short circuits in the AC grid may result in generator overheating and poor lubrication may lead to the bearing fatigue [15].

2.5 Yaw System

The yaw system controls the nacelle rotation to achieve the maximum efficiency under the uncertain wind direction. The main rotary components of yaw consist of drive motor, reducer and gear ring. Motor vibration, yaw gear tooth wear, abnormal noise, yaw limit switch fault, lubricant leakage, the cracking of yaw drive shafts, pitting of the yaw bearing race and bearing damage are all categorised as yaw system failures [20, 21]. The root causes are severe weather and highly vibration on overload.

3   WT Diagnostic Knowledge Model

Currently, some computer systems for WT condition monitoring and fault diagnosis have been developed to monitor and control the early faults. In addition to some sorts of failures that can be detected in an intuitive way such as oil leaking, corrosion, the physical condition of a WT can also be indicated by the sound coming from the bearings. However, the process of diagnosis is very complicated, and there are various reasons causing WT failures that will result in unplanned downtime and economic loss for the wind power industry. Therefore, more sophisticated and professional approaches are needed to diagnose faults and improve WT maintenance. Building a diagnostic knowledge model is necessary for improving the accuracy of fault diagnosis. This section proposes a knowledge model used to classify, capture and store WT fault diagnosis knowledge with a specifically devised structure. Constructing a diagnostic knowledge model is a complicated systematic process which requires a range of extensible knowledge classification methods in the research area and collection of a large amount of fault records [22]. An understanding of the failures, operation parameters, root causes, characteristics and diagnosis methods is important for realising knowledge capture and representation based on the WT diagnostic model. As WT is a complex electromechanical equipment that operates in a harsh environment, fault diagnosis of WT becomes even more intricate. The process of fault diagnosis is to identify the types of failure by using diagnostic methods that take account of fault features, operating parameters and vibration characteristics of WT. Based on an analysis of the fault diagnosis information of WT, the diagnostic knowledge model is composed of six elements as follows:

Issue: Identifying the fault location in subsystems based on WT structure.

Answer: Determining the specific failure in the subsystems of WT.

Characteristic: When the failure occurs in WT, there are a lot of symptoms will come out like abnormal Vibration, noise, leaking oil, and overheating, etc.

Parameter: The running state of WT can be indicated by the operational parameter. And each condition matches a consistent value, if the value changes, meaning the component is being in the failure status. In addition, the data of WT are acquired by SCADA system and vibration sensors.

Cause: There are various reasons lead to WT fault by independent or coupled reasons caused.

Argument: Diagnosis methods for supporting the corresponding failures could be monitoring, and diagnosis methods have been applied in the real-wind farms, or hypothesis experts proposed.

The diagnostic knowledge model of WT is a foundation for implementing quick information and knowledge sharing among users. The structure map of the diagnostic knowledge model is shown in Fig. 2. There are five parts (issues) in this model: Blades, Auxiliary System, Generator, Gearbox and Tower, as well as other four elements (Answers, Characteristics, Parameters, Causes) and their sub-classes. Besides information aggregation, the argument element is another critical section of this model that collection a lot of diagnosis methods which experts proposed for different faults. Therefore, following this knowledge model that the corresponding diagnostic method can be adopted to detect specific failure quickly. Because the process of finding diagnosis method is based on the fault characteristic, root causes and running parameter of the moment when the failure occurs in WT.

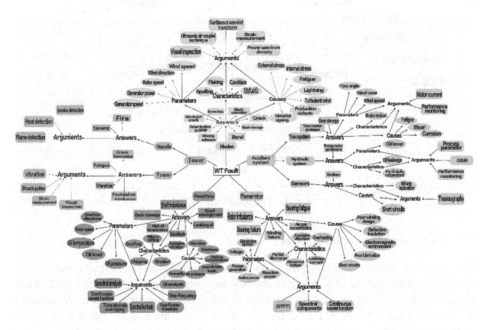

Fig. 2. Structure map of the WT diagnostic knowledge model

3.1 Blades Sub-model

The typical faults of WT are discussed in Sect. 2. Figure 3 shows the fault characteristics in blades represented by spalling, flaking, cavities and blades out-of-control rotation. The root causes of blades fault come from external and internal stress, fatigue, turbulent wind, lightning, unstable load, blades material ageing and production defects, etc. In the meanwhile, related parameters used to reveal blades condition status include wind speed, wind direction, rotor speed, generator power and generator speed. The arguments provide a series of diagnosis methods by specialists. For some outside failures such as blades crack and bend, utilise visual inspection to detect by high-powered telescope. Jeffries and Chambers [23] proposed a method, without additional sensors on blades, to measure the power spectrum density at the small WT generator terminals to identify rotor unbalance and defects. The continuous wavelet transform and strain measurement approaches are applied to blades. In addition, the ultrasonic air-coupled technique has been used to recognise the possible geometric figure, and similar dimension of failure through the ultrasonic images by obtained [24].

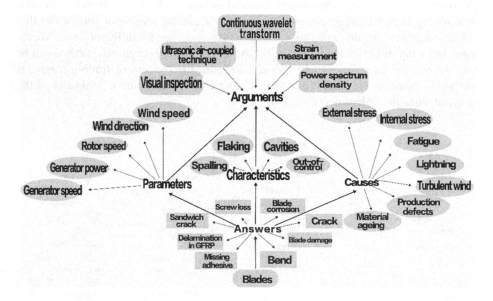

Fig. 3. The blade sub-model

3.2   Auxiliary System Sub-model

The auxiliary system is divided into yaw system, hydraulic system and sensors. Yaw motor fault and gear damage occur in the yaw system, and express in brake locked due to gear fatigue, wear, and corrosion, etc. Usually, utilisation the approach monitors operational parameter of WT to determine the WT condition state, the parameter are including yaw angle, wind vane data, wind speed, motor current, etc. Performance monitoring has been applied to WT systems for condition monitoring and fault diagnosis. The problems about the hydraulic system are oil leakage and hydraulic retardant. The same as yaw system approach, this involves monitoring oil level to detect as well as operation and maintenance of the whole system. Moreover, thermography analysis is also used as a way to recognise sensor's failure. IR thermography has already been identified as one of the most effective tools for condition monitoring in the WT for diagnosis of electric equipment [25]. This method is based on the sensor in the working environment that emits heat over its normal temperature (Fig. 4).

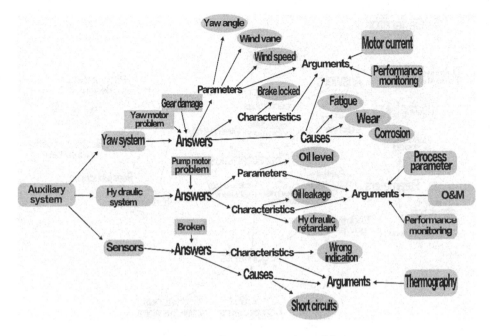

Fig. 4. Auxiliary system sub-model

3.3 Generator Sub-model

Because the generator operates in the electromagnetic environment, short circuits in the AC grid, defective insulation, poor lubrication and poor winding design are reasons lead to generator failures. The familiar features of a generator include overheating, excessive vibration, abnormal noises, leakage current increasing and partial discharge, etc. In addition, generator speed, voltage, power generation, generator active and reactive power are also applied to detect health status. Therefore, in the steady-state of induction generator, diagnosing the turn-to-turn fault, bearing failures and air gap eccentricities can be realised by the steady-state spectral components of the stator quantities [26, 27]. However, the wind speed is variable with unstable generator signals and thus the short time Fourier transform (STFT) was put forward to analyse rotor unbalance and short-circuits in the stator winding of WT with the heavy load in transient conditions [28]. The other method, i.e. continuous wavelet transform, is also used to detect rotor electrical unbalance and stator turn faults in a DFIG [29, 30] (Fig. 5).

3.4 Gearbox Sub-model

The gearbox is a component of the main drive train and is also the most frequently damaged subsystem that may lead to non-direct drive WT downtime. It has spawned more and more attention from experts to develop effective diagnosis methods. The gearbox fault is caused suffering harsh weather conditions, frequent stoppage and staring, corrosion and also particle contaminations result in micro-pitting. Parameters

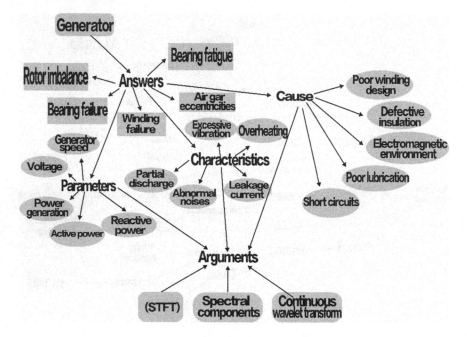

Fig. 5. Generator sub-model

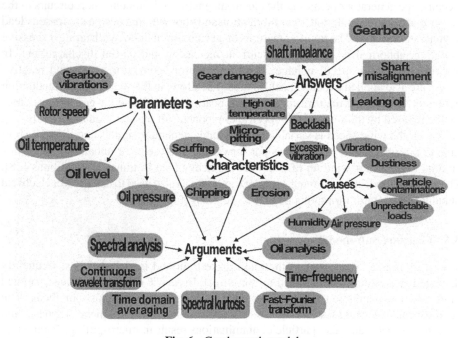

Fig. 6. Gearbox sub-model

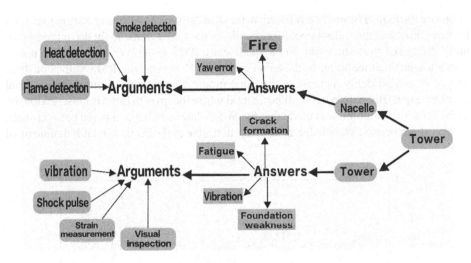

Fig. 7. Tower sub-model

monitoring of gearbox contain gearbox vibrations, the temperature of gearbox oil, rotor speed, oil level in gearbox sump and gearbox oil pressure. The vibration analysis, time domain averaging, spectral kurtosis and time-frequency analysis are common methods for non-stationary gearbox faults [31]. Wavelet and Fourier transformation are the most popular methods to detect gearbox failures [32]. When vibration has no obvious change, oil analysis also can be considered as a diagnostic means (Fig. 6).

3.5 Tower Sub-model

The tower is the foundation of a WT, and the whole tower system consists of nacelle and tower. Fire and yaw error usually happen in the nacelle. Crack formation, fatigue, vibration and foundation weakness are categorised as tower failure. In order to ensure the nacelle security and normal running need to carry out smoke, heat and flame detection. The detection techniques on the tower are not widespread such as shock pulse method, strain measurement and vibration analysis [7] except visual inspection (Fig. 7).

3.6 Evaluation

To evaluate the proposed methods, an example is used through developing condition monitoring and fault diagnosis model for a wind farm in China. A number of WTs are monitored using a pre-established data collection system. One WT is detected with the fault in its gearbox through the operation parameters collected by sensors on drive train of the gearbox. The fault characteristic is mainly in vibration signal abnormalities and amplitude increasing. The points on gearbox measured from the motor to the main shaft include axial direction of high-speed shaft, high-speed shaft Radial and parallel shaft horizontal. Figure 8 shows the RMS tendency of each measuring point. According to the record of diagnosis knowledge model, a vibration analysis method was applied to diagnose the fault. The method is based on the distribution of gearbox spectrum to extract feature index and the failure evolution trend. As the speed of the main drive train gradually decreased from the motor to the main shaft, RMS values of three measure points also showed declined trend in the same direction. However, the RMS values of three sets of points suddenly increase at the same moment before the 250th sampling serial number, especially in parallel shaft horizontal where the growth trend is most obviously. Therefore, a speculation was made that the WT failure may happen in the parallel shaft. Hence, the proposed knowledge model is practicable and effective in fault diagnosis of WT.

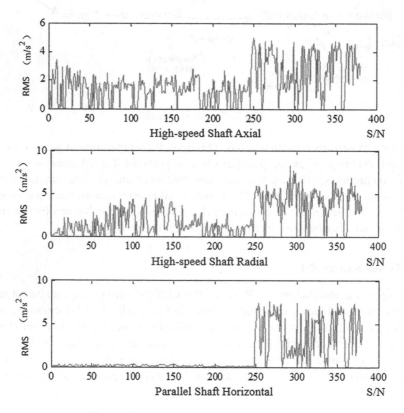

Fig. 8. The RMS trend of each measure point

4 Conclusion

In recent years, the wind power industry has paid more and more attention that reducing the cost of operation and maintenance while improving the WT reliability. This paper summarised typical failures in WTs and a diagnosis knowledge model was constructed on six elements: issue, answer, fault characteristic, running parameter, root causes and argument. The result of evaluation reveals the diagnostic knowledge model is efficient. At the same time, the fault diagnosis methods and knowledge model could provide technical supports, references and information sharing for users in different fields. In the future works should be focused on the expert collaboration improving the content of diagnosis knowledge model and increasing knowledge expressivity. Moreover, establishing a fault diagnosis database to realise application flexibility of detectable methods is necessary for sharing and reuse of diagnostic knowledge.

References

1. Global Wind Energy Council: Wind in power 2016 European statistics, pp. 6–7 (2016)
2. Hatch, C.: Improved wind turbine condition monitoring using acceleration enveloping. Orbit **61**, 58–61 (2004)
3. Kusiak, A., Li, W.: The prediction and diagnosis of wind turbine faults. Renew. Energy **36**, 16–23 (2011)
4. García Márquez, F., Tobias, A., Pinar Pérez, J., Papaelias, M.: Condition monitoring of wind turbines: techniques and methods. Renew. Energy **46**, 169–178 (2012)
5. Milborrow, D.: Operation and maintenance costs compared and revealed. Windstats Newslett. **19**, 1–3 (2006)
6. Lau, B.C.P., Ma, E.W.M., Pecht, M.: Review of offshore wind turbine failures and fault prognostic methods. In: 2012 IEEE Conference on Prognostics and System Health Management (PHM), pp. 1–5. IEEE (2012)
7. Takoutsing, P., Wamkeue, R., Ouhrouche, M., Slaoui-Hasnaoui, F., Tameghe, T., Ekemb, G.: Wind turbine condition monitoring: state-of-the-art review, new trends, and future challenges. Energies **7**, 2595–2630 (2014)
8. Birolini, A.: Reliability Engineering, vol. 5. Springer, Heidelberg (2007)
9. International Electrotechnical Commission. http://www.electropedia.org/. Accessed 21 Apr 2017
10. Nie, M., Wang, L.: Review of condition monitoring and fault diagnosis technologies for wind turbine gearbox. Procedia CIRP **11**, 287–290 (2013)
11. Liu, W., Tang, B., Han, J., Lu, X., Hu, N., He, Z.: The structure healthy condition monitoring and fault diagnosis methods in wind turbines: a review. Renew. Sustain. Energy Rev. **44**, 466–472 (2015)
12. Amirat, Y., Benbouzid, M., Al-Ahmar, E., Bensaker, B., Turri, S.: A brief status on condition monitoring and fault diagnosis in wind energy conversion systems. Renew. Sustain. Energy Rev. **13**, 2629–2636 (2009)
13. Kithil, R.: Case study of Lightning Damage to wind turbine blade. National Lighting Safety Institute (NLSI) (2008)
14. Yang, M., Chengbing, H., Xinxin, F.: Institutions function and failure statistic and analysis of wind turbine. Phys. Procedia **24**, 25–30 (2012)
15. Major failures in the wind turbine components and the importance of periodic inspections. http://www.windaction.org/posts/44617-major-failures-in-the-wind-turbine-components-and-the-importance-of-periodic-inspections#.WPlXF4jyvIU. Accessed 20 Apr 2017
16. Purarjomandlangrudi, A., Nourbakhsh, G., Esmalifalak, M., Tan, A.: Fault detection in wind turbine: a systematic literature review. Wind Eng. **37**, 535–548 (2013)
17. Babu, J.R., Jithesh, S.V.: Breakdown risks in wind energy turbines. Pravartak J. Insur. Risk Manag. National Insur. Acad. Pun **3**(3) (2008)
18. The Confederation of Fire Protection Association CFP A Europe.: Wind turbines fire protection guideline (2010)

19. Benbouzid, M.: Bibliography on induction motors faults detection and diagnosis. IEEE Trans. Energy Convers. **14**, 1065–1074 (1999)
20. Ghaedi, A., Abbaspour, A., Fotuhi-Firuzabad, M., Moeini-Aghtaie, M.: Toward a comprehensive model of large-scale dfig-based wind farms in adequacy assessment of power systems. IEEE Trans. Sustain. Energy **5**, 55–63 (2014)
21. Stubkier, S., Dyrbye, E.: AVN energy designs a new hydraulic yaw system and reduces the fatigue loads of the wind turbine. Mod. Energy Rev. **3**(2), 45 (2011)
22. Chen, R., Zhou, Z., Liu, Q., Pham, D.T., Zhao, Y., Yan, J., Wei, Q.: Knowledge modeling of fault diagnosis for rotating machinery based on ontology. In: 2015 International Conference on Industrial Informatics (INDIN), pp. 1050–1055. IEEE (2015)
23. Jeffries, W.Q., Chambers, J.A., Infield, D.G.: Experience with bicoherence of electrical power for condition monitoring of wind turbine blades. IEE Proc.-Vis. Image Sig. Process. **145**(3), 141–148 (1998)
24. Jasiūnienė, E., Raišutis, R., Šliteris, R., Voleišis, A., Jakas, M.: Ultrasonic NDT of wind turbine blades using contact pulse-echo immersion testing with moving water container. Ultragarsas "Ultrasound" **63**(3), 28–32 (2016)
25. Liu, W., Tang, B., Jiang, Y.: Status and problems of wind turbine structural health monitoring techniques in China. Renew. Energy **35**, 1414–1418 (2010)
26. Benbouzid, M.E.H., Kliman, G.B.: What stator current processing-based technique to use for induction motor rotor faults diagnosis? IEEE Trans. Energy Convers. **18**(2), 238–244 (2003)
27. Benbouzid, M.E.H.: A review of induction motors signature analysis as a medium for faults detection. IEEE Trans. Industr. Electron. **47**(5), 984–993 (2000)
28. Wilkinson, M.R., Spinato, F., Tavner, P.J.: Condition monitoring of generators & other subassemblies in wind turbine drive trains. In: IEEE International Symposium on Diagnostics for Electric Machines, Power Electronics and Drives, pp. 388–392. IEEE (2007)
29. Yang, W., Tavner, P.J., Crabtree, C.J., Wilkinson, M.: Research on a simple, cheap but globally effective condition monitoring technique for wind turbines. In: 2008 18th International Conference on Electrical Machines, ICEM 2008, pp. 1–5. IEEE (2008)
30. Kim, K.: Parlos, A.: Induction motor fault diagnosis based on neuropredictors and wavelet signal processing. IEEE/ASME Trans. Mechatron. **7**, 201–219 (2002)
31. Ahadi, A.: Wind turbine fault diagnosis techniques and related algorithms. Int. J. Renew. Energy Res. (IJRER) **6**(1), 80–89 (2016)
32. Zhang, Z., Verma, A., Kusiak, A.: Fault analysis and condition monitoring of the wind turbine gearbox. IEEE Trans. Energy Convers. **27**, 526–535 (2012)

Robust Object Tracking via Structure Learning and Patch Refinement in Handling Occlusion

Junwei Li[1], Xiaolong Zhou[1($\boxtimes$)], Shengyong Chen[2], Sixian Chan[1], and Zhaojie Ju[3]

[1] College of Computer Science and Technology, Zhejiang University of Technology,
Hangzhou, China
{junwei.li,zxl}@zjut.edu.cn
[2] School of Computer Science and Engineering, Tianjin University of Technology,
Tianjin, China
csy@tjut.edu.cn
[3] School of Computing, University of Portsmouth, Portsmouth, UK

Abstract. Object tracking is a challenging task especially when occlusion occurs. In this paper, we propose a robust tracking method via structure learning and patch refinement to handle occlusion problem. First, we pose the tracking task as a structured output learning problem to mitigate the gap between pattern classification and the objective of object tracking. Contrary to the random target candidates selection method, we utilize the object independent proposal strategy to generate high quality training and testing samples in structured learning. Second, we over-segment the tracked target to a set of superpixel patches, and then train a background/foreground binary classifier to remove the background patches within the tracked object rectangle area for refining the tracking precision. The objective of target refining is to mitigate tracking model degradation and enhance model robustness for adapting our tracker for long-term and accurate tracking. Experimental results conducted on publicly available tracking dataset demonstrate that the proposed tracking method achieves excellent performance in handling target occlusion.

Keywords: Object tracking · Structure learning · Appearance model · Superpixel · Classification

1 Introduction

Visual tracking has been a fundamental research focus in computer vision and robotics because of its wide applications in intelligent video surveillance, transportation monitoring and robot-human interaction, etc. Although lots of excellent algorithms have been proposed for visual tracking, it remains a challenging problem for a tracker to handle occlusions, abrupt motion, large appearance variations and background clutters.

© Springer International Publishing AG 2017
Y. Huang et al. (Eds.): ICIRA 2017, Part III, LNAI 10464, pp. 449–459, 2017.
DOI: 10.1007/978-3-319-65298-6_41

Currently, most of the tracking methods can be categorized into either generative or discriminative model based methods. Generative based tracking methods evaluate target likelihood by searching for the most probable target candidates as tracking result. In [1], Ross *et al.* propose to use principal component analysis for measuring the distance between target candidates and the subspace. Sparse representation is also used to object tracking. In [2,3], authors introduce to employ dictionaries learning or regulation rules to reconstruct target candidates by sparse combination of templates, and the target candidates with minimize reconstruction error are selected as tracking result. Contrary to generative tracking model, discriminative model formulates tracking task as a binary classification problem to distinguish target from background. To achieve better target location estimate, lots of powerful online learning classifiers including support vector machine, structured output support vector machine, multiple instance leaning, and correlation filter are used to make foreground/background prediction. Avidan *et al.* [4] propose to use SVM classifier combining optical flow for target classification. Lately, Hare *et al.* [5] present a structured output SVM to predict target location. Recently, correlation filters [6,7] have been used for efficient and effective visual tracking. The aforementioned discriminative trackers maily perform tracking in Haar-like [8] or color [9–11] feature space to track single or multiple targets. In recent years, convolutional neural network based method [12–14] has been widely employed in object tracking task by learning discriminative target representation from raw pixels, and achives state-of-the-art performance. However, lacking of training data has been a bottleneck to train efficient and robust convolutional neural network for tracking. A simple and efficient way to alleviate the problem is to use pre-trained model to extract target feature, but this method cannot handle the occasion when target experiences heavy occlusion, small scale, or fast appearance variation. In [12], convolutional features from multiple layers are combined to correlation filters to predict target location jointly. In [13],Wang proposes to train a stacked auto-encoder network and uses the learned feature vector to separate the target from background. Nam *et al.* [14] propose to utilize the offline trained deep network model for online tracking.

However, the aforementioned trackers cannot perform robust object tracking for video sequence with occlusion challenge. In this paper, we propose a new framework to handle occlusion. The proposed method includes two modules: initial tracking estimation by structured output leaning, and tracking refinement to improve tracking precision and prevent target drift. The pipeline of the proposed method is shown in Fig. 1. First, we train a structured output SVM from the initial annotated frames, and the training sample is generated by object independent proposal. When a test frame arrives, we predict the confidence of each target candidates and select the maximum one as initial tracking result. Second, we over-segment the tracked target patch by SLIC [15] algorithm to a set of superpixels, and then assign a label to each superpixel patch indicating background/foreground. The refined tracking result is achieved from the background/foreground map. Compared with the most related two trackers,

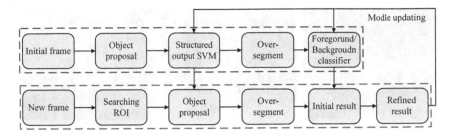

Fig. 1. Pipeline of the proposed tracking method. The section covered by purple bounding box indicates training module, and those covered by blue boundingbox indicates the tracking module when a new frame arrives. (Color figure online)

Struck [5] tracker and SPT tracker [16], the proposed tracker owns the ability to detect target occlusion and prevents model degradation by stopping tracker updating when heavy occlusion occurs. On the other hand, our method takes advantage of high-level target representation to achieve intial tracking result by structured output learning instead of merely using middle-level superpixel cue to estimate target location.

The rest of the paper is organized as follows. First, we give a brief description of the proposed tracker in Sect. 2. Then, the experimental setup and evaluation are detailed in Sect. 3. We concluded this paper in Sect. 4.

2   The Proposed Tracking Method

The proposed tracking method mainly consists of two individual modules: structured learning and predicting with object independent proposal, and target refining with superpixel matching for accurate and long-term tracking. In [5], Hare *et al.* propose to use structure output support vector machine (Struct) to learn a discriminative tracking model to mitigate the gap between label predicting and accurate position estimation. However, Struct tracker lacks the capability in handling object occlusion, which results in tracking model degradation and target drift. The main reason of model degradation is resulted from false model updating. To further improve the performance of structured learning in object tracking, we employ patch based object matching to remove those patches both belonging to target rectangle and part of background (as shown in Fig. 2).

In this section, we first introduce the motion model of our tracking method, and then give a brief presentation of structured learning. Third, we describe how to perform tracking refining to improve tracking accuracy.

2.1   The Motion Model of the Proposed Tracker

The motion model is used to define which candidate proposals should be tested in the current frame to locate target. The currently proposed tracking motion model can be divided into three categories: particle-based search, searching for

(a) Woman   (b) Lemming   (c) Tiger1

Fig. 2. Demonstration of three occlusion occasions. The yellow rectangle stands for groundtruth, and the area with blue mask represent the patch within target rectangle and belonging to background. (Color figure online)

the hard negatives, and object independent proposal. The particle-based motion model has been extensively studied in recent years, which performs importance sampling on the target location of previous frame as target candidates, but it is unable to perceive structural information and visual features. Searching for the hard negatives is becoming partly popular in tracking-by-detection framework, the motion model allows to train and update a classifier online thanks to the dense sampling. However, the computation complex will increase significantly when the search area is large. Object proposal method has been a successful motion model used in modern discriminative trackers because of its low computation complex and global search property, but object proposal cannot well handle scale variation resulting from the local structure objectness.

In this paper, we propose to use a weighted object proposal as target motion model. First, we employ a high efficient proposal method proposed by Bing [17] to generate a set of object candidates at current frame t, $\mathcal{B}_t = \{B_1, B_2, ..., B_n\}$, around the target location in previous frame. The proposal method exploits the edge cue to capture object structure information, so the proposed target candidates prone to be with a high objectness. However, the main challenge lies in the distraction of full structure information. To improve the proposal confidence, we penalize those candidates with either large scale variations or transformation by weighting the candidates with w.

$$w_{t,i} = \lambda_1 w_{t,i}^s + \lambda_2 w_{t,i}^p \tag{1}$$

and

$$w_{t,i}^s = 1 - \frac{\triangle width_{(t,i)} \times \triangle height_{(t,i)}}{width_{t-1} \times height_{t-1}} \tag{2}$$

$$w_{t,i}^p = 1 - \frac{\triangle p_{(t,x)} \times \triangle p_{(t,y)}}{width_{t-1} \times height_{t-1}} \tag{3}$$

where λ_1 and λ_2 in Eq. (1) are parameters to trade off the importance of scale variation and position transformation. The $w_{t,i}^s$ and $w_{t,i}^p$ in Eqs. (2) and (3) represent the weightings assigned to the i-th target candidate in the t-th frame. The symbol $\triangle(\bullet)$ stands for differential operations. Without losing generality,

$\triangle width_{(t,i)} = |width_{t-1} - width_{(t,i)}|$, in which $width_{t-1}$ and $width_{t,i}$ corresponding to target width in previous frame and the candidate width in t frame, respectively. The $p_{(t,x)}$ and $p_{(t,y)}$ in Eq. (3) correspond to the x-coordinate and y-coordinate of target candidate.

2.2 Structure Output Support Vector Machine

Compared with traditional tracking-by-detection tracking framework by learning binary classifier from the initial annotated frame, we employ structure output support vector machine to learn a regression function (Eq. (4)) directly to estimate target transformation from previous frame to the next.

$$f : \mathcal{X} \longrightarrow \mathcal{Y} \tag{4}$$

where $\mathcal{X}$ stands for the training sample set, and $\mathcal{Y}$ represents the output space (all possible target transformations).

The main difference from traditional classifier is that the training set $\{(x_1, y_1), ..., (x_n, y_n)\}$ are annotated by continuous labels which demonstrate the distance between candidate targets and groundtruth. Lots of distance definitions have been proposed in structured learning, such as Gaussian distribution distance which takes a maximum value one for the candidate located in the target center and smoothly decreases to zero for large transformation. Another popular and successful distance is overlap measurement as shown in Eq. (5).

$$sp_{p^t}^o(y_t^i, y_t^j) = \frac{(p_t \circ y_t^i) \bigcap (p_t \circ y_t^j)}{(p_t \circ y_t^i) \bigcup (p_t \circ y_t^j)} \tag{5}$$

Our goal is to learn a discriminant function $F : \mathcal{X} \times \mathcal{Y} \longrightarrow \mathbf{R}$ to measure the compatibility of the given training dataset $(x_1, y_1, ..., (x_n, y_n))$. In this paper, the training samples are generated by an object independent proposal method. The label set $\{y_1, ..., y_n\}$ corresponding to target candidates image patches $(x_1, ..., x_n)$ are the desired transformation of the candidates. Different from traditional post probability, we weight every target candidate with a penalization in Eq. (1) to constraint object scale and transformation variation. The tracking problem then can be formulated to

$$\mathbf{y}_t = f(\mathbf{x}_t^{p_{t-1}}) = arg \max_{y \in \mathcal{Y}} w_{t,i} * F(\mathbf{x}_t^{p_{t-1}}, \mathbf{y}) \tag{6}$$

It should be noted that the formulation in Eq. (6) includes label $\mathbf{y}$ explicitly, which means it considers the label into account in the optimization process. One of advantages is that the we do not treat all the training examples equally weighted, and the theoretical support lies in it is unreasonable to treat a negative training example with large overlaps with groundtruth and one negative example which overlaps very little. This learning strategy works out a new way on how to define the boundary of negative and positive training example.

Given a training example pair (x, y), the discriminative function F measures its compatibility. The output confidence is high if the example is well matched, otherwise, the confidence is low. The form of $F(\mathbf{x}, \mathbf{y})$ can be expressed as

$$F(\mathbf{x}, \mathbf{y}) = \langle \mathbf{w}, \varPhi(\mathbf{x}, \mathbf{y}) \rangle \tag{7}$$

where w is the parameters to be learn, and $\varPhi(\mathbf{x}, \mathbf{y})$ represents a joint kernel map. Then the learning process is converted to solve a convex objective function

$$\min_{\mathbf{w}} \frac{1}{2} \|\mathbf{w}\|^2 + C \sum_{i=1}^{n} \xi_i$$

$$s.t. \quad \forall i : \xi_i \geq 0$$

$$\forall i, \forall \mathbf{y} \neq \mathbf{y}_i : \langle \mathbf{w}, \delta\varPhi(\mathbf{x}, \mathbf{y}) \rangle \geq \triangle(\mathbf{y}_i, \mathbf{y}) - \xi_i \tag{8}$$

where $\xi = \xi_1, ..., \xi_n$ represents the slack variables vector corresponding to each training examples, and C stands for the regularization parameter to control trade-off between margin maximization and training error minimization. The $\delta\varPhi(\mathbf{x}, \mathbf{y}) = \varPhi(\mathbf{x}_i, \mathbf{y}_i) - \varPhi(\mathbf{x}_i, \mathbf{y})$. The $\triangle(\mathbf{y}_i, \mathbf{y})$ is used to quantify the loss associated between a prediction and the true output. The loss function is defined as Eq. (9).

$$\triangle(\mathbf{y}_i, \mathbf{y}) = 1 - sp_{p^t}^o(y_t^i, y_t^j) \tag{9}$$

The learning formulation in Eq. (8) aims to ensure the output of function $F(\mathbf{x}_i, \mathbf{y}_i)$ is bigger than $F(\mathbf{x}_i, \mathbf{y})$ for any $\mathbf{y} \neq \mathbf{y}_i$ by a margin $\triangle(\mathbf{y}_i, \mathbf{y}) - \xi_i$. The difference of traditional classifier and structured learning lies in the loss representation. In the structured learning, the loss function is vary from one to another which is determined on the quantitative similarity between training examples and groundtruth. The loss variation avoids to treat all the training examples equally.

By introducing kernel trick and standard Lagrangian duality techniques, the optimization is converted to its dual form

$$\max_{\beta} - \sum_{i, \mathbf{y} \neq \mathbf{y}_i} \triangle(\mathbf{y}, \mathbf{y}_i) \beta_i^{\mathbf{y}} - \frac{1}{2} \sum_{\substack{i, \mathbf{y} \neq \mathbf{y}_i \\ j, \mathbf{y} \neq \mathbf{y}_j}} \beta_i^{\mathbf{y}} \beta_i^{\bar{\mathbf{y}}} \langle \varPhi(\mathbf{x}_i, \mathbf{y}), \varPhi(\mathbf{x}_j, \bar{\mathbf{y}}) \rangle$$

$$s.t. \quad \forall i, \forall \mathbf{y} : \beta_i^{\mathbf{y}} \leq \delta(\mathbf{y}, \mathbf{y}_i) C$$

$$\forall i : \sum_{\mathbf{y}} \beta_i^{\mathbf{y}} = 0 \tag{10}$$

where the kernel function is defined as $k(\mathbf{x}, \mathbf{y}, \bar{\mathbf{x}}, \bar{\mathbf{y}}) = \langle \varPhi(\mathbf{x}, \mathbf{y}), \varPhi(\bar{\mathbf{x}}, \bar{\mathbf{y}}) \rangle$, and $\beta_i^{\mathbf{y}} = -\alpha_i^{\mathbf{y}}$ if $\mathbf{y} \neq \mathbf{y}_i$, otherwise $\beta_i^{\mathbf{y}} = \sum_{\bar{\mathbf{y}} \neq \mathbf{y}_i} \alpha_i^{\bar{\mathbf{y}}}$. Those example features with $\beta_i^{\mathbf{y}} \neq 0$ corresponding to support vectors, and the discriminative function F can be written as $F(x, y) = \sum_{i, \bar{y}} \beta_i^{\bar{\mathbf{y}}} \langle \varPhi(\mathbf{x}_i, \mathbf{y}), \varPhi(\mathbf{x}_j, \bar{\mathbf{y}}) \rangle$.

2.3 Tracking Refining

The tracking results achieved by tracking-by-detection based methods are represented by a rectangle. However, the results are mainly determined from two aspects, one is the object independent proposal performance, and another is the label predicted by a classifier. The former focuses on extracting object with independent structure, and the classifier is to predict a label to each object candidate. The latter is unrelated to tracking tasks, which is employed to precisely estimate the target position. To mitigate the gaps between object detection and object tracking, this paper proposes to refine the tracking results (labeling a candidate in detection framework) by superpixel patch classification. Our goal is to improve the tracking precision and prevent object drift. The refined tracking result T_i^r is used to compute an occlusion coefficient

$$occ = 1 - \frac{T_t \bigcap T_t^r}{T_t \bigcup T_t^r} \tag{11}$$

where T_t and T_t^r represent the initial tracking result and refinement, respectively.

The tracking refinement is proposed based on superpixel by over-segmentation instead of raw pixel considering that superpixel encodes both structure and spatial information effectively. It has been widely used in variety applications as a robust middle level feature representation. In this paper, we employ two kinds of features (HSI color and relative position) to encode a superpixel.

HSI color: HSI color space is robust to the effect of lighting changes and shows more discriminative ability. We use the mean value in $(H,\ S,\ I)$ to represent a superpixel patch. We then encode a superpixel based on another color based distribution, the mean value of four superpixels (left, right, up and bottom) around $sp_(i,j)$. The same feature extracting method is performed in RGB color space. A total of thirty dimension feature is used to encode a superpixel both considering structure information and spatial color distribution.

Position: Position provides reliable information for refining tracking result. In this paper, we use the centroid $c = (sp_x, sp_y)$ of each superpixel as spatial feature mainly considering a hypothesis that those superpixel patches with foreground label is continuous distribution.

When a new frame comes, we firstly perform object independent proposal and structure output SVM to predict the confidence of each candidate target, and search for the maximum confidence based on Eq. (6) to get the initial tracking result (denoted as T_t). The simple linear iterative clustering (SLIC) [15] is used to oversegment the tracked target patch T_t. To learn a robust and discriminative classifier, we track object in the first 5 frames by structured output SVM merely, and then extract the color based and position based feature vector of each superpixel. We denote the feature vector of the i-th superpixel in the t-th frame as $f_{t,i}$. Each superpixel corresponds to a label $y_{t,i}$ and $y_{t,i} \in \{+1, -1\}$. We employ neural network to learn a discriminative classifier, when a test image uses the learned neural network to regress the label confidence being part of target. If the superpixel confidence is greater than *threshold*, it is labeled as a target, otherwise, it is labeled as background. The initial tracked result T_t is mapped to a binary label at superpixel level, and then a more accurate tracking result is achieved from the map.

3   Experiments and Validation

We evaluate the proposed tracking method on 29 video sequences in OTB-50 [18] with occlusion challenges. SLIC algorithm is used to over-segment the inital tracking result, and the spatial weight and superpixel numbers are set to 10 and 200, respectively. In the training stage, the first 5 tracking results from structured output SVM are employed as training set. The object is determined as heavy occlusion when occlusion coefficient *occ* is greater than 0.3, and the structured SVM will not be updated to prevent model degradation. The proposed tracking method is implemented in Matlab and runs at 6.8 frames per second. We compare our method with some state-of-the-art tracking methods such as SPT [16], HDT [19], CT [20], MIL [21], TLD [22], Struck [5], and SRDCF [23]. Among them the SPT tracker is the most advanced and similar superpixel based tracking method. SRDCF is state-of-the-art KCF based tracking algorithm, HDT is based on hierarchical convolutional neural network and correlation filter, and the other trackers are selected due to their excellent performance in OTB benchmark. To fully assess our method, we use one-pass evaluation (OPE) to evaluate our tracking method. The *precision* scores indicate the percentage of frames in which the estimated locations are within 20 pixels compared to the ground-truth positions. The success scores are defined as the area under curve (AUC) of each success plot, which is the average of the success rates corresponding to the sampled overlap threshold.

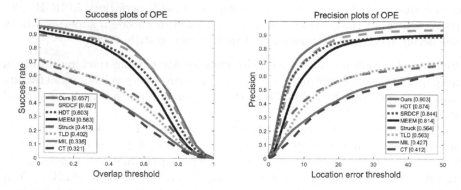

Fig. 3. Average success plot (left) and precision plot (right) of OPE on 29 video sequences with occlusion challenges in OTB-50 datasets. The performance of each tracker is shown in the legend. The proposed method performs best in both precision and success plot.

Figure 3 illustrates the success plot and precision plot with OPE evaluation. The proposed method performs the best in both success and precision rates. While Struct tracker ranks 5 place, which is a popular and advanced tracking algorithm based on structured learning. The initial tracking result is achieved by Struct, and improved with refinement module by 24.4% and 43.9% in average success rate and precision rate, respectively. The deep feature based tracker HDT ranks the second and third places in precision and success plot which employ the pre-trained VGG-19 model to extract target appearance representation and then combine correlation filter to estimate target transformation. However, HDT only tracks targets with fixed-scale and it is time consuming. SRDCF is the state-of-the-art correlation filter based tracker that considers spatially regularizing in training process to mitigate the unwanted boundary effects. Our method outperforms the SRDCF 3.0% and 5.9%, respectively for average success plot and precision plot. Figure 4 presents some tracking results on four video sequences. It is obvious that our method performs significantly better than other methods especially in handling occlusion.

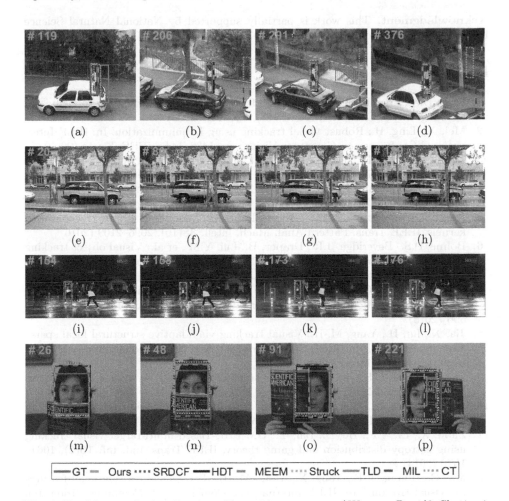

(a)   (b)   (c)   (d)

(e)   (f)   (g)   (h)

(i)   (j)   (k)   (l)

(m)   (n)   (o)   (p)

— GT — Ours ···· SRDCF — HDT — MEEM ···· Struck — TLD — MIL ···· CT

Fig. 4. Tracking result visualization of four video sequences (*Woman, David3, Skating1, FaceOcc1*)

4   Conclusion

In this paper, we have proposed a robust tracking method for handling occlusion challenge. The method was divided into two modules. We employed a structure output learning to estimate initial target location and scale, and then performed tracking refinement to handle target occlusion. The refinement process was based on over-segmentation, and a background/foreground map was achieved by classifying the superpixel patches to background or foreground. Extensive evaluations on benchmark dataset with occlusion challenge showed that the proposed method outperformed the state-of-the-art methods in both success plot and precision plot.

Acknowledgment. This work is partially supported by National Natural Science Foundation (Grant No. 61403342, U1509207, 61325019).

References

1. Ross, D.A., Lim, J., Lin, R.-S., Yang, M.-H.: Incremental learning for robust visual tracking. Int. J. Comput. Vision **77**(1), 125–141 (2008)
2. Mei, X., Ling, H.: Robust visual tracking using L1 minimization. In: IEEE International Conference on Computer Vision, pp. 1436–1443. IEEE, Kyoto (2009)
3. Zhong, W., Lu, H., Yang, M.: Robust object tracking via sparse collaborative appearance model. IEEE Trans. Image Process. **23**(5), 2356–2368 (2004)
4. Avidan, S.: Support vector tracking. IEEE Trans. Pattern Anal. Mach. Intell. **26**(8), 1064–1072 (2004)
5. Hare, S., Golodetz, S., Saffari, A., et al.: Struck: Structured output tracking with kernels. IEEE Trans. Pattern Anal. Mach. Intell. **38**(10), 2096–2109 (2016)
6. Bolme, D.S., Beveridge, J.R., Draper, B., Lui, Y.M., et al.: Visual object tracking using adaptive correlation filters. In: IEEE International Conference on Computer Vision, pp. 2544–2550. IEEE, Istanbul (2010)
7. Henriques, J.F., Caseiro, R., Martins, P., Batista, J.: High-speed tracking with kernelized correlation filters. IEEE Trans. Pattern Anal. Mach. Intell. **37**(3), 583–596 (2015)
8. Jia, X., Lu, H., Yang, M.-H.: Visual tracking via adaptive structural local sparse appearance model. In: IEEE Computer vision and pattern recognition, pp. 1822–1829. IEEE, Providence (2012)
9. Zhong, W., Lu, H., Yang, M.: Robust object tracking via sparse collaborative appearance model. IEEE Trans. Image Process. **23**(5), 2356–2368 (2014)
10. Zhou, X., Yu, H., Liu, H., Li, Y.F.: Tracking multiple video targets with an improved GM-PHD tracker. Sensors **15**(12), 30240–30260 (2015)
11. Zhou, X., Li, Y.F., He, B., Bai, T.: GM-PHD-Based multi-target visual tracking using entropy distribution and game theory. IEEE Trans. Ind. Inf. **10**(2), 1064–1076 (2014)
12. Ma, C., Huang, J.-B., Yang, X., Yang, M.-H.: Hierarchical convolutional features for visual tracking. In: IEEE International Conference on Computer Vision, pp. 3074–3082. IEEE, Boston (2015)
13. Wang, N., Yeung, D.-Y.: Learning a deep compact image representation for visual tracking. In: Advances in Neural Information Processing Systems, pp. 809–817. Lake Tahoe (2013)

14. Nam, H., Han, B.: Learning multi-domain convolutional neural networks for visual tracking. In: 2016 IEEE Conference on Computer Vision and Pattern Recognition, pp. 4293–4302. IEEE, Las Vegas (2016)
15. Achanta, R., Shaji, A., Smith, K., Lucchi, A., Fua, P., Susstrunk, S.: Slic superpixels compared to state-of-the-art superpixel methods. IEEE Trans. Pattern Anal. Mach. Intell. **34**(11), 2274–2282 (2012)
16. Yang, F., Lu, H., Yang, M.H.: Robust superpixel tracking. IEEE Trans. Image Process. **23**(4), 1639–1651 (2014)
17. Cheng, M.-M., Zhang, Z.B., et al.: Binarized normed gradients for objectness estimation at 300fps. In: IEEE Conference on Computer Vision and Pattern Recognition, pp. 3286–3293. IEEE, Columbus (2014)
18. Wu, Y., Lim, J., Yang, M.H.: Online object tracking: a benchmark. In: IEEE Conference on Computer Vision and Pattern Recognition, pp. 2411–2418. IEEE, Portland (2013)
19. Qi, Y., Zhang, S., Qin, L., et al.: Hedged deep tracking. In: IEEE Conference on Computer Vision and Pattern Recognition, pp. 4303–4311. IEEE, Las Vegas (2016)
20. Zhang, K., Zhang, L., Yang, M.-H.: Real-time compressive tracking. In: Fitzgibbon, A., Lazebnik, S., Perona, P., Sato, Y., Schmid, C. (eds.) ECCV 2012. LNCS, vol. 7574, pp. 864–877. Springer, Heidelberg (2012). doi:10.1007/978-3-642-33712-3_62
21. Babenko, B., Yang, M.H., Belongie, S.: Visual tracking with online multiple instance learning. In: IEEE International Conference on Computer Vision and Pattern Recognition, pp. 983–990. IEEE, Miami (2009)
22. Kalal, Z., Matas, J., Mikolajczyk, K.: P-N learning: bootstrapping binary classifiers by structural constraints. In: IEEE International Conference on Computer Vision and Pattern Recognition, pp. 49–56. IEEE, California (2010)
23. Danelljan, M., Hager, G., Shahbaz Khan, F., et al.: Learning spatially regularized correlation filters for visual tracking. In: IEEE International Conference on Computer Vision and Pattern Recognition, pp. 4310–4318. IEEE, Boston (2015)

Graspable Object Classification with Multi-loss Hierarchical Representations

Zhichao Wang, Zhiqi Li, Bin Wang$^{(\boxtimes)}$, and Hong Liu

Harbin Institute of Technology, Harbin 150001, China
wbhit@hit.edu.cn

Abstract. To allow robots to accomplish manipulation work effectively, one of the critical functions they need is to precisely and robustly recognize the robotic graspable object and the category of the graspable objects, especially in data limited condition. In this paper, we propose a novel multi-loss hierarchical representations learning framework that is capable of recognizing the category of graspable objects in a coarse-to-fine way. Our model consists of two main components, an efficient hierarchical feature learning component that combines kernel features with the deep learning features and a multi-loss function that optimizes the multi-task learning mechanism in a coarse-to-fine way. We demonstrate the power of our proposed system to data of graspable and ungraspable objects. The results show that our system has superior performance than many existing algorithms both in terms of classification accuracy and computation efficiency. Moreover, our system achieves a quite high accuracy (about 82%) in unstructured real-world condition.

Keywords: Hierarchical feature · Multi-loss · Kernel feature · Deep learning

1 Introduction

In the past 10 years, robotic grasping has been attracting great interests in robotic community, which aims to accomplish autonomous manipulation tasks. Although the performances of robotic grasping tasks have been improved significantly, the robot is still hard to accomplish even a daily housework, such as "clean a table" or "catch an apple for me". The fundamental problem of robotic grasping is to perceive the environment of the robot, such as graspable object discrimination, graspable object classification and grasping area detection.

Recently, with the rise of the deep learning (DL) method, massive DL based approaches were proposed to detect the robotic grasping area and grasping pose of an object [1,2], which have been achieved tremendous progress. These DL based methods need to be trained on tremendous data. However, the graspable characteristic of an object is based on the affordance of the objects and the robotic gripper, which means the robotic grasping data is difficult to generate. According to [3], many methods show overfitting problems. Moreover, these

© Springer International Publishing AG 2017
Y. Huang et al. (Eds.): ICIRA 2017, Part III, LNAI 10464, pp. 460–471, 2017.
DOI: 10.1007/978-3-319-65298-6_42

grasping detection methods ignored the graspable characteristic and the category of the object in the scene, which is a prerequisite condition for fine manipulation tasks such as providing specify object for senior people in household service.

The graspable object classification in data-limited robotic manipulation domains is still an open problem in robotic community. Due to the tremendous gap of data amount between ungraspable object and one specific graspable object, it is inappropriate to classify the objects to $K+1$ categories (K graspable object category and one special ungraspable object category). Therefore, there are two tasks in this problem, graspable object discrimination and graspable object classification. The first task aims to avoid the interference of ungraspable objects in unstructured environment. And, the second task estimate the category of the graspable object.

In this paper, we propose and study a novel multi-loss hierarchical representation learning approach for the graspable object classification in a coarse-to-fine way. Our model consists of two main components, a hierarchical feature learning component and a coarse-to-fine multi-task optimization approach. Specifically, in the hierarchical feature learning component, the data-efficient kernel descriptors are extracted from raw image information directly and the expressive high-level features are learned from the kernel descriptors by using a neural networks. Inspired by the Fast Region-based Convolutional Network method (Fast R-CNN) [4], we adopt a jointly learning mechanism to simultaneously accomplish the graspable discrimination and category classification. Furthermore, we use a modified loss function to ensure that the model is optimized on multi-tasks rather than one particular task. An overview of our system is shown in Fig. 1.

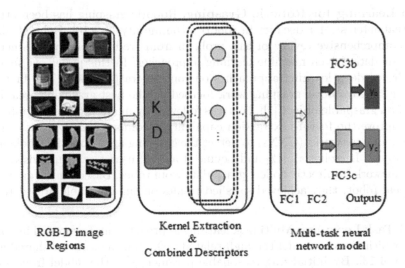

RGB-D image   Kernel Extraction   Multi-task neural
Regions   &   network model
                 Combined Descriptors

Fig. 1. The overview of our hierarchical feature learning framework for multi tasks. Our model is stacked up by kernel feature extractor, one fully connected layer neural networks, and two sibling outputs: graspable probabilities and the probabilities of categories. The architecture is end-to-end trainable with a multi-loss function.

The key contributions of this paper are two-fold. (1) A novel hierarchical feature learning architecture is developed to learn robotic graspable object classification in data-limited condition. (2) We propose coarse-to-fine multi-task jointly learning strategy to improve the performance of graspable object classification. The results show that our system has superior performance than many existing algorithms both in terms of classification accuracy and computation efficiency.

The remainder of this paper is organized as follows. Some related works are presented in Sect. 2. In Sect. 3, we presents the algorithmic details of our model, including the hierarchical presentation learning component and the multi-task jointly learning component. Section 4 describes a series of experiments to validate our approach. Finally, Sect. 5 concludes the paper.

2   Related Work

Hierarchical Feature Learning: Hierarchical feature learning (HFL) methods have demonstrated substantial improvements over the conventional hand-designed local features in recent years. Bo et al. [5] presented a hierarchical hand-designed matching pursuit approach to estimate the category of the objects by fusing color, texture, geometry and 3D shape features. Recently, deep neural networks (DNN) architecture [6], one typical automatic hierarchical feature learning approach, has achieved impressive progresses in the field of object recognition and semantic scene understanding [7,8]. Rather than manually extracting shadow features, DNN approach automatically learns expressive hierarchical features from large amounts of labeled raw data.

Deep Learning for Robotic Grasping: Robotic grasping has been attracting great interests for decades in robotics community [3,9,10]. Bohg et al. [11] gives comprehensive review of this problem from tradition analytical methods to recent data-driven machine learning approaches. In this paper, we focus on data-driven deep learning approaches on robotic grasping task. The first deep learning based robotic grasping work focused on the best graspable area detection of an graspable object [1]. This was followed by many variations approaches that improve the feature extraction capability. Redmon et al. [2] used convolutional neural networks (CNNs) to detect the grasping area from the whole image of an object. Lerrel et al. [12] and Levine et al. [10] used a self-supervised learning framework to directly collect grasp data from robot. Based on the grasp data from real robot, they achieved satisfied results on the graspable area detection task.

Multi Task Learning: Multi task learning approach is also proposed in rencent years, which aims to model the high-related task by using same or shared architecture [4,13]. By initializing reasonable parameters of the model from related task model, the multi-task learning approach presents superior performance in deep learning approach based task [14].

In this paper, we focus on developing a multi-loss hierarchical feature learning model to tackle grasping object classification problem, which is an prerequisite for graspable area detection task or further human-robot collaboration task. It is an extension of our former graspable area detection work [3].

3 Approach

In this section, we describe the detail of our coarse-to-fine graspable object classification approach, including the hierarchical feature learning structure and the end-to-end trainable multi-task optimization.

3.1 Hierarchical Feature Learning

Due to the constrain of training data amount, we use a hierarchical feature learning method to describe the objects, which integrates data-efficient low-level kernel descriptors and automatic high-level features. Firstly, the kernel descriptor attempt to extract efficient kernel features from raw RGB-D images. Then, a deep neural networks architecture is utilized to learn more expressive high-level representations from the kernel representations for the coarse-to-fine graspable object classification.

Low-level Kernel Feature Extraction: Kernel descriptor is one kind of efficient local feature which transfers the pixel attributes to useful local features [15] from limited amount data. We adopt a set of kernel features, including gradient and shape kernels over color images, size and shape kernels over depth images. In this paper, we just describe the computation method of the gradient kernel descriptor for brevity. Other three types of kernel descriptors can be extracted in the same way. As for the detail of the kernel feature extraction, we direct the readers to [5] and our former work [3].

To obtain the gradient kernel descriptors, we first compute the corresponding gradient match kernel descriptors to approximate the similarity of two sets of features points, followed by Eq. 1.

$$K_g(P1, P2) = \sum_{z \in P1} \sum_{z' \in P2} \hat{m}(z)\hat{m}(z')k_v\left(\hat{\theta}(z), \hat{\theta}(z')\right)k_p(z, z') \tag{1}$$

where P are rectangle patches of the image; $\hat{m}(z)$ is normalized gradient magnitude which weights the contribution of each pixel; $\hat{\theta}(z) = (\sin(\theta(z)), \cos(\theta(z)))$ is gradient orientation vector where $\theta(z)$ is gradient orientation at pixel z; $k_v\left(\hat{\theta}(z), \hat{\theta}(z')\right) = \exp(-\lambda_v||\hat{\theta}(z) - \hat{\theta}(z')||^2)$ and $k_p(z, z') = \exp\left(-\lambda_p||z - z'||^2\right)$ are Gaussian kernel over orientation and position to describe the distance between two pixels.

Then, we extract kernel features from the matching kernel descriptor by sampling sufficient basis vector from support region [15] to mitigate the computation load of kernel descriptor matching, and use the kernel principal component analysis (KPCA) method [16,17] to reduce the dimension of the kernel features. The gradient kernel features are extracted following Eq. 2.

$$G_g(P) = \sum_{z \in P} \hat{m}(z)\sigma\left(\varphi_v\left(\hat{\theta}(z)\right), \varphi_p(z)\right) \tag{2}$$

where $\varphi_v(\hat{\theta}(z))$ and $\varphi_p(z)$ are the orientation and position components of the feature descriptor; σ is the Kronecker product function. The gradient kernel descriptors of two corresponding image patches satisfies $K_g(P1, P2) = G_g(P1)^T G_g(P2)$.

Other kernel descriptors are extracted in the same way, shown in Eq. 3.

$$
\begin{cases}
G_{sc}(P) = \sum_{z \in P} \hat{n}(z)\sigma\left(\varphi_{sc}\left(b(z)\right), \varphi_p(z)\right) \\
G_{sd}(P) = \sum_{z \in P} \sigma\left(\varphi_{od}\left(\hat{\gamma}(z)\right), \varphi_{sd}\left(c(z)\right)\right) \\
G_{ld}(P) = \sum_{l_p \in P} \varphi_{ld}\left(l_p\right)
\end{cases}
\tag{3}
$$

where $G_{sc}(P)$, $G_{sd}(P)$ and $G_{ld}(P)$ are the shape kernel descriptor on color image, shape kernel descriptor and size kernel descriptor on depth image. Specifically, $\hat{n}(z)$ is normalized standard deviation; $b(z)$ is a comparison indicator vector; $\varphi_{sc}\left(b(z)\right)$ and $\varphi_p(z)$ are the shape and position components of the shape kernel descriptor. $\hat{\gamma}(z)$ is the orientation parameter; $c(z)$ is the parameter to measure the distance of the spin parameters; φ_{od} and φ_{sd} are the components of the 3D shape kernel descriptor on depth image. $\varphi_{ld}\left(l_p\right)$ is the parameter to measure the size of the 3D object; l_p is the $L2$ distance between the points z and a reference point $\hat{z}$.

The combined kernel features are $G_{all} = \left(G_g^{(m)}, G_{sc}^{(m)}, G_{sd}^{(m)}, G_{ld}^{(m)}\right)$, $G_{all} \in \mathbb{R}^D$, D is the dimension of the kernel features. However, due to the intrinsic constrain of the local visual feature, the low-level kernel feature are unable to describe the global information and the relation between different channels. Therefore, the deep neural networks is utilized to extract more expressive high-level features.

High-level Feature Learning: After the extraction of the low-level kernel descriptors, the kernel descriptors are transferred to our network parametrized by Φ to learn high-level representations. We incorporate the graspable object discrimination task and the graspable object classification task into our neural networks to accomplish the coarse-to-fine graspable object classification. The structure of the deep neural networks is shown in Fig. 2. And the implementation detail of our networks follows Eq. 4. It is worth noting that the bias terms are suppressed in the equation for readability.

$$
\begin{aligned}
h_j^{[1](m)} &= \delta\left(\sum_{i=1}^{D} \bar{x}_i^{(m)} W_{i,j}^{[1]}\right) \\
h_j^{[2](m)} &= \delta\left(\sum_{i=1}^{K1} h_i^{[1](m)} W_{i,j}^{[2]}\right) \\
h_{bj}^{[3](m)} &= \delta\left(\sum_{i=1}^{K2} h_i^{[2](m)} W_{b,i,j}^{[3]}\right) \\
h_{cj}^{[3](m)} &= \delta\left(\sum_{i=1}^{K2} h_i^{[2](m)} W_{c,i,j}^{[3]}\right) \\
P\left(\hat{y}_b^{(m)} = 1 | \bar{x}^{(m)}; \Phi\right) &= \Delta_1\left(\sum_{i=1}^{K3b} h_{b,i}^{[3](m)} W_{b,i}^{[4]}\right) \\
P\left(\hat{y}_c^{(m)} = l | \bar{x}^{(m)}; \Phi\right) &= \Delta_2\left(\frac{\sum_{i=1}^{K3c} h_{c,i}^{[3](m)} W_{c,i,l}^{[4]}}{\sum_{j=1}^{L} \sum_{i=1}^{K3c} h_{c,i}^{[3](m)} W_{c,i,j}^{[4]}}\right)
\end{aligned}
\tag{4}
$$

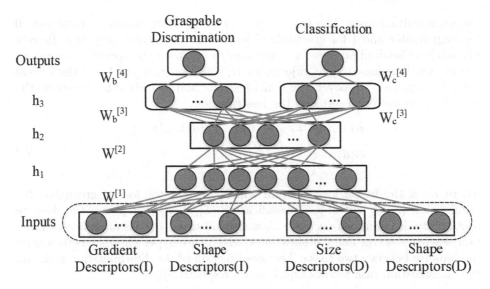

Fig. 2. The architecture of our networks with two outputs. The two output vectors are logistic graspable probabilities and softmax category probabilities of the graspable objects.

where the kernel features are represented by $\bar{x}^{(m)} = G_{all}$. $y_b^{(m)} \in \{0, 1\}$ is the graspable label of $\bar{x}^{(m)}$. If $y_b^{(m)} = 1$, $y_c^{(m)}$ is the corresponding category label of L graspable object. The logistic classifier (Δ_1) and softmax classifier (Δ_2) are utilized to respectively discriminate the graspable characteristic binary label y_b and classify the category y_c of the graspable object. $W^{[1]}$ and $W^{[2]}$ are the parameters of hidden layer $h1$ and $h2$. Moreover, $W_b^{[3]}$, $W_b^{[4]}$ and $W_c^{[3]}$, $W_c^{[4]}$ are the parameters of binary output layer and category output layer. The corresponding units number of each layer are $K1$, $K2$, $K3b$, and $K3c$. δ is the sigmoid activation function.

We maximized the output probability $\hat{y}_b$ and $\hat{y}_c$ to optimize the model:

$$\Phi^* = \underset{\Phi}{\text{argmax}} \sum_{m=1}^{M} \gamma_1 \log P\left(\hat{y}^{(m)} | \bar{x}^{(m)}; \Phi\right) + \gamma_2 \log P\left(\hat{y}^{(m)} | \bar{x}^{(m)}; \Phi\right) \quad (5)$$

where γ_1 and γ_2 are hyper-parameters to control the impact of two classifiers on the result. Φ^* is the final optimization parameters of the model.

3.2 Jointly Learning Multi-task

Inspired by the Fast Region-based Convolutional Network method (Fast R-CNN) [4], we adopt a multi-loss function to jointly train the graspable discrimination and category classification tasks. Rather than learning features separately, the tasks shared the same kernel features and first two layers of the neural networks. Specifically, the features extracted by the hierarchical feature learning architecture are duplicated and fed into two sibling discrete probability distribution output simultaneously. The first outputs a binary distribution of an object (0 for ungraspable and 1 for graspable object). And the second outputs a discrete probability distribution over K categories to estimate the specific category of the object. For ungraspable objects, there is no category label and the second output is ignored. Moreover, the multi-loss function is utilized to encourage the model to use both two terms in the loss function.

$$L_f = G(d)L_{dis}(p_d, d) + \lambda_2 \sigma(d)L_{cls}(p_c, c)$$
$$G(d) = \begin{cases} \lambda_1, d = 0 \\ 1, d = 1 \end{cases} \tag{6}$$

where L_f is the fusion loss function which contains the loss of graspable discrimination (L_{dis}) and object classification (L_{cls}). p_d and p_c are the discrete probability distributions over two classification tasks. $\sigma(d)$ is an indicator function, 1 when d is true and 0 otherwise. λ_1 and λ_2 are hyper-parameters to control the balance between two tasks. The loss function of the discrimination task and the classification task follows Eqs. 7 and 8 respectively.

$$L_{dis} = -\frac{1}{M}\left[\sum_{i=1}^{M} y^{(i)} \log h_{\theta_1}(x^{(i)}) + (1 - y^{(i)})\log(1 - h_{\theta_1}(x^{(i)}))\right] + \gamma_1 L(\theta_1) \tag{7}$$

where θ_1 are the parameters of the discrimination task. $x^{(i)}$ and $y^{(i)}$ are the training data and corresponding ground-truth graspable label. γ_1 is hyper-parameter to control the effect of weight penalty term $L(\theta_1)$.

$$L_{cls} = -\frac{1}{M}\left[\sum_{i=1}^{M}\sum_{j=1}^{K} 1\{y^{(i)} = j\} \log \frac{e^{\theta_{2j}^T x^{(i)}}}{\sum_{l=1}^{K} e^{\theta_{2l}^T x^{(i)}}}\right] + \gamma_2 L(\theta_2) \tag{8}$$

where θ_2 are the parameters of the classification task. K is the total number of ground-truth category.

3.3 Jointly Training

In the jointly training procedure, we first computed the parameter gradients of graspable discrimination task on corresponding discrimination training data batch. Then the parameter gradients of graspable object classification task are updated on corresponding classification training data batch. Finally, the gradients of shared parameters in the network are updated. The update follows Eq. 9.

$$W_b \leftarrow RMS\left(\frac{\partial L_{dis}}{\partial W_b}\right)$$
$$W_c \leftarrow RMS\left(\frac{\partial L_{cls}}{\partial W_c}\right) \tag{9}$$
$$W_f \leftarrow RMS\left(\frac{\partial L_f}{\partial W_{all}}\right)$$

where W_f are the parameters in the shared network, containing $W^{[1]}$ and $W^{[2]}$; W_b are the parameters in the graspable object discrimination subnet, containing

$W_b^{[3]}$ and $W_b^{[4]}$; W_c are the parameters in the graspable object classification subnet, containing $W_c^{[3]}$ and $W_c^{[4]}$. RMS is a gradient descent method [18].

Trained in this manner, our model explicitly to infer the category of the graspable object in a coarse-to-fine way. We summarize the multi-loss hierarchical representation learning approach in Algorithm 1.

Algorithm 1. Hierarchical Feature learning

Input: A pair of RGB-D images, I_{src} and D_{src}
Output: Inferred graspable discrimination $\hat{y}_b$ or category $\hat{y}_c$
for $m = 1; m \leq M; m + +$ **do**
  /*Hierarchical Feature extraction*/
  $\bar{x}^{(m)} = K_{all}^{(m)} \leftarrow K(I_{src}^{(m)}, D_{src}^{(m)})$;
  $\tilde{x}^{(m)} = \Phi_{DNN}(\bar{x}^{(m)})$;
  /*Graspable object discrimination*/
  $P(\hat{y}_b^{(m)} = 1 | \tilde{x}^{(m)}, \Phi^*)$;
  $\hat{y}_b \leftarrow Stack(\hat{y}_b^{(m)})$;
  if $(P(\hat{y}_b^{(m)}) \geq \eta)$ **then**
    /*Graspable object classification*/
    $P(\hat{y}_c^{(m)} = l | \tilde{x}^{(m)}, \Phi^*)$;
    $\hat{y}_c \leftarrow Stack(\hat{y}_c^{(m)})$;
  end
end
return $\hat{y}_b$ and $\hat{y}_c$;

---

4   Experiments

In this section, we conduct several experiments to verify: (1) whether our coarse-to-fine multi-task graspable object classification approach outperforms traditional step-by-step approach, (2) whether our multi-loss hierarchical feature learning approach is data efficient.

4.1   Experimental Setup and Procedures

In the experiment, we evaluate our approach on a hybrid dataset which contains RGB-D object dataset [19] and our ungraspable object dataset. Our dataset mainly contains color and depth images of ungraspable objects from different orientations and distances. We manually labelled the whether the object is graspable or not based on our robotic gripper whose maximal opening length is around 100 millimeters. A desktop computer, with an i7-4790CPU 3.6GHZ and GeForce GTX 980 graphics card, was used to implement the algorithm of our intention recognition system.

The detail of the experimental procedure is arranged as following:

- Preprocessing the data. The color images are transformed to grayscale images, and the intensities are normalized into $[0, 1]$.
- Generating training patches. The size of patches used for kernel feature extraction is set to 16×16. And, the KPCA [16] ratio is set to 0.2.
- Normalizing kernel features. We normalize the kernel feature of each channel and combine the final kernel descriptor vector $\bar{x} \in R^{4000}$.
- Pretraining our model. We use unsupervised stack sparse auto-encoder (SSAE) to pre-train the parameters of our neural networks.
- Jointly training the model. We jointly train the neural networks to fine-tune the parameters. The sparse hyper-parameter is set to 0.5 to reduce the over-fitting risk of model. And the hyper-parameters λ_1 and $lambda_2$ in Eq. 6 are set to 1.5 and 1, respectively.

4.2   Results and Comparison

We compare the performances of our multi-loss hierarchical feature learning method (M-HFL) against some existing algorithms which take all the ungraspable object as one particular category.

1 SIFT, in which case SIFT feature is used [20];
2 KF, in which case the kernel feature is used [15];
3 DNN, in which case a four-layer neural network algorithm is used [6];
4 HFL, in which case the hierarchical representation of our model is used.
5 M-HFL, in which case our multi-loss hierarchical representation is used.

Table 1 shows the comparison results of different algorithms. The color, depth, and combined columns in the table mean the result are obtained from color, depth images, and both two images, respectively. It can be observed that our M-HFL approach distinctly outperforms the other algorithms, which reaches an accuracy of 86.78%. The table also shows that the combination of the two channels improves the performance of our system, as well as SIFT, KF and HFL

Table 1. Classification results on different algorithms

Model	Accuracy (%)		
	Color	Depth	Combined
SIFT	64.77	62.29	**68.52**
KF	72.60	67.15	**77.89**
DNN	**74.18**	71.53	72.64
HFL	76.56	74.51	**78.49**
M-HFL	84.93	80.42	**86.78**

methods. Furthermore, the color information is more important than depth information for the purpose of graspable object classification. The major reason is that the diversities of the depth are much bigger than color channel.

From the comparison between DNN and HFL, we can observe that it is more efficient and effective to learn high-level features from the low-level features than simply learn from the raw images when the training data are inadequate. The comparison between the HFL and M-HFL shows the benefit of using multi-task learning mechanism, which gives a significant improvement (8.29%) in terms of accuracy.

Table 2. Coarse-to-fine vs Step-by-step

Model	Accuracy (%)			Speed (fps)
	Graspable	Category	Final	
KF-2	92.32	86.20	79.58	20.4
DNN-2	84.51	76.83	64.92	**56.6**
HFL-2	94.84	90.25	85.59	17.2
M-HFL	-	-	**86.78**	32.3

The experiment results in Table 2 show the advantages of our coarse-to-fine MHL approach compared to step-by-step approaches. The KF-2, DNN-2 and HFL-2 in Table 2 are step-by-step approaches, meaning the graspable discrimination and graspable object classification are successively conducted by two similar algorithms. We observe that the M-HFL shows the best performance in the case of accuracy. Compared to KF-2 and HFL-2 approaches, the improvement of M-HFL is obvious in the term of computational efficiency. Furthermore, compared to two step-by-step HFL algorithm, the accuracy of M-HFL increases 1.19% and the computation speed boosts almost 1 time. The DNN-2 approach obtains the best speed performance. However, it shows the worst performance in the case of accuracy due to the overfitting problem. And, the comparison between DNN-2 and HFL-2 confirms the data efficiency of our hierarchical feature learning method again.

In order to verify the performance of our model in real-world condition, we conducted a ground experiment in which our model is utilized to discriminate the graspability of several objects and classify the graspable objects. Firstly, we use the RANSAC plane-fitting algorithm to find the plane of table by using depth information. Then, the grouped points above the plane are viewed as segmented objects. Finally, our coarse-to-fine model is utilized to classify the segmented objects. Our model achieves 82.33% (494/600) accuracy on an enlarged 50 objects image set in real world environment, containing 600 ($50 \times 6 \times 2$, each object contains 12 images in different orientation and distance) pairs of image patches of the objects. Figure 3 shows an example of the ground experiment.

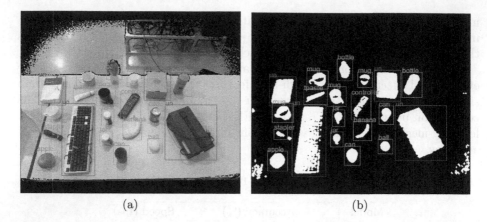

(a)   (b)

Fig. 3. Graspable object classification in real-world environment. (a) shows the re-projected color image in depth world coordinates. (b) shows the valid image whose points are above the plane. The red rectangles and green rectangles show the ungraspable objects and the categories of graspable objects.

5   Conclusions

In this work, we presented a novel multi-loss hierarchical feature learning approach for graspable object classification in limited training data domain. The data efficient hierarchical feature learning approach is utilized to extract expressive features from limited training data. Furthermore, a multi-task learning mechanism is utilized to tackle the graspable object classification problem in a coarse-to-fine way. Specifically, we proposed a multi-loss function to jointly optimize the parameters of the multi-task model. Experiments demonstrated the superior performance of our method both in the cases of accuracy and computation speed. Our approach can be easily extended to other multi-task object recognition problem, simply requiring new training data and corresponding labels.

References

1. Lenz, I., Lee, H., Saxena, A.: Deep learning for detecting robotic grasps. Int. J. Robot. Res. **34**(4–5), 705–724 (2015)
2. Redmon, J., Angelova, A.: Real-time grasp detection using convolutional neural networks 2015, pp. 1316–1322 (2015)
3. Wang, Z., Li, Z., Wang, B., Liu, H.: Robot grasp detection using multimodal deep convolutional neural networks. Adv. Mech. Eng. **8**(9) (2016). doi:10.1177/1687814016668077
4. Girshick, R.: Fast R-CNN. In: International Conference on Computer Vision (ICCV) (2015)
5. Bo, L., Lai, K., Ren, X., Fox, D.: Object recognition with hierarchical kernel descriptors. In: 2011 IEEE Conference on Computer Vision and Pattern Recognition (CVPR), pp. 1729–1736. IEEE (2011)

6. Hinton, G.E., Salakhutdinov, R.R.: Reducing the dimensionality of data with neural networks. Science **313**(5786), 504–507 (2006)
7. Krizhevsky, A., Sutskever, I., Hinton, G.E.: Imagenet classification with deep convolutional neural networks. In: Advances in Neural Information Processing Systems, pp. 1097–1105 (2012)
8. Karpathy, A., Fei-Fei, L.: Deep visual-semantic alignments for generating image descriptions. In: Proceedings of the IEEE Conference on Computer Vision and Pattern Recognition, pp. 3128–3137 (2015)
9. Saxena, A., Driemeyer, J., Ng, A.Y.: Robotic grasping of novel objects using vision. Int. J. Robot. Res. **27**(2), 157–173 (2008)
10. Levine, S., Pastor, P., Krizhevsky, A., Quillen, D.: Learning hand-eye coordination for robotic grasping with deep learning and large-scale data collection (2016)
11. Bohg, J., Morales, A., Asfour, T., Kragic, D.: Data-driven grasp synthesis a survey. IEEE Trans. Rob. **30**(2), 289–309 (2014)
12. Pinto, L., Gupta, A.: Supersizing self-supervision: Learning to grasp from 50k tries and 700 robot hours (2015)
13. Dai, J., He, K., Sun, J.: Instance-aware semantic segmentation via multi-task network cascades, pp. 3150–3158 (2016)
14. Wang, K., Lin, L., Zuo, W., Gu, S., Zhang, L.: Dictionary pair classifier driven convolutional neural networks for object detection. In: IEEE Conference on Computer Vision and Pattern Recognition, pp. 2138–2146 (2016)
15. Bo, L., Ren, X., Fox, D.: Kernel descriptors for visual recognition. In: Advances in Neural Information Processing Systems, pp. 244–252 (2010)
16. Schölkopf, B., Smola, A., Müller, K.-R.: Kernel principal component analysis. In: Gerstner, W., Germond, A., Hasler, M., Nicoud, J.-D. (eds.) ICANN 1997. LNCS, vol. 1327, pp. 583–588. Springer, Heidelberg (1997). doi:10.1007/BFb0020217
17. Wang, Q.: Kernel principal component analysis and its applications in face recognition and active shape models (2012). arXiv preprint arXiv:1207.3538
18. Dauphin, Y., De Vries, H., Chung, J., Bengio, Y.: RMSprop and equilibrated adaptive learning rates for non-convex optimization. arxiv preprint (2015). arXiv preprint arXiv:1502.04390
19. Lai, K., Bo, L., Ren, X., Fox, D.: A large-scale hierarchical multi-view RGB-D object dataset. In: 2011 IEEE International Conference on Robotics and Automation (ICRA), pp. 1817–1824. IEEE (2011)
20. Lowe, D.G.: Distinctive image features from scale-invariant keypoints. Int. J. Comput. Vision **60**(2), 91–110 (2004)

A Robotized Data Collection Approach for Convolutional Neural Networks

Yiming Liu[1]($\boxtimes$), Shaohua Zhang[1], Xiaohui Xiao[2], and Miao Li[2]

[1] Wuhan Cobot Technology Co., Ltd., Wuhan, China
sjtulym@outlook.com
[2] Wuhan University, Wuhan, China
miao.li@whu.edu.cn

Abstract. Convolutional Neural Networks are powerful tools in object classification which are widely used in Robot Vision. One of the basic requirements of this approach is the demand for a massive data set. However, in many scenarios, it is either economically expensive or difficult (impossible) to collect many valid data with few samples. To this end, in this paper we proposes an automatic approach to collecting data for food industry. First, a robotized data collection system is introduced which uses an industry robot with 6 Degree of Freedoms (DOF). Second, we analysis the key parameters of the proposed system in order to improve the quality of the training model. Finally, the effectiveness of our approach is demonstrated on real experimental platform.

Keywords: Convolutional Neural Networks · Data collection

1 Introduction

Object detection and classification is of great importance in many scenarios, ranging from house-ware service robots to industrial robot applications. Convolution Neural Networks (CNNs) has been recently considered as one of the main tools in this area. However, the performance of CNNs is highly determined by the training data set and the structure of the network. While there are a large number of studies on the design of novel network structures, few works have been investigated on the collection of the valid training data set.

There are some popular methods for data collection, which include 'virtual' data from Internet or sensory data from physical sensors. The biggest labeled data set ImageNet [3] was organized according to the WordNet [7] hierarchy which was collected from Internet and annotated by humans, with around 1000 pictures for each classes. MNIST Database of handwritten digits is the first database applied to convolutional neural networks which were created by data from American Post Letters. These data are in fact evolved from a very long time period, e.g., MNIST includes the postcode data from several years. However, for many scenarios, it is impractical to collect data from the past several years and a more effective approach to data collection is required in this case.

© Springer International Publishing AG 2017
Y. Huang et al. (Eds.): ICIRA 2017, Part III, LNAI 10464, pp. 472–483, 2017.
DOI: 10.1007/978-3-319-65298-6_43

2   Related Works

CNNs are widely used in vision fields like classification, object detection, pose estimation, image segmentation and face recognition. All these applications require a large number of data to train the network so as to obtain a reasonable generalization performance. Popular research area like face recognition has established specialized databases, among which Labeled faces in the wild (LFW [10]), Deep Face Recognition (DFR [15]), industrial database form Face-Book and Google are widely used for benchmark testing. There are also databases for general deep learning tasks such as ImageNet [3] and Pascal Voc [6]. ImageNet has over 21000 subclasses, each of which has an average of 1000 images [3]. That makes ImageNet the biggest database for computer vision. Eventhough ImageNet has a large variety of classes, the data distribution of each class is not even. In another word, for some classes, there are more images than others. For example, while the class person and class tree include 2035 and 993 subclasses respectively, the class of fungus only includes 308 subclasses. Moreover, for some cases, there is no available dataset at all, especially in industrial area where the products are usually newly designed.

To solve the problem, Sapp et al. have introduced a data collection and augmentation method for object recognition [18]. Enenkel et al. proposed a method that trains workers to use mobile for food security data collection [5]. Elmasry et al. collected food data from camera over the conveyor belt for food quality recognition [4]. However, all these methods are time-consuming and expensive, which are not suitable for our cases.

In this work, we proposed a systematic approach to collect data for food industry. First, an automatic data collection approach is introduced to gather enough valid data for the CNN. Second, we compare different system parameters during the data collection and preprocessing to improve the performance of the trained CNN. The rest of this paper is organized as follows. In next section, the system architecture of data collection is introduced. Then we discussed the strategy for key parameters setting in motion control and data processing. Finally we demonstrated the effectiveness of our approach from real experimental platform, with a discussion on possible extension of this work.

3   System Architecture for Data Collection

In this section, we will first give a brief introduction to the system architecture and also the pipeline for data collection.

3.1   System Architecture

We choose a Universal Robots 3 (UR3) as camera carrier. The robot is controlled through Robot Operating System (ROS [16]) at 125 Hz and the path of the end-effector is pre-defined. PC communicates with the robot and camera through TCP/IP and USB respectively. The camera is RealSense R200 which

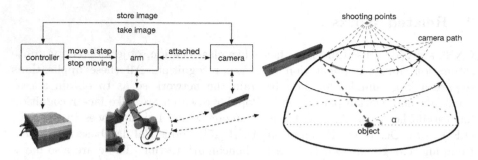

Fig. 1. This figure shows the architecture of the system. Computer controlled the arm carried with camera to take photos of the objects from a set of angles. The camera takes photos in the shoot point which is depicted in the right picture. The length between two shooting points is the step size which is labeled red with the right picture. The right picture shows the predefined path to the robot arm to traverse, which is shown as semicircle with different height and shooting angles α. After a set of shooting, we changed the objects' side for another shooting. (Color figure online)

runs on RGB mode. The resolution ratio is set to 1080×1080, while white balance and exposure were set to auto mode. As shown in Fig. 1, the camera carried by the arm goes along the cycles on the half ball and the camera takes photos in the breaks between two steps. The arm goes a semicircle on each height and changes angle α after finishing each semicircle. The step size is shown as red arc in Fig. 1. Considering the lens angle of the camera above the conveyor, seeing Fig. 3, we should chose angle α ranging from $45°$ to $80°$. As shown in the right picture in Fig. 1, the robot only take the path of the solid line on the quarter of the ball and the object is manually rotated $180°$ after the semicircle to photo another side. After finishing one surface, we rotate the object $180°$ to take another surface. So each photo capturing procedure needs four loops. Two of them are for front and two for bottom.

3.2   Pipeline for Data Processing

In this section, we will discuss the procedure for data processing. Providing proper images of target objects need to reduce the noise and background textureless. To ensure the cropping quality, the object detection should be accurate. For the training set, as the environment changing during data collection procedure, it's hard to guarantee the consistency of the collected images. Therefore we designed different data processing strategies for bin-picking camera system and data collection system.

We designed a Convolutional Neural Network for classification which is inspired by AlexNet [14]. Although deeper networks such as Inception [22] and VGG [20] performance better on classification coming out later. The computational overhead of these networks is very large. As shown in Fig. 2 We added batch normalization layers after data input layer and convolution layers [11]. These layers reduce the workload on adjust parameters when using SGD

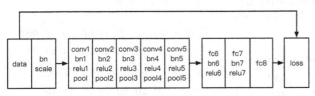

Fig. 2. Our network design which is based on AlexNet [14]. In our network batch normalization layers are added after data input layer and convolution layers. These layers reduce the workload on adjusting parameters when using SGD (Stochastic Gradient Descent [1]) strategy thus reducing computational resources required. Bn: Batch Normalization layer; conv: convolution layer; relu: Rectified Linear Units layer; scale: Scale Parameter; pool: Pooling Parameter.

Fig. 3. This is the conveyer camera system. The range of the red rectangle represents the camera's field of view.

(Stochastic Gradient Descent [1]) strategy thus reducing computational resources required. All our experiments were conducted on this network which is trained on a GTX1050Ti platform.

4   Parameters Selection

Many parameters may affect the quality of the network training. In this section, we introduce some of the key parameters and explain how they influence the network.

4.1   Hardware

The Robot we choose is an UR3 which communicates with the PC at 125 Hz through TCP/IP. We chose the RealSense R200 camera which uses soft trigger to communicate with the PC through USB. The camera is set to 1080×1080 resolution ratio in RBG mode with auto exposure and white balance.

4.2   Motion Planning

We are equally concerned with efficiency and quality of the model which is highly impacted by several parameters in the motion plan.

To train the model we have to collect thousands of pictures which takes a lot of time. Reducing the number of photos taken per winding can save time but may affect model quality. Longer the step size is, larger the differences between the adjacent photos are. So we tried 4 different step sizes to find out how the step size influences the network. As Table 1 showing, increasing the spacing between images can save time. The greater the spacing, the more time it spends on moving

Table 1. The accuracy of CNN model trained by data that uses different step sizes and the number of photos taken for each surface

Step length	Accuracy	Photos	Time(s)
Basic size ×1	90.9%	676	580
Basic size ×2	89.3%	336	460
Basic size ×3	89.7%	224	420
Basic size ×4	87.3%	168	412

Table 2. Number of steps on each degrees of angle α.

Degree	45	47	49	51	53	55	57	69
Steps	28	28	28	24	24	24	24	24
Degree	62	65	68	71	74	77	80	
Steps	20	20	20	20	18	18	18	

rather than the time it spends to stop taking pictures. The model, which is trained by images shoot with original step size, is most accurate. So the original step should be used in the case of high accuracy requirements, while longer steps should be used in time-sensitive situations. For each shiitake, we take 676 pictures from 15 different angles for each surface. In Table 2, it shows the number of steps on each angle.

4.3 Edge Recognition

In order not to lose too much detail, we need to cut the target object from the image which needs correct detection and edge recognition.

We detected the shiitakes by finding their contours which means finding a way to tell the shiitakes apart from the background. Figure 4 shows the typical appearance of the four kinds. Some types of textures make it difficult for edge recognition. One example is the contours of white texture shiitakes which are difficult to distinguish from the background. We need an algorithm to eliminate the influence of the background without affecting the edge recognition.

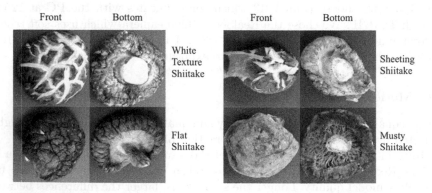

Fig. 4. The four kinds of shiitakes are shown in the figures. The classifier is required to tell the four kinds of surfaces and bottoms apart. The challenge occurred on the shiitakes with smooth surface which is hard to be detected.

One of the most commonly used edge recognition algorithm is proposed by Canny [2] in 1986. Canny's method recognizes the contour by calculating the gradient. This algorithm is well performed in good lighting condition but it enhances the noise when the bottom noise is high. So we found out the strategy of choosing a proper threshold, which is using high threshold in good photo condition while using low threshold when noise is high. Comparing the advantages and disadvantages, we used Canny edge detection with low threshold when capturing data on the conveyor belt. The conveyor belt is equipped with light source and better camera which guarantees stable and good image quality. However the training set contains some pictures that the noise is so strong that can not use Canny's method. So we didn't use the method when processing training data set. If Canny's method is not used, the contours of some pictures may not be fully recognized. Detecting images with half-baked contours may crop pictures in improper place. Considering random crop is a data augmentation method, some of the images that are not just cut on the edge of shiitakes have little influence on the model. Thus not using Canny's method in training image set is acceptable. Meanwhile it's supposed to use Canny's method in pictures taken by conveyor camera system. As Fig. 5 shows, pictures taken on conveyor belt have better quality, in which Canny's method has sound effects. The problem of background noise still remained unsolved. So we used median blur combining with eroding operation to remove the noise. Erode operation might make the contours incomplete. So we used adaptive threshold binarization operation [12] to make the contour line stronger. The method of adaptive threshold reserves few noises and a clear but discontinuous boundary.

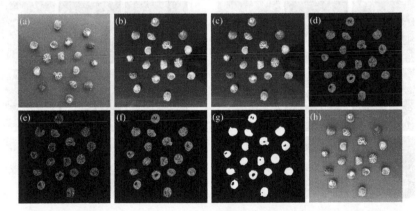

Fig. 5. This set of pictures shows how we process images taken on conveyor belt Picture (a) is the original image taken on the conveyor belt. Picture (b) is the red channel of the original picture. Picture (c): median blur; Picture (d) (e) are processed separately from picture (c) that uses Canny's method and adaptive threshold binaryzation. Picture (f) is added by picture (d) and picture (e). Picture (g): morphology method(erosion and dilation several times). In picture (h), we found the minimum bonding box for each object.

Another problem is shadow. We use the red channel of image to avoid the influence of shadow. Shadow may blur the edge of target object which may lead the contour hard to be recognized. Since the conveyor belt is green and shiitakes don't contain any green color. Red channel will filter out all the green backgrounds. This method makes the edge of target object sharper and easier to be recognized. Despite the advantages of using the red channel, brightness bottom noise will stand out from the background. So we only use this method on the images taken on the conveyor belt.

The kind of flat shiitake which has a plain and relatively smooth black umbrella is hard to be detected on internal texture. Even changing different parameters of adaptive threshold would fail to recognize the edge completely. So we use morphology operation [19] to bold the contour line and fill the blanks between contours.

After we got the proper contour, we can find a bonding box to determine the location of shiitakes. Then pictures would be cropped along the edge of bonding box. As it is presented in Fig. 5 shiitakes would be divided into singles to be classified. Both bin-picking system and data collection system use this procedure. In Fig. 6 these pictures were selected from training set. Our detection method works good in both the two situations.

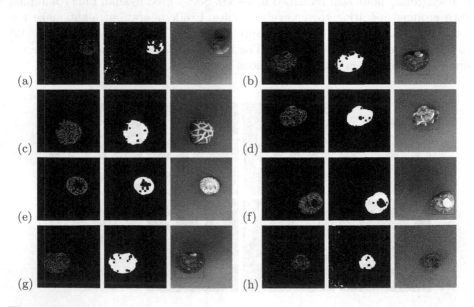

Fig. 6. These group of pictures is the result of detection captured by data collection system. The first pictures in each sets show the contours of the shiitakes. Second ones are the result of morphology operation. Last pictures are detection result. Our detection method works well in the training set. Picture (a)(h) shows our method deals with noise well. In (a)(f)(g)(h) detection did not affected by shadow. (Color figure online)

5 Experiment Results

In this section we compared our classification result to the result from rotation method and discuss why our approach better. We took 12692 pictures from White Shiitake, 12708 from Flat Shiitake, 14050 from Sheeting Shiitake, 9878 from Musty Shiitakes as training set and validation set. we took another 6828 pictures which are around one eighth of the training set as testing set. Those numbers above don't contain broken files and overexposure pictures. For rotation method we used data augmentation method to create 66064 images from 74 samples, 4404 images.

5.1 Implementation Details

The classification result is shown in Fig. 7. We first detect the position and edge of the objects through our detection method. The shiitakes will then be cut out from the original pictures. After that the classifier calculates the possibility of each shiitake on each category, thus to find out the most possible type.

5.2 Rotation Experiment

From the previous works we know that Regularization Dropout [21] Unsupervised pre-training and Data Augmentation could help with training model

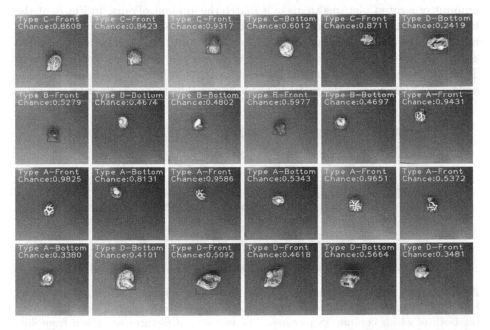

Fig. 7. These are some classification results. Type A: white shiitake; Type B: sheeting shiitake. Type C: flat shiitake. Type D: musty shiitake. Chance of the type is printed on the second line.

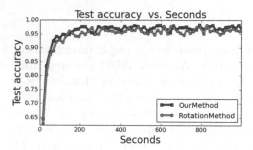

Fig. 8. Accuracy-iteration plot between our method and rotation method during training

Table 3. A: rotation method and B: our method. Comparison between the rotation method and our method

Iteration	A	B
5000	74.7%	89.3%
10000	75.0%	91.0%

with few data. Among those methods, Data Augmentation is the only one that processes data rather than changing the networks. Data Augmentation improves the robustness and reduces overfitting, some of the which includes rotation, flip, zoom, shift, scale, contrast, noise, Principle Component Analysis (PCA [13]). While our data collecting system takes photo of the target object from 0 to 360°, it raises the question that whether our method could replace rotation operation. So we compared the models which were trained from data collected by our method and data created by rotation. Figure 9 shows the comparison between rotated pictures and pictures taken by our system. We controlled the amount of rotated pictures to make it match the amount of ours. In Fig. 8, at the first 800 iteration our method and rotation method have similar trend on accuracy during training. But our method leads the accuracy than rotation method almost all the training procedure. The result that ran on the test set is shown in Table 3. Looking into the data of rotation method in Table 4, we found rotation method accuracy drops dramatic in the bottoms. The reason is shiitake feet grow

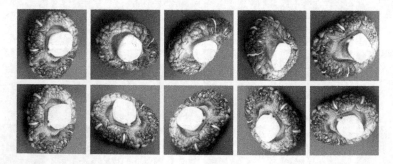

Fig. 9. The comparison between Our method (first column) and rotation method (second column). Both the two sets of pictures are selected from their original picture set both of which contain 28 pictures. The two sets of pictures are taken from 45°, bottom of a same flat shiitake.

Table 4. Comparison between the rotation method and our method in different kinds. Type A: white shiitake; Type B: sheeting shiitake. Type C: flat shiitake. Type D: musty shiitake. Method A: our method. Method B: rotation method

Type	Method A	Method B
Type A front	99.4%	98.6%
Type A bottom	83.5%	34.4%
Type B front	90.7%	87.8%
Type B bottom	73.7%	13.3%
Type C front	99.9%	95.7%
Type C bottom	86.2%	92.1%
Type D front	100.0%	97.3%
Type D bottom	97.3%	85.3%
Total	91.0%	75.0%

on the bottom side. Thus making the shiitakes bottom side look different from different view points. Apparently our method is better than doing rotation.

5.3   Step Size Experiment

Different choices of arm step sizes and processing strategy lead to different classification result. As the Fig. 10 showing, if not considering accuracy difference step sizes do not affect training speed. Accuracy went up fast at the beginning, then raised slowly and steady. Different methods followed similar trend during all the training procedure, which means the choice of step sizes should mainly consider the requirement of accuracy and data capturing time. Those results has been shown in Table 1. Shorter step size provide higher accuracy while longer step size saves time. So it could be said that in the small sample situation, our method is better in accuracy no matter how much iteration it takes in training.

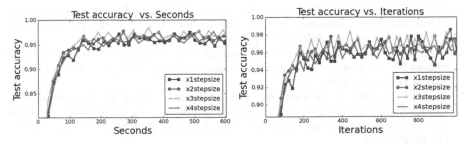

Fig. 10. The 2 figures show the comparison between the four step sizes. The fist picture shows plot of time and accuracy. All the accuracy is test on validation set. Second picture shows plot of iteration and accuracy. Accuracy changes few after dramatically going up at the beginning of training.

Although our method doesn't keep us away from using other data augmentation methods like scaling, mirror, shift, zoom etc. It still remained unknown what other method would influence the network.

We used the same way to create test set with training set, but the sample size is even smaller. For the reason that the individuality of some kinds of shiitakes is very different and the small sample size of the test set, the accuracy rate can not be fully reflected. Solving this problem needs large number of samples which is not in line with our situation. How to use a small number of testing data to test the model performance would be a challenging direction of our future work.

6   Conclusion

In this paper, we introduced a robotized approach to collect data for CNNs. This method is mainly used for the situation where data can not be collected by database. Moreover, this approach can quickly create a data set from a small number of samples with few manual intervention, which is quite suitable for industrial applications. The experiment shows that Comparing to using rotation to do data augmentation, our method can obtain a better performance.

For the future work, one of the possible direction is to use R-CNN (Rich Feature Hierarchies Convolutional Neural Network) [8,9,17] to detect, segment and classify object in one shot. However, this method requires training data including the category information as well as a bounding box indicating the position of the object. Our future work is to collect this kind of training dataset and extend our method to more general cases.

References

1. Bottou, L.: Large-scale machine learning with stochastic gradient descent. In: Lechevallier, Y., Saporta, G. (eds.) Proceedings of COMPSTAT 2010. pp. 177–186. Springer, Heidelberg (2010). doi:10.1007/978-3-7908-2604-3_16
2. Canny, J.: A computational approach to edge detection. IEEE Trans. Pattern Anal. Mach. Intell. **6**, 679–698 (1986)
3. Deng, J., Dong, W., Socher, R., Li, L.J., Li, K., Fei-Fei, L.: Imagenet: a large-scale hierarchical image database. In: 2009 IEEE Conference on Computer Vision and Pattern Recognition, CVPR 2009, pp. 248–255. IEEE (2009)
4. Elmasry, G.M., Nakauchi, S.: Image analysis operations applied to hyperspectral images for non-invasive sensing of food quality a comprehensive review. Biosyst. Eng. **142**, 53–82 (2016)
5. Enenkel, M., See, L., Karner, M., lvarez, M., Rogenhofer, E., Baraldsvallverd, C., Lanusse, C., Salse, N.: Food security monitoring via mobile data collection and remote sensing: results from the central African Republic. Plos One **10**(11), e0142030 (2015)
6. Everingham, M., Gool, L.V., Williams, C.K.I., Winn, J., Zisserman, A.: The pascal visual object classes (VOC) challenge. Int. J. Comput. Vision **88**(2), 303–338 (2010)
7. Fellbaum, C., Miller, G.: WordNet: an electronic lexical database. MIT Press, Cambridge (1998)

8. Girshick, R.: Fast R-CNN. In: IEEE International Conference on Computer Vision, pp. 1440–1448 (2015)
9. Girshick, R., Donahue, J., Darrell, T., Malik, J.: Rich feature hierarchies for accurate object detection and semantic segmentation. In: Computer Vision and Pattern Recognition, pp. 580–587 (2014)
10. Huang, G.B., Ramesh, M., Berg, T., Learned-Miller, E.: Labeled faces in the wild: a database for studying face recognition in unconstrained environments. Technical report, 07–49, University of Massachusetts, Amherst (2007)
11. Ioffe, S., Szegedy, C.: Batch normalization: accelerating deep network training by reducing internal covariate shift. In: International Conference on Machine Learning, pp. 448–456 (2015)
12. Jain, A.K.: Fundamentals of Digital Image Processing. Prentice-Hall Inc., Upper Saddle River (1989)
13. Jolliffe, I.: Principal Component Analysis. Wiley, Hoboken (2002)
14. Krizhevsky, A., Sutskever, I., Hinton, G.E.: Imagenet classification with deep convolutional neural networks. In: Advances in neural information processing systems, pp. 1097–1105 (2012)
15. Parkhi, O.M., Vedaldi, A., Zisserman, A.: Deep face recognition. In: BMVC, vol. 1, p. 6 (2015)
16. Quigley, M., Conley, K., Gerkey, B., Faust, J., Foote, T., Leibs, J., Wheeler, R., Ng, A.Y.: ROS: an open-source robot operating system. In: ICRA workshop on open source software, vol. 3, p. 5. Kobe (2009)
17. Ren, S., He, K., Girshick, R., Sun, J.: Faster R-CNN: towards real-time object detection with region proposal networks. In: International Conference on Neural Information Processing Systems, pp. 91–99 (2015)
18. Sapp, B., Saxena, A., Ng, A.Y.: A fast data collection and augmentation procedure for object recognition. In: National Conference on Artificial Intelligence, pp. 1402–1408 (2008)
19. Serra, J.: Image Analysis and Mathematical Morphology, V. 1. Academic press, Cambridge (1982)
20. Simonyan, K., Zisserman, A.: Very deep convolutional networks for large-scale image recognition. arXiv preprint arXiv:1409.1556 (2014)
21. Srivastava, N., Hinton, G.E., Krizhevsky, A., Sutskever, I., Salakhutdinov, R.: Dropout: a simple way to prevent neural networks from overfitting. J. Mach. Learn. Res. 15(1), 1929–1958 (2014)
22. Szegedy, C., Liu, W., Jia, Y., Sermanet, P.: Going deeper with convolutions, pp. 1–9 (2015)

Virtual Simulation of the Artificial Satellites Based on OpenGL

Yang Liu(iD), Yikun Gu, Zongwu Xie$^{(\boxtimes)}$, Haitao Yang,
Zhichao Wang, and Hong Liu

State Key Laboratory of Robotics and System, Harbin Institute of Technology,
Harbin 150001, Heilongjiang, People's Republic of China
xiezongwu@hit.edu.cn

Abstract. In order to generate the images of the satellites in outer space, a virtual simulation platform was build up based on Visual C++ and OpenGL API. Take the Sinosat-2 Satellite as an example, which is a failure satellite of China. Its projection model was established using computer-graphics technology. Its kinematics formulas were described using quaternions to avoid singularity. A spinning model was set up to demonstrate its motion in 3D space. The platform could simulate the synthetic images of the satellite and then send them to the virtual serve system through TCP/IP in real time.

Keywords: Binocular vision · Virtual reality · OpenGL · SINOSAT-2 · DFH-4 · Spinning model

1 Introduction

Satellites can be separated into four categories: remote sensing, space science, communication, and global navigation. Most operational satellites fall into the communications category (57%). Besides, communications satellites are also expensive which can cost millions of dollars.

According to [1], 43 percent of the satellites fall to work in one year and the number increases to 59 percent in three years, leading to plenty of economic cost. Among these failures, only 18 percent are beyond retrieving, which means most of these satellites could be rescued by auxiliary orbits maintain, or simple repairs. Meanwhile, a failed satellite in orbit occupies an orbital resource and increases the probability of collisions, which poses a serious potential threat to space activities.

Thus, on-orbit service has been recognized as the most potential space application of the future space missions and drawn the attention of many countries and organizations. Related studies and demonstrations have been recently developed, such as the Orbital Express (OE) demonstration mission, the Micro-satellite Technology Experiment (MiTEx) and the Spacecraft for the Universal Modification of Orbits (SUMO) project.

After rendezvous with the target to a relative distance of approximately 50 m, the chaser spacecraft maintains a constant distance measured using images taken from the onboard cameras. During this station-keeping phase the chaser captures images of the

Y. Huang et al. (Eds.): ICIRA 2017, Part III, LNAI 10464, pp. 484–494, 2017.
DOI: 10.1007/978-3-319-65298-6_44

target to allow visual inspection. Images are also used in the grapple and retreat phase to provide the relative pose of the targets. Therefore, machine vision is an important subsystem in the on-orbit service system, which can reduce the possibility of collision.

The sunlight in space environment is intense and directional, besides there is no diffusing background other than the earth [2]. Thus, the grabbed images always have very high image contrast and the only diffuse light source comes from the earth's albedo [3].

In this paper, a virtual simulation platform is proposed to generate images of the satellite in outer space. It can provide synthetic images that is useful in the virtual serve system with cheap cost. Take the Sinosat-2 Satellite as an example, the image-forming principle is introduced to illustrate how to project a 3D model onto the image plane.

2 The Sinosat-2 Satellite

Communications satellites play an important role in the development of the Chinese space program. In 1984, China launched its first communication satellite that then brought television and modem communications to the whole Chinese landmass. Moreover, it marks the beginning of the Dong Fang Hong series of communications satellites.

As shown in Fig. 1, DFH-4 is the third generation of China-built large geostationary satellite platform with larger output power, payload capacity and longer service lifetime [5]. It could be used in many services such as high capacity broadcast communication, direct TV broadcasting and digital audio broadcasting. The main body of the DFH-4 satellite is a box-form structure with the size of 2.36 m × 2.1 m × 3.6 m and the wingspan can reach 33.0 m with the height of 6.4 m. More specific parameters can be found in Table 1.

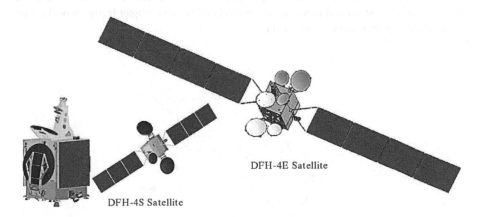

DFH-4E Satellite

DFH-4S Satellite

Fig. 1. DFH-4 satellite configurations [4]

The first satellite based on DFH-4 is Sinosat-2, China's first direct-to-home satellite. It was designed for domestic communications in China, which would enable

Table 1. The technical Specifications of the DFH-4 Platform

Name	DFH-4
Platform mass	5000–5200 kg
Payload mass	500–700ke
Attitude and orbit control mode	3-axis Stabilization
Pointing accuracy	$\pm 0.06°$(R), $\pm 0.06°$(Pi), $\pm 0.2°$(Y)
Solar array power output	11 kW
Power available for payload	8 kW
Design lifetime	15 Years
Station-keeping accuracy	$\pm 0.05°$(E/W), $\pm 0.05°$(N/S)

indigenous broadcasters to disseminate live TV to the nation, rather than relying on foreign broadcasts during the 2008 Beijing Olympic Games. Unfortunately, it was revealed a month later that its solar wing failed to deploy, depriving it of power.

As described in [1], the solar wing is the most easily damaged component. Thus, we take the SINOSAT-2 satellite for example in this paper. A 1/1 scale ProE model is established and then used as the target satellite in the virtual simulation platform.

3   Image-Forming Principle

The pinhole camera model describes the mathematical relationship between the coordinates of a 3D point and its projection onto the image plane. The model does not take the geometric distortions or blurring of unfocused objects caused by lenses into consideration. It's used as a first order approximation of the mapping from a 3D scene to a 2D image.

As shown in Fig. 2, the central perspective-imaging model is widely used in computer vision system and there are several related frames: object frame, world frame, camera frame, screen frame and so on.

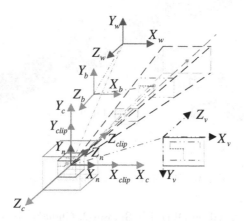

Fig. 2. The camera imaging mechanism

The relation of one point between the coordinates in the physical world and on the projection screen is called a projective transformation, the homogeneous coordinates are widely used in such transforms.

Figure 3 shows how to convert the viewing frustum (truncated pyramid) into a unit cube using projection matrices. You can see the process as if the viewing frustum being normalized. The viewing frustum which defined by the near and far clipping planes is not a cube at all, is indeed "warped" into a cube. Which makes it easier to operate on points as a cube is much easier geometrical form to work with than a truncated pyramid.

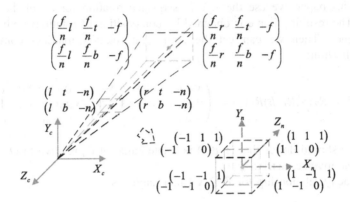

Fig. 3. Transformation between the frustum and normalized cube

In OpenGL, we use the function gluPerspective() to define the frustum, which creates a matrix for a symmetric perspective-view frustum. Using the function glViewPort() you can define a pixel rectangle in the window onto which the final image is mapped [6–8].

The calibration algorithm calculates the camera matrix using the extrinsic and intrinsic parameters. The extrinsic parameters represent a rigid transformation from 3-D world coordinates frame to the 3-D camera's coordinate frame and the intrinsic parameters represent a projective transformation matrix from the 3-D camera's coordinates to the 2-D image coordinates.

The intrinsic matrix contains five intrinsic parameters. These parameters encompass focal length, skew coefficient, and principal point. The relationship between the intrinsic parameters and the frustum is:

$$\begin{cases} f_x = Wn/(r-l), f_y = -Hn/(t-b) \\ c_x = Wl/(r-l), c_y = -Ht/(t-b) \end{cases} \tag{1}$$

where W and H are the width and height of the output image, r and l represent the x coordinates of the right and left corners of on the near plane, besides, t and b represent the y coordinates of the top and bottom corners of on the near plane.

4   Kinematics Description of the Satellites

To describe the position of the target, the body fixed frame is established at the center of mass (CM) of the tumbling target spacecraft, where the x-axis is located on the centerline of the solar panels, the y-axis points towards the orientation of the exhaust nozzle, and the z-axis is along the direction of the antenna.

Euler angles are widely used to describe the attitude of a reference frame relative to another through three successive rotation angles about an axis of a moving frame. As a minimum representation, it is an effective tool for visualization as it is intuitively easier to understand. There are twelve different conventions to represent rotations via Euler angles. In this paper, we use the Z-Y-X sequence rotation for all of the presented examples. The resulting frame is obtained by composition of rotations with respect to current frames. Then we can get the equivalent rotation matrices computed via post-multiplication:

$$^j R_i = R_Z(\alpha)R_Y(\beta)R_X(\gamma) = \begin{pmatrix} c_\alpha c_\beta & c_\alpha s_\beta s_\gamma - s_\alpha c_\gamma & c_\alpha s_\beta c_\gamma + s_\alpha s_\gamma \\ s_\alpha c_\beta & s_\alpha s_\beta s_\gamma + c_\alpha c_\gamma & s_\alpha s_\beta c_\gamma - c_\alpha s_\gamma \\ -s_\beta & c_\beta s_\gamma & c_\beta c_\gamma \end{pmatrix} \tag{2}$$

Here you should notice that $(\alpha\ \beta\ \gamma)$ is rotate angle of O_i relative to O_j, otherwise you'll get the inverse matrix.

The velocity relationship for Z-Y-X Euler angles is:

$$\begin{pmatrix} \dot\alpha \\ \dot\beta \\ \dot\gamma \end{pmatrix} = \begin{pmatrix} -c_\beta & 0 & 1 \\ c_\beta s_\gamma & c_\gamma & 0 \\ c_\beta c_\gamma & -s_\beta & 0 \end{pmatrix} \begin{pmatrix} \omega_x \\ \omega_y \\ \omega_z \end{pmatrix} \tag{3}$$

Singularity occurs when the first and last rotations are about the same axis, so you should take care that $\beta \neq \pm 90°$, known as "gimbal lock". Thus, quaternions are proposed since they do not suffer from singularities as Euler angles do. Which means that all the reachable orientations in the workspace of robot manipulators can be expressed by unit quaternion.

The uniform expression of unit quaternion is as follow:

$$q = q_w + q_x i + q_y j + q_z k = (\cos(\theta/2)\quad \vec n \sin(\theta/2)) \tag{4}$$

Unit quaternion has a clear physical significance: rotation of objects is about the axis $\vec n$ with θ, so the equivalent rotation matrix can be written as follow:

$$^j R_i = R_Z(\phi)R_Y(\theta)R_X(\psi) = \begin{pmatrix} \cos\phi/2 \\ 0 \\ 0 \\ \sin\phi/2 \end{pmatrix} \begin{pmatrix} \cos\theta/2 \\ 0 \\ \sin\theta/2 \\ 0 \end{pmatrix} \begin{pmatrix} \cos\psi/2 \\ \sin\psi/2 \\ 0 \\ 0 \end{pmatrix} \tag{5}$$

For velocity analysis, the time derivative of the quaternion could be related to the angular velocity vector as:

$$\begin{pmatrix} \dot{q}_w \\ \dot{q}_x \\ \dot{q}_y \\ \dot{q}_z \end{pmatrix} = \frac{1}{2} \begin{pmatrix} -q_x & -q_y & -q_z \\ q_w & -q_z & q_y \\ q_z & q_w & -q_x \\ -q_y & q_x & q_w \end{pmatrix} \begin{pmatrix} \omega_x \\ \omega_y \\ \omega_z \end{pmatrix} \tag{6}$$

5 The Spinning Model

In the study of motion analysis, we regard the artificial satellite as a rigid body despite it may contain stiffness components, cables or instrumentations. A spinning model is built up in this section and its motion is analyzed in detail [9].

Suppose that all the axes pass through the centroid and the inertia parameters are I_x, I_y and I_z. Based on Euler method, the dynamics equation of the satellite can be written as (Here we assume the satellite is symmetric around Z-axis and the external torque is equal to zero):

$$\begin{cases} I\dot{\omega}_x + (I_z - I)\omega_z\omega_y = 0 \\ I\dot{\omega}_y + (I - I_z)\omega_z\omega_x = 0 \\ \dot{\omega}_z = 0 \end{cases} \tag{7}$$

Here $I = I_x = I_y$ and let $\Omega = (I_z/I - 1)\omega_z$, then Eq. (7) can be expressed as:

$$\begin{cases} \dot{\omega}_x + \Omega\omega_y = 0 \\ \dot{\omega}_y - \Omega\omega_x = 0 \\ \dot{\omega}_z = 0 \end{cases} \tag{8}$$

Assume that the initial state are as follows

$$\begin{cases} \omega_x = \omega_0 \cos(\varphi) \\ \omega_y = \omega_0 \sin(\varphi) \\ \omega_z = \omega_s \end{cases} \tag{9}$$

Here ω_0 and ω_s are constant. Then Eq. (8) can be expressed as:

$$\omega = \begin{bmatrix} \omega_x \\ \omega_y \\ \omega_z \end{bmatrix} = \begin{bmatrix} \omega_0 \cos(\Omega t + \varphi) \\ \omega_0 \sin(\Omega t + \varphi) \\ \omega_s \end{bmatrix} \tag{10}$$

As shown in Fig. 4, the spin axis is rotating around the angular momentum vector at the speed of Ω. The angle between the spin axis and the angular momentum vector is named as nutation angle.

Here you should notice that $I < I_x$, otherwise the control of the satellite cannot achieve spinning stability.

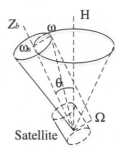

Fig. 4. Speed of the satellite

6   Virtual Simulation Platform

As a real-time visual simulation platform, this environment is a practical application of the visual reality technology. It adopts the computer-graphics technology to establish the kinematics model of the satellite and takes the advantage of cheap costs, reliable technology, flexibility and convenience [10, 11]. Thence it is a useful tool in the field of virtual serve system.

The virtual simulation platform is implemented by C++ programming language and tested on a PC (I3-3240 at 3.4 GHz, 4 GB of Ram) with Win10. As an interface about graphics and hardware [12], OpenGL is used to establish 3D models of the Sinosat-2 satellite. Different from other graphic program designs, it offers very clear functions with high performances.

As shown in Fig. 5, the SINOSAT-2 satellite is illustrated on the virtual simulation platform. We have made a 3D model using the software of ProE and convert it into obj format. The platform reads the data from the obj file and demonstrates the satellite using three cameras: global camera, left camera and right camera. In addition, the average time to generate images is 28.63 ms.

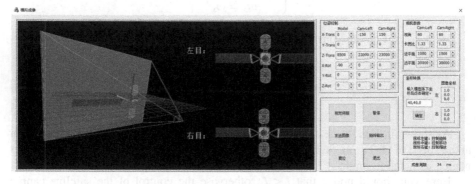

Fig. 5. Virtual simulation platform

With a black background, these cameras are listed in two columns. On the left, the global camera is used to show the relationship between the other two cameras and the satellite. Since the distance between the satellite and the cameras is much larger than the offset of the two related cameras, the viewing frustums in red and green are much closed. The light gray color means the far clipping planes, objects become invisible when they cross the plane. On the right, the left and right cameras are used to grab images that will be sent to the virtual serve system later. You can change the position of the global camera for more details using the computer mouse: rotating, moving or scaling. Moreover, the pose of the satellite and the cameras could be changed through the spin control (The global camera is fixed). Parameters about the cameras can also be changed to get more specific images.

The satellite model can carry out movement using data from txt files. It describes the kinematics of the satellite and has already been precomputed. The target satellite is moving during the task of virtual serve and the relative pose is changing. Besides, this platform is also a client that can send the images out using TCP/IP. As shown in Table 2, the movement is demonstrated through consecutive images with related pose of the camera.

A 3D point can be defined in the Cartesian space with three coordinates, and then the platform can figure out the corresponding coordinates on the image plane. The extrinsic and intrinsic parameters are calculated and you can check it through camera calibration.

To evaluate this platform, a checkerboard calibration method is introduced. As we can obtain the intrinsic parameters through manual setting or stereo calibration, comparison between these two results can prove whether our algorithm is true or not. First, we generate synthetic images with checkerboard. Then we detect all the corners in the images and reject the images with wrong detection. After that, calibration is carried out to get the results and the reprojection error is analyzed as well. As shown in Fig. 6, the corners are all detected as there is little noise. The green circles represent the detected corners while the reprojected points are printed in red. In addition, the yellow square means the origin of the checkboard and it should follow the consistency between two relative images.

The parameters about the virtual simulation platform are as follow:

$$
\left\{
\begin{aligned}
M_{pl} = M_{pr} &= \begin{bmatrix} 478.345 & 0 & 276.5 \\ 0 & 478.345 & 276.5 \\ 0 & 0 & 1.0 \end{bmatrix} \\
R = \begin{bmatrix} 1 & 0 & 0 \\ 0 & 1 & 0 \\ 0 & 0 & 1 \end{bmatrix} \quad & T = \begin{bmatrix} -300 \\ 0 \\ 0 \end{bmatrix}
\end{aligned}
\right.
\tag{11}
$$

And the results from stereo calibration is:

Table 2. Consecutive Images during the Movement

ID	Pose	Translation (mm)	Rotation (degree)	Left Image	Right Image
1	Left	-247.2, 87.1, 3974	5, -10, 5		
	Right	47.2, 112.9, 4026	5, -10, 5		
	Satellite	100, 100, 300	-86, 2, -20		
2	Left	-199.5, 145.6, 3710	4.29, -4.54, 1.69		
	Right	99.5, 154, 3733	4.29, -4.54, 1.69		
	Satellite	200, 308, 110.6	-87.9, -0.565, -6.55		
3	Left	-199.9, 154, 3430	3.27, 0.89, -1.677		
	Right	99.9, 146, 3425	3.27, 0.89, -1.677		
	Satellite	200, 308, 110.6	-90.3, -2.6, 7.05		
4	Left	66.4, 122, 3136	2.09, 0.90, -0.935		
	Right	366.4, 118, 3131	2.09, 0.90, -0.935		
	Satellite	126.7, 260, 20	-90.2, -2.87, 6.27		
5	Left	346.8, 120, 2842	0.91, 0., 3, -0.19		
	Right	647, 119, 2837	0.91, 0., 3, -0.19		
	Satellite	100, 148, -49.9	-89.7, -2.94, 3.87		
6	Left	360, 119, 2548	-0.28, 0.96, 0.549		
	Right	659, 121, 2543	-0.28, 0.96, 0.549		
	Satellite	100, 40.5, -50	-89.3, -2.98, 1.464		

$$
\begin{cases}
M'_{pl} = \begin{bmatrix} 482.25 & 0 & 284.25 \\ 0 & 486.56 & 277.30 \\ 0 & 0 & 1.0 \end{bmatrix} & M'_{pr} = \begin{bmatrix} 483.80 & 0 & 277.49 \\ 0 & 488.70 & 277.01 \\ 0 & 0 & 1.0 \end{bmatrix} \\
R' = \begin{bmatrix} 1.00 & 0 & -0.02 \\ 0 & 1 & 0 \\ 0.02 & 0 & 1 \end{bmatrix} & T' = \begin{bmatrix} -302.85 \\ 0.46 \\ 7.77 \end{bmatrix}
\end{cases}
\tag{12}
$$

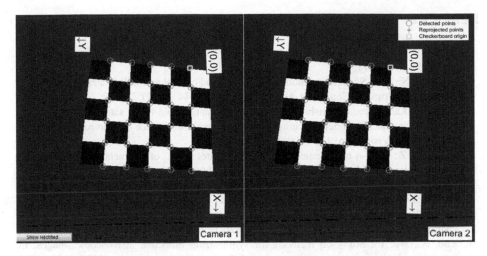

Fig. 6. Corner detection of the synthetic images (Color figure online)

The vector angle method is used to evaluate the results and the relative accuracy is $S = 99.97\%$.

In this paper, the adapter ring (also called interface ring) is chosen as the image feather, also a good candidate to be grasped. It is a high-strength torus with a radius of either 1194 mm or 1666 mm, which is used to connect the satellite to the launch vehicle. The global camera points to the adapter ring, which is actually the back of the satellite (opposite to the antenna).

7   Conclusion

Based on the three-dimensional modeling techniques, we proposed a virtual simulation platform. It can generate the synthetic images of the satellites in outer space. The Sinosat-2 satellite is taken as an example to describe its kinematics formulas. Using the central perspective-imaging model, we then map the satellite from 3D scene to 2D image. The platform can calculate the extrinsic and intrinsic parameters of the camera and then send the output images to the virtual serve system using TCP/IP.

Acknowledgement. The research is supported by the Foundation for Innovative Research Groups of the National Natural Science Foundation of China (Grant No. 51521003 and NO. 61603112) and the Self-Planned Task (NO. SKLRS201721A) of State Key Laboratory of Robotics and System (HIT).

References

1. Zhao, H., Zhang, Y.: Statistical analysis of DFH- 3 serial satellites failure. Spacecraft Eng. **16**(1), 33–37 (2007)
2. Terui, F., Nishida, S.: Relative motion estimation and control to a failed satellite by machine vision. in: Automatic Control in Aerospace, pp. 639–644 (2007)
3. Terui, F., Kamimura, H., Nishida, S.: Motion estimation to a failed satellite on orbit using stereo vision and 3D model matching. In: International Conference on Control, Automation, Robotics and Vision, pp. 1–8 (2006)
4. Zhicheng, Z.: DFH-4 based communications satellites: series of products gradually entering into the international aerospace market. Aerosp. China **14**(3), 8–14 (2013)
5. Zhou, Z., Li, F., Li, Z., Zhang, B.: Development of DFH-4, the third generation of China GEO platform. Febs Lett. **318**(2), 139–144 (1993)
6. Zhang, H., Xu, K., Zhao, M., et al.: Development of semi-physical simulation system for vision servo control of robot. J. Syst. Simul. **20**(1), 68–69 (2008)
7. Yao, X., Wu, P.: Study on methods for synthesizing binocular stereoscopic images. In: International Conference on Electrical and Control Engineering, pp. 6127–6130. IEEE (2011)
8. Fu, G., Huang, P., Chen, K., et al.: Design of semi-physical simulation sys-tem for binocular vision servo control. Comput. Meas. Control **17**(12), 2494–2496 (2009)
9. Liu, H., Liang, B., Wang, X., et al.: Autonomous path planning and experiment study of free-floating space robot for spinning satellite capturing. In: International Conference on Control Automation Robotics and Vision, pp. 303–331 (2015)
10. Li, R.: Flight environment virtual simulation based on OpenGL. In: International Conference on Information and Computing Science, pp. 141–144 (2009)
11. Li, R.: Study on the virtual simulation of flight environment based on OpenGL. Comput. Inf. Sci. **2**(2), 92 (2009)
12. Bin, W., Shao, Y., Zhang, H.: OpenGL based collision avoidance technology in a virtual environment for robot path planning simulation. J. Inf. Computat. Sci. **10**(7), 2133–2139 (2013)

Wearable Rehabilitation Training System for Upper Limbs Based on Virtual Reality

Jianhai Han[1,2(✉)], Shujun Lian[1], Bingjing Guo[1,2], Xiangpan Li[1,2], and Aimin You[3]

[1] School of Mechatronics Engineering, Henan University of Science and Technology, Luoyang 471003, China
jianhaihan@haust.edu.cn

[2] Henan Provincial Key Laboratory of Robotics and Intelligent Systems, Luoyang 471003, China

[3] The First Affiliated Hospital, Rehabilitation Center, Henan University of Science and Technology, Luoyang 471003, China
18238832685@163.com

Abstract. In this paper, wearable rehabilitation training system for the upper limb based on virtual reality is designed for patients with upper extremity hemiparesis. The six-axis IMU sensor is used to collect the joint training angles of the shoulder and elbow. In view of the patient's shoulder and elbow joint active rehabilitation training, the virtual rehabilitation training games based on the Unity3D engine are designed to complete different tasks. Its purpose is to increase the interest of rehabilitation training. The data obtained from the experiment showed that the movement ranges of the shoulder and elbow joint reached the required ranges in the rehabilitation training game. The basic function of the system is verified by the experiments, which can provide effective rehabilitation training for patients with upper extremity hemiparesis.

Keywords: Virtual reality · Upper limb · Hemiplegia · Six-axis IMU sensor · Rehabilitation training

1 Introduction

The human upper limb is an important part and one of the most complex parts of the activities, and it connects the brain to accomplish complex movement through the nervous system. With the intensification of the aging problem in the world, the amount of patients with stroke hemiplegia has shown a growing trend, especially in China, where the population is large [1]. In addition, frequent traffic accidents have also become one of the common causes of hemiplegia in recent years. Limb hemiplegia that is caused by stroke or traffic accidents not only gives the patient's life, work and learning a great deal of inconvenience and makes the patient suffered great mental pain, but also brings a heavy spirit stress and medical burden on the patient's family and society. It has been

Henan scientific and technological research program (172102210036).
Natural Science Foundation of Henan Province (162300410082).

Y. Huang et al. (Eds.): ICIRA 2017, Part III, LNAI 10464, pp. 495–505, 2017.
DOI: 10.1007/978-3-319-65298-6_45

one of the urgent problems in the society how to help the patients to recover the function of the body health movement and bring them return to normal life [2]. At present, rehabilitation therapy of upper limb is still using traditional rehabilitation therapy, that is, rehabilitation therapists perform rehabilitation training on patients one by one. With the development of robot, the rehabilitation of robot-assisted training is also rising up, but the patients are treated with low therapeutic initiative and with the single and dull treatment's process, while the treatment cost is expensive and the treatment effect is not obvious [3, 4]. According to the research, if we can increase the design with games that are based on the virtual reality to mobilize the subjective initiative of patients, the rehabilitation effect will be greatly improved [5]. Currently, the automatic rehabilitation equipment on the market are mostly complex and expensive, which are bought by the major hospitals and clinics and are not afforded for ordinary patients. Therefore, it is necessary to study the upper limb rehabilitation training system with minimal structure, low cost and suitable for popularization in the family. But in China, the researches in this field just begin, most of which are in the experimental stage and not yet market-oriented [6, 7]. In 2014, Xiling Xiao designed a hand function rehabilitation system based on virtual reality technology. In 2015, Hong Wang who comes from Shanghai University made a study about design and implementation of sEMC virtual training system based on android. The research of virtual reality system which is applied to rehabilitation training was initiated a few years ago in foreign countries. In 2013, Mulroy S made a study about the effect of rehabilitation training of patients with shoulder joints in a virtual environment [8]. In 2014, Fischer H.C. conducted a preliminary study that stroke patients in the virtual environment assist themselves in training their hand [9]. These overseas studies are mainly designed for the single joint of the upper limb rehabilitation training and are also relatively little about together with shoulder and elbow joints rehabilitation training. Therefore, it is necessary to carry out research on shoulder and elbow joints rehabilitation training in this field.

This paper will introduce a wearable rehabilitation training system for the upper limb based on virtual reality, which has some advantages of simple structure and low price and can mobilize the initiative of patients to carry out rehabilitation training so that the rehabilitation effect of the upper limbs of the patients is obviously improved.

2   The Overall Design of the System

The main function of the wearable rehabilitation training system for the upper limb based on virtual reality is rehabilitation training for extension/flexion and abduction exercises of upper arms and the internal rotation/external rotation and flexion movements of the forearm. It is easy to improve the interests of the patients by adding the games in the training processes so that patients actively carry out rehabilitation training and will greatly improve the efficiency of the treatments. It can effectively address the problem of the medical resources shortage of upper limb rehabilitation field.

2.1   Overall Scheme of the System

The system is composed of two parts: the upper limb posture detection system and the virtual rehabilitation training scene, as shown in Fig. 1. The upper limb posture detection system based on six-axis IMU sensor is worn on the patient's arm. The angle measured by six-axis IMU sensor is changed according to the attitude of the arm, and the angle value is transmitted to PC by wireless Bluetooth. The system uses Visual Studio programming platform to collect the angle value of rotation and swing of the upper arm and the forearm. The virtual rehabilitation training scene achieves human-computer interaction by the right rehabilitation game designed with Unity3D game engine and allows patients to take more active rehabilitation training. At the same time, it can display the changes of the parameters of the patients in real time in the scene and show the state of rehabilitation training of the patients more clearly, which forms the basis for the training effect assessment for the physical therapist.

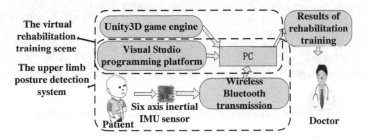

Fig. 1. Schematic diagram of wearable rehabilitation training system for the upper limb based on virtual reality

2.2   Design and Analysis of Upper Limb Posture Detection System

According to the analysis of the movement of the human upper limb shoulder and elbow based on The Physiology of the Joints written by Dr. Adalbert Kapandji, the different game scenes are designed from easy to difficult level and from single action to multiple action combination training in the system. The game scene is designed for the extension/flexion $(-50°{\sim}0°{\sim}180°)$ and abduction $(0°{\sim}180°)$ of the upper arm and the internal rotation/external rotation $(-90°{\sim}0°{\sim}90°)$ and flexion $(0°{\sim}145°)$ of the patient's the forearm [10]. There are two six-axis IMU sensors are placed on the patient's the upper arm and the forearm, as shown in Fig. 2 (the coordinates defined as that of Fig. 3). Wherein, the upper and lower swing angle of the upper arm/forearm is collected by the rotation of the X1/X2 axis of the sensor on the upper arm, and the left and right swing angle of the upper arm/forearm is collected by the Z1/Z2 axis of the sensor on the upper arm, and the rotation angle of the upper arm/forearm is collected by the Y1/Y2 axis of the sensor on the upper arm.

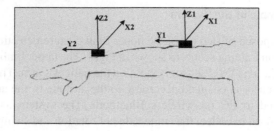

Fig. 2. Schematic diagram of two six-axis IMU sensors on the upper limb

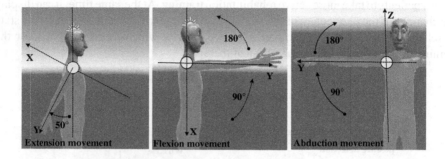

(a) Schematic diagram of the upper arm's single training action

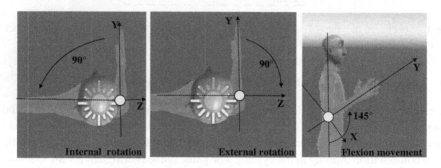

(b) Schematic diagram of the forearm's single training action

Fig. 3. Schematic diagram of the upper limb single training

The single patient training action is shown in Fig. 3(a) and (b) (The coordinates in the Fig. 2 are the coordinates of the corresponding sensors). The patient's upper limb posture in the side view of Fig. 3(a) is the position of the upper arm extension/flexion movement at the angle of 0 degrees. The patient's upper limb posture in the forward view of Fig. 3(a) is the position of the upper arm abduction movement at the angle of 0 degrees. The patient's upper limb posture in the top view of Fig. 3(b) is the position of the forearm's internal/external rotation at the angle of 0 degrees. The patient's upper limb posture in the side view of Fig. 3(b) is the position of the forearm flexion movement at the angle of 0°.

3   Design and Analysis of Virtual Rehabilitation Training Scene

Compared with the traditional upper limb rehabilitation therapy, the design of virtual rehabilitation scene based on Unity3D engine is added in the system. It can give patients a real interactive experience so that patients unknowingly immersed in training. At the same time, it can display the movement angle of the patient's joints in all directions in time, which is convenient to correct the posture of the patients during the rehabilitation training, so that the patients can achieve the effect of rehabilitation treatment in playing the game [11]. The soft is including two parts: (1) the 2D rehabilitation training scene is aimed at the extension/flexion and abduction of the upper arm; (2) the 2D rehabilitation training scene is aimed at the internal rotation/external rotation and flexion action of the forearm.

3.1   Upper Arm Rehabilitation Training Scene

The rehabilitation of the upper arm in the virtual game scene is a fishing game for big fish eating small fish. Depending on the severity and the rehabilitation of the patient's condition, four-game scenes were designed. The movement angle of the patient is displayed in real time in the scene, which is convenient for the medical staff to observe and correct the patient's rehabilitation, thus the patient can feel relaxed and the effect of rehabilitation training will be more obvious. Four upper arm rehabilitation game scenes are shown in Fig. 4.

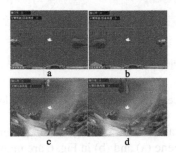

Fig. 4. The game scene of the upper arm's rehabilitation (Color figure online)

As shown in Fig. 5, rehabilitation training for the patient's arm is mainly divided into two parts. The scene a and b in Fig. 4 are mainly aimed at the extension/flexion movement of the upper arm. With the swing of the patient's the upper arm, the red fish will move around to eat the still fish. The scene c and d in Fig. 4 are mainly aimed at the abduction of the upper arm. As the patient's the upper arm swing up and down, the red fish moves up and down to eat the still fish.

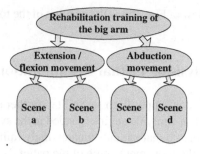

Fig. 5. Analysis of the upper arm's rehabilitation training scene

3.2 Forearm Rehabilitation Training Scene

The forearm rehabilitation scene is angry birds rescue partner. According to the different condition and rehabilitation of patient, the game is designed with four game scenes. The game is full of comfortable music for patients with physical and mental relaxation, which is more conducive to the rehabilitation of patients. Four forearm rehabilitation game scenes are shown in Fig. 6.

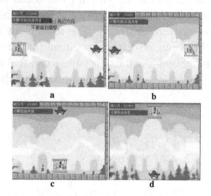

Fig. 6. Game scene for forearm rehabilitation (Color figure online)

As shown in Fig. 7, rehabilitation training for the patient's the forearm is divided into two major parts. The scene (a) and (b) in Fig. 6 are mainly aimed at the training of the internal/external rotation of the forearm. With the patient's the forearm swinging, the red bird moves around to save the bird in the cage. The scene (c) and (d) in Fig. 6 are mainly aimed at the training of the forearm flexion. With the patient's arm swings up and down, the red bird moves up and down to rescue the caged bird.

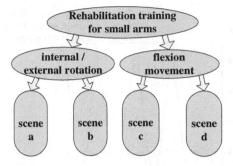

Fig. 7. Analysis of the forearm's rehabilitation training scene

4 Experiment and Data Analysis

In order to make patients more comfortable during training with the rehabilitation system, it is allowed that patients sit in front of the virtual scene. Then, the rehabilitation training experiment was carried out for extension/flexion and abduction of the upper arm and the internal rotation/external rotation and flexion of the forearm. The system is mainly used for the training of joint movement, so the rehabilitation posture of the patients is not evaluated. The initial zero coordinate of the sensor is fixed in the absolute coordinate system. In this case, the initial value of the sensor is cleared in the programming so that the synchronization and accuracy of the patient's movement and the virtual scene can still be guaranteed when the patient is wearing in any direction or not accurate enough.

4.1 Experiment and Data Analysis of the Patient's Upper Arm

Seen in Fig. 8(a) and (b), the curves indicate the relationship between time and movement angles when the upper arm is extension/flexion and abduction in the virtual environment. According to the curves, it is clear that the extension/flexion angle range of the upper arm is $-50°{\sim}0°{\sim}180°$ and the abduction angle is in the range of $0°{\sim}180°$ in the process of rehabilitation exercise training. Therefore, the patient's arm extension/flexion and abduction angle also achieved the desired effect, which is further proved that the system is suitable for upper limb rehabilitation training.

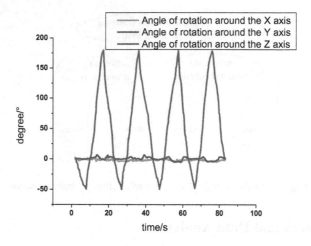

(a) Extension / flexion angle

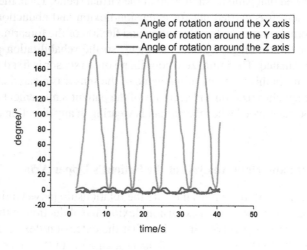

(b) Abduction angle

Fig. 8. Curves of the movement angle and time in the rehabilitation training of the upper arm

4.2 Experiment and Data Analysis of the Patient's Forearm

Based on the collection of the angle of internal rotation and external rotation and flexion exercise training of the forearm in virtual rehabilitation training, the relationship between the angle of the corresponding training and the time was drawn. According to the data curves in Fig. 9(a) and (b), it can be seen that the angles of the internal/external rotation of the patient's the forearm are in the range of −90°~0°~90° and the angle of the forearm's flexion movement swings in the range of 0°~145°. The training action of the forearm reaches the corresponding standard angles.

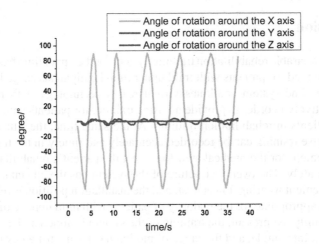

(a) Angle of the internal rotation and external rotation

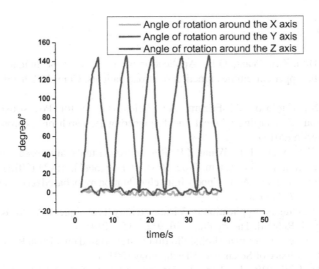

(b) Angle of flexion

Fig. 9. Curves of the movement angle and time in the rehabilitation training of the forearm

Through the above experimental data, we can analysis that the training range of the upper arm and the forearm of the patient have reached the range of the corresponding joint in the reference book of "Physiology of the Joints". From the experimental data in the Figs. 8 and 9, it can be seen that the stability and accuracy of the angle in each training action acquisition are relatively high in the rehabilitation training of the patient's arm and the forearms.

5   Conclusion

In this paper, a wearable rehabilitation training system for the upper limb based on virtual reality was designed for patients with mild upper limb hemiparesis caused by stroke or traffic accident. The system combines virtual reality technology with rehabilitation medicine effectively in order to provide a virtual training for patients and increased the interests of patients in rehabilitation training. At the same time, the situation of rehabilitation exercise training can be recorded accurately and timely in the training scene, which is convenient for the medical staff to correct the patient's rehabilitation training posture immediately. The overall structure of the system has the advantages of simple structure, convenient wearing, lower cost, and the installation position is relatively free. Therefore, it is appropriate for the majority of patients with mild upper extremity hemiplegia in the family. At present, the training of the system is aimed at the rehabilitation training of the joints' angle, and the range of motion has been verified by experiments. The function of the system is also extended so that the design of the wrist and hand training can be added in the later stage in order to make the upper limb active rehabilitation training system more perfect.

References

1. Liang, M., Dou, Z.L., Wang, Q.H.: Application of virtual reality technique in rehabilitation of hemiplegic upper extremities function of stroke patients. Chin. J. Rehabil. Med. **02**, 114–118 (2013)
2. Valencia, N., Cardoso, V., Frizera, A.: Serious Game for Post-stroke Upper Limb Rehabilitation. Converging Clinical and Engineering Research on Neurorehabilitation II. Springer, Berlin (2017)
3. Lei, Y., Yu, H.L., Wang, L.L., Wang, Z.P.: Research on virtual reality-based interactive upper-limb rehabilitation training system. Prog. Biomed. Eng. **36**(1), 21–24 (2015)
4. Xu, B.G., Peng, S., Song, A.G.: Upper-limb rehabilitation robot based on motor imagery EEG. Robot **33**(3), 307–313 (2011)
5. Wang, H.T.: Status of Application of Virtual Reality Technique in Motor Rehabilitation in Stroke. Chin. J. Rehabil. Theory Pract. **10**, 911–915 (2014)
6. Zhang, J.L.: Research of Finger Rehabilitation System Based on Virtual Reality Technology. Huazhong University of Science and Technology (2012)
7. Mei, Z., He, L.W., Wu, L., Jian, Z.: Design and test of a portable exoskeleton elbow rehabilitation training device. Chin. J. Rehabil. Med. **11**, 1155–1157 (2015)
8. Mazzone, B., Haubert, L.L., Mulroy, S.: Intensity of shoulder muscle activation during resistive exercises performed with and without virtual reality games. In: International Conference on Virtual Rehabilitation, pp. 127–133. IEEE (2013)
9. Fischer, H.C., Stubblefield, K., Kline, T.: Hand rehabilitation following stroke: a pilot study of assisted finger extension training in a virtual environment. Topics Stroke Rehabil. **14**(1), 1–12 (2014)

10. Kapandji, A.I.: The Physiology of the Joints, 6th edn. People's Military Medical Press, Beijing (2011)
11. Gu, Y., Tian, L.H., Chen, H.: Application of virtual reality training system and rehabilitation therapy in the treatment in hemiplegic patients with upper limb dysfunction. Chin. J. Rehabil. Med. 26(6), 579–581 (2011)

Active Gait Rehabilitation Training System Based on Virtual Reality

Bingjing Guo[1,2(✉)] ⓘ, Wenxiao Li[1], Jianhai Han[1,2], Xiangpan Li[1,2], and Yongfei Mao[1]

[1] School of Mechatronics Engineering, Henan University of Science and Technology,
Luoyang 471003, China
bingjing@haust.edu.cn
[2] Henan Provincial Key Laboratory of Robotics and Intelligent Systems, Luoyang 471003, China

Abstract. Aiming at improving the low participation and inactive motion intention of patients in traditional gait rehabilitation training, an active gait rehabilitation training system is designed based on Virtual Reality (VR) technology. This project focuses on the design of the gait parameters real-time detecting system and the virtual reality rehabilitation training scene system. Based on an analysis of gait rehabilitation medicine theory, the lower-limb joints range of motion (ROM) and the plantar pressure of the affected limb are selected as the important gait parameters, thus a built-in sensing system is constructed with three inertial measurement units (IMU) and the multi-point force sensing resistors (FSR). Through the wireless Bluetooth communication interface, the lower-limb motor parameters of patients are transmitted into the virtual training games as the motion control signals for character driven in games and the scientific evidence for rehabilitation assessment. Error analysis and compensation method of sampled data are elaborated in this paper. The experiments are carried out about data acquisition, man-machine interaction, and functions of the rehabilitation training scenes. The results show that the active rehabilitation training system is able to assist patients with real-time interaction and immersive sensing and provides better visual feedback information to patients. It improves the training initiative as well as provides an effective means for nerve remodeling.

Keywords: Virtual reality · Active gait rehabilitation · Wearable sensing system · Joint range of motion · Plantar pressure

1 Introduction

With the aggravation of the aging society, people with lower-limb dysfunction caused by nerve injury grow with years. This motor deficit seriously affects the patient's quality of life and brings a heavy burden on family and society [1, 2]. The traditional rehabilitation training is often complied with the physiotherapist's manual assistance, which is often the low efficiency and tedious process [3]. This physiotherapist has usually been conducted on limited populations, with little objective measurement of its effects on outcomes, and with few standardized assessments over the course of rehabilitation.

Based on the neural plasticity and kinesiotherapy theory, rich training environment can promote the process of rehabilitation [4]. Therefore, the establishment of rich virtual rehabilitation scenes with VR technology is an effectively method to improve the patient's training interest, enthusiasm, thereby the efficiency of rehabilitation. Haptic

© Springer International Publishing AG 2017
Y. Huang et al. (Eds.): ICIRA 2017, Part III, LNAI 10464, pp. 506–516, 2017.
DOI: 10.1007/978-3-319-65298-6_46

Walker robot system [5] simulates a variety of training scenarios, such as walking on the ground, up and down stairs, which are developed by the Furlong Hofer Institute in Germany. Multi-degree of freedom lower limb trainer provides a real-life street scene for patients in rehabilitation training, which is presented by the McGill University in Canada. A virtual scene of aircraft flight is integrated into the active and passive training ankle rehabilitation system of the Hebei University of Technology. The researchers developed different kinds of VR-based training systems and used for the stroke patients with hemiparesis in different stages. The results demonstrated that compared with the routine rehabilitation, VR-based training paradigms contributed in motor function and activities of daily living. Overall, these studies had important clinical implications for the development of future VR training protocols.

However, VR-based training system usually combines with robot and/or partial body weight-supported treadmill training equipment in medical institutions. In the case of Rutgers Ankle, the ankle joint motion controls the driving of the virtual plane and the small boat to complete the rehabilitation training. Most virtual reality research facilities are complex, expensive, and require specialized assistance of physiotherapists [6]. This situation is not conducive to the patient's autonomy training, especially for the patient who leaves hospital and lives at home or lives in a remote rural area, due to the lack of family portable intelligent rehabilitation training equipment.

As discussed above, meeting with the rehabilitation therapy requirements of family and individuals, we designed a portable wearable lower-limb active rehabilitation training equipment based on VR technology, which administers to repair the damaged neural pathways by repetitive motor exercise training and the visual feedback from the virtual scene. This active training system can effectively improve the immersion, interest, and initiative of patients in rehabilitation training to achieve the goal of remodeling the motor function.

2   Overall Scheme of Active Gait Rehabilitation Training System

From the perspective of modern nerve injury rehabilitation theory, the repeated random exercise training for the limbs with motor dysfunction can cause the corresponding cortical area to expand and improve the transmission efficiency of neural circuits [7].

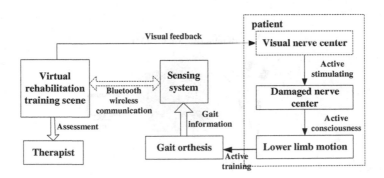

Fig. 1. Schematic diagram of the system

Therefore, limb motor training is conducive to the reconstruction of neural circuits. Active rehabilitation training mode provides appreciate damping force for patients to strenuously overcome. During the training, they have a certain degree of independent participation. It is helpful for the physical sensation recovery and the stability of hip, knee and ankle joint being enhanced. Then the physical strength and endurance are gradually strengthened. Combined with the correct and repeated visual feedback, the brain cortex can be stimulated to expand related brain activity area, so as to form the corresponding conditioned reflex and establish the normal movement pattern [8].

The active gait rehabilitation training system consists of four parts: a joint orthosis, an embedded sensing system, a Bluetooth wireless communication interface and a virtual rehabilitation training scene system, as shown in Fig. 1. The joint orthosis can perform a single joint movement or multi-joint complex movement for hip, knee, and ankle joint. The sensor system is embedded in the orthosis, which integrates microcontroller unit (MCU), inertial measurement unit (IMU), and force sensing resistor (FSR). The joint angles of the lower-limb and the plantar pressures are sampled through the analog-digital converter built in MCU.

Virtual rehabilitation training scenes are programmed with Unity3D software platform on PC. In order to meet the portable requirements, the sampled gait parameters are transferred from the sensing system to the PC through the wireless Bluetooth module, and power module adopts mini mobile power supply to the system. The patient actives the gait orthosis to provide motion control signals for the game characters in the scene to walk, rotate and other actions. Meanwhile, the patient acquires the visual feedback information, so as to maintain or adjust the gait motion in order to achieve the best training effect. The patient and the therapist are able to aware of the gait assessment results promoted by the rehabilitation evaluation module in real time. The module is developed with the Unity3D plug-in Next-Gen UI kit.

3   The Construction of Sensing System

The sensing system is integrated as a whole with MCU, IMU, FSR, and Bluetooth modular, as showed in Fig. 2. There are three IMUs installed on the thigh, calf, and ankle

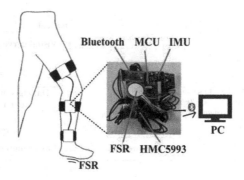

Fig. 2. Hardware composition of wearable sensing system

of the ipsilateral affected lower-limb. They are used to capture lower extremity kinematics information [9, 10]. Four FSRs are embedded in the insole to capture contact information between the plantar and the ground. The motion data of the lower-limb are sent to the scene by wireless Bluetooth to match the characters and to determine the situation of rehabilitation.

3.1 Joint Angle Acquisition

The hip, knee and ankle joints of the lower-limb have their own range of physiological activities [11], as summarized in Table 1. These ranges are the basis of lower-limb rehabilitation training equipment design, joint angle sensor selection, and virtual scene planning.

Table 1. The physiological range of the joints of the lower limbs

Joint name	Movement	Joint activity angle
Hip	Bend, stretch	$-120° \sim 65°$
	Adduction, abduction	$(-30°\sim -35°) \sim 40°$
	Internal rotation, external rotation	$(-15°\sim-30°) \sim 60°$
Knee	Bend, stretch	$(-120°\sim -160°) \sim 0°$
Ankle	Dorsiflexion, plantar flexion	$-20° \sim (40°\sim 50°)$
	Internal rotation, external rotation	$-15° \sim (30°\sim 50°)$
	Varus, eversion	$(-30°\sim -35°) \sim (15°\sim 20°)$

Human motion coordinate system is defined in Fig. 3. IMU (MPU6050) is an attitude sensor to measure joint angles in this system. If the sensor axes are placed parallel to the human coordinate (o-xyz), the roll angle is defined as the rotation of the sagittal axis, the pitch angle of the coronal axis and the yaw angle of the vertical axis (Fig. 3). Three single-axis accelerometers and three gyroscopes are integrated into the IMU to measure the angular velocity and acceleration in a three-dimensional space. Then the posture of

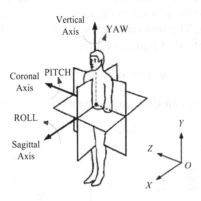

Fig. 3. Definition of joint angles

the object can be calculated by the algorithm. The accelerometer detects the acceleration signal in the direction of gravity. Detection requires a reference value, so the gyroscope can accurately correct the roll and pitch angle, but the yaw angle cannot be amended. Therefore a magnetometer HMC5883 is added to get the module horizontal rotation angle based on the data fusion.

3.2   Measurement Error Analysis of IMU

Because of the inherent error and installation error of IMU, it is necessary to carry out error analysis through experiment. The bend motion of the knee causes the maximum joint angle range in Table 1, so the pitch angle of the knee is selected as an example to illustrate the error analysis and compensation. According to the human lower-limb joint angle range, the pitch angle changes from 0° to 150°.

From Figs. 4 and 5, it can be seen that the pitch angle is within the operating range of [0°, 150°] and the error range of (–3°, 2°). The error is relatively larger than the accuracy of joint angle in rehabilitation training.

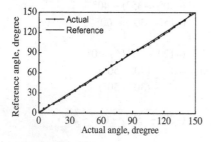

Fig. 4. Curves of the pitch angle of knee   **Fig. 5.** Error curve of the pitch angle of knee

3.3   Error Compensation

The polynomial fitting method is used to calibrate and compensate the error, so as to reduce the measurement error of the IMU unit. According to the least squares fitting principle, the measured value is used as the independent variable and the actual value of the dependent variable. The two variables are linearly fitted.

Linear fitting model of Least Squares Fitting is

$$\varphi(x) = a_0 + a_1 x \tag{1}$$

According to m pairs of raw data, a corresponding fitting matrix can be formed.

$$Ga = y \tag{2}$$

where $a = (a_o, a_1)^T$, $y = (\sum_{i=0}^{m} y_i, \sum_{i=0}^{m} x_i y_i)^T$, $G = \begin{bmatrix} m+1 & \sum_{i=0}^{m} x_i \\ \sum_{i=0}^{m} x_i & \sum_{i=0}^{m} x_i^2 \end{bmatrix}$

In Fig. 5, measurement error varies in different intervals, so the data is dealt with piecewise fitting. The pitch angle is divided into five intervals: [0°, 10°], (10°, 60°], (60°, 90°], (90°, 140°], (140°, 150°].

$$y = \begin{cases} 1.0362x - 1.4762, x \in [0°, 10°] \\ 1.0091x + 1.325, x \in (10°, 60°] \\ 0.9521x + 3.302, x \in (60°, 90°] \\ 0.9894x + 3.0812, x \in (90°, 140°] \\ 1.0157x - 3.8779, x \in (140°, 150°] \end{cases} \tag{3}$$

where x is the raw angle value and y is the fitting value.

The angular error distribution of the fitting pitch angle is shown in Fig. 6. It can be seen that the pitch angle error after error compensation reduces to $(-1°, 1°)$.

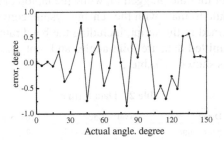

Fig. 6. Error curve of pitch angle after fitting

3.4 Plantar Pressure Data Acquisition

For the leg lesions, the plantar pressure will be a corresponding change [12]. By detecting and analyzing changes and distributions in plantar pressure, we can obtain some of the body's physiological or pathological information to provide an important reference for the identification of gait obstruction. The thin film pressure sensor is adopted in the system, the production of Interlink Electronics Company. This sensor is made of an ultra-thin flexible material, suitable for flat acquisition, widely used to measure the point, line or surface contact pressure distribution.

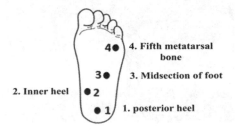

Fig. 7. Distribution of plantar pressure sensors

According to the distribution characteristics of plantar pressure in hemiplegic patients, four sensors were arranged in the posterior heel, inner heel, midsection of the foot, and the fifth metatarsal bone, as the feature markers [13], as shown in Fig. 7.

3.5 Data Wireless Transmission

Taking into account a large number of sensors in the system, as well as the convenience of walking in gait training, the wireless communication is applied for data transmission [14]. Wireless data transmission makes use of virtual rehabilitation system flexibly. Patients with lower-limb disorders can get rid of the shackles of data transmission lines, so that the virtual rehabilitation more convenient and easy to configure.

SPP-CA Bluetooth module is selected with the advantages of low power consumption, small size, and high sensitivity of transceiver data. The module communicates with the single chip of the built-in serial interface. The PC is paired with the sensing system by the Bluetooth adapter module plugged in, while the laptop or mobile phone uses the built-in Bluetooth module to mate with the sensing system conveniently. The communication protocol is defined in the script, including the baud rate, data format and data check mode. The transmitted data are joint angles and foot pressures. Each is 8 bytes long. The data format is shown in Table 2.

Table 2. Data format

Start	Data1	Interval	Data2	Interval	Data3	Interval	Data4	Stop
!	Hip angles	#	Knee angles	#	Ankle angles	#	pressures	*

4 Virtual Rehabilitation Training Scene

The gait rehabilitation virtual environment is allowed for adaptation to a wide range of patients with respect to different hip, knee and ankle training demands. The development of virtual reality scene, including the terrain, the surrounding environment, the role controller and the driven script, is based on the powerful and multi-platform releasing Unity3D software [15]. The first scene is designed for the patients to perform single joint training or the walking training preparatory exercises (Fig. 8). The muscles to be trained are selected according to the corresponding action in Table 3.

Fig. 8. Virtual training scene 1

Table 3. Rehabilitation training targeted muscle

Muscle position	Human action
Upper of anterolateral thigh	Thigh bending
Front side of thigh	Hip or knee joint stretching
Back side of thigh	Calf bending
Front side of calf	Ankle Stretching, foot varus
Back side of calf	Ankle plantar flexion, knee flexion

When a patient wears the joint orthosis, he repeats the joint rehabilitation training in accordance with the setting angle range. The character in the scene 1 will follow the movements synchronously and the training data show on the interface immediately.

The second scene is designed for patients to perform walking training as shown in Fig. 9. Patients are divided into different rehabilitation grades according to Bruustrom's six stage evaluations of hemiplegia recovery [16]. Patients or therapists choose the difficulty level of the game to ensure the suitability of rehabilitation training. The easy level scene is for patients with lower extremity paralysis of I~III. The scene is designed with a flat road, where the patient walks, as showed in Fig. 9(a). The scene is also accompanied by bird calls, the sound of running water, warning tone and other auditory feedback to give the patients a pleasant feeling. The difficulty level scene is for the patients with lower extremity paralysis of IV~VII. The scene creates with a realistic forest atmosphere. The patient's way forward needs to avoid all kinds of obstacles, such as hills, ditches, and trees, sometimes he needs making a turn, as showed in Fig. 9(b). The attitude detection unit IMU collects the

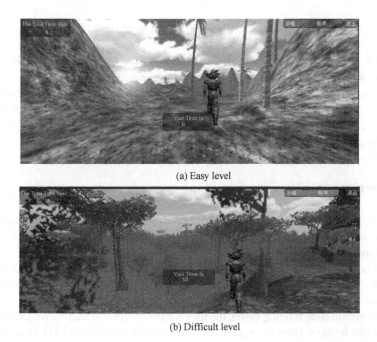

(a) Easy level

(b) Difficult level

Fig. 9. Virtual training scenes of different levels (a) Easy level (b) Difficult level

Fig. 10. Experimental system for the knee joint

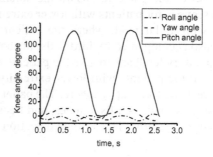

Fig. 11. Curves of knee joint angle

patient's lower-limb posture, and then the data is sent by the wireless Bluetooth to the scene, as the parameters to control the game.

Built-in rehabilitation evaluation system is constructed according to the completion time of the games, joint range of motion and the plantar pressure value in the training. In addition, the scene can display gait parameters in real time, the trainer knows his gait by watching the data and the game character movement, so as to adjust and correct the action. This visual feedback is helpful to the never plasticity.

5   Experimental Verification

At present, there is no clinical trial for the experimental prototype, so the feasibility of the whole system design is verified by the experiments with healthy people. The prototype is shown in Fig. 10.

The flexion and extension angle of the knee is the pitch angle around the coronal axis, as defined in Fig. 3. When the rehabilitation game is running, the detection curve of the knee is shown in Fig. 11.

In the rehabilitation training game, the user's knee movement angle setting range is [0°, 120°]. As can be seen from the curve, IMU obtained the measured PITCH angle corresponds to the setting range. The error value is within the range of [−0.6°, +0.8°] to verify the

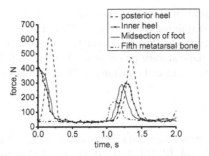

Fig. 12. Curves of plantar pressure

effectiveness of the error compensation method. The gait cycle of the experimenter is about 1.4 s. As a result of the knee joint shake, roll and yaw angle are also acquired.

In Fig. 12, it can be seen that the test values of the plantar pressure correspond to the force points where the four FSRs arranged in the shoes. And the curves correctly reflect the normal human gait when he walks. The support phase is estimated at about 58% as well as the swing phase of 42%. Through the test of the active gait rehabilitation training system, the accuracy of rehabilitation data acquisition and the validity of gait parameter detection in the training process are verified.

Since the entire system has not yet become a product, no patient clinical trials have been carried out. If the experimental subject is the patient, during the VR rehabilitation game running, his or her joint mobility curve can be obtained as that of Fig. 11 and compared with the set point to evaluate the rehabilitation of the motion of the joint. The gait cycle and gait rehabilitation status were also assessed by plantar pressure test as that of Fig. 12. The character in the game can reflect and match with the movement of patients while the virtual scene of the rehabilitation process runs continuously and smoothly. During the experiments, the experimenter concentrates on the virtual scene and the task pattern, which effectively increases the interest of rehabilitation and the initiative of training.

6   Conclusion

A wearable sensing system and an active rehabilitation system based on VR technology are elaborated in this paper, with the merits of portability, simple operation, low cost, and immersion experience. The sensing system includes MCU, IMU and foot pressure sensors to collect posture information of patients and send them to the virtual scene wirelessly to control the motion of the character in the game. The sensor error is compensated by the polynomial fitting algorithm to improve accuracy. Real-time detection of the joint range of motion and plantar pressure provides an effective basis for gait analysis and achieves the assessment of rehabilitation. Virtual reality technology provides real-time audio-visual feedback, giving a good incentive for patients to increase their attention, initiative, and enthusiasm.

The gait information obtained by the system can be used to analyze the patient's gait cycle, gait phase switching time, joint range of motion, and foot pedal force in walking. Accordingly, the system provides support for gait analysis for further clinical rehabilitation.

Acknowledgments. This study was supported by the "Research on key generic technologies of pneumatic gait rehabilitation training robot" project (172102210036) granted from "Project of science and technology of the Henan Province", and the "Research on bionic driving mechanism and control strategy of non-skeletal waist power assisted robot" project (162300410082) granted from "Program for the Natural Science Foundation of Henan Province".

References

1. Kawamoto, H.: Pilot study of locomotion improvement using hybrid assistive limb in chronic stroke patients. BMC Neurol. **13**, 141–148 (2013). doi:10.1186/1471-2377-13-141
2. Wang, J.: Application and research progress of lower limb rehabilitation robot in stroke patients with walking disorder. Chin. J. Rehabil. Med. **29**(8), 784–788 (2014). doi:10.3969/j.issn.1001-1242.2014.08.025
3. Chen, B.: Recent developments and challenges of lower extremity exoskeletons. J. Orthop. Transl. **5**, 26–37 (2016). doi:10.1016/j.jot.2015.09.007
4. Guo, X.H.: Active and passive training system of lower limb rehabilitation based on virtual reality. J. Xi'an Jiaotong Univ. **50**(2), 124–131 (2016). doi:10.7652/xjtuxb201602021
5. Meng, W.: Recent development of mechanisms and control strategies for robot-assisted lower limb rehabilitation. Mechatronics **31**, 132–145 (2015). doi:10.1016/j.mechatronics.2015.04.005
6. Liu, R.S.: Design and simulation of a lower limb rehabilitative robot. China Mech. Eng. **27**(20), 2722–2727 (2016). doi:10.3969/j.issn.1004-132X.2016.20.005
7. Liu, H.L.: Effect of multi-position lower limb rehabilitation robot on motor function in stroke patients with hemiplegia. Chin. J. Rehabil. Theory Pract. **19**(8), 722–724 (2013). doi:10.3969/j.issn.1006-9771.2013.08.003
8. Li, J.Q.: Research on exoskeleton remote rehabilitation system based on virtual reality technology. Mach. Des. Res. **27**(4), 35–38 (2011)
9. Hassan, M.: Wearable gait measurement system with an instrumented cane for exoskeleton control. Sensors **14**(1), 1705–1722 (2014). doi:10.3390/s140101705
10. Bergmann, J.H.: A portable system for collecting anatomical joint angles during stair ascent: a comparison with an optical tracking device. Dyn. Med. **8**(3), 1–7 (2009). doi:10.1186/1476-5918-8-3
11. Zhang, J.F., Chen, Y., Yang, C.J.: Flexible Exoskeleton Human-Robot Intelligent System, 1st edn. Science Press, Beijing (2011)
12. Deng, X.: Wearable plantar pressure detecting system based on FSR. Transducer Microsyst. Technol. **32**(2), 81–86 (2013). doi:10.3969/j.issn.1000-9787.2013.02.024
13. Cao, H.: Structure optimization analysis for exoskeleton foot. J. Eng. Des. **17**(1), 35–39 (2010). doi:10.3785/j.issn.1006-754X.2010.01.006
14. Jin, Z.H.: Design of intelligent measurement and control system based on Bluetooth wireless interface. J. Donghua Univ. **30**(1), 72–75 (2004). doi:10.3969/j.issn.1671-0444.2004.01.017
15. Vahe, K.: Introduction to Game Programming: Using C# and Unity 3D, 1st edn. Noorcon Inc., Los Angeles (2016)
16. Brunnstrom, S.: Motor testing procedures in hemiplegia: based on sequential recovery stages. Phys. Ther. **46**(4), 357–375 (1966)

A Realtime Object Detection Algorithm Based on Limited Computing Resource

Fei Liu[✉], Yanbin Wang, Yimin Wei, Chunxue Li, and Li Tang

Ludong University, Yantai, Shandong, China
liufeildu@163.com

Abstract. Aiming at the problem of realtime object detection in humanoid robot, this paper presents a method that combines the Support Vector Machine (SVM) classifier with the Histograms of Oriented Gradients (HOG) feature of local area in the image. In order to reduce time consumption, the paper uses one image segmentation and scan policy to get candidate local area in the image. To evaluate the proposed method, we have established a data set of black and white ball images from RoboCup SPL games. Experimental results have demonstrated that recognition efficiency has been improved greatly and the algorithm can be executed in realtime on the NAO robot that has only a 1.6 GHz CPU.

Keywords: Object detection · HOG · SVM · Limited computing

1 Introduction

Reliable object detection in static image is essential to image understanding [1]. Feature extraction and classifier designing are two key steps. HOG [2] plus SVM [3] is one of the most successful detection algorithms, especially in human detection field [1,2,4–8]. But it is too time-consuming to be applied in realtime object detection because of complex environment and the limited computing hardware. For example, in RoboCup SPL competition, robots are required to identify the black and white ball in a very short time and under limited computing power (1.6 GHz CPU). In this area people have made many researches [9–13] and have got great progress. For instance, NAO Devils team [9] proposed a feature matching method to find the black area of the image and then to identify whether it is consistent with the characteristics of the ball. TJArk team [10] proposed a method to fit the candidate region by the 24 boundary points firstly, and then use the Otsu algorithm [14] to identify the characteristics of the ball. However, the improvement of recognition efficiency is still difficult to achieve.

In order to reduce time consumption, the paper firstly uses one image segmentation and scan policy to get candidate local area in the image, and then presents a method that combines the SVM classifier with the HOG feature of candidate local area in the image. The experimental results prove that recognition efficiency has been improved greatly and the algorithm can be executed in realtime on the NAO robot that has only a 1.6 GHz CPU.

© Springer International Publishing AG 2017
Y. Huang et al. (Eds.): ICIRA 2017, Part III, LNAI 10464, pp. 517–523, 2017.
DOI: 10.1007/978-3-319-65298-6_47

The remainder of this paper is organized as follows: in Sect. 2, the proposed approach is discussed. The experimental results are presented in Sect. 3. Finally, we provide a brief summary in Sect. 4.

2   Proposed Approach

Our method is divided into three phases. In the first phase, the algorithm scans the image from upper or lower camera of the robot and locates the non-green regions because the filed's carpet is green but the ball is black and white. And then it computes the radius and center of the region. If the computed radius is greater than the minimum acceptable radius, the center is regarded as one candidate ball point marked by one black cross in Fig. 1. In the second phase, the algorithm uses all of the candidate ball points in the first phase to compute the expected radius, fits the rectangle from the ball center in the expected radius and then chooses the rectangle as the candidate region corresponding to the two black rectangles in Fig. 2. The last phase is to select the real ball (see the green circle in Fig. 3) from candidate regions using some classification algorithms.

Fig. 1. Finding the candidate ball points in the first phase of the ball detection process

In this phase, we use grayscale images of candidate regions to extract the HOG feature as the descriptor and introduce the SVM classifier to evaluate HOG of the candidate regions and improve ball recognition efficiency. The size of grayscale images of candidate regions is normalised as 32×32 pixels. The size of the detection window also is 32×32. Because the algorithms have found

Fig. 2. Fitting the candidate regions in the second phase of the ball detection process

Fig. 3. Selecting the real ball in the third phase of the ball detection process (Color figure online)

candidate regions in the second phase, it don't need to scan the whole image with sliding detection window and it can increase the efficiency of computing the HOG features. The first step of HOG extraction is to compute the gradient of every pixel as follows:

$$
\begin{aligned}
G_x(x,y) &= H(x+1,y) - H(x-1,y) \\
G_y(x,y) &= H(x,y+1) - H(x,y-1) \\
G(x,y) &= \sqrt{G_x(x,y)^2 + G_y(x,y)^2} \\
\alpha(x,y) &= \tan^{-1}\left(\frac{G_y(x,y)}{G_x(x,y)}\right)
\end{aligned}
\tag{1}
$$

where $G_x(x,y)$, $G_y(x,y)$, $H(x,y)$ respectively represent the horizontal gradient, vertical gradient and the value of the pixel (x,y) in the images. The gradient magnitude and orientation of the pixel (x,y) is $G(x,y)$ and $\alpha(x,y)$.

The second step of HOG extraction is that the input grayscale image is divided into 3×3 overlapping blocks (size: 16×16). Each block includes 4 cells and the size of every cell is 8×8 pixels. In each cell, the orientation histogram has 9 bins, which correspond to orientations $\frac{z\pi}{9}, z = 0,1,2,\ldots,8$. Thus each block contains 4×9 features and each candidate region contains $3 \times 3 \times 36 = 324$ features. Trilinear interpolation [4] is employed to compute HOG features in this step. Finally, a 324-dimensional vector is achieved and is used as the input data of SVM model.

Next we employ SVM algorithm from reference [3]. The first step is offline training. We divide the candidate ball regions into the positive samples and negative samples. In this phase, the images that include ball are regarded as the positive samples, while others are regarded as the negative samples. We extract HOG features from two type samples and train SVM model. Finallly, we get the SVM classifier as follows:

$$
\begin{aligned}
(x_i, y_i), i &= 1,2,\ldots,N \\
yi &\in \{+1, -1\} \\
f(x) &= w^T \cdot x + b \\
f(x_i) &\geq 0, y_i = +1 \\
f(x_i) &< 0, y_i = -1
\end{aligned}
\tag{2}
$$

where (x_i, y_i) is one sample and x_i is the vector of the HOG feature. $f(x)$ defines a hyperplane to divide all the samples into two categories. w and b are the tuned parameter set.

3 Experimental Results and Analysis

The proposed algorithms have been evaluated on our own data set. Our own ball data set has been collected with upper and lower camera installed above the robot's head. The positive images are shown in Fig. 4. Negative training images are shown in Fig. 5. The size of every image is 32×32 pixels. In our experiment, the testing data set includes 2502 negative samples and 9242 positive samples.

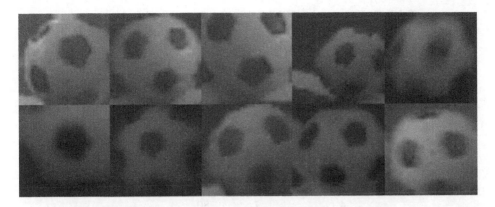

Fig. 4. Ten examples of positive training images of our data set

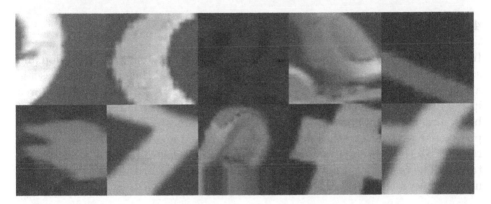

Fig. 5. Ten examples of negative training images of our data set

The experimental result of using the trained SVM classifier to filter out the candidate regions is obtained as Table 1. We adopted recall, precision and accuracy [15] to measure the performances of the proposed method. According to their definitions and Table 1, we can get that the recall is 89%, the precision is 94.6% and the accuracy is 87.4%. It takes 3–4 ms in every detection and our program runs at 60 Hz, so it can analyze the pictures in real-time without losing any images.

We define the recognition rate as the number of true positives (i.e. they are correctly labeled as belonging to the positive class) divided by the total number of candidate regions. We adopt recognition rate to compare the performances of the proposed method with CNS approach [13] on data set of the images taken by the robot from a range of distances. We compare the results of two experiments as Table 2. From Table 2, it is noted that the recognition rate of the CNS approach will drop rapidly as the distance increases, but the recognition rate of SVM classifier remains relatively stable, especially for a long distant target.

Table 1. The result of SVM classifier on our testing data set

Test samples	True	False	Total
Positive	8225	1017	9242
Negative	466	2036	2502

Table 2. Accuracy of CNS and SVM

Distance (m)	CNS (%)	SVM (%)
1.5	90	87.4
3	32	87.5
4	7	80

4   Conclusions

In order to reduce time consumption, the paper uses one image segmentation and scan policy to get candidate local area in the image. And then the paper presents a method that combines the SVM classifier with the HOG feature of local area in the image. To evaluate the proposed method, we have established a data set of black and white ball images from RoboCup SPL games. The experimental results prove that recognition efficiency has been improved greatly, and the recognition rate of SVM classifier remains relatively stable, especially for a long distant target. The results also have demonstrated that the algorithm can be executed in realtime on the NAO robot that has only a 1.6 GHz CPU.

Acknowledgement. This work has been funded by Program for Shandong Science and Technology (2012YD03111), Laboratory of Robotics in Ludong University, Multi-agent Systems Laboratory and Information Center in the University of Science and Technology of China. In addition, we would like to thank Shu Li and other members from Tongji University because the initial idea was inspired from our discussion with them.

References

1. Pang, Y., Yuan, Y., Li, X., Pan, J.: Efficient HOG human detection. Signal Process. **91**, 773–781 (2011). doi:10.1016/j.sigpro.2010.08.010
2. Dalal, N., Triggs, B.: Histograms of oriented gradients for human detection. In: 2005 IEEE Computer Society Conference on Computer Vision and Pattern Recognition, vol. 1, pp. 886–893. IEEE, San Diego, CA, USA (2005). doi:10.1109/CVPR.2005. 177
3. Burges, C.J.: A tutorial on support vector machines for pattern recognition. Data Mining Knowl. Discov. **2**(2), 121–167 (1998). doi:10.1023/A:1009715923555
4. Dalal, N.: Finding people in images and videos. Institut National Polytechnique de Grenoble-INPG (2006). https://tel.archives-ouvertes.fr/tel-00390303/document

5. Wang, X., Han, T.X., Yan, S.: An HOG-LBP human detector with partial occlusion handling. In: 2009 IEEE 12th International Conference on Computer Vision, pp. 32–39. IEEE, Kyoto, Japan (2009). doi:10.1109/ICCV.2009.5459207
6. Zhu, Q., Yeh, M.-C., Cheng, K.-T., Avidan, S.: Fast human detection using a cascade of histograms of oriented gradients. In: 2006 IEEE Computer Society Conference on Computer Vision and Pattern Recognition, vol. 2, pp. 1491–1498. NY, USA (2006). doi:10.1109/CVPR.2006.119
7. Schwartz, W.R., Kembhavi, A., Harwood, D., Davis, L.S.: Human detection using partial least squares analysis. In: 2009 IEEE 12th International Conference on Computer Vision, pp. 24–31. IEEE, Kyoto, Japan (2009). doi:10.1109/ICCV.2009.5459205
8. Chen, Y.T., Chen, C.S.: Fast human detection using a novel boosted cascading structure with meta stages. IEEE Trans. Image Process. **17**(8), 1452–1464 (2008). doi:10.1109/TIP.2008.926152
9. Hofmann, M., Schwarz, I., Urbann, O., Rensen, F., Larisch, A., Moos, A., Hemmers, J.: NAO Devils Dortmund team report 2016. Technical report (2016). https://github.com/NaoDevils/CodeRelease2016/blob/master/TeamReportNaoDevils2016.pdf
10. Li, S., et al.: TJArk team description paper and research report 2016. Technical report (2002). https://github.com/TJArk-Robotics/coderelease_2016/blob/master/TJArk%20Team%20Research%20Report%202016.pdf
11. Müller, J., Frese, U., Röfer, T.: Grab a mug-object detection and grasp motion planning with the Nao robot. In: 12th IEEE-RAS International Conference on Humanoid Robots, pp. 349–356. IEEE, Osaka, Japan (2012). doi:10.1109/HUMANOIDS.2012.6651543
12. Hall, B., Harris, S., Hengst, B., Liu, R., Ng, K., Pagnucco, M., Pearson, L., Sammut, C., Schmidt, P.: RoboCup SPL 2015 champion team paper. In: Almeida, L., Ji, J., Steinbauer, G., Luke, S. (eds.) RoboCup 2015. LNCS (LNAI), vol. 9513, pp. 72–82. Springer, Cham (2015). doi:10.1007/978-3-319-29339-4_6
13. Röfer, T., Laue, T., Kuball, J., Lübken, A., Maaß, F., Müller, J., Post, L., Richter-Klug, J., Schulz, P., Stolpmann, A., Stöwing, A., Thielke, F.: B-Human team report and code release 2016. Technical report (2016). https://github.com/bhuman/BHumanCodeRelease/raw/master/CodeRelease2016.pdf
14. Vala, H.J., Baxi, A.: A review on Otsu image segmentation algorithm. Int. J. Adv. Res. Comput. Eng. Technol. **2**(2), 387–389 (2013)
15. Junker, M., Hoch, R., Dengel, A.: On the evaluation of document analysis components by recall, precision, and accuracy. In: the Fifth International Conference on Document Analysis and Recognition, pp. 713–716. IEEE, Bangalore, India (1999). doi:10.1109/ICDAR.1999.791887

Aerial and Space Robotics

Linearity of the Force Leverage Mechanism Based on Flexure Hinges

Jihao Liu, Enguang Guan, Peixing Li, Weixin Yan[(✉)],
and Yanzheng Zhao

State Key Lab of Mechanical System and Vibration,
Shanghai Jiao Tong University, Shanghai 200240, China
Xiaogu4524@sjtu.edu.com

Abstract. This paper proposes development of a force leverage mechanism based on the flexure hinges. The primary function of this leverage mechanism is to transform an objective unbalance force/moment to a force sensor in the static unbalance measure system. The measure precision is dependent on the linearity of the force transmission of the force leverage mechanism. The kinematics of the force leverage mechanism is modeled based on the elastic model. The finite element method is used to verify the analytical solutions. Moreover, the effect of the initial external load on the linearity is investigated. Further, the virtual experiment is carried on to verify the linearity and sensitivity. The static unbalance measure system employing the proposed leverage mechanism has the advanced sensitivity of less than 0.03 gcm and performs excellent linearity.

Keywords: Force leverage mechanism · Flexure hinge · Force transmission · Linearity · Elastic model

1 Introduction

The compliance mechanisms with flexure hinges is defined as monolithic structures that rely on the small deformation to produce the smooth force and motion [1–4]. The compliance mechanism is the optimal option, which has no backlash, no lubrication requirement, repeatable motion and vacuum compatibility, which has been widely in the precision engineering applications including the micro/nano positioning mechanism [5, 6], micro manufacture [7–9], medical devices [10] and measurement systems [11].

The kinematics analysis, Castigliano's second theorem and the pseudo-rigid-body model (PRBM) are available to model and design the compliance mechanisms [12–16]. Yao [13] takes use of the forward and reverse position and velocity kinematics to analyze and optimize the workspace of a multi-axis nano-positioning stage. Li [14] and Luo [15] prove that the PRBM method is one optimal method to analyze the displacement of the compliance mechanisms. In the PRBM method, the flexure hinge is regarded as a pivot with a torsional spring. However, the elastic model based on the elastic beam theorem performs a high precision superior to aforementioned methods [16].

For the unmanned Aerial vehicles, the gimbaled seekers are crucial components to guarantee the precision and reliability characteristics during the track procedure [17]. Space Electronics company proposes a static unbalance measure method for the

© Springer International Publishing AG 2017
Y. Huang et al. (Eds.): ICIRA 2017, Part III, LNAI 10464, pp. 527–537, 2017.
DOI: 10.1007/978-3-319-65298-6_48

gimbaled seekers based on the centroid of gravity projection (CGP) [18]. So far their commercial products qualify sensitivity of 0.1 gcm.

Based on Zhan's work in the early stage [19], a force leverage mechanism with flexure hinges is developed for a static unbalance measure system. The measure precision is dependent on the performance of the force transmission of the proposed mechanism. This leverage mechanism is modeled to investigate the linearity characteristics using the elastic model. The finite element method (FEM) and the experimental investigation are employed to verify the elastic analysis and the linearity performance.

2   Static Unbalance Measurement System

2.1   Measurement Instrument

Figure 1 demonstrates the static unbalance measurement system that is composed of a load platform, force leverage mechanisms, the force sensors, a computer, a pneumatic isolating platform and an outer cover. The seeker is mounted on the load platform. The force leverage mechanism connects the load platform with the force sensor. The computer is used to read results from the sensors and calculate unbalance moments occurring at a seeker. The pneumatic isolating platform and the outer cover are used to prevent the leverage mechanisms from the low-frequency vibration sources including the winds, sounds and so on.

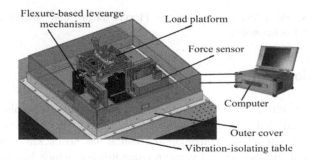

Fig. 1. View of static unbalance measurement system

2.2   Force Leverage Mechanism

The function of the proposed leverage mechanism is to transport force from the load platform to the force balance sensor. Therefore, the linearity characteristics in the force transmission of the proposed leverage mechanism has significant effects on the measuring precision.

Figure 2 shows the schematics of the force leverage mechanism. In terms of the mechanical configuration, it is divided into a parallel four-linkage and a two-stage leverage mechanism. The flexure hinges a_{L1}, a_{L2}, a_{f1} and a_{f2} exist in the parallel four-linkage, three flexure hinges b_i, b_o and b_f exist in the first-stage leverage, and the second consists of flexure hinges c_i, c_o and c_f. The rigid body A_i in Part 1 is connected

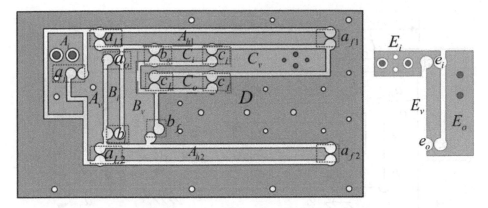

Fig. 2. Schematics of the flexure-based leverage mechanism

with the load platform. The rigid body C_v in Part 1 is assembled with the rigid body E_i in Part 2. Part 2 is designed to be easy to assemble with the force sensor by the rigid body E_o.

3 Modeling Force Transmission

According to the principle of CGP method, since the rotation of an unbalanced gimbal produces the moment change, then it is sensed by force sensors via the flexure-based leverage mechanisms. The unbalance moment causes the small deformation at the flexure hinges of the leverage mechanism during the force transmission. It is essential to model the proposed leverage mechanism considering the deformation of the flexure hinges and self-weights of the rigid bodies.

In order to simplify calculations and analysis, several assumptions exist as follows:

(1). The deformation only occurs in the flexure hinges. Flexure hinges are regarded as flexible components, while the other are rigid;
(2). The elastic deformation consists of two parts: small angle deflection (rotation angle of less than 1°) and the tension in the axial direction.

Different from other compliant mechanisms, the proposed leverage mechanisms are assembled vertically in the measuring system. Hence, the self-weight of rigid bodies is taken into account in the analytical progress.

Figure 3 shows the schematic of the leverage mechanism. The parallelogram linkage transports the force to the first-stage lever. It functions to eliminate the parasitical deformation and coupling force in the horizontal direction. Considering the bending deformation occurring at the flexure hinges, the equation that express the moment equilibrium in the parallel four linkage is derived as:

$$[F_{in} + (m_{Av} + m_{Ah})g - F_a]L_{al-af} = 4K_M\theta_a \tag{1}$$

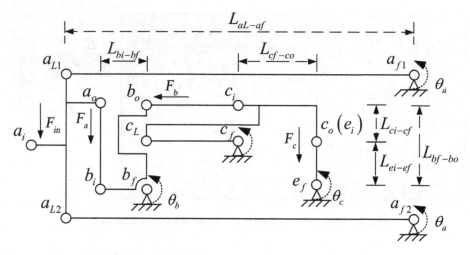

Fig. 3. Schematic of flexure-based leverage mechanism

Where

F_{in} is the input force for the force leverage mechanism;
F_a is the transmitted force into the first-stage lever;
θ_a is bending angles at four flexible hinges in the parallelogram mechanism;
K_M is rotational stiffness of the flexure hinge;
L_{al-af} is the length of the swing arm of the parallelogram mechanism;
m_{Av}, m_{Ah} is the mass of the rigid bodies in the parallelogram mechanism;

The parallel four linkage produces a vertical force, the flexure hinge b_f correspondingly rotates a small angle in the first-stage leverage. The moment equilibrium occurring at the first-stage leverage is derived based on a small deformation as:

$$\left[F_a + \left(m_{Bi} + \frac{1}{2} m_{Bv} \right) g \right] L_{bi-bf} + \frac{1}{2} m_{bo-bf} g L_{bf-bo} \theta_b = F_b L_{bo-bf} + 3 K_M \theta_b \quad (2)$$

Where

L_{bi-bf} is the length of the input arm of the first-level lever;
L_{bf-bo} is the output arm of the first-level lever;
m_{Bi}, m_{Bv} represents the mass of B_i, B_v respectively;
θ_b is the deformation angle occurring at the first-stage leverage;
F_b is the horizontal output force of the first-stage leverage;

Similarly, the moment equilibrium in the second-stage lever is formulated as:

$$F_b L_{ci-cf} + \frac{1}{2} m_{Co} g L_{cL-cf} = 3K_M \theta_c + F_c L_{cf-co} + \left(\frac{1}{2} m_{Cv} + m_{Ei} + m_{Ev} + m_{Eo} \right) g L_{cf-co}$$

(3)

Where

L_{ci-cf} is the input moment arm in the second-stage lever;
L_{cf-co} is the output moment arm in the second-stage lever;
θ_c is the bending angle occurring at the second-stage leverage;
F_c is the output force;
$m_{Co}, m_{Cv}, m_{Ei}, m_{Ev}$ and m_{Eo} represent the mass of C_o, C_v, E_i, E_v and E_o respectively;

The equation that expresses the bending angle at the two-stage leverage are derived as:

$$L_{al-af} \theta_a = -\frac{2F_a + m_{Bi} g}{K_x} + L_{bi-bf} \theta_b$$

(4)

$$\theta_b = \frac{2F_b}{K_x L_{bo-bf}} + \theta_c \frac{L_{ci-cf}}{L_{bo-bf}}$$

(5)

$$\theta_c = \frac{2F_b + m_{Ev} g}{L_{co-cf}}$$

(6)

Combining Eqs. (1)– (6), the force amplification ratio η of the force leverage mechanism is defined as:

$$\eta = \frac{F_c}{F_{in}} = \frac{L_{al-af} F_c}{FF_c + G}$$

(7)

where

$$A = \frac{1}{4} m_{Co} g L_{cl-cf} - \left(m_{Ev} + m_{Ei} + m_{Eo} + \frac{1}{2} m_{Cv} \right) g L_{cf-co} - \frac{3K_M}{K_x L_{cf-co}} m_{Ev} g$$

(8)

$$B = \frac{K_x L_{cf-co}^2 + 6K_M}{K_x L_{cf-co} L_{ci-cf}}$$

(9)

$$C = \frac{6K_M - m_{Bv} g L_{bf-bo}}{2K_x L_{bo-bf}}$$

(10)

$$D = \frac{B(L_{bo-bf} + 2C)}{L_{bi}} + \frac{2L_{ci-cf}}{L_{bi} L_{cf-co}} C$$

(11)

$$E = \frac{L_{ci-cf}}{L_{bi}L_{cf-co}} Cm_{Ev}g - \frac{A}{L_{bi}L_{ci-cf}}\left(L_{bo-bf}+2C\right) - \left(m_{Bi} + \frac{1}{2}m_{Bv}\right)g \qquad (12)$$

$$F = DL_{al-af} + \frac{4K_M}{K_x}\left(\frac{2L_{bi-bf}B}{L_{bo-bf}L_{cf-co}} + \frac{2L_{bi-bf}L_{ci-cf}}{L_{bo-bf}L_{cf-co}^2} - \frac{2D}{L_{cf-co}}\right) \qquad (13)$$

$$G = \frac{4K_M}{K_x}\left(\frac{L_{bi-bf}L_{ci-cf}}{L_{bo-bf}L_{cf-co}^2}m_{Ev}g - \frac{2E+m_{Av}g}{L_{cf-co}} - \frac{2L_{bi-bf}A}{L_{bo-bf}L_{ci-cf}L_{cf-co}}\right)$$
$$+ [E - (m_{Av}+m_{Ah})g]L_{al-af} \qquad (14)$$

The ideal force amplification ratio is assumed to be η_{ideal} that is derived based on the geometric relationships:

$$\eta_{ideal} = \frac{L_{bi-bf}L_{ci-cf}}{L_{bf-bo}L_{cf-co}} \qquad (15)$$

4   Simulation and Experiment

The FEM is employed to investigate static kinematics of the proposed leverage mechanism. Because the linearity characteristics has significant impact on the measuring error and precision, it is essential to carry on the static structural simulation on the proposed leverage mechanism subject to various forces. The mechanical parameters of the force leverage mechanism are listed in Table 1.

Table 1. The mechanical parameters

Material	Alloy
Density	2.78 gcm^3
Young modulus E	69 GPa
t	0.2 mm
R	5 mm
b	5 mm
L_{al-af}	140.8 mm
L_{bi-bf}	21.51 mm
L_{bf-bo}	99.51 mm
L_{ci-cf}	22.18 mm
L_{cl-cf}	18.99 mm
L_{cf-co}	63.56 mm

In the simulation procedure, the standard earth gravity is applied while the identical external loads are applied step by step. As shown in Fig. 4, the results of the elastic analysis are closed to that from FEM. Figure 4(a) depicts the curve of the output force when the weight of 0.04 g is added into the leverage mechanism every step. Figure 4 (b) and 4(c) depict the curves of the output force of the proposed mechanism subject to every step of 0.469 g and 1.41 g respectively. In terms of the results in Fig. 4, the force amplification ratio of the developed leverage mechanism is 0.082 in the ANSYS simulation, and the ratio is 0.0775 based on the elastic analysis. According to the comparison, the analytical solutions is verified and is one valid method to investigate the force transmission of the proposed force leverage mechanism.

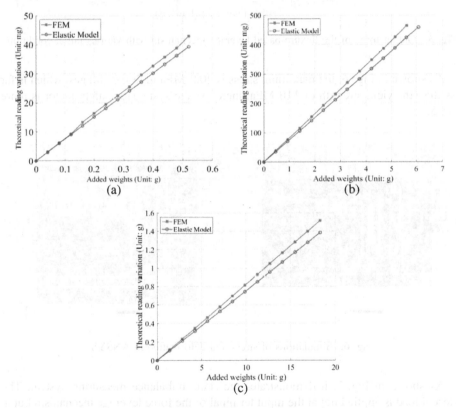

Fig. 4. Output force of the flexure-based leverage mechanism subject to various loads

During the actual static unbalance measurement, the initial unbalance results from the random location. Hence, the effect of various initial external forces on the linearity is evaluated by FEM in ANSYS. Figure 5 shows three curves that represent every step of 0.047 g with the initial unbalance is zero, 300 g and 600 g respectively. It indicates that the linearity of the force transmission is not affected by the applied force.

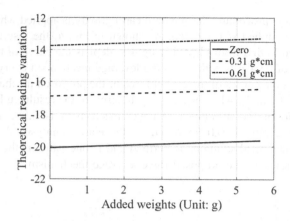

Fig. 5. Output force of the flexure-based leverage mechanism with various initial condition

As shown in Fig. 6, the maximum stress is 20.8 MPa in flexure hinges, which is far less than the yield strength of 110 MPa, therefore the systematic safety factor is more than 5.

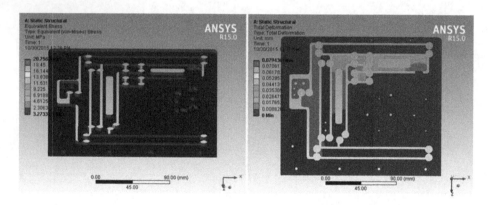

Fig. 6. Simulations of stress and deformation in ANSYS

As shown in Fig. 7, it demonstrates the static unbalance measuring system. The external load is applied not at the input terminal of the force leverage mechanism but at the load platform with double-frame. A weights of 1 g on the center of edge of the load platform is equal to 8 gcm unbalance moment applied into this system. While the reading difference of 1 g on the force sensor means 0.6 gcm unbalance moment occurring at loading platform.

As shown in Fig. 8(a), it indicates the max hysteresis error is 0.15 g and the amplification ratio is about 0.0652 in the case that one weighs of 5.52 g is used every step. In Fig. 8(b), the weights of 3.62 g is used, the max hysteresis error is 0.074 g while the displacement ratio is 0.0647; Fig. 8(c) indicates the max hysteresis error is 0.015 g and the ratio is 0.0613 using the weights of 2.84 g; Fig. 8(d) indicates the max

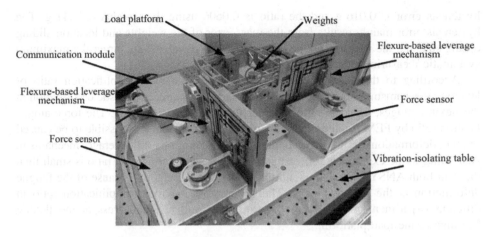

Fig. 7. Test system for static unbalance measuring platform

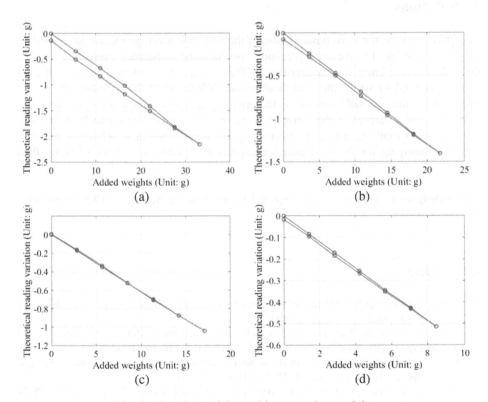

Fig. 8. Experimental data subject to various weights

hysteresis error is 0.016 g and the ratio is 0.0606 using the weights of 1.41 g. The hysteresis error mainly results from the mass error of the weights and location sliding during stacking the weights. So the heavier the added weighs, the larger the maximum systematic measuring error is 0.12 gcm.

According to the aforementioned investigation, the ideal amplification ratio of leverage mechanism is about 1:12.12. Due to the motion loss and elastic deformation in the flexible hinges, the ideal ratio is larger than the elastic analysis. The force amplification ratio by FEM is a little larger than the analytical one. It's possible to be caused by the deformation of the rigid bodies of the flexure-based leverage mechanism. According to the experimental investigation, the force amplification ratio is small than those in both ANSYS simulation and the theoretical calculation because of the fatigue deformation at the flexure hinges. The different of the force amplification ratio in different experiments results from the force loss and residual stress at the flexure bearings as the load platform.

5   Solutions

This paper models the force transmission of the proposed leverage mechanism based on the elastic model. The measure precision for the static unbalance measure system is dependent on the linearity characteristics of the leverage mechanism. The finite element method is used to verify the analytical model. We conduct the investigation of the effects the initial external loads on the linearity characteristics of the force transmission. A conclusion is arrived at the force leverage mechanism performs excellent linearity characteristics that guarantees the measuring precision. The static unbalance measure system employing the flexure-based leverage could qualify sensitivity of better than 0.03 gcm.

Acknowledgment. This work was supported by the National Science of China Foundation (No. 51475305).

References

1. Meng, Q., Berselli, G., Vertechy, R., Castelli, V.P.: An improved method for designing flexure-based nonlinear springs (2015)
2. Beroz, J., Awtar, S., John Hart, a: Extensible-link kinematic model for characterizing and optimizing compliant mechanism motion. J. Mech. Des. **136**(3), 31008 (2014)
3. Sun, X., Chen, W., Fatikow, S., et al.: A novel piezo-driven microgripper with a large jaw displacement. Microsyst. Technol. **21**(4), 931–942 (2015)
4. Lee, V.D., Gibert, J.M., Ziegert, J.C.: Hybrid bi-directional flexure joint. Precis. Eng. **38**(2), 270–278 (2014)
5. Lum, G.Z., Teo, T.J., Yang, G., Yeo, S.H., Sitti, M.: Integrating mechanism synthesis and topological optimization technique for stiffness-oriented design of a three degrees-of-freedom flexure-based parallel mechanism. Precis. Eng. **39**, 125–133 (2015)

6. Xu, Q.: Design of a large-range compliant rotary micropositioning stage with angle and torque sensing. IEEE Sens. J. **15**(4), 2419–2430 (2015)

7. Xu, Q.: Robust impedance control of a compliant microgripper for high-speed position/force regulation. IEEE Trans. Ind. Electron. **62**(2), 1201–1209 (2015)

8. Zhang, D., Zhang, Z., Gao, Q., Xu, D., Liu, S.: Development of a monolithic compliant SPCA-driven micro-gripper. Mechatronics **25**, 37–43 (2015)

9. Jayanth, G.R., Menq, C.H.: Design and modeling of an active five-axis compliant micromanipulator. J. Mech. Robot. **6**(4), 41014 (2014)

10. Rubbert, L., Caro, S., Gangloff, J., Renaud, P.: Using singularities of parallel manipulators to enhance the rigid-body replacement design method of compliant mechanisms. J. Mech. Des. **136**(5), 051010 (2014)

11. Clark, L., Shirinzadeh, B., Tian, Y., et al.: Development of a passive compliant mechanism for measurement of micro/nanoscale planar 3-DOF motions. IEEE/ASME Trans. Mechatron. **21**(3), 1222–1232 (2016)

12. Kim, K., Lee, J., Ju, J., Kim, D.-M.: Compliant cellular materials with compliant porous structures: a mechanism based materials design. Int. J. Solids Struct. **51**(23–24), 3889–3903 (2014)

13. Yao, Q., Dong, J., Ferreira, P.M.: A novel parallel-kinematics mechanisms for integrated, multi-axis nanopositioning. Part 1. Kinematics and design for fabrication. Precis. Eng. **32**(1), 7–19 (2008)

14. Li, Y., Wu, Z.: Design, analysis and simulation of a novel 3-DOF translational micromanipulator based on the PRB model. Mech. Mach. Theory **100**, 235–258 (2016)

15. Luo, Y., Liu, W., Wu, L.: Analysis of the displacement of lumped compliant parallel-guiding mechanism considering parasitic rotation and deflection on the guiding plate and rigid beams. Mech. Mach. Theory **91**, 50–68 (2015)

16. Friedrich, R., Lammering, R.: Flexure hinge mechanisms modeled by nonlinear euler-bernoulli-beams. PAMM **15**(1), 193–194 (2015)

17. Hebert, M.H., Thorpe, C.E., Stentz, A. (eds.): Intelligent Unmanned Ground Vehicles: Autonomous Navigation Research at Carnegie Mellon, vol. 388. Springer Science & Business Media, Heidelberg (2012)

18. Boynton, R., Wiener, P.K., Kennedy, P., Rathbun, B., Engineer, A.C.: Static balancing a device with two or more degrees of freedom–(The Key to Obtaining High Performance On Gimbaled Missile Seekers). In: The 62nd Annual Conference of Society of Allied Weight Engineers, Inc. (2003)

19. Yan, W.X., Zhan, S.T., Qian, Z.Y., Fu, Z., Zhao, Y.Z.: Design of a measurement system for use in static balancing a two-axis gimbaled antenna. Proc. Inst. Mech. Eng. Part G: J. Aerosp. Eng. **228**(13), 2530–2541 (2014)

A Kind of Large-Sized Flapping Wing Robotic Bird: Design and Experiments

Erzhen Pan[1], Lianrui Chen[1], Bing Zhang[1], and Wenfu Xu[1,2(✉)]

[1] Shenzhen Graduate School, Harbin Institute of Technology,
Shenzhen 518055, China
wfxu@hit.edu.cn
[2] The State Key Laboratory of Robotics and System,
Harbin Institute of Technology, Harbin, China

Abstract. A flapping wing aerial vehicle (FAV) has advantages in the aerodynamic performance, flight flexibility and flight efficiency than the fixed wing and rotary wing flyers. It can be used for military investigation, environment exploration and life rescue. Considering long-distance flight and large payload ratio, we developed a large-sized flapping wing robotic bird. The structure and size comes from that of large-sized birds. The mechanism is first designed by bionics. It has a relatively simple mechanical configuration. Then, the kinematics model is derived to analyze the flapping properties. The sensor system, the control system and the software are also designed according to the requirement of flying control. Then, a prototype was developed. It has a span of 1.1 m, and only weighs 455 g. Finally, a lot of outdoors experiments are performed under nature weather conditions. The continuous flight time is more than 15 min. By analyzing the experimental data, we have found out some interesting phenomena and results.

Keywords: Flapping wing · Flapping flight · FAV

1 Introduction

With the development of society and the advancement of technologies, whether in the field of public security administration, the monitoring and reconnaissance of our environment, resources exploration or the rescue and evaluation of the scene of natural disasters have put forward even higher demand for the tools to implement these tasks. The unmanned aerial vehicle(UAV), as a kind of aircraft that can flexibly and freely manoeuver in the air, which is characterized by a wide field of view and of low costing, has been widely used in both civilian and military cases. At the same time, research for UAV technologies has also became a new hot issues for many years.

There are two kinds of aircrafts for traditional UAVs, fixed wing aircrafts and rotorcrafts. Generally speaking, fixed wing UAVs are more suitable for high-speed and long distance tasks scenarios, and rotorcrafts are then more preferable for occasions that require hovering flight and low speed cruise flight. However, compared to flapping wing flights, these two types of aircrafts both have the drawbacks of high energy consumption rate and low lift production efficiency while flying at low speed. However, aerodynamics

© Springer International Publishing AG 2017
Y. Huang et al. (Eds.): ICIRA 2017, Part III, LNAI 10464, pp. 538–550, 2017.
DOI: 10.1007/978-3-319-65298-6_49

for flapping flight is more complicated, and by far there are still quite a lot of challenges exist in this area [1–3].

The flapping wing flight of birds can generate the requiring lift and propulsion for flying at the same time by flapping their wings regularly, and to maneuver and change direction by transforming and twisting their wings in a certain way. We can safely come to the conclusion that the flapping wing flight has some merits such as good aerodynamic performance, high flight efficiency and notable maneuverability and flexibility. These arguments can be testified by the facts that many a lot birds can continuously fly hundreds of kilometers one time after having their meal, and some certain raptorial birds can precisely and rapidly prey other birds or insects when flying.

After we reviewed literatures published in the past several years, we knew that researches for flapping wing flight hasn't been becoming a hot topic for a long time. Except that some scientists and engineers has been in this field for a lifetime, the project that was initiated by Defense Advanced Research Projects Agency (DARPA) in some 20 years ago aiming to develop some sophisticated Unmanned reconnaissance systems by using certain kinds of small and micro scaled UAVs to perform indoor tasks or to scout in confined spaces has played an important role in promoting the development of both theory researches and fabrication of ornithopter prototypes. Up to now, much of far-reaching works have been done by researchers and scientists. However, in the past, the majority of works have been focused on the realm of small and micro FAVs and are specifically about mimicking the flapping flight motion of these flying animals, which are typically like drosophila, mosquitos, hummingbirds and bats [4–15]. Although ornithopters of small sized have high lift generation efficiency and can execute tasks secretly and in occasions like confined spaces, they are on the other hand featured with low-payload capability and short flight range. So when we want to employ some kind of vehicles that satisfies requirements of efficient low speed cruise flight and of long flight ranges, long endurance time and low cost, we do not seem to have too many choices for the moment. But since flapping wing flight has some advantages over the traditional aircrafts, so maybe to develop a FAV of large and medium sized that can exploit its advantages to the full, i.e. low energy-consumption and long endurance time.

Among those FAVs which mimics large and medium sized birds and has good flight performance, Smartbird [16] from the German company Festo and RoboRaven [17, 18] from Maryland University are the two outstanding ones.

By using two symmetrical and synchronized crank and rocker mechanisms, Smartbird's left and right wings can flap ups-and-downs simultaneously. And by adding a link mechanism with two links and three revolute pairs, the outer wing can fold up to a certain angle relative to the inner wing during up-stroke which will finally reduce the negative lift produced in up-stroke, without adding more degrees of freedom. The tail is controlled by two servos, which are used to adjust the roll and pitch motion of the body correspondingly. The main body is structured with carbon fiber material, covered by extruded polyurethane foam skin, and with specially designed wings with wing section. Smartbird do have reproduced the motion of bird's wings in a good way when flapping by folding its outer wing.

From literatures that we can refer, RoboRaven have been made for more than five generations, and the most prominent feature which have been passed down from the right start of this research is that the wings of the FAV were driven by two independent servos. This characteristics enables RoboRaven can perform various kinds of acrobatic flight by independently control the two servos, which result in that these prototypes possess great maneuverability. Generally speaking, RoboRaven is a rather successful FAV. In RoboRaven III, by putting some flexible solar cells on its wings can slightly increase its flight endurance. In RoboRaven V, by adding two propellers to its rear part, the FAV can easily switch between flapping mode and fixed wing mode, which allows RoboRaven to fly in a larger speed range.

To conclude, although Smartbird and RoboRaven are both very successful, they still haven't exploited the advantages of flapping fly to the full. Or, in other words, they just reproduce the motion of flapping in a certain degree, rather than have deeply revealed the principle of flapping flight and have make full use of this mode of motion. In order to improve the practicability and pay-load capability of FAVs, we have fabricated our flapping wing vehicle our FAV prototype on the basis of the current theoretical and empirical findings. Our FAV prototype imitates the flapping motion of large and medium sized birds, which means flaps at a relative low frequency and the flapping mechanism is simplified to only one DOF, resulting in the motion of the wing endpoint is featured with planar trajectory. We will explain the prototype from the following aspects, the mechanical design, kinematics analysis, avionics and sensory system and experimental analysis.

2   Mechanical Design

Statistically, birds are all born with complex and special bone structure and muscular system, which endows their wings with more than ten DOF for each wing. They can on the one hand perform the basic flapping motion when flying, and on the other hand are able to deliberately adjust their muscles and feather to change their morphology, thus to maneuver and to change direction and attitudes at the same time. And that's why birds are superior in deforming their wings and have remarkable flying performance.

At the time when we first get started with developing our FAV prototype, considering that the power that can be generated by the current MEMS technology, artificial chemical muscle and piezoelectric actuator is quite limited, which have been used by other scientists in their researches in the making of their ornithopters [3, 19], we finally decided to use the more traditional means to drive the whole system, i.e. to use the more popular and mature technology, motor and mechanical transmission mechanism, to produce the required power and to achieve the payload of more than 100 grams. Since the schematic of these kind are mostly bulky, so in order to reduce weight and at the same time just to focus on exploring the inherent working principle of flapping flight, we simplified the motion of bird's wings by adopting only one degree of freedom, that is, we made the wings of our prototype can only flap ups-and-downs. The flapping motion of the two wings are known as the planar flapping with which the wing endpoints have only planar trajectory, without considering the twisting and folding motion of real birds' wings and without the typical spatial motion and adjustment. After determining the final arrangement of aerodynamic layout, transmission chain and methods to adjust our prototype's attitude, we finished the designing and assembling process of our FAV prototype. The experimental results verified its feasibility. Figure 1 shows the FAV prototype.

Fig. 1. The flapping wing aerial vehicle prototype we built.

There are 3 DOFs in total for this prototype. One is particularly used for the generation of flapping motion, of which the power is originated from the brushless motor, and the other two are for the adjustment movement of the pitch angle and roll angle of the tail provided by two servo motors mounted in the tail part, which will further directly affect the attitude of the body during flight. The 3D model of the prototype's flapping mechanism and tail attitude adjustment mechanism is shown in Fig. 2.

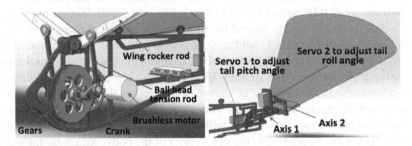

Fig. 2. The transmission and flapping mechanism (left) and the tail attitude adjustment mechanism of the FAV prototype (right).

From Fig. 2 we can see that the whole flapping system is comprised of brushless motor, mechanical transmission mechanism, spatial crank rocker mechanism and the wings. The movement and power will be transmitted in the following way: firstly we remotely control the rotational speed of the brushless motor, and the double reduction gear unit will slow down the revolving speed and at the same time increase torque acted on the spatial crank rocker mechanism. The ball head tension rod connecting the crank and the wing rocker rod will transform the rotational motion of the crank to be the reciprocating swing of the wings. The rocker that swings to and fro acts as the radius and humerus of birds' wings. When the rocker swings, the wings flap ups-and-downs. The wing surface is covered with a kind of cloth membrane, which are usually used to make kites. When flapping, the wing interacts with the air, and the air then will be pushed backward, producing a certain amount of thrust and lift. On each side of the wing, we glued spars (see as Fig. 3) that are used to strengthen the wing plane. The function of these spars mounted on the wing surface play a quite important role in generating aerodynamic forces, both their amplitude and direction, known as the wing flexibility determines the efficiency of flapping aerodynamics.

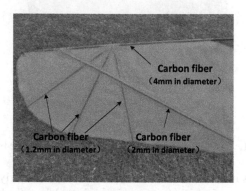

Fig. 3. Distribution of wing spars.

On the other hand, from the 3D model of the tail part we can see that the tail has the same plane structure as the wings. It also use the same kind of membrane cloth as its surface, and use two servo motors to adjust the movement of the tail part. The motion of the two servos is transmitted to the tail by using two parallelogram mechanisms. Among which servo 1 is used to control the pitch motion of the tail plane, which results in the tail revolves around axis 1. And servo 2 is used to control the roll motion of the tail plane, which will result in the tail revolves around axis 2, which is also the output axis of servo 2. While the wings of the FAV are responsible for generating thrust and lift, the empennage is then mainly in charge of generating the pitch torque and yaw torque relative to the center of body mass. The principle of the generation of these torques is that when the tail surface are applied with wind and air, the forces generated on the tail will produce a torque relative to the mass center of the body. And these torques will in turn effect the attitude and direction of the flight trajectory and the attitude of the body.

The flapping frequency of the wings is directly controlled by adjusting the rotational speed of the motor, the requiring value is about 3~7 Hz. And the flapping amplitude and stroke range is fixed in our model, which is about −11°~45°.

The material of the main body is mostly glass fiber (GF), which is cut from a large and complete glass fiber board by using CNC machining. The main GF body frame is used to support and mount all other parts. Speaking of the membrane cloth covered on the wing surface, of which the thickness and softness is of great significance. We've tried various clothes of different stiffness, and we found that although mostly they are capable of generating enough lift and propulsion for flight, relatively, the kind that are thin and soft have better performance. This also verifies the theory that flexibility of an ornithopter wing is rather decisive, and the stiffness distribution of the wing will directly affect the efficiency of lift and propulsion generation. The spars glued to the wings (see as Fig. 3) together with the wing cloth, they change and determine the flexibility of the whole wing. After conducting many times of experiments, we choose the layout form as shown in Fig. 3. What need to be stressed here is that the layout that we are using now is just the result of experience, and we still haven't thoroughly research about this. But we believe the wing flexibility problem is a rather complicated and deterministic one, and we will try to do more in-depth study in this aspect in the near future. Interested readers can refer to some relating papers and findings in reference when the RoboRaven team doing their researches.

What needs to be emphasized is that the flapping mechanism design is the most difficult and important part. After many times of trials and modifications, we came to the conclusion that the selection of motor and gear reduction units and their combinations are of decisive importance for the problem whether the FAV will fly. We choose the appropriate motor by computing the maximum required power output according to the intended maximum weight and payload and the specified flapping frequency. And then compute the suitable reduction ratio and the battery voltage according to the required torque needed to drive the spatial crank and rocker mechanism. Last but not the least, the position of the mass center of the whole body is also quite important, which will directly determine the stability of the prototype when flying. Since when flapping, aerodynamic forces generated on the wings surfaces and tail surface is changing all the time, both the working point and the direction. So the torques of these forces relative to the center of body mass is also changing. How to make these two torques to equilibrate is then become a problem how to arrange components' positions, especially those bulky ones. Our experience is that to simplify the prototype as a fixed wing airplane with a certain anhedral, since the FAV has a larger positive extreme position than negative extreme position. And then we can use the simplified analysis method that the resultant aerodynamic forces generated in wings are roughly located in the position about one third of the chord length away from the leading edge. We should guarantee that the working point of this force is closer to the prototype's head then the center of body mass (CM), and is higher than CM along z axis, while the CM can be obtained by using tools of any 3D designing software. Our experience demonstrate that this simplification idea is quite useful. However, there are still quite a lot of trivial details and factors that will make the result not so desirable.

After determining all components' size, we assembled our prototype of an early version, and some important configuration is listed as follows (Table 1):

Table 1. Parameters of the FAV prototype.

Parameters	Values
Total mass	455 g
Stroke range	$-11° \sim 45°$
Flapping frequency	3~7 Hz
Kv value of brushless motor	Kv 3600
Battery	3S LiPo 11.1 V 800 mAh
Flight endurance	15 min

3   Kinematics

Since we adopted the scheme that the left and right wings are sharing the same power resource and using the same pair of transmission mechanism, so the wing flapping motion of each side is completely symmetrical. Which means, in any time, they both have the exact opposite and antisymmetrical angular position and angular velocity. Meanwhile, in one cycle the flapping motion of the up-stroke and down-stroke is also antisymmetrical, for each of them accounting for half the time of the whole flapping period, and for wings of each side it has opposite characteristics at the same position of down stroke and up stroke. Figure 4 is the schematic diagram of the flapping mechanism.

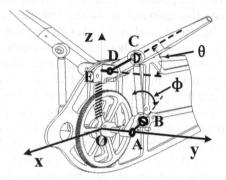

Fig. 4. The schematic diagram of the flapping mechanism.

From the schematic diagram above we can see that OABCDE is then the spatial link mechanism which realizes the flapping motion. While AB is the crank which transmits rotational motion of motor, and CD is the rocker which acts as the wing radius. B and C are two spherical bearing pairs. Link BC transfer the rotational motion of crank AB to the reciprocating swing of the rocker CD. φ is the angle between crank AB and z axis, which refers to the angular position of the crank. And θ is the angle between the rocker and y axis, which is used to denote the angular position of the wing.

When crank AB revolves at a constant speed, we specifically analyzed the angular position and angular velocity of the rocker CD. Figure 5 shows the two limit angles of the rocker during strokes and Fig. 6 shows the variation of angular velocity the rocker CD when the flapping frequency is normalized to 1 Hz. We can see that the positive extreme angular position is 45°, and the lowest extreme angular position is −11°. During our fabrication of prototypes, we found that the length of the spatial link BC determines the stroke ranges of the rocker CD, and this range will greatly affect the pitching stability when flying.

According to the above simulation result, we can see that when crank AB rotates at a constant speed rocker CD swings circularly and accordingly, and at the mean time CD does not swing at a constant angular velocity. During down stroke, the angular velocity of the rocker first increases to a maximum value from a minimum velocity, and then decreases to a minimum speed. For up stroke, the changing process is to the opposite, increases first and then decreases. What calls for special attention is that the time elapse

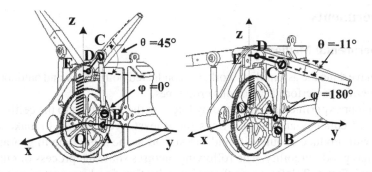

Fig. 5. Angular position of two extreme position of rocker CD.

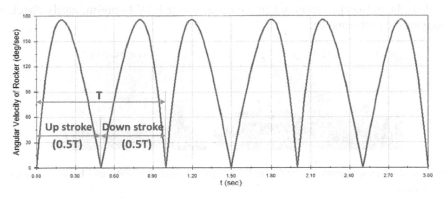

Fig. 6. The angular velocity of rocker CD when flapping frequency is normalized to 1 Hz.

for the angular velocity increasing process of up stroke is the same as the time elapse spent in the decreasing process of down stroke, and the time elapse for the decreasing process of up stroke is the same as the time elapse spent in the increasing process of up stroke. The up stroke and down stroke process of the whole flapping period is completely antisymmetrical. Rocker CD reaches its highest angular velocity at the position of roughly the center of each stroke, which is about 17°.

The flapping motion in up stroke and down stroke are completely the same, which means theoretically speaking, the positive lift generated during the down stroke and the negative lift generated during the up stroke should be balanced out, however prototypes of this kind can still fly, which means the resulting lift forces arc still positive. And this phenomenon shows that lift produced by flapping motion cannot be explained in a traditional way. Maybe theories like leading-edge vortex [20] effect put forward by scientists can illustrate why our ornithopters can still make its flight.

4   Experiments

4.1   Experimental Methods

After we have done the assembling work, we conducted both indoor and outdoor experiments to verify the performance of our prototypes.

The indoor experiments conducted by hitching the prototype to the ceiling to flap at a certain frequency and usually flaps for at least five minutes, to make sure the structure will produce enough lift and thrust, and at the same time to guarantee the structure has good durability. The following pictures show the process of our indoor experiments. Figure 7 (left) shows that we usually first fixed our prototype to a resilient mount which can beat up and down circularly when the FAV flaps its wings, and Fig. 7 (right) shows the scene when we hanging our FAV to preliminarily check its aerodynamic performance.

Fig. 7. Indoor experiments to check the FAV's mechanical durability and basic aerodynamic performance

After we have preliminarily examined our FAV's mechanical and aerodynamic performances, we then conducted the outdoor experiments (see as Fig. 8), i.e. free flight, to examine the systematical performance. Usually, the performance of free flight is affected by various aspects. Except from the mechanical and aerodynamic layout itself, the wind conditions and the operator's proficiency of remote controlling also play an important role. The following pictures show scenes when we were flying our FAV prototype.

Fig. 8. Outdoor experiments

4.2 Experimental Analysis

By using hall sensors installed in the large gear which is used to transmit power and torque to the flapping mechanism, we can obtain the flapping frequency at any time it flaps. The onboard chip and circuit is equipped with inertial measurement unit, which includes accelerometer, gyroscope and magnetometer. Besides, it also includes signal transceiver unit, which on one hand receives signals sent from the upper computer or from remote controller, on the other hand it sends back information collected by various transducers to the upper computer for further analysis. The picture of onboard chip is displayed as follows. Figure 9 are the avionics circuit board (left), and the installation result (middle and right).

Fig. 9. Onboard avionics circuit.

When conducting free flight experiments, we use a computer to receive information transmitted from the onboard avionics circuit and then carry out analysis by using a software developed by Labview. In this software, we can get the complete information collected by the onboard sensors. Figure 10 are respectively the Zigbee module (left) we utilized to transmit and receive signals and the connection of the Zigbee module and the upper computer (right).

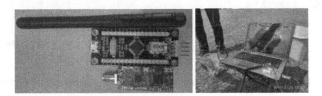

Fig. 10. Transceiver module for upper computer.

4.3 Experimental Analysis

After applied procedures like smoothing and filtering to the raw data collected by sensors, we obtained some results about the free flight. We specifically paid attention to the flapping frequency when flying. Figure 11 (left) shows the change of flapping frequency for a period of free flight time, and Fig. 11 (right) is the statistical information (number of times for each flapping frequency) for this period. From pictures above we can see that for our FAV prototype, the flapping frequency is mainly in the range of 2–3 Hz and 4–6 Hz. When the FAV flaps at 4–6 Hz it can generate enough lift and thrust for flight, and when the airflow is turbulent the FAV then should flap at a relatively low frequency to stabilize itself and to remain a right posture.

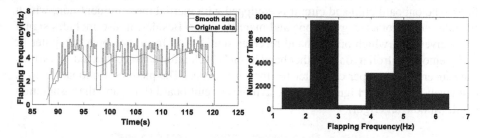

Fig. 11. Variation of flapping frequency & the corresponding flapping frequency distribution

When flying, the pitch angle of the FAV also changes all the time (see as Fig. 12), and this is the typical characteristics and result of flapping flight. According to the following picture we can see that the range of our FAV is mostly from 40° to 80°. Generally speaking, when flapping at a low frequency, the extent of the pitch angle fluctuation is more drastic than flaps at a higher frequency. This phenomenon is in some way has the same feature like flying insects, they flap rapidly to stabilize their body posture and even to perform hovering and acrobatic flight.

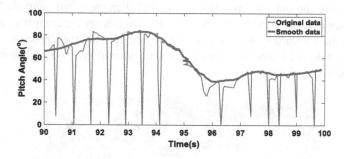

Fig. 12. The fluctuation of pitch angles for the above flapping period.

5   Conclusions

We fabricated our FAV prototype, and have done some preliminary analysis of both the transmission mechanism and flight data. Through experiments we have gained a lot of experience to make our FAV better. However, we still haven't research deeply in how parameters like flapping frequency, stroke amplitude and the range of strokes affect dynamics of the flight. We need to model our prototype and base on that work out the corresponding control strategies to make sure it flies as expected. We will direct at these problems and give some more detailed and in-depth analysis in future works.

Acknowledgements. This work was supported by the National Natural Science Foundation of China (U1613227), and the Basic Research Program of Shenzhen (JCYJ201604271835532-03, CKFW2016033016372515).

References

1. Platzer, M.F., Jones, K.D., Young, J., Lai, J.C.: Flapping wing aerodynamics: progress and challenges. AIAA J. **46**(9), 2136–2149 (2008)
2. Shyy, W., Aono, H., Chimakurthi, S.K., Trizila, P., Kang, C.K., Cesnik, C.E., Liu, H.: Recent progress in flapping wing aerodynamics and aeroelasticity. Progress in Aerospace Sciences 2010, vol. 46, pp. 284–327. Elsevier, Armstrong (2010)
3. Chin, D.D., Lentink, D.: Flapping wing aerodynamics: from insects to vertebrates. J. Exp. Biol. **219**(7), 920–932 (2016)
4. Keennon, M., Klingebiel, K., Won, H.: Development of the nano hummingbird: a tailless flapping wing micro air vehicle. In: 50th AIAA Aerospace Sciences Meeting Including the New Horizons Forum and Aerospace Exposition (2012)
5. Wood, R.J.: Design, fabrication, and analysis of a 3DOF, 3 cm flapping-wing MAV. In: Conference IROS, pp. 1576–1581 (2007)
6. Zdunich, P., Bilyk, D., MacMaster, M., Loewen, D., DeLaurier, J., Kornbluh, R., Low, T., Stanford, S.: Dennis Holeman.: Development and testing of the mentor flapping-wing micro air vehicle. J. Aircr. **44**(5), 1701–1711 (2007)
7. Fenelon Michael, A.A., Furukawa, T.: Design of an active flapping wing mechanism and a micro aerial vehicle using a rotary actuator. Mech. Mach. Theory **45**(2), 137–146 (2010)
8. Whitney, J.P., Wood, R.J.: Conceptual design of flapping-wing micro air vehicles. Bioinspiration Biomimetics 7(3), 036001–036010 (2012)
9. Teoh, Z.E., Fuller, S.B., Chirarattananon, P., Pérez-Arancibia, N.O., Greenberg, J.D., Wood, R.J.: A hovering flapping-wing microrobot with altitude control and passive upright stability. In: Conference 2012, IROS, pp. 3209–3216. IEEE (2012)
10. Landgraf, T.: RoboBee: A Biomimetic Honeybee Robot for the Analysis of the Dance Communication System. Doctoral dissertation, Freie Universität Berlin (2013)
11. Fong, D., Dorothy, M., Chung, S.-J.: Robobat: dynamics and control of a robotic bat flapping flying testbed. AIAA Infotech @ Aerospace Conference 2011, p. 1435. MO, St. Louis (2011)
12. Ramezani, A., Shi, X., Chung, S.-J., Hutchinson, S.: Bat Bot (B2), an Articulated-Winged Bat Robot (2015)
13. Ramezani, A., et al.: Bat Bot (B2), a biologically inspired flying machine. In: Conference 2016, ICRA, pp. 3219–3226. IEEE (2016)
14. Teoh, Z.E., Wood, R.J.: A flapping-wing microrobot with a differential angle-of-attack mechanism. In: Conference 2013, ICRA, pp. 1381–1388. IEEE (2013)
15. Mackenzie, D.: A flapping of wings. Science **6075**(335), 1430–1433 (2012)
16. Smartbird Homepage. https://www.festo.com/group/en/cms/10238.htm
17. John, G., Alex, H., Ariel, P.-R., Luke, R., Adrian, G., Eli, B., Johannes, K., Deepak, L., Chen-Haur, Y., Bruck, H.A., Gupta, S.K.: Robo Raven: a flapping-wing air vehicle with highly compliant and independently controlled wings. Soft Robot. **4**(1), 275–288 (2014)
18. Holness, A.E., Bruck, H., Gupta, S.K.: Design of propeller-assisted flapping wing air vehicles for enhanced aerodynamic performance. In: Conference 2015, ASME, p. V05BT08A017. ASME (2015)

19. Wood, R.J., Avadhanula, S., Menon, M., Fearing, R.S.: Microrobotics using composite materials: the micromechanical flying insect thorax. In: Conference 2003, ICRA, vol. 2, pp. 1842–1849. IEEE (2003)
20. Birch, J.M., Dickinson, M.H.: Spanwise flow and the attachment of the leading-edge vortex on insect wings. Nature **6848**(412), 729–733 (2001)

A Trajectory Planning and Control System for Quadrotor Unmanned Aerial Vehicle in Field Inspection Missions

Gang Chen[1], Rong Wang[2], Wei Dong[1], and Xinjun Sheng[1(✉)]

[1] State Key Laboratory of Mechanical System and Vibration,
School of Mechanical Engineering, Shanghai Jiao Tong University,
Shanghai 200240, China
xjsheng@sjtu.edu.cn
[2] Shanghai Pujiang Bridge and Tunnel Management Co. Ltd.,
Shanghai 200231, China

Abstract. In this work, a trajectory planning and control system designed for quadrotor unmanned aerial vehicles (UAVs) in automatic field inspection missions is presented. This system controls the UAV by two steps. First, a ground station laptop generates a complete coverage flight path regarding the specific working field. Subsequently, an onboard computer controls the UAV in real-time with position, velocity and acceleration setpoints calculated by a trajectory planning algorithm. The algorithm considers both the dynamic property of quadrotor UAVs and the acceleration and velocity limitations in real working condition. UAV controlled by this system performs a rapid and stable flight in windy environment and can be widely utilised in various domains.

Keywords: Quadrotor UAV · Complete coverage path · Trajectory planning · Trajectory tracker

1 Introduction

The quadrotor UAV is a popular type of flying robot with compact mechanical structure, high load capacity, capability of hovering and low cost. Due to its unique advantages, the quadrotor UAV is utilized in various application scenarios. In industrial inspection field, it is applied in power line inspection [1], bridge defect detection [2] and solar power plant measurement [3]. With UAVs, inspection work can be safer, lower-cost and much more efficient. Also, there are great demands for quadrotor UAVs in precision agriculture. Loading appropriate equipment, they can monitor and analyse the quality and health condition of growing crops [4,5]. Furthermore, quadrotor UAVs are utilized in traffic monitoring [6] and public safety [7] domains.

In some of the scenarios mentioned above, such as bridge defect detection, solar power plant measurement and crop health monitoring, the UAV is desired

© Springer International Publishing AG 2017
Y. Huang et al. (Eds.): ICIRA 2017, Part III, LNAI 10464, pp. 551–562, 2017.
DOI: 10.1007/978-3-319-65298-6_50

to fulfil an inspection mission in specified field. The whole field should be covered with minimal time cost. Stable and smooth flight is also required in order to achieve continuous inspection result. Currently, most UAVs are manually controlled to perform such kind of missions, which requires professional UAV pilots and causes high manpower cost. Therefore, the corresponding automatic flight control system is demanded.

The main algorithm in this control system is a trajectory planning algorithm. First a complete coverage path should be generated to cover the working field. Coverage path planning algorithm research starts from ground robots, such as cleaning robots [8] and farming robots [9]. Later, Valente J et al. designed a coverage path planning algorithm for quadrotor UAVs working on irregular fields measured by images. The algorithm divides the field into squares and treat the issue as a travel salesman problem [10]. Franco et al. proposed an energy-aware coverage path planning method for UAVs [11]. These methods concern the general path to cover the field, but ignore the real-time control of the quadrotor UAV. Y. Bouktir et al. researched on the dynamic properties of quadrotor UAVs and raised a trajectory planning method based on B-spline functions [12]. Mark W. Mueller et al. proposed a computationally efficient trajectory generation algorithm for quadrotor UAVs flying along a straight line [13]. These algorithms perform excellently in simulation or indoor experiment environment. In view of the state-of-art, the control of quadrotor UAVs in real working condition is still required to be thoroughly considered to improve the performance.

Motivated by this consideration, a trajectory planing and control system for quadrotor UAV is proposed. The system has competence to generate complete coverage flight path and calculate real-time setpoints to control the UAV. The hardware of this system is composed of a flight control unit (FCU), an onboard computer and a ground station laptop. The ground station laptop generates coverage path based on the Global Position System (GPS) geography information of the working field in a practical approach. The onboard computer calculates real-time trajectory setpoints through an improved algorithm and controls the UAV to realize a stable and efficient flight.

The rest of this paper is arranged as follows. Section 2 introduces how the whole system functions. Section 3 describes the complete coverage polyline path generation method. Section 4 presents a trajectory planning algorithm on each line and a trajectory tracker for real-time control. Section 5 shows the experiment results and Sect. 6 states the conclusion.

2   System Overview

The basic flight of a quadrotor UAV is controlled by FCU. An open-source FCU named PX4 is chosen in this program [14]. With OFFBOARD control mode in PX4, the UAV can be controlled by acceleration setpoints received from serial port. One approach is to transmit the real-time setpoints calculated by the ground station laptop with wireless transmitter, such as XBee. However, smooth flight can not be achieved because of the terrible packet dropout phenomenon. Therefore, an onboard computer is utilized to calculate the real-time setpoints.

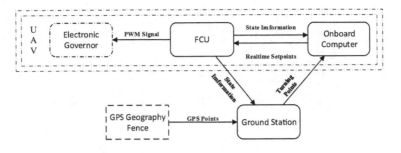

Fig. 1. System framework

The whole system is operated as Fig. 1 shows. Firstly, a GPS geography fence of the working field is measured and imported to the ground station laptop. Then the ground station generates a complete coverage polyline path based on the working field and the current position of the UAV. The generated turning points, namely the endpoints of each line, are transmitted to the onboard computer through wireless transmitter. After that, the onboard computer calculates the real-time trajectory setpoints on the basis of the path generated by ground station and the flight state information given by FCU. Finally, the FCU receives the setpoints and controls the UAV to fly as expected. When the mission changes, the ground station generates another coverage path and sends to the onboard computer to proceed new mission. This system allows the UAV operators to change missions on the ground anytime they want and realize smooth automatic flight control.

3   Complete Coverage Ployline Path Generation

To cover the selected working field, a complete coverage polyline path is generated on the ground station laptop with geography fence. Geography fence is usually measured by a smartphone or a professional GPS device. The fence consists of GPS points and forms a polygon after converting from global position system to local position system. In Fig. 2, the blue polygon is the imported fence. The flight height of the UAV is always settled in particular mission, thus only a two dimensional path is considered in this paper.

Firstly, the line equation defined by two adjacent corner points on the fence polygon can be calculated. Denote corner point $A = (x_1, y_1)$ and $B = (x_2, y_2)$. To avoid infinite result during calculation, the equation of Line AB is described as the standard form

$$ax + by = c \tag{1}$$

where $a = (y_2 - y_1)$, $b = (x_1 - x_2)$ and $c = x_1 y_2 - x_2 y_1$. The endpoints of each line are also recorded in a matrix in the same order as lines parameters.

Denote reference working width d_r and the flight direction $\vec{\omega} = (x_\omega, y_\omega)$, where d_r is often given by the property of the UAV and $\vec{\omega}$ is given according to the requirement in specific situation. For example, solar panels inspection mission

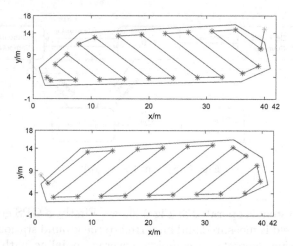

Fig. 2. Complete coverage polyline path. Blue polygon in the figure is the geography fence for the working field and red polylines are the generated flight path. Fly direction is $(-1, 1)$ in the upper figure and $(1, 1)$ in the below figure. Green star represents the position of the UAV at this moment. Red stars are the turning points. (Color figure online)

in solar power plant requires the UAV to fly along the direction the panels are ranged and d_r is width that a image of the inspection camera could cover. In crop inspection or fertilization mission, fly direction is along the direction the crops are planted.

Any line along direction $\vec{\omega}$ can be written as

$$y_\omega x - x_\omega y = m \tag{2}$$

Substitute corner points of fence polygon into Eq. 2 and find the maximum and minimum value of variable m, namely m_{max} and m_{min}. The number of the parallel lines required in the ployline flight path n can be calculated by

$$n = \left\lceil \frac{m_{max} - m_{min}}{\sqrt{y_\omega^2 + x_\omega^2} \times d_r} \right\rceil + 1 \tag{3}$$

From small to large value of m with interval $\delta m = (m_{max} - m_{min}) \div n$, crossover points of the parallel lines and fence polygon are given by solving the following equation for each parallel path line and boundary line of the polygon.

$$\begin{pmatrix} a & b \\ y_\omega & -x_\omega \end{pmatrix} \begin{pmatrix} x \\ y \end{pmatrix} = \begin{pmatrix} c \\ m_{min} + \frac{\delta m}{2} + k\delta m \end{pmatrix} \tag{4}$$

where a and b are the variables in Eq. 1, $k \in \{1, 2, \cdots, n\}$ and valid result (x, y) should be on the boundary of the fence polygon. The precondition of solving Eq. 4 is

$$\begin{vmatrix} a & b \\ y_\omega & -x_\omega \end{vmatrix} \neq 0 \tag{5}$$

which indicates that the flight path line and the boundary line can not be parallel. After the crossover points are acquired, move those points a distance of $\frac{\delta m}{2}$ from boundary to cut the overlap area. Finally, rearrange the sequence of crossover points based on the current position of the UAV. The result is shown in Fig. 2.

4 Trajectory Planning and Real-Time Control

After the ground station generates polylines flight path and transmits to the onboard computer, trajectory planning and real-time control threads start in sequence. These threads are realized in the onboard computer. Trajectory planning is an off-line work to generate the desired acceleration, velocity and position of the UAV on the whole flight path, while real-time control adjust and send setpoints to FCU to control the actual behaviour of the UAV.

4.1 Trajectory Planning on Line

The polyline flight path can be divided into lines. On each line, the UAV should fly through three periods, acceleration period, uniform velocity period and deceleration period. In a real working situation, uniform velocity is always expected to achieve a excellent inspection result. This part focus on the algorithm to realize a smooth and rapid flight with long uniform velocity period on each line.

The algorithm Mark W. Mueller et al. proposed is established on the dynamic characteristic of quadrotor UAVs by employing Pontryagin's minimum principle [13]. Three parameters α, β and γ are derived to acquire position, velocity and acceleration setpoints on a single line path as follows

$$\begin{pmatrix} \alpha \\ \beta \\ \gamma \end{pmatrix} = \frac{1}{T^5} \begin{pmatrix} 60T^2 & -360T & 720 \\ -24T^3 & 168T^2 & -360T \\ 3T^4 & -24T^3 & 60T^2 \end{pmatrix} \begin{pmatrix} \delta a \\ \delta v \\ \delta p \end{pmatrix} \tag{6}$$

where T is the expected flight time, a, v and p represent acceleration, velocity and position respectively. Denote initial state $s_0 = \{p_0, v_0, a_0\}$ and final state $s_f = \{p_f, v_f, a_f\}$. δa, δv and δp in Eq. 6 are given by

$$\begin{pmatrix} \delta a \\ \delta v \\ \delta p \end{pmatrix} = \begin{pmatrix} a_f - a_0 \\ v_f - v_0 - a_0 T \\ p_f - p_0 - v_0 T - \frac{1}{2} a_0 T^2 \end{pmatrix} \tag{7}$$

Then the flight jerk, namely the derivative of acceleration, is calculated by

$$j(t) = \frac{1}{2}\alpha + \beta t + \gamma, t \in [0, T] \tag{8}$$

Furthermore, position, velocity and acceleration setpoints at time t can be expressed as

$$\begin{pmatrix} p(t) \\ v(t) \\ a(t) \end{pmatrix} = \begin{pmatrix} \frac{\alpha}{120}t^5 + \frac{\beta}{24}t^4 + \frac{\gamma}{6}t^3 + \frac{a_0}{2}t^2 + v_0 t + p_0 \\ \frac{\alpha}{24}t^4 + \frac{\beta}{6}t^3 + \frac{\gamma}{2}t^2 + a_0 t + v_0 \\ \frac{\alpha}{6}t^3 + \frac{\beta}{2}t^2 + \gamma t + a_0 \end{pmatrix}, t \in [0, T] \tag{9}$$

The algorithm realizes a smooth and rapid accelerating and decelerating process, but it is not really suitable for the inspection mission. The flight time T in this algorithm is calculated by dividing flight distance by average velocity. If the average velocity increases, T would be smaller and a large maximum velocity would be generated. For example, when $s_0 = \{0, 0, 0\}$, $s_f = \{20, 0, 0\}$ and $T = 5$ s, the average velocity is 4 m/s while the maximum velocity is up to 7.5 m/s. However, in the real working condition, there are always limitations on both velocity and acceleration due to the safety and specific mission requirements in outdoor flight environment, such as stability in wind, emergency brake dealing with unexpected obstacles, and the maximum detection speed determined by real-time image processing methods. Under velocity and acceleration limitations, we hope to achieve a high average velocity to increase working efficiency. Therefore, a piecewise trajectory generation algorithm is proposed and discussed in this paper.

When the length of a single line on the path is short, with acceleration limitation a_{max}, the quadrotor UAV would not reach the allowed maximal velocity v_{max}. Therefore, suppose the length threshold is s_{th}, when the real line length is shorter than s_{th}, the algorithm described from Eqs. 6 to 9 can be applied simply by calculating time T, which will be given in Eq. 15. If the real line length is longer than s_{th}, the newly proposed piecewise trajectory generation algorithm is utilized.

In our mission, the UAV flies along the path line by line, thus the initial and final acceleration and velocity are all zero. Denote the initial position p_0 and the final postion p_f, Eqs. 8 and 9 can be respectively transformed into

$$j(t) = 360 \frac{d}{T^5} t^2 - 360 \frac{d}{T^4} t + 60 \frac{d}{T^3}, t \in [0, T] \tag{10}$$

$$\begin{pmatrix} p(t) \\ v(t) \\ a(t) \end{pmatrix} = \begin{pmatrix} \frac{6d}{T^5} t^5 - \frac{15d}{T^4} t^4 + \frac{10d}{T^3} t^3 1ex \\ \frac{30d}{T^5} t^4 - \frac{60d}{T^4} t^3 + \frac{30d}{T^3} t^2 1ex \\ \frac{120d}{T^5} t^3 - \frac{180d}{T^4} t^2 + \frac{60d}{T^3} t 1ex \end{pmatrix}, t \in [0, T] \tag{11}$$

In order to get the maximum acceleration on a line flight, let the jerk, the derivative of acceleration be zero. Eq. 10 can be rewritten as

$$\frac{360 s_{th}}{T^5} (t - Tt + \frac{T^2}{6}) = 0 \tag{12}$$

Solve Eq. 12 and take accelerating period into consideration, the answer is

$$t_m = \frac{3 - \sqrt{3}}{6} T \tag{13}$$

Substitute t_m into acceleration equation in Eq. 11, the maximum acceleration can be described as

$$a(t_m) = \frac{120 s_{th}}{T^5} (t_m)^3 - \frac{180 s_{th}}{T^4} (t_m)^2 + \frac{60 s_{th}}{T^3} t_m \approx 5.77 \frac{s_{th}}{T^2} \tag{14}$$

$a(t_m)$ should be smaller than a_{max}, thus we can get the minimum flight time T_s to satisfy the acceleration limitation.

$$T_s = \sqrt{\frac{5.77s_{th}}{a_{max}}} \tag{15}$$

According to symmetry of velocity in Eq. 11, the maximum velocity exists at time $t = \frac{T}{2}$.

$$v\left(\frac{T}{2}\right) = \frac{30d}{T^5}\frac{T^4}{2} - \frac{60d}{T^4}\frac{T^3}{2} + \frac{30d}{T^3}\frac{T^2}{2} = \frac{15}{8}\frac{s_{th}}{T} \le v_{max} \tag{16}$$

where $\frac{s_{th}}{T}$ represents the average velocity. Hence the maximum velocity is $\frac{15}{8}$ times of the average velocity. Replace T with T_s given by Eq. 15 and consider the threshold state, the relational expression between a_{max}, v_{max} and s_{th} is expressed as

$$s_{th} = \frac{8}{15}v_{max}\sqrt{\frac{5.77s_{th}}{a_{max}}} \tag{17}$$

Then the distance threshold s_{th} can be calculated by

$$s_{th} = \frac{64}{225}v_{max}^2\frac{5.77}{a_{max}} \approx 1.64\frac{v_{max}^2}{a_{max}} \tag{18}$$

Therefore, given the limitations v_{max} and a_{max} by experience in real flight environment, distance threshold s_{th} can be derived. When the coverage polyline path in Sect. 3 is generated, compare the length of each line between two turning points. If the length s is smaller than s_th, we use Eq. 11 and T is acquired from Eq. 15 by replace s_{th} with actual length s. Otherwise, the flight trajectory is calculated separately in three periods as the following Period 1 to Period 3. The total fight time is $T_s + T'$. An example state curve is shown in Fig. 3.

Period 1 Accelerating period. The position, velocity and acceleration setpoints are calculated by

$$\begin{pmatrix} p(t) \\ v(t) \\ a(t) \end{pmatrix} = \begin{pmatrix} \frac{\alpha}{120}t^5 + \frac{\beta}{24}t^4 + \frac{\gamma}{6}t^3 + p_0 \\ \frac{\alpha}{24}t^4 + \frac{\beta}{6}t^3 + \frac{\gamma}{2}t^2 \\ \frac{\alpha}{6}t^3 + \frac{\beta}{2}t^2 + \gamma t \end{pmatrix}, \quad t \in [0, \frac{T_s}{2}] \tag{19}$$

where T_s is derived by Eq. 18 and α, β and γ are given by

$$\begin{pmatrix} \alpha \\ \beta \\ \gamma \end{pmatrix} = \frac{1}{T_s^5}\begin{pmatrix} 720s_{th} \\ -360s_{th}T_s \\ 60s_{th}T_s^2 \end{pmatrix} \tag{20}$$

Period 2 Uniform velocity period. In this period, the acceleration remains zero and velocity remains v_{max}. The state values are

$$\begin{pmatrix} p(t) \\ v(t) \\ a(t) \end{pmatrix} = \begin{pmatrix} p_0\frac{s_{th}}{2} + v_{max}(t - \frac{T_s}{2}) \\ v_{max} \\ 0 \end{pmatrix}, \quad t \in [\frac{T_s}{2}, \frac{T_s}{2} + T'] \tag{21}$$

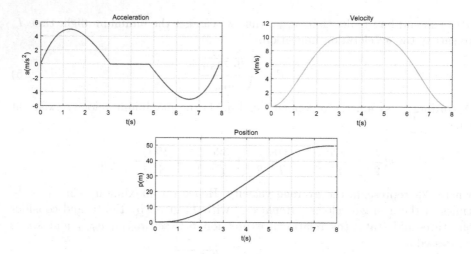

Fig. 3. A trajectory curve on a line 50 m in length. The red curve, green curve and blue curve respectively represent acceleration, velocity and position. Set $v_{max} = 10\,\text{m/s}$ and $a_{max} = 5\,\text{m/s}^2$. The derived s_{th} is 32.8 m and T_s is 6.15 s. (Color figure online)

where T' is the time for this period and is described as

$$T' = \frac{s - s_{th}}{v_{max}} \tag{22}$$

Period 3 Decelerating period. The velocity decelerate from v_{max} to 0. With the same α, β and γ in Eq. 20, the state values are expressed as

$$\begin{pmatrix} p(t) \\ v(t) \\ a(t) \end{pmatrix} = \begin{pmatrix} \frac{\alpha}{120}(t-T')^5 + \frac{\beta}{24}(t-T')^4 + \frac{\gamma}{6}(t-T')^3 + p_0 + v_{max}T' \\ \frac{\alpha}{24}(t-T')^4 + \frac{\beta}{6}(t-T')^3 + \frac{\gamma}{2}(t-T')^2 \\ \frac{\alpha}{6}(t-T')^3 + \frac{\beta}{2}(t-T')^2 + \gamma(t-T') \end{pmatrix},$$

$$t \in [\frac{T_s}{2} + T', T_s + T'] \tag{23}$$

4.2   Trajectory Tracker for Real-Time Control

When the flight trajectory is derived, the position, velocity and acceleration setpoints are saved in the onboard computer. The FCU receives only acceleration setpoints, which is directly related to the attitude of a quadrotor UAV. Thus we need a trajectory tracker to control the UAV to follow the desired position, velocity and acceleration regardless of wind effect. The output of the tracker is only acceleration setpoints. The tracker is designed as Fig. 4 shows.

5   Experiment

To verify the algorithm in real environment, a quadrotor UAV with onboard computer Raspberry Pi is utilized, as is shown in Fig. 5. The weight of the UAV

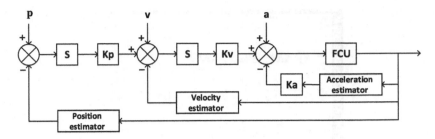

Fig. 4. Trajectory tracker. p, v and a are respectively position, velocity and acceleration setpoint. kp, kv and ka are amplification coefficients.

is 2.1 kg and the axle base is 400 mm. Raspberry Pi runs the algorithm introduced in Sect. 4. Besides, a ground station software is programmed to generate the complete coverage path introduced in Sect. 3 and to monitor the real-time status of the UAV. The software is shown in Fig. 6. In a GPS available outdoor environment, a piece of grass land is chosen as the test field. In real working condition, the UAV is always loaded with particular equipment and becomes less flexible. Therefore, 1.0 kg load is mounted on the UAV and the thrust weight ratio is reduced from 2.38 to 1.61. Considering the flight safety in the wind, the velocity limitation is set to be 5 m/s and acceleration limitation is set to be 3 m/s^2. Firstly, the geography fence is measured by a GPS gauge and transmitted into the ground station software. Then the path and trajectory is generated as Sect. 2 introduced. Finally, the UAV is controlled by the setpoints received from onboard computer.

The experiment results are shown in Fig. 7, which are the position maps drawn in ground station software. Due to the manual take-off, the start position is a little different from the planned position. In Fig. 7(a), the average position error in gentle breeze condition (3 °) is less than 0.2 m. The maximum position error is about 0.3 m when a gust of wind occurred. In Fig. 7(b), the maximum position error is about 0.4 m in strong wind condition(5–6 °). The error is acceptable in application scenarios.

(a) The experiment quadrotor UAV

(b) The onboard computer Raspberry Pi 2B+

Fig. 5. Experiment hardware

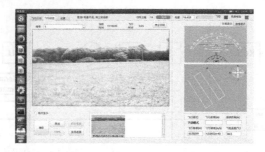

Fig. 6. Ground station software

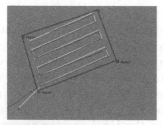

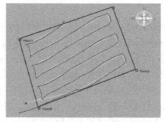

(a) Position in gentle breeze   (b) Position in strong wind

Fig. 7. Fly position map in experiments. Blue lines are the GPS fence. Yellow lines are the expected path and red lines are the real flight path. (Color figure online)

The current UAV products with automatic field inspection control system are very few. Here, $MG-1$ designed by the famous multirotor UAV company DJI is chosen as the contrast. $MG-1$ is utilized in agricultural domain. It has an automatic flight mode which flies polylines based on a given fixed length of parallel lines. Hence it only fits rectangle fields. It can fly as fast as 8 m/s. Choose the identical maximum velocity and apply in our test UAV. The size of the test field is 24 m × 55 m. Set the working width to 4 m and 6 m, the working speed result of $MG-1$ and our system is presented in Table 1. The thrust weight ratio of $MG-1$ is 1.78, which is higher than our UAV. However, with this control system, the working speed of our UAV are respectively 22.2% and 25.0% higher than that of $MG-1$.

Table 1. Working speed comparison

UAV system	Working width	Working speed	Working width	Working speed
MG-1	4 m	2.7 ha./h	6 m	4.0 ha./h
MG-1	4 m	3.3 ha./h	6 m	5.0 ha./h

6   Conclusion

In this paper, a trajectory planning and control system for quadrotor UAVs working in field inspection missions is proposed. The hardware of this system is composed of a FCU, a ground station laptop and an onboard computer. In order to achieve an efficient inspection work, a complete coverage path generation method operating in the ground station laptop and a piecewise trajectory planning algorithm running in the onboard computer are respectively designed. With this system, the UAV could fly stably and rapidly to cover the whole working field. The maximum position error in windy environment is less than 0.5 m and the working efficiency is higher than that of commercial UAV products. In future, further researches will be carried to reduce the position error and the time cost. More experiments and improvements will be made to acquire more excellent performance in real working condition.

References

1. Luque-Vega, L.F., Castillo-Toledo, B., Loukianov, A., Gonzalez-Jimenez, L.E.: Power line inspection via an unmanned aerial system based on the quadrotor helicopter. In: Mediterranean Electrotechnical Conference, pp. 393–397 (2014)
2. Chen, J., Wu, J., Chen, G., Dong, W., Sheng, X.: Design and development of a multi-rotor unmanned aerial vehicle system for bridge inspection. In: Kubota, N., Kiguchi, K., Liu, H., Obo, T. (eds.) ICIRA 2016. LNCS (LNAI), vol. 9834, pp. 498–510. Springer, Cham (2016). doi:10.1007/978-3-319-43506-0_44
3. Matsuoka, R., Nagusa, I., Yasuhara, H., Mori, M., Katayama, T., Yachi, N., Hasui, A., Katakuse, M., Atagi, T.: Measurement of large-scale solar power plant by using images acquired by non-metric digital camera on board uav. ISPRS - Int. Arch. Photogram., Remote Sens. Spat. Inf. Sci. **39**, 435–440 (2012)
4. Tabanlioglu, A., Yucedag, A.C., Tuysuz, M.F., Tenekeci, M.E.: Multicopter usage for analysis productivity in agriculture on gap region. In: Signal Processing and Communications Applications Conference, pp. 1102–1105 (2015)
5. Malveaux, C., Hall, S.G., Price, R.: Using drones in agriculture: unmanned aerial systems for agricultural remote sensing applications. In: 2014 Montreal, Quebec Canada, 13 July–16 July 2014, p. 1. American Society of Agricultural and Biological Engineers (2014)
6. Bizin, I., Vlasov, V., Demidov, A., Kanatnikov, N.: The use of multicopters for traffic monitoring in dense urban areas. In: Proceedings of the 2014 Conference on Electronic Governance and Open Society: Challenges in Eurasia, pp. 42–44. ACM (2014)
7. Baker, R.E.: Combining micro technologies and unmanned systems to support public safety and homeland security. J. Civil Eng. Archit. **6**(10), 1399 (2012)
8. De Carvalho, R.N., Vidal, H., Vieira, P., Ribeiro, M.: Complete coverage path planning and guidance for cleaning robots. In: 1997 Proceedings of the IEEE International Symposium on Industrial Electronics, ISIE 1997, vol. 2, pp. 677–682. IEEE (1997)
9. Zuo, G., Zhang, P., Qiao, J.: Path planning algorithm based on sub-region for agricultural robot. In: 2010 2nd International Asia Conference on Informatics in Control, Automation and Robotics (CAR), vol. 2, pp. 197–200. IEEE (2010)

10. Valente, J., Sanz, D., Del Cerro, J., Barrientos, A., de Frutos, M.Á.: Near-optimal coverage trajectories for image mosaicing using a mini quad-rotor over irregular-shaped fields. Precision Agric. **14**(1), 115–132 (2013)
11. Di Franco, C., Buttazzo, G.: Energy-aware coverage path planning of UAVs. In: 2015 IEEE International Conference on Autonomous Robot Systems and Competitions (ICARSC), pp. 111–117. IEEE (2015)
12. Bouktir, Y., Haddad, M., Chettibi, T.: Trajectory planning for a quadrotor helicopter. In: 2008 16th Mediterranean Conference on Control and Automation, pp. 1258–1263. IEEE (2008)
13. Mueller, M.W., Hehn, M., D'Andrea, R.: A computationally efficient algorithm for state-to-state quadrocopter trajectory generation and feasibility verification. In: 2013 IEEE/RSJ International Conference on Intelligent Robots and Systems (IROS), pp. 3480–3486. IEEE (2013)
14. Meier, L., Honegger, D., Pollefeys, M.: PX4: a node-based multithreaded open source robotics framework for deeply embedded platforms. In: 2015 IEEE International Conference on Robotics and Automation (ICRA), pp. 6235–6240. IEEE (2015)

Attitude and Position Control of Quadrotor UAV Using PD-Fuzzy Sliding Mode Control

Jong Ho Han[1(✉)], Yi Min Feng[2], Fei Peng[2], Wei Dong[1],
and Xin Jun Sheng[1]

[1] School of Mechanical Engineering, Shanghai Jiao Tong University,
Dongchuan Road 800, Shanghai 200240, China
{jonghohan, chengquess, xjsheng}@sjtu.edu.cn
[2] Shanghai Pujiang Bridge and Tunnel Operation Management Co. Ltd.,
Shanghai, China
fengyimin@dp600.com, rocfly@163.com

Abstract. In this paper designs PD-fuzzy sliding mode control for the quadrotor UAV to implement the trajectory-tracking mission. Firstly, dynamic model is introduced for quadrotor UAV. Secondly, the PD-fuzzy sliding mode control is proposed to make the real value of the UAV reach the desired value command, although the UAV is even with system uncertainties and disturbances. The convergence of the complete equations of motion of the UAV is proved by the Lyapunov stability theory. Computer simulation results illustrate the effectiveness of the proposed control schemes.

Keywords: Quadrotor · PD-Fuzzy · Sliding mode · Disturbance · Lyapunov

1 Introduction

Unmanned Aerial Vehicles (UAVs) have been developed for performing various missions in the military and civil areas. Quadrotors are one of UAVs which consist of two rods and four actuators as shown in the Fig. 1. Even though its structure is simple, the quadrotor is a VTOL (Vertical Task off and Landing) and can perform most of missions that helicopters can do. The dynamic model of quadrotor UAV has six degree-of freedom (DOF) with only four independent thrust forces generated by four rotors. It is difficult to control all these six outputs with only four control inputs. Moreover, uncertainties associate with dynamic model also bring more challenge for control design. To solve the quadrotor UAV tracking control problem, many techniques have been proposed [1–5]. In [6], PID and LQR controller were proposed to stabilize the attitude. The PID controller showed the ability to control the attitude in the presence of minor perturbation and the LQR controller provided average results, due to model imperfections Madani et al. it studied a full-state backstepping technique based on the Lyapunov stability theory and backstepping control [7, 8], another backstepping control method was proposed by P. Castillo et al. they used this controller with a saturation function and it performed well under perturbation [8]. The sliding mode control has been applied extensively to control quadrotors. The advantage of this approach is its insensitivity to the model errors, parametric uncertainties, ability to globally stabilize the system and other disturbances [9–12].

© Springer International Publishing AG 2017
Y. Huang et al. (Eds.): ICIRA 2017, Part III, LNAI 10464, pp. 563–575, 2017.
DOI: 10.1007/978-3-319-65298-6_51

In [3] the authors presents a continuous sliding mode control method based on feedback linearization applied to a quadrotor UAV. In [17], this paper represents fuzzy logic control and sliding mode control techniques based on backstepping approach for an under-actuated quadrotor UAV system under external disturbances and parameter uncertainties. In [18], wide flight envelope span experiencing large parametric variations in the presence of uncertainties, a fuzzy adaptive tracking controller (FATC) is proposed. In [19], presents self-tuning PID based on fuzzy logic system. the system is able to adjust from uncertain parameters. In this paper, PD-fuzzy sliding mode control strategy to solve the attitude, altitude and position problem of the quadrotor UAV is proposed and applied to the dynamic model of the quadrotor UAV.

The control system is to guarantee all system state variables converge to their own desired value. Finally, simulations are carried out to verify the model and the control strategy. This study is organized as follows. The proposed robust control algorithms are described in Sect. 2. Section 3 is devoted to the presentation and the discussion of simulation results, when the proposed scheme is applied to the quadrotor. Finally, conclusions and futures advances are provided in Sect. 4.

2 Design of a PD-Fuzzy Sliding Mode Controller

In this section, we developed PD-Fuzzy Sliding mode controller [17–19]. Basically, the sliding mode control is a variable structure control(VSC). It has been included the several different continuous functions which has sliding surface. We considered the sliding surface and select to it. Firstly, we designed the PD control system to reduce the input error. The benefit of this approach is that each controller can compensate the unknown bounded terms and disturbance [13–15].

2.1 Altitude Control for the Quadrotor

To design the sliding mode controller, the sliding surface was set as follows:

$$s_z(t) = c_i e(t) + \dot{e}(t) \tag{1}$$

where c_1 is the gain constant to be determined, $e(t)$, the difference between the real altitude angle, z_{eq}, and the reference angle, z_{ref}, is given as:

$$e(t) = z_{eq}(t) - z_{ref}(t) \tag{2}$$

To obtain an equivalent input, u_{eq}, the $\dot{s}$ term can be obtained from Eq. (1) as:

$$
\begin{aligned}
\dot{s}_z &= (c_1 \dot{e} + \ddot{e}) \\
&= \left(c_1 \dot{e} - \ddot{z}_{ref} + \ddot{z}_{eq}\right) \\
&= c_1 \dot{e} - \ddot{z}_{ref} + \left(g + \frac{c\phi c\theta}{m} u_1\right)
\end{aligned}
\tag{3}
$$

where $\ddot{z}_{eq}$ can be obtained from $\ddot{z} = \frac{1}{m}(\cos\phi \cos\theta)u_1 - g - \frac{k_z \dot{z}}{m}$.

The control input, z_{eq}, can be selected to satisfy the condition $\dot{s}_z = 0$ such that the control state variables of $e(t)$ and $\dot{e}(t)$ gradually approaches to zero [11].

$$\ddot{z}_{eq} = c_1\dot{e} - \ddot{z}_{ref} + (g + \frac{c\phi c\theta}{m}u_1 + \frac{k_z\dot{z}}{m}) \tag{4}$$

To keep the system states on the sliding surface, the control input u can be set as:

$$u = z_{eq} + \gamma_z \cdot \text{sgn}(s_z) \tag{5}$$

where $\gamma_z > 0$ and $\text{sgn}(s_z)$ is defined as follows:

$$\text{sgn}(s_z) = \begin{cases} -1 & if s_z < 0 \\ 1 & f s_z > 0 \end{cases} \tag{6}$$

For the stability condition, $\dot{V} = s_z\dot{s}_z \leq 0$, Eqs. (1) and (3) are substituted into s_z and $\dot{s}_z$. When the u_{eq} is substituted for u in Eq. (5), $\dot{V}$ is derived as follows:

$$
\begin{aligned}
V &= \frac{1}{2}s_z^2 \\
\dot{V} &= s_z\dot{s}_z \\
&= s_z[c_1\dot{e} - \ddot{z}_{ref} + (g + \frac{c\phi c\theta}{m}u_1 + \frac{k_z\dot{z}}{m}) \\
&\quad - \{c_1\dot{e} - \ddot{z}_{ref} + (g + \frac{c\phi c\theta}{m}u_1 + \frac{k_z\dot{z}}{m})\} \\
&\quad - \gamma_z\text{sgn}(s_z) - u] \\
&= s_z \cdot -\gamma_z\frac{|s_z|}{s_z} \leq 0 \\
&= -\gamma_z|s_z| \leq 0
\end{aligned} \tag{7}
$$

By selecting u using Eq. (5), $\dot{V} = s_z\dot{s}_z \leq 0$ condition has been satisfied such that the system states exists on the sliding surface [3, 13–15, 18, 19].

(1) The design of Fuzzy system for the quadrotor

To obtain the equivalent control input, z_{eq}, in the sliding mode algorithm, fuzzy logic was utilized in this research [17–19]. Even though the z_{eq} has already been determined in Eq. (4) for the sliding mode control of the frictions, disturbance and uncertainties in the system. This limits the accuracy of the equivalent control input for the optimal control. Fuzzy logic resolves these time-varying and nonlinear effects in real time in determining z_{eq}.

The fuzzy rules are represented as:

$$\text{Rule i : } \textbf{IF} \text{ error is } F_e^i \textbf{ AND } \text{change - of - error is } F_c^i \textbf{ THEN } \text{ouput is } \delta_i \tag{8}$$

where $\delta_i, i = 1, 2, \ldots, m$ are singleton values, F_e^i and F_c^i are fuzzy sets for the error and the derivative of the error, respectively.

The triangular membership functions are used for both the IF and THEN parts. The fuzzy rules are summarized in Table 1.

Table 1. Two input one output fuzzy rule for δ

δ		Change of error										
		NH	NL	NB	NM	N S	Z	P S	PM	P B	P L	P H
Error	NH	PH	PH	PH	PH	PH	PH	PL	PB	PM	PS	Z
	NL	PH	PH	PH	PH	PH	PL	PB	PM	PS	Z	NS
	NB	PH	PH	PH	PH	PL	PB	PM	PS	Z	NS	NM
	NM	PH	PH	PH	PL	PB	PM	PS	Z	NS	NM	NB
	NS	PH	PH	PL	PB	PM	PS	Z	NS	NM	NB	NL
	Z	PH	PL	PB	PM	PS	Z	NS	NM	NB	NL	NH
	PS	PL	PB	PM	PS	Z	NS	NM	NB	NL	NH	NH
	PM	PB	PM	PS	Z	NS	NM	NB	NL	NH	NH	NH
	PB	PM	PS	Z	NS	NM	NB	NL	NH	NH	NH	NH
	PL	PS	Z	NS	NM	NB	NL	NH	NH	NH	NH	NH
	PH	Z	NS	NM	NB	NL	NH	NH	NH	NH	NH	NH

In Table 1, N and P imply negative and positive, respectively, while H, L, B, M, S, and Z represent Huge, Large, Big, Medium, Small, and Zero, in that order.

For the de-fuzzification, the center of gravity method was used as:

$$u_{fz} = \frac{\sum_{i=1}^m w_i \cdot \delta_i}{\sum_{i=1}^m w_i} \qquad (9)$$

where w_i is the weight for the ith rule.

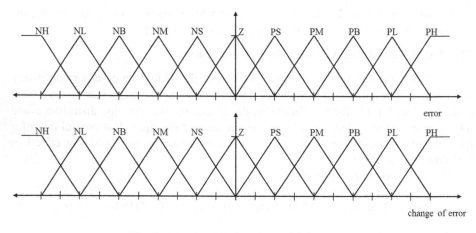

Fig. 1. Membership functions of fuzzy rules.

To determine the stability of this fuzzy-sliding mode algorithm, the altitude dynamics can be defined as

$$\ddot{z}_{eq} = f_{altitude}(q) - u \tag{10}$$

where $f_{atitude}(q)$ includes the uncertain time varying and nonlinear terms in the altitude dynamics. From Eq. (5), for the control input u, z_{eq} is replaced by u_{fz} and the result is substituted into Eq. (7) for $\dot{V} = s_z \dot{s}_z$

$$
\begin{aligned}
\dot{V} &= s_z \dot{s}_z \\
&= s_z \left(c_1 \dot{e} - \ddot{z}_{ref} + f_{altitude}(q) - u \right) \\
&= s_z \left(c_1 \dot{e} - \ddot{z}_{ref} + f_{altitude}(q) - u_{fz} - \gamma_1 \mathrm{sgn}(s_z) \right)
\end{aligned}
\tag{11}
$$

where the condition $u_{fz} \leq c_1 \dot{e} - \ddot{z}_{ref} + f_{altitude}(q)$ is satisfied by the fuzzy rules, which guarantee the stability condition for $\dot{V} = s_z \dot{s}_z \leq 0$.

(2) Reduction of chattering problem

Chattering phenomenon occurs in the sliding mode control due to the signum switching function. When γ is decreased, in an attempt to reduce the chattering, the settling time of the state variables become longer. To counter this phenomenon, in this study, a fuzzy tuning algorithm was developed to eliminate the time delay while reducing the chattering. γ is reduced near the sliding surface to reduce the boundary layer, and it is kept high at the initial stage to shorten the settling time.

The fuzzy rules for this are as follows:

$$\text{Rule i}: \textbf{IF } s \text{ is } F_s^i \textbf{ THEN } \text{ouput is } \gamma_i \tag{12}$$

where s represents the sliding surface, $\gamma_i, i = 1, 2, \ldots, m$ is a singleton value, and F_s^i is the fuzzy set for s. The fuzzy rules are summarized in Table 3.

In Table 2, N and P represent negative and positive, respectively, while H, L, B, M, S, and Z represent huge, large, big, medium, small, and zero, respectively.

Table 2. Fuzzy rule for γ fuzzy-tuning

S	NH	NL	NB	NM	NS	Z	PS	PM	PB	PL	PH
γ	H	L	B	M	S	Z	S	M	B	L	H

The center of gravity method is used for the de-fuzzification. Note that γ is kept as a positive constant to satisfy the following condition: $\dot{V} = s\dot{s} \leq 0$.

Figure 2 shows the chattering phenomenon between signum function and application of the fuzzy rules. As shown in solid line, the chattering phenomenon disappears from the sliding surface input when fuzzy rules are applied to tune gains.

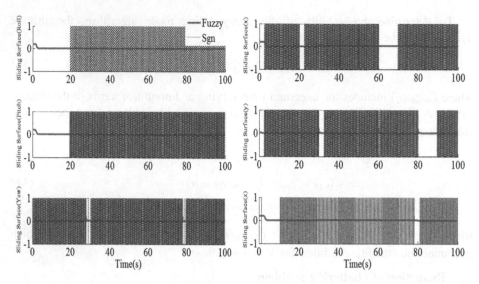

Fig. 2. The results of fuzzy-tuning for Chattering

Finally, the synthesized stabilizing altitude control laws are following:

$$u_1 = \frac{m}{\cos\theta\,\cos\phi}\left\{u_{fz} - g - \frac{k_z\dot{z}}{m} - \gamma_{fz}sgn(s_z)\right\}$$ (13)

where u_{fz} includes $c_1\dot{e} - \ddot{z}_{ref} + f_{altitude}(q)$.

3 Simulation Results

In this section, the proposed algorithm is implemented in Matlab/Simulink simulation environment. The computer simulation results are given to demonstrate the effectiveness of all the proposed control schemes. Moreover, we have been compared with sliding mode and backstepping controller. The initial position and angle values of the quadrotor for simulation test are [0, 0, 0] m and [0, 0, 0] rad. In addition, the quadrotor model variables are listed in Table 3 and the controller parameters are listed in Table 4.

Table 3. Quadrotor model parameters

Variable	Value	Units
m	2	kg
$I_x = I_y$	1.25	Ns^2/rad
I_z	2.2	Ns^2/rad
l	0.2	m
J_r	1	Ns^2/rad
b	2	Ns^2
d	5	$N\,ms^2$
g	9.806	N/s^2
$k_x = k_y = k_z$	0.01	Ns/m
$k_\phi = k_\theta = k_\psi$	0.012	Ns/m
C	1	

Table 4. The each of controller parameters

	PD controller	
	K_p	K_d
z	80	15
ϕ	80	15
θ	100	5
ψ	80	15
x	80	15
y	80	5
	Backstepping controller	
z	$\alpha_1 = 3, \alpha_2 = 15$	
ϕ	$\alpha_3 = 3, \alpha_4 = 15$	
θ	$\alpha_5 = 3, \alpha_6 = 15$	
ψ	$\alpha_7 = 3, \alpha_8 = 3$	
x	$\alpha_9 = 3, \alpha_{10} = 3$	
y	$\alpha_{11} = 3, \alpha_{12} = 15$	
	Sling mode controller	
z	$c_1 = 10, k_3 = 15$	
ϕ	$c_2 = 10, k_2 = 15$	
θ	$c_3 = 10, k_3 = 15$	
ψ	$c_4 = 10, k_4 = 15$	
x	$c_5 = 10, k_5 = 15$	
y	$c_6 = 10, k_6 = 15$	

Figures 3 and 4 represent the results of the roll, pitch, yaw (ϕ, θ, ψ) and x, y, z error data, respectively. The results of each error show the proposed method is more better than others.

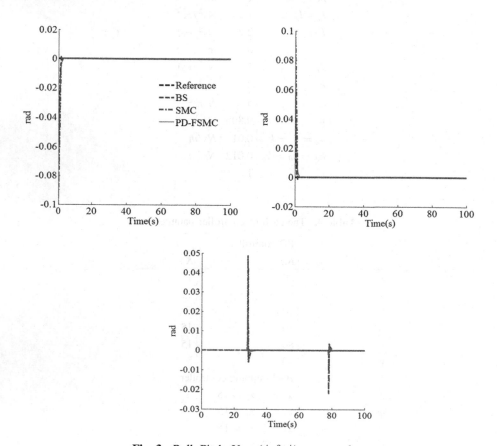

Fig. 3. Roll, Pitch, Yaw (ϕ, θ, ψ) error results.

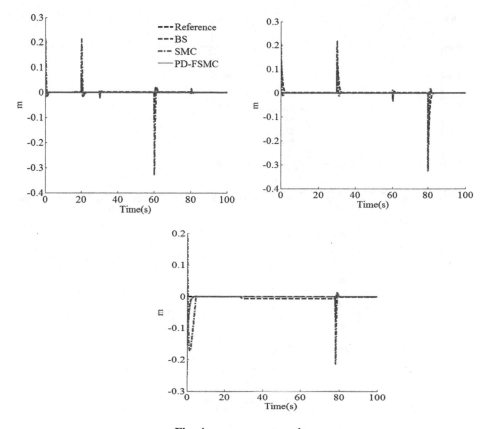

Fig. 4. x, y, z error results.

Here, Figs. 5 and 6 represent the results of the roll, pitch, yaw and x, y, z input data. As seen the result, The proposed method is stable than others and the desired values reach to be faster than the others.

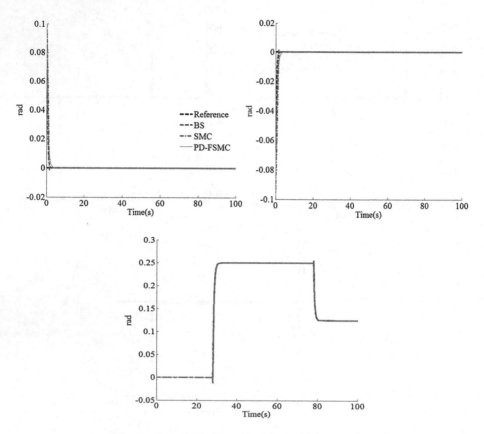

Fig. 5. Roll, Pitch, Yaw (ϕ, θ, ψ) input data results.

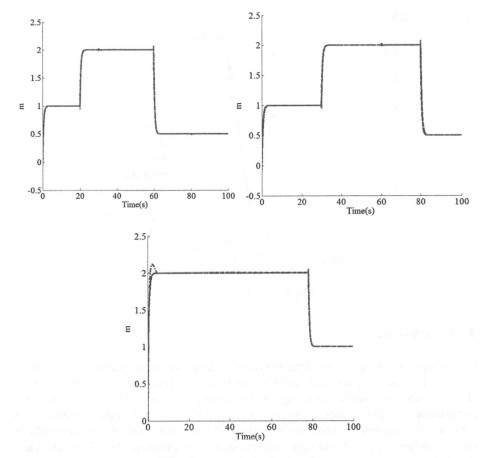

Fig. 6. x, y, z input data results.

The position results is shown in Fig. 7. The Fig. 7 shows the tracking of desired trajectory by proposed method and the evolution of the quadrotor in space and its stabilization. The result shows improvement than others. Finally, the robustness of the proposed overall control method is performed to evaluate the simulation tests and is compared with others controller which are based on attitude, altitude and position tracking of the quadrotor UAV.

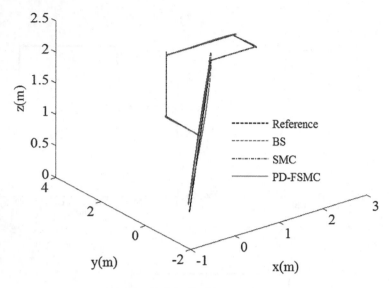

Fig. 7. The Position results.

4 Conclusions

In this paper, we have proposed the PD fuzzy sliding mode controller for the quadrotor UAV with unknown parameter variations and external disturbances, which allows the UAV to achieve complete tracking of the desired trajectory. The position tracking performance of the proposed control scheme compared with other methods was achieved by simulation. Also, the PD fuzzy sliding mode control system results in robust performance with strongly reduced chattering problem. All of the computer simulations results demonstrate that the proposed scheme is indeed feasible and effective.

References

1. Samir, Z.: Sliding mode control strategy for a 6 DOF quadrotor helicopter. J. Electr. Eng. **10**, 1–7 (2002)
2. Erginer, B., Altug, E.: Design and implementation of a hybrid fuzzy logic controller for a quadrotor VTOL vehicle. Int. J. Control Autom. Syst. **10**(1), 61–70 (2012)
3. Lee, D., Jin Kim, H., Sastry, S.: Feedback linearization vs. adaptive sliding mode control for a quadrotor helicopter. Int. J. Control Autom. Syst. **7**(3), 419–428 (2009)
4. Aguilar-Ibanez, C., Humberto Sossa-Azuela, J., Suarez-Castanon, M.S.: A backstepping-based procedure with saturation functions to control the PVTOL system. Nonlinear Dyn. **83**, 1247–1257 (2016)
5. Lin, H., Yu, Y., Zhong, Y.: Robust trajectory control for a laboratory". Nonlinear Dyn. **77**, 621–634 (2014)

6. Bouabdallah, S., Noth, A., Siegwart, R.: PID vs LQ control techniques applied to an indoor micro quadrotor. In: Proceedings of 2004 IEEE/RSJ International Conference Intelligent Robots and Systems, pp. 2451–2456 (2004)
7. Madani, T., Backstepping, A.B.: Control for a quadrotor helicopter. In: Proceedings of the 2006 IEEE/RSJ International Conference on Intelligent Robots and Systems, Beijing, China (2006)
8. Castillo, P., Albertos, P., Garcia, P., Lozano, R.: Simple real-time attitude stabilization of a quadrotor aircraft with bounded signals. In: Proceedings of the 45th IEEE Conference on Decision and Control, pp. 1533–1538 (2006)
9. Mokhtari, A., Benallegue, A., Mian, B.D.A.A., Daobo, W.: Modeling and backstepping-based nonlinear control strategy for a 6 DOF quadrotor helicopter. Chin. J. Aeronaut. **21**, 261–268 (2008)
10. Utkin, V.I.: Sliding Modes in Control and Optimization. Spinger, Berlin (1992)
11. Fang, Z., Zhi, Z., Jun, L., Jian, W.: Feedback linearization and continuous sliding mode control for a quadrotor UAV. In: Proceedings of the 27th Chinese Control Conference, pp. 16–18 (2008)
12. Bouabdallah, S., Siegwart, R.: Backstepping and sliding-mode techniques applied to an indoor micro quadrotor. In: Proceedings of the 2005 IEEE International Conference on Robotics and Automation (2005)
13. Rong, X., Özgüner, Ü.: Sliding mode control of a class of underactuated systems. Automatica **44**, 233–241 (2008)
14. Lee, K., Back, J., Choy, I.: Nonlinear disturbance observer based robust attitude tracking controller for quadrotor UAVs. Int. J. Control Autom. Syst. **12**(6), 1266–1275 (2014)
15. Wang, X., Shirinzadeh, B., Ang, M.H.: Nonlinear doubleintegral observer and application to quadrotor aircraft. Ind. Electron. IEEE Trans. **62**(2), 1189–1200 (2015)
16. Zheng, E.-H., Xiong, J.-J., Luo, J.-L.: Second order sliding mode control for a quadrotor UAV. ISA Trans. **53**, 1350–1356 (2014)
17. Khebbache, H., Tadjine, M.: Robust fuzzy backstepping sliding mode controller for a quadrotor unmanned aerial vehicle. Control Eng. Appl. Inf. **15**(2), 3–11 (2013)
18. Zhi, L., Youn, W.: Fuzzy adaptive tracking control within the full envelope for an unmanned aerial vehicle. Chin. J. Aeronaut. **27**(5), 1273–1287 (2014)
19. Novak, D., Cermak, P.: Fuzzy-based method of detecting the environment character for UAV optical stabilization. Adv. Electr. Electron. Eng. **13**(3), 255–261 (2015)

Integrated Design and Analysis of an Amplitude-Variable Flapping Mechanism for FMAV

Peng Nian[✉], Bifeng Song, Wenqing Yang, and Shaoran Liang

Micro Aerial Vehicle Research Laboratory, School of Aeronautics,
Northwestern Polytechnical University, Xi'an 710072, China
2015100276@mail.nwpu.edu.cn

Abstract. Bird is one of natural flying masters. It can take advantage of limited energy to fly a long distance. However, the performance of FMAV (Flapping-wing Micro Air Vehicle) is far from the birds. Zoologists have showed that birds can dynamically adjust the flapping amplitude and frequency of its wings. This means that the energy consumption of birds in a variety of flight conditions is minimum. The development of a novel flapping mechanism for FMAV was introduced. The mechanism can achieve independently controllable left and right wings' flapping-amplitude, both symmetric and asymmetric. The kinematics equations and the static equilibrium equations are deduced for the design and optimization of the AVFM (Amplitude-Variable Flapping Mechanism). The kinetic character of the mechanism is evaluated through computation and simulation. The result shows that the flapping-amplitude of the wings can be changed symmetrically as well as asymmetrically without affecting in-phase flapping motions.

Keywords: FMAV · Flapping mechanism · Amplitude-variable · Mechanism optimization · Kinematics simulation

1 Introduction

Flapping-wing micro air vehicle (FMAV) is a kind of special micro air vehicles which use the flapping-wing to produce thrust and lift at the same time. Besides due to its flapping flying bionic characteristics, it has a unique advantage than conventional aircraft. Its potential military/civilian value has been the concern of many national research institutions. In recent years, a variety of FMAV successfully carried out a test flight. Such as AeroVironment's Nano Hummingbird [1, 2], Delft's Delfly [3], Festo's Smart Bird [4, 5], NPU's Pigeon [6], etc. It is noteworthy that the overall bionic degree of the current FMAV is still low, in terms of manipulation, flight strategy, wing structure, etc. is still difficult to achieve complete bionic.

Flapping mechanism as the main power source of FMAV, the bionic degree of FMAV is highly dependent on the form of motion that the mechanism can provide [7]. By observing the flight of nature birds, it is not difficult to find that birds can choose different flapping amplitude and frequency in different stages of flight. Such as the take-off stage with large flapping amplitude and high frequency combined to produce high lift, the

© Springer International Publishing AG 2017
Y. Huang et al. (Eds.): ICIRA 2017, Part III, LNAI 10464, pp. 576–588, 2017.
DOI: 10.1007/978-3-319-65298-6_52

cruise stage medium amplitude and low frequency to ensure a lower energy consumption, while the landing stage is to take a large angle of gliding approach. At the same time as the flight directional control, many flying creatures rely on their wings to achieve horizontal navigation [8], rather than the use of control surface deflections. Researchers from the perspective of bionics, developed a series of bionic mechanism, Park et al. used two-stage mechanism design to achieve asymmetric in-phase flap of the wings [9], Iqbal et al. designed an experimental device which can provide asymmetric frequency flapping [10], etc. These driving devices solve the problem of bionic manipulation of flapping wing aircraft to a certain extent, but they cannot provide the most important flapping amplitude symmetric control in bird flight, which restricts the performance of FMAV.

This paper presents an Amplitude-Variable Flapping Mechanism(AVFM), inspired by creatures, to realize the independent amplitude variation of the wings, which has the function to provide symmetric amplitude control as well as asymmetric amplitude control. The mechanism has a single motor to generate flapping motion and two servos to adjust the flapping amplitude of each side of the wings separately. A kinematics model is established and simplified. The characteristics of the mechanism are optimized, combined with static equilibrium equations. The kinematic characteristics of the digital mock-up are analyzed and a prototype is designed to verify the design functions.

2 Design of Amplitude-Variable Flapping Mechanism

In order to obtain the independent and in-phase adjustable amplitude of the flapping wings, the transmission scheme of the flapping mechanism should be determined firstly. At present, the flapping mechanism is used to make crank-rocker mechanism principle, among which the three typical transmission schemes are shown in Fig. 1 [11]. The three transmission schemes can be divided into two categories by the number of front main drive gears: (1) The left and right links share the single drive gear, (2) The left and right links have independent drive gears. The scheme (c) in Fig. 1 is a more successful configuration driven by a single gear. It has the advantages of simple structure, few parts, light weight and reliable transmission. However, due to the existence of short-circuit characteristics of the four-bar mechanism, in the process of drive gear running a circle, the left and right rockers flap asymmetrically. And there is a certain phase difference, which will lead to aerodynamic differences among left and right flapping wings. So, the link parameters need to be carefully optimized to reduce this adverse phase difference [12, 13]. The scheme (a) in Fig. 1 uses a pinion gear to drive the large gears on both sides at the same time. It is easy to know that the rotation direction of the two large gears are the same, which means the motion of the left and right rockers are not mirrored symmetrically. Its rocker's movement also exists phase difference. The scheme (b) in Fig. 1 uses a pinion gear to drive the right large gear and the left large gear is driven by the right large gear then. This type of transmission scheme ensures that the motion of the rockers on both sides are completely mirror symmetrical. Besides the changing of one-side parameters does not affect the kinematic characteristics of the other side. Therefore, the transmission scheme like scheme (b) can meet the needs of in-phase, independent movement of the two sides wings in flapping mechanism design goals.

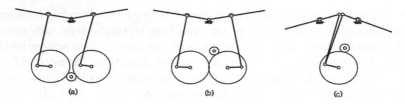

Fig. 1. Three different flapping-wing transmission schemes [11].

The paper presents an AVFM in the form of multi-links as shown in Fig. 2. Take the right side as an example where *FCGH* is a double-rocker mechanism, when the *GH* position is fixed, point C will have a definite position. When point C has a definite position, *ABCD* is a crank and swing slider mechanism. *AB* can rotate around point A. The rotation of the *AB* drives the *BD* to slide in the slider at point C, and *BD* also swings around point C. The trajectory of point is similar to the spindle shape. Point D connect rocker *EI* with a slider. Rocker *EI* produce flap movement with the driving by point D. As *GH* is rotated clockwise to a certain angle, the position of point C changed, which result in the relative distance between point D's trajectory and point E gets longer. Then the flap amplitude of *EI* will reduce. Similarly, when *GH* is rotated counterclockwise, the flap amplitude of *EI* will increase. The motion principle of the left side is same to the right side, their motions are in-phase, independent and do not interfere with each other.

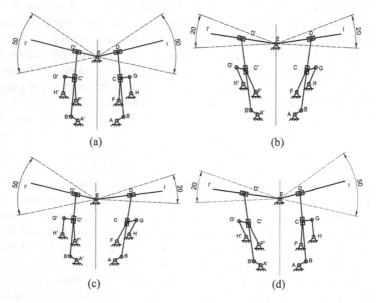

Fig. 2. AVFM schematic diagram. (a) *GH* and *G'H'* mirror rotate a same angle, causing both sides of the wings flapping amplitude to increase symmetrically. (b) *GH* and *G'H'* mirror rotate a same angle, causing both sides of the wings flapping amplitude to decrease symmetrically. (c) *GH* and *G'H'* rotate a different angle, causing the left wing flapping amplitude larger than the right wing's flapping amplitude. (d) *GH* and *G'H'* rotate a different angle, causing the left wing flapping amplitude smaller than the right wing's flapping amplitude.

3   Kinematic Analysis

For the sake of facilitating the parametric design of the entire mechanism, it is necessary to simplify the mathematical model of the mechanism. As the geometric parameters of the left side and the right side are exactly the same, so take the right side as an example, the paper creates a simplified model and a cartesian coordinate system as shown in Fig. 3. The simplified model contains 6 design parameters (L_1, L_2, L_3, α, xA, yE). In order to facilitate the solution of kinematic equations, the model is decomposed into crank and swing slider mechanism $ABCD$ and crank and swing guide-bar mechanism $EDCB$.

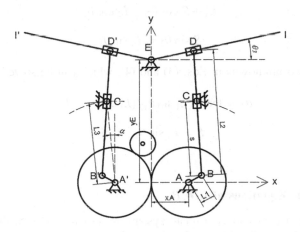

Fig. 3. AVFM simplified model.

3.1   Kinematic Equations of $ABCD$

Before applying the method of complex vector to establish the position equation of $ABCD$, it is necessary to use the vector to represent the links and build the closed vector polygon of the mechanism. As shown in Fig. 4, the link vector and azimuth are marked. When given L_1, L_2, L_3 and θ_1, there are two unknown variables S and θ_2.

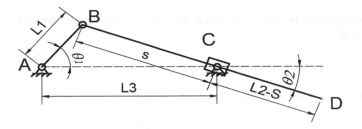

Fig. 4. Crank and swing slider mechanism $ABCD$.

The closed vector equation can be obtained from the closed vector polygon:

$$\vec{L} + \vec{S} = \vec{L_1} \tag{1}$$

Rewrite Eq. (1) as a complex vector form:

$$L_3 + S e^{i\theta_2} = L_1 e^{i\theta_1} \tag{2}$$

Applying the Euler formula $e^{i\theta} = \cos\theta + i\sin\theta$ to separate the real and imaginary parts of Eq. (2):

$$L_3 + S \cos\theta_2 = L_1 \cos\theta_1 \tag{3}$$

$$S \sin\theta_2 = L_1 \sin\theta_1 \tag{4}$$

There are two unknowns in Eqs. (3) and (4), which can be solved in parallel:

$$\theta_2 = \tan^{-1}(L_1 \sin\theta_1 / (L_1 \cos\theta_1 - L_3)) \tag{5}$$

$$S = L_1 \sin\theta_1 / \sin\theta_2 \tag{6}$$

3.2   Kinematic Equations of *EDCB*

Firstly, creating a cartesian coordinate system as shown in Fig. 5, the variables are marked on the figure. Combined with the previous results, we can get the following coordinate relationships:

$$xC = xA + L_3 \sin\alpha \tag{7}$$

$$yC = L_3 \cos\alpha \tag{8}$$

$$xD = xC + (L_2 - S)\sin(\alpha - \theta_2) \tag{9}$$

$$yD = yC + (L_2 - S)\cos(\alpha - \theta_2) \tag{10}$$

According to the obtained coordinate relationship, the flapping angle of *EI* (θ_3) can be calculated:

$$\theta_3 = \tan^{-1}((yD - yE)/(xD - xE)) \tag{11}$$

And the flap amplitude (A_{flap}) can be expressed as:

$$A_{flap} = \max(\theta_3) - \min(\theta_3) \tag{12}$$

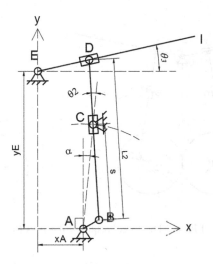

Fig. 5. Crank and swing guide-bar mechanism *EDCB*.

3.3 Angular Velocity/Acceleration of *EI*

The angular velocity and angular acceleration of the wing are obtained indirectly by the time derivative of the real-time flapping angle θ_3. The actual calculated instantaneous flapping angle θ_3 is not a continuous function, so the method of discrete mathematics is used to calculate the flapping angular velocity and angular acceleration. The paper uses the second order central method:

Angular velocity:

$$\omega_3(i) = \frac{1}{2dt}(\theta_3'(i+1) - \theta_3'(i-1)) \tag{13}$$

Angular acceleration:

$$\omega_3(i) = \frac{1}{2dt}(\theta_3'(i+1) - \theta_3'(i-1)) \tag{14}$$

Where $\theta_3'(-1)$ and $\theta_3'(n)$ are equal to the value of the first point of the original data θ_3, $\omega_3'(-1)$ and $\omega_3'(n)$ are equal to the value of the first point of the original data ω_3, and f represent the frequency of the flapping.

4   Dynamic-Static Analysis

In order to analyze the mechanical characteristics of the AVFM, it is necessary to carry out the dynamic static analysis. Finding the force/torque to be applied to the link, as well as the reaction of the joints to balance the aerodynamic load of the flapping wing using the static method [14]. The force/torque of the AVFM is shown in Fig. 6,

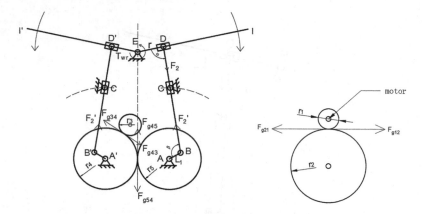

Fig. 6. The force/torque of the AVFM.

where T_{wr} is the aerodynamic torque of the right wing and J represents the moment of inertia of the rocker EI. As the other links' movement range is small, so their inertial force is ignored. The following equilibrium equations are obtained:

$$T_{wri} = r_i \sin \alpha_i F_{2i} + J\alpha_{3i} \tag{15}$$

$$F_{2i}L_1 \sin \beta_i = F_{g45i}r_5 \tag{16}$$

$$F_{g45i} = \frac{L_1 \sin \beta_i}{r_5 r_i \sin \alpha_i}(T_{wri} - J\alpha_{3i}) \tag{17}$$

$$F_{g54i}r_4 + F'_{2i}L'_1 \sin \beta'_i = F_{g34i}r_4 \tag{18}$$

$$F_{g54i}r_4 + F'_{2i}L'_1 \sin \beta'_i = F_{g34i}r_4 \tag{19}$$

$$F_{g34i} = F_{g43i} = 2F_{g45i} \tag{20}$$

It can be seen from the Eq. (20) that the maximum force on the front of the mechanism is the part of gear No. 3 meshing with the gear No. 4, thus the paper defines a load transfer coefficient w to characterize the force of the point.

$$w = \frac{L_1 \sin \beta_i}{r_5 r_i \sin \alpha_i} \tag{21}$$

The force on the back of the mechanism is shown in Fig. 6. The gear No. 3 and the gear No. 2 transfer torque through an axis.

$$F_{g43i}r_3 = F_{g12i}r_2 \tag{22}$$

$$T_{motori} = F_{g21i}r_1 = \frac{2r_1 r_3 L_1 \sin \beta_i}{r_2 r_4 r_i \sin \alpha_i}(T_{wri} - J\alpha_{3i}) \tag{23}$$

Gear transmission ratio:

$$i = \frac{r_2 \, r_4}{r_1 \, r_3} \tag{24}$$

$$T_{motor_i} = \frac{2L_1 \sin \beta_i}{i r_i \sin \alpha_i} (T_{wri} - J\alpha_{3i}) \tag{25}$$

When input T_{wr}, T_{motor} can be calculated, it can be used as a reference to select the motor, which can provide enough balance torque.

5 Multi-objective Optimization

From the dynamic static analysis in Sect. 4, it can be seen that the load at the meshing point of the gear No. 3 and the gear No. 4 is larger than anywhere. So, the design parameters must meet the minimum w target, and to reduce the impact of the swing of the BD link on point C, the amplitude of θ_2 (A_{02}) should be the smallest. The wind tunnel experiment results show that the amplitude of the flapping is the dominant factor in the aerodynamic force of the FMAV [15]. So, the amplitude A_{flap} should be designed within a reasonable range. Also, the flapping up and down angle of deviation dA should be limited to a small angle to ensure the stability of the FMAV. The design parameters should also meet some geometric constraints. In order to optimize multi objectives, the paper uses MATLAB and Isight to solve the global optimal parameters satisfying the constraint conditions by applying multi-island genetic algorithm. A MATLAB program based on the equations derived previously calculate the required data and Isight call the program to optimize the design parameters. The optimization algorithm settings and the optimization results are shown in Tables 1 and 2.

Table 1. Algorithm settings.

Option	Value
Sub-population size	200
Number of islands	2
Number of generations	600
Scale factor of θ_2	10
Weight factor of θ_2	0.35
Scale factor of w	0.1
Weight factor of w	0.65

Table 2. Optimization results.

Parameters	Initial	Optimized
L_1	4	4.279
L_2	36	37.039
L_3	25	27.750
α	-0.0698	-0.0648
xA	10	10
yE	36	36.190

Objectives: $f(x) = \min(w(x), A_{\theta_2}(x))$, $x = (L_1, L_2, L_3, \alpha, xA, yE)^T$.

Constraints:
$$\begin{cases} L_1 - L_3 \leq 0 \\ -L_2 + L_3 \leq 0 \\ L_1 - L_2 + L_3 \leq -5 \\ L_1 - xA \leq -3 \\ L_3 - yE \leq 0 \\ xA - L_3 \cos\alpha \leq 0 \\ 59 \leq A_{flap} \leq 70 \\ 0 \leq dA \leq 15 \end{cases}$$
Parameters:
$$\begin{cases} 3.5 \leq L_1 \leq 10 \\ 30 \leq L_2 \leq 50 \\ 20 \leq L_3 \leq 50 \\ -0.0873 \leq \alpha \leq 0.1745 \\ 5 \leq xA \leq 15 \\ 20 \leq yE \leq 45 \end{cases}$$

Getting a set of appropriate parameters through repeated optimization, using MATLAB programming to plot the following curves. The curves show that flapping amplitude has increased, which is conducive to the generation of aerodynamic. The changes of angular velocity and angular acceleration are not significant. The load transfer coefficient w which is defined before, its peak value is significantly reduced,

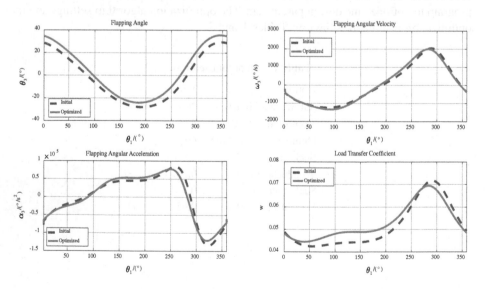

Fig. 7. The curves of optimization result.

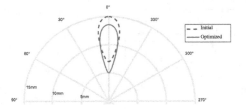

Fig. 8. The trajectory of point D in polar coordinate system.

which will significantly improve the mechanical properties of the AVFM (Fig. 7). The trajectory of point D in polar coordinate system is shown in Fig. 8. As can be seen from the figure, the amplitude of θ_2 ($A_{\theta 2}$) has a significant reduction, which will improve the stability of the AVFM.

6 Modeling, Kinematic Simulation, and Fabrication

Based on the optimization parameters, a detailed digital mock-up is built in UG (Fig. 9). Several joints and drive motions are added to the digital, and then carried out a series of simulations. The simulation results are showed in Fig. 10. And can be used to verify the design goals of the mechanism.

Fig. 9. The digital mock-up and the physical prototype of AVFM.

To further test the functions of the mechanism, a physical prototype is fabricated with aluminum, titanium alloy and other materials. The procedure of the test with flapping wings is showed in Fig. 11.

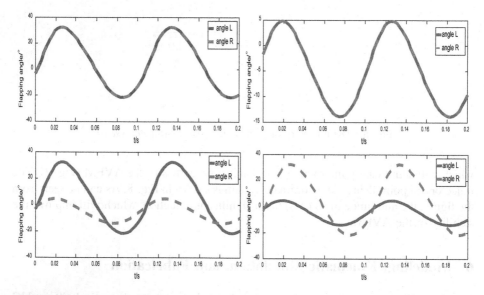

Fig. 10. Kinematics simulation results. (a) Both side of the wings flapping at high amplitude symmetrically. (b) Both sides of the wings flapping at low symmetrically. (c) The left wing's flapping amplitude larger than the right wing. (d) The left wing's flapping amplitude smaller than the right wing.

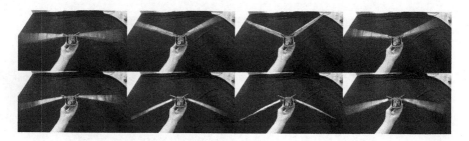

Fig. 11. Physical prototype test with flapping wings.

7   Conclusion

In this paper, a novel Amplitude-Variable Flapping Mechanism(AVFM) is presented along with its kinematic analysis, dynamic-static analysis, multi-objective optimization and kinematic simulation. The inspiration of this mechanism comes from creatures in nature. The mechanism implements the independent motion control of the left and right flapping wings with a single motor and two servos. It can make the amplitude of the flapping wings increase or decrease simultaneously without phase delay, and can also make the flapping wings have different amplitude. Real-time amplitude adjustment capability can increase the energy efficiency of FMAV, and the independent control of left and right flapping amplitude will improve the maneuverability of FMAV. A set of optimized design parameters have been obtained through kinematics modeling and optimization design. The kinematics simulation and physical prototype testing result show that the AVFM can meet the above functional requirements, and can be applied to high performance FMAV.

The work of this paper focuses on the conceptual design, modeling, optimization and simulation of a novel AVFM, and does not fully consider the influence of aerodynamics on the mechanism during flapping. The physical prototype is only built for testing and does not take into account the size and the weight constraints. In future work, wind tunnel experiments based on the mechanism will be carried out to obtain the actual aerodynamic force and bring into the digital mock-up to study the dynamic performance with the aerodynamic force.

Acknowledgments. The authors gratefully acknowledge the funding from the National Natural Science Foundation of China (11402208, 11572255, U1613227) for the support of this work.

References

1. Keennon, M., Klingebiel, K.: Development of the Nano Hummingbird: a tailless flapping wing micro air vehicle. In: 50th AIAA Aerospace Sciences Meeting including the New Horizons Forum and Aerospace Exposition, pp. 588–615. AIAA, Nashville (2012)

2. Kadam, S., Gajbhiye, S., Banavar, R.: A geometric approach to the dynamics of flapping wing micro aerial vehicles: modelling and reduction. Comput. Sci. **10**(10), 1219–1220 (2015)

3. Croon, G., Clercq, K., Ruijsink, R., Remes, B., Wagter, C.: Design, aerodynamics, and vision-based control of the DelFly. Int. J. Micro Air Veh. **1**(2), 71–97 (2009)

4. Wong, B.: New robot designs are for the birds. Electron. Des. **59**(6), 14 (2011)

5. Send, W., Fischer, M., Jebens, K., Mugrauer, R.: Artificial hinged-wing bird with active torsion and partially linear kinematics. In: 28th International Congress of the Aeronautical Sciences, pp. 23–28. ICAS, St. Maarten (2012)

6. Wang, L.: Research on the System Design and Development Approach for Biomimetic Flapping-Wing Micro Air Vehicle. Doctor, Northwestern Polytechnical University (2013)

7. Fenelon, M., Furukawa, T.: Design of an active flapping wing mechanism and a micro aerial vehicle using a rotary actuator. Mech. Mach. Theor. **45**(2), 137–146 (2010)

8. Takagi, M., Nishimura, T., Miyoshi, T.: Development of a roll and yaw moment generation mechanism with flapping amplitude control. J. Abmech **4**(1), 56–62 (2015)

9. Park, J., Yang, E., Zhang, C., Agrawal, S.: Kinematic design of an asymmetric in-phase flapping mechanism for MAVs. In: IEEE International Conference on Robotics and Automation, pp. 5099–5104. IEEE, River Centre (2012)

10. Iqbal, S., Gilani, H., Azim, R., Malik, A., Zafar, S.: Design and analysis of a flapping wing mechanism with asymmetric frequencies for a micro air vehicle. Iran. J. Sci. Technol. Trans. A Sci. **40**(2), 75–89 (2016)

11. Shao, L., Song, B., Li, W.: Synthesized designing method on flapping mechanism of minityped flapping wing aircraft based on fuzzy judgement and optimization. J. Mach. Des. **24**(9), 28–32 (2007)

12. Zhou, K., Fang, Z., Cao, X., Zhang, M.: Optimization design for single-crank and double-rocker kind of driving mechanism of FMAV. J. Aerosp. Power **23**(1), 184–188 (2008)
13. Park, J., Agrawal, S.: Dynamic effects of asymmetric in-phase flapping (AIF) on forward flight. In: IEEE International Conference on Robotics and Automation, pp. 3550–3555. IEEE, Hong Kong Convention and Exhibition Center (2014)
14. Liu, C.: Mechanism Design and Dynamic Simulation Research on Bionics Flapping-Wing Air Vehicles. Master, Harbin Institute of Technology (2010)
15. Shao, L., Song, B., Song, W.: Experimental research of flapping-wing propulsion for micro air vehicles. J. Exp. Fluid Mech. **23**(1), 1–6 (2009)

Modeling and Hover Control of a Dual-Rotor Tail-Sitter Unmanned Aircraft

Jingyang Zhong$^{(\boxtimes)}$, Bifeng Song, Wenqing Yang, and Peng Nian

Micro Aerial Vehicle Research Laboratory, School of Aeronautics,
Northwestern Polytechnical University, Xi'an 710072, China
zjynwpu@mail.nwpu.edu.cn

Abstract. A tail-sitter unmanned aircraft is capable of transition between horizontal and vertical flight. This paper highlights topics of interest to developing a more accurate model to make the simulation results more reasonable. The modeling processes are presented in details. Aerodynamics of this unmanned aircraft is obtained by wind tunnel tests associated with aerodynamic estimation software. Characters of propeller slipstream are analyzed and the mathematical relationships among slipstream velocity, propeller speed, radial location and axial location of propeller plane are deduced from the experiment data. Besides, separate consideration of the propeller slipstream on wings and control surfaces gives better estimations on the dynamic pressure. Models of actuators and motors are also obtained through some tests to make the results reliable. Furthermore, a simple controller is designed to implement the hover attitude control.

Keywords: Dual-rotor Tail-sitter · Reliable model · Hover · Slipstream · Actuator · Control

1 Introduction

As a result of advances in communication, computation and sensing, unmanned aerial vehicles are into the stage of rapid development. Most common unmanned aerial vehicles are multirotor while the drawback is obvious whose forward speed is relatively low [1–3]. Although the fixed wing aircraft can remedy the deficiency of low forward speed, appropriate runway for takeoff and landing has also limited its use. A tail-sitter aircraft could not only takeoff and landing vertically, but also fly forward at high speed like conventional fixed wing aircraft.

The University of Sydney has been studied a tail-sitter aircraft named T-wing for many years. The characters of the aerodynamics have been analyzed extensively. Through simulation and flight test results, the effectiveness of the controller is validated both at hover, transition and forward flight stage [4, 5]. The Brigham Young University focuses the study on a delta wing tail-sitter aerial vehicle named Pogo and quaternion based controllers are their main method [6, 7]. Drexel university takes advantage of a fixed wing aircraft with high thrust-weight ratio to implement hovering and two small propellers mounted in the wing tip are used to counter the effects of motor torque [8, 9]. Georgia Tech has also developed the UAV named GTEdge to perform aggressive aerobatic maneuver including transition [10]. Most of the aircrafts discussed above are

© Springer International Publishing AG 2017
Y. Huang et al. (Eds.): ICIRA 2017, Part III, LNAI 10464, pp. 589–601, 2017.
DOI: 10.1007/978-3-319-65298-6_53

using propellers to generate thrust and making use of the control surfaces within the slipstream to hold attitude. However, when evaluating the impact of slipstream on aerodynamic forces and moments for wings and control surfaces, many people use momentum theory to simplify model which will definitely affect the accuracy of the model especially the evaluation for efficiency of the rudder. Besides, the bandwidth of actuators plays an important role in the performance of system and should also be considered when building the model.

The outline of this paper is as follows. In Sect. 2, the design of aircraft is presented. In Sect. 3, the mathematical model of aircraft is derived by experiments and basic aerodynamics knowledge. Besides, the slipstream model and the bandwidth of the actuator are also studied through experiments. In Sect. 4, simulation results are presented and conclusions are provided in Sect. 5.

2 Aircraft Design

TBS Caipirinha is a flying wing which is easy to assemble and robust against hard crashes. This makes it a perfect platform for flight testing. By equipping with dual rotors, a propeller rotating clockwise and the other rotating counter clockwise, the Caipirinha could be modified for vertical takeoff and landing as in Fig. 1. Two Sunnysky-2208 brushless motors with a pair of 8045 propellers are installed on each wing by using two 3D printing motor cabinets. The UAV comes with two elevons for roll and pitch control and two rotors for yaw control.

The guidance, navigation and control operations on the UAV are implemented with the Pixhawk autopilot system. The system is integrated with a 168 MHz stm32f427 as main processor and a 24 MHz stm32f100 as co-processor for fail-safe. Three axis accelerometers, a three-axis gyro, a three-axis magnetometer, a barometer, GPS and a digital airspeed sensor are also integrated to constitute a complete autopilot system.

Fig. 1. Prototype of UAV.

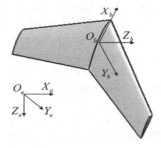

Fig. 2. Definition of earth and body coordinate.

3 Modeling

The 6 DOF model of the vehicle consists of five primary subsystems: rigid-body dynamic, aerodynamic model of wings and control surfaces, motor model, slipstream model and actuator model.

3.1 Coordinate Definition

The coordinates used in this paper follow the conventional fixed-wing aircraft. It is assumed that the Earth reference frame (X_e, Y_e, Z_e) is the right handed inertial coordinates and (X_b, Y_b, Z_b) denotes frame attached to the aircraft body whose origin is located at the center of gravity with X_b pointing to the nose of the vehicle. Figure 2 shows the direction of each axis.

3.2 Rigid-Body Dynamic Model

The Newton's second law provides the overall equations of motion of a rigid body. The translational motion of the body-fixed coordinate frame is given below, where the applied forces $[F_x, F_y, F_z]$ are in the body-fixed frame and the mass of the body m is assumed constant.

$$\overrightarrow{F}_b = \begin{bmatrix} F_x \\ F_y \\ F_z \end{bmatrix} = m\left(\dot{\overrightarrow{V}}_b + \overrightarrow{\omega} \times \overrightarrow{V}_b\right) \tag{1}$$

$$\overrightarrow{V}_b = \begin{bmatrix} u_b \\ v_b \\ w_b \end{bmatrix}, \quad \overrightarrow{\omega} = \begin{bmatrix} p \\ q \\ r \end{bmatrix} \tag{2}$$

Where $\overrightarrow{V}_b$ denotes the linear velocity in body frame and $\overrightarrow{\omega}$ denotes the angular velocity.

The rotation dynamics of the body-fixed frame are given below, where the applied moments are $[L, M, N]$ and the inertia tensor I is with respect to the center of gravity.

$$\overrightarrow{M}_b = \begin{bmatrix} L \\ M \\ N \end{bmatrix} = I\dot{\overrightarrow{\omega}} + \overrightarrow{\omega} \times \left(I\overrightarrow{\omega}\right) \tag{3}$$

The relationship between the body-fixed angular velocity $\overrightarrow{\omega}$ and the quaternion $\overrightarrow{\eta}$ can be determined by resolving the follows equations,

$$\dot{\overrightarrow{\eta}} = \frac{1}{2}\left(\overrightarrow{\eta}^X + \eta_4 I_3\right)\overrightarrow{\omega}, \quad \dot{\eta}_4 = -\frac{1}{2}\overrightarrow{\eta}^T\overrightarrow{\omega} \tag{4}$$

Where the notation of I_3 is identity matrix and ξ^x denotes the skew-symmetric matrix given by:

$$\vec{\xi}^X = \begin{pmatrix} 0 & -\xi_3 & \xi_2 \\ \xi_3 & 0 & -\xi_1 \\ -\xi_2 & \xi_1 & 0 \end{pmatrix} \tag{5}$$

The total $\vec{F}_b$ consists of the thrust $\vec{F}_{rotor}$ generated by two counter rotating rotors, aerodynamic forces on the fixed wings $\vec{F}_{aerodynamic}$, aerodynamic force on the control surfaces $\vec{F}_{control}$, gravity of the aircraft $\vec{F}_G$ and forces due to external disturbances $\vec{F}_{disturbance}$,

$$\vec{F}_b = \vec{F}_{aerodynamics} + \vec{F}_{control} + \vec{F}_{rotor} + \vec{F}_G + \vec{F}_{disturbance} \tag{6}$$

The total moment $\vec{M}_b$ consists of the moments $\vec{M}_{rotor}$ created by two rotors, aerodynamic moments on the fixed wing $\vec{M}_{aerodynamic}$, aerodynamic moments on the control surfaces $\vec{M}_{actuator}$ and moments due to external disturbances $\vec{M}_{disturbance}$,

$$\vec{M}_b = \vec{M}_{aerodynamic} + \vec{M}_{control} + \vec{M}_{rotor} + \vec{M}_{disturbance} \tag{7}$$

3.3   Aerodynamic Model of Wings and Control Surfaces

The aerodynamic model of the fixed wings and control surfaces are obtained from both wind tunnel tests [11] for static derivatives and aerodynamic estimation software for dynamic derivatives as well as the derivatives that wind tunnel tests could not get. Considering the large flight envelope which means 90 degrees' attitude change, using linearized model to describe the aerodynamics of fixed wings is not appropriate, so the coefficients should be calculated as follows

$$C_L = C_{L\,exp}(\alpha, V, \delta_e) + \frac{c}{2V} C_{Lq} q \tag{8}$$

$$C_Y = C_{Y\,exp}(\beta, V) + \frac{b}{2V} \left(C_{Cp} p + C_{Cr} r \right) \tag{9}$$

$$C_D = C_{D\,exp}(\alpha, V, \delta_e) + \frac{c}{2V} C_{Dq} q \tag{10}$$

$$C_l = \frac{b}{2V} \left(C_{lp} p + C_{lr} r \right) + f(\delta_e)\delta_a \tag{11}$$

$$C_m = C_{m\,\exp}(\alpha, V, \delta_e) + \frac{c}{2V}C_{mq}q \qquad (12)$$

$$C_n = \frac{b}{2V}(C_{np}p + C_{nr}r) \qquad (13)$$

$C_{L\,\exp}(\alpha, V, \delta_e)$, $C_{Y\,\exp}(\beta, V)$, $C_{D\,\exp}(\alpha, V, \delta_e)$ and $C_{m\,\exp}(\alpha, V, \delta_e)$ can be found in the database from wind tunnel experiments and dynamic derivatives such that C_{mq} are obtained using the open-source Tornado VLM (Vortex Lattice Method) code [12]. The Fig. 3 shows the Geometry plot of this aircraft used in the aerodynamic estimation calculation and Fig. 4 shows the distribution of pressure for each panel.

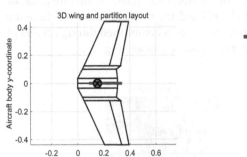

Fig. 3. Geometry definition in Tornado.   **Fig. 4.** Pressure distribution.

The lift equation can then be used after transformation from wind to body frame which yield the final aerodynamic forces and moments as

$$\vec{F}_{aerodynamics} = \frac{1}{2}\rho V^2 S \begin{bmatrix} \cos(\alpha)C_L - \sin(\alpha)C_D \\ -C_Y \\ \sin(\alpha)C_D - \cos(\alpha)C_D \end{bmatrix} \qquad (14)$$

$$\vec{M}_{aerodynamic} = \frac{1}{2}\rho V^2 S \begin{bmatrix} -bC_l \\ cC_m \\ -bC_n \end{bmatrix} \qquad (15)$$

3.4 Motor Model

The $\vec{F}_{rotor}$ and $\vec{M}_{rotor}$ terms in (6) and (7) represent the thrust and moment generated by the propellers

$$\vec{F}_{rotor} = \begin{bmatrix} F_{rotor\_l} + F_{rotor\_r} \\ 0 \\ 0 \end{bmatrix} \qquad (16)$$

$$\vec{M}_{rotor} = \begin{bmatrix} M_{rotor\_l} + M_{rotor\_r} \\ 0 \\ (F_{rotor\_l} - F_{rotor\_r})dr \end{bmatrix} \tag{17}$$

dr is the distance between the thrust axis and the x axis of the body frame.

Due to the focus of this paper is on the hover stage, the thrust and moment tests are under the condition of no free stream. So F_{rotor_r} and M_{rotor_r} are only the functions of propeller rotations. By least square fitting of the experiment results, mathematical expression for $F_{rotor\_r} = k_1 \bullet rpm^2$ and $M_{rotor\_r} = k_2 \bullet rpm^2$ could be obtained.

The experiments were carried out with the rotation speeds of the propeller ranging from 2000 rpm to 10000 rpm and 1000 rpm was set as measurement interval. DC regulated power supply instead of lithium-ion battery was used to guarantee the favorable voltage level during whole tests. The measurement range of the thrust sensor is 1 kg with the relative resolution of $\pm 0.1\%$ $F.S$ and the moment sensor of these are 0.5Nm and $\pm 0.1\%$ $F.S$ separately. Data collection was accomplished through NI-9215 acquisition card with a 2000 Hz sampling rate of analog input (Fig. 5).

Fig. 5. Force and moment testbed.

For ensuring the reliability of the tests, repetitive experiments were done for 5 times. The Fig. 6 shows one of these test results.

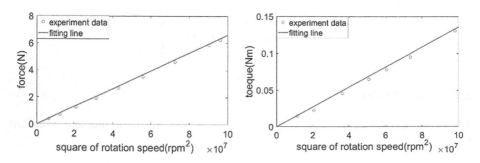

Fig. 6. Force and moment fitting line vs experiment data.

The mathematical models are $F_{rotor} = 6.61 \times 10^{-8}$ rpm^2 and $M_{rotor} = 1.36 \times 10^{-9}$ rpm^2 respectively and the coefficients of k_1 and k_2 are calculated according to the five time tests.

3.5   Slipstream Model

In the hovering stage, propeller slipstream plays a crucial role for maintaining roll and pitch attitude as it creates the dynamic pressure on the control surface. Although the momentum theory could give relatively reasonable results, lacking of the consideration of the diffusion [13] within the slipstream reduces the accuracy of model in some degree. An experiment about testing the slipstream behind propeller was done to offer more accurate model. An 8045 propeller(203.2-mm-diam) is mounted on a 0.5 m stand which keeps the propeller away from the ground disturbance as Fig. 7 below.

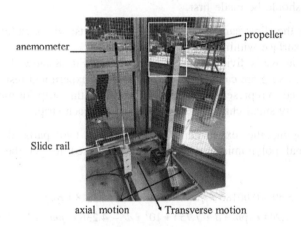

Fig. 7. Slipstream experiment setup.

Five sets of experiments were performed at different rotation speed near the hovering rotation speed (about 7700 rpm): 6200 rpm, 7000 rpm, 7700 rpm, 8000 rpm and 8500 rpm. Induced velocity was measured downstream of the propeller at different axial and radial locations using a TESTO hot-wire anemometer, which has a sampling rate of 2 Hz and a resolution of 0.01 m/s. By taking advantage of the slide rail, the hot-wire anemometer could be traversed in the axial and radial direction within the slipstream region. A measurement grid was set from 35 mm to 515 mm at 20 mm interval in the axial location measured from the propeller plane and radial location from 200 mm to 0 mm at 10 mm interval from the center of the propeller. Besides, due to the rapid change of slipstream speed and low sampling rate of the hot-wire anemometer, recording the average value for a continue time of 10 s as final results in a certain position would be reasonable.

To establish the mathematical model among slipstream speed (m/s), propeller revolution(rpm) and the radial location(m) of the propeller (d = 0.2 m represents the

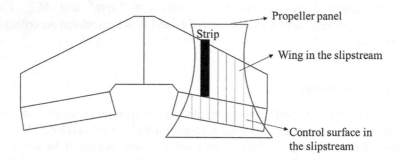

Fig. 8. Strip in the wing and control surface.

propeller center and 0.04 m are away from the propeller in radial direction), some simplifications should be made first.

1. The areas of the slipstream are divided into two parts: wings within the slipstream and control surface within the slipstream
2. Each part of above is divided into different strip as Fig. 8 showed below and mean values of each strip are calculated according to the experiment results. These mean values are used to represent the slipstream speed of this strip for modeling because of the relatively small change of the speed within each strip.

After averaging the experiment results within two parts discussed above, multi-dimensional polynomial fitting method was used to get the following two functions

$$Slipspdw = -15.59 + 0.00186 * rpm + 821.9 * d - 0.08818 * rpm * d - 1.368 * 10^4 * d^2$$
$$+ 1.244 * rpm * d + 7.754 * 10^4 * d^3 - 4.236 * rpm * d^3 - 1.333 * 10^5 * d^4$$
(18)

$$Slipspdcs = 5.043 - 0.00011 * rpm - 204.2 * d + 0.002482 * rpm * d + 930.8 * 10^4 * d^2$$
$$+ 0.277 * rpm * d + 308.4 * 10^4 * d^3 - 1.319 * rpm * d^3 - 537.2 * d^4$$
(19)

d is the radial location(m) of the propeller, *Slipspdw* and *Slipspdcs* represent the speed of the slipstream on the wing and control surface separately.

The R-square of the above fittings are 0.9852 and 0.988 which means good matches among experiment data as in Fig. 9. By using the above two nonlinear functions, calculation of term $1/2\rho V^2 S$ could be distributed to each strip within the slipstream region and sum up the results as the final value to calculate aerodynamic force and moment. Compared to the momentum theory, this experiment based method could offer more accurate results. Different location and size of the control surface may produce different gap between this method and momentum theory results.

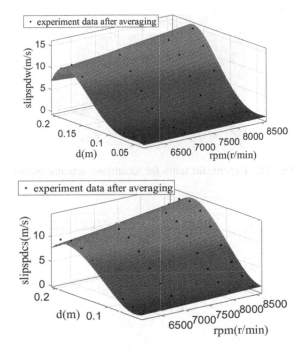

Fig. 9. Slipstream airspeed vs rpm, d within the wing and control surface region.

3.6 Actuator Model

The response of the actuator has a direct impact on the whole control system because of the finite bandwidth of the actuator, an experiment was made below to acquire the model of the actuator. An Arduino board was used to send pulse-width signals to the actuator to follow the desired motion law and the actual response angles were measured through a potentiometer Fig. 10. The expected motion of the actuator is the following cosine function

$$A = 30\cos(2\pi ft + \frac{\pi}{2}) + 90 \tag{20}$$

f is the frequency and A is the amplitude in degree. The 90 degrees' bias is to make the actuator work near the neutral point which is the same as the real working state. Making use of the measurement results in different frequencies, the model of the actuator could be obtained through system identification method.

After mathematical computation, the bandwidth of this actuator near the specified working condition is identified as 26 rad/s and the damping ratio is 0.69 with the undamped natural frequency of 25 rad/s. The Fig. 11 shows the results of the response of the actuator and the identified model under the command of 2 Hz cosine input.

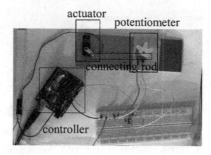

Fig. 10. Experiment setup for identifying actuator model.

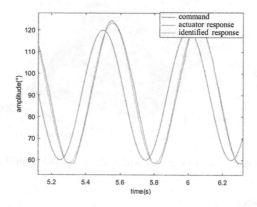

Fig. 11. Comparison between command, actual and identified model response of the actuator.

4   Flight Control and Simulation Results

The system mainly consists of 5 parts: controller model, slipstream model, aerodynamic model, actuator model and rigid-body dynamic model as in Fig. 12. In order to test the whole system, an attitude controller (PID) was designed and tested in the simulation.

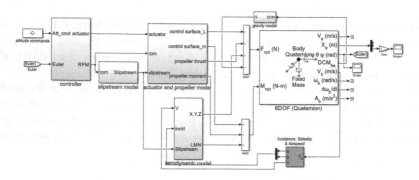

Fig. 12. Simulink model of the whole system.

From the simulation results, it can be seen that under the non-zero initial condition, the aircraft could follow the command and perform well because of the low oscillation and fast response time which means the appropriate controller design Fig. 13.

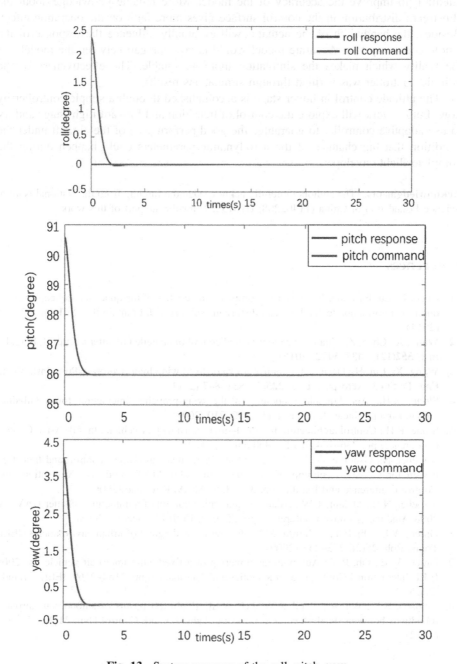

Fig. 13. System response of the roll, pitch, yaw.

5 Conclusion

This paper discussed the modeling of a dual-rotor tail-sitter aircraft and paying more attention to improve the accuracy of the model. More reliable knowledge about the slipstream distribution in the control surface gives more faith on the operation effect. Besides, the bandwidth of the actuator will eventually influence the response of the whole control system. Accurate model would narrow the gap between the model and the reality which makes the simulation more reasonable. The effectiveness of the attitude controller was verified through simulations results.

The attitude control in hover stage is accomplished through a simple controller by now. Future work will explore the control of transition and forward flight stage and try to use adaptive controller to guarantee the good performance of the aircraft under the condition that big changes of the aerodynamic parameters could happen during the complete flight envelope.

Acknowledgments. The authors gratefully acknowledge the funding from the National Natural Science Foundation of China (11402208, U1613227) for the support of this work.

References

1. Chen, F., Lu, F., Jiang, B.: Adaptive compensation control of the quadrotor helicopter using quantum information technology and disturbance observer. J. Franklin Inst. **351**(1), 442–455 (2014)
2. Wang, X., Chen, Z., Yuan, Z.: Modeling and control of an agile tail-sitter aircraft. J. Franklin Inst. **352**(12), 5437–5472 (2015)
3. Wang, X., Lin, H.: Design and control for rotor-fixed wing hybrid aircraft. Proc. Inst. Mech. Eng. Part G J. Aerospace Eng. **225**(7), 831–847 (2011)
4. Stone, R.H.: Aerodynamic modeling of the wing-propeller interaction for a tail-sitter unmanned air vehicle. J. Aircr. **45**(1), 192–210 (2008)
5. Stone, R.H.: Control architecture for a tail-sitter unmanned air vehicle. In: 5th Asian Control Conference, pp. 736–744. IEEE, Melbourne (2004)
6. Knoebel, N., Osborne, S., Snyder, D., et al.: Preliminary modeling, control, and trajectory design for miniature autonomous tailsitters. In: 2006 AIAA Guidance, Navigation, and Control Conference and Exhibit, pp. 345–356. AIAA, Keystone (2006)
7. Knoebel, N.B., Mclain, T.W.: Adaptive quaternion control of a miniature tailsitter UAV. In: 2008 American Control Conference, pp. 2340–2345. IEEE, Seattle (2008)
8. Green, W.E., Oh, P.Y.: A hybrid MAV for ingress and egress of urban environments. IEEE Trans. Rob. **25**(2), 253–263 (2009)
9. Green, W.E., Oh, P.Y.: Autonomous hovering of a fixed-wing micro air vehicle. In: 2006 IEEE International Conference on Robotics and Automation, pp. 2164–2169. IEEE, Orlando (2006)
10. Johnson, E.N., Wu, A., Neidhoefer, J.C., et al.: Flight-test results of autonomous airplane transitions between steady-level and hovering flight. J. Guid. Control Dyn. **31**(2), 358–370 (2012)

11. Sun, J., Li, B., Shen, L., et al.: Dynamic modeling and hardware-in-loop simulation for a tail-sitter unmanned aerial vehicle in hovering flight. In: 2017 AIAA Modeling and Simulation Technologies Conference, pp. 874–886. AIAA, Grapevine (2017)
12. Tornado Vortex Lattice Method. http://www.redhammer.se/tornado. Accessed 16 Mar 2017
13. Khan, W., Nahon, M.: Development and validation of a propeller slipstream model for unmanned aerial vehicles. J. Aircr. **52**(6), 1985–1994 (2015)

Experimental Study on Dynamic Modeling of Flapping Wing Micro Aerial Vehicle

Shaoran Liang$^{(\boxtimes)}$, Bifeng Song, Wenqing Yang, and Peng Nian

School of Aeronautics, Northwestern Polytechnical University,
Xi'an 710072, China
lsr910906@126.com

Abstract. Flapping-wing Micro Aerial Vehicle (FWMAV) is a kind of aircraft which can produce lift and thrust by flapping the wings just like birds or insects. Wind tunnel experiment played an important role in the Micro Aerial Vehicle (MAV) research. This paper introduces wind tunnel experiments of a pair of flexible flapping wings, and analyzes mean and instantaneous characteristics of aerodynamic force and torque. According to the characters of force and torque, they are divided into average signals and periodic alternating signals. Average signals are related to motion characteristics of aircraft, like fixed wing aerial vehicle. Meanwhile, periodic alternating signals are related to periodic flapping up and down, and the frequency is the same as flapping frequency. Finally, the article establishes an available longitudinal flight dynamic model of FWMAV using small perturbation theory to deal with average signals and the model was verified by the real flight tests.

Keywords: Wind tunnel experiment · Flapping wing · Dynamic modeling · Average signals · Period alternating signals

1 Introduction

There has been an upsurge interest in the development of FWMAV, which can generate lift and thrust at the same time by flapping the wings just like birds and insects. The flapping flight is believed to have several advantages such as better maneuverability, lower power cost, better bio-imitability and so on, compared with the same scale fixed-wing aircraft and rotorcraft [1]. Inspired by nature's flyers, researchers have conducted a number of studies on the important aspects of flapping flight in nature. Actually, the biological flying birds have flexible flapping wings, which can change the motion law and passive or active deformation law real-timely during the up and down motion to achieve the optimal aerodynamic performance. The wings produce lift to resist gravity and thrust to balance drag, meanwhile, the tail provides control moment during take-off, flight and landing process. The effective cooperation and coordination between wings and tail is the guarantee of flight stability and maneuverability.

Being different from birds, insects possess marvelous flight skills including high-speed forward flight, steady hovering flight, quick stop and so on, as a result of adjusting the three degree of freedom wing kinematic parameters in various flight modes. However, nearly all of the research of insect flight mostly focuses on the

© Springer International Publishing AG 2017
Y. Huang et al. (Eds.): ICIRA 2017, Part III, LNAI 10464, pp. 602–612, 2017.
DOI: 10.1007/978-3-319-65298-6_54

theoretical analysis and system simulation, but lacks test verification and actual application because of the limits of miniaturization and low energy consumption of sensor system, drive mechanism, etc.

At present, recent research activities in the area of ornithopter mainly focus on the following several aspects: (1) the flowing mechanism of the flow field and flow control, (2) the coupling between deformation of wing and aerodynamic force, (3) the efficient three-dimensional flapping mechanism, (4) the problem of flight stability and control at low Reynolds numbers. Deep understanding of instantaneous and mean characteristics for aerodynamic force and moment is the key to solve problems of ornithopter show as above.

During the flight of flying creatures in nature, they use complicated wing motions consisting of flapping up and down, which is the main motion form, active and passive twisting along spanwise and chordwise, folding along spanwise and forward and backward sweeping, etc. [2].

The calculation of aerodynamic force and moment, considering satisfactory calculation precision, acceptable time and computational consumption at the same time, is still one of the most sophisticated problems needed to be solved urgently. The calculation methods of aerodynamic characteristics are mainly classified as the estimation method based on blade-element theory, the numerical simulation method using computational fluid dynamic technology and the wind tunnel test.

Most of them have a common requirement that the need of a complete aerodynamic model of ornithopter, which describes aerodynamic force generation mechanism. This quasi-steady assumption has been used by many researchers, including Declarer [3] and Larijani and DeLaurier [4], to analyze existing ornithopters and to design a human-piloted ornithopter. However, in Haithem E Taha's work [5], the longitudinal flight dynamic of FWMAV and insects is considered. Results show that direct averaging is not sufficient to assess the flight characteristics of the relatively low flapping frequency systems.

The aerodynamic models are mainly classified as the blade-element theory models, the computational fluid dynamic models and the experimental models. The blade-element theory is an estimation method for qualitative analysis of the aerodynamic characteristics, and its general use is under the assumption of rigid wings, neglecting time-changing wing flexible deformation [6]. On the other hand, when calculating force and torque during the flight, the application of CFD method usually means relatively long time cost, which is unacceptable for real-time simulation. As mentioned above, in the aspect of establishing mathematical model of ornithopter with the capacity of passive deformation and complicated wing motion forms, wind tunnel test method is more accurate and direct, compared with the former two methods.

Based on the wind tunnel test system of laboratory, this paper introduced wind tunnel experiments of a pair of flexible flapping wings, analyzed mean and instantaneous characteristics for aerodynamic force and moment. According to its own characters, force and torque signals are divided into average signals and periodic alternating signals. Average signals are related to motion characteristics of aircraft, like fixed wing aerial vehicle. Meanwhile, periodic alternating signals are related to periodic flapping up and down, and the frequency is the same as the flapping frequency. Finally, the article established an available simulation model and the model is verified by the real flight tests.

2 Wind Tunnel Experiment

2.1 Purpose of Experiment

The purpose of wind tunnel experiment is to research the relationship between dynamic aerodynamic characteristics (force and moment) and flight parameters (angle of attack, airspeed and flapping frequency).

2.2 The Wind Tunnel and Measuring Balance

Wind tunnel utilized in the flapping wing tests is the low turbulence wind tunnel in Northwestern Polytechnical University (NWPU) (as shown in Fig. 1), which is an absorption-type closed one. The turbulence of test section is almost as low as atmospheric environment, and the wind tunnel is very suitable for experimental study of low-speed aircrafts. The physical performance parameters of testing wind tunnel are shown as follows:

Fig. 1. Low turbulence wind tunnel of NWPU.

- Length of wind tunnel: 39.5 m
- Size of three-dimensional test section: 2.8 m long × 1.2 m wide × 1.05 m high respectively
- Range of airspeed of three-dimensional test section: 3 m/s to 22 m/s
- Turbulence: ≤ 0.02%
- Range of angle of attack: −10° to 36°

In order to meet the requirement of flapping wing measurement, F/T Sensor Nano SI-12-0.12, made by American corporation ATI, is chosen as measuring balance. Its merit is satisfactory dynamic performance, large measurement range, high sensitivity, small volume and so on. The performance parameters of measuring balance are showed in Table 1.

Table 1. Performance parameters of measuring balance

Component	Fx Fy	Fz	Tx Ty	Tz
Range	12 N	17 N	0.12Nm	0.12Nm
Precision	1.00%	1.00%	1.50%	1.75%

2.3 Experiment Model

Flapping Wing Model. The experiment flapping wing model is showed in Fig. 2., which is confirmed feasible and reliable in our successful flapping flight. The model is a kind of flexible flapping wing with special designed camber. It is composed of carbon fiber wing frame and polyether thin films. The wing frame is composed of a front spar, an inclined beam and five ribs of airfoil. The film is a flat covering on the surface of wing frame. A specially designed and constructed junction device is used in the inner end of beam, by which the flapping wing and flapping mechanism can be linked. The geometric size of flapping wing model is showed in Fig. 3.

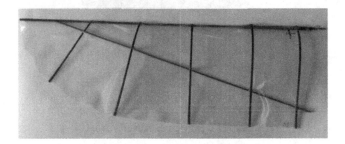

Fig. 2. Flapping wing model.

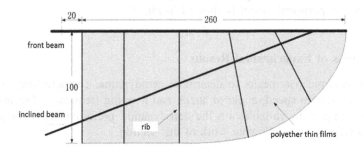

Fig. 3. Size of flapping wing model.

Flapping Mechanism Model. The flapping mechanism model is showed in Fig. 5, which is designed based on the principle of four-bar. Left and right rocker motion simultaneously through gear transmission and circular disk plays the same role as crank. By changing the position of connecting hole, six different flapping amplitudes

Fig. 4. Experiment model.

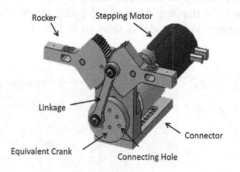

Fig. 5. Flapping mechanism model.

can be selected and they are 35.2°, 47.7°, 54.2° 61.0° and 76.1°, respectively. By adjusting the speed of stepping motor, a range of 1 Hz-10 Hz flapping frequency can be generated Experiment model is showed in Fig. 4.

2.4   Analysis of Experimental Results

The main dynamic parameters to determine aerodynamic characteristics of flapping wing are free stream speed, angle of attack and flapping frequency. The influence of upper three important variables on the aerodynamic performance of flapping wing model is studied in the following work of this section.

Influence of free stream speed. Figure 6 shows the relationship between airspeed and average lift in the case of 12° angle of attack and 8 Hz flapping frequency. Linear fitting to the airspeed-average lift curve was carried out and their correlation coefficient is 0.9979. Results show that there is a good linear relation between 6 m/s and 10 m/s.

The relationship between airspeed and average pitch torque is showed in Fig. 7, and result of linear fitting shows that the linear relation (r = 0.9972) is excellent within the range from 6 m/s to 10 m/s.

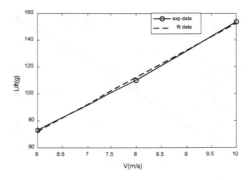

Fig. 6. Relationship between airspeed and average lift.

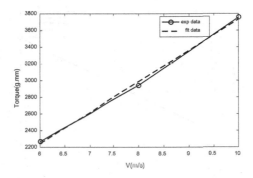

Fig. 7. Relationship between airspeed and average pitch torque.

Influence of angle of attack. The relationship between angle of attack and average lift is given in Fig. 8 in the case of 8 m/s airspeed and 8 Hz flapping frequency. By using linear fitting method, a linear relationship (r = 0.9952) holded between average lift and the range of angle of attack between 9° to 18°.

Under the condition of 8 m/s airspeed and 8 Hz flapping frequency, influence relation between angle of attack and average pitch torque has been displayed in Fig. 9. Then, linear fitting for the four points measured when α = 9°, 12°, 15°and 18°. Results show that their correlation coefficient r = 0.0996, a linear relation between angle of attack and average pitch torque concentration is found in the range of 9° to 18°.

Influence of flapping frequency. Figure 10 contains two curves, one is experiment data for average lift and flapping frequency, the other is fitting curve for experimental data. Results show that there is a good linear relation in the range of 4 Hz to10 Hz, the correlation coefficient r = 0.9962. Experimental data is measured under the condition of airspeed V = 8 m/s and angle of attack α = 12°.

Using the same data processing methods, the influence relation between flapping frequency and average pitch torque was investigated. As shown in Fig. 11, their correlation exhibits favorable linear characteristics with r = 0.9901.

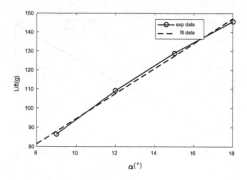

Fig. 8. Relationship between angle of attack and average lift.

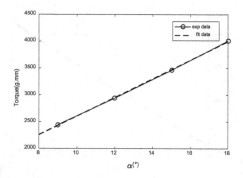

Fig. 9. Relationship between angle of attack and average pitch torque.

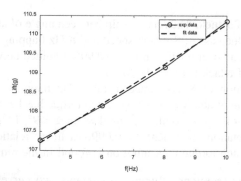

Fig. 10. Relationship between average lift and flapping frequency.

Average Features of Flapping Flight. The average aerodynamic characteristics of flapping flight is analyzed using averaging theory. The relation between mean flight force and moment and experimental variables, airspeed, angle of attack and flapping frequency, is established from experimental results. Based on the analysis above, a strong linear correlation is observed among these flight states. It means that average features during steady flight can be dealt with linearization analysis technology.

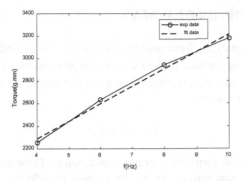

Fig. 11. Relationship between average pitch and flapping frequency.

Instantaneous Features of Flapping Flight. Figure 12 shows the change of instantaneous lift and pitch torque in one flapping period. Dashed line represents the result of fitting instantaneous experimental data, using Fourier fitting method. Solid line represents instantaneous experiment data. Increasing the order number of lift fitting to second, correlation coefficient reaches to r = 0.9931. When torque fitting order number is 3, correlation coefficient reaches to r = 0.99367. Fitting results are the superposition of a constant and several trigonometric functions. The aerodynamic force and torque, which are the important components of real-time simulation, are calculated as the form of Fourier fitting as Formulas 1 and 2.

$$f(t) = A_0 + \sum_{n=1}^{i} a_n \cos(nwt) + b_n \tag{1}$$

$$w = \frac{2\pi}{T} \tag{2}$$

In the above formula, constant term stands for mean characteristics while trigonometric function terms expresses the periodic alternating characteristics during flapping flight.

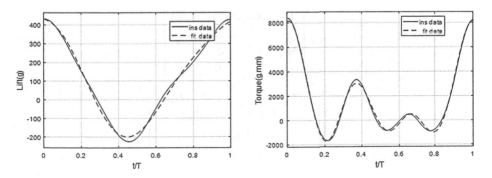

Fig. 12. Instantaneous lift and Instantaneous pitch torque during one flapping period.

3 Modeling Method of FWMAV

Based on the above analysis, force and torque signals are divided into average signals and periodic alternating signals. In Formula 3, $\bar{F}$ represents average signals, which are independent of time t, and $\tilde{F}(t)$ represents periodic alternating signals, whose frequency is equal to flapping frequency.

$$F_t = \bar{F} + \tilde{F}(t) \tag{3}$$

By using the linear form of average aerodynamic force and torque, shown as Formula 3, real force and torque can be determined as follows.

$$L = L_0 + L_V \Delta V + L_\alpha \Delta \alpha + L_{\delta_e} \Delta \delta_e + \tilde{L}(t) \tag{4}$$

$$M = M_0^a + \left(\frac{\partial M_a}{\partial V}\right)_0 \Delta V + \left(\frac{\partial M_a}{\partial \alpha}\right)_0 \Delta \alpha + \left(\frac{\partial M_a}{\partial \delta_e}\right)_0 \Delta \delta_e + \left(\frac{\partial M_a}{\partial q}\right)_0 \Delta q + \tilde{M}(t) \tag{5}$$

$$T = T_0 + T_V \Delta V + T_{\delta_p} \Delta \delta_p + \tilde{T}(t) \tag{6}$$

Because of the coupling relationship between real thrust and drag, the force measured by balance is resultant force in the direction of X. Drag is estimated by Tornado calculation software, which is an estimation program using vortex lattice method.

Based on the small perturbation theory, equilibrium equation of longitudinal flight dynamic of FWMAV can be simplified. Periodic alternating signals are kept invariant. Using the preceding representation, the equations of the longitudinal body motion are written similarly to those of a conventional aircraft [7]; that is

$$\dot{x} = Ax + Bu + C \tag{7}$$

$$A = \begin{bmatrix} X_u + X_{Tu}\cos(\alpha_0 + \varphi_T) & X_\alpha - \frac{1}{m}\tilde{T}(t)\sin(\alpha_0 + \varphi_T) & 0 & -g\cos\gamma_0 \\ \frac{(Z_u - Z_{Tu}\sin(\alpha_0 + \varphi_T))}{V_0} & \frac{Z_\alpha + \frac{1}{m}\tilde{T}(t)\cos(\alpha_0 + \varphi_T)}{V_0} & \frac{V_0 + Z_q}{V_0} & \frac{-g\sin\gamma_0}{V_0} \\ M_u + M_{Tu} & M_{\dot{\alpha}} + M_{Tu} & M_q & 0 \\ 0 & 0 & 1 & 0 \end{bmatrix} \tag{8}$$

$$B = \begin{bmatrix} X_{\delta e} & X_{\delta p}\cos(\alpha_0 + \varphi_T) \\ \frac{Z_{\delta e}}{V_0} & \frac{-X_{\delta p}\cos(\alpha_0 + \varphi_T)}{V_0} \\ M_{\delta e} & M_{\delta p} \\ 0 & 0 \end{bmatrix} \tag{9}$$

$$C = \begin{bmatrix} \frac{1}{m}(\tilde{T}(t)\cos(\alpha_0 + \varphi_T) - \tilde{D}(t)) \\ \frac{\frac{1}{m}(\tilde{T}(t)\sin(\alpha_0 + \varphi_T) - \tilde{D}(t))}{V_0} \\ \frac{1}{I_y}\tilde{M}(t) \\ 0 \end{bmatrix} \tag{10}$$

$$x = \begin{bmatrix} \Delta V \\ \Delta \alpha \\ \Delta q \\ \Delta \theta \end{bmatrix} \tag{11}$$

$$u = \begin{bmatrix} \Delta \delta e \\ \Delta \delta p \end{bmatrix} \tag{12}$$

As can be seen in Formulas 7, 8, 9, 10, 11 and 12, alternating signals change the coefficient of state variables. The last term is introduced due to periodic flapping motion.

4 Simulation Results

In order to verify this model, some simulations need to be conducted. Based on the above analysis, aerodynamic model is made up of average value and periodic alternating value, respectively. Under the limitation of experimental conditions and experimental design method, some simulation data cannot be obtained from experimental results such as real drag and several dynamic derivatives. To solve this problem, estimation software is used to supplement data needed by simulation. Considering the

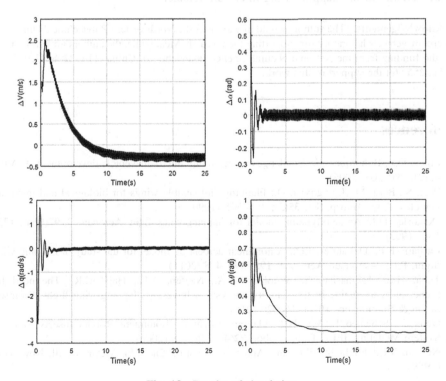

Fig. 13. Results of simulation.

alternating signal is composed of sinusoidal function and cosine function, whose average value is zero, mean balance point can be obtained by solving equilibrium equation.

To simulate the FWMAV, the initial conditions are set as follows: 10 m/s forward flying speed, 100 altitude and 13.2° angle of attack.

The simulation results obviously show that state vatiables fluctuate around average value. Like the real flight test, they are in good agreement with the simulation output and the real flight test results (Fig. 13).

5   Conclusion

The article puts forward an available simulation model of an ornithopter with a pair of flexible flapping wings by testing its aerodynamic characteristics in a low turbulence wind tunnel. Experimental results are analyzed by treating the aerodynamic force and torque as the superposition of a constant and several trigonometric functions. Through linear fitting method, the average parts are investigated. Alternating parts are expressed in the form of Fourier series. On the basis of small perturbation theory, average values are expanded at the mean balance point. However, periodic alternating signals are kept invariant. With these assumption, longitudinal flight dynamic model of FWMAV is given, and simulation results evidence that the established model can reflect the basic characteristics of the flapping wing micro air vehicle.

Acknowledgements. The authors would like to thank the Northwestern Polytechnical University Micro air vehicles laboratory for supporting the research. Meanwhile, the authors gratefully thank for the funding from the National Natural Science Foundation of China (11402208, U1613227, 11572255) for the support of this work.

References

1. Pines, D.J., Bohorquez, F.: Challenges facing future micro-air-vehicle development. J. Aircr. **43**(2), 290–305 (2006)
2. Wei, S., Berg, M., Ljungqvist, D.: Flapping and flexible wings for biological and micro air vehicles. Prog. Aerosp. Sci. **35**(5), 455–505 (1999)
3. Delaurier, D.J.: An aerodynamic model for flapping-wing flight. Aeronaut. J. **97**(964), 125–130 (1993)
4. Larijani, R.F., DeLaurier, J.D.: A nonlinear aeroelastic model for the study of flapping wing flight. Prog. Astronaut. Aeronaut. **195**, 399–428 (2011)
5. Taha, H.E., Tahmasian, S., Woolsey, C.A., Nayfeh, A.H., Hajj, M.R.: The need for higher-order averaging in the stability analysis of hovering, flapping-wing flight. Bioinspiration & Biomimetics **10**(1), 16002 (2015)
6. Deng, X., Schenato, L.: Wu, W.C: Flapping Flight for Biomimetic Robotic Insects: Part I — System Modeling. IEEE Trans. Robot. **22**(4), 776–788 (2006)
7. Nelson, R.C.: Flight Stability and Automatic Control, 2nd edn. McGraw-Hill, New York (1989)

A Novel Low Velocity Robotic Penetrator Based on Ampere Force

Jingkai Feng[1,2], Jinguo Liu[1(✉)], and Feiyu Zhang[1,2]

[1] The State Key Laboratory of Robotics, Shenyang Institute of Automation,
Chinese Academy of Sciences, Shenyang 110016, China
liujinguo@sia.cn
[2] University of the Chinese Academy of Science, Beijing 100049, China

Abstract. The subsurface access technology is of great significance to study evolution of stars and probe landing, and the low velocity robotic penetrator (LVRP) is a commonly used detection technique. In this study, a novel LVRP based on ampere force (AFRP) is proposed, which uses the ampere force of the coils located in high intensity magnetic field as the driving force. The AFRP is composed of stator and mover. In the working state, the mover is subjected to a linear reciprocating motion, colliding with the stator, and displacement occurs. Firstly, parametric modeling of AFRP is carried out and the internal magnetic field distribution is analyzed. The results show that the magnetic induction between the two poles can be increased by the relative mounting of the permanent magnets, and the magnetic induction is affected by the gap and the diameter of the permanent magnets. Then, ISIGHT® software is used for integrated optimization, whose purpose is to optimize the design variables and make the output force of each unit reaches the maximum when the casing size of the AFRP is certain. Furthermore, to achieve the optimization design goal that the overall output force is maximum. According to the optimization results, the prototype was developed. The experimental results show that the AFRP operating frequency can be controlled in the range of 1 ~ 10 Hz, the average displacement of each motion cycle is 2 mm. The proposed AFRP can output a larger driving force by changing the current and the number of units. In addition, the AFRP can be used as a novel actuator for robots.

Keywords: Low velocity robotic penetrator · Subsurface access technologies · Electromagnetic actuator · Ampere force

1 Introduction

The subsurface access technology is considered to be of great significance in space exploration. The development of subsurface access technology has increased the human understanding of the evolution of stars and the solar system, provides a basis for the lander to choose the appropriate landing mode and landing location. Moreover, this technology is used to seek new energy and extraterrestrial civilization.

The most frequently used subsurface access technology can be divided into the following categories: drills, penetrators, hypervelocity impactors, and excavating

© Springer International Publishing AG 2017
Y. Huang et al. (Eds.): ICIRA 2017, Part III, LNAI 10464, pp. 613–623, 2017.
DOI: 10.1007/978-3-319-65298-6_55

machines [1]. The drilling method is mature and widely used, but in the microgravity environment, it is necessary to provide sufficient positive pressure acting on the drill bit, and the cutting heat generated by the drill cuttings is likely to cause the drill pipe failure, which happened to the Apollo 15 mission [2]. Penetrators use only axial force generated by some type of linear actuator to push beneath the surface, and there is no gripping force required. Hypervelocity impactors are mounted on the on-orbit detector and launched, which can obtain a high enough speed and impact the surface at an angle of 90° as much as possible [3]. The excavating machines can scoop or dig into subsurface due to high efficiency and no environmental limitation. But there are also some disadvantages such as large volume and difficulty in launching. Therefore, considering that the penetrator has the advantages of compact-sized, small volume and low cost, it has a good application prospect in stellar surface detection.

Over the past few years, several robotic penetrators have been used in space exploration missions. T. Spohn et al. [4–8] designed a penetrator called MUPUS for the Rosetta Lander Philae to detect the geological characteristics of comet. MUPUS is driven by an electromagnet (EM) drive, and the energy of the system is stored in foil capacitors. The accumulated energy is discharged through a coil to achieve the attraction and release of the hammer, and accelerate the hammer to hit the penetrator to move into the ground. But it requires the corresponding release mechanism to place it in the target area. On the basis of MUPUS, a new generation of mole penetrator called KRET was designed and successfully tested at the Space Research Center PAS in Poland [9, 10]. KRET employs a latch to catch the hammer, pull it and release it to hit the casing when the driving spring is fully compressed. As a result, the casing is inserted. Then, the support mass is decelerated by the return spring and hits the casing once again which causes its additional displacement. But it has a relatively long operation period, the work area of latch will appear varying degrees of fatigue damage. DLR designed the HP^3 Mole penetrator for the ExoMars mission. The hammering mechanism contains a cylindrical cam mechanism driven by a motor, which can apply a periodic load to the spring. But the proposed driven mechanism leads to complex stroke process [11, 12]. The three types of robotic penetrators mentioned above are mainly composed of three parts: the casing, the hammer, and the support mass [13]. They work in a similar principle, where the intrusion in the vertical direction is achieved by the interaction between three masses. However, there are some disadvantages as follows: the complicated mechanism, relatively long operation period, non-adjustable driving force and operating frequency. Also, with the increase of operating, some components have fatigue damage.

In view of the mentioned problems, this paper presents the so-called AFRP, the same magnetic poles are mounted oppositely to form a strong magnetic gap. According to Ampere's Law, by changing the direction of the current in each coil, the coils will be subjected to ampere force in the same direction when they are located in the magnetic gap with different polarities. Changing the magnitude of all currents in the coils and the number of units, the driving force can be adjusted. In addition, the adjustment of the operating frequency can be achieved by the time interval of the coil current commutation. The simulation and optimization of the magnetic field show that the designed AFRP can output a large driving force. The prototype test verifies its feasibility.

2   Principle of Operation

AFRP is composed of stator and mover. The mover consists of a series of oppositely mounted magnets sandwiched a magnetic conductive gasket in the middle, thus forming a strong magnetic gap. The stator includes the casing, the coils and the magnetic conductive gaskets. When the coils are energized, the stator is subjected to the ampere force, and the mover is forced in the opposite direction, moving in the axial direction and interacting with the stator. The spring is used to reduce the rebound. Figure 1 shows the AFRP's working principle.

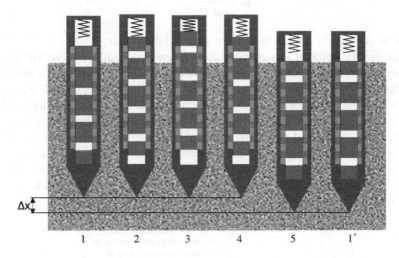

Fig. 1. Schematic principle of operation of AFRP: the casing is shown as black, orange is for coil, green is the gasket, blue represents the magnet S pole, red is for the N pole and the white block is for the magnetic disk. In the order from bottom to top, the odd coil is marked as winding 1, the even coil is winding 2 (Color figure online)

In a single work cycle of AFRP six phases can be discriminated:

PHASE 1: the initial state, no current flows through winding 1 and winding 2.
PHASE 2: the winding 1 is energized, and the winding 2 is not. The current direction in the coil is controlled to be subjected the downward ampere force, so that the motion of the mover is reversed.
PHASE 3: the winding 2 is energized, and the winding 1 is not. The ampere force is still controlled to be downward, and the mover moves on to reach the highest position and compresses the spring.
PHASE 4: changing the current direction of the winding 2, the winding 1 is de-energized, the mover moves downward by the action of the downward ampere force and the spring restoring force.
PHASE 5: changing the current direction of the winding 1, the winding 2 is de-energized, the mover continues to move downward and hits the casing to cause the insertion Δx.

PHASE 1′: restore the initial state, and start the next cycle.

3 Magnetic Field Analysis

The third generation of rare-earth permanent magnetic material (Nd-Fe-B) has the most excellent magnetic properties in the existing permanent magnets [14]. The B-H curve is a straight line, the relative permeability can be obtained from Eq. (1).

$$\mu_r = \frac{B_r}{\mu_0 H_c} \tag{1}$$

Where, $\mu_0 = 4\pi \times 10^{-7}$, is the relative permeability of air.

The magnetic fields of a single permanent magnet and magnets mounted oppositely are analyzed respectively [15]. Figure 2 shows the distribution of magnetic induction in the X direction on the same path of the magnetic field. It can be seen: (1) The permanent magnets mounted oppositely can increase the magnetic induction intensity and form a strong magnetic gap. (2) The magnetic induction of the strong magnetic gap is symmetrical about the middle plane of the gap and reaches the maximum at the plane of symmetry, which is about 2 times the maximum value of the single magnet. (3) The magnetic induction at other locations is not twice the magnetic induction of a single magnet. Therefore, through the design proposed in this paper, magnetic induction intensity increased significantly and can provide a larger output force for the AFRP.

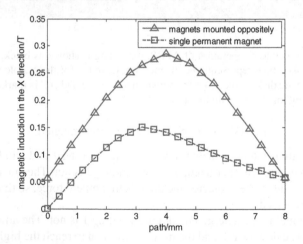

Fig. 2. The distribution of magnetic induction intensity on the same path

The AFRP is an axisymmetric structure, which consists of multiple identical units, and its final output force is the product of the output force of each unit and the number of units. The two-dimensional axisymmetric model of the AFRP's unit is established and the geometric parameters of a single unit are defined in Fig. 3(a). Numerical analysis was made on magnetic field distribution of a single unit based on the axisymmetric element type in ANSYS® software, and the result is presented in Fig. 3(b). According to the analysis, it is found that the magnetic field distribution of adjacent permanent magnets is symmetrical to the middle of the *gap*, and the magnetic field produced by the coil is far smaller than that of the permanent magnets. Therefore, the influence of the coil on the magnetic field can be ignored.

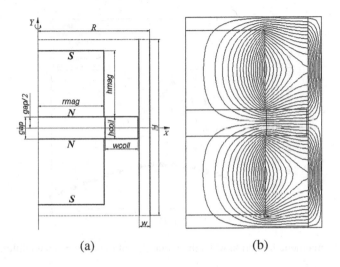

(a)   (b)

Fig. 3. (a) Geometric parameter definition of a single unit, (b) Magnetic field distribution

The magnitude of magnetic induction at any point is affected by the size of magnetic gap and the permanent magnet. When the radial size of the permanent magnet (*rmag*) is fixed and the gap width (*gap*) between the adjacent permanent magnets is taken different values, the distribution of magnetic induction in the X direction is shown in Fig. 4. It can be seen that with the decrease of the *gap*, the magnetic induction in the gap

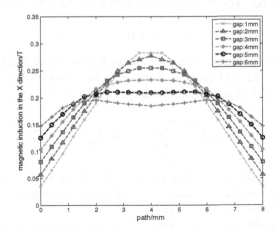

Fig. 4. The magnetic induction when the *gap* is taken different values

within a certain range is increasing. When the *gap* is fixed and the *rmag* is taken different values, the results are shown in Fig. 5. The magnetic induction in the *gap* decreases as the *rmag* of the permanent magnet increases. Besides, when the *rmag* is continuously reduced to the minimum value, the magnetic induction does not reach the maximum value at the middle point.

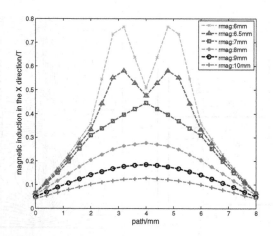

Fig. 5. The magnetic induction when the radial size of the *rmag* is taken different values

According to the above analysis, it can be seen that the distribution of the magnetic induction of AFRP unit in the space is complex. To maximize the output force, the parameters of the model need to be optimized.

4   Parameter Optimization Design

When the number of units is constant, the maximum force of each unit can maximize the overall output force. The AFRP's working principle follows Ampere's Law [16]:

$$d\vec{F} = I\vec{dl} \times \vec{B} \tag{2}$$

From the above analysis, we can see that the distribution of magnetic induction in the field is complicated, and the overall magnetic force is along the Y direction. The value of the force can be calculated according to Eq. (3).

$$F_Y = I \sum_{(rmag,hcoil/2)}^{(rmag+wcoil,hcoil/2)} B_{i,j}L_{i,j} \tag{3}$$

Where, I is the current of the winding, $L_{i,j}$ is the length of the single turn winding coil located at the coordinate system *(i,j)*, and $B_{i,j}$ is the magnetic induction of the component in the X direction at the winding.

Therefore, the problem of maximizing the output force under the condition of the size of the AFRP in the diameter direction is constant can be converted into the optimal design process by changing the value of the design variable to maximize the output force. And the optimization problem can be described as follows:

Design variables:

$$var = \{rmag, hmag, gap/2, hcoil, wcoil\}$$

Constraints:

$$rmag + wcoil + w = R$$
$$2hmag + 2gap = H$$

(4)

Optimization objective function:

$$Maximize: F_Y = I \sum_{(rmag,hcoil/2)}^{(rmag+wcoil,hcoil/2)} B_{i,j}L_{i,j}$$

In this study, ANSYS® software is integrated into ISIGHT® platform, and Hooke-Jeeves algorithm is used to optimize the design variables of AFRP. Each winding is supplied with a current of 100 mA. The specific optimization process is shown in Fig. 6.

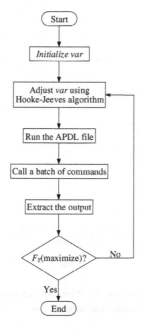

Fig. 6. Optimization design process

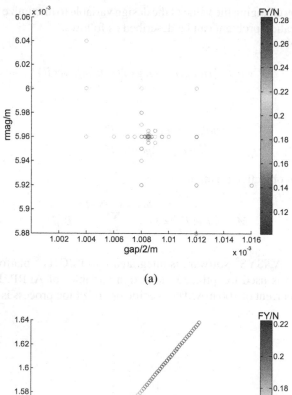

(a)

(b)

Fig. 7. The optimization result: (a) the relationship between F_Y and *rmag* and *gap*, (b) the relationship between F_Y and *rmag/hmag* and *wcoil/hcoil*

Figure 7 shows the optimization result, which reflects the relationship between the unit output force F_Y of AFRP and the design variables *var*. Figure 7(a) presents the relationship between F_Y and the permanent magnet radial dimension (*rmag*) and the gap width (*gap*). The relationship between F_Y and permanent magnet parameter ratio of *rmag* to *hmag* and coil parameter ratio of *wcoil* to *hcoil* is shown in Fig. 7(b). From the optimization result, we can see that even if the change of design variables is very small, the influence on the output force is large. When F_Y obtains a larger value, design variables should satisfy the equation set (5).

$$
\begin{cases}
\dfrac{rmag}{hmag} \in [0.98, 1] \\
\dfrac{gap}{2} \in [1, 1.01] \\
\dfrac{wcoil}{hcoil} \in [1.47, 1.57]
\end{cases}
\tag{5}
$$

Indeed, in the above figures, each point represent a group data including *gap*, permanent magnet parameters *rmag* and *hmag*, coil parameters *wcoil* and *hcoil*. When F_Y reaches the maximum and the minimum, the corresponding values for each variable are listed in Table 1, respectively.

Table 1. Variable optimization results

Variable		Value		Unit
Permanent magnet radial dimension	*rmag*	5.96	5.96	*mm*
Permanent magnet height	*hmag*	6	5.995	*mm*
Gap width	*gap*	1.0085	1.00825	*mm*
Coil width	*wcoil*	3.04	3.04	*mm*
Coil thickness	*hcoil*	1.96	1.96	*mm*
Unit output force	*FY*	*FYmax*	*FYmin*	*N*
		0.28231	0.10154	

5 Principle Prototype

Based on the previous optimization results, a prototype was developed. When the maximum and minimum values of F_Y are obtained, the values of each variable are not significant. In order to make F_Y relatively large, the variables of AFRP principle prototype are *rmag/hmap* = 1, *gap* = 2 *mm* and *wcoil/hcoil* = 1.5. The parameters of the prototype are shown in Table 2.

Table 2. The AFRP prototype parameters

Outer diameter [mm]	Length [mm]	Number of units	rmag [mm]	hmag [mm]	gap [mm]	wcoil [mm]	hcoil [mm]
20	140	5	6	6	1	3	2

In the experiment, the current direction of the two windings is controlled by the different input signals of the H-bridge driving circuit. By changing the time interval of the current commutation, the operating frequency of the AFRP is controlled. The processor outputs the *PWM* waveform to adjust the current in the coil, further changing the output force. In addition, changing the number of units and the corresponding coils, the output force will change accordingly. To simulate the motion in a low-gravity environment, the AFRP was placed horizontally on the test bed for testing, as shown in Fig. 8. Through the test, it is found that the frequency range of the AFRP is 1 ~ 10 Hz, and the average displacement of each work cycle is 2 mm by reading the scale. We have also found that when the coil current is 100 mA, the output force is about 1.2 N.

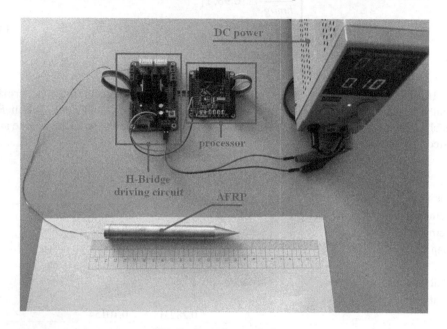

Fig. 8. Experiment platform

6   Conclusion

In this paper, a novel low velocity robotic penetrator based on ampere force (AFRP) is proposed for the study of penetrator for exploration missions of celestial bodies. Moreover, the application of the AFRP principle of operation may be promising, it can be used as a linear actuator to provide power for self-driven equipment. In order to maximize the output force F_Y, the ANSYS® software is integrated into the ISIGHT® platform to optimize the design variables. According to the optimization results, the developed prototype was tested. The results show the novel low velocity robotic penetrator's feasibility and can be used as a valid actuator for robots.

The future work will focus on studying the kinematics and dynamics of the prototype, and to improve the experimental platform to test the relationships between the output force of the AFRP and the operating frequency, displacement and power.

Acknowledgments. This work is supported by Research Fund of China Manned Space Engineering (050102), the Key Research Program of the Chinese Academy of Sciences

(Y4A3210301), the National Science Foundation of China (51175494, 61128008, and 51575412), and the State Key Laboratory of Robotics Foundation.

References

1. Glaser, D.L., Ball, A.J., Zacny, K.A.: A review of penetrometers for subsurface access on comets and asteroids. Meteorit. Planet. Sci. **43**, 1021–1032 (2008)
2. Gouache, T., Yang, G., Gourinat, Y., et al.: Wood Wasp Inspired Planetary and Earth Drill. Research gate Net (2010)
3. Davies, P., Phipps, A., Taylor, M., et al.: UK Lunar Science Missions: Moonlite & Moonraker, Bremen, pp. 774–779 (2007)
4. Banaszkiewicz, M., Grygorczuk, J., Marczewski, W., et al.: Rosetta, a cornerston mission of ESA: objectives, instruments and the Polish contribution. In: International Conference on Microwaves, Radar, and Wireless Communications, pp. 1–4 (2014)
5. Grygorczuk, J., Wisniewski, Ł., et al.: High energy and efficiency penetrator – HEEP. In: 15th European Space Mechanisms & Tribology Symposium-ESMATS 2013, Noordwijk, The Netherlands (2013)
6. Grygorczuk, J., Seweryn, K., Przybyla, R., et al.: Priorities of the SMRL SRC PAS involvement in exploration of the planetary bodies' surface based on the existing and planned to be developed technologies. In: International Conference on Microwave Radar and Wireless Communications, pp. 422–425. IEEE (2012)
7. Spohn, T., Seiferlin, K., Hagermann, A., et al.: Mupus – a thermal and mechanical properties probe for the rosetta lander Philae. Space Sci. Rev. **128**(1), 339–362 (2007)
8. Grygorczuk, J., Banaszkiewicz, M., Seweryn, K., et al.: MUPUS insertion device for the Rosetta mission. Space Technol. **17**(6), 59–64 (2007)
9. Seweryn, K., Banaszkiewicz, M., Bednarz, S., et al.: Mole penetrator 'KRET' for Lunar Exploration (2011)
10. Grygorczuk, J., Seweryn, K., et al.: Technological features in the new mole penetrator "KRET". In: 13th European Space Mechanisms and Tribology Symposium– ESMATS 2009 (2009)
11. Lichtenheldt, R., Schäfer, B., Krömer, O.: Hammering beneath the surface of Mars - modeling and simulation of the impact-driven locomotion of the HP^3-Mole by coupling enhanced multibody dynamics and discrete element method. In: 58th Ilmenau Scientific Colloquium Technische Universität Ilmenau (2014)
12. Spohn, T., Grott, M., Knollenberg, J., et al.: InSight: measuring the martian heat flow using the heat flow and physical properties package (HP^3). In: Lunar and Planetary Science Conference, p. 1916 (2012)
13. Grygorczuk, J., Banaszkiewicz, M., Seweryn, K., et al.: Space penetrators — Rosetta case study. In: International Conference on Methods and MODELS in Automation and Robotics, pp. 441–447 (2013)
14. Dempsey, N.M., Walther, A., May, F., et al.: High performance hard magnetic NdFeB thick films for integration into micro-electro-mechanical systems. Appl. Phys. Lett. **90**(9), 535 (2007)
15. Alqadi, M.K., Alzoubi, F.Y., Saadeh, S.M., et al.: Force analysis of a permanent magnet and a superconducting hollow cylinder. J. Supercond. Novel Magn. **25**(5), 1469–1473 (2012)
16. Harrold, W.: Calculation of equipotentials and flux lines in axially symmetrical permanent magnet assemblies by computer. IEEE Trans. Magn. **8**(1), 23–29 (1972)

Research of the Active Vibration Suppression of Flexible Manipulator with One Degree-of-Freedom

Luo Qingsheng$^{(\boxtimes)}$ and Li Xiang$^{(\boxtimes)}$

Beijing Institute of Technology, Beijing 100081, China
xiangzi201102@126.com

Abstract. A flexible manipulator with concentrated mass at its end is simplified to be a Bernoulli-Euler beam module in this paper. The oscillatory differential equation and system decoupling equation of the one DOF flexible manipulator is deduced according to the module simplified from practical engineering. And then, based on the establishment of dynamic model, we get the state of space expression through the Lagrange equation. Take the first order modal parameters of the flexible manipulator to track the angular displacement of the manipulator and the deflection of its end. According to the established state space expression, a comparison is done between the PD controller and the integral separation PID controller in the active control of flexible manipulator through simulation in Simulink. Result shows that the integral separation PID controller has obvious advantage than the PD controller in the control of flexible manipulator. The former is better to quickly arrive at the specified position and both its static error and overshoot are smaller.

Keywords: Flexible manipulator · Active vibration suppression · PD controller · Integral separation PID controller

1 Introduction

As a device of automation, flexible manipulator is widely used in aviation, aerospace and engineering. In field of aerospace, under disturbance, the free vibration with large amplitude of space manipulator will last for a long time. It will affect the stability and the accuracy of position of the mechanical arm, especially when in need of highly accuracy of the position. Active vibration suppression is a kind of control strategy, through real-time calculation, based on the detected vibration to drive the actuator and finally to eliminate the vibration. Various kinds of control methods in the fields of automatic control have been introduced into the active vibration suppression of the flexible manipulator system [1–3]. The search of the linear control strategy on the vibration suppression of manipulator begins in 1980s, Cannon [4] uses PID feedback control strategy to carry out a series of experiment on the flexible robot. Cai Guoping uses an approximate model to study the linear control strategy of the flexible manipulator, and results show that the linear control strategy can make the system reach the designated position, but it is poor in the time of arrival. Because of its characteristics, the active

© Springer International Publishing AG 2017
Y. Huang et al. (Eds.): ICIRA 2017, Part III, LNAI 10464, pp. 624–636, 2017.
DOI: 10.1007/978-3-319-65298-6_56

control of flexible manipulator only relies on P and D controller. Ozen figen [5] proposes a new control strategy to control the trajectory of flexible manipulator. This control law uses the parameter, easily to be obtained, such as the joint angle, the angular velocity, the deformation and the speed of each end of the manipulator. Talebi [6] and Khorsani [7] use the PD controller to control the flexible manipulator. Yigit [8] study the robustness of PD controller of the independent joint, the result show that the stability of PD controller is not depend on the parameters of the system, but on the Non discretized or linearized equations of the system. Kelly [9] use the PD controller, based on the approximate differential method, to control the flexible manipulator. In this paper, a comparison between the PD controller and integral separation PID controller is done in the field of active vibration suppression of manipulator with one degree-of-freedom.

2   Introduction

The oscillatory differential equation and system decoupling equation of the one DOF flexible manipulator is deduced according to the module simplified from practical engineering. And then, based on the establishment of dynamic model, we get the state of space expression through the Lagrange equation.

2.1   Physical Model

As is shown in the Fig. 1, a flexible arm, with a mass concentrated at its end, driven by a motor rotates on a horizontal plane with the ignorance of the gravity. Due to the large transmission ratio of the harmonic gear in the motor, we can consider the boundary condition as the cantilever as the torsional stiffness is much larger than the beam stiffness.

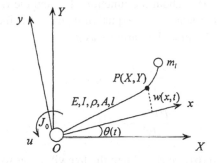

Fig. 1. Physical model of flexible manipulator

XOY is the fixed coordinate system, xoy is the coordinate system fixed with rotation shaft. OX axis coincident with the shaft when the shaft is not deformed. $w(x, t)$ is the elastic deformation of the point of the OX axis in the position of x and at the time of t. J_0 is the sums of the inertia of motor shaft and mechanical chuck related to rotation shaft. l is the length of flexible manipulator. E is the Elastic modulus of materials.

I is the inertia moment of the manipulator to the neutral axis. ρ is the density. θ is the angle of turn of the manipulator. m_l is the mass of the mass block. u is the torque that input by the motor. A is the cross-sectional area of the manipulator.

2.2 Oscillatory Differential Equation

The equation of free vibration modal analysis of beam is [10]

$$\rho A \frac{\partial^2 w(x,t)}{\partial t^2} + EI \frac{\partial^4(x,t)}{\partial x^4} = 0 \tag{1}$$

Assume that

$$w(x,t) = W(x)q(t) \tag{2}$$

$W(x)$ is the vibration amplitude of the beam in the position of x, $q(t)$ is the function of time t.

Substitute (2) into (1),

$$\rho A W(x) \frac{d^2 q(t)}{dt^2} + EI \frac{d^4 W}{dx^4} q(t) = 0 \tag{3}$$

The Eq. (3) can also be written as

$$\frac{W^{(4)}(x)}{W(x)} = -\frac{\rho A}{EI} \cdot \frac{\ddot{q}(t)}{q(t)} \tag{4}$$

The left side of Eq. (4) is about the function of x, and the right side is the function of t. So, only both of the side of the equation equal to an independent variable the equation can be set up. Assume the variable as p^2.

$$\frac{d^2 q(t)}{dt^2} + p^2 q(t) = 0 \tag{5}$$

$$\frac{d^4 W(x)}{dx^4} - \lambda^4 W(x) = 0 \tag{6}$$

Solve the Eq. (5) and (6), we can obtain the free vibration solution of flexible beam.

$$W(x) = a_1 \sinh \lambda x + a_2 \cosh \lambda x + a_3 \cos \lambda x + a_4 \sin \lambda x \tag{7}$$

$$q(t) = b_1 \cos \omega t + b_2 \sin \omega t \tag{8}$$

Eigenvalue λ and mode function $W(x)$ are all determined by boundary condition of the beam.

According to Mechanics of materials, the boundary condition of the cantilever beam is that the deflection and the rotation angle of the fixed part are zero and the bending moment and shear force are zero at the free part.

At the fixed part,

$$w(x, 0) = 0 \tag{9}$$

$$\frac{\partial w(0, t)}{\partial x} = 0 \tag{10}$$

At the free part,

$$EI \frac{\partial^2 w(l, t)}{\partial x^2} = 0 \tag{11}$$

$$EI \frac{\partial^3 w(l, t)}{\partial x^3} = m_l \frac{\partial^2 w(l, t)}{\partial t^2} \tag{12}$$

The natural frequency of the system is

$$\omega_i = \left(\frac{\lambda_i}{l}\right)^2 \sqrt{\frac{EI}{\rho A}}, i = 1, 2 \ldots \tag{13}$$

The vibration function of the system is

$$W_i(x) = K[(ch \lambda_i x - \cos \lambda_i x) - \frac{sh \lambda_i l - \sin \lambda_i l}{ch \lambda_i l + \cos \lambda_i l}] \tag{14}$$

The relative elastic deformation of the flexible manipulator can be expressed as

$$w(x, t) = \sum_{i=1}^{\infty} W_i(x) q_i(t) \tag{15}$$

$W_i(x)$ is the vibration mode function of the flexible manipulator, $q_i(t)$ is the corresponding modal coordinates. In practice, we generally take the finite sums

$$w(x, t) = \sum_{i=1}^{N} W_i(x) q_i(t) \tag{16}$$

N is the modal order that we take.

2.3   System Decoupling Equation

The natural vibration mode of beam bending vibration is orthogonal. From the Eq. (6) we can obtain,

$$\rho A w_i^2 W_i(x) = EI \frac{d^4 W(x)}{dx^4} \tag{17}$$

Both sides of Eq. (17) multiply $W_j(x)$ and take the integral of x along the beam length,

$$\rho A w_i^2 \int_0^l W_i(x) W_j(x) dx = EI \int_0^l \frac{d^4 W(x)}{dx^4} W_j(x) dx \tag{18}$$

Substitute the Eqs. (9), (10), (11), (12) into Eq. (18),

$$EI \int_0^l \frac{d^2 W_i(x)}{dx^2} \frac{d^2 W_j(x)}{dx^2} dx = w_i^2 \left[\rho A \int_0^l W_i(x) W_j(x) dx + m_l W_i(l) W_j(l) \right] \tag{19}$$

We can also write the equation as follows, because i and j are taken random.

$$EI \int_0^l \frac{d^2 W_i(x)}{dx^2} \frac{d^2 W_j(x)}{dx^2} dx = w_j^2 \left[\rho A \int_0^l W_i(x) W_j(x) dx + m_l W_i(l) W_j(l) \right] \tag{20}$$

Take the Eq. (19) to subtract the Eq. (20),

$$(\omega_i^2 - \omega_j^2) \left[\rho A \int_0^l W_i(x) W_j(x) dx + m_l W_i(l) W_j(l) \right] = 0 \tag{21}$$

When $i \neq j$ and $\omega_i \neq \omega_j$, we can obtain

$$\rho A \int_0^l W_i(x) W_j(x) dx + m_l W_i(l) W_j(l) = 0 \tag{22}$$

And

$$EI \int_0^l \frac{d^2 W_i(x)}{dx^2} \frac{d^2 W_j(x)}{dx^2} dx = 0 \tag{23}$$

The Eqs. (22) and (23) are the orthogonality condition of the natural mode shape function of the flexible manipulator with concentrated mass at the end.

When $i = j$ we can obtain from the Eq. (19) that

$$w_i^2 \left[\rho A \int_0^l W_i^2(x) dx + m_l W_i^2(l) \right] = EI \int_0^l \left[\frac{d^2 W_i(x)}{dx^2} \right]^2 dx \tag{24}$$

Define

$$\rho A \int_0^l W_i^2(x)dx + m_l W_i^2(l) = M_i \tag{25}$$

$$EI \int_0^l \left[\frac{d^2 W_i(x)}{dx^2}\right]^2 dx = K_i \tag{26}$$

From Eq. (26) we can obtain that

$$K_i = \omega_i^2 M_i \tag{27}$$

M_i, K_i, respectively, is called the generalized mass and generalized stiffness corresponding to the i modal of the flexible manipulator.

2.4 Dynamic Model

The total kinetic energy of flexible manipulator T comprises the kinetic of motor shaft and clamp T_1, the kinetic of the manipulator itself T_2, and the kinetic of mass at the end T_3.

$$T = T_1 + T_2 + T_3 \tag{28}$$

And

$$T_1 = \frac{1}{2}J_0\dot{\theta}^2 \tag{29}$$

The position coordinates of P in Fig. 1 is (X, Y)

$$X = x\cos\theta - w(x, t)\sin\theta \tag{30}$$

$$Y = x\sin\theta + w(x, t)\cos\theta \tag{31}$$

Seeking t derivative is

$$\dot{X} = -(x\sin\theta + w(x, t)\cos\theta)\,\dot{\theta} + \dot{x}\cos\theta - \dot{w}(x, t)\sin\theta \tag{32}$$

$$\dot{Y} = (x\cos\theta - w(x, t)\sin\theta)\,\dot{\theta} + \dot{x}\sin\theta + \dot{w}(x, t)\cos\theta \tag{33}$$

In the Eqs. (32), (33), $\dot{x} = 0$. The square of the velocity of the P point is

$$v^2 = \dot{X}^2 + \dot{Y}^2 = (x^2 + w^2(x, t))\,\dot{\theta}^2 + \dot{w}^2(x, t) + 2\dot{\theta}x\dot{w}(x, t) \tag{34}$$

So the kinetic energy of the manipulator is

$$v^2 = \dot{X}^2 + \dot{Y}^2 = (x^2 + w^2(x,t))\,\dot{\theta}^2 + \dot{w}^2(x,t) + 2\dot{\theta}x\dot{w}(x,t) \tag{35}$$

The kinetic energy generated by the mass concentration at the end is

$$T_3 = \frac{1}{2}m_l(l^2 + w^2(l,t))\,\dot{\theta}^2 + \dot{w}^2(l,t) + 2\dot{\theta}l\dot{w}(l,t) \tag{36}$$

Substitute the Eq. (16) into the Eqs. (35), (36), and use the orthogonality condition (22), (23), The total kinetic energy of flexible manipulator is

$$T = T_1 + T_2 + T_3 = \frac{1}{2}(J_0 + \rho A \int_0^1 x^2 dx + m_l l^2)\,\dot{\theta}^2 + \frac{1}{2}\sum_{i=1}^N M_i \left[\frac{dq_i(t)}{dt}\right]^2$$
$$+ \frac{1}{2}\dot{\theta}^2 \sum_{i=1}^N M_i[q_i(t)]^2 + \dot{\theta}\sum_{i=1}^N \left[\rho A \int_0^1 xW_i(x)dx + m_l lW_i(l)\right]\frac{dq_i(t)}{dt} \tag{37}$$

The potential energy of elastic deformation is

$$V = \frac{1}{2}EI \int_0^l \left[\frac{\partial w(x,t)}{\partial x^2}\right]^2 dx \tag{38}$$

Substitute the Eq. (17) into the Eq. (38), and use the Eq. (26), we can obtain the potential energy of elastic deformation is

$$V = \frac{1}{2}EI \int_0^l \left[\sum_{i=1}^N \frac{d^2W_i(x)}{dx^2}q_i(t)\right]^2 dx = \frac{1}{2}\sum_{i=1}^N K_i[q_i(t)]^2 \tag{39}$$

Substitute the Eqs. (37) and (39) into the Lagrange function,

$$L = T - V \tag{40}$$

$$L = \frac{1}{2}J\,\dot{\theta}^2 + \frac{1}{2}\sum_{i=1}^N M_i \left[\frac{dq_i(t)}{dt}\right]^2 + \frac{1}{2}\dot{\theta}^2 \sum_{i=1}^N M_i[q_i(t)]^2$$
$$+ \dot{\theta}\sum_{i=1}^N [\rho A\sigma_1 + m_l lW_i(l)]\frac{dq_i}{dt} - \frac{1}{2}\sum_{i=1}^N K_i[q_i(t)]^2 \tag{41}$$

And

$$J = J_0 + \rho A \int_0^l x^2 dx + m_l l^2 \tag{42}$$

$$\sigma_i = \int_0^l xW_i(x)dx \tag{43}$$

Lagrange equation

$$\frac{d}{dt}\left(\frac{\partial L}{\partial \dot{\theta}}\right) - \frac{\partial L}{\partial \theta} = u \tag{44}$$

$$\frac{d}{dt}\left(\frac{\partial L}{\partial \dot{q}}\right) - \frac{\partial L}{\partial q_i} = 0 \tag{45}$$

The dynamic equation of the flexible arm with concentrated mass is

$$J\ddot{\theta} + \sum_{i=1}^{N}[\rho A\sigma_i + m_l l W(l)]q_i = u(t) \tag{46}$$

$$[\rho A\sigma_i + m_l l W_i(l)]\ddot{\theta} + M_i\ddot{q}_i + K_i q_i = 0 \tag{47}$$

Turn the Eqs. (46) and (47) into matrix

$$M\ddot{z} + Kz = Fu \tag{48}$$

$$z = [\theta, q_1, \ldots, q_N]^T \tag{49}$$

$$F = [1, 0, \ldots, 0]^T \tag{50}$$

Generalized mass matrix M is

$$M = \begin{bmatrix} J & \rho A\sigma_1 + m_l l W_1(l) & \cdots & \rho A\sigma_N + m_l l W_N(l) \\ \rho A\sigma_1 + m_l l W_1(l) & M_1 & \cdots & 0 \\ \cdots & \cdots & \cdots & \cdots \\ \rho A\sigma_N + m_l l W_N(l) & 0 & \cdots & M_N \end{bmatrix} \tag{51}$$

Generalized mass matrix K is

$$K = \begin{bmatrix} 0 & 0 & \cdots & 0 \\ 0 & K_1 & \cdots & 0 \\ \cdots & \cdots & \cdots & \cdots \\ 0 & 0 & \cdots & K_N \end{bmatrix} \tag{52}$$

Use the state space equation to express the kinetic equation

$$\dot{x}(t) = Ax(t) + Bu(t) \tag{53}$$

$$y = Cx(t) \tag{54}$$

Chose the variable $x(t) = [z^T, \dot{z}^T]^T$, The corresponding coefficient matrix is

$$A = \begin{bmatrix} 0 & I \\ -M^{-1}K & 0 \end{bmatrix}$$ (55)

The state space model of the rotating flexible manipulator is

$$\dot{x}(t) = \begin{bmatrix} 0 & I \\ -M^{-1}K & 0 \end{bmatrix} x(t) + \begin{bmatrix} 0 \\ M^{-1}F \end{bmatrix} u(t)$$ (56)

Equation (56) did not consider the damping of flexible manipulator, the matrix form of the kinetic equation that consider the damping is

$$M\ddot{z} + C\dot{z} + Kz = Fu$$ (57)

$$C = \begin{bmatrix} 0 & 0 & \dots & 0 \\ 0 & C_1 & \dots & 0 \\ \dots & \dots & \dots & \dots \\ 0 & 0 & \dots & C_N \end{bmatrix}$$ (58)

$C_i(i = 1, 2, \dots, N)$ is the modal damping coefficient, the relationship between modal damping coefficient and modal damping factor ζ_i is

$$\zeta_i = \frac{C_i}{\sqrt{2K_iM_i}}$$ (59)

The state space model of the flexible arm with damping obtained by formula (1.57) is

$$\dot{x}(t) = \begin{bmatrix} 0 & I \\ -M^{-1}K & -M^{-1}C \end{bmatrix} x(t) + \begin{bmatrix} 0 \\ M^{-1}F \end{bmatrix} u(t)$$ (60)

Theoretical analysis and numerical simulation demonstrates that the second modal response of the one DOF flexible manipulator is very small. Generally, the former mode can satisfy the requirement of accuracy.

The structural parameters of flexible manipulator is shown in the Table 1.

Table 1. Structural parameters of flexible manipulator.

Density ρ	$7.8 \times 10^3 \, \text{kg/m}^3$
Elastic modulus E	$2.0 \times 10^{11} \, \text{N/m}^2$
Section width b	40 mm
Depth of Section h	3 mm
length l	1.5 m
Concentrated mass m_l	0.1 kg
Moment of inertia J_O	$0.8 \, \text{kg} \cdot \text{m}^2$

According to the structure parameters of flexible manipulator in Table 1, the natural frequency, modal mass and modal stiffness, before the three order, of the flexible manipulator is calculated as shown in Table 2.

Table 2. The first three order modal parameters of flexible manipulator

Modal order	Natural frequency	Modal mass	Modal stiffness
1	1.2518	0.4245	9.649
2	7.6173	0.3967	23.016
3	39.9060	0.3621	143.467

We take the first mode to track the movement of flexible manipulator. Set state variable as $x = \left[\theta, q_1, \dot{\theta}, \dot{q}_1\right]^T$, the state space model of flexible manipulator is

$$\dot{x} = \begin{bmatrix} 0 & 0 & 1 & 0 \\ 0 & 0 & 0 & 1 \\ 0 & 19 & 0 & 6 \\ 0 & -105 & 0 & -2 \end{bmatrix} + \begin{bmatrix} 0 \\ 0 \\ 5 \\ -2 \end{bmatrix} u$$

We chose θ', the angular displacement of the end of the flexible manipulator, as the output variable. And $\theta' = Cx, C = [1, \frac{W_1(l)}{l}, \ldots, \frac{W_N(l)}{l}, 0, 0, \ldots 0]$,

$$\dot{x} = \begin{bmatrix} 0 & 0 & 1 & 0 \\ 0 & 0 & 0 & 1 \\ 0 & 19 & 0 & 6 \\ 0 & -105 & 0 & -2 \end{bmatrix} x + \begin{bmatrix} 0 \\ 0 \\ 5 \\ -2 \end{bmatrix} u$$

$$y = \theta' = [1, -0.030, 0, 0]x$$

The eigenvalue of the system matrix is $\lambda_1 = \lambda_2 = 0, \lambda_3 = -1 + 10.198i, \lambda_4 = -1 - 10.198i$, The real part of eigenvalue is not bigger than zero. From the control theory, we can get that the system is stable [11].

3 Comparison of Two Control Methods

According to the state space expression, we build two kinds of simulation model based on different control method, PD controller and integral separation PID controller. We set the initial condition as $x = \begin{bmatrix} 0 & 0 & 0 & 0 \end{bmatrix}^T$ and the driver as unit step response.

Then compare the strengths and weaknesses of those methods in the control of flexible manipulator.

3.1   PD Controller

In PD controller, the introduction of K_P to reduce the raise time and steady error of the system, and the introduction of K_D to enhance the stability and reduce the overshoot [12]. According to the principle and method of parameter tuning of PID controller [13]. Initial condition is $x = \begin{bmatrix} 0 & 0 & 0 & 0 \end{bmatrix}^T$, the driver is unit step response. Fig. 2 is the corresponding simulation structure.

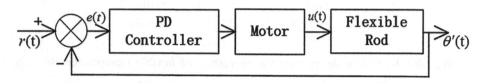

Fig. 2. The PD controller simulation structure

3.2   Integral Separation PID Controller

The basic method of the integral separation PID controller is that when the offset between measured quantity and set value is large, we cancel the integral controller to avoid the increase of overshoot, and that when the offset between measured quantity and set value is small, we add the integral controller to improve the accuracy of the control [14]. Figure 3 is the corresponding simulation structure.

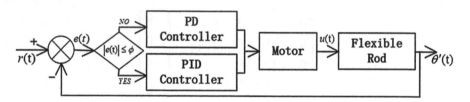

Fig. 3. The integral separation PID controller simulation structure

3.3   Comparison Results

In order to compare the two above mentioned control methods in the control of one DOF flexible manipulator, we take the angular displacement of the flexible manipulator θ' and the deflection of the flexible manipulator $w(l, t)$ as the out of our simulation. As shown in Fig. 4 is the displacement of the end of the end of the flexible manipulator θ', in Fig. 5 is the deflection of the end of the flexible manipulator $w(l, t)$.

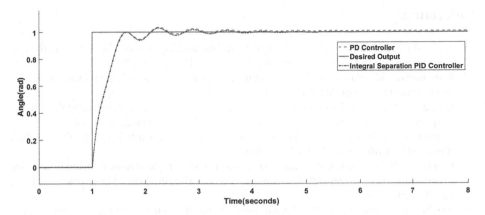

Fig. 4. The angular displacement of the end of the flexible manipulator

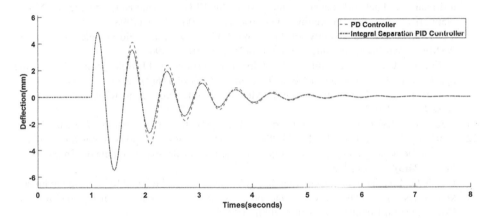

Fig. 5. The deflection of the end of the flexible manipulator

3.4   Conclusion

From the result we can see that the integral separation PID controller has obvious advantage than the PD controller in the control of flexible manipulator. In the control of the integral separation PID controller, the manipulator can quickly arrive at the specified position and both its static error and overshoot are small. However, with PD controller control the manipulator performance, static error exists, so the manipulator cannot accurately reach the specified position, and the deflection of the end of the flexible manipulator is larger.

References

1. Kerr, M.L., Asokanthan, S.F., Jayasuriya, S.: Robust tracking control of a single-link flexible manipulator. In: IFAC Proceedings, vol. 35(1), pp. 241–246 (2002)
2. Rios-Bolívar, M., Navas, A.: Output feedback regulation of a flexible joint manipulator. In: IFAC Proceedings, vol. 40(12), pp. 342–347 (2007)
3. Qiu, Z.-c., Zhao, Z.-l.: Vibration suppression of a pneumatic drive flexible manipulator using adaptive phase adjusting controller. J. Vib. Control 21(15), 2959–2980 (2014)
4. Cannon, R.H., Schmitz, E.: Initial experiments on end-point control of a flexible one-link robot. Int. J. Robot. Res. 3(3), 62–75 (1984)
5. Ozen, F: New non-linear stabilizing control law for flexible-link manipulator. In: Proceedings of the Japan/USA Symposium on Flexible-Link Manipulator, vol. 1, pp. 291–298 (1996)
6. Talebi, H.A., Khorasani, K.: Neural network based control schemes for flexible-link manipulators: simulations and experiments. Neural Netw. 11(7–8), 1337–1357 (1998)
7. Talebi, H.A., Khorasani, K: Tip-position tracking for flexible-link manipulator using artificial neural networks: experimental results. In: IEEE International Conference on Neural Networks Conference Proceedings, vol. 3(4–9), pp. 2063–2068 (1998)
8. Yigit, A.S.: On the stability of PD control for a two-link rigid-flexible manipulator. ASME J. Dyn. Syst. Measur. Control 116(2), 208–2015 (1994)
9. Kelly, R., Omega, R., Ailton, A.: Global regulation of Flexible joint robots using approximate differentiation. IEEE Trans. Autom. Control 116(2), 208–215 (1994)
10. Robert, F.: Steidel: An Introduction to Mechanical Vibrations. John wiley and sons Press, New Jersey (1989)
11. Ogata, K.: Modern Control Engineering, 5th edn. Prentice Hall Press, Hong Kong (2009)
12. Lou, J.-q., Wei, Y.-d., et al.: Hybrid PD and effective multi-mode positive position feedback control for slewing and vibration suppression of a smart flexible manipulator. Smart Mater. Struct. 24(3), 1–14 (2015)
13. Dai, Y.Q., Loukianov, A., Uchyama, M.: A hybrid numerical method for solving the inverse kinematics of a class flexible manipulators. In: Proceedings of IEEE International Conference on Robotics and Automation (1997)
14. Jahed, A., Piltan, F., et al.: Design computed torque controller with parallel fuzzy inference system compensator to control of robot manipulator. Int. J. Inf. Eng. Electron. Bus. (IJIEEB) 5(3), 66–77 (2013)

Space Robotic De-Tumbling of Large Target with Eddy Current Brake in Hand

Jiayu Liu[1,2(✉)], Baosen Du[2], and Qiang Huang[1]

[1] School of Mechatronical Engineering, Intelligent Robotics Institute,
Beijing Institute of Technology, 5 Nandajie, Zhongguancun,
Haidian, Beijing, China
jyliu_bit@hotmail.com
[2] Beijing Institute of Precise Mechatronics and Controls (the 18th Institute),
China Academy of Launch Vehicle Technology (CALT), 1 Nandahongmenlu,
Fengtai, Beijing, China

Abstract. Space debris has increased with recent launch missions greatly. Former active debris removal tests using space robot mainly focused on the fundamental technology of target recognition, motion control and path planning. However, robot contacts directly with the surface of targets with large mass and angular momentum will cause severe collision problems. A target de-tumbling strategy is proposed in this paper by using two manipulators. Each arm is equipped with magnetic coil, which can generate eddy current in conductive targets and gradually de-tumble rotation without contact. The three-dimension rotation model of a discarded satellite and upstage is established based on its distribution of the moment of inertia and the safe working space of the robots is calculated. By analyzing the point of application and direction of the magnetic force, an optimized de-tumbling trajectory for the robot is presented to minimize the de-tumbling time by reducing the targets' angular momentum. At last, a simulation is processed to verify the optimized de-tumbling method.

Keywords: Space robot · Eddy current brake · On-orbit · Tumbling target · Large debris capture

1 Introduction

Satellites in orbit have greatly increased in recent years with the unceasing space development. With the limited space resources, space debris removal should be taken into action to make better use of the space.

1.1 Typical Target Capture Method

The debris removal method includes active debris removal and passive debris removal method. Passive debris removal usually puts a damper on the debris and drag it to the atmospheric layer by earth magnetic field or pneumatic forces. Active debris removal needs a space craft to perform the capture of the debris and then remove it with the spacecraft's energy. And this removal procedure often equips with manipulator.

© Springer International Publishing AG 2017
Y. Huang et al. (Eds.): ICIRA 2017, Part III, LNAI 10464, pp. 637–649, 2017.
DOI: 10.1007/978-3-319-65298-6_57

1.2   Target Manipulation with Manipulator

Robot manipulator, also named robotic arm, has been used in on-orbit refueling, module replacement and assisted space docking. These robot arms have different D-H parameters and joint numbers [1–3]. And researches on robot arms have extended to multi-robot and dedicated operation from single robot. Some research work focused on the eddy current brake have got wonderful results [4–7].

Chaser with a manipulator de-tumbles a large tumbling debris with an eddy current brake is shown in Fig. 1, and this paper is focused on the challenge.

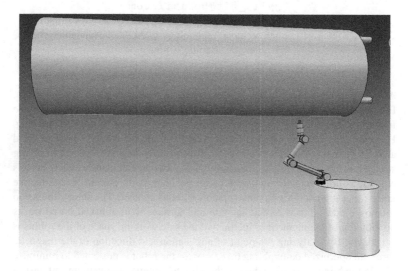

Fig. 1. De-tumbling of large target with Eddy Current Brake in Hand

1.3   Problems of Debris with High Angular Momentum

Former debris target to be captured by robot arm often has such characters: a, the target's attitude is limited, it has a relatively constant pose to the spacecraft. b, the target is small and little in weight, it is easy to be handled by the robot arm's gripper. c, the target's angular velocity is negligible, the spacecraft can keep still to the target by orbital maneuver.

However, debris generated in recent years are mainly satellites and abandoned rocket's upper stage which are large and weight. And these debris also have angular velocity which lead to large angular momentum.

Traditional method using rigid gripper to directly contact with the debris has following problems: a, the gripper may break at this moment when the arm first contact with the debris. b, the enormous torque transferred from the debris after the manipulator has linked to the debris may cause the joint motor to break down or even break the transmission gear. c, the sudden impact may cause the spacecraft out of control. d, the impact may drive the debris away and out of sight.

In this paper, a contactless debris removal strategy is proposed to accomplish the de-tumbling procedure of debris with large angular momentum. And avoid the problem of rigid contact by using eddy current brake. The kinematic model of the debris in spin state is established to analyze the de-tumbling procedure. Then an optimized de-tumbling method is proposed by calculation referring time consuming. At last, a simulation is designed to illustrate the efficiency of optimized strategy.

2 Target Recognition and De-Tumbling

The capture system consists of two parts: the target to be captured, meaning the space debris in orbit with large mass; the performer of the capture mission, meaning the space craft with manipulator on board.

The chaser keeps the target in the robot arm's manipulation space by attitude adjustment after it reaches the target's zone by orbital maneuver.

The eddy current brake in the hand of the manipulator starts to work after the manipulator has moved to the target coordinate by path planning. It will then gradually decrease the target's spin movement. At the same time, the controller system maintains the chaser-target's relative position and the manipulator's action pose by real-time calculation.

2.1 Target Approach and Coordinates Building

The target's orbit is assessed by observation, then the chaser can fly to the target's zone through orbital maneuver Fig. 2.

Typical capturing method as described here is called active debris removal strategy. Debris in the orbit is the target of the chaser-target system while the capture vehicle acting as the chaser. First, the orbit of the target debris is observed.

When the chaser reaches the same orbit with the target. It slowly changes its relative pose to the target while maintaining their position until the end effector of the onboard manipulator is in flexible manipulation space.

The spinning motion of the target rotating around the coordinate axes can be reconstructed by vision camera. Two different frames are built here to explain the

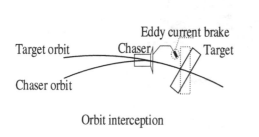

Orbit interception

Fig. 2. Capturing of large tumbling target

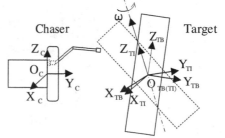

Fig. 3. Coordinates of target and chaser

target's angular velocity and angular momentum. The body frame's origin of coordinate is the center of mass of the debris. And set the minimum inertia spindle as the Z_{TB} axis. And X_{TB}, Y_{TB} axes are mutually perpendicular in the transection of the cylinder. Then the complex spin kinematic of the target can be described as rotation movement around three perpendicular axes OX_{TB} OY_{TB} OZ_{TB} with the angular velocity ω_X ω_Y ω_Z.

The origin of the inertia frame is also the target's center of mass. Set ω as the resultant vector of angular velocity's component ω_X ω_Y ω_Z. Then the inertia frame's Z_{IB} axis is parallel to ω.

In the same way, the chaser's body frame can be built based on the center of its load's setting surface. Y_C axis is perpendicular to the manipulator's fixing panel Fig. 3.

First, when the captor reaches the target's spin zone, maintaining the relative position between the spacecraft and the target.

$$P_{O_C}^{O_{TI}} = const \tag{1}$$

Then, keeping the three axes of the captor's body frame parallel to that of the target's inertia frame by adjusting the spacecraft's spatial attitude.

$$R_{O_C XYZ}^{O_{TI} XYZ} = \begin{bmatrix} 1 & 0 & 0 \\ 0 & 1 & 0 \\ 0 & 0 & 1 \end{bmatrix} \tag{2}$$

$$T_{O_C XYZ}^{O_{TI} XYZ} = \begin{bmatrix} R_{O_C XYZ}^{O_{TI} XYZ} & P_{O_C}^{O_{TI}} \\ 0 & 1 \end{bmatrix} = const \tag{3}$$

By this way, the position and orientation matrix between the target's inertia frame and the captor's body frame is established. The manipulator onboard can calculate the target's position through coordinate translation.

From [8–11], the dynamics motion of target with external force from action point between hand and target can be described as:

$$H_t \ddot{r}_t + C_t = J_t^T f_{hd} \tag{4}$$

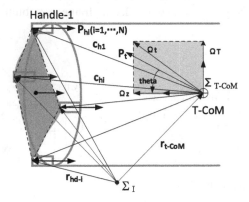

Fig. 4. The model of rocket upstage

Where, $\ddot{r}_t = \mathrm{col}(p_t, \varphi_t) \in \Re^6$ is the generalized velocity vector consisting of the position of the center of mass $p_t \in \Re^3$ and the attitude $\varphi_t \in \Re^3$, respectively. Figure $4C_t \in \Re^6$ represents the centrifugal and Coriolis terms. The vector $f_{hd} \in \Re^6$ is the force and moment exerted by the arms. $H_t \in \Re^{6 \times 6}$ is the generalized mass matrix which can be described as:

$$H_t = \begin{bmatrix} m1 & 0 \\ 0 & I_c \end{bmatrix} \tag{5}$$

Where, m and I_c denote the mass and the inertia tensor of the target, respectively.

$$c_t = \begin{bmatrix} m\tilde{\omega}_t v_0 \\ \tilde{\omega}_t I_c \omega_0 \end{bmatrix} \tag{6}$$

Generalized Jacobin matrix A^k can be described following with $\rho \in \Re^3$ being the position vector of the kth action point with respect to its center of mass which detailed in 8. The operator $\sim$ hat a vector denotes the cross product with the following vector.

$$J_t^k = \begin{bmatrix} 1_3 & \tilde{\rho}_k \\ 0 & 1_3 \end{bmatrix} \tag{7}$$

At this point, the dynamic model of the tumbling target has been established, and the follow-up will carry out the study of the trajectory optimization of de-tumbling planning with eddy current brake based on the motion characteristics of the tumbling target.

2.2   Contactless De-Tumbling

The relative pose between the target and the manipulator's end effector can be determined when the pose between the end and the robot's base frame, the pose between the robot's base frame and the captor's body frame, the pose between the target's inertia frame and the captor's body frame. And the target's pose in its inertia frame are determined.

A constant magnetic field will be generated when the eddy current brake at the robot's end is powered on. During the de-tumbling procedure, make sure the brake is close to the external surface of the target's spin state with the magnetic field's direction parallel to the external surface's normal direction. Current is produced when the target moves cutting the magnetic induction lines. And the direction of the force F_B produced by the brake is opposed to the target's movement Fig. 5.

By keeping the magnetic induction lines' direction perpendicular to the direction of the target's resultant vector ω and the brake close to the target, an electromagnetic torque will continually decrease the target's angular momentum. And eventually makes the target captured by the captor.

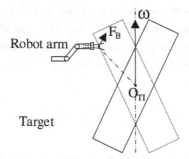

Fig. 5. Brake force by eddy current

3   Optimization of De-Tumbling Strategy

As the brake force's direction is always opposed to the target's movement and the action time is relative to the brake's shape size and the relative speed between the target and the manipulator. The major factor that affects the de-tumbling torque's performance is the brake force's point of action. In this section, the de-tumbling procedure of the target by cutting down its angular momentum through eddy current brake is analyzed. And an optimized de-tumbling strategy is proposed referring to time consuming.

3.1   Dynamics of Target De-Tumbling

The relationship between the target's angular velocity vector and angular momentum is shown as below when it is in spin state Fig. 6.

The target's Euler equations of spin movement in orbit are:

$$
\begin{aligned}
J_x\dot{\omega}_x + \omega_x \times J_x\omega_x &= \sum T_x \\
J_y\dot{\omega}_y + \omega_y \times J_y\omega_y &= \sum T_y \\
J_z\dot{\omega}_z + \omega_z \times J_z\omega_z &= \sum T_z
\end{aligned}
\tag{8}
$$

The angular velocity around X Y Z axes remain unchanged when the target's resultant momentum is zero, and can be expressed as below:

$$
\begin{aligned}
H_x &= J_x\omega_x \\
H_y &= J_y\omega_y \\
H_z &= J_z\omega_z
\end{aligned}
\tag{9}
$$

Usually, the target's angular velocity vector is not parallel to its angular momentum vector. The eddy current brake generates the angular momentum H_F by applying the brake force F_B on the target to decrease the target's angular momentum H Fig. 7.

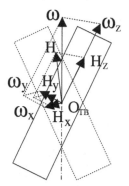

Fig. 6. Rotation model

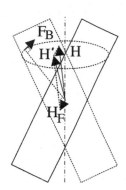

Fig. 7. Angular momentum of target

The eddy-current-brake angular momentum applies once when the target circles one time around its angular velocity vector. And the target's angular momentum is decreased from H to H'. Which also means a decrease in the target's angular velocity. Its angular velocity turns out to be zero by continuous de-tumbling method.

3.2   Calculation of Optimized De-Tumbling Strategy

The eddy-current-brake angular momentum vector's value and direction depends on the brake force's point of action.

The total time to finish the de-tumbling procedure is set as the principle to judge the strategy's efficiency. The less the remained angular momentum of the target is after one revolution, the better the de-tumbling strategy is.

So, the optimized de-tumbling strategy is to find out the minimum value of H'.

$$\text{Min}\{H'\} \tag{10}$$

Assuming the value of F_B is constant, L_C is the width of the eddy current brake's magnetic field.

θ' is the angle between the target's angular velocity vector and the arm of F_B.

θ is the angle between the target's angular velocity vector and the Z axis of the target's body frame.

δ is the angle between the target's angular velocity vector and its angular momentum, calculated as the angle between H and the target's Z axis minus θ (Fig. 8).

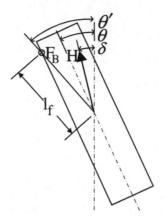

Fig. 8. Calculation of H'

The eddy-current-brake angular momentum is expressed as:

$$H_F = F_B \cdot l_f \cdot t = F_B \cdot l_f \cdot \left(\frac{L_C}{\omega \cdot l_f \cdot \sin \theta'} \right)$$
$$= \frac{F_B \cdot L_C}{\omega \cdot \sin \theta'} = \frac{K_F}{\omega} \cdot \frac{1}{\sin \theta'} \tag{11}$$

Where, $K_F = F_B \cdot L_C$. Apply the law of cosines in the triangle built by H_F, H', H. The detailed process is

$$
\begin{aligned}
|H'|^2 &= |H|^2 + |H_F|^2 - 2 \cdot |H| \cdot |H_F| \cdot \cos \gamma \\
&= |H|^2 + |H_F|^2 - 2 \cdot |H| \cdot |H_F| \cdot \sin(\theta' + \delta) \\
&= |H|^2 + \left(\frac{K_F}{\omega} \cdot \frac{1}{\sin \theta'} \right)^2 - 2 \cdot |H| \cdot \left(\frac{K_F}{\omega} \cdot \frac{1}{\sin \theta'} \right) \cdot \sin(\theta' + \delta) \\
&= |H|^2 + \left(\frac{K_F}{\omega} \right)^2 \cdot \frac{1}{\sin^2 \theta'} - 2 \cdot |H| \cdot \frac{K_F}{\omega} \cdot \frac{\sin \theta' \cdot \cos \delta + \cos \theta' \cdot \sin \delta}{\sin \theta'} \\
&= |H|^2 + \left(\frac{K_F}{\omega} \right)^2 \cdot \frac{1}{\sin^2 \theta'} - 2 \cdot |H| \cdot \frac{K_F}{\omega} \cdot \frac{\cos \theta' \cdot \sin \delta}{\sin \theta'} - 2 \cdot |H| \cdot \frac{K_F}{\omega} \cdot \cos \delta
\end{aligned}
\tag{12}
$$

In the above equation, the change of the brake force's point of action will lead to the change of θ', which will affect the value of H'. Set $|H'|^2$ as the function with θ' as the independent variable: $f(\theta')$.

Calculate the function's minimum value by its first derivative:

$$f'(\theta') = -2 \cdot \left(\frac{K_F}{\omega}\right)^2 \cdot \frac{\cos\theta'}{\sin^3\theta'} - 2 \cdot |H| \cdot \frac{K_F}{\omega} \cdot \sin\delta \cdot \frac{-\sin\theta' \cdot \sin\theta' - \cos\theta' \cdot \cos\theta'}{\sin^2\theta'}$$

$$= 2 \cdot \left(\frac{K_F}{\omega}\right)^2 \cdot \frac{\cos\theta'}{\sin^3\theta'} + 2 \cdot |H| \cdot \frac{K_F}{\omega} \cdot \sin\delta \cdot \frac{1}{\sin^2\theta'} \tag{13}$$

$$= 2 \cdot \frac{K_F}{\omega} \cdot \frac{1}{\sin^2\theta'} \left(|H| \cdot \sin\delta - \frac{K_F}{\omega} \cdot \frac{\cos\theta'}{\sin\theta'}\right)$$

$$\theta' \in \left(\theta'_{min}, \frac{\pi}{2}\right); \theta'_{min} = \theta + \tan^{-1}\left(\frac{2R}{h}\right) \tag{14}$$

Set $f'(\theta') = 0$, then:

$$\theta' = \tan^{-1}\left(\frac{K_F}{\omega|H| \cdot \sin\delta}\right) \tag{15}$$

It can be concluded that:

$$f'(\theta') > 0, 0 < \theta' < \tan^{-1}\left(\frac{K_F}{\omega|H| \cdot \sin\delta}\right)$$
$$\tag{16}$$
$$f'(\theta') < 0, \tan^{-1}\left(\frac{K_F}{\omega|H| \cdot \sin\delta}\right) < \theta' < \frac{\pi}{2}$$

Function $f(\theta')$ has its minimum value when $\theta' = \tan^{-1}\left(\frac{K_F}{\omega|H| \cdot \sin\delta}\right)$.

Summing up the above, the optimized de-tumbling strategy performs best when $\theta' = \tan^{-1}\left(\frac{K_F}{\omega|H| \cdot \sin\delta}\right)$ in the target's every spin cycle.

4 Simulation of Optimized De-Tumbling Method

4.1 Simulation

A simulation is performed to estimate the optimized method using MATLAB tools. The target's model is based on the abandoned rocket's upper stage with a launch weight of 2154 kg. The sizes are shown below, where R = 2.6 m, r = 2.5 m, h = 7.4 m Fig. 9.

Its moment of inertia around each axis are:

$$J_X = \frac{m}{12} \cdot [3 \cdot (R^2 + r^2) + h^2] = \frac{16835 \, m^2}{kg}$$

$$J_Y = \frac{m}{12} \cdot [3 \cdot (R^2 + r^2) + h^2] = \frac{16835 \, m^2}{kg} \tag{17}$$

$$J_Z = \frac{m}{2} \cdot (R^2 + r^2) = \frac{14012 m^2}{kg}$$

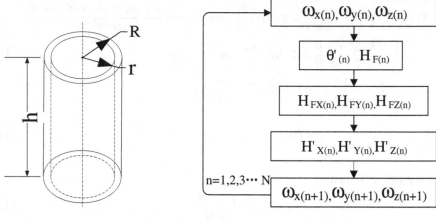

Fig. 9. Shape of target   **Fig. 10.** Iteration cycle

Also, set the target's initial angular velocity $\omega = \frac{\pi}{18}\left(10^\circ\right)$ and the angle between the angular velocity and the Z axis of the target's body frame $\theta = \frac{\pi}{6}\left(30^\circ\right)$.

The simulation is carried out as following process, the parameters' value of every cycle is calculated by iteration Fig. 10. The angular parameters θ, θ'_{min}, δ are calculated using angular velocity data first.

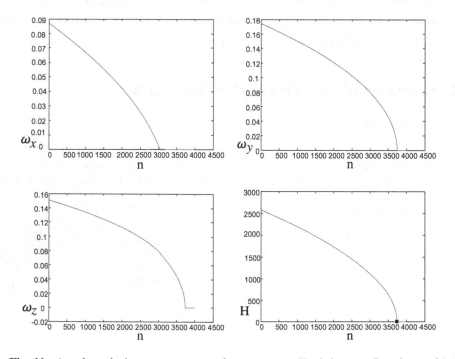

Fig. 11. Angular velocity ω_x, ω_y, ω_z, angular momentum H relative to n (iteration cycle)

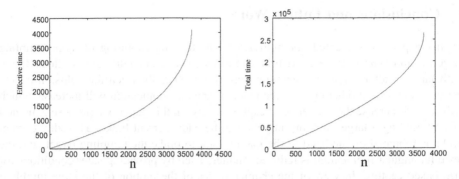

Fig. 12. Effective time of force in each cycle, Total time relative to n

Then put the angular parameters in to the Equation to calculate θ'. Then the remained angular momentum of target H' of present cycle is computed, which can be used to determine the initial angular velocity of the target's next spinning cycle.

Finish the iteration by using revolution numbers n as the iteration times. The de-tumbling process will end when H' is close to zero.

Above figures show that the target's angular velocity around each axis is gradually decreased cycle by cycle Fig. 11. But the de-tumbling rate becomes slow as the target's angular velocity slows down.

Eventually, the process stops at n = 3740 when H' is close to zero. The effective time of force in each cycle and the total time of the de-tumbling procedure can be calculated Fig. 12.

On the other hand, if the brake force's point of action is not optimized, but constantly acting at the middle part of the target Fig. 13. The simulation results as below Fig. 14:

Put the angular parameters into iteration, the total time can be computed relative to cycle time. Comparing to the total time in optimized de-tumbling, it can be concluded that the time consuming in debris removal can be spared by optimized de-tumbling strategy based on the debris' kinematic parameters of spin movement.

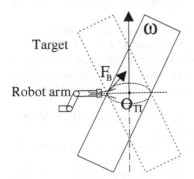

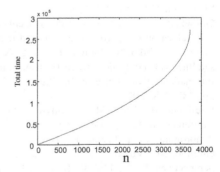

Fig. 13. Un-optimized method of de-tumbling   **Fig. 14.** Total time of un-optimized method

5   Conclusions and Future Work

In this paper, we presented our motivation for studying capturing of large tumbling target. In most cases, the space debris who loses power and control always floats freely with large residual angular momentum on-orbit. Especially when the collision occurs between debris and other objects, the residual angular momentum will increase, which will further increase the challenge of capture safety. In this paper, we propose a method of de-tumbling a large tumbling target with the eddy current brake. The eddy current brake is mounted on the hand of the manipulator to brake the tumbling target, and the residual angular momentum is reduced smaller. Thereby providing safer conditions for arm-based capture. In view of the characteristics of the motion of the large tumbling target, this paper presents an optimized method of braking, and obtains the optimized planning trajectory by simulation. Then, the validity of the above method is verified by simulation.

The effectiveness of large debris de-tumbling based on eddy current brake in hand of manipulator is verified by the study of the noncontact de-tumbling technology by the simplified model. The ground experiment based on the 6 DoF space manipulator [12] and air-bed experimental system [13] will be carried out in the future and the results should be applied to practice.

Acknowledgment. Thanks for the support of the research work by China Academy of Launch Vehicle Technology (CASC). The authors would like to acknowledge the support of the team members for their outstanding contributions to the paper.

References

1. Greaves, S., Boyle, K., Doshewnek, N.: Orbiter boom sensor system and shuttle return to flight: operations analyses. In: AIAA Guidance Navigation, and Control Conference and Exhibit, San Francisco, p. 5986 (2005)
2. Liu, J., Fan, Q., Yang, T., Li, K., Huang, Q.: A space robot manipulator system: designed for capture. Int. J. Mechatron. Autom. **5**(2/3), 125–132 (2015)
3. Shoemaker, J., Wright, M.: Orbital express space operations architecture program. Space systems technology and operations. In: Tchoryk Jr., P., Shoemaker, J. (eds.) Proceedings of SPIE, vol. 5088, pp. 1–9 (2003)
4. Gomez, N.O., Walker, S.J.I.: Eddy currents applied to de-tumbling of space debris: analysis and validation of approximate proposed methods. Acta Astronaut. **114**, 34–53 (2015)
5. Sugai, F., Abiko, S., Tsujita, T., Jiang, X., Uchiyama, M.: De-tumbling an uncontrolled satellite with contactless force by using an eddy current brake. In: Proceedings of IEEE International Conference on Intelligent Robots and Systems, Tokyo, 3–7 November 2013, pp. 783–788 (2013)
6. Reinhardt, B., Peck, M.A.: Eddy-current space tug. In: Proceedings of AIAA SPACE Conference & Exposition, Long Beach, 27–29 September 2011
7. Ambroglini, F., Battiston, R., Burger, W.J., Calvelli, V., Musenich, R., Spillantini, P., Giraudo, M., Vuolo, M.: Active magnetic shielding for manned space missions present perspectives. In: Sgobba, T., Rongier, I. (eds.) Space Safety is No Accident, pp. 151–160. Springer, Cham (2015). doi:10.1007/978-3-319-15982-9_18

8. Yoshida, K.: Engineering test satellite VII flight experiment for space robot dynamics and control: Theories on laboratory test beds ten years ago, now in orbit. Int. J. Rob. Res. **22**(5), 321–335 (2003)
9. Murray, R.M., Li, Z., Sastry, S.S.: A Mathematical Introduction to Robotic Manipulation. CRC Press, Boca Raton (1994)
10. Featherstone, R.: Robot Dynamics Algorithms. Kluwer, Norwell, MA (1987)
11. Landzettel, K., Brunner, B., Hirzinger, G., Lampariello, R., Schreiber, G., Steinmetz, B.-M.: A unified ground control and programming methodology for space robotics applications-demonstrations on ETS-VII. In: Proceedings of International Symposium on Robotics, Montreal pp. 422–427 (2000)
12. Liu, J., Fan, Q., Wang, Y., et al.: A space robot hand arm system: designed for capture. In: IEEE International Conference on Mechatronics and Automation. pp. 1247–1252 (2015)
13. Liu, J., Huang, Q., Wang, Y., Deng, T.: A method of ground verification for energy optimization in trajectory planning for six DOF space manipulator. In: Proceedings of IEEE International Conference on Fluid Power and Mechatronics, Harbin, 5–7 August 2015, pp. 791–796 (2015)

Development of Modular Joints of a Space Manipulator with Light Weight and Wireless Communication

Liang Han[1], Can Luo[1], Xiangliang Cheng[1], and Wenfu Xu[1,2(✉)]

[1] Shenzhen Graduate School, Harbin Institute of Technology, Shenzhen 518055, China
wfxu@hit.edu.cn
[2] The State Key Laboratory of Robotics and System, Harbin Institute of Technology, Harbin 150001, China

Abstract. Due to a large number of electrical cables in the joints of traditional space manipulator, the assembly, testing and maintenance are very inconvenient. In this paper, a modular joint with light weight and wireless communication link is developed to solve the above problem. The performance of the joint is firstly determined according to task requirement. Then, the mechanical structure is designed to meet the performance requirement. It is mainly constructed by a brushless DC motor, harmonic reducer, bearing, and supporting structure. The electrical sub-system includes sensors, servo controller and WIFI communication link. Multiple sensors, including three hall sensors, an incremental magnetic encoder, an absolute magnetic encoder, and a torque sensor are installed in the joint. The vector control is used for motor. The WIFI communication link is designed for the communication between the central controller and the servo controller. Therefore, the task command can be sent through the WIFI link. Electrical cables between the central controller and joint servo controller are not required. Finally, the developed joint is tested on a test platform. The experimental results verify the performance of the joint.

Keywords: Space manipulator · Modular joint · On-orbital servicing · Wireless communication · Light weight

1 Introduction

Space manipulators are expected to play an increasingly important role in space activities. One broad area of application is in the servicing, construction, and maintenance of satellites and large space structures in orbit. Therefore, space robotic technologies have been emphasized by many countries [1–3]. Such applications often require that a space manipulator should have high load-to-weight ratio, simple and convenient test interface. Moreover, the manipulator itself may need repairing on orbit to assure that the on-orbital servicing task be completed successfully. For the space manipulators used previously, such as ETS-VII [4], Orbital Express [5], SSRMS [6], and so on [7], there are a large number of electrical cables connecting each joint and the central controller. Therefore, it is very inconvenient for assembly, testing and maintenance.

© Springer International Publishing AG 2017
Y. Huang et al. (Eds.): ICIRA 2017, Part III, LNAI 10464, pp. 650–661, 2017.
DOI: 10.1007/978-3-319-65298-6_58

NASA and Canada developed the MSS space station service system [8], which includes a removable base, a seven degrees of freedom arm SSRMS and the special purpose dexterous manipulator (SPDM). SPDM can be connected to the end of the SSRMS by the adapter on the body for orbit operations. DLR has completed the development of the third generation light arm called LWR [9]. The joint has a big load-to-weight ratio and a compact structure by using a DC brushless motor and a harmonic reducer. The Reconfigurable Brachiating Space Robot (called BRB) [10, 11] developed in Japan belongs to the typical space reconfigurable manipulator. The most important feature of this manipulator is that it can reconstruct the arm by the adapter of the manipulator. Robonaut2 developed by NASA includes head, arms and the end dexterous hand. With a mount of sensors integrated into the arms and fingers, it is possible to perform compliance and detailed manipulation [12, 13]. In 2004, Mars landing Rovers, courage and opportunity, are installed in the light arm for the sampling and analysis of soil and rock [14]. The light weight manipulator has five degrees of freedom, with the length of 1 m. It can reach the range of 0.75 m and load 4.4 kg with weight 2 kg.

Recently, wireless communication technology is extended for aerospace field. For example, SuperBot [15] uses wireless remote control to communicate. The National Defense Advanced Research Projects Agency (DARPA) has constructed a new structure system of spacecraft called F6. One of the cores is the establishment of the wireless communication network between satellite clusters [16]. The application of wireless communication in space is firstly used in the rapid response of satellite. The world's first satellite which used the wireless communication technology is Delfi-C3 which designed by the University of Delft [17]. Autonomous wireless Sun Sensor (AWSS) of the Delfi-C3 satellite communicates with other devices in the satellite through the wireless communication. The tactical satellite called TacSat [18] developed by U.S. verified that WIFI communication can transmit data stably through the transformation of the communication protocol. At present, the application of wireless communication in space manipulator is less, however, wireless communication has great research significance and application prospect for shortening the development cycle of the space manipulator, reducing the weight of space robot system and making the manipulator operation more flexible.

In this paper, we developed a type of joints with light weight and WIFI communication for space manipulators. The main contributions include performance index analysis, mechanical design, and electrical design. The communication link is based on WIFI. The motor is controlled under vector control mode. A joint test platform is also developed to test and verify the joint prototype. Wireless communication realization can shorten the development cycle of the space manipulator, reduce the weight of the space robot system, expand the scope of the manipulator, and make the manipulator more flexible. Wireless communication network is helpful to realize the rapid response of space vehicles.

2   The Design of Mechanical System

2.1   Design Index of the Joint

The light weight space manipulator developed in this paper has 7 degrees of freedom which contained the shoulder with three degrees of freedom, the elbow with one degrees of freedom and the wrist with three degrees of freedom. Both three joints in the shoulder and the wrist are respectively vertical. Moreover, the shoulder is configured in roll-yaw-pitch, the elbow is configured in pitch and the wrist is used the Pitch-Yaw-Roll configuration. The joint configuration and dimensions of the manipulator are shown in Fig. 1.

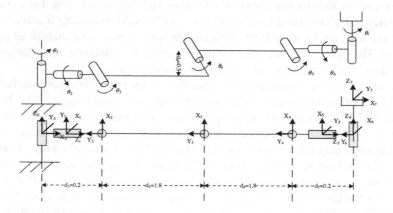

Fig. 1.   Configuration and dimensions of the manipulator

The main technical indicators of the manipulator:

The absolute position and attitude accuracy of the manipulator: better than 8 mm, better than 1°;

The repetitive position and attitude accuracy of the manipulator: superior to 2 mm and better than 0.5°;

The maximum acceleration of the manipulator joint: $2°/s^2$;

By modeling in the 3D software, we can estimate mass of the joint, link and manipulator. These parameters will be used to calculate the joint torque. According to Fig. 1, we can determine the design index of the joint as shown in Table 1.

Table 1.   The design index of the joint

Parameters	Value
Mass	5 kg
Torque	100 NM
Accuracy	0.06°

2.2   Modular Joint Design

Based on the typical design of the robot joint, a highly integrated modular joint is designed in this paper. The joint uses a DC brushless motor as driving mechanism. The harmonic reducer is used as the transmission mechanism to increase the output torque and reduce the joint error. The input shaft and the output shaft of the joint are both supported by a pair of deep groove ball bearings. On the one hand, the joint can compensate for thermal expansion through the outer ring of the bearing. On the other hand, the joint has high rigidity and precision. Joint has a wealth of sensor resources to improve the environmental adaptability and high precision. The joint sensors include an incremental encoder for the control of the speed loop, current hall sensors for the current loop control, the absolute encoder for the position loop control, and the torque sensor for the torque loop control. The joint installs an electromagnetic brake which is used for the braking of the manipulator when the system is powered down. The structure of the joint is shown in Fig. 2.

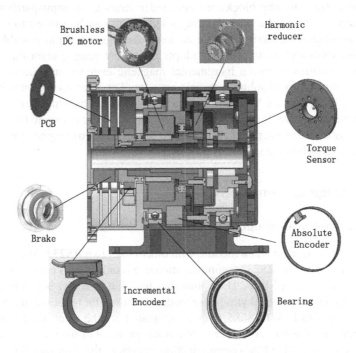

Fig. 2. The structure of the joint

The overall structure of the joint. The modular joint is composed of four shafts. Firstly, the movable block, the movable block end cover, the torque sensor, the output shaft and the flexible wheel form a shaft through rabbets and screws. Secondly, the static block, the stator sleeve, the steel wheel fixing part, the steel wheel, the output shaft and the flexible wheel form a shaft through rabbets and screws. Thirdly, the rotating shaft, the wave generator, the brake armature and the incremental encoder form a shaft through rabbets and screws. Finally, the end cover of static block, the sleeve and the magnetic yoke of the brake form the fourth shafts. These four shafts are supported by two pairs of bearings. The concentricity of the shaft which determines the performance of the joint is guaranteed by rabbets and bearings. The magnetic grid of the absolute encoder is arranged on the inner wall of the movable block. The output parts are matched with a taper pin to eliminate the empty back after fixed with screws.

Transmission principle. In order to reduce the size of joint, the joint uses separating and ultrathin type harmonic reducer. The steel wheel is fixed and flexible wheel is output by an output shaft. The brushless DC motor drives the wave generator to rotate as input part, and the reduced motion is output by the flexible wheel.

Technical features. The static block is a cylinder, and the movable block extends the interface on the cylinder for the connection of the next joint; A brushless DC motor and a wave generator are connected in series as an input part, and the flexible wheel, the output shaft and the movable block are connected in series as an output part; One bearing is completely fixed and the other bearing's outer ring can be moved to compensate thermal expansion of the shaft. Distance between bearings is as far as possible to ensure transmission accuracy; Motor control adopts the vector control algorithm. The joint speed loop is controlled by an incremental magnetic encoder and the position loop control is controlled by absolute magnetic encoder; The joint servo controller is integrated in the joint, and the WIFI is used for wireless transmission; Harmonic reducer using separate three parts to reduce the size of joints; In order to ensure the accuracy of joint, four shafts are respectively combined to process; Electromagnetic brake is used to avoid accident when the system is powered down.

3   The Design of Control System

3.1   WIFI Wireless Communication

Generally, WIFI will be made a module with other buses like RS232, SPI, and SDIO to simplify the use of WIFI. RS232 communication rate is only 20 KB per second while SPI communication rate is 2 Mbps. The communication rate of SDIO is much higher than SPI. It can reach tens of megabytes per second and be used for real-time image transmission. The MARVELL WIFI8686 chip is integrated on the joint servo controller to communicate with the host computer. The transfer protocol of the module is 802.11b/g. When the joint uses SDIO as communication interface, the data transfer rate can be 54 Mbps. Once WIFI communication is dropped, it will be automatically connected. WIFI communication could use check bits to prevent packet loss.

This paper uses a PC as the WIFI debug interface to send commands to the joint servo controller. WIFI interface is as shown in Fig. 3. The interface includes the WIFI parameter configuration, the joint selection, the different modes of joint, the instruction transmission, the controller reset and the other status bar. For example, we enter the corresponding number of instructions in the position mode, the joint controller will control the motor to drive the corresponding movement. Through the check of the frame head 0XFF and the frame tail 0X00, loss of the packets are avoided. When errors occur at the frame head or the frame tail, the data will be automatically retransmitted.

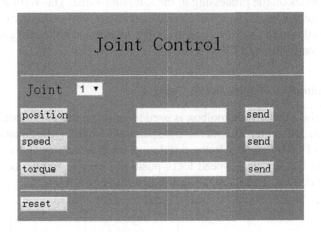

Fig. 3. WIFI interface

3.2 Joint Servo Controller

The motor adopts closed-loop vector control method. The joint servo controller drives the joint according to the command of host control including position, speed, torque, and braking signals. According to the requirements of the light space manipulator, the joint control system based on STM32 F103ZET6 and WIFI8686 module is designed in this paper. The hardware system of joint is shown in Fig. 4.

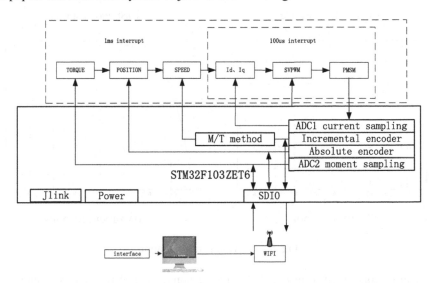

Fig. 4. The hardware system of joint

The joint servo controller is designed according to the joint size, which is divided into two boards. The control board mainly integrates the signal sampling circuit of sensors and carries on the processing to the sampling value. The WIFI module is integrated on the control board, which is connected with the SDIO interface of ARM. Drive and power board is used for converting the power to the rated voltage of each component and drive the motor through PWM wave.

3.3   The Simulation of Joint Servo Control

In this paper, the SIMULINK toolbox is used to simulate the vector control algorithm of the brushless DC motor. In order to transplant the program, the control algorithm is written by MATLAB function. In the simulation, we give the step signal to observe the response of the current loop, speed loop, position loop and torque loop respectively under the PID control. The results of simulation are shown in Fig. 5.

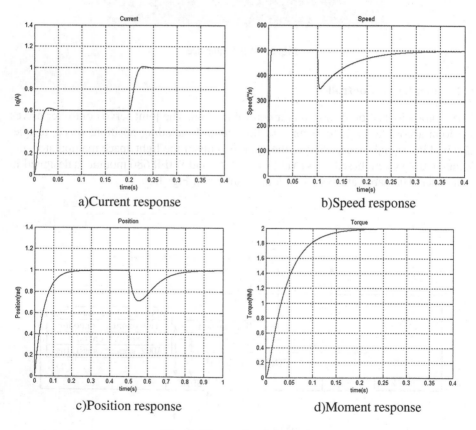

a)Current response   b)Speed response

c)Position response   d)Moment response

Fig. 5. The results of simulation

It can be concluded from Fig. 5 that the response time of four loops is short. The current and the moment can follow a given value with a small fluctuation. The speed loop and the position loop have a good anti-jamming performance. When the motor has a load disturbance, the speed and the position can quickly return to the given value. The position loop has no overshoot In order to prevent the manipulator from moving over.

Prototype and Experiment

In order to verify the performance of modular joint and control system, we develop the joint test platform to carry out joint performance test. The joint test platform is shown in Fig. 6.

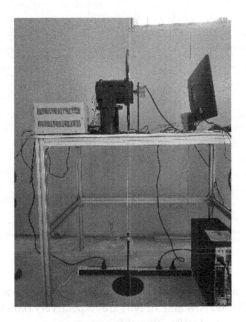

Fig. 6. The joint test platform

The joint test platform includes weights, the table, the photoelectric encoder, the joint, the controller and so on. The joint test platform can carry on the position precision test, the speed precision test and the torque precision test.

3.4 Position Precision Experiment

The central controller PC sends position command to joint servo control through WIFI communication. The joint turns 30° respectively for 6 times under load 0 NM, 30 NM, 60 NM, and 80 NM. The classical PID control algorithm is used for position control. K_p, K_i and K_d are set to 0.2, 0.02 and 0.08 respectively. Record the value of the external encoder when the joint rotates clockwise and counter clockwise, the position precision curve and error value can be obtained as shown in Fig. 7.

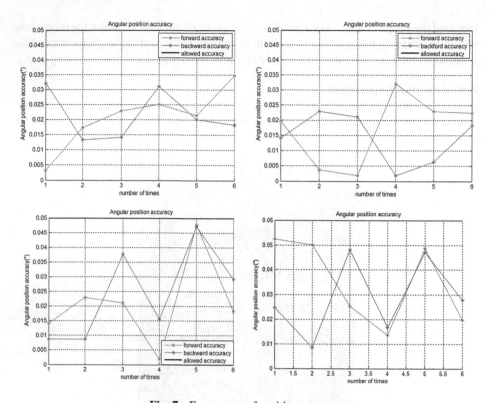

Fig. 7. Error curve of position accuracy

Take the maximum position error as the position precision, position precision under the load 0 NM, 30 NM, 60 NM and 80 NM are respectively 0.0348°, 0.0320°, 0.0477° and 0.050° which are all under the allowable error range.

3.5   Angular Velocity Accuracy and Stability Test

The central controller PC sends angular velocity command to joint servo control through WIFI communication. The joint angle data will be recorded every second for 10 times under the uniform motion. Differences are worth to calculate the average as the angular velocity of the joint.

$$\Delta \bar{\theta} = \frac{1}{10} \sum_{i=1}^{10} \Delta \theta_i \tag{1}$$

Angular velocity accuracy is:

$$\delta_{\omega} = \frac{1}{\dot{\theta}_g} (\Delta \bar{\theta} - \dot{\theta}_g) \tag{2}$$

$\dot{\theta}_g$ is expected angular velocity in the formula.

The joint angular velocity stability σ_ω can be obtained by the following formula:

$$\sigma_\omega = \frac{1}{\Delta\bar{\theta}} \sqrt{\frac{1}{N-1} \sum_{i=1}^{N} (\Delta\theta_i - \Delta\bar{\theta})^2} \tag{3}$$

The classical PID control algorithm is used for speed control. Kp, Ki and Kd are set to 0.4, 0.02 and 0.12 respectively. Under the 0.1 degrees per second, 0.5 degrees per second, 1 degrees per second, 5 degrees per second and 10 degrees per second respectively, the joint angular velocity accuracy and angular velocity stability are shown in Fig. 8.

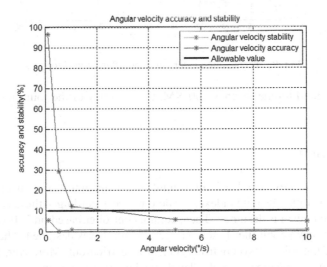

Fig. 8. Error curve of angular velocity and stability

In the range of 0.1 degrees per second to 10 degrees per second, the angular velocity accuracies are within the allowable range, however, the stabilities of the angular velocity do not meet the requirements when the angular velocity is too low. That is mainly because the joint appears the phenomenon of low-speed crawling.

3.6 Load Torque Test

By adding weights to increase the load torque of the joint, the position error of the joint are measured under the forward and reverse rotation. We measured the position accuracy of the joint under 0 NM, 30 NM, 60 NM, 80 NM and 100 NM. The position error curves of different load torques are shown in Fig. 9.

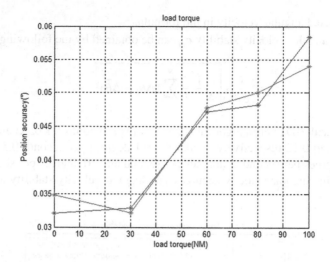

Fig. 9. The position error curve of different load torques

When the load torque reaches 100 NM, the accuracy of the joint position reaches 0.06°.

4 Conclusion

In this paper, we developed a light weight space manipulator joint using wireless communication link. The joint indexes are determined according to the task requirement. Then, the mechanism is designed. Then, central controller based on WIFI communication and joint servo controller based on vector control was designed. The closed loop simulation of the joint servo control system was carried out. Moreover, the joint test platform was also developed, and the developed joint was tested on it. The experimental results show that the designed joint has high control precision and large load torque. In the future, we will make the wireless communication more stable and develop the reconfigurable interface and the whole manipulator system.

Acknowledgements. This work was supported by the National Natural Science Foundation of China (Grant No. 61573116), the Foundation for Innovative Research Groups of the National Natural Science Foundation of China (Grant No.51521003), and the Basic Research Program of Shenzhen (JCYJ20160427183553203).

References

1. Shan, M., Guo, J., Gill, E.: Review and comparison of active space debris capturing and removal methods. Prog. Aerosp. Sci. **80**, 18–32 (2016)
2. Huang, P., Zhang, F., Cai, J., Wang, D., Meng, Z., Guo, J.: Dexterous tethered space robot: design, measurement, control and experiment. IEEE Trans. Aerosp. Electron. Syst. **99**, 1–14 (2017)

3. Xu, W., Peng, J., Liang, B., Mu, Z.: Hybrid modeling and analysis method for dynamic coupling of space robots. IEEE Trans. Aerosp. Electron. Syst. **52**(1), 85–98 (2016)
4. Oda, M., Kibe, K., Yamagata, F.: ETS-VII, space robot in-orbit experiment satellite. In: Proceedings of the IEEE International Conference on Robotics and Automation, Minneapolis (1996)
5. Friend, R.: Orbital express program summary and mission overview. In: Proceedings of the SPIE, Sensors and Systems for Space Applications II, Orlando (2008)
6. Laryssa, P., Lindsay E., Layi O., Marius O., Nara K., Aris L.: International space station robotics: a comparative study of ERA, JEMRMS and MSS. In: Proceedings of the 7th ESA Workshop on Advanced Space Technologies for Robotics and Automation, Noordwijk, The Netherlands, pp. 1–8 (2002)
7. Liang, B., Li, C., Xue, L.J., Qiang, W.Y.: A Chinese Small Intelligent Space Robotic System for On-Orbit Servicing. In: Proceedings of the IEEE/RSJ International Conference on Intelligent Robots and Systems, Beijing (2006)
8. Stieber, M.E., Hunter, D.G., Abramovici, A.: Overview of the mobile servicing system for the international space station. European Space Agency-Publications-ESA SP (1999)
9. Hirzinger, G., Butterfass, J., Fischer, M.: A mechatronics approach to the design of light-weight arms and multifingered hands. In: IEEE International Conference on Robotics & Automation. IEEE (2000)
10. Matunaga, S.: Advanced space robot for in-orbit servicing missions. JSME News, 6–8 (2002)
11. Shibata, T., Ohkami Y.: Development of brachiating control system for reconfigurable brachiating space robot. In: Proceedings of the Third International Workshop on RoMoCo. IEEE (2002)
12. Ambrose, R.O., Aldridge, H., Askew, R.S.: Robonaut: NASA's space humanoid. IEEE Intell. Syst. (2000)
13. Diftler, M.A., Mehling, J.S., Abdallah, M.E.: Robonaut 2-the first humanoid robot in space. In: 2011 IEEE International Conference on Robotics and Automation. IEEE (2011)
14. Billing, R.: Caging mechanisms for the Mars Exploration Rover instrument deployment device (2003)
15. Salemi, B., Moll, M., Shen, W.M.: SUPERBOT: a deployable, multi-functional, and modular self-reconfigurable robotic system. In: 2006 IEEE/RSJ International Conference on Intelligent Robots and Systems. IEEE (2006)
16. Qin, W.: General situation of the system F6. In: Space International, pp. 31–35 (2008)
17. Boom, C.W.D., Leijtens, J.A.P., Heiden, N.V.D.: Micro Digital Sun Sensor: "a Matchbox Miracle" (2006)
18. Doyne, T., Wegner, P., Riddle, R.: A TacSat and ORS update including TacSat-4. In: The 4th Responsive Space Conference, Los Angeles (2005)

Accurate Dynamics Modeling and Feedback Control for Maneuverable-Net Space Robot

Yakun Zhao[1,2], Panfeng Huang[1,2(✉)], and Fan Zhang[1,2]

[1] National Key Laboratory of Aerospace Flight Dynamics,
Northwestern Polytechnical University, Xian 710072, China
[2] Research Center for Intelligent Robotics, School of Astronautics,
Northwestern Polytechnical University, Xian 710072, China
pfhuang@nwpu.edu.cn

Abstract. Tethered Space Net (TSN) has been proposed since there is an increasing threat of space debris to spacecraft and astronauts in recent years. In this paper, we propose an improved TSN, named Maneuverable-Net Space Robot (MNSR), which has four maneuvering units in its four corners (square net). The four maneuverable units make the MNSR controllable. Because of autonomous maneuverability, the attitude dynamics of the platform, master tether and flexible net are strongly coupled. In order to design an effective controller to maintain the configuration of the maneuverable net, an accurate dynamics model of MNSR based on the Lagrangian method is derived. In our model, we consider the three-dimensional attitude of the platform, master tether and maneuvering-net as well. Due to the vibration of the in-plane and out-of-plane angles of the net tethers, feedback control is employed for MNSR. The simulation results demonstrate that the proposed control is efficient and suitable for the MNSR system.

Keywords: Space debris · Maneuverable-net space robot · Dynamics modeling · Feedback control

1 Introduction

There is an increasing threat of space debris to spacecraft and astronauts since the first satellite was launched in October 4th, 1957 [1]. Some researchers proposed Tethered Space Capturing System to complete the capturing task which aimed at space debris removal [2]. Tethered Space Capturing System contains Tethered Space Robot (TSR) [3] (shown in Fig. 1), Tethered Space Harpoon (TSH) [4] (shown in Fig. 2) and Tethered Space Net (TSN) [5] (shown in Fig. 3). Nowadays, Tethered Space Net (TSN) has attracted much attention [5,6]. TSN consists of platform satellite, tether, space net and four flying weights in each corner of the net, which is shown in Fig. 3. TSN is a flexible system and it converts traditional point-to-point capture into surface-to-point capture, which lowers the requirement on capture precision. Besides, it can be used for uncooperative target capture and long distance capture. So Tethered Space Net (TSN) is one of the most promising solutions for active space debris removal.

© Springer International Publishing AG 2017
Y. Huang et al. (Eds.): ICIRA 2017, Part III, LNAI 10464, pp. 662–672, 2017.
DOI: 10.1007/978-3-319-65298-6_59

Fig. 1. Tethered Space Robot (TSR)

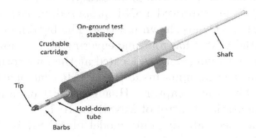

Fig. 2. Tethered Space Harpoon (TSH)

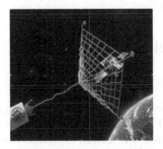

Fig. 3. Tethered Space Net (TSN)

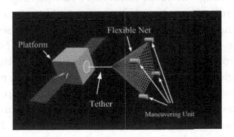

Fig. 4. Maneuverable-Net Space Robot (MNSR)

A lot of studies have been focused on the TSN. Based on the previous studies of space webs, various dynamics model of TSN have been derived, such as rigid model [7], mass-spring model [8], absolute nodal coordinate formulation (ANCF) model [9]. Rigid model is most commonly used in tethered space system. It can be better used for analyzing in-plane and out-of-plane angles of the tether. Mass-spring model is always used for establishing a quadrangular mesh net, while absolute nodal coordinate formulation (ANCF) model can better describe the flexibility between two nodes on the net and reflect the dynamics of TSN. However, rigid model always treats the platform satellite as point mass, while the literatures about ANCF model only demonstrate the dynamics of the net and neglect the dynamics of platform. Thus, an accurate model for TSN, which considered the dynamics of platform, is necessary.

Based on the TSN, an improved TSN is proposed, named Maneuverable Net Space Robot (MNSR), which is shown in Fig. 4. The big difference between TSN and MNSR is that the MNSR has four maneuvering units in four corners (square net) instead of the four flying weights located at the four corners of traditional TSN. The four maneuvering units make the MNSR controllable. So the MNSR is more preferable for orbital capture. Huang et al. have already studied the dynamics and configuration control of MNSR in [10,11].

In this paper, an accurate dynamic model of MNSR is derived in Sect. 2. Then in Sect. 3, a feedback control scheme is proposed for maintaining the configuration of maneuverable net. Some numerical simulations are shown to verify the control scheme in Sect. 4. Finally, the contribution of this paper is briefly summarized in Conclusion Sect. 5.

2   Dynamics Model

2.1   Description of the System

The schematic figure of the MNSR and generalized coordinates used to describe the motion are shown in Fig. 5. The position of the centre of the mass of the system C in its orbit around the Earth is defined by the true anomaly γ and the orbit radius R_c. The coordinate system $C - x_o y_o z_o$ has z_o axis along the orbit normal, y_o axis radially outward away from the Earth along the local vertical and x_o axis along the local horizontal completing the right hand triad. The six rotating coordinate system ($C_p - x_p y_p z_p, C_t - x_t y_t z_t, C_k - x_k y_k z_k, k = 1, 2, 3, 4$) for platform, master tether and net tether are used with their origins at the center of mass of each of them respectively. The coordinate $C_p - x_p y_p z_p$ is obtained by the rotation θ (pitch angle of the platform satellite) about z_o axis, and ψ (roll angle of the platform satellite) about y_o axis. The coordinate system $C_t - x_t y_t z_t$ is obtained via the rotation α (the in-plane angle of master tether) about z_o axis, and β (the out-of-plane angle of master tether) about y_o axis. The coordinate $C_k - x_k y_k z_k$ is obtained via the rotation α_k (the in-plane angle of net tether) about z_o axis, and β_k (the out-of-plane angle of net tether) about y_o axis. m_0, m_t, m are the masses of the platform, master tether

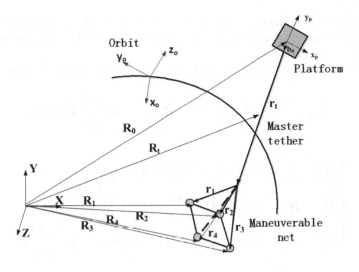

Fig. 5. Generalized coordinates for MNSR

and maneuverable net tether respectively. l, lw are the length of master tether and tether net respectively.

2.2 Motion Equations of the System

The following assumptions are made for the dynamic model of MNSR.

1. Four maneuvering units are considered as mass points. Thus, the maneuverable net can be considered as four tethers (net tether) and four mass points (maneuvering unit) (shown in Fig. 5). The effect of the variation of maneuvering unit's attitude and material damping of the net are ignored. The platform is considered as rigid body, and tethers are considered as rigid, inextensible and remaining straight.
2. Gravity is treated to be the only external force acting on the system.
3. The center of mass of the system is assumed to follow a circular Keplerian orbit.

The position vectors of platform, an elemental mass of master tether and net tether with respect to the center of the Earth, are respectively

$$\begin{cases} \boldsymbol{R}_t = \boldsymbol{R}_0 + \boldsymbol{b}_0 + \frac{1}{2}\boldsymbol{r}_t \\ \boldsymbol{R}_1 = \boldsymbol{R}_0 + \boldsymbol{b}_0 + \boldsymbol{r}_t + \boldsymbol{r}_1 \\ \boldsymbol{R}_2 = \boldsymbol{R}_0 + \boldsymbol{b}_0 + \boldsymbol{r}_t + \boldsymbol{r}_2 \\ \boldsymbol{R}_3 = \boldsymbol{R}_0 + \boldsymbol{b}_0 + \boldsymbol{r}_t + \boldsymbol{r}_3 \\ \boldsymbol{R}_4 = \boldsymbol{R}_0 + \boldsymbol{b}_0 + \boldsymbol{r}_t + \boldsymbol{r}_4 \end{cases} \qquad (1)$$

where, $\boldsymbol{R}_0, \boldsymbol{R}_t, \boldsymbol{R}_k(k = 1, 2, 3, 4)$ are the position vectors of the mass of platform, master tether and net tethers respectively. $\boldsymbol{b}_0$ is the offset from center of mass of platform to the attachment points on it. $r_t, r_k (k = 1, 2, 3, 4)$ denote the position vectors of an elemental mass of master tether and net tether with respect to the center of mass of the system.

The kinetic energy due to translation of the whole system is derived as

$$T_{rans} = \tfrac{1}{2} m_0 \dot{R}_0 \dot{R}_0 + \tfrac{1}{2} m_t \dot{R}_t \dot{R}_t + \tfrac{1}{2} m \sum_{k=1}^{n=4} (\dot{R}_k \dot{R}_k) \tag{2}$$

where

$$
\begin{cases}
\dot{R}_0 = \dot{R}_c - \frac{m_t + 4m}{M} \dot{b}_o - \frac{\frac{1}{2} m_t + 4m}{M} \dot{r}_t - \frac{m}{M} \sum_{k=1}^{n=4} \dot{r}_k \\
\dot{R}_t = \dot{R}_c + \frac{m_0}{M} \dot{b}_o + \frac{\frac{1}{2}(m_0 - 4m)}{M} \dot{r}_t - \frac{m}{M} \sum_{k=1}^{n=4} \dot{r}_k \\
\dot{R}_k = \dot{R}_c + \frac{m_0}{M} \dot{b}_o + \frac{m_0 + \frac{1}{2} m_t}{M} \dot{r}_t - \frac{m - M}{M} \dot{r}_k - \frac{m}{M} \sum_{\substack{q \neq k}}^{1,2,3,4} \dot{r}_q
\end{cases}
\tag{3}
$$

and $M = m_0 + m_t + 4m$.

The rotational kinetic energy of the whole system is determined via

$$T_{rot} = \tfrac{1}{2} \omega_0{}^T I_0 \omega_0 \tag{4}$$

where I_0 is the moments inertia of platform, $I_0 = diag(I_{oxx}, I_{oyy}, I_{ozz})$

The potential energy of the system is given by

$$V = -\mu m_0 \frac{1}{|R_0|} - \mu m_t \frac{1}{|R_t|} - \mu m \sum_{k=1}^{n=4} \frac{1}{|R_k|} \tag{5}$$

where μ is the gravitational constant of the Earth.

The Lagrangian equation is used to obtain the equations of motion of the system from the kinetic and potential energy expressions

$$\frac{d}{dt} \left(\frac{\partial T}{\partial \dot{q}_i} \right) - \frac{\partial T}{\partial \dot{q}_i} + \frac{\partial V}{\partial \dot{q}_i} = Q_i \tag{6}$$

where T and V are the total kinetic and potential energies of the system; q_i are the generalized coordinates, which are chosen to be $q_i = \{\theta \ \ \psi \ \ \alpha \ \ \beta \ \ l \ \ \alpha_k \ \ \beta_k\}^T, (k = 1, 2, 3, 4)$. Consequently, generalized forces vector is Q_i, which is also the control force and chosen to be $Q_i = \{Q_\theta \ \ Q_\psi \ \ Q_\alpha \ \ Q_\beta \ \ Q_l \ \ Q_{\alpha_k} \ \ Q_{\beta_k}\}^T$.

A set of dimensionless quantities are defined as follows

$$\Lambda = l/Lr, \ \ \tau = \dot{\gamma}t, \ \ ()' = d\,()/d\tau \tag{7}$$

where Lr is the reference length of tether, Λ is the nondimensional tether length, and τ is the non-dimensional time.

The non-dimensional equations of MNSR are then given by

$$
\theta''[M_A \cos^2\psi + Ioxx \cdot \sin^2\psi + Iozz \cdot \cos^2\psi] + \alpha'' M_D \Lambda Lr \cos\psi \cos\beta
$$
$$
+ \sum_{k=1}^{n=4}[\alpha_k'' M_E lw \cos\psi \cos\beta_k]
$$
$$
-\psi' \left[
\begin{array}{l}
2M_A(1+\theta')\cos\psi\sin\psi + M_D\Lambda Lr(1+\alpha')\sin\psi\cos\beta \\
+ \sum_{k=1}^{n=4}[M_E lw(1+\alpha_k')\sin\psi\cos\beta_k] + 2Ioxx(1+\theta')\sin\psi\cos\psi \\
- 2Iozz(1+\theta')\sin\psi\cos\psi
\end{array}
\right]
$$
$$
-\beta' M_D\Lambda Lr(1+\alpha')\cos\psi\sin\beta - \sum_{k=1}^{n=4}[\beta_k' M_E lw(1+\alpha_k')\cos\psi\sin\beta_k]
$$
$$
+\Lambda' M_D Lr(1+\alpha')\cos\psi\sin\beta + 3M_A\cos^2\psi\cos\theta\sin\theta
$$
$$
+ M_D\Lambda Lr(2\sin\theta\cos\psi\cos\alpha\cos\beta + \cos\theta\cos\psi\sin\alpha\cos\beta)
$$
$$
+ \sum_{k=1}^{n=4}[M_E lw(2\sin\theta\cos\psi\cos\alpha_k\cos\beta_k + \cos\theta\cos\psi\sin\alpha_k\cos\beta_k)]
$$
$$
-Q_\theta/\Omega^2 = 0 \tag{8}
$$

$$
\psi''[M_A + I_{oyy}] + \beta'' M_D\Lambda Lr + \sum_{k=1}^{n=4}[\beta''_k M_E lw] + M_D Lr \Lambda'\beta' + M_A(1+\theta')^2\cos\psi\sin\psi
$$
$$
+ M_D\Lambda Lr(1+\theta')(1+\alpha')\cos\beta\sin\psi + \sum_{k=1}^{n=4}[M_E lw(1+\theta')(1+\alpha'_k)\cos\beta_k\sin\psi]
$$
$$
+ 3M_A\cos^2\theta\cos\psi\sin\psi + M_D\Lambda Lr(2\sin\psi\cos\theta\cos\alpha\cos\beta
$$
$$
- \sin\psi\sin\theta\sin\alpha\cos\beta + \cos\psi\sin\beta)
$$
$$
+ \sum_{k=1}^{n=4}[M_E lw(2\sin\psi\cos\theta\cos\alpha_k\cos\beta_k - \sin\psi\sin\theta\sin\alpha_k\cos\beta_k + \cos\psi\sin\beta_k)]
$$
$$
- I_{oxx}(1+\theta')^2\sin\psi\cos\psi + I_{ozz}(1+\theta')^2\sin\psi\cos\psi
$$
$$
-Q_\psi/\Omega^2 = 0 \tag{9}
$$

$$
\alpha'' M_B\cos^2\beta\Lambda^2 Lr^2 + \theta'' M_D\Lambda Lr \cos\psi\cos\beta + \sum_{k=1}^{n=4}[\alpha_k'' M_F lw\Lambda Lr\cos\beta\cos\beta_k]
$$
$$
-\beta'\left[2M_B\Lambda^2 Lr^2\cos\beta\sin\beta\cdot(1+\alpha') + M_D\Lambda Lr(1+\theta')\cos\psi\sin\beta \right.
$$
$$
\left. + \sum_{k=1}^{n=4}[M_F lw\Lambda Lr(1+\alpha'_k)\sin\beta\cos\beta_k]\right]
$$
$$
-\psi' M_D\Lambda Lr(1+\theta')\sin\psi\cos\beta - \sum_{k=1}^{n=4}[\beta'_k M_F lw\Lambda Lr(1+\alpha'_k)\cos\beta\sin\beta_k]
$$
$$
+ Lr\Lambda'\left[2M_B\Lambda Lr(1+\alpha')\cos^2\beta + M_D(1+\theta')\cos\psi\cos\beta \right.
$$
$$
\left. + \sum_{k=1}^{n=4}[M_F lw(1+\alpha'_k)\cos\beta\cos\beta_k]\right]
$$
$$
+ 3M_B\Lambda^2 Lr^2\cos^2\beta\cos\alpha\sin\alpha + M_D\Lambda Lr(2\sin\alpha\cos\theta\cos\psi\cos\beta + \cos\alpha\sin\theta\cos\psi\cos\beta)
$$
$$
+ \sum_{k=1}^{n=4}[M_F lw\Lambda Lr(2\sin\alpha\cos\beta\cos\alpha_k\cos\beta_k + \cos\alpha\cos\beta\sin\alpha_k\cos\beta_k)]
$$
$$
-Q_\alpha/\Omega^2 = 0 \tag{10}
$$

$$\beta'' M_B \Lambda^2 L r^2 + \psi'' M_D \Lambda L r + \sum_{k=1}^{n=4} \left[\beta_k'' M_F l w \Lambda L r \right]$$

$$+ \Lambda' L r \left[M_D \psi' + \sum_{k=1}^{n=4} \left[M_F l w \beta_k' \right] + 2 M_B \beta' \Lambda L r \right]$$

$$+ M_B \Lambda^2 L r^2 (1 + \alpha')^2 \cos \beta \sin \beta + M_D \Lambda L r (1 + \theta')(1 + \alpha') \cos \psi \sin \beta$$

$$+ \sum_{k=1}^{n=4} \left[M_F l w \Lambda L r (1 + \alpha')(1 + \alpha_k') \cos \beta_k \sin \beta \right] + 3 M_B \Lambda^2 L r^2 \cos^2 \alpha \cos \beta \sin \beta \qquad (11)$$

$$+ M_D \Lambda L r (2 \sin \beta \cos \theta \cos \psi \cos \alpha - \sin \beta \sin \theta \cos \psi \sin \alpha + \sin \psi \cos \beta)$$

$$+ \sum_{k=1}^{n=4} \left[M_F l w \Lambda L r (2 \sin \beta \cos \alpha \cos \alpha_k \cos \beta_k - \sin \beta \sin \alpha \sin \alpha_k \cos \beta_k + \cos \beta \sin \beta_k) \right]$$

$$- Q_\beta / \Omega^2 = 0$$

$$\Lambda'' L r M_B - M_B \Lambda L r \left[(1 + \alpha')^2 \cos^2 \beta + \beta'' \right] - M_D \left[(1 + \theta')(1 + \alpha') \cos \psi \cos \beta + \psi' \beta' \right]$$

$$- \sum_{k=1}^{n=4} \left[M_F l w \left[(1 + \alpha')(1 + \alpha_k') \cos \beta \cos \beta_k + \beta' \beta_k' \right] \right] - 3 M_B \Lambda L r \cos^2 \beta \cos^2 \alpha$$

$$+ M_D \left[-2 \cos \theta \cos \psi \cos \alpha \cos \beta + \sin \theta \cos \psi \sin \alpha \cos \beta + \sin \psi \sin \beta \right] \qquad (12)$$

$$+ \sum_{k=1}^{n=4} \left[M_F l w \left[-2 \cos \alpha \cos \beta \cos \alpha_k \cos \beta_k + \sin \alpha \cos \beta \sin \alpha_k \cos \beta_k + \sin \beta \sin \beta_k \right] \right]$$

$$- Q_\Lambda / \Omega^2 = 0$$

$$\alpha_k'' M_C l w^2 \cos^2 \beta_k + \theta'' M_E l w \cos \psi \cos \beta_k + \alpha'' M_F l w \Lambda L r \cos \beta \cos \beta_k$$

$$+ \sum_{q \neq k}^{1,2,3,4} \left[M_G l w^2 \cos \beta_k \cos \beta_q \cdot \alpha_q'' \right]$$

$$+ \beta_k' \left[\begin{array}{l} M_C l w^2 (1 + \alpha_k') 2 \cos \beta_k (-\sin \beta_k) + M_E \cdot l w (1 + \theta') \cos \psi (-\sin \beta_k) \\ + M_F l w \Lambda L r (1 + \alpha') \cos \beta (-\sin \beta_k) + \sum_{q \neq k}^{1,2,3,4} \left[M_G l w^2 (1 + \alpha_q')(-\sin \beta_k) \cos \beta_q \right] \end{array} \right]$$

$$+ \psi' M_E l w (1 + \theta')(-\sin \psi) \cos \beta_k + \Lambda' L r M_F l w (1 + \alpha') \cos \beta \cos \beta_k$$

$$+ \beta' M_F l w \Lambda L r (1 + \alpha')(-\sin \beta) \cos \beta_k + \sum_{q \neq k}^{1,2,3,4} \left[M_G l w^2 (1 + \alpha_q') \cos \beta_k (-\sin \beta_q) \beta_q' \right]$$

$$+ 3 M_C l w^2 \cos^2 \beta_k \cos \alpha_k \sin \alpha_k + M_E l w (2 \sin \alpha_k \cos \theta \cos \psi \cos \beta_k + \cos \alpha_k \sin \theta \cos \psi \cos \beta_k)$$

$$+ M_F l w \Lambda L r (2 \sin \alpha_k \cos \alpha \cos \beta \cos \beta_k + \cos \alpha_k \sin \alpha \cos \beta \cos \beta_k)$$

$$+ \sum_{q \neq k}^{1,2,3,4} \left[M_G l w^2 (2 \sin \alpha_k \cos \beta_k \cos \alpha_q \cos \beta_q + \cos \alpha_k \cos \beta_k \sin \alpha_q \cos \beta_q) \right]$$

$$- Q_{\alpha_k} / \Omega^2 = 0$$

$$(13)$$

$$\beta_k'' M_C l w^2 + \psi'' M_E l w + \beta'' M_F l w \Lambda L r$$

$$+ \sum_{q \neq k}^{1,2,3,4} \left[M_G l w^2 \beta_q'' \right] + M_C l w^2 (1 + \alpha_k')^2 \cos \beta_k \sin \beta_k$$

$$+ M_E l w (1 + \theta')(1 + \alpha_k') \cos \psi \sin \beta_k + M_F l w \Lambda' L r \beta'$$

$$+ M_F l w \Lambda L r (1 + \alpha')(1 + \alpha_k') \cos \beta \sin \beta_k$$

$$+ \sum_{q \neq k}^{1,2,3,4} \left[M_G l w^2 (1 + \alpha_k')(1 + \alpha_q') \cos \beta_q \sin \beta_k \right] + 3 M_C l w^2 \cos^2 \alpha_k \cos \beta_k \sin \beta_k$$

$$+ M_E l w (2 \sin \beta_k \cos \theta \cos \psi \cos \alpha_k - \sin \beta_k \sin \theta \cos \psi \sin \alpha_k + \sin \psi \cos \beta_k)$$

$$+ M_F l w \Lambda L r (2 \sin \beta_k \cos \alpha \cos \beta \cos \alpha_k - \sin \beta_k \sin \alpha \cos \beta \sin \alpha_k + \sin \beta \cos \beta_k)$$

$$+ \sum_{q \neq k}^{1,2,3,4} \left[M_G l w^2 (2 \sin \beta_k \cos \alpha_k \cos \alpha_q \cos \beta_q - \sin \beta_k \sin \alpha_k \sin \alpha_q \cos \beta_q + \cos \beta_k \sin \beta_q) \right]$$

$$- Q_{\beta_k} / \Omega^2 = 0$$

$$(14)$$

where $M_A = \frac{m_0(m_t+4m)}{M}$, $M_B = \frac{m_0(\frac{1}{2}m_t+4m)^2+\frac{1}{4}m_t(m_0-4m)^2+4m(m_0+\frac{1}{2}m_t)^2}{M^2}$, $M_C = \frac{m(M-m)}{M}$, $M_D = \frac{m_0(\frac{1}{2}m_t+4m)}{M}$, $M_E = \frac{mm_0}{M}$, $M_F = \frac{m(m_0+\frac{1}{2}m_t)}{M}$, $M_G = -\frac{m^2}{M}$.

3 Control Scheme

In many studies about the dynamics analysis of tethered space system, we can see that the in-plane and out-of-plane angles of the tether vibrate with the frequencies of $\sqrt{3}$ and 2 times the orbital frequency respectively [12], which may lead to tangle of the master tether with platform and the chaos of the four net tethers, and lead to the failure of the mission. An appropriate controller for maintaining the configuration of the net is necessary.

It is clear that the dynamics of MNSR is complex and the state variables are strongly coupled. Some simplifications are made for the dynamic equations for the convenience of controller design. Assume that the length of master tether keeps constant during the station-keeping phase, that is, $\Lambda = 1$ and $\Lambda'' = \Lambda' = 0$. And the platform is controllable and keeps constant, and the in-plane and out-of-plane angles of master tether can be kept in the desired values. Then, we only need to control the in-plane and out-of-plane angles of net tether, namely, the configuration of the maneuverable net.

The feedback control laws for MNSR are designed as

$$Q_{c\alpha_k} = kp_{\alpha k} \cdot (\alpha_{kd} - \alpha_k) + kd_{\alpha k} \cdot (\alpha'_{kd} - \alpha'_k)$$
$$Q_{c\beta_k} = kp_{\beta k} \cdot (\beta_{kd} - \beta_k) + kd_{\beta k} \cdot (\beta'_{kd} - \beta'_k) \tag{15}$$

where α_{kd}, β_{kd} and α'_{kd}, β'_{kd} are the desired values of in-plane and out-of-plane angles and derivative of in-plane and out-of-plane angles respectively. Thus, $\alpha_{kd} - \alpha_k$, $\beta_{kd} - \beta_k$ and $\alpha'_{kd} - \alpha'$, $\beta'_{kd} - \beta'_k$ are the error of angles and derivative of error of angles respectively. $kp_{\alpha k}$, $kp_{\beta k}$ and $kd_{\alpha k}$, $kd_{\beta k}$ are the parameters of error and derivative of error.

Then the control inputs $Q_{\alpha k}, Q_{\beta k}(k = 1, 2, 3, 4)$ can be derived as

$$Q_{\alpha_k} = (M_C lw^2\cos^2\beta_k \cdot Q_{c\alpha k} + \sum_{q\neq k}^{1,2,3,4} \left[M_G lw^2 \cos\beta_k \cos\beta_q \cdot Q_{c\alpha q}\right]$$
$$+\beta'_k \left[\begin{array}{l} M_C lw^2(1+\alpha'_k)2\cos\beta_k(-\sin\beta_k) + M_E \cdot lw(-\sin\beta_k) \\ +M_F lwLr(-\sin\beta_k) + \sum_{q\neq k}^{1,2,3,4}\left[M_G lw^2(1+\alpha'_q)(-\sin\beta_k)\cos\beta_q\right] \end{array}\right]$$
$$+ \sum_{q\neq k}^{1,2,3,4}\left[M_G lw^2(1+\alpha'_q)\cos\beta_k(-\sin\beta_q)\beta'_q\right] + 3M_C lw^2\cos^2\beta_k \cos\alpha_k \sin\alpha_k$$
$$+M_E lw(2\sin\alpha_k\cos\beta_k) + M_F lwLr(2\sin\alpha_k\cos\beta_k)$$
$$+ \sum_{q\neq k}^{1,2,3,4}\left[M_G lw^2(2\sin\alpha_k\cos\beta_k\cos\alpha_q\cos\beta_q + \cos\alpha_k\cos\beta_k\sin\alpha_q\cos\beta_q)\right]) \cdot \Omega^2 \tag{16}$$

$$Q_{\beta_k} = (M_C lw^2 \cdot Q_{c\beta k} + \sum_{q \neq k}^{1,2,3,4} M_G lw^2 \cdot Q_{c\beta q} + M_C lw^2 (1 + \alpha'_k)^2 \cos \beta_k \sin \beta_k$$

$$+ M_E lw(1 + \alpha'_k) \sin \beta_k + M_F lw Lr(1 + \alpha'_k) \sin \beta_k$$

$$+ \sum_{q \neq k}^{1,2,3,4} \left[M_G lw^2 (1 + \alpha'_k)(1 + \alpha'_q) \cos \beta_q \sin \beta_k \right]$$

$$+ 3 M_C lw^2 \cos^2 \alpha_k \cos \beta_k \sin \beta_k + M_E lw(2 \sin \beta_k \cos \alpha_k) + M_F lw Lr(2 \sin \beta_k \cos \alpha_k)$$

$$+ \sum_{q \neq k}^{1,2,3,4} \left[M_G lw^2 (2 \sin \beta_k \cos \alpha_k \cos \alpha_q \cos \beta_q - \sin \beta_k \sin \alpha_k \sin \alpha_q \cos \beta_q + \cos \beta_k \sin \beta_q) \right]) \cdot \Omega^2$$

$$(17)$$

4 Simulations

4.1 Simulation Environment

The simulations about controlled MNSR are demonstrated as follows and the simulation parameters are shown in Table 1.

Table 1. Simulation parameters

Property	Value
Radial coordinate for center of mass of the system, R_c (km)	6470
Mass of platform, m_1 (kg)	5000
Mass of maneuverable net tether, m (kg)	2
Mass of the master tether, m_t (kg)	5
Moments inertia of platform, $I_0 (\mathrm{kg} \cdot \mathrm{m}^2)$	$diag(46, 50, 50)$
Length of maneuverable net tether, lw (m)	5
Reference length of master tether, Lr (m)	100

The initial values of in-plane and out-of-plane angles of net tethers are $\alpha_k = \pi/18,\ \pi/18,\ -\pi/18,\ -\pi/18$ and $\beta_k = \pi/18,\ -\pi/18,\ -\pi/18,\ \pi/18$, and the desired values of in-plane and out-of-plane angles of net tethers are $\alpha_{kd} = \pi/6,\ \pi/6,\ -\pi/6,\ -\pi/6$ and $\beta_{kd} = \pi/6,\ \pi/6,\ -\pi/6,\ \pi/6$ respectively.

4.2 Simulation Results

All the simulation results are shown in Figs. 6 and 7. Figure 6 shows that the in-plane and out-of-plane angles can be controlled in the desired values and the configuration of maneuverable net can be maintained in the desired state. Figure 7 demonstrates that control inputs of maneuvering units. The values of the control forces are in the acceptable range for spacecraft.

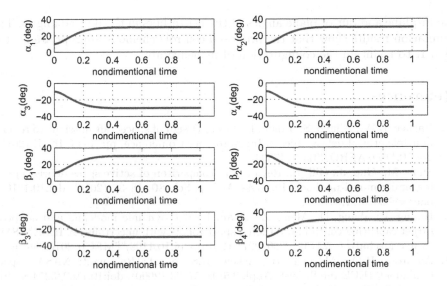

Fig. 6. Variation of in-plane and out-of-plane angles of net tethers

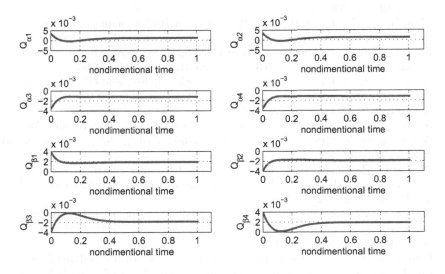

Fig. 7. Variation of control forces of MNSR

5   Conclusions

In this paper, an improved Tethered Space Net (TSN) System, named Maneuverable-Net Space Robot (MNSR), have been proposed for active space debris removal. In order to maintain the configuration of maneuverable net, it is necessary to employ a control strategy for MNSR. First, the 3-D dynamics model of MNSR is derived. Then, a feedback control is employed for maintaining the in-plane and out-of-plane angles of the net tethers in the desired angles. The simulation results show that the proposed control scheme is appropriate for the MNSR to maintain the configuration of maneuverable net.

References

1. Barbee, B.W., Alfano, S., Pinon, E., et al.: Design of spacecraft missions to remove multiple orbital debris objects. In: Aerospace Conference, pp. 1–14. IEEE (2011). doi:10.1109/AERO.2011.5747303
2. Shan, M., Guo, J., Gill, E.: Review and comparison of active space debris capturing and removal methods. Progress Aerosp. Sci. **80**, 18–32 (2016). doi:10.1016/j.paerosci.2015.11.001
3. Huang, P., Wang, D., Meng, Z., et al.: Impact dynamic modeling and adaptive target capturing control for tethered space robots with uncertainties. IEEE/ASME Trans. Mechatron. **21**(5), 2260–2271 (2016). doi:10.1109/TMECH.2016.2569466
4. Ambrose, R.O., Aldridge, H., Askew, R.S., et al.: Robonaut: NASA's space humanoid. IEEE Intell. Syst. Appl. **15**(4), 57–63 (2000). doi:10.1109/5254.867913
5. Liu, H.T., Zhang, Q.B., Yang, L.P., et al.: Dynamics of tether-tugging reorbiting with net capture. Sci. China Technol. Sci. **57**(12), 2407–2417 (2014). doi:10.1007/s11431-014-5717-8
6. Zhai, G., Qiu, Y., Liang, B., et al.: Research of attitude dynamics with time-varying inertia for space net capture robot system. J. Astronaut. **4**, 011 (2008)
7. Zhai, G., Zhang, J., Yao, Z.: Circular orbit target capture using space tether-net system. Math. Probl. Eng. **2013** (2013). doi:10.1155/2013/601482
8. Benvenuto, R., Salvi, S., Lavagna, M.: Dynamics analysis and GNC design of flexible systems for space debris active removal. Acta Astronaut. **110**, 247–265 (2015). doi:10.1016/j.actaastro.2015.01.014
9. Shan, M., Guo, J., Gill, E.: Deployment dynamics of tethered-net for space debris removal. Acta Astronaut. **132**, 293–302 (2017). doi:10.1016/j.actaastro.2017.01.001
10. Zhang, F., Huang, P.: Releasing dynamics and stability control of maneuverable-net space robot. IEEE/ASME Trans. Mechatron. **22**(2), 983–993 (2016). doi:10.1109/TMECH.2016.2628052
11. Huang, P., Zhang, F., Ma, J., et al.: Dynamics and configuration control of the maneuvering-net space robot system. Adv. Space Res. **55**(4), 1004–1014 (2015). doi:10.1016/j.asr.2014.11.009
12. Kumar, K.D.: Review on dynamics and control of nonelectrodynamic tethered satellite systems. J. Spacecraft Rockets **43**(4), 705–720 (2006). doi:10.2514/1.5479

Research on Space Manipulator System Man Machine Cooperation On-Orbit Operation Mode and Ground Test

Dongyu Liu[1,2], Hong Liu[1(✉)], Bainan Zhang[2], Yu He[2], Chao Luo[2], and Yiwei Liu[1]

[1] State Key Laboratory of Robotics and System, Harbin Institute of Technology,
Harbin 150080, China
hong.liu@hit.edu.cn
[2] Institute of Manned Spacecraft System Engineering, CAST, Beijing 100094, China

Abstract. In this paper, space manipulator man machine cooperation system structure and operation mode of Mir Space Station and International Space Station are reviewed briefly. Based on this, system structure, control mode, operation mode of Chinese first space manipulator and dexterous hand man machine cooperation experiment are introduced. Mathematics models and ground tests methodology of man machine cooperation are given. Via on orbit experiment, flight scheme and ground validation of space manipulator and dexterous hand man machine cooperation are reasonable. Through Chinese first on orbit space manipulator and dexterous hand man machine cooperation technology experiments, not only the technology foundation of man machine cooperation is established, but also it is accumulated experience for Chinese space robot.

Keywords: Space manipulator · Man machine cooperation · Operation mode · Ground test

1 Introduction

During the construction and operation of the Mir Space Station and the International Space Station, the man machine cooperation mode of the space manipulator has played an important role.

Twice extravehicular activities of the Mir Space Station in 1991 and 1996, the Russian astronauts respectively installed 14 m long, 45 kgs weighs of telescopic Boom "Strela Boom" in the Mir Space Station core module (Base block) column section of one side and Kvant another side, as shown in Fig. 1. Each boom is able to support 700 kgs weighs, used to move a large solar array, and the astronauts can be moved from one place to another place. The boom manipulator is manually operated and the astronaut will move the manipulator around the base by turning a pair of cranks [1, 2].

The Canada arm and Canada arm 2, known as the Space Shuttle and International Space Station have played an irreplaceable role in the process of construction, operation and maintenance of the International Space Station. The astronauts can observe the operation and operating environment of the cabin through the three LCD displays on the Robot Workstation or directly through the porthole of the space shuttle. RWS unit has 2 sets of joystick: a set of rotary manual joystick, a set of translational manual

© Springer International Publishing AG 2017
Y. Huang et al. (Eds.): ICIRA 2017, Part III, LNAI 10464, pp. 673–685, 2017.
DOI: 10.1007/978-3-319-65298-6_60

Fig. 1. Mir space station strela boom

joystick. The space shuttle Canada arm mainly includes autonomous mode, astronaut manual mode and single joint mode. Autonomous mode mainly includes pre-programmed trajectories and direct control of astronaut directives. The astronaut manual control mode allows the astronaut to adjust the position and pose of the end of the robot via the hand controller, mainly used for crawling, capturing non-parked loads; tracking and capturing free flight loads; coarse positioning. In the case of manipulator joints drive hardware failure, the control mode can be switched to direct drive and passive drive derating using fail-safe mode [3] (Fig. 2).

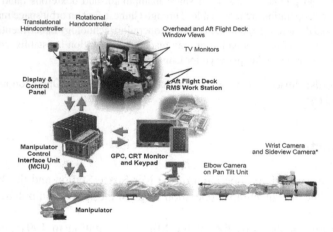

Fig. 2. Shuttle Remote Manipulator System preliminary system structure [3]

Compared to the United States, Russia, Germany and Japan [4], Space manipulator technology in China starts late, but develops rapidly in recent years. In Tiangong-2 Space Laboratory mission, Chinese first space manipulator and dexterous hand man machine cooperation on orbit experiments have completed successfully. The experiments have been carried out to verify the operation of the space manipulator and the dexterous hand in the Space Laboratory, local use of commercial products, with the smallest resources and funding to obtain on-orbit data and technical accumulation. This paper gives a brief introduction to the basic system composition, control mode and working mode of the space manipulator and dexterous hand man machine cooperation experiments, the mathematical model of man machine cooperation and the method of ground verification, The correctness of the man machine cooperation scheme design and the ground verification system of the space manipulator proposed in this paper is verified through the implementation on orbit. It has accumulated some experiences for Chinese space robot technology through this experiments on orbit.

2   System Composition and Man Machine Cooperation Mode

2.1   System Composition

Space manipulator and dexterous hand man machine cooperation experiments are equipped with a manipulator, a dexterous hand, a hand-eye camera, a controller, global cameras, orbit replaceable units. The system block diagram is shown in Fig. 3.

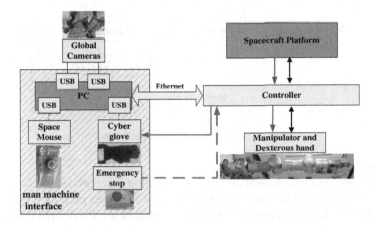

Fig. 3. Manipulator and dexterous hand system structure

The manipulator, dexterous hand and hand-eye camera are powered by the controller. The controller is the information center of the system. Control instructions and data acquisitions of manipulator, dexterous hand, hand-eye camera are emitted and incepted by the controller. The controller controls the operation of system actions. Global cameras measurement data are collected and processed through the PC, transmission by cable to controller for the use of system closed-loop control.

The personal computer acts as a man machine interface to the astronauts, and is connected to the controller via Ethernet. The astronauts can send control commands via the interaction software on the personal computer or monitor the system state. In the astronauts manual mode of operation, the astronauts can teleoperate the manipulator and dexterous hand respectively by space mouse and cyberglove.

The controller is set an emergency stop button for the astronauts for system emergency stop.

2.2   Control Mode

The system consists of three basic control modes: position control, impedance control and visual servo control.

Position control

Position control block diagram is shown in Fig. 4. In ground test, in order to unload the manipulator's own gravity, to improve the dynamic control performance, to control accuracy, it is set the gravity term $G(q)$ in the manipulator control loop. The gravity compensation term $G(q)$ is set to zero during orbital operation. The control mode is the basic control mode for the manipulator operation.

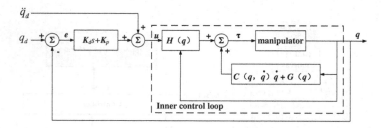

Fig. 4. Position control of manipulator

Impedance control

The manipulator is used an position control inner loop based impedance controller [5, 6], as shown in Fig. 5. The outer impedance control loop uses each joint torque sensor data for end effector force computed through force Jacobian, the desired position of the end of the manipulator is calculated by the admittance link. At last, the amount of change in each joint to the closed-loop feedback is calculated by inverse kinematics. This mode is mostly used in contact tasks.

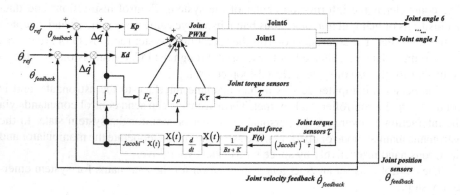

Fig. 5. Impedance control of manipulator

When the stiffness K in the impedance link is set to zero, the impedance control can be set to zero force control. This mode is used astronauts teaching playback.

Visual servo control
The visual servo controller receives the camera-identified position and pose information for closed-loop control, combined position control or impedance control. In the screwing screw task, if there is also a slight error between the end of the power tool and the screw hole, this small error makes the power tool head cannot accurately be put into the screw hole, but be on the edge of the screw hole. At this time, the impedance controller of the manipulator is adjusted according to the condition of the end force feedback, and the control variable of the desired position from the visual servo controller is corrected, so that the manipulator terminal will be inserted into the screw hole compliantly. The control block diagram is shown in Fig. 6. The control mode is used in the autonomous task of the manipulator system.

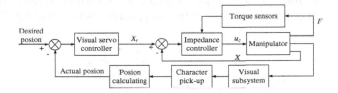

Fig. 6. Vision servo control fused impedance control

2.3   Operating Mode

Normal conditions
The astronauts command manipulator pre-programming instructions by operating the interaction software button on the personal computer, manipulator acts according to the pre-set action. Astronauts observe the entire experiment process, confirm each step to ground at any time. Based on telemetry and astronaut feedback, ground confirms the on-orbit experiment status [7, 8].

Backup conditions
When an unexpected failure occurs during the experiment, astronauts immediately press the emergency stop button. When the ground sends instructions cannot complete the test, astronauts also press the emergency stop button or ground send the "system standby" command. At this time, astronauts operate space mouse and cyberglove to operating manipulator and dexterous hand system.

After astronauts have adjusted the position and posture of the manipulator, the operator and ground will confirm the current position and posture of the manipulator terminal. The joint angle data will be saved in the PC control software by the astronauts. The data are sent via PC to the controller, to complete the experiment task.

Fail safe conditions
When an astronaut judges that the manipulator will harm his personal safety, or if the astronaut judges that the manipulator is about to intervene with the equipment in the cabin, the astronaut can stop the system by pressing the emergency stop button.

When faced with ground path planning difficult task, the ground will switch the manipulator control mode to zero force control mode, manipulator will be driven directly by the astronauts on orbit to adjust the terminal to proper position and posture. Ground records and analyzes the position and posture by telemetry parameters.

3 Man Machine Cooperation Operation Mathematical Model

3.1 Kinematic Model of Manipulator

The manipulator and dexterous hand are modeled and controlled separately. The manipulator has six joints, each joint corresponds to a joint coordinate system.

The joint coordinate system of the manipulator uses the modified D-H method to establish the joint coordinate system of the manipulator, as shown in Fig. 7.

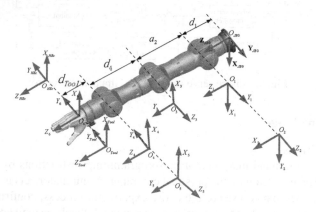

Fig. 7. Manipulator D-H reference frame definition

Substituting the link parameters into the link transformation general formula (1), the relative transformation matrix $^{i-1}_{i}T$ of each link can be obtained, $i = 1, 2, \cdots 5, 6$.

$$^{i-1}_{i}T = \begin{bmatrix} c\theta_i & -s\theta_i & 0 & a_{i-1} \\ s\theta_i c\alpha_{i-1} & c\theta_i c\alpha_{i-1} & -s\alpha_{i-1} & -d_i s\alpha_{i-1} \\ s\theta_i s\alpha_{i-1} & c\theta_i s\alpha_{i-1} & c\alpha_{i-1} & d_i c\alpha_{i-1} \\ 0 & 0 & 0 & 1 \end{bmatrix} \tag{1}$$

The transformation matrix is multiplied by right in turn, it can be obtained the transformation matrix of the end of the manipulator relative to the origin of the coordinate system, as shown in Eq. (2).

$$^{0}_{6}T = {}^{0}_{1}T(\theta 1){}^{1}_{2}T(\theta 2){}^{2}_{3}T(\theta 3){}^{3}_{4}T(\theta 4){}^{4}_{5}T(\theta 5){}^{5}_{6}T(\theta 6) \tag{2}$$

Sufficient conditions for the existence of closed solutions for the inverse kinematics of the manipulator (Pieper criterion): the three joints of the end of manipulator intersect at one point, Paul's inverse transformation method can be used [9]. When the end posture is given by the RPY angle, the corresponding rotation matrix is obtained. Combined with the end position input, that is, the relative transformation matrix of the given manipulator end relative to the base coordinate system. By giving the above matrix, closed solution of joint 1 to 6 can be solved by the inverse transform method.

3.2 Kinematic Model of Dexterous Hand

The dexterous hand has five fingers, each finger structure is exactly the same. Every single finger can be seen as a joint robot with 4 joints, 3 degrees of freedom. The fourth joint and the third joint have a similar 1:1 coupling relationship, satisfying the following equation. Single finger coordinate system is shown in Fig. 8 [10]. Substituting the link parameters into the link transformation matrix general formula (1), a transformation matrix between the links can be obtained.

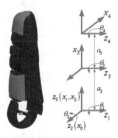

Fig. 8. Dexterous hand fingers reference frame definition

As a finger has only three degrees of freedom, and the range of joint movement of the fingers is limited in mechanical dimensions, there is only one finger gesture at the same fingertip position. If the position of the fingertip in the base coordinate system (x, y, z) is known, according to the principle of inverse kinematics in robotics, the only inverse solution in the joint space of the finger can be obtained.

3.3 Man Machine Cooperation Mapping Model

Man machine cooperative manipulator mapping model
In this experiment, space mouse is used to control three directions of the translation and three directions of rotation of end tip of the manipulator, as shown in Fig. 9. Space mouse movement and rotation range are tiny, so in this experiment space mouse output is set to the end increment of the manipulator [8].

Fig. 9. Space mouse

Space mouse and personal computer exchange data using USB2.0 port. Astronauts use the space mouse to control the end position and posture of the manipulator, the astronauts - manipulator interface logic as shown in Fig. 10.

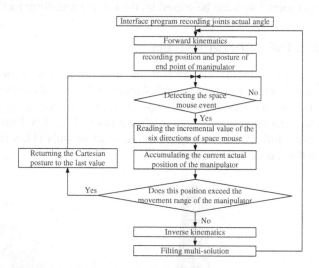

Fig. 10. Space mouse cartesian space control logistic chart

When the astronauts operate manipulator and dexterous hand, the interface program detects the space mouse event, reads the incremental value of the six directions of space mouse, to accumulate the current actual position of the manipulator. If this position does not exceed the movement range of the manipulator and is not singular, the human-machine interface program will perform inverse kinematics calculations. Through filting multi-solution, the control variables of the joint space of the manipulator are finally generated.

If the control data of the space mouse will make the manipulator out of the range of motion, the program returns the Cartesian posture to the last value, until the data entered by the space mouse again causes the manipulator to be in the range of motion.

Man machine cooperative dexterous hand mapping model

The commercial cyber gloves are selected for dexterous hand man-machine cooperative operating. The information exchange interface between the cyber glove and the personal computer are USB 2.0. At each finger, there are 3 joint buckling sensors, 4 joint outreach sensors, 1 palm sensor, and 2 measurements of joint flexion and joint abduction sensors. The 19 joint angle sensors which are operated by man-machine cooperation. During the ground test, the joint sensors measurement angle of cyber gloves are calibrated according to the astronauts's palm. The control of the dexterous hand uses the position mapping of the joint space. For each joint of the dexterous hand, a first order linear function is used for mapping as shown in Eq. (3) [11].

$$G_0 = AGi + B \tag{3}$$

In Eq. (3):

G_i ——Cyber glove sensors measurement matrix (5×3);
G_0 ——Calibrated dexterous hand control matrix (5×3);
A ——Calibration gain coefficient matrix (3×5);
B ——Calibration bias matrix (5×3).

The motion relationship of the astronaut controlling dexterous hand by cyber glove is shown in Fig. 11. Cyber gloves measure the angle of the joints of the astronaut's fingers, control the movement of the dexterous hand, through the joint position mapping to dexterous hand joints space.

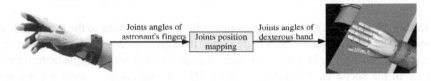

Fig. 11. Astronaut and dexterous hand cooperation mapping relation based cyber glove

4   Ground Test

In this paper, two kinds of ground test to complete the space manipulator system man machine on-orbit cooperation experiment verification work. First of all, through simulation verification to ensure that the manipulator system trajectory planning is reasonable. Through the ground flight system 1:1 physical test to ensure that experiment of the flight procedures are reasonable, the failure schemes are feasible.

4.1   Simulation Verification

The simulation system is worked by VC++. Using the 3D model of the design phase to transform into the 3D model of Open Inventor.

The root of the simulation environment is located at the center of the base joint of the robot arm, which is the origin of the kinematic base coordinates of the manipulator. And then the various parts of the model are loaded in program with the Open Inventor class library function, placed into the scene for assembly. The relationship between the two models follows the parent-child relationship, that is, the movement of the parent node affects his child nodes, and the movement of the child nodes cannot affect the parent node. Such as between any joints, as shown in Fig. 12.

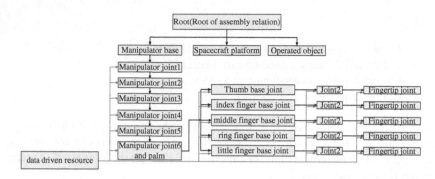

Fig. 12. Simulation models assembly relation and data drive relation

In this relationship, it is formed a control chain from the base to the dexterous hand fingertip. The control chain is as a node joined the virtual scene. The center point of the base joint is as the origin point, the aircraft platform, the orbital replacable units are located in the scene by the actual position [12]. A simulation system for the space manipulator system based on graphical predictive emulation is shown in Fig. 13.

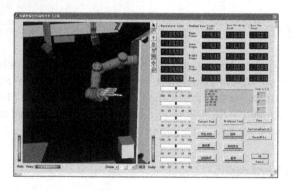

Fig. 13. Manipulator and dexterous hand system man machine cooperation simulation

A pre-programmed trajectory instruction is used to drive the simulation system, or the man-machine interaction device cyber gloves and a space mouse to drive the manipulator and dexterous hand directly in the simulation model.

4.2   Test Verification

Before on orbit implementation, 1:1 test verification is carried out using ground flight system. As shown in Fig. 14, The ground technicians were operating the manipulator and dexterous hand through space mouse and cyber gloves.

Fig. 14. Manipulator and dexterous hand system man machine cooperation test validation

Besides further verifying the manipulator and dexterous hand operating trajectory, in test verification, it is more important to verify the real dynamic response and tracking performance of the manipulator and dexterous hand hardware system during manual operating, the correctness of the fault plan switching, the ergonomic performance.

5   On-Orbit Implementation

Space manipulator system man machine cooperation has been tested successfully in Tiangong-2 Space Laboratory Missions on-orbit operation display task.

The astronauts operated the dexterous hand by the cyber gloves, and successfully completed the gestures specified in the flight procedure as shown in Fig. 15, the trajectory following by dexterous hand is smooth and stable as shown in Fig. 17a.

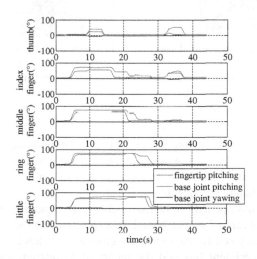

Fig. 15. Joint angle data of dexterous hand man machine cooperation on-orbit experiment

As shown in Fig. 17b, the astronauts operated the manipulator by the space mouse, and successfully completed the movement requirements specified in the flight procedure. As shown in Fig. 16, the joins error of manipulator is not greater than 0.01°, the tracking performance of manipulator is well.

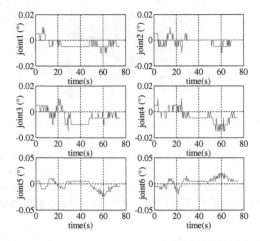

Fig. 16. Joints following error of manipulator man machine cooperation on-orbit experiment

a)Astronauts operationing manipulator   b)Astronauts operationing dexterous hand

Fig. 17. Telemetry images of manipulator man machine cooperation on-orbit experiment

6   Conclusion

Through the successful implementation of the on-orbit experiment, it is proved that the system configuration and control mode presented in this paper are reasonable, and the man-machine cooperation mode is reasonable, and the ground verification method combining simulation and physical test is feasible.

The experiment has opened up the manipulator man machine cooperation mode in our country, and has carried on the beneficial exploration to the system configuration structure, the necessary control mode and the ground verification technology system of this model. The experiment has laid the technical foundation for the man-machine cooperation mode of astronauts and manipulators in the China's space station mission. The overall layout of the experiment is in the cabin, it has carried on space manipulator technology verification. The subsequent part of the manipulator transplanted outside the cabin for operation missions, space manipulator composition structure, operation mode, control mode and ground test methods can be inherited.

In subsequent missions, when the astronauts in cabin control the manipulator outside the cabin, it is noteworthy to control the time delay in the closed-loop system and set up the necessary cabin cameras to meet the astronauts' observation requirements.

References

1. Portree, D.S.F.: Mir hardware heritage. NASA Reference Publication 1357, pp. 146–159 (1995)
2. Sue, M.: Mir mission chronicle. NASA/TP-98-207890, pp. 20–54 (1998)
3. Glenn, J., Elizabeth, B.: SRMS history, evolution and lessons learned. In: AIAA SPACE Conference and Exposition, vol. 9, pp. 1–53 (2011)
4. Fukuda, T., Fujisawa, Y., Kosuge, K., et al.: Manipulator for man-robot cooperation. In: International Conference on Industrial Electronics, Control and Instrumentation, vol. 10, pp. 996–1001 (1991)
5. Liu, Y., Jin, M., Liu, H.: Joint torque-based Cartesian impedance control. J. Southeast Univ. **12**, 492–497 (2008)
6. Jiang, Z., Zhao, J., Liu, H.: Haptic flexible virtual fixture for teleoperation. Robot **11**, 685–687 (2011)
7. Jiang, Z., Xie, Z., Wang, B., et al.: A teleprogramming method for internet-based teleoperation. In: IEEE International Conference on Robotics and Biomimetics, pp. 322–326 (2006)
8. Liu, D.: Research on Robot Arm/Hand Teleoperation Toward Satellite on Orbit Servicing. Harbin Institute of Technology, Harbin (2010)
9. Liu, H.: Space Robot and Teleoperation. Harbin Institute of Technology Press, Harbin (2012)
10. Liu, Y., Jin, M., Fan, S., Lan, T., Chen, Z.: Five-finger dextrous robot hand DLR/HIT hand II. J. Mech. Eng. **11**, 10–16 (2009)
11. Hu, H., Gao, X., Li, J., Wang, J., Liu, H.: Calibrating human hand for teleoperating the HIT/DLR hand. IEEE Int. Conf. Robot. Autom. **5**, 4571–4576 (2004)
12. Zhao, M., Zhang, H., Chen, J.: Technology of VR teleoperation and its development. J. Syst. Simul. **19**, 3248–3252 (2007)

Collaborative Optimization of Robot's Mechanical Structure and Control System Parameter

Yuanchao Cheng[1(✉)], Ke Li[1], Fan Yang[1], Songbo Deng[1], Yuanyuan Chang[2], Yanbo Wang[1], and Xuman Zhang[1]

[1] 18th Institute of CALT, Beijing, China
cheng_yuanchao@163.com
[2] Science and Technology on Space Physics Laboratory, Beijing, China

Abstract. In this paper, an inverse kinematics analysis algorithm of a robot is proposed based on the analysis of the configuration and kinematics of the robot. With the inverse kinematics method, the working path of the robot's joints is planned. Based on multi-body dynamics, finite element analysis, topology optimization and PID control, using co-simulation, this paper presents the method including Structural Optimization, Mechanical System Optimization and Controller Optimization. The main bearing part (large arm) is selected as the optimization object, in order to improve the repeated positioning accuracy of the robotic robot and verify the collaborative optimization methods, the effectiveness of the method is improved by comparing the error of the end position with the same working path by comparing the robotic system before and after the collaborative optimization.

Keywords: 6-DOF lightweight robot · Kinematics analysis · Topology optimization · Collaborative optimization · Co-simulation

1 Introduction

The mechanical system affects the control system parameters; conversely, changes of the parameters of the control system also affect the dynamic performance of the mechanical structure. Mechanical system and control system are closely related, the performance of the two systems together determine the overall performance of the robot. It is important to optimize the overall performance of the robot by considering the structural parameters of the mechanical system and the controller parameters in the robot optimization design process [1].

The mechanical system fundamentally limits the optimal degree of repeatability of the end of the robot [2]. The servo control system of the joint directly determines the follower error at the end of the robot. The combined effect of them together determines the repeatability of the end of the robot.

In order to optimize the mechanical structure to improve the mechanical system performance, this paper uses the structural topology optimization design, in the range of design variables to find the optimal design to achieve optimal design of mechanical system [3].

© Springer International Publishing AG 2017
Y. Huang et al. (Eds.): ICIRA 2017, Part III, LNAI 10464, pp. 686–697, 2017.
DOI: 10.1007/978-3-319-65298-6_61

The robot topology optimization design mainly includes two aspects [4]: for the robot, in the initial design of the robot, in the initial design space, according to the design specification, the entire structure of the robot topology optimization design; for the robot's parts, under the specified load condition, its topology is optimized. Topology optimization changes shape and size parameters while optimizing topology structure [5].

The control system is one of the main factors that determine the function and performance of the robot, which restricts the development of robot technology to a certain extent. The advantages and disadvantages of the robot arm control system directly affect the speed, accuracy and reliability of the robot. The parameter adjustment process of robot control system is a basic work of optimizing the control system.

In this paper, a collaborative optimization method is proposed for mechanical structure and control system parameters of mechanical arm based on joint simulation.

2 Kinematic Analysis

2.1 Forward Kinematics Analysis

According to the actual configuration of 6-DOF lightweight robot, the mechanism schematic diagram of robot is established, and the standard D-H parameter method is used to establish the D-H coordinates of the robot, as shown in Fig. 1. The origin of coordinate system 5 coincide with the origin of coordinate system 6. The corresponding mechanical arm D-H parameters are shown in Table 1.

The problem of the forward kinematics of the lightweight robot is known as the angle of each joint of the robot to solve the pose of the coordinate system 6 which is related to the coordinate system 0.

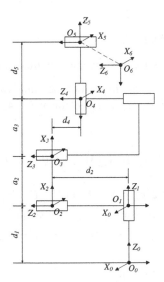

Fig. 1. 6-DOF lightweight robot's D-H coordinates.

Table 1. 6-DOF lightweight robot's D-H parameters.

i	$\alpha_{i-1}(°)$	a_{i-1} (mm)	θ_i (°)	d_i (mm)	Joint motion range
1	0	0	$\theta_1(0)$	d_1 (123.5)	$-180 \sim +180°$
2	90	0	$\theta_2(90)$	d_2 (197.5)	$-180 \sim +180°$
3	0	a_2 (436)	$\theta_3(0)$	0	$-180 \sim +180°$
4	0	a_3 (386)	$\theta_4(90)$	d_4 (−28.61)	$-180 \sim +180°$
5	90	0	$\theta_5(0)$	d_5 (136.39)	$-180 \sim +180°$
6	−90	0	$\theta_6(0)$	0	$-180 \sim +180°$

Table 1 lists all D-H parameters of the lightweight robot as α_i, a_i, θ_i, d_i, only θ_i are variables and the other parameters are constants. The homogeneous transformation matrix of coordinate system i relative to coordinate system i-1 can be determined by these D-H parameters.

Use D-H parameters and (1), we can figure out six homogeneous transformation matrices. And then we can figure out the pose of the coordinate system 6 relative to the coordinate system 0, as shown in (2).

$$
{}_i^{i-1}A = \begin{bmatrix} \cos\theta_i & -\sin\theta_i & 0 & a_{i-1} \\ \sin\theta_i \cos\alpha_{i-1} & \cos\theta_i \cos\alpha_{i-1} & -\sin\alpha_{i-1} & -d_i \sin\alpha_{i-1} \\ \sin\theta_i \sin\alpha_{i-1} & \cos\theta_i \sin\alpha_{i-1} & \cos\alpha_{i-1} & d_i \cos\alpha_{i-1} \\ 0 & 0 & 0 & 1 \end{bmatrix} \tag{1}
$$

$$
{}_6^0T = {}_1^0A {}_2^1A {}_4^3A {}_5^4A {}_6^5A = \begin{bmatrix} r_{11} & r_{12} & r_{13} & p_x \\ r_{21} & r_{22} & r_{23} & p_y \\ r_{31} & r_{32} & r_{33} & p_z \\ 0 & 0 & 0 & 1 \end{bmatrix} \tag{2}
$$

2.2 Inverse Kinematics Analysis

The analytical solution of this robot's inverse kinematics is as follows.

1. Using (3) element (2, 4), we can figure out the value of θ_1, using (3) element (2, 3) to figure out the value of θ_5, using (3) element (1, 4) and (3, 4) to figure out the value of θ_6, as (4) shown.

$$
\left[{}_1^0A\right]^{-1}{}_6^0T = {}_2^1A(\theta_1){}_3^2A(\theta_1){}_4^3A(\theta_1){}_5^4A(\theta_1){}_6^5A(\theta_1) \tag{3}
$$

$$
\begin{cases} \theta_1 = A\tan 2(p_x, p_y) - A\tan 2\left(-E_1, \pm\sqrt{1-E_1^2}\right) \\ \theta_5 = \pm arc\cos[\sin(\theta_1)\cdot r_{13} - \cos(\theta_1)\cdot r_{23}] \\ \theta_6 = A\tan 2\left(\frac{-\sin(\theta_1)\cdot r_{12} + \cos(\theta_1)\cdot r_{22}}{\sin(\theta_5)}, \frac{\sin(\theta_1)\cdot r_{11} - \cos(\theta_1)\cdot r_{21}}{\sin(\theta_5)}\right) \end{cases} \tag{4}
$$

Where, $E_1 = (d_2 + d_4)/\sqrt{p_x^2 + p_y^2}$.

2. Using (5) element (1, 4) and (3, 4), we can figure out the value of θ_2, as (6) shown.

$$\left[{}_1^0 A\right]^{-1} {}_6^0 T \left[{}_6^5 A\right]^{-1} \left[{}_5^4 A\right]^{-1} = \tfrac{1}{2} A(\theta_1){}_3^2 A(\theta_1){}_4^3 A(\theta_1) \tag{5}$$

$$\theta_2 = A \tan 2 \left(\frac{G_4}{\sqrt{E_4^2 + F_4^2}}, \pm\sqrt{1 - \frac{G_4^2}{E_4^2 + F_4^2}} \right) - A \tan 2(E_4, F_4) \tag{6}$$

Where, $E_4 = d_5 \cdot s_6 \cdot c_1 \cdot r_{11} + d_5 \cdot s_6 \cdot s_1 \cdot r_{21} + d_5 \cdot c_6 \cdot c_1 \cdot r_{12} + d_5 \cdot c_6 \cdot s_1 \cdot r_{22} + c_1 \cdot p_x + s_1 \cdot p_y$

$F_4 = d_5 \cdot s_6 \cdot r_{31} + d_5 \cdot c_6 \cdot r_{32} + p_z - d_1;\ G_4 = -(a_3^2 - E_4^2 - F_4^2 - a_2^2)/2a_2.$

3. Using (7) element (1, 3) and (2, 3) figure out the value of $(\theta_2 + \theta_3)$, using (7) element (1, 4) and (2, 4) figure out the value of θ_3 and finally figure out the value of θ_4, as (8) shown.

$$\left[{}_2^1 A\right]^{-1} \left[{}_1^0 A\right]^{-1} {}_6^0 T = {}_3^2 A(\theta_1){}_4^3 A(\theta_1){}_5^4 A(\theta_1){}_6^5 A(\theta_1) \tag{7}$$

$$\theta_3 = A \tan 2(H_6, G_6)$$

$$\theta_4 = A \tan 2 \left(\frac{F_5}{-\sin(\theta_5)}, \frac{E_5}{-\sin(\theta_5)} \right) - \theta_3 \tag{8}$$

Where,

$E_5 = c_2 \cdot c_1 \cdot r_{13} + c_2 \cdot s_1 \cdot r_{23} + s_2 \cdot r_{33};\ F_5 = -s_2 \cdot c_1 \cdot r_{13} - s_2 \cdot s_1 \cdot r_{23} + c_2 \cdot r_{33};$
$E_6 = c_2 \cdot c_1 \cdot p_x + c_2 \cdot s_1 \cdot p_y + s_2 \cdot p_z - s_2 \cdot d_1;\ F_6 = -s_2 \cdot c_1 \cdot p_x - s_2 \cdot s_1 \cdot p_y + c_2 \cdot p_z - c_2 \cdot d_1;$
$G_6 = ((E_6 - a_2 + d_5 \cdot F_5)/\sin(\theta_5))/a_3;\ H_6 = ((F_6 - d_5 \cdot E_5)/\sin(\theta_5))/a_3.$

In the above process of solving the inverse kinematics, the joint angles θ_1, θ_2 and θ_5 have two groups of analytical solutions. Consequently, the inverse kinematics of this 6-DOF lightweight robot has eight groups of analytical solutions.

3 Trajectory Planning

3.1 Workspace Analysis

All the joints' motion of this lightweight robot can reach the range of -180 degrees to 180 degrees. The workspace of the robot is shown in Fig. 2.

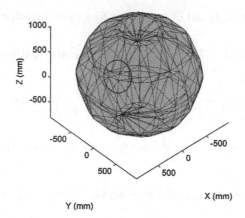

Fig. 2. Workspace and trajectory of the robot.

3.2 Circular Trajectory Planning in Cartesian Space

The end of the robot is projected to move along the circular trajectory in Cartesian coordinates. The centre of the circular trajectory is chosen as (500, 50, 400), the radius is 250 mm, and the normal vector of the circle plane is (0, 4, 3). Then we can figure out the circular trajectory equation, as (9) shown.

$$\begin{cases} (x - 500)^2 + (y - 50)^2 + (z - 400)^2 = 250^2 \\ 4(y - 50) + 3(z - 400) = 0 \end{cases} \tag{9}$$

The planning of circular trajectory in the working space of the robot shown in Fig. 2.

3.3 Joints Trajectory Planning

Converse the circular trajectory planning in Cartesian space into joint trajectory by means of the robot's inverse kinematics solving method that mentioned above. The joint trajectories shown in Fig. 3, 4, 5, 6, 7, 8 and 9. The curves of 0-5 s are the joint trajectory curves that let the robot move from the initial state to the start point.

The interpolation point of each joint trajectory angles are shown in Table 2.

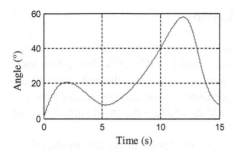

Fig. 3. Joint 1 trajectory curve.

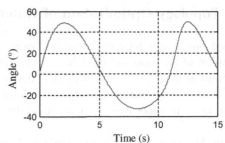

Fig. 4. Joint 2 trajectory curve.

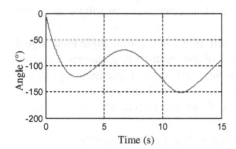

Fig. 5. Joint 3 trajectory curve.

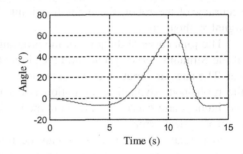

Fig. 6. Joint 4 trajectory curve.

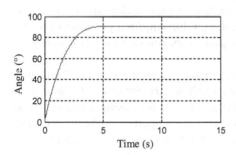

Fig. 7. Joint 5 trajectory curve.

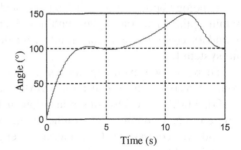

Fig. 8. Joint 6 trajectory curve.

Table 2. Joint trajectories' angle-time interpolation point.

t(s)	θ_1	θ_2	θ_3	θ_4	θ_5	θ_6
0	0	90	0	90	0	0
5.0	8.03	95.12	−88.84	83.71	90	98.03
5.4	7.61	86.63	−81.49	84.85	90	97.61
......						
14.6	9.94	104.10	−97.13	83.00	90	99.94
15	8.03	95.12	−88.84	83.71	90	98.03

4   Topology Optimization of Robot Arm

Based on the topology optimization design of the boom of the robot and the establishment of the rigid and flexible hybrid dynamic model in ADAMS, the system joint simulation model is established. In order to accurately describe the influence of the servo control system of the large arm drive joint on the positioning accuracy of the robot arm, the other five joints are ideal drive and can follow the input signal without error.

In this paper, based on the topological optimization design of the arm of the robot, the joint simulation model of the system electromechanical joint is established on the basis of ADAMS rigid and soft hybrid dynamic model. In order to accurately describe the influence of the servo control system of the large arm drive joint on the positioning accuracy of the robot, the other five joints are ideal drive and can follow the input signal without error.

The joint curve of the joint of the programmed joint 2 is the joint angle input signal of the Simulink servo system, and the simulation of the multi-body dynamics is carried out. The angular variation curve of the joint 2 is Simulink's joint angle input signal, and the multi-body dynamics simulation is carried out to obtain the end position of the mechanical arm caused by the combined effect of the arm flexure and the joint servo control system error.

Structural stiffness is a significant design factor in engineering design, which can affect dynamic performance and steady-state error in control system. In this paper, to improve dynamic performance of a robot, a topology optimization method based on variable density theory has been applied.

Topology optimization is a mathematical method that use optimization iteration to obtain optimal material layout within a given design domain, for a given set of loads, boundary conditions and constraints with the goal of maximising the performance of the system [6].

There are two main approaches for solving topology optimization [7]: Solid Isotropic Material with Penalisation (SIMP) and Evolutionary Structural Optimization (ESO). SIMP is a continuous relaxation of the problem solved using a mathematical programming technique and so inherits the convergence properties of the optimization method. By contrast, ESO is based on engineering heuristics and has no proof of optimality [8].

As for SIMP, variable density method is the most suitable and common method for establishing mathematic model based finite element concerning structural payload and modal, which can be operated in some CAE software, such as Hyperworks, Ansys and Abaqus. In this project, Hyperworks was selected as operation software because of its user-friendly operation environment and variable density method working better.

The main principle of variable density method: the material density of each element is selected as the design variable. During optimization process, intermediate density is penalised to force the design variables to approach 0 or 1. If the material density of an element is close to 0 at the end of optimization, the element remains void, which can also be seen in (10). In the variable density method, continuous functions of Young's modulus matrix, stiffness matrix and mass matrix depict the relation between original values and optimized values based on different density of each element, which can be seen below 69.23:

$$
\begin{cases}
E = (\rho_i)^q \cdot E_0 \\
K = (\rho_i)^q \cdot K_0 \\
m = \rho_i \cdot m_0
\end{cases}
\tag{10}
$$

Where, E element equivalent Young's modulus matrix; K element equivalent stiffness matrix; m element equivalent mass matrix; E_0 initial element Young's modulus matrix; K_0 initial element stiffness matrix; m_0 initial element mass matrix; q penalty factor.

In order to eliminate intermediate density, penalty factor contributes to penalising the design variable in the interval (0, 1), making topology model with continuous variable approaching traditional 0-1 discrete model. Larger penalty factor enhances penalisation effect. However extensively large q can cause system matrix singular. Therefore, it is practically considered to be 3. This method is simple to come true and of high efficiency.

For the minimum compliance (maximum global stiffness) problem, it can be written in the form:

$$
\min_{uE_e} f^T u
$$
$$
s.t. \quad K(E_e)u = f; \; E_e \in E_{ad}.
\tag{11}
$$

Where u and f are the displacement and load vector respectively. The global stiffness matrix K depends on stiffness E_e in each element and E_{ad} is the admissible stiffness matrix.

$$
K = \sum_{e=1}^{N} K_e(E_e)
\tag{12}
$$

In the process of topology optimization, the stress, strain, modal, mass and displacement of the structure are taken into account, and the stiffness of the whole

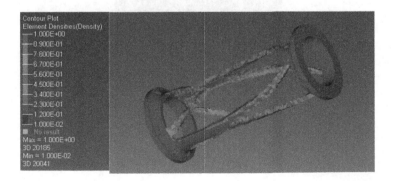

Fig. 9. Topology optimization results.

Table 3. Static analysis results.

	Initial	Optimized
Miss (Kg)	1.708	1.711
Maximum stress (Mpa)	2.408	1.797
Maximum strain (mm)	2.981e-5	2.225e-5
Maximum displacement (mm)	1.811	1.809

Table 4. Model analysis results.

	Initial (Hz)	Optimized (Hz)
Mode 1	940.657	1053.892
Mode 2	940.935	1054.197
Mode 3	2631.028	2758.495
Mode 4	3883.735	3820.265
Mode 5	3884.852	3871.489

structure is chosen as the main optimization target. The optimized topology results is shown in Fig. 9, the static analysis results is shown in Table 3 and the model analysis results is shown in Table 4.

5 Collaborative Optimization of Control Parameters

For a robot joint, the following robot joints, arm, end effector and workload are all the effective load, which is a variable related to the robot's position, posture, velocity and acceleration. It is more complicated and will cost more time to consider the dynamic characteristics of the robot in Matlab compared with Adams. Therefore, this paper use the professional dynamic analysis software-Adams, to solve the dynamic problem of the robot, then use Matlab co-simulate the robot's working process.

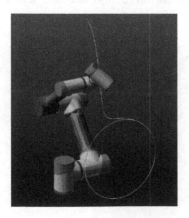

Fig. 10. Robot's Adams model.

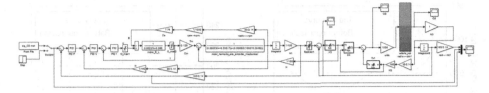

Fig. 11. Matlab model of robot's joint 2 servo control system.

Establish the Adams flexible model based on the optimization of robot arm. Then start the Matlab-Adams co-simulation, as shown in Figs. 10 and 11.

Optimize the parameters of the joint servo control system. Compare the parameters, as shown in Table 5.

Table 5. Joint servo control system parameters.

Parameters	Current loop		Velocity loop	Position loop
	P	I	P	P
Initial	6	600	30	4
Optimized	6	1000	50	10

6   Optimization Effect Verification

Compare the position error of the two robots, before and after the collaborative optimization, at the same working path. In order to verify the optimization effect of the collaborative optimization method to improve the robot's positioning accuracy, as shown in Fig. 12, 13, 14, 15, 16, 17, 18 and 19.

Table 6 shows the comparison of the maximum error of O6, which proves that after the collaborative optimizing the robot's maximum error in each direction reduces greatly.

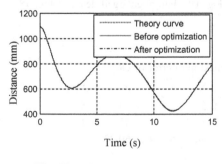

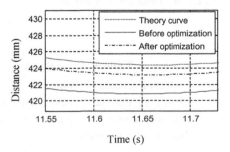

Fig. 12. Distance from O0 to O6.

Fig. 13. Partial enlarged detail of Fig. 12.

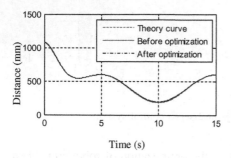

Fig. 14. Distance in Z direction.

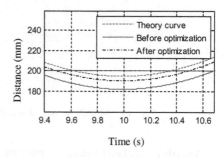

Fig. 15. Partial enlarged detail of Fig. 14.

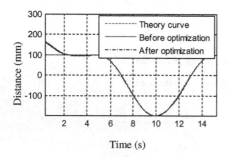

Fig. 16. Distance in Y direction.

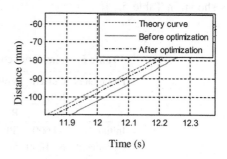

Fig. 17. Partial enlarged detail of Fig. 16.

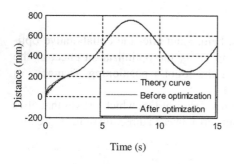

Fig. 18. Distance in X direction.

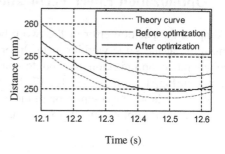

Fig. 19. Partial enlarged detail of Fig. 18.

Table 6. Comparison of the maximum error of O6.

Direction (coordinate system 0)	X	Y	Z	Mag
Maximum error before optimization (mm)	5	7	13	4
Maximum error after optimization (mm)	3	3	4	1
Rate of change gradient	40%	57.14%	69.23%	75%

Especially in major stress direction (i.e. vertical direction), the reduction achieved 69.23%, that indicate the effectiveness of the collaborative optimization method to improve the positioning accuracy of the robot.

Table 6 shows the comparison of the maximum error of O6, which proves that after the collaborative optimizing the robot's maximum error in each direction reduces greatly. Especially in major stress direction (i.e. vertical direction), the reduction achieved 69.23%. Show the effectiveness of the collaborative optimization method to improve the positioning accuracy of the robot.

7 Conclusion

The collaborative optimization method is based on the interaction between the mechanical and control system in the electromechanical system, and the optimization of the mechanical parameters of the mechanical and electrical system is realized in the optimization process. The optimization of the mechanical parts is optimized by the multi-body dynamics optimization and the controller parameters.

In this paper, the proposed method and process are verified by a robot with flexible arm. The results show that the collaborative optimization method can optimize the steady-state and transient performance of the system and improve the dynamic characteristics of the control system. The parameters of the mechanical system can be made more reasonable and the mechanical system can be operated more smoothly, and can significantly improve the positioning accuracy of the robot.

Acknowledgment. Thanks for all the team members for their outstanding contributions to the project. Special thanks to Anping Chen Engineer for his guidance during the research.

References

1. Grujicic, M.F., Arakere, G.S., Pisu, P.T.: Application of topology, size and shape optimization methods in polymer metal hybrid structural lightweight engineering. Multidiscipline Model. Mater. Struct. **4**(4), 305–330 (2008)
2. Lee, S.H.F., Kim, J.S., Park, F.C.T.: Newton-type algorithms for dynamics-based robot movement optimization. IEEE Trans. Robot. **21**(4), 657–667 (2005)
3. Liu, L.F., Wang, G.D.S., Xiao, R.Y.T.: Optimization of the method to palletize firebricks by robot based on pareto genetic algorithm. In: Applied Mechanics and Materials 2014, vol. 620, pp. 337–342. Trans Tech Publications (2014)
4. Lim, K.B.F., Junkins, J.L.S.: Robustness optimization of structural and controller parameters. J. Guidance Control & Dyn. **12**(12), 89–96 (2012)
5. Deng, K.F., Pan, P.S., Sun, J.T.: Shape optimization design of steel shear panel dampers. J. Constr. Steel Res. **99**, 187–193 (2014)
6. Bourdin, B.F., Kohn, R.V.S.: Optimization of structural topology in the high-porosity regime. J. Mech. Phys. Solids **56**(3), 1043–1064 (2004)
7. Tai, K.F., Wang, S.S., Akhtar, S.T.: Structural topology optimization using a genetic algorithm and a morphological representation of geometry. Dissertations & Theses – Gradworks **599**, 319–323 (2003)
8. Sigmund, O.F., Maute, K.S.: Topology optimization approaches. Struct. Multi. Optim. **48**(6), 1031–1055 (2013)

Nonlinear MPC Based Coordinated Control of Towed Debris Using Tethered Space Robot

Bingheng Wang[1,2], Zhongjie Meng[1,2], and Panfeng Huang[1,2(✉)]

[1] Research Center for Intelligent Robotics, School of Astronautics,
Northwestern Polytechnical University,
Xi'an 710072, Shaanxi, People's Republic of China
pfhuang@nwpu.edu.cn
[2] National Key Laboratory of Aerospace Flight Dynamics,
Northwestern Polytechnical University,
Xi'an 710072, Shaanxi, People's Republic of China

Abstract. Using tethered space robot (TSR) for active debris removal (ADR) is promising but subject to collision and entanglement due to the debris tumbling. To detumble the towed debris, this paper proposes the nonlinear model predictive control (NMPC) based coordinated control strategy. The TSR consists of a gripper for capture, thrusters and a tethered manipulator (TM) with variable length to which the tether is attached. The proposed strategy works in the way that the TM coordinates with the thrusters for de-tumbling by changing its length accordingly so that the tension torque can be adjusted. The attitude model of the debris is first established, followed by the definition of attitude equilibrium. The NMPC is then designed with the prediction model discretized by 4-order Runge-Kutta method. Simulation results validate this strategy and show that the debris attitude can maneuver to the equilibrium smoothly in the presence of the constraints on TM and thrusts.

Keywords: Active debris removal · Tethered space robot · Model predictive control

1 Introduction

To reduce the debris population, the use of space tether for active debris removal (ADR) proves to be promising compared to the rigid manipulator and has attracted a great deal of research interests [1–3]. This technology, also termed as 'towing removal', is achieved by employing an active maneuverable platform, attaching the tether to the debris and towing it to the disposal orbit.

However, the towing removal involves two main technical challenges, namely collision between the two end bodies and the entanglement with tether. They are mainly attributable to the flexible structure of tether and the debris tumbling [4]. Further, many factors can actually cause debris to tumble, such as residual angular velocities, off-centered capture [5] and flexible appendages [6].

© Springer International Publishing AG 2017
Y. Huang et al. (Eds.): ICIRA 2017, Part III, LNAI 10464, pp. 698–706, 2017.
DOI: 10.1007/978-3-319-65298-6_62

As a result, debris de-tumbling is imperative for towing mission success. Taking advantages of the torque generated by the tether tension is effective and economic for de-tumbling the tethered debris, but the tether fails to yield the tension torque along itself. To achieve 3-axis detumbling, the coordinated control was proposed by Huang, et al. [7,8] which combines the varying tension with the gripper thrusters.

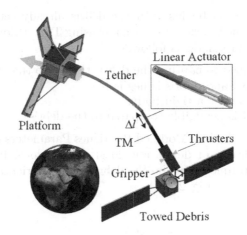

Fig. 1. Illustration of towing removal and TSR

In this paper, the coordinated control is applied to the towed debris detumbling. For towing removal, the tether tension should be stabilized to make debris keep pace with the platform for collision avoidance. For this respect, moving tether attachment point (TAP) is the only way to obtain the desired tension torque. Thus, the tethered space robot (TSR) is proposed that consists of a gripper for capture, thrusters and a tethered manipulator (TM), as shown in Fig. 1. The TM is a linear actuator with variable length Δl to which the tether is attached. During de-tumbling, the TM coordinates with thrusters by changing the length so that the tension torque acting on debris is adjusted.

However, the limits on thrusts and TM length change as well as the velocity of TAP pose severe challenges on controller design. Fortunately, nonlinear model predictive control (NMPC) which features feedback control and receding horizon enables the control systems to perform well in the presence of control constraints. And its many successful applications to aerospace has demonstrated the great effectiveness [9–11]. For this reason, the main contribution of this paper is to apply NMPC to the coordinated control of towed debris using TSR.

The paper is organized as follows. In Sect. 2, the attitude model of debris towed with off-centered capture by TSR is developed and the attitude equilibrium is defined. The design procedure of NMPC is presented in Sect. 3. Section 4 gives the simulation results and discussions. The conclusion is presented in Sect. 5.

2   Attitude Model and Equilibrium Definition

In this section, the attitude model of towed debris is first derived taking into account the tension torque provided by TM, then the attitude equilibrium to be tracked during detumbling is defined.

2.1   Attitude Model

The detumbling scenario begins with the debris already tethered by the off-centered capture on the bracket of solar panel using TSR. Prior to the modelling, several assumptions are made as follows.

1. The TSR is assumed to capture the debris tightly so that the relative movement between TSR and debris is negligible.
2. The debris is viewed as a rigid body.
3. The mass of TSR is negligible compared to the debris.

The attitude described by Modified Rodrigues Parameters (MRPs) is defined between the body frame of debris and local vertical local horizontal (LVLH) frame. The tension torque acting on the debris and the orientation of TSR after capture are illustrated in Fig. 2.

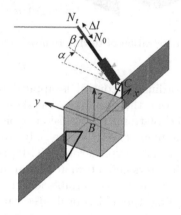

Fig. 2. Illustration of tension torque and orientation of TSR

In Fig. 2, B is the debris center of mass (CoM), C is the off-centered capture point, N_0 denotes the initial TAP when $\Delta l = 0$ and N_t represents the TAP during detumbling. The α and β are the orientation angles of TSR, according to assumption 1, they are constant when the capture is complete. As shown in Fig. 2, the tension torque $\boldsymbol{\tau}_T$ includes two parts, namely the induced torque $\boldsymbol{\tau}_{Ti}$ caused by the off-centered arm $\overrightarrow{BN_0}$ and the control torque $\boldsymbol{\tau}_{Tc}$ caused by the length change $\overrightarrow{\Delta l}$.

$$\boldsymbol{\tau}_T = \boldsymbol{\tau}_{Ti} + \boldsymbol{\tau}_{Tc}$$
$$= \left(\overrightarrow{BN_0} + \overrightarrow{\Delta l}\right) \times \boldsymbol{T}|_D \tag{1}$$

where $\overrightarrow{\Delta l} = [\Delta l \cos \beta \sin \alpha, \Delta l \cos \beta \cos \alpha, \Delta l \sin \beta]^T$ is length change vector in body frame, $T|_D = RT$ is tension vector in body frame, R denotes the transformation matrix and T is tension vector in LVLH frame. Therefore, we can obtain the attitude model of the form.

$$\begin{cases} \dot{\sigma} = G(\sigma)\,\omega \\ J\dot{\omega} = -\omega^\times J\omega + \tau_T + \tau_F \end{cases} \tag{2}$$

where $\sigma = [\sigma_x, \sigma_y, \sigma_z]^T \in \Re^3$ is MRPs of debris attitude, $\omega = [\omega_x, \omega_y, \omega_z]^T \in \Re^3$ is debris angular velocity. $(\cdot)^\times \in \Re^{3\times3}$ is a skew-symmetric operator, $J \in \Re^{3\times3}$ is principal inertia tensor of the debris, $\tau_F \in \Re^3$ is thruster torque, and $G(\sigma) = 1/4\left[(1 - \sigma^T\sigma)\,I_3 + 2\sigma^\times + 2\sigma\sigma^T\right]$. The transformation matrix used in Eq. 1 is defined as $R(\sigma) \overset{\Delta}{=} I_3 - \frac{4(1-\sigma^2)}{(1+\sigma^2)^2}[\sigma^\times] + \frac{8}{(1+\sigma^2)^2}[\sigma^\times]^2$.

2.2   Equilibrium Definition

The control objective is to steer the debris towards the equilibrium where the tension force acts through the debris CoM so that the attitude in this case can be stable and maintained. The equilibrium is determined by the TM length and we define that the length returns to its initial value l_0 after detumbling. As a result, the equilibrium can be obtained by the following equation.

$$\overrightarrow{BN_0} \times R(\sigma_{\mathrm{eq}})\,T = 0 \tag{3}$$

where σ_{eq} is the MRPs in equilibrium.

3   Nonlinear MPC Design

As a effective feedback optimal control, the NMPC can optimize a control sequence over a future horizon using the prediction model in order to minimize a cost function subject to constraints [9]. The first elements of the optimized control is applied to the plant over the first sampling interval. And the optimization horizon subsequently recedes and the process is repeated again. In this paper, the application of NMPC is shown in Fig. 3 where the thruster torque τ_F and TM length change Δl are the control variables to be optimized, the errors of MRPs and angular velocity (σ_e and ω_e) are the state x to be controlled.

Note that the navigation system is outside the scope of this paper. Therefore, the error attitude model is first derived from Eq. 2 and then is discretized by 4-order Runge-Kutta method. The next step is to design the cost function,and the system constraints should be considered and implemented in controller.

3.1   Error Model and Discretization

Define $\sigma_e = [\sigma_{ex}, \sigma_{ey}, \sigma_{ez}]^T$ and $\omega_e = [\omega_{ex}, \omega_{ey}, \omega_{ez}]^T$ as:

$$\begin{cases} \sigma_e = \sigma \otimes \sigma_{\mathrm{eq}}^{-1} \\ \omega_e = \omega - R(\sigma_e)\,\omega_{\mathrm{d}} \end{cases} \tag{4}$$

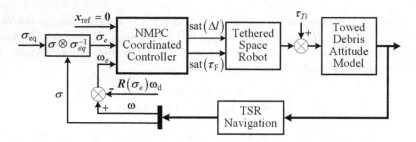

Fig. 3. The structure of NMPC coordinated controller

where $\boldsymbol{\omega}_d$ is the desired angular velocity and set to be zeros, The operator $\otimes$ denotes the MRP multiplication defined as below.

$$\boldsymbol{\sigma} \otimes \boldsymbol{\sigma}_{eq}^{-1} = \frac{\left(1 - \boldsymbol{\sigma}_{eq}^T \boldsymbol{\sigma}_{eq}\right) \boldsymbol{\sigma} + \left(\boldsymbol{\sigma}^T \boldsymbol{\sigma} - 1\right) \boldsymbol{\sigma}_{eq} - 2\boldsymbol{\sigma}_{eq} \times \boldsymbol{\sigma}}{1 + \left(\boldsymbol{\sigma}_{eq}^T \boldsymbol{\sigma}_{eq}\right) \left(\boldsymbol{\sigma}^T \boldsymbol{\sigma}\right) + 2\boldsymbol{\sigma}_{eq}^T \boldsymbol{\sigma}} \tag{5}$$

Therefore, the error model with the new states $\boldsymbol{x}$ can be formed as:

$$\dot{\boldsymbol{x}} = \boldsymbol{f}\left(\boldsymbol{x}, \Delta l, \boldsymbol{\tau}_F\right) \tag{6}$$

where $\boldsymbol{x} = [\boldsymbol{\sigma}_e, \boldsymbol{\omega}_e]^T$ and $\boldsymbol{f} = \left[\boldsymbol{G}\left(\boldsymbol{\sigma}_e\right)\boldsymbol{\omega}_e, -\boldsymbol{J}^{-1}\boldsymbol{\omega}_e^\times \boldsymbol{J}\boldsymbol{\omega}_e + \boldsymbol{J}^{-1}\left(\boldsymbol{\tau}_T + \boldsymbol{\tau}_F\right)\right]^T$.

The above continuous-time error model is discretized using the 4-order Runge-Kutta method of the form.

$$\boldsymbol{x}\left(n+1\right) = \boldsymbol{x}\left(n\right) + \frac{1}{6}\left(k_1 + 2k_2 + 2k_3 + k_4\right) \tag{7}$$

where $\boldsymbol{x}\left(n\right)$ denotes the state $\boldsymbol{x}$ at n moment and k_1, k_2, k_3, k_4 are defined as:

$$\begin{cases} k_1 = \Delta t_s f\left(\boldsymbol{x}\left(n\right), \Delta l\left(n\right), \boldsymbol{\tau}_F\left(n\right)\right) \\ k_2 = \Delta t_s f\left(\boldsymbol{x}\left(n\right) + 0.5k_1, \Delta l\left(n\right), \boldsymbol{\tau}_F\left(n\right)\right) \\ k_3 = \Delta t_s f\left(\boldsymbol{x}\left(n\right) + 0.5k_2, \Delta l\left(n\right), \boldsymbol{\tau}_F\left(n\right)\right) \\ k_4 = \Delta t_s f\left(\boldsymbol{x}\left(n\right) + k_3, \Delta l\left(n\right), \boldsymbol{\tau}_F\left(n\right)\right) \end{cases} \tag{8}$$

where $\Delta t_s = \frac{\Delta T}{m}$ is the step size, ΔT is the MPC sampling time and m is the discretization number.

3.2   Cost Function and Constraints

The cost function should be designed to penalize the state $\boldsymbol{x}$ in order to make the attitude track the equilibrium. The increments of control variables should also be penalized as the velocity of length change $\dot{\Delta l}$ is limited and a stead control is of significance. The terminal cost function F_t and constraints Ω_t are absent due to two reasons [12]. First, designing F_t and Ω_t to achieve a asymptotical stability is still an open problem. Second, including F_t and Ω_t will give rise to nonconvex optimization problems. As a result, the cost function is defined as below.

$$\bar{J}\left(\boldsymbol{x}, \Delta l, \boldsymbol{\tau}_F\right) = \sum_{n=0}^{N-1} \left(\boldsymbol{x}(n)^T \boldsymbol{Q} \boldsymbol{x}(n) + \widehat{\Delta l}(n) R_l \widehat{\Delta l}(n) + \widehat{\boldsymbol{\tau}_F}(n)^T \boldsymbol{R}_\tau \widehat{\boldsymbol{\tau}_F}(n)\right) \quad (9)$$

where N is the optimization horizon, $\boldsymbol{Q}$, R_l and $\boldsymbol{R}_\tau$ are appropriate weighting matrices, and $\widehat{\Delta l}(n)$ and $\widehat{\boldsymbol{\tau}_F}(n)$ denote the control increments.

The system constraints on states and control variables are defined as follows.

$$\begin{cases} 0 \leq \Delta l(n) \leq \Delta l_{\max} \\ \boldsymbol{\tau}_{F\min} \leq \boldsymbol{\tau}_F(n) \leq \boldsymbol{\tau}_{F\max} \\ \boldsymbol{x}(0) = \boldsymbol{x}(t) \\ \boldsymbol{x}(N) = \boldsymbol{0} \end{cases} \quad (10)$$

where $\Delta l_{\max}$, $\boldsymbol{\tau}_{F\min}$ and $\boldsymbol{\tau}_{F\max}$ are the bounds of control variables, and $\boldsymbol{x}(t)$ is the current state.

4   Simulation and Discussion

The simulation begins with the debris already captured by TSR and being towed on the way to the disposal orbit. The inertia parameters of debris and the geometrical size of TSR are set as $\boldsymbol{J} = diag\left(1500, 2000, 3000\right) \text{kgm}^2$ and $\left\|\overrightarrow{CN_0}\right\| = 1.7\,\text{m}$. The coordinate of capture point U in body frame is defined as $C = [2, 0, 0.577]^T$. The tension vector $\boldsymbol{T}$ in LVLH frame is assumed to be constant as $\boldsymbol{T} = [0, 75, 0]^T N$. In this case, according to Eq. 3, the equilibrium is

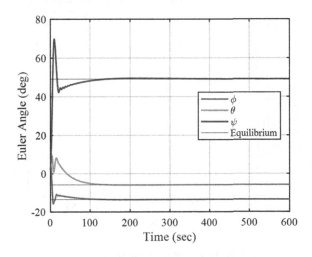

Fig. 4. Euler angles of debris

therefore obtained and can be expressed using Euler angles with 1-2-3 rotation sequence as $[\phi_{eq}, \theta_{eq}, \psi_{eq}] = [-13.48, -5.78, 49.11]^T$ deg.

The bounds of TM length change and thruster torque are set as 1m and $\pm 15\,\mathrm{Nm}$ respectively. The ΔT is 0.1 s, m is 5 and N is 2. The state weighting matrix is $\boldsymbol{Q} = diag\left(1.2 \times 10^4, 1.5 \times 10^4, 1 \times 10^4, 2500, 2500, 2500\right)$ and weighting matrices for control are chosen as $R_l = 5$ and $\boldsymbol{R}_\tau = diag\left(0.001, 0.001, 0.001\right)$ respectively since the aggressive control of thruster is feasible but the quick change of the TM length should be avoided.

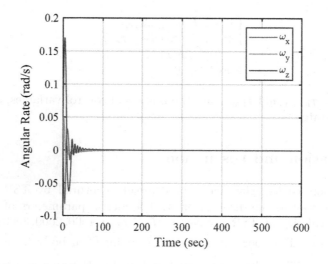

Fig. 5. Angular velocity of debris

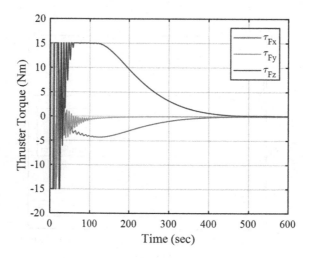

Fig. 6. Thruster control torque

Figure 4 shows that the attitude varies dramatically during the first 20 s then converge to the equilibrium smoothly. In Fig. 5, the angular velocity present the dumping oscillations and gradually decay to zeros, demonstrating the system is asymptotically stable. Figure 6 shows that the thruster torque components fluctuate within the bounds initially and begin to converge towards zeros after 120 s. Figure 7 presents the TM length change whose velocity is feasible thanks to the large weighting parameter. In Fig. 8, the total tension torque can converge to zero as the debris attitude maneuvers to the equilibrium, validating the equilibrium definition and the effectiveness of the proposed control strategy.

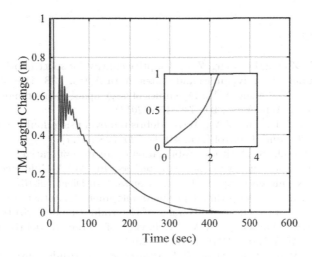

Fig. 7. TM length change

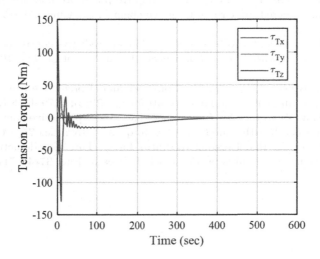

Fig. 8. The tension torque

5   Conclusion

The NMPC based coordinated control of towed debris using tethered space robot is studied in this paper. Simulation results illustrate that the TSR with the proposed control algorithm enables the debris to reorient to the equilibrium. This result is encouraging since it ensures the avoidance of the collision and entanglement, which is of significance for the safe conduction of ADR mission. In the future, a more novel TSR that uses tether only to implement detumbling control is worth discussion, and the robustness to dynamics uncertainty should be considered in controller design.

References

1. Huang, P.F., Zhang, F., Meng, Z.J.: Adaptive control for space debris removal with uncertain kinematics, dynamics and states. Acta Astro. **128**, 416–430 (2016)
2. Aslanov, V.S., Yudintsev, V.V.: Dynamics of large space debris removal using tethered space tug. Acta Astro. **91**, 149–156 (2013)
3. Qi, R., Misra, A.K., Zuo, Z.Y.: Active debris removal using double-tethered space-tug system. J. Guid. Control Dyn. **40**(3), 720–728 (2017)
4. Hovell, K., Ulrich, S.: Attitude stabilization of an uncooperative spacecraft in an orbital environment using visco-elastic tethers. In: AIAA Guidance, Navigation, and Control Conference, pp. 1–16. AIAA press, San Diego (2016)
5. Jasper, L.E.Z.: Open-loop thrust profile development for tethered towing of large space objects. Ph.D. Thesis, University of Colorado, Boulder (2014)
6. Aslanov, V.S., Yudintsev, V.V.: Behavior of tethered debris with flexible appendages. Acta Astro. **104**(1), 91–98 (2014)
7. Huang, P.F., Wang, D.K., Meng, Z.J.: Post-capture attitude control for a tethered space robot-target combination system. Robotica **33**(4), 898–919 (2015)
8. Huang, P.F., Wang, D.K., Meng, Z.J.: Adaptive postcapture backstepping control for tumbling tethered space robot target combination. J. Guid. Control Dyn. **39**(1), 150–156 (2015)
9. Petersen, C.D., Leve, F., Kolmanovsky, I.: Model predictive control of an Underactuated spacecraft with two reaction wheels. J. Guid. Control Dyn. **40**(2), 320–332 (2016)
10. Slegers, N., Kyle, J., Costello, M.: Nonlinear model predictive control technique for unmanned air vehicles. J. Guid. Control Dyn. **29**(5), 1179–1188 (2006)
11. Li, H.P., Yan, W.S., Shi, Y.: Continuous-time model predictive control of under-actuated spacecraft with bounded control torques. Automatica **75**, 144–153 (2017)
12. Yan, Z., Wang, J.: Model predictive control for tracking of underactuated vessels based on recurrent neural networks. IEEE J. Oceanic Eng. **37**(4), 717–726 (2012)

Design and Experimental Study on Telescopic Boom of the Space Manipulator

Shicai Shi$^{(\boxtimes)}$, Qingchao He, and Minghe Jin

State Key Laboratory of Robotics and System, Harbin Institute of Technology,
Harbin 150001, China
sschit@hit.edu.cn, heqingcao123@163.com

Abstract. This paper achieves the design of the key component of space manipulator – the telescopic boom and develops the principle prototype, to solve the contradiction between large working space, small load and fine operation, high rigidity, high precision in space manipulator handling operations. The stiffness test platform is developed for the telescopic boom, and through the test the boom meets the design requirements. The control strategy used to stretch the boom is also discussed in detail. Relevant experiments show that the telescopic boom can successfully achieve the telescopic movement.

Keywords: Space manipulator · Telescopic boom · Test platform · Telescopic strategy

1 Introduction

Space environment with high vacuum, large temperature difference, strong radiation and other characteristics, will pose a threat to the safety of astronauts working in space [1]. By space robots to complete the above operations can effectively avoid the problem, which can improve the safety, economic benefit, efficiency and quality of the task [2]. The most widely used space robot system is the space manipulator, such as the Canadarm and Canadarm2 [3, 4].

Nowadays, the mature space manipulators all use fixed length arm, which brings the problem of large occupied space, high cost and fixed working space [5]. Although the Canadarm2 uses a collapsible structure that reduces the need for launch space and improves operational flexibility, it needs the astronaut's extravehicular operations to assist its development, which brings some risks. Therefore, the retractable space manipulator is the future direction of development. The main advantages of retractable space manipulator mainly include: (1) it occupies a small size when shrank, and has a large working space after being stretched. (2) it is more flexible and can be competent for more complex tasks. (3) it reduces the cost of launch.

© Springer International Publishing AG 2017
Y. Huang et al. (Eds.): ICIRA 2017, Part III, LNAI 10464, pp. 707–716, 2017.
DOI: 10.1007/978-3-319-65298-6_63

2 Design of the Telescopic Boom

2.1 Design Specifications

The telescopic boom has the characteristics of light weight and small space occupation, and it also needs to have enough rigidity, strength, reliability and security, to meet the needs of the complex space mission. In this paper, the design index of telescopic boom of space manipulator is shown in Table 1.

Table 1. Design index

Serial Number	Property	Index
1	Weight	≤ 8 kg
2	Torsional stiffness	3×10^4 Nm/rad
3	Bending stiffness	7×10^4 Nm/rad
4	Tensile stiffness	2.5×10^5 Nm/rad
5	Telescopic ratio	$\geq 3 : 2$

2.2 Overall Design of the Telescopic Boom

In this paper, several telescopic structures are compared. It can be seen from the results of Table 2 that the sleeve telescopic mast has the advantages of moderate length, high carrying capacity, high reliability and low complexity [6, 7]. The telescopic scheme of the boom is divided into two types: the active telescopic scheme and the passive telescopic scheme according to the power source. Although the active telescopic scheme has the advantages of strong versatility, large proportion of boom extension and easy change of arm length, it has the characteristics of complex structure, large volume and heavy weight. As for passive scheme, the arm is not driven by a separate drive source, but is rotated by the rotation of the joint, mounted on both ends of the boom, to drive the boom to achieve a change in the length of the arm. Therefore, the passive scheme is simpler and lighter. After comparison, the telescopic boom developed in this paper adopts the sleeve structure, using the passive telescopic scheme.

Table 2. Comparisons of the telescopic strategies

Strategy	Stretching length	Carrying capacity	Reliability	Complexity
Collapsible Tubular mast	Short	Small	Low	Low
Telescopic Sleeve mast	Moderate	Large	High	Low
Coilable mast	Long	Larger	High	High

Figure 1 shows the overall design of the telescopic boom. The telescopic arm system consists of two sets of telescopic boom, two active joints and one driven joint, arranged symmetrically. An active joint is mounted at the end of each boom, and two

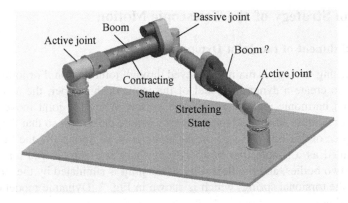

Fig. 1. The overall structure of the boom

sets of booms are connected by a driven joint. In the initial state, the booms are contracted and locked. The stretching process is as follows: positions of two active joint are fixed, unlock the arm to be expanded, the joint at the end of another boom rotates, another active joint follows the movement, passive extension starts, stop joint's operation after reaching the designed position and then lock booms again. The constriction process is similar to the above procedure, with the reverse rotation of the joint.

2.3 Internal Structure of the Boom

The structural design of the telescopic mechanism is shown in Fig. 2, consisting mainly of internal and external booms. In order to ensure smooth sliding between the inner and outer booms, the outer boom is equipped with a pulley, and the inner boom has a corresponding guide rail. The inner part of the boom is provided with a locking mechanism which is driven by a locking motor, to lock booms after the telescopic motion. A stroke switch is arranged at the limit position of the extension and contraction of the boom to control the start and stop of the joint motor and the locking motor.

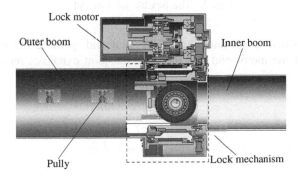

Fig. 2. The inner structure of the boom

3   Control Strategy of the Telescopic Motion

3.1   Establishment of the Joint Dynamic Model

Passive stretching strategy is mainly achieved through joint motion. For joint control, it is necessary to create a dynamic model of the joint. In this paper, the active joint is equipped with harmonic reducer and torque sensor. When the joint movement, harmonic reducer flexible wheel and torque sensor will be deformed, so that the joint with a certain degree of flexibility. Based on the assumption of Spong [8], the flexible joint can be regarded as a simple two body system: there is a zero inertia torsion spring between the two bodies, and the flexibility of the joint is simulated by the stiffness and damping of the torsional spring, which is shown in Fig. 3. Dynamic model of flexible joint is as follows:

$$\tau_m - \tau_{f1} = J_1\ddot{q}_1 + D(\dot{q}_1 - \dot{q}_2) + K(q_1 - q_2)$$
$$\tau = K(q_1 - q_2) \tag{1}$$
$$K(q_1 - q_2) + D(\dot{q}_1 - \dot{q}_2) + \tau_{ext} = J_2\ddot{q}_2 + \tau_{f2}$$

Where q_1 and q_2 denote the rotation angle of the motor and the load, J_1 and J_2 are the inertias of the motor and the load separately. τ_{f1} and τ_{f2} represent the friction torque respectively applied on motor and load, K and D are the stiffness and damping coefficients of the sensor. τ_m and τ_{ext} respectively represent the input torque of the motor and the torque applied to the load. τ is the measured value of the torque sensor.

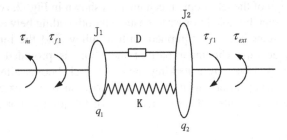

Fig. 3. The flexible joint model

Without considering the damping of the joint and ignoring the influence of the friction torque at the motor end and the load, the joint dynamics model can be simplified as:

$$\tau_m = J_1\ddot{q}_1 + K(q_1 - q_2)$$
$$\tau = K(q_1 - q_2) \tag{2}$$
$$K(q_1 - q_2) + \tau_{ext} = J_2\ddot{q}_2$$

3.2   Position Control of the Active Joint

It is the basis for the telescopic motion of the boom to carry out the accurate position control of the active joint, which makes it follow the planned path precisely. When the joint position control is concerned, it often involves the rapid start and stop or the large change of the joint setting value. The introduction of integral controller will cause the system overshoot, and even the system of shock. Therefore, this paper uses the PD control strategy. By the Lyapunov stability theorem and LaSalle theorem, PD control has the asymptotic stability, and its control law is:

$$\tau = k_P(q_d - q) + k_D(\dot{q}_d - \dot{q}) \tag{3}$$

Where k_p and k_p are scale factor and derivative coefficient. q_d and $\dot{q}_d$ denote the desired position and speed of the joint angle. q and $\dot{q}$ represent the actual position and speed.

3.3   The Impedance Control of the Joint

When an active joint is moved by position control, the other active joint follows the movement by impedance control. By establishing the relationship between the force and the position, the impedance control makes the joint flexibility [9], the expression is:

$$M_\theta(\ddot{\theta} - \ddot{\theta}_r) + D_\theta(\dot{\theta} - \dot{\theta}_r) + K_\theta(\theta - \theta_r) = \tau \tag{4}$$

Where M_θ, D_θ and K_θ represent the inertia matrix, damping matrix and stiffness matrix of the desired impedance model.

Since the joint in this paper does not have the acceleration sensor, and the acceleration information obtained by the second derivative of the position information will produce very strong noise. Therefore, the inertia term is neglected in this paper. The impedance control expression becomes:

$$D_\theta(\dot{\theta} - \dot{\theta}_r) + K_\theta(\theta - \theta_r) = \tau \tag{5}$$

In order to obtain smaller ideal stiffness and damping, this paper adopts the impedance control strategy based on joint force, which consists of impedance control outer ring and force control inner ring [10], and uses internal force controller to compensate the nonlinear dynamic influence of manipulator.

Because of the flexibility of joints, the stiffness of joints and the effective moment of inertia of the rotor will affect the accuracy of the mechanical model. For this reason, this paper introduces the output torque of the torque sensor as the feedback of the force controller, so that the effective inertia of the rotor of the force closed loop system is smaller, and the model is shown in Fig. 4. The motor output torque is:

$$\tau_m = J_1 J_{1\theta}^{-1} \tau_1 + (1 - J_1 J_{1\theta}^{-1})\tau \tag{6}$$

Where $J_{1\theta}$ represents the effective moment of rotation of the motor, and τ_1 is a new control input whose expression is $\tau_1 = K_\theta\theta + D_\theta\dot{\theta}$. K_θ and D_θ mean the ideal stiffness

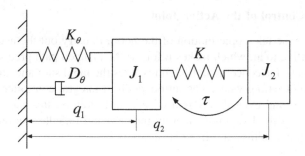

Fig. 4. Introducing moment negative feedback

and ideal damping of joints in impedance control respectively. According to (2), the impedance control strategy is:

$$K_\theta \theta + D_\theta \dot{\theta} = J_{1\theta} \ddot{q}_1 + K(q_1 - q_2)$$
$$\tau = K(q_1 - q_2) \quad\quad (7)$$
$$K(q_1 - q_2) + \tau_{ext} = J_2 \ddot{q}_2$$

4 Experimental Verification of the Telescopic Boom

4.1 The Stiffness Test of the Boom

In order to test the actual operation of the telescopic boom and its stiffness, this paper has worked on the principle prototype of the boom and developed the corresponding rigidity test platform, which is shown in Fig. 5.

The experiment platform can fix the boom in the vertical direction, and exert force on the boom in all directions through the pulley, cable and weight. Apply the vertical and horizontal forces on the boom to obtain the corresponding deformation. Through the linear fitting of the experimental data by MATLAB, we can get the stiffness data of

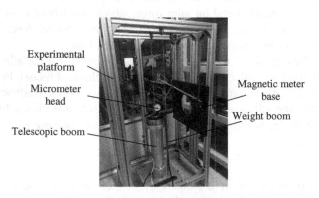

Fig. 5. The stiffness test platform of the boom

Table 3. Stiffness index of the boom

	Contraction	Stretching	Units
Compression stiffness	2.75×10^6	2.48×10^6	N/m
Tensile stiffness	3.38×10^6	2.22×10^6	N/m
Bending stiffness	9.83×10^4	7.86×10^4	Nm/rad
Torsional stiffness	–	3.05×10^4	Nm/rad

different states of the boom, as shown in the Table 3. By comparing with Table 2, we can see that the stiffness of the boom can reach the design target.

4.2   The Test of the Position Control of the Active Joint

The whole telescopic motion of the boom contains the following stages, as shown in Table 4.

Table 4. Boom movement stage

	Joint 1	Joint 2	Boom 1	Boom 2
Phase 1	Active	Passive	Locked	Stretch
Phase 2	Passive	Active	Stretch	Locked
Phase 3	Passive	Active	Shrink	Locked
Phase 4	Active	Passive	Locked	Shrink

In order to test the actual effect of the PD and impedance control strategy applied in the telescopic boom, the experimental system is set up in Fig. 6. The joints mounted on both ends of the telescopic arm are secured to the horizontal test platform.

This paper first adjusts the PD controller parameters, to ensure the accuracy of the boom telescopic movement. At the same time to ensure the stability of the system, we should minimize the steady-state error of the system and speed up the system. In this paper, different values of k_p and k_D are selected, and the reasonable parameters are obtained by comparing the actual trajectories with the given trajectories. For the active

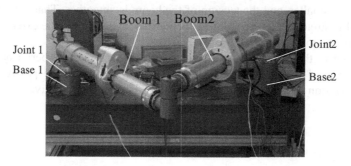

Fig. 6. The experimental system of the boom

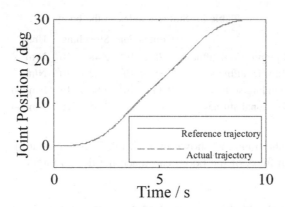

Fig. 7. Reference trajectory and actual trajectory of joint

joint, the actual trajectory can track the target trajectory well, as shown in Fig. 7, the maximum position error does not exceed 0.1°.

4.3   The Test of the Impedance Control of the Following Joint

For the joint that follows the movement, the stiffness and the damping coefficient in the impedance parameters are set to zero. The signal measured by the torque sensor is multiplied by the amplification factor as feedforward and input to the motor control terminal so that the effect of zero force tracking is as good as possible. Because the torque sensor has a certain noise, the amplification factor can not be taken too large, otherwise it will cause instability of the system.

For the four movement stages in the previous section, the moments of the two joints are shown in Fig. 8. It can be seen from the image that the moment curves of the two joints also show some symmetry due to the symmetry of the four stages of motion: the stage one is similar to the stage four, and the phase two and the phase are similar. Therefore, we discuss only the joint moment curves of stage one and two.

The output torque of the active joint is both large at the beginning of stage 1 and stage 2, for the reason that the anti-twist block gets stuck in the card slot, causing a great friction resistance. Compared with stage two, the output torque of the active joint in stage one is smaller, and the tension of the boom is smoother. That is because, after the boom II is stretched, its arm of force becomes longer, so that the extension boom I requires a larger joint torque.

According to the experimental results, it can be seen that the performance of the impedance control strategy based on the joint torque is well, the passive joint can follow the rotation of the active joint and reflect the better flexibility.

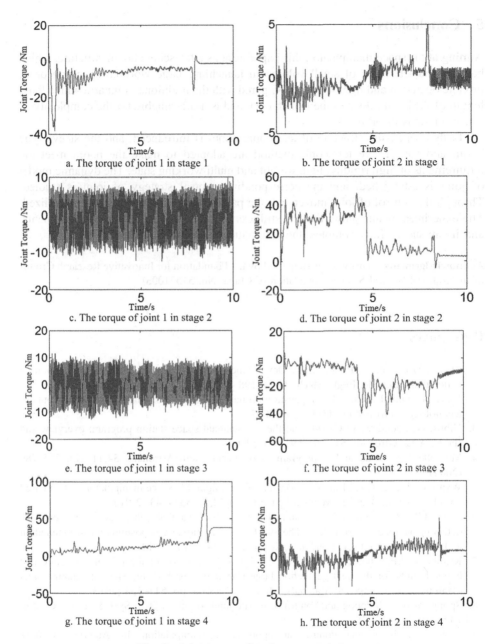

Fig. 8. Joint torque of each movement stage

5 Conclusions

Aiming at the space manipulator, this paper designs a telescopic boom structure, which has the characteristics of small volume in launching, large working space on orbit, flexible operation and light weight. Compared with the traditional joint arm with a fixed length of the boom, it saves the launch cost, and is more suitable for the complexity of space missions in orbit.

In this paper, the structure of telescopic boom is introduced, and the sleeve type boom and the passive telescopic method are adopted to make the boom meet the requirements of high rigidity, light weight and multi working state. The dynamic model of joint is established, and the corresponding control strategy of joint is studied. Through the control of joint movement, the passive extension of the boom is realized. The experiment result shows that the stiffness of the boom can meet the requirements, and it can successfully complete the automatic telescopic movement.

Acknowledgements. This work is supported by the Foundation for Innovative Research Groups of the National Natural Science Foundation of China (No. 51521003)

References

1. Zhang, W.H., Ye, X.P., Ji, X.M.: Development summarizing of space robot technology national and outside. Flight Mech. **31**(3), 198–202 (2013)
2. Yu, D.Y., Sun, J., Ma, X.R.: Suggestion on the development of Chinese space manipulator technology. Aerosp. Eng. **16**(4), 1–8 (2007)
3. Gibbs, G., Sachdev, S.: Canada and the international space station program: overview and status. Acta Astronaut. **51**(1–9), 591–600 (2002)
4. Marcotte, B.: Canadian ISS program involvement. Acta Astronaut. **54**(11–12), 785–786 (2004)
5. Wang, K., Liang, C.H., Lin, Y.C.: Design of reconfigurable space manipulator system based on telescopic mechanism. Manned Space Flight **22**(5), 537–543 (2016)
6. Li, D., Chu, Z.Y., Cui, J.: Design and adaptive control of a deployable manipulator for space detecting payload supporting. In: The 8th IEEE International Symposium on Instrumentation and Control Technology, London, UK, pp. 229–234. IEEE (2012)
7. Krimbalis, P.P., Djokic, D., Hay, G.: Design and validation of the primary structure and bonded joints for the next generation large Canadarm test bed. In: The 19th International Conference on Composite Materials, Montréal, Canada, pp. 544–554 (2013)
8. Spong, M.W.: Modeling and control of elastic joint robots. J. Dyn. Syst. Measure. Control **109**, 310–319 (1987)
9. Hogan, N.: Impedance control: an approach to manipulation. In: American Control Conference, pp. 304–313. IEEE Xplore (2009)
10. West, H., Asada, H.: A method for the design of hybrid position/Force controllers for manipulators constrained by contact with the environment. In: Proceedings of the IEEE International Conference on Robotics and Automation, pp. 251–259. IEEE (1985)

Trajectory Planning of Space Robot System for Reorientation After Capturing Target Based on Particle Swarm Optimization Algorithm

Songhua Hu[✉], Ping He, Zhurong Dong, Hongwei Cui,
and Songfeng Liang

School of Automotive and Transportation Engineering,
Shenzhen Polytechnic, Shenzhen, China
husonghua@szpt.edu.cn

Abstract. A typical servicing operation in space mainly includes three phases: capturing the target, re-orientating of the whole system with the target, and repairing the target. A method is proposed to achieve re-orientating the space robot system, planning the manipulator configuration and spacecraft orientation at the same time. Firstly, angular momentum preloaded in the manipulator will provide a favorable condition for the capturing and reorientation after capturing, so the manipulator moves with some initial velocity after capturing. The constraints on the manipulator and the objective function are defined according to the planning problem. Then the joint trajectories are parameterized by using sinusoidal polynomials functions and the cost function is proposed according to the accuracy requirements. Finally, Particle Swarm Optimization (PSO) is used to search for the global optimal resolution of the parameters. When the parameters are found, each joint trajectory can be determined. The simulation results show that this method is better than other approaches both in convergence rate and accuracy and the joints trajectories are very smooth and suitable for control.

Keywords: Space robot · Trajectory planning · Particle swarm optimization

1 Introduction

The space robotic system is becoming one of the most attractive areas of On-Orbit Servicing (OOS) in recent years [1]. In general, a typical servicing operation mainly includes the following three phases: capturing the target by space robots, reorientation of the entire space robot system with the target, and repairing the target. During the capturing, the attitude control system will be turned off in order to avoid the collision with the target and save energy. Therefore, the attitude of the coupled satellite pair may change much and needs stabilizing and re-orientating after capturing. In other words, there are two tasks to be completed. One is to berth the target with suitable manipulator configuration, and the other is to stabilize the attitude of the whole robot system for communication and power supply. In the paper, a method to complete the two tasks at the same time is proposed. The characteristic of the method is that only the manipulator

Y. Huang et al. (Eds.): ICIRA 2017, Part III, LNAI 10464, pp. 717–728, 2017.
DOI: 10.1007/978-3-319-65298-6_64

joints are activated and traditional attitude control devices such as reaction wheels and the propulsion system are not required, which is useful in reducing fuel consumption of the spacecraft and extend its operational life on the orbit.

Many literatures focus on the target capturing of the space robot system. A capturing operation can be divided into three phases: the pre-impact phase, the impact phase and the post-impact phase. A comprehensive discussion about the usage of the so-called reaction null space is made by Nechev and Yoshida [2]. They showed that obtaining joint velocities using this approach does not influence the momentum distribution whatsoever. D. N. Dimitrov and K. Yoshida [3] introduce a new satellite capturing strategy with bias momentum. They pointed out that the pre-impact angular momentum distribution is closely related with the attitude profile after the contact with the target.

R. Lampariello et al. [4] applied the method to a system with rotational joints and the method is used for point-to-point maneuvers and attitude variation maneuvers. E. Papadopoulos [5] introduce a new method to realize smooth planning for free-floating space robots using polynomials. Wen-Fu Xu [6] employed a method for target berthing and reorientation after capturing. They solved the problem by Genetic Algorithm. In this paper, a method is proposed, for planning the trajectory of joint of the manipulator and simultaneous the attitude of the base vehicle arrive the expectation, so the purpose of reorientation has achieved. The trajectory planning problem is solved by Particle Swarm Optimization (PSO).

The paper is organized as follows: In section two, we derive the equations used to plan the path. The idea of preloaded angular momentum in manipulator is proposed in section three. Then, the trajectory planning problem is discussed in section four. And the problem is solved based on PSO in Section five. In section six, computer simulation study and results will be shown. Section seven is the discussion and conclusion of the work.

2   Equations of Motion

The equation of motion of a free-flying space robot is generally expressed in the following form:

$$
\begin{bmatrix} H_b & H_{bm} \\ H_{bm}^T & H_m \end{bmatrix} \begin{bmatrix} \ddot{x}_b \\ \ddot{q}_m \end{bmatrix} + \begin{bmatrix} c_b \\ c_m \end{bmatrix} = \begin{bmatrix} F_b \\ \tau \end{bmatrix} + \begin{bmatrix} J_b^T \\ J_m^T \end{bmatrix} F_e \tag{1}
$$

where,

$\dot{x}_b \in R^6$: the inertia linear velocity and angle velocity of the base, i.e. $\dot{x}_b = \left[v_0^T, \omega_0^T \right]^T$,

$q_m \in R^n$: the joint angle vector (n is the degree of freedom of the manipulator) i.e. $q_m = [q_1,...,q_n]^T$),

$H_b \in R^{6\times6}$: inertia matrix of the base,

$H_m \in R^{n\times n}$: inertia matrix for the manipulator arms (the links except the base),

$H_{bm} \in R^{6\times n}$: coupling inertia matrix,

$c_b \in R^6$: velocity dependent non-linear term for the base,

$c_m \in R^6$: velocity dependent non-linear term for the arm,

$F_b \in R^6$:   force and torque exert on the centroid of the base,
$F_h \in R^6$:   force and torque exert on the manipulator hand,
$\tau \in R^n$:   torque on the manipulator joints,

Since there are no external forces and torques acted on the free-floating system, with the assumption made of linear and angular momentum initially zero, the system momentum keeps zero according to the conservation law, i.e.:

$$H_b \left[v_0^T, \omega_0^T \right]^T + H_{bm} \dot{q}_m = 0 \tag{2}$$

From (2), v_0 and ω_0 are solved as:

$$\left[v_0^T, \omega_0^T \right]^T = -H_b^{-1} H_{bm} \dot{q}_m = H \dot{q}_m \tag{3}$$

For the attitude of the base, the relationship exists:

$$\omega_0 = J_{\omega q}(Q_0, q_m) \dot{q}_m \tag{4}$$

where $J_{\omega q}(Q_0, q_m) \in R^{3 \times 6}$ is the sub-matrix of H and it represents the relationship between the base attitude motion and the manipulator joint motion. $Q_0(= [\alpha, \beta, \gamma]^T)$ is the attitude of the satellite and it is expressed in terms of z-y-x Euler angles. Therefore, the attitude transformation matrix is

$$A = Rot(Z, \alpha) Rot(Y, \beta) Rot(X, \gamma) = \lfloor n \quad o \quad a \rfloor$$
$$= \begin{bmatrix} c\alpha c\beta & -s\alpha c\gamma + c\alpha s\beta s\gamma & s\alpha s\gamma + c\alpha s\beta c\gamma \\ s\alpha c\beta & c\alpha c\gamma + s\alpha s\beta s\gamma & -c\alpha s\gamma + s\alpha s\beta c\gamma \\ -s\beta & c\beta s\gamma & c\beta c\gamma \end{bmatrix} \tag{5}$$

where, $c\alpha = cos(\alpha)$, $s\alpha = sin(\alpha)$, $c\beta = cos(\beta)$, $s\beta = sin(\beta)$, $c\gamma = cos(\gamma)$, $s\gamma = sin(\gamma)$. And the relationship between the attitude velocity and the Euler angle velocity are given by:

$$^I\omega_0 = \begin{bmatrix} \omega_{0x} \\ \omega_{0y} \\ \omega_{0z} \end{bmatrix} = \begin{bmatrix} 0 & -s\alpha & c\alpha c\beta \\ 0 & c\alpha & s\alpha c\beta \\ 1 & 0 & -s\beta \end{bmatrix} \begin{bmatrix} \dot{\alpha} \\ \dot{\beta} \\ \dot{\gamma} \end{bmatrix} = K_r \dot{Q}_0 \tag{6}$$

$$\dot{Q}_0 = K_r^{-1} {}^I\omega_0 = K_r^{-1} J_{\omega q} \dot{q}_m \tag{7}$$

The following equations are obtained by integrating the Eq. (7):

$$Q_0(t) = \int_{t_0}^t K_r^{-1} J_{\omega q} \dot{q}_m dt \tag{8}$$

where t_0 is the initial time of the maneuver. The integral cannot be obtained analytically, but it can be calculated numerically.

3 Angular Momentum Distribution Before Capture

It is very important to keep the base attitude motion within some limits during the approach to the target, the impact and the post-capture motion.

One of the main characteristics of a capturing operation in orbit is the momentum conservation if there are no external forces. Our aim is to capture a target without changing the chaser's base orientation. In other words, during the three phases of the capturing operation we want the smallest possible amount of angular momentum to be stored in the base rotational motion.

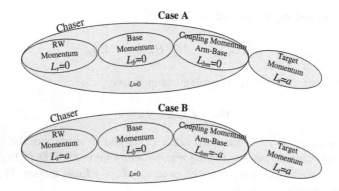

Fig. 1. Two cases of angular momentum distribution in the approach phase

The angular momentum in the entire system consisted of both the chaser and target can be sufficiently defined by four variables (Fig. 1):

(1) L_{bm}: coupling angular momentum between the manipulator and the base
(2) L_{br}: coupling angular momentum between the reaction wheels and the base.
(3) L_b: angular momentum stored in the base as a result of its angular motion
(4) L_t: angular momentum in the target.

$$L_{bm} = \tilde{H}_{bm}\dot{q}_m; L_{br} = \tilde{H}_{br}\dot{q}_r; L_b = \tilde{H}_b\omega_b \qquad (9)$$

Two typical possible distributions are depicted in Fig. 1.

3.1 Case a: Non-bias Case

In Case A, during the approach to the target, $L_{bm} = 0$, we call it a non-bias case. As a result of the maximum torque restriction, the reaction wheels will most likely fail to accommodate the angular momentum transferred to the base in a short time. Hence, obvious base rotational motion will occur.

3.2   Case B: Bias Case

After the contact, L_t "entering" the chaser could be canceled out with the one preloaded in the manipulator L_{bm}. Therefore, the remaining amount of angular momentum in the base, manipulator and target should be redistributed in order the system to come to a complete stop in the post-impact phase.

Case B is a "full" bias distribution and obtaining it is not always possible. A "partial" bias distribution is realizable, the follow conditions should be satisfied.

$$\begin{cases} |L_{bm}^k| \leq |L_t^k| \\ L_{bm}^k L_t^k < 0 \end{cases} \tag{10}$$

where $k = \{x, y, z\}$ stand for the x, y and z components of a three dimensional vector. The momentum that should be preloaded in the manipulator has to be with smaller or equal magnitude and opposite direction to the one in the target.

4   Trajectory Planning Problem

4.1   Constraints of the Path Planning

The initial and final joint angles are:

$$q_i(t_0) = q_{i0}, \; \dot{q}_i(t_0) = \dot{q}_{i0}, \; \ddot{q}_i(t_0) = 0 \tag{11}$$

$$q_i(t_f) = q_{if}, \; \dot{q}_i(t_f) = 0, \; \ddot{q}_i(t_f) = 0 \tag{12}$$

where t_f is the final time of the maneuver and q_{if} is determined according to the docking configuration. Furthermore, the path planning solution has to satisfy kinematics and dynamic constraints. The constraints can be expressed as

$$q_{i\_min} \leq q_i(t) \leq q_{i\_max} \tag{13}$$

In order to satisfy the need for communication and power supply, the satellite attitude must change to desired values.

$$Q_{0f} = \int_{t_0}^{t} K_r^{-1} J_{\omega q} \dot{q}_m dt = Q_{0d} \tag{14}$$

Where Q_{0f} is the final attitude and Q_{0d} is the desired attitude. Correspondingly, the transformation matrixes are $A_f ([n_f, o_f, a_f])$ and $A_d([n_d, o_d, a_d])$ respectively. A choice of the object function can be

$$J = \|\Delta Q_0\| = \|0.5(n_f \times n_d + o_f \times o_d + a_f \times a_d)\| \tag{15}$$

4.2   Joint Function Parameterization

In order to satisfy the kinematical constraints, sinusoidal functions, whose arguments are the polynomial functions, are used to parameterize the joint angle functions:

$$q_i(t) = \Delta_{i1}\ \sin\left(\begin{array}{c} a_{i7}t^7 + a_{i6}t^6 + a_{i5}t^5 + a_{i4}t^4 \\ + a_{i3}t^3 + a_{i2}t^2 + a_{i1}t + a_{i0} \end{array}\right) + \Delta_{i2} \tag{16}$$

where, $0 < t < t_f$ and a_{i0}-a_{i7} are the coefficients of the polynomial. Δ_{i1} and Δ_{i2} are defined as

$$\Delta_{i1} = (q_{i\,max} - q_{i\,min})/2,\ \Delta_{i2} = (q_{i\,max} + q_{i\,min})/2 \tag{17}$$

The corresponding angle rates and accelerates are

$$\begin{aligned} \dot{q}_i =& \Delta_{i1}\ cos\left(a_{i7}t^7 + a_{i6}t^6 + a_{i5}t^5 + a_{i4}t^4 + a_{i3}t^3 + a_{i2}t^2 + a_{i1}t + a_{i0}\right) \\ &\cdot \left(7a_{i7}t^6 + 6a_{i6}t^5 + 5a_{i5}t^4 + 4a_{i4}t^3 + 3a_{i3}t^2 + 2a_{i2}t + a_{i1}\right) \end{aligned} \tag{18}$$

$$\begin{aligned} \ddot{q}_i =& - \Delta_{i1}\ sin\left(a_{i7}t^7 + a_{i6}t^6 + a_{i5}t^5 + a_{i4}t^4 + a_{i3}t^3 + a_{i2}t^2 + a_{i1}t + a_{i0}\right) \\ &\cdot \left(7a_{i7}t^6 + 6a_{i6}t^5 + 5a_{i5}t^4 + 4a_{i4}t^3 + 3a_{i3}t^2 + 2a_{i2}t + a_{i1}\right)^2 \\ &+ \Delta_{i1}\ cos\left(a_{i7}t^7 + a_{i6}t^6 + a_{i5}t^5 + a_{i4}t^4 + a_{i3}t^3 + a_{i2}t^2 + a_{i1}t + a_{i0}\right) \\ &\cdot \left(42a_{i7}t^5 + 30a_{i6}t^4 + 20a_{i5}t^3 + 12a_{i4}t^2 + 6a_{i3}t + 2a_{i2}\right) \end{aligned} \tag{19}$$

Applying the equality constraints of (11) and (12) to (16) (18) (19), the following results are found:

$$a_{i0} = sin^{-1}[(q_{i0} - \Delta_{i2})/\Delta_{i1}] \tag{20}$$

$$a_{i1} = \dot{q}_{i0}/\Delta_{i1}\ cos\ a_{i0} \tag{21}$$

$$a_{i2} = \frac{1}{2}a_{i1}^2\ tan\ a_{i0} \tag{22}$$

$$a_{i3} = -\frac{3a_{i7}t_f^7 + a_{i6}t_f^6 + 3a_{i2}t_f^2 + 6a_{i1}t_f + 10a_{i0} - 10\Delta}{t_f^3} \tag{23}$$

$$a_{i4} = \frac{8a_{i7}t_f^7 + 3a_{i6}t_f^6 + 3a_{i2}t_f^2 + 8a_{i1}t_f + 15a_{i0} - 15\Delta}{t_f^4} \tag{24}$$

$$a_{i5} = -\frac{6a_{i7}t_f^7 + 3a_{i6}t_f^6 + a_{i2}t_f^2 + 3a_{i1}t_f + 6a_{i0} - 6\Delta}{t_f^5} \tag{25}$$

where

$$\Delta = sin^{-1}((q_{id} - \Delta_{i2})/\Delta_{i1}) \tag{26}$$

Therefore, there are only two parameters (a_{i6} and a_{i7}) in each joint function after parameterization. If $n = 6$, let $a = [a_{16}, a_{17}, a_{26}, a_{27},..., a_{66}, a_{67}] \in \mathbf{R}^{12}$.

5 Solution to the Planning Problem

Particle Swarm Optimization (PSO) is a population based stochastic optimization technique developed by Dr. Eberhart and Dr. Kennedy, inspired by social behavior of bird flocking or fish schooling [7].

In PSO, the potential solutions, called particles, fly through the problem space by following the current optimum particles. Compared to GA, the advantages of PSO are that PSO is easy to implement and there are few parameters to adjust. PSO has been successfully in application to wide range of engineering and computer science problems. PSOt is a PSO Toolbox for use with Matlab. Here, PSOt is used to solve the problem of the trajectory planning proposed above. The procedure is:

(1) The particles are chosen as

$$a = [a_{16}, a_{17}, a_{26}, a_{27}, \cdots, a_{66}, a_{67}] \tag{27}$$

(2) Define the Parameters for the PSO.

We define the fitness function according to the object function (15) and constraints, and define the range of particles, VarRange, the maximum change one particle can take during one iteration, Vmax.

$$\begin{cases} a_{16} \in [-10^{-4}, 10^{-4}], a_{17} \in [-10^{-4}, 10^{-4}] \\ a_{26} \in [-10^{-4}, 10^{-4}], a_{27} \in [-10^{-4}, 10^{-4}] \\ a_{36} \in [-10^{-4}, 10^{-4}], a_{37} \in [-10^{-4}, 10^{-4}] \\ a_{46} \in [-10^{-4}, 10^{-4}], a_{47} \in [-10^{-4}, 10^{-4}] \\ a_{56} \in [-10^{-3}, 10^{-3}], a_{57} \in [-10^{-4}, 10^{-4}] \\ a_{66} \in [-10^{-4}, 10^{-4}], a_{67} \in [-10^{-4}, 10^{-4}] \end{cases} \tag{28}$$

$$V \max_{ij} = 0.1(a \max_{ij} - a \min_{ij}) \tag{29}$$

where, $a \max_{ij}$ and $a \min_{ij}$ is the upper limit and the low limit of a_{ij}.

(3) Search the optimal parameters use PSOt.

6 Simulation Study

A space robot system is composed of a chaser spacecraft (Base), a 6 DOFs manipulator mounted on the spacecraft and an object as a target to be captured by the manipulator, as shown in Fig. 2. Let B_0 be the satellite, B_i ($i = 1,...,n$) the ith link of the manipulator in order. J_i is the joint which connects body i-1 with body i. The body fixed frames of the free-flying multi-body system are defined by the convention of the reference 0, as shown in Fig. 3. The frame of the base satellite (i.e. B_0) and the ith body frame are represented by {0} and {i} respectively. The vector a_i is the vector from J_{i-1} to the CM (center of mass) of B_{i-1} and b_i is the vector from the CM of B_i to J_i. a_i and b_i are expressed in {i}. ^iI_i is the inertia matrix of B_i respect to its CM, expressed in {i}. These values are shown in Table 1.

After capturing, the base spacecraft orientates at [0.5°, 9°, 0.5°], the joint angle of manipulator is q_{m0} = [2, –30, 30, –1, –225, 1] degree. The desired base attitude is [0, 0, 0] after the reorientation, and the desired joint angle of the manipulator is arrived at q_{mf} = [0, –55, 40, 0, –215, 0] degree at the same time. Because of the angular momentum preloaded in the manipulator during the approach the target, the manipulator moves with a joint rate $\dot{q}_{m0}$ = [0, –2, 2, 0, 3, 0] °/s after capturing. Using the method proposed above, the aim can be attained. The mechanical limits on the joints are given as:

$$-210^o \leq q_1 \leq 90^o, -240^o \leq q_2 \leq 60^o, -60^o \leq q_3 \leq 240^o$$
$$-150^o \leq q_4 \leq 150^o, -270^o \leq q_5 \leq 30^o, -150^o \leq q_6 \leq 150^o$$

The limits on the joint rates are:

$$|\dot{q}_1| \leq 50^o/s^2, |\dot{q}_2| \leq 50^o/s^2, |\dot{q}_3| \leq 50^o/s^2$$
$$|\dot{q}_4| \leq 50^o/s^2, |\dot{q}_5| \leq 50^o/s^2, |\dot{q}_6| \leq 50^o/s^2$$

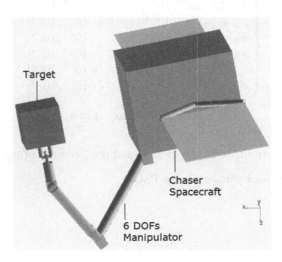

Fig. 2. A space robot system

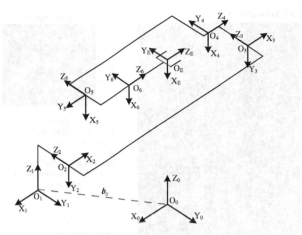

Fig. 3. The body fixed frame

Table 1. The mass properties of each body.

	B_0	B_1	B_2	B_3	B_4	B_5	B_6 (Before capture)	B_6 (After capture)
Mass	400	2.5	10	2.5	10	2.5	8	30
a_{ix}	0.5	0	0.5	0	0	0	0	0.0367
a_{iy}	0	0	0	0	0.1	0	0	0
a_{iz}	0	0.1	−0.2	0.1	−0.4	−0.1	0.1	0.5400
b_{ix}	0.5	−0.1	0.5	−0.1	0	0	0	−0.0367
b_{iy}	0	0	0	0	0	−0.1	0	0
b_{iz}	0.5	0	0.2	0	−0.4	0	0.2	−0.2400
Ixx	60	0.01	0.05	0.01	1	0.01	0.2	4.812
Iyy	70	0.01	2	0.01	1	0.01	0.2	5.3267
Izz	90	0.01	2	0.01	0.2	0.01	0.1	2.6147
$-Ixy$	0	0	0	0	0	0	0	0
$-Ixz$	0	0	0	0	0	0	0	−0.1760
$-Iyz$	0	0	0	0	−0.2	0	0	0

Particle Swarm Optimization (PSO) is used to search the optimal parameters. We set psoParams = [40 1600 24 2 2 0.9 0.4 1500 1e-25 250 NaN 0 0];

After the algorithm converges, the optimal parameters and object function are:

$$a^* = [-0.0371, -0.0010, 0.0262, 0.0082, 0.0197, -0.0444,$$
$$- 0.0439, -.0020, 0.3596, -0.0689, -0.0119, -0.0267] \times 10^{-3} \tag{30}$$

$$\hat{J}^* = 7.5432 \times 10^{-4} \tag{31}$$

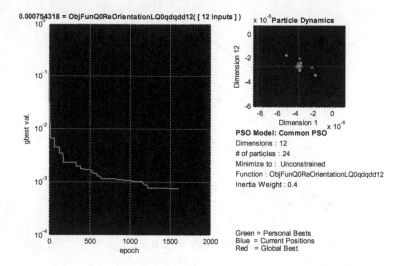

Fig. 4. Best values of the object function (PSOt)

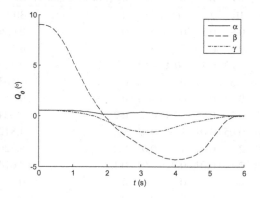

Fig. 5. The process of base attitude variation

According to the simulation result, Fig. 4 shows the best value of objective function (optimized use PSOt). The base attitude angles (z-y-x Euler angle) vary as shown in Fig. 5. We can see they change to the desired values ([0, 0, 0]). The planned joint trajectories and joint rates are shown as Figs. 6 and 7 respectively. We can see they are very smooth and applicable to the control of the manipulator. They are not beyond the domains of their limits.

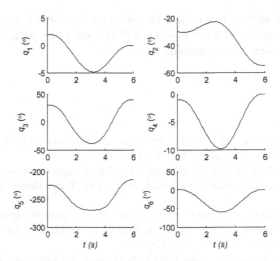

Fig. 6. The manipulator joint angles

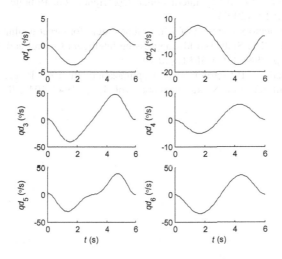

Fig. 7. The manipulator joint rates

7   Conclusion

A perfectible condition for reorientation after capturing is provided by the use of the capturing strategy with bias momentum. A new method is proposed to plan manipulator configuration and the base attitude at the same time using manipulator motion only based on the nonholonomic characteristics of free-floating space system. Optimal path planning for the reorientation of the entire system is the key issues of the paper. The parameters are solved based on Particle Swarm Optimization. After the trajectory is planned, a control strategy may be designed to obtain the desired result.

When the target is unknown, a certain identification algorithm is needed to determine the target mass property. Our future work will be focused on parameter identification and the stability issues of the method.

References

1. Liang Bin, X., Wenfu, L.C., et al.: The status and prospect of orbital servicing in the geostationary orbit. J. Astronaut. **31**(1), 1–13 (2010)
2. Nenchev, D.N., Yoshida, K.: Impact analysis and post-impact motion control issues of a free-floating space robot subject to a force impulse. IEEE Trans. Robot. Autom. **15**(3), 548–557 (1999)
3. Dimitrov, D.N., Yoshida, K.: Momentum distribution in a space manipulator for facilitating the post-impact control. In: IEEE/RSJ International Conference on Intelligent Robots and Systems, vol. 4, pp. 3345–3350. IEEE (2004)
4. Lampariello, R., Deutrich, K.: Simplified Path Planning for Free-Floating Robots, Internal Report DLR 515-99-04 (1999)
5. Papadopoulos, E., Tortopidis, I., Nanos, K.: Smooth planning for free-floating space robots using polynomials. In: IEEE International Conference on Robotics and Automation, pp. 4272–4277. IEEE (2005)
6. Xu, W., et al.: Trajectory planning of space robot system for target berthing and reorientation after capturing. In: The Sixth World Congress on Intelligent Control and Automation, 2006, WCICA 2006, pp. 8981–8985. IEEE (2006)
7. Kennedy, J., Eberhart, R.: Particle swarm optimization. In: 1995 Proceedings of the IEEE International Conference on Neural Networks, vol. 4, pp. 1942–1948. IEEE (2002)

An Iterative Calculation Method for Solve the Inverse Kinematics of a 7-DOF Robot with Link Offset

Shaotian Lu, Yikun Gu, Jingdong Zhao$^{(\boxtimes)}$, and Li Jiang

State Key Laboratory of Robotics and System,
Harbin Institute of Technology, Harbin 150080, China
llb908029@hit.edu.cn

Abstract. This paper presents an inverse kinematics solution method for a 7-degrees-of-freedom (DOF) redundant robot with offset shoulder and wrist joints. Aiming at the plan-points of the 7-DOF redundant robot, we propose an iterative method to solve the inverse kinematics for a specific application scenario, that is, the orientation matrices and arm angle keep the same values during the movement of robot which can be found in some specific operations, like painting on a plane. The position, orientation and arm angle of current plan-point are used to calculate the corresponding values of next plan-point. Through iterative calculation, the inverse kinematics solutions can be obtained with the characteristic that the orientation matrices and arm angle maintain the same values for all plan-points. The comparison and analytical results show this iterative solving method is effective.

Keywords: Redundant robot · Inverse kinematics · Arm angle

1 Introduction

Compared with non-redundant robot, redundant robot possesses many advantages, for instance, it can avoid singularity, avoid obstacle, carry out fault-tolerant operation and so on. The 7-DOF robot is one of the most attractive redundant robots. Researchers generally divide the 7 DOFs into three joints, the shoulder joint (3 DOFs), the elbow joint (1 DOF) and the wrist joint (3 DOFs). Some robots contained spherical joints, or the joint axis of the shoulder/wrist joint are parallel. There is also a kind of 7-DOF robot has offset configuration; the Space Station Remote Manipulator System (SSRMS) belongs to this type.

At present, there are many studies on the inverse kinematics of the non-offset Spherical-Roll-Spherical (SRS) 7-DOF robot. Typical examples are shown in [1–3]. However, the studies on the inverse kinematics for the offset configuration (offset mainly at the shoulder, elbow, and wrist joints) robot are relative less. Xu Wenfu et al. proposed two methods, the joint angle parameterized and arm angle parameterized method, to solve the inverse kinematics for SSRMS type 7-DOF redundant robot with offset configuration. The former inferred some expressions by using one joint angle as parameter according to the configuration characteristics and obtained 8 possible solutions.

© Springer International Publishing AG 2017
Y. Huang et al. (Eds.): ICIRA 2017, Part III, LNAI 10464, pp. 729–739, 2017.
DOI: 10.1007/978-3-319-65298-6_65

The latter firstly constructed a zero-offset SRS manipulator corresponding to the SSRMS type manipulator, then applied the arm angle as parameter to solve the inverse kinematics of SRS manipulator, and gained 8 inverse kinematics solutions for the offset configuration SSRMS type 7-DOF redundant robot according to the relationship between the SRS and SSRMS manipulators. The effectiveness of the proposed method was verified by an instance [4]. Yu Chao et al. proposed a method using a virtual joint to replace the offset-wrist of a 7-DOF manipulator, and formed a virtual joint with a sphere-wrist in order to solve the inverse kinematics of the 7-DOF redundant manipulators with offset-wrist. They supposed the joint angles of the virtual manipulator and the real manipulator keep the same all the time. The kinematic simulations verified the proposed method [5]. Singh et al. presented a method of computing the joint-variables for any geometric pose in order to gain the inverse kinematics for the 7-DOF Barrett Whole Arm Manipulator with link offsets. They gave a method to compute the joint variables for any geometric pose (position and orientation), and established a set of feasible poses for some joint angle constraints [6]. Crane et al. presented an efficient reverse analysis for three 6-DOF subchains of the 7-DOF SSRMS. There were at most eight different arm configurations in each of the three subchains. It was verified through choosing the orientation of the plane containing two longest links (the upper arm and forearm) to avoid collisions with obstacles [7]. Luo et al. proposed an analytical inverse kinematics method for a 7-DOF manipulator with offset at the shoulder and wrist joint. The key point of this method was using different joint as redundancy based on parameterization method to obtain closed-form inverse kinematic solution. Then a proper redundant joint was chosen as redundancy at different tip position and orientation by setting redundant rotational joint and observing elbow displacement per degree. The effectiveness of the proposed method was verified by both simulation and experiments on a modularized 7-DOF manipulator [8]. Jiang Li et al. presented an integrated method in order to obtain the inverse kinematics of the 7-DOF humanoid arm with offset wrist. The simulation results demonstrated the integrated method has better precision and optimization during trajectory tracking [9].

In the aforementioned inverse kinematics studies, many researchers choose the arm angle as a parameter because of its evident geometric significance. It can fully reflect the redundancy characteristics of the 7-DOF series robot. In addition, previous studies mainly focus on the non-offset SRS robot; the studies on the inverse kinematics of the offset configuration robot are still relatively less, so it is necessary to carry out in-depth study in this area.

When carrying out some specific operation tasks, like painting on a plane, the orientation matrices and arm angle do not need to change during the movement of a 7-DOF robot. For this application scenario, an iterative method for solving the inverse kinematics of a 7-DOF redundant robot with offset at the shoulder, elbow and wrist joints is proposed in this paper. For the plan-points in Cartesian space, when the inverse kinematics of the former plan-point is known, the position, orientation and arm angle of former plan-point can also be obtained. This information is utilized to calculate the corresponding values of next plan-point. By these iterative processes, the inverse kinematics for the next plan-point can be uniquely obtained while the orientation matrices and arm angle of the plan-points maintain the same values. This method provides a new idea for real-time motion planning and control for 7-DOF offset type robot.

The remainder of this paper is organized as follows: A 7-DOF offset configuration robot is presented in Sect. 2. Section 3 describes the configuration control and arm angle methods used to solve the inverse kinematics of the 7-DOF robot. Section 4 presents the inverse kinematics results. Section 5 presents the conclusions of this study.

2 The Kinematics Modeling of the 7-DOF Redundant Robot

A 7-DOF offset configuration robot is shown in Fig. 1. The parameters of the robot are shown in Table 1. This robot has 7 modular joints. It resembles the configuration of the Space Station Remote Manipulator System (SSRMS). In order to calculate the transformation matrix more conveniently, the coordinate system $(x_0'\ y_0'\ z_0')$ is added. Firstly, it is necessary to calculate the transformation matrix $^0_{0'}T$ from the base coordinate system $(x_0\ y_0\ z_0)$ to the coordinate system $(x_0'\ y_0'\ z_0')$. Thus, the subsequent transformation matrix can be calculated based on $^0_{0'}T$. The pose transformation matrix between adjacent coordinate systems is defined as follows:

$$
^{i-1}_iT = \begin{bmatrix} c\theta_i & -s\theta_i & 0 & a_{i-1} \\ s\theta_i c\alpha_{i-1} & c\theta_i c\alpha_{i-1} & -s\alpha_{i-1} & -d_i s\alpha_{i-1} \\ s\theta_i s\alpha_{i-1} & c\theta_i s\alpha_{i-1} & c\alpha_{i-1} & d_i c\alpha_{i-1} \\ 0 & 0 & 0 & 1 \end{bmatrix} \tag{1}
$$

where $c = cos$, $s = sin$.

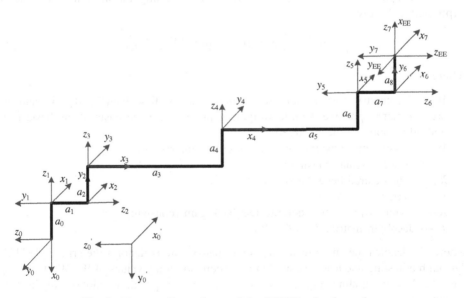

Fig. 1. The coordinate frame of the 7-DOF redundant robot

Table 1. Parameters of the 7-DOF redundant robot

Number	a_{i-1}	α_{i-1}	d_i	θ_i
1	0	90°	a_0	θ_1 (0°)
2	0	90°	a_1	θ_2 (0°)
3	0	−90°	a_2	θ_3 (−90°)
4	a_3	0°	a_4	θ_4 (0°)
5	a_5	0°	a_6	θ_5 (90°)
6	0	90°	a_7	θ_6 (0°)
7	0	−90°	a_8	θ_7 (0°)
8	0	90°	0	θ_8 (90°)

3   The Inverse Kinematics Analysis of the 7-DOF Redundant Robot

3.1   Configuration Control Method

The configuration control method is an effective method for solving the inverse kinematics of redundant robot [10, 11]. It can ensure the inverse kinematics solution is unique and especially suitable for the robot performing cyclic motion. In addition, this method can be calculated fast and it is particularly suitable for real-time control of redundant robots. In this paper, this method is used to resolve the inverse kinematics of the 7-DOF redundant robot.

The damped-least-squares (DLS) method of the configuration control scheme is expressed as follows:

$$\dot{q} = W_v^{-1} J^T [J W_v^{-1} J^T + \lambda^2 W^{-1}]^{-1} (\dot{X}_d + K E_e) \tag{2}$$

where

W——symmetric positive-definite weighting matrix, $W = \text{diag}[W_e, W_c]$, W_e and W_c are symmetric positive-definite weighting matrices for the basis of the basic task and additional task, respectively;
W_v——symmetric positive-definite weighting matrix;
λ——positive scalar constant;
X_d——the desired behavior of the robot;
E_e——error, $E_e = X_d - X$;
K——symmetric positive-definite feedback gain (constant);
J——Jacobian matrix, $J = (J^{ee}, J^\psi)^T$.

where J^{ee} denotes the end-effector Jacobian matrix, and ψ denotes the arm angle [12]. Through adjusting and test we can obtain the corresponding values of W_e, W_c, W_v, λ, X_d and K. Via integrating Eq. (2) we can obtain the position solution of inverse kinematics.

3.2   Solving Arm Angle

The arm angle is an important parameter which is often used to solve the inverse kinematics of redundant robot with offset configuration. It is defined as the angle between the reference plane and the current plane, as shown in Fig. 2.

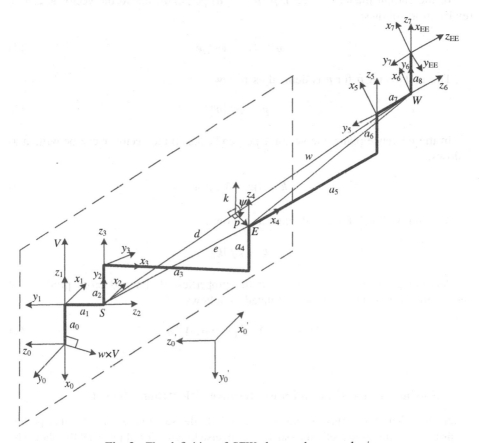

Fig. 2. The definition of SEW plane and arm angle ψ

In Fig. 2, we define the intersection between the axes of joint 2 and joint 3 as the point S, the coordinate origin of the joint 4 as the point E, and the intersection between the axes of joint 6 and joint 7 as the point W. The plane formed by points S, E and W is called the current plane. And the position vectors from the reference coordinate origin S to points E and W are e and w, respectively. The unit vector V ($V = [0, 0, 1]^T$) is parallel with the axis of the joint 1. The plane formed by vector V and point W is defined as the reference plane. According to the right hand rule, the reference plane rotates around the vector w until it coincides with the current plane. The obtained rotation angle ψ is defined as the arm angle. Let $\hat{w} = w/\|w\|$, and the projection of the vector e on the vector w is expressed as follows:

$$d = \hat{w}(\hat{w}^{T}e) \tag{3}$$

In the current plane, the vector p ($p = e-d$) perpendicular to the vector w can be rewritten as follows:

$$p = (I - \hat{w}\hat{w}^{T})e \tag{4}$$

The unit vector $\hat{p}$ for p is defined as follows:

$$\hat{p} = p/\|p\| \tag{5}$$

In the reference plane, the vector k perpendicular to the vector w can be written as follows:

$$k = (w \times V) \times w \tag{6}$$

The unit vector $\hat{k}$ for k is defined as follows:

$$\hat{k} = k/\|k\| \tag{7}$$

According to the Eqs. (3)–(7) and the properties of dot and cross products of vector, the arm angle ψ can be calculated as follows:

$$\Psi = \mathrm{atan2}(\hat{w}^{T}(V \times p), V^{T}p) \tag{8}$$

3.3   Solving Inverse Kinematics by Iterative Calculation Method

When the orientation matrices and arm angle hold the same values for all plan-points, we propose an iterative method to solve the inverse kinematics of the 7-DOF robot. The flowchart for describing the proposed inverse kinematics method is shown in Fig. 3. When applying Eq. (2) to solve the inverse kinematics, the error e can be obtained as follows:

$$E_e = \left[e_o; e_p; e_\Psi\right]^{T}, e_o = 1/2(n \times n_d + o \times o_d + a \times a_d) \tag{9}$$

where $e_p = p_d - p$, $e_\Psi = \psi_d - \psi$; n, o and a denote the rotary vectors; h denotes the position vector; ψ denotes the arm angle; n_d, o_d and a_d denote the desired rotary vectors; h_d denotes the desired position vector; ψ_d denotes the desired arm angle.

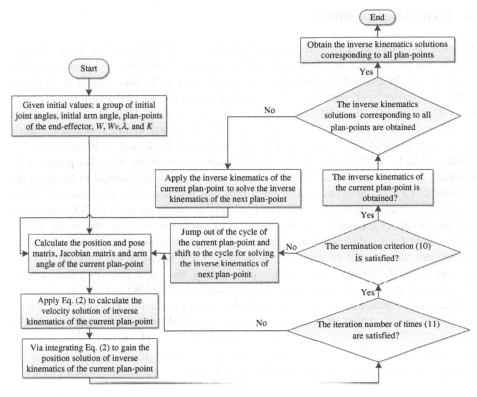

Fig. 3. The flowchart for describing the proposed inverse kinematics method

The termination criterion is introduced as follows:

$$\|Ee\| \leq \varepsilon_e \tag{10}$$

where ε_e denotes the constraint error.

In addition, in order to ensure the inverse kinematics of each plan-point can be obtained as soon as possible when it satisfies the termination criterion, to prevent falling into dead loop or wasting too much time, the maximum iteration number of times $inter_{max}$ is introduced. Namely, when $inter_{max}$ is reached but the termination criterion is not satisfied during solving certain planned point, output the current result directly, jump out of the cycle for solving the inverse kinematics of current point, and shift to the cycle for solving the inverse kinematics of next point. The constraints for $inter_{max}$ are defined as follows:

$$i \leq inter_{max} \tag{11}$$

where i denotes the current number of times of iteration.

4 Experimental Study

In order to verify the algorithm proposed in Sect. 3, a test task is set up in experiment, that is, the end-effector tracks a circle with a constant orientation and arm angle. The robot will be in a singular situation when we directly use the joint angles as shown in Table 1. In order to prevent this, the initial joint angles are selected as follows: $\theta_1 = 0°$, $\theta_2 = 0°$, $\theta_3 = -90°$, $\theta_4 = 60°$, $\theta_5 = 90°$, $\theta_6 = 0°$ and $\theta_7 = 0°$. Thus, the initial position of the end-effector is $(-2679.2, -2173.7, -3765.0)$ (mm) in the coordinate frames $(x_0\ y_0\ z_0)$. The test task in experiment is to track a circle. Its radius R and center are 1000 mm and $(-2679.2, -2173.7, -2765.0)$ (mm), respectively. The plan-points at the tracking circle are calculated from the central angle, which is assumed to be along the counter-clockwise direction and divided into uniform acceleration section $(0°-60°)$, uniform velocity section $(60°-300°)$ and uniform deceleration section $(300°-360°)$ for a total of three segments. At the uniform acceleration section, the robot begins accelerating from initial position with the angular acceleration of $120/(1/80 \times n)2$ °/s2 (n denotes the amount of planned via-points). Then it enters the uniform velocity section when the central angle equals $60°$. Finally the robot goes into the uniform deceleration section with the angular deceleration of $120/(1/80 \times n)2$ °/s2 when the central angle equals $300°$. It decelerates until returning to the initial point. Here, we set via-points $n = 8$. Other parameters are defined as follows: $W_e = \text{diag}[1,1,1,1,1,1]$, $inter_{max} = 10000$, $W_c = 1$, $\lambda = 1$, $W_v = \text{diag}[1,1,1,1,1,1,1]$, $\varepsilon_e = 10^{-10}$, $K = \text{diag}[1,1,1,1,1,1,1]$, $\dot{X}_d = (0,0,0,0,0,0,0)^T$. The pose matrix of the end-effector of the 7-DOF robot at the initial position is given as follows:

$$
{}^0_8T = \begin{bmatrix}
-1.0000 & -0.0000 & -0.0000 & -2.6792 \\
0 & 0.5000 & -0.8660 & -2.1737 \\
0.0000 & -0.8660 & -0.5000 & -3.7650 \\
0 & 0 & 0 & 1.0000
\end{bmatrix}
\tag{12}
$$

The arm angle of the robot at the initial point is $87.0751°$.

When Eq. (2) is applied to solve the inverse kinematics of the robot, the obtained inverse kinematics solutions for each plan-point are shown in Table 2. Point 1 coincides with point 9 when the end-effector of the robot finishes a cyclic motion. Hence,

Table 2. Inverse kinematics solutions of the proposed iterative calculation method

Joint	Plan-points (°)								
	1	2	3	4	5	6	7	8	9
1	0	-9.64	-32.66	-37.74	-15.44	4.21	20.26	9.80	-0.14
2	0	0.00	-0.00	-0.00	-0.00	0.00	0.00	0.00	0.00
3	-90.00	-85.80	-79.19	-79.02	-77.37	-79.99	-95.51	-94.99	-89.78
4	60.00	65.70	88.73	116.13	112.24	81.75	55.89	55.14	58.95
5	90.00	91.35	91.92	84.66	75.24	79.18	87.45	93.33	95.52
6	0	-0.00	0.00	0.00	0.00	0.00	-0.00	-0.00	-0.00
7	0	-1.61	-8.79	-24.03	-34.67	-25.15	-8.08	-3.29	-4.56

Table 3. Data of plan-points and the results of forward kinematics verification according to relevant points

Circular central angle (rad)	Plan-points of the end effector (m)			Forward kinematics verification (m)		
	X axis	Y axis	Z axis	X axis	Y axis	Z axis
0	−2.6792	−2.1737	−3.7650	−2.6792	−2.1737	−3.7650
0.2618	−2.6792	−1.9149	−3.7309	−2.6792	−1.9149	−3.7309
1.0472	−2.6792	−1.3077	−3.2650	−2.6792	−1.3077	−3.2650
2.0944	−2.6792	−1.3077	−2.2650	−2.6792	−1.3077	−2.2650
3.1416	−2.6792	−2.1737	−1.7650	−2.6792	−2.1737	−1.7650
4.1888	−2.6792	−3.0397	−2.2650	−2.6792	−3.0397	−2.2650
5.2360	−2.6792	−3.0397	−3.2650	−2.6792	−3.0397	−3.2650
6.0214	−2.6792	−2.4325	−3.7309	−2.6792	−2.4325	−3.7309
6.2832	−2.6792	−2.1737	−3.7650	−2.6792	−2.1737	−3.7650

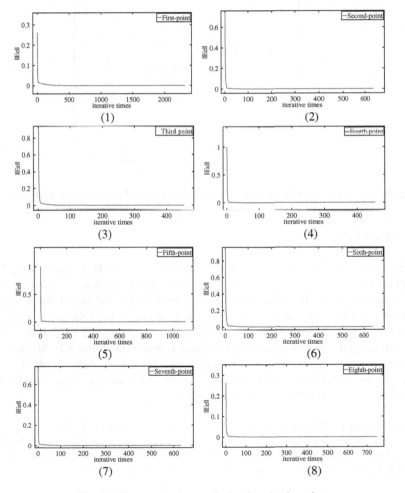

Fig. 4. The termination analysis of each via-point

there are nine plan-points while the number of via-points is eight. Through bringing the inverse kinematics solutions into the corresponding transformation matrix, we can find that the orientation matrices at all the plan-points are exactly the same as that at the initial point. And the arm angles at all plan-points are 87.0751° after implementing relational calculation, which is also exactly the same as that at the initial point.

The data about plan-points and the results of forward kinematics verification according to the obtained inverse kinematics at the plan-points are shown in Table 3. In order to demonstrate the comparative effect more directly, the comparison between plan-points and forward kinematics verification results are shown in Fig. 4. The termination analysis of each via-point is shown in Fig. 5. The norm of error $\|E_e\|$ converges to the threshold quickly for all plan- points, which verify the efficiency of the proposed method.

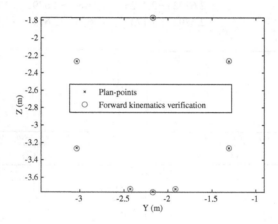

Fig. 5. The comparison between plan-points and forward kinematics verification results

5   Conclusions

This paper proposes an iterative method to solve the inverse kinematics of the 7-DOF redundant robot with link offset. The objective of this method is through iterative calculation to realize the inverse kinematics solution for redundant robot when the orientation matrices and arm angle of all the plan-points hold the same values. The simulation results show that the inverse kinematics solution can be obtained quickly when satisfying the termination criterion (error is under the threshold). Meanwhile the orientation matrices and arm angle are maintained the same values for all plan-points. Moreover, this method has some advantages, such as it is simple, it has clear geometric significance, and it can be easily implemented.

Acknowledgments. This work is supported in part by the Foundation for Innovative Research Groups of the National Natural Science Foundation of China (Grant No. 51521003), the National Natural Science Foundation of China (NO. 61603112) and the Self-Planned Task (NO. SKLRS201721A) of State Key Laboratory of Robotics and System (HIT).

References

1. Xu, W.F., Lei, Y., Wang, Z.Y.: Dual arm-angle parameterisation and its applications for analytical inverse kinematics of redundant manipulators. Robotica, 1–20 (2015)
2. Shimizu, M., Kakuya, H., Yoon, W.K., Kitagaki, K., Kosuge, K.: Analytical inverse kinematic computation for 7-DOF redundant manipulators with joint limits and its application to redundancy resolution. IEEE Trans. Robot. 24(5), 1131–1142 (2008)
3. Zu, D., Wu, Z.W., Tan, D.L.: Effecient inverse kinematic solution for redundant manipulators. Chin. J. Mech. Eng. 41(6), 71–75 (2005)
4. Xu, W.F., Zhang, J.T., Yan, L., Wang, Z.Y.: Parameterized inverse kinematics resolution method for a redundant space manipulator with link offset. J. Aeronaut. 36(1), 33–39 (2015)
5. Yu, C., Jin, M.H., Liu, Y.: An analytical solution for inverse kinematic of 7-DOF redundant manipulators with offset-wrist. In: Proceedings of the 2012 International Conference on Mechatronics and Automation (ICMA), Chengdu, pp. 92–97 (2012)
6. Singh, G.K., Claassens, J.: An analytical solution for the inverse kinematics of a redundant 7DOF manipulator with link offsets. In: IEEE/RSJ International Conference on Intelligent Robots and Systems (IROS), TaiPei, pp. 2976–2982 (2010)
7. Crane III, C.D., Duffy, J.: A kinematic analysis of the space station remote manipulator system (SSRMS). J. Robot. Syst. 8(5), 637–658 (1990)
8. Luo, R.C., Lin, T.W., Tsai, Y.H.: Analytical inverse kinematic solution for modularized 7-DOF redundant manipulators with offsets at shoulder and wrist. In: IEEE/RSJ International Conference on Intelligent Robots and Systems (IROS), Chicago, pp. 516–521 (2014)
9. Jiang, L., Huo, X.J., Liu, Y.W., Liu, H.: An integrated inverse kinematic approach for the 7-DOF humanoid RRM with offset wrist. In: Proceeding of the IEEE International conference on Robotics and Biomimetics (ROBIO), Shenzhen, pp. 2737–2742 (2013)
10. Seraji, H.: Configuration control of redundant manipulators: theory and implementation. IEEE Trans. Robot. Autom. 5(4), 472–490 (1989)
11. Glasst, K., Colbaught, R., Lim, D., Seraji, H.: On-line collision avoidance for redundant manipulators. In: Proceedings of the IEEE International Conference on Robotics and Automation, Atlanta, pp. 36–43 (1993)
12. Kenneth, K.D., Long, M., Seraji, H.: Kinematic analysis of 7-DOF manipulators. Int. J. Robot. Res. 11(4), 469–481 (1992)

Kinematic Nonlinear Control of Aerial Mobile Manipulators

Víctor H. Andaluz$^{(\boxtimes)}$, Christian P. Carvajal, José A. Pérez, and Luis E. Proaño

Universidad de las Fuerzas Armadas ESPE, Sangolquí, Ecuador
vhandaluz1@espe.edu.ec, chriss2592@hotmail.com,
joansll@hotmail.com, luis.e.proa@gmail.com

Abstract. This work proposes a kinematic modeling and a kinematic nonlinear controller for an autonomous aerial mobile manipulator robot that generates saturated reference velocity commands for accurate waypoint path following. The dynamic compensation controller is considered through of a helicopter-inner-loop system to independently track four velocity commands: forward, lateral, up/downward, and heading rate; and arm-inner-loop system to independently track angular velocity commands. Stability and robustness of the complete control system are proved through the Lyapunov method. Finally, simulation results are presented and discussed, which validate the proposed controller.

Keywords: Aerial mobile manipulators · Kinematic modeling · Nonlinear controller · Lyapunov · Non-linear systems

1 Introduction

At present the area of robotics has presented numerous advances in literature and in the area of development, with the incursion of new techniques and technology that allow to improve the mobility of robots. Among the main subjects of academic research are mobile manipulators, since they allow sophisticated tasks and access to places of difficult access, highlighting unmanned aerial manipulators (UAVs), mobility in the three axes, managing to expand the functionalities as: (i) construction [1]; (ii) freight transport [2]; (iii) electric laying applications [3]; (iv) dangerous tasks that affect the human being, among others [4, 5].

Mobile robots have been developed for terrestrial, aerial and aquatic environments [6], in literature and in practice the combination of mobile platforms with robotic arms denominated as mobile manipulators is performed which perform complex tasks and its applications are innumerable in the military, commercial, domestic, mobility, among others [7]. To perform the study and control of these systems two methods are obtained: (i) kinematic model separately, i.e. the kinetic model and the control of the mobile platform and the manipulator is performed independently; (ii) kinematic study together, i.e. modeling and kinematic control is performed from the attached system, for the model and control is performed with respect to the final effector of the mobile manipulator.

© Springer International Publishing AG 2017
Y. Huang et al. (Eds.): ICIRA 2017, Part III, LNAI 10464, pp. 740–749, 2017.
DOI: 10.1007/978-3-319-65298-6_66

Different ways to control mobile manipulators have been studied and some control strategies have been developed to solve the path-tracking problem, which has resulted in a wide field of research and varied range of applications [9]. Being $\mathcal{P}(s)$ a path to follow during the path to be followed is expressed by a non-parameterized equation in the time defined in the 3 axes, it is necessary to implement controls that solve this problem, in the literature there are several control algorithms for the path tracking where υ_P is the desired speed To follow the path, this is considered as one of the inputs for control [10]. It is also worth mentioning algorithms based on sampling which are based on the random generation of the states to calculate the path, by sampling the configuration space [1]. Other form for the road tracking of the aerial mobile manipulators is based on the detection of images of the environment through artificial vision sensors that detect objects that can be differentiated by their forms, this allows the tracking of the destination through the captured visual information [11].

This paper presents a robust non-linear control strategy for resolving the path following problem of a aerial mobile manipulator. Which is constituted by an helicopter mounting a robotic arm of 3 degrees of freedom mounted on back of base. For the design of the controller, the kinematic model of the aerial mobile manipulator is used which has as input the speed and orientation, this controller is designed in two blocks: (i) the first one is a kinematic controller which uses saturation commands of speed based on the aerial mobile manipulator's kinematic and (ii) the second one is a dynamic compensation controller for the independent monitoring of the seven speed orders, 4 corresponding to the aerial platform: forward, lateral, up/downward and orientation, the last 3 are those who command the Manipulator. It is also pointed out that the workspace has a single reference that is located in the operative end of the aerial mobile manipulator $<R(x\,y\,z)>$. The stability of the controller is analyzed by the Lyapunov's method and to validate the proposed control algorithm, experimental processes are presented and discussed in this paper.

The paper is organized in 5 sections including de Introduction, in Sect. 2 is presented the kinematic modeling of the aerial mobile manipulator; the design of the kinematic controller, stability and robustness analyses is presents in Sect. 3; while that the results are presented and discussed in Sect. 4 and finally, conclusions are given in Sect. 5.

2 Helicopter and Robotic Arm

The kinematic model of the aerial mobile manipulator is composed by a set of seven velocities represented at the spatial frame $<\Upsilon>$. The displacement of the aerial mobile manipulator is guided by the three linear velocities v_x, v_y and v_z defined in a rotating right-handed spatial frame $<\Upsilon>$, and the angular velocity ω, as shown in Fig. 1.

Each linear velocity is directed as one of the axes of the frame $<\Upsilon>$ attached to the center of gravity of the helicopter: v_x points to the frontal direction; v_y points to the left-lateral direction, and v_z points up. The angular velocity ω rotates the referential system $<\Upsilon>$ counterclockwise, around the axis Υ_Z (considering the top view). While the maneuverability of the robotic arm is defined by three angulars velocities with

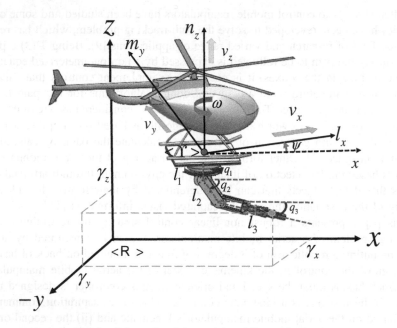

Fig. 1. Schematic of the aerial mobile manipulator.

respect to the reference system $<\Upsilon>$, *i.e.*, the Cartesian motion of the aerial mobile manipulator is defined by, γ_x, γ_y, γ_z and ψ are all measured with respect to the inertial frame $<\mathcal{R}>$; the point of interest (whose position is being controlled) is the end/effector of the aerial mobile manipulator model can be written in compact form as $\dot{\gamma} = f(\mathbf{q}, \psi)\mathbf{v}$, *i.e.*,

$$\dot{\gamma}(t) = \mathbf{J}(\mathbf{q}, \psi)\mathbf{v}(t) \tag{1}$$

where, $\mathbf{J}(\mathbf{q}, \psi) \in \Re^{m \times n}$ with $m = 3$ and $n = 7$ represents the Jacobian matrix that defines a linear mapping between the velocity vector of the aerial mobile manipulator $\mathbf{v} \in \Re^n$ where $\mathbf{v} = [v_x\, v_y\, v_z\, \omega\, \dot{q}_1\, \dot{q}_2\, \dot{q}_3]^T$ and the velocity vector of the operative end $\dot{\gamma} \in \Re^m$ where $\dot{\gamma} = [\dot{\gamma}_x\, \dot{\gamma}_y\, \dot{\gamma}_z]^T$.

3 Controller Design and Stability Analysis

As represented in Fig. 2, the path to be followed is denoted as $\mathcal{P}(s)$; the actual desired location $P_D = (x_\mathcal{P}(s_D), y_\mathcal{P}(s_D), z_\mathcal{P}(s_D))$ is defined as the closest point on $\mathcal{P}(s)$ to the end-effector of the aerial mobile manipulator, with s_D being the curvilinear abscissa defining the point P_D; $\tilde{\gamma}_x = x_\mathcal{P}(s_D) - \gamma_x\psi$ is the position error in the $\mathcal{X}$ direction; $\tilde{\gamma}_y = y_C(s_D) - \gamma_y\psi$ is the position error in the $\mathcal{Y}$ direction; $\tilde{\gamma}_z = z_C(s_D) - \gamma_z\psi$ is the position error in the $\mathcal{Z}$ direction; ρ represents the distance between the end-effector position of the aerial mobile manipulator $\gamma(x, y, z)$ and the desired point P_D on inertial frame $<\mathcal{R}(\mathcal{X}, \mathcal{Y}, \mathcal{Z})>$.

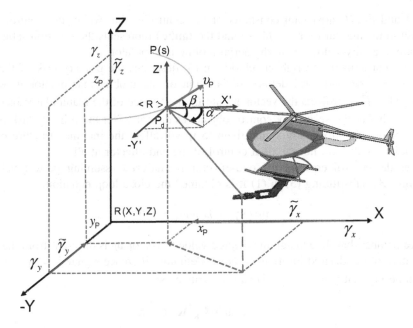

Fig. 2. Problem of control.

The controller proposed to solve the path following problems of the aerial mobile manipulator aims to calculate at every time $\tilde{\gamma}_x(t)$, $\tilde{\gamma}_y(t)$ and $\tilde{\gamma}_z(t)$ and use these measures to drive both the helicopter and the robotic arm in a direction which decreases the control errors. The proposed kinematic controller is based on the kinematic model of the aerial mobile manipulator (1). Hence following control law is proposed,

$$
\begin{bmatrix} v_x \\ v_y \\ v_z \\ \omega \\ \dot{q}_1 \\ \dot{q}_2 \\ \dot{q}_3 \end{bmatrix} = \mathbf{J}^{\#} \left(\begin{bmatrix} |\upsilon_P| \cos(\alpha)\cos(\beta) \\ |\upsilon_P| \cos(\alpha)\sin(\beta) \\ |\upsilon_P| \sin(\alpha) \end{bmatrix} + \begin{bmatrix} l_x \tanh\left(\frac{k_x}{l_x}\tilde{\gamma}_x\right) \\ l_y \tanh\left(\frac{k_y}{l_y}\tilde{\gamma}_y\right) \\ l_z \tanh\left(\frac{k_z}{l_z}\tilde{\gamma}_z\right) \end{bmatrix} \right) + \left(\mathbf{I} - \mathbf{J}^{\#}\mathbf{J}\right) \begin{bmatrix} k_{\eta v_x}\tanh(v_{xobs}) \\ k_{\eta v_y}\tanh(v_{obs}) \\ k_{\eta v_z}\tanh(v_{zobs}) \\ k_{\eta\omega}\tanh(\omega_{obs}) \\ k_{\tilde{q}_1}\tanh(\tilde{q}_1) \\ k_{\eta\tilde{q}_2}\tanh(\tilde{q}_2) \\ k_{\eta\tilde{q}_3}\tanh(\tilde{q}_3) \end{bmatrix}
\tag{2}
$$

Equation (2) can be represented in compact form by:

$$
\mathbf{v} = \mathbf{J}^{\#}\left(\upsilon_P + \mathbf{L_K}\tanh\left(\mathbf{L_K^{-1}K}\,\tilde{\gamma}\right)\right) + \left(\mathbf{I} - \mathbf{J}^{\#}\mathbf{J}\right)\mathbf{K}_{\eta}\tanh(\boldsymbol{\eta})
\tag{3}
$$

where υ_P is the reference velocity input of the aerial mobile manipulator for the controller; $\mathbf{J}^{\#}$ is the matrix of pseudoinverse kinematics for the aerial mobile manipulator where $\mathbf{J}^{\#} = \mathbf{W}^{-1}\mathbf{J}^{\mathrm{T}}\left(\mathbf{J}\mathbf{W}^{-1}\mathbf{J}^{\mathrm{T}}\right)^{-1}$ with $\mathbf{W}$ being a definite positive matrix that weighs the control actions of the system; while that $l_x > 0$, $k_x > 0$, $l_y > 0$, $k_y > 0$,

$l_z > 0$ and $k_z > 0$ area gain constants of the controller that weigh the control error respect to the inertial frame $<\mathcal{R}>$; and the **tanh**(.) represents the function saturation of maniobrability velocities in the aerial mobile manipulator.

The first term of the right hand side in (3) describes the primary task of the end effector. The second term defines self-motion of the mobile manipulator in which matrix $\left(\mathbf{I} - \mathbf{J}^{\#}\mathbf{J}\right)$ projects the vector $\boldsymbol{\eta}$ onto the null space of the manipulator Jacobian $\mathcal{N}(\mathbf{J})$ such that the secondary control objectives do not affect the primary task of the end-effector. Therefore, any value given to $\boldsymbol{\eta}$ will affect the internal structure of the manipulator only, but not the final control of the end-effector at all.

For the control error of the end-effector is analyzed assuming perfect velocity tracking. By substituting (3) in (1) it is obtained the close loop equation,

$$(\upsilon_P - \dot{\gamma}) + \mathbf{L}\tanh(\tilde{\gamma}) = \mathbf{0} \tag{4}$$

Remember that, in general, the desired velocity vector υ_P is different from the time derivative of the desired location $\dot{\gamma}_{\mathbf{d}}$. Now, defining difference signal $\boldsymbol{\xi} = \dot{\gamma}_{\mathbf{d}} - \upsilon_P$ and remembering that $\dot{\tilde{\gamma}} = \dot{\gamma}_{\mathbf{d}} - \dot{\gamma}$, (2) can be written as:

$$\dot{\tilde{\gamma}} + \mathbf{L_K}\tanh\left(\mathbf{L_K^{-1}K}\tilde{\gamma}\right) = \boldsymbol{\xi}. \tag{5}$$

Remark 1: υ_P is collinear to $\dot{\gamma}_{\mathbf{d}}$ (tangent to the path), then $\boldsymbol{\xi}$ is also a collinear vector to υ_P and $\dot{\gamma}_{\mathbf{d}}$.

For the stability analysis the following Lyapunov candidate function is considered $V(\tilde{\gamma}) = \frac{1}{2}\tilde{\gamma}^{\mathbf{T}}\tilde{\gamma}$. Its time derivative on the trajectories of the system is, $\dot{V}(\tilde{\gamma}) = \tilde{\gamma}^{\mathbf{T}}\boldsymbol{\xi} - \tilde{\gamma}^{\mathbf{T}}\mathbf{L_K}\tanh\left(\mathbf{L_K^{-1}K}\tilde{\gamma}\right)$. A sufficient condition for $\dot{V}(\tilde{\gamma}) < 0$ to be negative definite is,

$$\left|\tilde{\gamma}^{\mathbf{T}}\mathbf{L_K}\tanh\left(\mathbf{L_K^{-1}K}\tilde{\gamma}\right)\right| > \left|\tilde{\gamma}^{\mathbf{T}}\boldsymbol{\xi}\right| \tag{6}$$

For large values of $\tilde{\gamma}$, the condition in (6) can be reinforced as,

$$\left\|\tilde{\gamma}^{\mathbf{T}}\mathbf{L_K'}\right\| > \|\tilde{\gamma}\|\|\boldsymbol{\xi}\|$$

with $\mathbf{L_K'} = \mathbf{L_K}\tanh\left(k_{aux}\mathbf{i}\right)$, where k_{aux} is a suitable positive constant and $\mathbf{i} \in \Re^m$ is the vector of unity components. Then, $\dot{V}$ will be negative definite only if

$$\|\mathbf{L}\| > \frac{\|\boldsymbol{\xi}\|}{\tanh\left(k_{aux}\right)} \tag{7}$$

hence, (7) establishes a design condition to make the errors $\tilde{\gamma}$ to decrease.

Now, for small values of $\tilde{\gamma}$, condition (6) will be fulfilled if (see Fig. 3)

$$\left\|\tilde{\gamma}^T\mathbf{K}\frac{\tanh\left(k_{aux}\right)}{k_{aux}}\tilde{\gamma}\right\| > \|\tilde{\gamma}\|\|\boldsymbol{\xi}\|$$

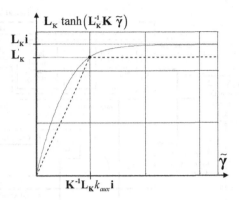

Fig. 3. Saturation function tanh(.) in solid line, and linear functions below tanh(.) in dash line.

which means that a sufficient condition for $\dot{V}(\tilde{\gamma}) < 0$ to be negative definite is,

$$\|\tilde{\gamma}\| > \frac{\|\xi\|}{\lambda_{\min}(\mathbf{K}) \tanh(k_{aux})}$$

thus implying that the error $\tilde{\gamma}$ is ultimately bounded by,

$$\|\tilde{\gamma}\| \leq \frac{k_{aux}\|\xi\|}{\varsigma\lambda_{\min}(\mathbf{K}) \tanh(k_{aux})} \text{; with } 0 < \varsigma < 1 \tag{8}$$

For the case of path following, the desired velocity is $\mathbf{v}_{\gamma d} = \dot{\gamma}_{\mathbf{d}} - \xi$. Once the control error is inside the bound (8), that is with small values of $\tilde{\gamma}$, $\mathbf{L_K} \tanh\left(\mathbf{L_K^{-1}K} \tilde{\gamma}\right) \approx \mathbf{K} \tilde{\gamma}$. Now, we prove by contradiction that this control error tends to zero. The closed loop Eq. (5) can be written as, $\dot{\tilde{\gamma}} + \mathbf{K} \tilde{\gamma} = \xi$ or after the transient, in Laplace transform,

$$\dot{\tilde{\gamma}}(s) = \frac{1}{s\mathbf{I} + \mathbf{K}}\xi(s) \tag{9}$$

According to (9) and recalling that $\mathbf{K}$ is diagonal positive definite, the control error vector $\tilde{\gamma}$ and the velocity vector ξ can not be orthogonal. Nevertheless both vectors are orthogonal by definition (see *Remark* 1 and remember the minimum distance criteria for $\gamma_{\mathbf{d}}$ on $\mathbf{P}$). Therefore the only solution for steady state is that $\tilde{\gamma}(t) \rightarrow \mathbf{0}$ asymptotically.

4   Results and Discussions

This section presents the simulation results of the waypoint tracking flight task in the 3D space using the kinematic nonlinear controller designed in the previous section. The goal of the simulations is to test the stability and performance of the proposed

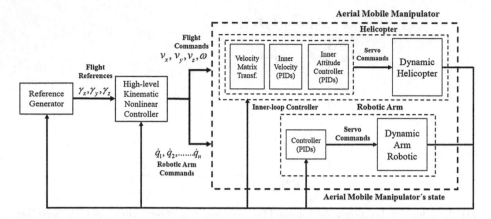

Fig. 4. Block diagram of the simulation system.

controller. Figure 4 represents the block diagram of the simulation system. The helicopter model used in the simulation is a very realistic nonlinear dynamic model of a small-scale helicopter found in [20]. The helicopter model considers not-ideal dynamics, such as flapping, drag, and actuator dynamics, and it describes accurately the system's dynamics both for hovering and for low speed translational flights. In the simulations, the helicopter's nominal parameters were used (which refer to *MIT's X-Cell .60* acrobatic helicopter); mientras que para el brazo robótico se considera la dinámica del *AX 12 A Smart Robotic Arm*.

In order to assess and discuss the performance of the proposed controller, it was developed a simulation platform for aerial mobile manipulators with Matlab interface, see the Fig. 5. This is an online simulator, which allows users to view three-dimensional environment navigation of the robot.

For the simulation presented below the path to be followed is a saddle described by, $\mathcal{P}_{xd} = \frac{1}{10}\cos(1.25t)$; $\mathcal{P}_{yd} = \frac{1}{10}\sin(1.25t)$ and $\mathcal{P}_{zd} = \frac{1}{2}+2$. The desired velocity of the end-effector of the aerial mobile manipulator will depend on the task, the control

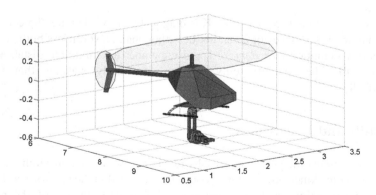

Fig. 5. Aerial mobile manipulator robot used by simulation platform developed

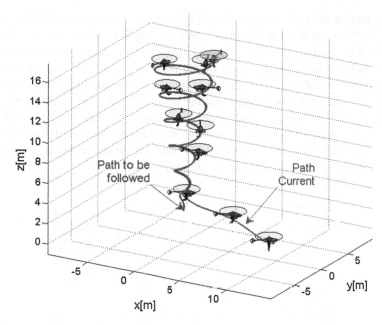

Fig. 6. Stroboscopic movement of the aerial mobile manipulator robot in the path following problem.

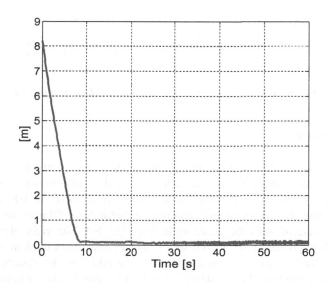

Fig. 7. Distance between the end-effector position and the closest point on the path.

error, the angular velocity, among others. In this case, it is considered that the reference velocity depends on the control errors and the angular velocity of the helicopter. It is defined as, $|\upsilon_P(s_D, \gamma)| = \upsilon_{P\max}/(1 + k_\rho \|\rho\|)$ where, $\upsilon_{P\max}$ represents the desired maximum velocity of the end-effector on the path $\mathcal{P}(s)$; and k_ρ is a positive constant that weigh the control error, respectively. Figure 6 shows the stroboscopic movement on the X–Y–Z space. It can be seen that the proposed controller works correctly. The evolution of the reference velocity and the actual velocity of the aerial mobile manipulator is illustrated in Fig. 7, where it can be seen that the reference velocity decreases in presence of large control errors; while Fig. 8 shows that $\rho(t)$ is ultimately bounded close to zero, considering that $\rho^2 = \tilde{\gamma}_x^2 + \tilde{\gamma}_y^2 + \tilde{\gamma}_z^2$.

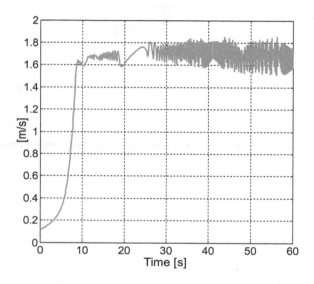

Fig. 8. End-effector velocity of the aerial mobile manipulator

5 Conclusions

A control system comprising two subsystems, a kinematic controller (responsible to accomplish the task of pathfollowing) and a dynamic compensator (responsible for dynamic compensations of the system) is here proposed to solve the 3D path following problem for a miniature aerial mobile manipulator. Saturation of the reference command to prevent the saturation of the actuators is here regarded. The main advantage of the control laws here proposed lies in their simplicity and easiness of implementation, when compared to other yet available in the literature. In addition, the system stability has been analytically proven. The simulation's results have proven the controller's ability to globally and asymptotically drive the controlled state variables to zero and simultaneously prevent any saturation in the flight commands. As future work, the implementation of such control system onboard a real aerial mobile manipulator will be tested, whose results are expected to confirm the effectiveness of the proposed control system.

References

1. Wang, Y., Lang, H., de Silva, C.W.: Visual servo control and parameter calibration for mobile multi-robot cooperative assembly tasks. In: IEEE International Conference on Automation and Logistics (2008)
2. Villareal, M., Cruz, C., Alvarez, E.: Robust structure-control design approach for Mechatronic system. IEEE/ASME Trans. Mechatron. **18**, 1592–1601 (2013)
3. Villareal, M., Cruz, J., Saldivar, L.: Dynamic coupling between a parallel manipulator and an omnidirectional mobile platform. In: XV Latinamerican Control Conference (2012)
4. Alisher, K., Alexander, K., Alexandr, B.: Control of the mobile robots with ROS in robotics courses. In: 25th DAAAM International Symposium on Intelligent Manufacturing and Automation (2014)
5. Díaz, C., Roa-Guerrero, E.: Development of mobile robotics platform for identification of land mines antipersonal in different areas of Colombia. IEEE (2015)
6. Andaluz, G., Andaluz, V., Terán, H., Arteaga, O., Chicaiza, F., Varela, J., Ortiz, J., Pérez, F., Rivas, D., Sánchez, J., Canseco, P.: Modeling dynamic of the human-wheelchair system applied to NMPC. In: Intelligent Robotics and Applications, pp. 179–190 (2016)
7. Yannick Morel, Y., Porez, M., Ijspeert, A.: Estimation of relative position and coordination of mobile underwater robotic platforms through electric sensing. IEEE (2012)
8. Brandao, A., Andaluz, V., Sarcinelli-Filho, M., Carelli, R.: 3-D path-following with a miniature helicopter using a high-level nonlinear underactuated controller. In: 2011 9th IEEE International Conference on Control and Automation (ICCA) (2011)
9. Andaluz, V., Carelli, R., Salinas, L., Toibero, J., Roberti, F.: Visual control with adaptive dynamical compensation for 3D target tracking by mobile manipulators. Mechatronics **22**(4), 491–502 (2012)
10. Boudjit, K., Larbes, C.: Detection and target tracking with a quadrotor using fuzzy logic. IEEE (2016)
11. From, P.J., Gravdahl, J.T., Pettersen, K.Y.: Vehicle-Manipulator Systems. Springer, London (2014)

Mechatronics and Intelligent Manufacturing

Design, Modeling and Analysis of a Magnetorheological Fluids-Based Soft Actuator for Robotic Joints

Daoming Wang[✉], Lan Yao, Jiawei Pang, and Zixiang Cao

School of Mechanical Engineering, Hefei University of Technology, Hefei 230009, China
cumtcmeewdm@hotmail.com, fanglanhf@163.com, pangjw2016@126.com,
884694387@qq.com

Abstract. Aiming at eliminating vibration generated during the motion state switch of robotic joints, this study proposes a magnetorheological fluids (MRFs) based soft actuator to achieve semi-active vibration control. In this paper, the configuration of the MRFs actuator is described firstly, followed by the theoretical modeling of the magnetic circuit and the transmitted torque. Then, the structural model of the actuator is designed and presented. After these, the influences of working induction and speed difference on both total transmitted torque and controllable coefficient are numerically calculated. Finally, an electromagnetic simulation is carried out with ANSYS 10.0® to verify the designed magnetic circuit of the actuator. The results indicate that the working induction holds a strong impact on both total transmitted torque and controllable coefficient; however, the influences of speed difference were relatively slight. Moreover, the designed circuit is proved to fulfill the requirements of both induction intensity and uniformity.

Keywords: Robotic joint · Magnetorheological fluids actuator · Structural design · Torque transmission · Electromagnetic simulation

1 Introduction

Robots are kinds of machine that rely on their own power and control ability to automatically realize various functions. Their mission is to assist or replace human to complete assigned works. Nowadays, robots are widely used in considerable works, such as automobile manufacturing [1], machinery processing [2, 3], palletizing [4, 5], automatic spraying [6], and heavy collaborative lifting [7–9]. Robotic joint is a basic component of the robot body. In practice, it should switch its motion frequently between active and passive states, which easily produces vibration during the motion to cause mechanical damage to joints and reduce the motion control accuracy [10, 11]. Consequently, robotic joints are required to have damping ability in the passive motion state. Traditional method is to access elastic elements, like dampers or springs, between the motor and shaft to absorb shock and vibration. However, this method is not adaptive to the damping requirements under various conditions, and also it is unable to realize semi-active vibration control.

© Springer International Publishing AG 2017
Y. Huang et al. (Eds.): ICIRA 2017, Part III, LNAI 10464, pp. 753–764, 2017.
DOI: 10.1007/978-3-319-65298-6_67

The emergence of intelligent materials greatly accelerates the development of novel engineering equipment. As a new smart material, magnetorheological fluids (MRFs) have the ability to rapidly and reversibly change their rheological behaviors between Newtonian fluids and quasi-solid structures with/without a magnetic field [12]. The phenomenon is generally known as the magnetorheological (MR) effect, which is characterized by smooth operation, rapid response and easy control [13, 14]. Thus, it has been extensively integrated in many mechanical devices, such as dampers [15], brakes [16, 17], clutches [18], vibration absorbers [19], robotics applications [20], and haptic devices [21].

Owing to the outstanding features of MRFs, they possess broad application prospects in the robotic advancement. Among the examples are assistive knee braces, exoskeletons and robot grippers. For instance, Chen and Liao [22] developed an MR actuator for assistive knee brace. Li et al. [23] proposed an MR brake for a prosthetic ankle joint to achieve a smooth walk for the user. Pettersson et al. [24] designed a MRFs-based robot gripper which had the ability to handle food products with varying shapes. Kikuchi et al. [25] utilized the MRFs clutch in the design of a leg-shaped robot for brain-injured patients.

The main focus of this study is on the design, modeling and analysis of a MRFs actuator for robotic joints. Figure 1 displays the application of the MRFs actuator in a flexible-joint robot arm. The actuator is installed between a servo motor and a reducer. Once vibration is generated due to the motion state switch of robotic joints, the MRFs actuator can effectively eliminate vibration to enable the joint to move smoothly owing to its soft transmission properties. In this paper, the configuration of the MRFs actuator is described firstly, followed by theoretical modeling of the magnetic circuit and the transmitted torque. Then, the structural model of the actuator is designed and presented. After these, the influences of working induction and speed difference on the total transmitted torque and controllable coefficient are numerically calculated. Finally, an electromagnetic simulation is carried out to verify the designed magnetic circuit of the actuator.

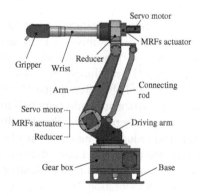

Fig. 1. Configuration of a flexible-joint robot arm utilizing the MRFs actuator

2   Design and Modeling of a Plate-Shaped MRFs Actuator

Figure 2 shows the configuration of MRFs actuator. In the actuator, a coil is enclosed in the shell cavity to produce an electromagnetic field. The isolation ring made of non-ferromagnetic material is installed to reduce flux leakage. An air gap with a width of 1 mm is formed between shell and plate. The input plate and output plate are respectively engaged with the input shaft and output shaft through eight bolts. The MRFs are confined between input and output plates by two O-rings.

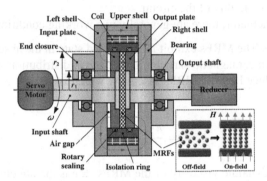

Fig. 2. Configuration of the MRFs actuator

The magnetic flux forms a closed loop after applying a coil current, as depicted in Fig. 3. Since the rheological effect only occurs to the MRFs within the magnetic circuit region, the actual working area of the actuator refers to the annular region formed between r_1 and r_2.

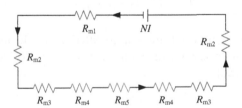

Fig. 3. Equivalent reluctance of the circuit for the MRFs actuator

Based on Ohm's law, magnetic reluctance of each circuit part is calculated by

$$R_{mi} = \frac{L_i}{\mu_i S_i} \tag{1}$$

where L is the length and S the cross-sectional area, μ is the magnetic permeability, and the subscript i is the serial number of each part.

Referring to Fig. 3, the magnetic flux forms a closed loop. Hence, the total magnetic reluctance of the circuit is derived as

$$R_{\text{total}} = R_{m1} + 2\left(R_{m2} + R_{m3} + R_{m4}\right) + R_{m5} \tag{2}$$

where R_{m1}, R_{m2}, R_{m3}, R_{m4}, R_{m5} are magnetic reluctances of upper shell, left and right shells, air gap, input and output plates and MRFs, respectively.

The magnetomotive force NI is a product of the coil turns N and the coil current I, which is expressed as

$$NI = \phi R_{\text{total}} \tag{3}$$

where ϕ is the magnetic flux of the circuit.

For the MRFs actuator, there are generally two working conditions:

(1) **Off-field State.** The MRFs exhibit Newton fluid state in the absence of a magnetic field. At this time, magnetic particles appear freely distribution state. Thus, a small torque is produced by the MRFs viscosity, which is known as viscous torque T_v given below

$$T_v = \frac{\pi \eta \Delta \omega r_2^4}{2h} \tag{4}$$

where η is the dynamic viscosity of the MRFs, $\Delta \omega$ is the angular speed difference between input and output plates, r_2 is the outer working radius, h is the distance between input and output plates (i.e., the working gap size).

(2) **On-field State.** After applying a magnetic field, an obvious change in rheological behavior occurs to the MRFs, forming particle chains along the field direction. Then, a magnetic torque TB is generated which is given below

$$T_B = \frac{2\pi \tau_B}{3}\left(r_2^3 - r_1^3\right) \tag{5}$$

where τ_B is the magnetic yield stress of the MRFs and r_1 the inner working radius.

The magnetic yield stress τ_B is a function of the magnetic flux intensity B and material parameters of MRFs [26].

$$\tau_B = kB^a \tag{6}$$

Hence, the total transmitted torque T_w is the sum of the magnetic torque T_B and the viscous torque T_v.

$$T_w = \frac{2\pi kB^a}{3}\left(r_2^3 - r_1^3\right) + \frac{\pi \eta \Delta \omega r_2^4}{2h} \tag{7}$$

Referring to design steps given in [27], structural model of MRFs actuator is designed in Fig. 4. Also, main parameters of the actuator are given in Table 1.

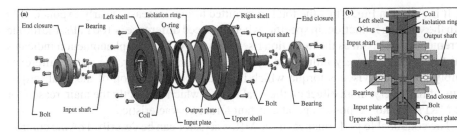

Fig. 4. Structural model of the MRFs actuator: (a) Exploded view; (b) Sectional view.

Table 1. Main Parameters of the Proposed MRFs Actuator

Parameter	Symbol	Value
Inner working radius	r_1	22 (mm)
Outer working radius	r_2	44 (mm)
Working gap size	h	1 (mm)
MRFs Viscosity (MRF-J01)	η	0.38 Pa·s
MRFs material parameters (MRF-J01)	k	182.2 kPa·T^{-1}
	a	1.506
Number of coil turns	N	250 (turns)
Maximum coil current	I	5.0 (A)

3   Performance Analysis of the MRFs Actuator

3.1   Torque Transmission and Controllable Performance

It is apparently in (7) that T_w is influenced only by B and $\Delta\omega$ once the basic parameters of MRFs and the actuator are given. The variation of total transmitted torque with working induction and speed difference is numerically calculated using Matlab 10.0®, as plotted in Fig. 5.

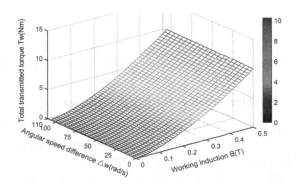

Fig. 5. Total transmitted torque versus working induction and angular speed difference

It is shown in Fig. 5 that the total transmitted torque increases nearly exponentially with the working induction. In specific, when the speed difference is 100 rad/s, the torque increment is 10.3 N·m as the induction rises from 0 to 0.5 T. The phenomenon indicates that the torque is largely influenced by the induction. However, the speed difference simply affects the viscous torque which holds a small proportion of the total torque. Thus, it has a very small effect on the total torque, which is exactly the main reason for the MRFs actuator to possess good constant torque characteristics.

The controllable performance of the MRFs actuator can be described by the controllable coefficient λ, which is represented as the ration of T_B and T_w. Note that the controllability of the transmitted torque becomes better as the controllable coefficient increases.

$$\lambda = \frac{T_B}{T_w} = \frac{4hkB^a\left(r_2^3 - r_1^3\right)}{4hkB^a\left(r_2^3 - r_1^3\right) + 3\eta\Delta\omega r_2^4} \tag{8}$$

Similarly, the variation of controllable coefficient with working induction and angular speed difference is obtained in Fig. 6. As the working induction increases, the controllable coefficient firstly rises rapidly, and then the increasing trend slows down gradually. In addition, there is no slip between the input and output plates once the speed difference is 0. At that time, no viscous torque is produced and the controllable coefficient maintains at a constant value of 1. With the increase of the speed difference, the controllable coefficient reduced slightly and the decreasing trend slows down with increasing working induction. On the whole, the working induction holds a strong impact on the controllable coefficient while the influence of speed difference is relatively slight.

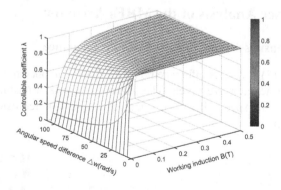

Fig. 6. Controllable coefficient versus working induction and angular speed difference

3.2 Electromagnetic Simulation of the MRFs Actuator

(1) *Finite Element Modeling:* The structure of the MRFs actuator is two-dimensional axisymmetric, and the magnetic circuit and boundary conditions are all consistent in the tangential direction. Thus, the electromagnetic simulation can be considered as a two-dimension problem. According to the configuration parameters of the

actuator, a PLANE13 element is used for finite element modeling with ANSYS 10.0®, as shown in Fig. 7.

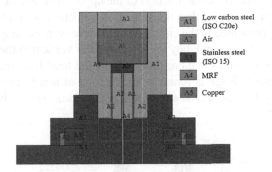

Fig. 7. Magnetic loop of the MRFs actuator

Grid meshing is required after the finite element modeling. The two-dimensional mapped meshing is utilized for regular quadrangles and the boundary meshing for irregular polygons. Additionally, the local refinement is carried out to the working gaps after the gird meshing.

(2) ***Material Property and Boundary Condition:*** Material properties of each component in the circuit should be endowed prior to the simulation. In the simulation, the relative permeabilities of the non-ferromagnetic materials, including stainless steel and air, are set to 1. The MRFs and the low-carbon steel belong to nonlinear ferromagnetic materials, whose relative permeabilities change with the magnetic field strength. Figure 8 shows the *B-H* curves of the nonlinear materials in the actuator.

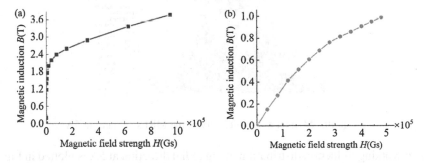

Fig. 8. *B-H* curves of the nonlinear materials in the MRFs actuator: (a) MRFs; (b) Low carbon steel (ISO C20e).

Assuming that the magnetic flux lines are completely confined in the internal of the model, the model frame is set as a parallel flux boundary. Also, the coil is applied on the coil domain as a current density load.

Simulation Results and Discussions: Figs. 9 and 10 separately show the magnetic flux lines and magnetic flux intensity of the MRFs actuator at 5 A. The highest magnetic induction intensity appears in the junction of the upper shell and the right shell with its value of 1.81 T, which is lower than the saturation induction of ISO C20e. It proves that a certain margin is presented in the circuit design. Moreover, the magnetic flux lines are strictly limited in the circuit and they are also basically vertical to the plate surface. The effective flux lines account for more than 90% of the total flux lines. Unavoidably, a small amount of flux leakage is found at the junction of non-ferromagnetic and ferromagnetic materials. In general, the simulation results indicate that the circuit meets the design requirements.

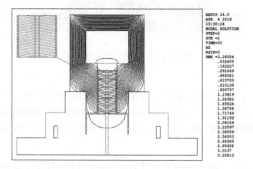

Fig. 9. Magnetic flux lines of the MRFs actuator at 5A

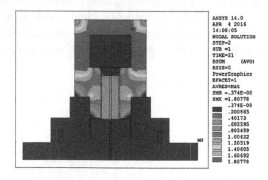

Fig. 10. Magnetic flux intensity of the MRFs actuator at 5A

The working induction distribution along redial direction at 5A is plotted in Fig. 11. As can be seen, the working induction increases from 0.551T to 0.558T from the inner radius to the outer radius, a variation ratio of nearly 1.3%. The phenomenon shows the magnetic inductions are basically distributed evenly along the radial direction.

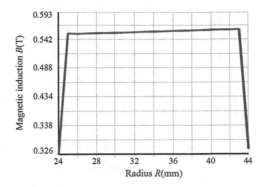

Fig. 11. Working induction distribution along radial direction at 5A

The working induction largely influences the torque transmission ability of the actuator. It is calculated by averaging the induction at each node of the working gap. Figure 12 is the variation of average working induction with coil current. Also, the relationship between transmitted torque and coil current is plotted in Fig. 13. It is obvious that both the working induction and transmitted torque increases with the coil current. However, the variation trend is not a linear relationship due to the effect of material nonlinearity. In specific, the increasing trend slows down as the current increases. At 5A, the working induction reaches about 0.561T that is greater than the saturation induction of MRF-J01® (0.5T). The results demonstrate that the designed circuit fulfills the induction intensity requirement. In addition, the maximum transmitted torque reaches about 11.9 N·m for a coil current of 5A.

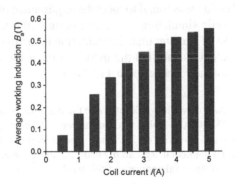

Fig. 12. Variation of average working induction with coil current

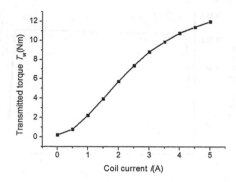

Fig. 13. Relationship curve between transmitted torque and coil current

4 Conclusions

This paper was concerned with the design, modeling and analysis of a MRFs actuator for application in the robotic joints to achieve semi-active vibration control during the motion state transformation. Configuration design, theoretical modeling and numerical simulation were performed to the MRFs actuator successively. The magnetic circuit and the transmitted torque in both off-field and on-field states were theoretically modeled. Numerical calculation results indicated that the working induction had a strong impact on both the total transmitted torque and controllable coefficient while the influences of speed difference were relatively slight. When the speed difference is 100 rad/s, the total torque increment is 408 N·m as the working induction increases from 0 to 0.5 T. In addition, the designed circuit was found to meet the requirement of induction uniformity through an electromagnetic simulation. For a coil current of 5A, the average working induction was about 0.561 T (greater than the saturation induction of the chosen MRFs). It demonstrated that the circuit fulfilled the induction intensity requirement. Also, the maximum transmitted torque reaches about 11.9 N·m.

Acknowledgment. The authors wish to acknowledge the National Natural Science Foundation of China (grant no. 51505114), the Anhui Provincial Natural Science Foundation (grant no. 1608085QE116), the China Postdoctoral Science Foundation funded project (grant no. 2015M571919), and the Fundamental Research Funds for the Central Universities (grant no. JZ2016HGTB0717) for their financial supports of this work.

References

1. Dang, Q.V., Nielsen, I., Steger-Jensen, K., et al.: Scheduling a single mobile robot for part-feeding tasks of production lines. J. Intell. Manuf. **25**, 1271–1287 (2014). doi:10.1007/s10845-013-0729-y
2. Shin, H., Kim, S., Jeong, J., et al.: Stiffness enhancement of a redundantly actuated parallel machine tool by dual support rims. Int. J. Precis. Eng. Manufact. **13**, 1539–1547 (2012). doi:10.1007/s12541-012-0203-3

3. Zi, B., Sun, H., Zhu, Z., et al.: The dynamics and sliding mode control of multiple cooperative welding robot manipulators. Int. J. Adv. Rob. Syst. **9**, 1–10 (2012). doi:10.5772/50641
4. Dzitac, P., Mazid, A.M.: An efficient control configuration development for a high-speed robotic palletizing system. In: IEEE Conference on Robotics, Automation and Mechatronics, pp. 140–145 (2008). doi:10.1109/RAMECH.2008.4681379
5. Liu, X.J., Li, J., Zhou, Y.: Kinematic optimal design of a 2-degree-of-freedom 3-parallelogram planar parallel manipulator. Mech. Mach. Theory **87**, 1–17 (2015). doi:10.1016/j.mechmachtheory.2014.12.014
6. Wang, G., Cheng, J., Li, R., et al.: A new point cloud slicing based path planning algorithm for robotic spray painting. In: IEEE International Conference on Robotics and Biomimetics, pp. 1717–1722 (2015). doi:10.1109/ROBIO.2015.7419019
7. Zi, B., Duan, B.Y., Du, J.L., et al.: Dynamic modeling and active control of a cable-suspended parallel robot. Mechatronics **18**, 1–12 (2008). doi:10.1016/j.mechatronics.2007.09.004
8. Qian, S., Zi, B., Ding, H.: Dynamics and trajectory tracking control of cooperative multiple mobile cranes. Nonlinear Dyn. **83**, 89–108 (2016). doi:10.1007/s11071-015-2313-9
9. Zi, B., Sun, H., Zhang, D.: Design, analysis and control of a winding hybrid-driven cable parallel manipulator. Rob. Comput.-Integr. Manufact. **48**, 196–208 (2017). doi:10.1016/j.rcim.2017.04.002
10. Kumagai, S., Ohishi, K., Shimada, N., et al.: High-performance robot motion control based on zero-phase notch filter for industrial robot. In: 11th IEEE International Workshop on Advanced Motion Control, pp. 626–630 (2010). doi:10.1109/AMC.2010.5464060
11. Zi, B., Lin, J., Qian, S.: Localization, obstacle avoidance planning and control of cooperative cable parallel robots for multiple mobile cranes. Rob. Comput. Integr. Manufact. **34**, 105–123 (2015). doi:10.1016/j.rcim.2014.11.005
12. Wang, D., Zi, B., Zeng, Y., et al.: Temperature-dependent material properties of the components of magnetorheological fluids. J. Mater. Sci. **49**, 8459–8470 (2014). doi:10.1007/s10853-014-8556-x
13. Jang, K.I., Nam, E., Lee, C.Y., et al.: Mechanism of synergetic material removal by electrochemomechanical magnetorheological polishing[J]. Int. J. Mach. Tools Manuf. **70**, 88–92 (2013). doi:10.1016/j.ijmachtools.2013.03.011
14. Wang, D., Zi, B., Zeng, Y., et al.: Measurement of temperature-dependent mechanical properties of magnetorheological fluids using a parallel disk shear stress testing device. Proc. Inst. Mech. Eng. Part C: J. Mech. Eng. Sci. **231**, 1725–1737 (2017). doi: 10.1177/0954406215621099
15. Xie, H.L., Liu, Z.B., Yang, J.Y., et al.: Modelling of magnetorheological damper for intelligent bionic leg and simulation of knee joint movement control. Int. J. Simul. Modell. **312**, 399–417 (2016). doi:10.2507/IJSIMM15(1)CO2
16. Wang, D.M., Hou, Y.F., Tian, Z.Z.: A novel high-torque magnetorheological brake with a water cooling method for heat dissipation. Smart Mater. Struct. **22**, 025019 (2013). doi: 10.1088/0964-1726/22/2/025019
17. Sarkar, C., Hirani, H.: Development of a magnetorheological brake with a slotted disc. Proc. Inst. Mech. Eng. Part D: J. Automobile Eng. **229**, 1907–1924 (2015). doi: 10.1177/0954407015574204
18. Wang, D., Zi, B., Zeng, Y., et al.: An investigation of thermal characteristics of a liquid-cooled magnetorheological fluid-based clutch. Smart Mater. Struct. **24**, 055020 (2015). doi: 10.1088/0964-1726/24/5/055020
19. Weber, F.: Semi-active vibration absorber based on real-time controlled MR damper. Mech. Syst. Signal Process. **46**, 272–288 (2014). doi:10.1016/j.ymssp.2014.01.017

20. Rossa, C., Lozada, J., Micaelli, A.: A new hybrid actuator approach for force-feedback devices. In: 2012 IEEE/RSJ International Conference on Intelligent Robots and Systems (IROS), pp. 4054–4059 (2012). doi:10.1109/IROS.2012.6385784
21. Blake, J., Gurocak, H.B.: Haptic glove with MR brakes for virtual reality. IEEE/ASME Trans. Mechatron. **14**, 606–615 (2009). doi:10.1109/TMECH.2008.2010934
22. Chen, J.Z., Liao, W.H.: Design, testing and control of a magnetorheological actuator for assistive knee braces. Smart Mater. Struct. **19**, 035029 (2010). doi: 10.1088/0964-1726/19/3/035029
23. Li, C., Tokuda, M., Furusho, J., et al.: Research and development of the intelligently-controlled prosthetic ankle joint. In: Proceedings of the 2006 IEEE International Conference on Mechatronics and Automation, pp. 1114–1119 (2006). doi:10.1109/ICMA.2006.257781
24. Pettersson, A., Davis, S., Gray, J.O., et al.: Design of a magnetorheological robot gripper for handling of delicate food products with varying shapes. J. Food Eng. **98**, 332–338 (2010). doi: 10.1016/j.jfoodeng.2009.11.020
25. Kikuchi, T., Oda, K., Furusho, J.: Leg-robot for demonstration of spastic movements of brain-injured patients with compact magnetorheological fluid clutch. Adv. Robot. **24**, 671–686 (2010). doi:10.1163/016918610X493534
26. Karakoc, K., Park, E.J., Suleman, A.: Design considerations for an automotive magnetorheological brake. Mechatronics **18**, 434–447 (2008). doi:10.1016/j.mechatronics. 2008.02.003
27. Wang, D., Hou, Y.: Design and experimental evaluation of a multidisk magnetorheological fluid actuator. J. Intell. Mater. Syst. Struct. **24**, 640–650 (2013). doi: 10.1177/1045389X12470305

Control of a Magnet-Driven Nano Positioning Stage with Long Stroke Based on Disturbance Observer

Letong Ma, Xixian Mo, Bo Zhang$^{(\boxtimes)}$, and Han Ding

State Key Laboratory of Mechanical System and Vibration,
School of Mechanical Engineering, Shanghai Jiao Tong University,
Shanghai 200240, China
{dictionary,wanjushengfang,b_zhang,hding}@sjtu.edu.cn

Abstract. In this paper, a proportional-derivative (PD) controller combined with a disturbance observer is adopted for a designed magnet-driven nano positioning stage. Firstly, the magnet-driven positioning stage with long stroke and nano positioning accuracy is introduced in detail. The stage adopts air bearings to eliminate complicated nonlinear friction effect. The actuator of the stage is a permanent magnet synchronous linear motor which uses ironless windings to eliminate the cogging force and attenuate the reluctance force. Then, a PD controller combined with a disturbance observer is used to control the stage. The validation experiment is carried out based on Matlab/Simulink Real-Time toolbox. The experimental results shows that the maximum travel range is 50 mm and positioning accuracy no bigger than 3 nm. The compared experiment with a conventional proportional-integral-derivative controller shows ten times better position accuracy is acquired by using proposed controller.

Keywords: Nano positioning stage · Disturbance observer · Long stroke · Magnet-driven actuator · Air bearing

1 Introduction

With the development of micro/nano technology, more and more high performance positioning systems with long stroke and micro/nano positioning accuracy are needed for fabricating and measuring micro structures on a large surface in the fields of lithography, micromachining, display technology and atomic force microscope (AFM)/scanning probe microscope (SPM). Frequently, the single degree of freedom (DOF) positioning stage attracts researchers more due to not only its relatively simple structure and widely application, but also its adaptation to be transformed or combined each other to form a multi-DOFs stage.

There is a dilemma between long stroke and nano positioning accuracy. The piezoelectric stack actuators (PSAs) is one of the most critical tools and is widely employed in nano positioning stages due to their ultrahigh resolution, fast response time and large stiffness [1–3], but their travel ranges are limited to

© Springer International Publishing AG 2017
Y. Huang et al. (Eds.): ICIRA 2017, Part III, LNAI 10464, pp. 765–776, 2017.
DOI: 10.1007/978-3-319-65298-6_68

several hundreds of micrometer only. An inchworm-type linear piezomotor can provide a travel range of several millimeters, however, its output force is very small. Another design of long stroke and nano positioning stage combines the two type stages together, which include a macro stage usually actuated using servo motor and a micro stage usually actuated by PSAs [4–6], but their structures are so sophisticated that it is difficult to design an efficient controller, also, the bandwidth of controller systems are usually low due to their high inertia.

Recently, more and more high performance permanent magnet synchronous linear motor (PMSLM) control systems are realized [7–11]. These control systems have an accuracy of micrometer or even sub micrometer. Though the accuracy level is somewhat lower compared to PACs, their travel ranges are tens of and thousands of longer than PACs and can be extended easily. There are two difficulties to further improve the positioning accuracy of PMSLM, one is external disturbances, and the other parameter variation. Nonlinear friction and ripple force are two mainly external disturbances. The two primary components of the force ripple are the cogging force and the reluctance force [12]. The cogging force arises as a result of the mutual attraction between the magnets and iron cores of the translator and the reluctance force is due to the variation of the self-inductance of the windings with respect to the relative position between the mover and the stator of the actuator. In [13], an adaptive robust control (ARC) method which combines the design techniques of adaptive control (AC) and those of deterministic robust control (DRC) [e.g. sliding mode control (SMC)] is adopted. The proposed ARC has a better tracking and transient performance in the presence of discontinuous disturbances, such as Coulomb friction. The experimental shows an accuracy less than $3\,\mu m$. In [14], another adaptive robust control method is proposed. The controller consists of three components: a simple feedforward compensator, a PID feedback controller and a radial-basis function (RBF) adaptive compensator used to compensate for uncertainties in the system, the final accuracy is less than $4\,\mu m$. In [15], Tomizuka and Lee demonstrated the robustness and accuracy of a motion controller that is composed of four modules: disturbance observer (DOB) as a velocity loop feedback controller, position loop feedback controller, feedforward tracking controller for the desired output and friction compensator, the $3\,\mu m$ accuracy is acquired. In [16], a variable structure control (VSC) method is proposed to achieve an overshoot smaller than $0.2\,\mu m$ and position precise positioning less than $0.03\,\mu m$. In [17], disturbance observer is adopted in controller, the accuracy less than $10\,\mu m$ is acquired. In these papers, it is no doubt that friction is very difficult to be compensated ideally even using sophisticated controller.

In this paper, a new single DOF magnet-driven nano positioning stage with long stroke is proposed. The actuator of the stage is a linear motor in essence. The stage is free from friction and is effected lightly by ripple force. A disturbance observer (DOB) controller and a proportional-derivative (PD) feedback controller are combined to get a high performance of the proposed positioning system.

The remainder of this paper is organized as follows. The design of the positioning stage is presented in Sect. 2. In Sect. 3, the system description and controller implementation are presented. The experimental setup and performance test are performed in Sect. 4, followed by the conclusion in Sect. 5.

2   Design of the Positioning Stage

Figure 1 shows the structure of the proposed nano positioning stage. The stage consists of a stage base and a moving platen. The actuator of the stage is a permanent magnet synchronous linear motor (PMSLM). Usually, there are two types of structure topologies for PMSLM, which are moving winding type and moving magnet type. In this paper, a moving magnet type is adopted in order

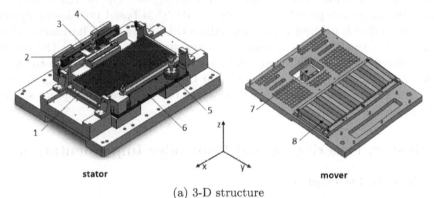

(a) 3-D structure
1. Stator windings; 2. Horizontal air bearings; 3. Horizontal granite guide way; 4. Linear encoder; 5. Vertical air bearings; 6. Vertical granite guide way; 7. Linear encoder read head; 8. Halbach magnet array

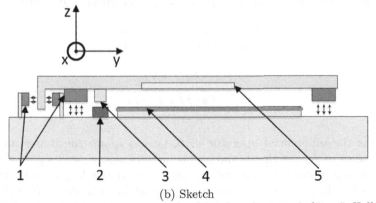

(b) Sketch
1. Air bearing; 2. Linear encoder; 3. Read head; 4. Stator winding; 5. Halbach magnetic array

Fig. 1. Structure of the stage

to avoid effect of thermal generation from windings, which thus can be conducted away through aluminum base. The Halbach magnet array, which can strengthen the amplitude of the magnetic flux density and reduce the higher-order harmonics of the magnetic flux at air gap side, is mounted underneath the mover (called rotor in rotary motor). The structure of stator is ironless, thus, the cogging force can be attenuated mostly. Ironless structure also reduce the windings inductance and improve the current response of the actuator, this is very favorable in improving system performance. The mover of the actuator is levitated using three air bearings in vertical direction. Since air bearings distribute their load over a relatively large and extremely flat area, they have built-in averaging effect that compensates for any minor irregularities in the reference surface. The motion in y direction is restricted by four horizontal air bearings. So the stage have only one DOF in x direction. Due to using of air bearings, the friction in the motion is nearly zero and is negligible. The designed stage has three advantages compared to conventional PMSLM-based positioning system. First, the Halbach magnet array can reduce the higher-order harmonics of the magnetic flux. Second, the ironless windings structure eliminates the cogging force, reduces reluctance force and improve the current response performance of actuator. Third, the complicated nonlinear friction is avoided by using air bearings. All of these design consideration makes the proposed stage very suitable for high precision positioning.

3   System Description and Controller Implementation

3.1   System Description

The control system of the proposed single DOF magnet driven nano positioning stage is described in this section.

The mathematical model of PMSLM can be described in the two axis d-q synchronously rotating frame by the following differential equations:

$$\begin{cases} u_q = Ri_q + p\psi_q + \frac{v\pi}{\tau}\psi_d \\ u_d = Ri_d + p\psi_d - \frac{v\pi}{\tau}\psi_q \end{cases} \tag{1}$$

$$\psi_d = L_d i_d + \psi_f \tag{2}$$

$$\psi_q = L_q i_q \tag{3}$$

$$v = \dot{x} \tag{4}$$

where p is the differential operator d/dt, i_d and i_q are the d-axis and q-axis currents, L_d and L_q are the d- and q-axis inductances, u_d and u_q are the d-axis and q-axis stator voltages, ψ_d and ψ_q are d- and q-axis flux linkage, ψ_f is the maximum flux linkage due to permanent magnet in each phase, R is the stator resistance, τ is the pole pitch, x is the position of the mover, v is the mover velocity. The elector-mechanical equation of the PMLSM is:

$$F = \frac{\pi}{\tau}[(L_d - L_q)i_d i_q + \psi_f i_q] \tag{5}$$

In this paper, the PMSLM is a surface mounted type, thus, L_d equals to L_q. Then, (5) can be simplified as:

$$F = \frac{\pi}{\tau}\psi_f i_q = K i_q \tag{6}$$

$$K = \frac{\pi}{\tau}\psi_f \tag{7}$$

where K is the force constant of the motor. The dynamic mechanical equation is:

$$m\ddot{x} = F \tag{8}$$

This equation is very simple due to absence of nonlinear friction.

Because the electrical time constant L/R is small (in this paper, $L = 0.002H$, $R = 7\Omega$), electrical transients decay very rapidly and $Ldi/dt \approx 0$, Thus, let $i_d = 0$, note $i = i_q$, $u = u_q$ the following simplified equation is given:

$$\ddot{x} + \frac{K^2}{mR}\dot{x} = \frac{K}{mR}u \tag{9}$$

so, the nominal transfer function from u to x is given:

$$P_n(s) = \frac{10^3 K/(mR)}{s^2 + K^2 s/(mR)} \tag{10}$$

where the gain of 10^3 is used to normalize the units of the system to 1 mm.

3.2 Controller Design

Figure 2 shows that the tracking controller consists of only two components: a feedback controller (Gc) which is a proportional derivative (PD) controller and a disturbance observer (DOB). Note that the DOB is applied in the position loop other than velocity loop, actually, there is no speed loop in this control system.

The DOB is designed based on linear system, this characteristic confine its performance in the usually motion control system because of the existence of nonlinear friction factor. In this paper, since friction in the system has been removed by using air bearings, the DOB is very suitable to compensate for external disturbance and plant uncertainties. The disturbance observer, shown demarcated within the dotted box in Fig. 2 estimates the disturbance based on the output x and the control signal u. P denotes the actual system, P_n denotes the nominal system, Q is a low-pass filter, d is the disturbance term including external disturbance and plant uncertainties, $\hat{d}$ represents the estimate of d, ξ is the measurement noise, x_d is the referent position, x is the actual position, e is the position error. Note that P_n^- is not realizable by itself but that QP_n^- can be made realizable by letting the relative degree of Q by equal to or greater than that of P_n^-. From the block diagram in Fig. 2, x is expressed as:

$$x(s) = G_{ux}u(s) + G_{dx}d(s) + G_{\xi x}\xi(s) \tag{11}$$

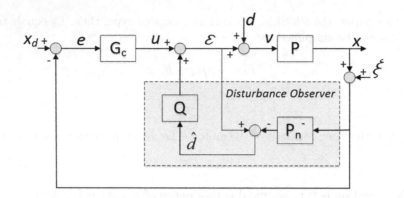

Fig. 2. Robust tracking control based on a disturbance observer

$$G_{ux} = \frac{PP_n}{P_n + (P - P_n)Q} \tag{12}$$

$$G_{dx} = \frac{PP_n(1 - Q)}{P_n + (P - P_n)Q} \tag{13}$$

$$G_{\xi x} = -\frac{PQ}{P_n + (P - P_n)Q} \tag{14}$$

If $Q(s) \approx 1$, the three transfer functions are $G_{ux} \approx P_n$, $G_{dx} \approx 0$, $G_{\xi x} \approx -1$. This implies that the disturbance makes the actual plant behave like the nominal plant, and this provides robustness to the control system. On the other hand, if $Q(s) \approx 0$, the three transfer functions are $G_{ux} \approx P$, $G_{dx} \approx P$, $G_{\xi x} \approx 0$, and the open dynamic is observed. Therefore, a sensible choice is to let the low frequency dynamics of $Q(s)$ close to 1 for disturbance rejection and model uncertainties, the high frequency dynamic close to 0 to reject measurement noise. So, $Q(s)$ is designed as a low-pass filter. A two order binomial filter which satisfies above stated properties has been chosen for this research

$$Q(s) = \frac{1}{(\tau s)^2 + 2\tau s + 1} \tag{15}$$

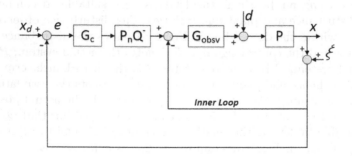

Fig. 3. Equivalent system to Fig. 2

The observation may be clearer by transforming Fig. 2 to the equivalent configuration of a two DOFs control system, as depicted in Fig. 3. Figure 3 shows that the disturbance observer is equivalent to an additional disturbance compensator G_{obsv}, which close a fast inner loop.

$$G_{obsv} = \frac{Q}{1-Q}P_n^-$$

(16)

4   Experiment Setup and Performance Test

In order to investigate the feasibility of the proposed magnet-driven nano positioning stage, a control system is contructed. A PD controller combined with DOB is adopted to handle the position step response control problem.

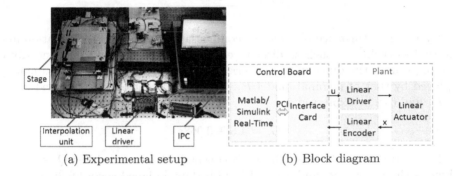

(a) Experimental setup   (b) Block diagram

Fig. 4. The experimental setup

The Fig. 4 shows the experimental setup. A prototype of the single DOF stage is fabricated and assembled. The control algorithm is implemented in Matlab/Simulink based on Real-Time toolbox, which can construct a real-time embedded operation system on industrial personal computer (IPC). This configuration actually is a simple hardware in loop like dSPACE but is more low-cost. A linear driver with a fix gain of 7.2 is used to provide excitation voltage (-36 V– 36 V) for the actuator. The high accurate linear incremental encoder L281 from HEIDENHAIN Corporation is employed as the position feedback sensor. The grating period of encoder is 2.048 μm. By 4 multiple frequency and 512 times interpolation with the interpolation unit GEMAC IPE16000-USB, a measurement resolution of 1 nm is acquired. The customized interface card plugged in IPC equipped with the 18-bit digital to analog converter (DACs) to output the excitation voltage (-5 V–5 V) for linear driver and encoder interface to capture the real-time displacement information from the linear encoder.

The parameter values of the actuator is shown in Table 1.

Table 1. The parameter values of the actuator

Parameter	Values
K(N/A)	43
m(kg)	6
L(H)	0.002
R(Ω)	7

The dominant linear model is computed to be:

$$P_n(s) = \frac{1023.81}{s^2 + 44.023s} \tag{17}$$

By using disturbance observer to compensate the parameters variation and external disturbance, such as discrete effects of DACs and low frequent vibration of ground, the dynamic characteristic of the system can be approximately replaced by that of nominal model P_n. So, the PD controller is designed based on P_n and can be expressed as:

$$G_c = 4.8 + 0.068s \tag{18}$$

For disturbance observer, the time constant of low-pass filter Q is 0.002s, which is chosen by trial and error. The Q is thus designed accordingly as:

$$Q(s) = \frac{1}{0.000004s^2 + 0.004s + 1} \tag{19}$$

Several different step response and a compared experiment with a conventional PID controller are planned to investigate the control performance and robustness of proposed controller. the steady state position error is selected as the performance indices of these controllers. The sampling frequency is chosen as 20 kHz for the following experiments. A third order trajectory planning method is used to get a smooth motion step point to point motion.

5   Step Response Experiments

In this section, two different continues displacements step response are specified to validate the performance of the DOB controller. The position response of 10 nm, 100 μm are shown in Fig. 5(a), (c) respectively, and corresponding process error are shown in Fig. 5(b), (d) respectively. Figure 5(a) shows the system have a 10 nm step performance, the stairs are very clear in the step motion. Figure 5(b) shows the positioning error are mostly in the range of −3 nm to 3 nm except for step periods that are not concerned about.

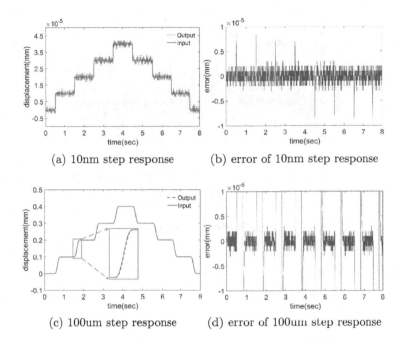

(a) 10nm step response

(b) error of 10nm step response

(c) 100um step response

(d) error of 100um step response

Fig. 5. Continues step response

6 Compared Experiment

In this section, the compared experiment between DOB and conventional PID is carried out to verify the effective of the proposed controller. A 1 mm step response experiment result is shown in Fig. 6(a). It can be seen that the DOB has a better track performance than conventional PID. Figure 6(b) shows the position error convergence speed of DOB is faster than that of PID. The steady state

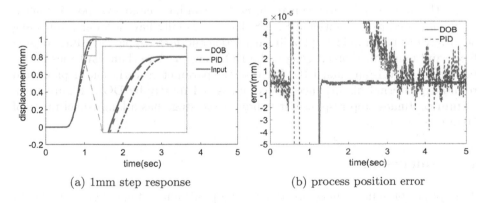

(a) 1mm step response

(b) process position error

Fig. 6. Compared step response

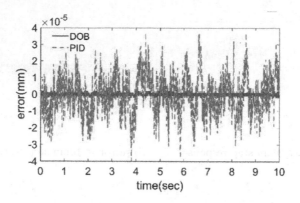

Fig. 7. Steady state error

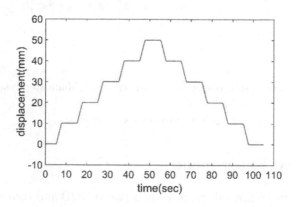

Fig. 8. 10 mm continues step response

positioning error is shows in Fig. 7. The error bound of DOB is within −3 nm to 3 nm, that of PID is −40 nm to 40 nm. There are two points needed to mention. First, this results show even conventional PID can have nano level position accuracy compared to conventional PMLSM which usually have a lower positioning accuracy due to complicated nonlinear friction and force ripple including cogging force and reluctance force; second, the DOB have a more than ten times better positioning precision than PID, because the parameters variation of plant and remained external disturbance are compensated mostly by DOB. Figure 8 is a 10 mm continues step response, it validates the stage has a long travel range of 50 mm.

7 Conclusion

This paper presents a magnet-driven nano positioning stage with long stroke. First, the detailed description of the stage is given. The stage is actually a permanent magnet synchronous linear motor (PMSLM), this is why it is called magnet-driven stage. The stage adopts air bearings to eliminate the complicated nonlinear friction, this is favor to linear control method of DOB. The ironless structure of the windings of actuator in stage avoid cogging force and reduce reluctance force existing in conventional PMSLM. By these design methods, the stage is very suitable for high precision positioning. Second, the paper proposed a very simple control method, which is a PD feedback controller plus a disturbance observer. The validation experiments shows the whole system have a total travel range of 50 mm and positioning error is with in −3 nm to 3 nm. Compared experiments shows even a conventional PID control method can get a nano level position accuracy due to absence of nonlinear friction and mostly force ripple included in conventional PMSLM. The paper gives a point that even a PMLSM can have nano level positioning accuracy.

References

1. Chu, C.-L., Fan, S.-H.: A novel long-travel piezoelectric-driven linear positioning stage. Precision Eng. **30**, 85–95 (2006)
2. Li, C.-X., Guo-Ying, G., Yang, M.-J., Zhu, L.-M.: Design, analysis and testing of a parallel-kinematic high-bandwidth XY nanopositioning stage. Rev. Sci. Instrum. **84**, 12522 (2013)
3. Li, C.-X., Gu, G.-Y., Yang, M.-J., Zhu, L.-M.: High-Speed tracking of a nanopositioning stage using modified repetitive control. IEEE Trans. Autom. Sci. Eng.
4. Liu, C.-H., Jywe, W.-Y., Jeng, Y.-R., Hsu, T.-H., Li, Y.-T.: Design and control of a long-traveling nanopositioning stage. Precision Eng. **34**, 497 506 (2010)
5. Jie, D., Sun, L., Liu, Y., Zhu, Y., Cai, H.: Design and simulation of a macro–micro dual-drive high acceleration precision XY-stage for IC bonding technology. In: 2005 6th International Conference on Electronic Packaging Technology
6. Elfizy, A.T., Bone, G.M., Elbestawi, M.A.: Design and control of a dual-stage feed drive. Int. J. Mach. Tools Manuf. **45**, 153–165 (2005)
7. Lin, C.-J., Yau, H.-T., Tian, Y.-C.: Identification and compensation of nonlinear friction characteristics and precision control for a linear motor stage. IEEE/ASME Trans. Mechatron. **18**(4), 1385–1396 (2013)
8. Liu, Z.Z., Luo, F.L., Rahman, M.A.: Robust and precision motion control system of linear-motor direct drive for high speed X-Y table positioning mechanism. IEEE Trans. Ind. Electron. **52**(5), 1357–1363 (2005)
9. Liu, T.-H., Lee, Y.-C., Chang, Y.-H.: Adaptive controller design for a linear motor control system. IEEE Trans. Aerosp. Electron. Syst. **40**(2), 601–616 (2004)
10. Cho, K., Kim, J., Choi, S.B., Oh, S.: A high-precision motion control based on a periodic adaptive disturbance observer in a PMLSM. IEEE/ASME Trans. Mechatron. **20**(5), 1083–1095 (2015)
11. Yao, B., Li, X.: Adaptive robust motion control of linear motors for precision manufacturing. Mechatronics **12**, 595–616 (2002)
12. Tan, K.K., Huang, S.N., Lee, T.H.: Robust adaptive numerical compensation for friction and force ripple in permanent-magnet linear motors. IEEE Trans. Magn. **38**(1), 221–228 (2002)
13. Yao, B., Al-Majed, M., Tomizuka, M.: High-performance robust motion control of machine tools: an adaptive robust control approach and comparative experiments. IEEE/ASME Trans. Mechatron. **2**(2), 63–76 (1997)

14. Tan, K.K., Huang, S.N., Dou, H.F., Lee, T.H., Chin, S.J., Lim, S.Y.: Adaptive robust motion control for precise trajectory tracking applications. ISA Trans. **40**, 57–71 (2001)
15. Lee, H.S., Tomizuka, M.: Robust motion controller design for high-accuracy positioning systems. IEEE Trans. Ind. Electron. **43**(1), 48–55 (1996)
16. Chang, S.B., Wu, S.H., Hu, Y.C.: Submicrometer overshoot control of rapid and precise positioning. Precision Eng. **20**(3), 161–170 (1997)
17. Tan, K.K., Lee, T.H., Dou, H.F., Chin, S.J., Zhao, S.: Precision motion control with disturbance observe for pulsewidth-modulated-driven permanent-magnet linear motors. IEEE Trans. Magnet. **39**(3) (2003)

Experimental Research of Loading Effect on a 3-DOF Macro-Micro Precision Positioning System

Lingbo Xie, Zhicheng Qiu, and Xianmin Zhang[✉]

Guangdong Provincial Key Laboratory of Precision Equipment and Manufacturing Technology,
School of Mechanical and Automotive Engineering, South China University of Technology,
Guangzhou 510640, China
lingboxie@163.com, {zhchqiu,zhangxm}@scut.edu.cn

Abstract. The precision positioning technology has developed for a long time and played a significant role in many fields. However, most precision positioning systems designed before only have one or two degrees of freedom, which greatly limits the application. In view of this, a kind of three-degrees-of-freedom (3-DOF) macro-micro precision positioning system is investigated and analyzed in the paper. The macro-micro precision positioning system is designed suitable for application in vacuum environment, which includes two main parts such as 3PRR (3 degrees of freedom, each branch consists of a prismatic pair (P) drive and two rotating pairs (R)) planar parallel platform and piezoelectric micro stage. Before conducting the experiment of macro-micro precision positioning system, it is necessary to investigate the loading effect on the 3PRR planar parallel platform. The loading experiment has showed that different loads have some effect on positioning accuracy and standard variance, but the positioning error is much less than the travel range of micro positioner. Then the experiment combining 3PRR planar parallel platform and piezoelectric micro stage has been carefully conducted by using laser interferometer as feedback control measurement. The experimental results demonstrate that the requirement of high positioning precision and large travel range can be simultaneously met by using the macro-micro precision positioning system.

Keywords: Precision positioning · 3-DOF · Macro-Micro · Vacuum · 3PRR · Loading

1 Introduction

Nowadays precision positioning technology requires larger travel range and higher precision, which plays an important role in the defense industry, aerospace, precision optical, biological and medical, precision processing and manufacturing fields [1]. For example, the key problem of high performance chip package positioning system is how to coordinate the contradiction between the high speed and high precision and how to solve the conflict between large travel and high precision. A number of domestic and foreign scientific institutions have carried out related research work and developed some high-speed and high-precision bonding machines [2, 3].

© Springer International Publishing AG 2017
Y. Huang et al. (Eds.): ICIRA 2017, Part III, LNAI 10464, pp. 777–787, 2017.
DOI: 10.1007/978-3-319-65298-6_69

Generally speaking, precision positioning technology is composed of the macro positioner and micro positioner. A. Sharon is the first scholar who proposed the concept of combining macro and micro stages together at Massachusetts Institute of Technology [4]. From then on, a lot of researchers have shown great interest in the macro-micro precision positioning system. Liu Hongzhong designed a kind of macro-micro ultra precision positioning system using double servo control and Chebyshev digital filter, which performed with travel range of 200 nm and positioning accuracy of 8 nm. The macro and micro positioning system was composed of ball screw guide rail and piezo-electric actuator, respectively [5]. In 2010, Chien-Hung Liu et al. developed a set of nanometer positioning platform with large travel range of 300 mm × 300 mm and positioning accuracy of 10 nm. The micro stage of piezoelectric actuator was installed in the traditional ball screw system with PID closed-loop control and measuring instrument of laser interferometer [6]. László Juhász et al. also developed a kind of macro-micro positioning platform, which combined the conventional servo motor of precision ball screw guide with the embedded piezoelectric actuator. It used the grating scale as closed-loop control feedback test with maximum travel range of 100 mm, maximum speed of 125 mm/s, and resolution of 2 nm [7]. The ultra precision platform composed of precise ball screw drive and voice coil motor with aerostatic slide was proposed by Hidenori Shinno et al. The laser interferometer was used as feedback measurement system. The maximum stroke of the platform could reach 150 mm, the positioning accuracy was 0.3 nm, and the maximum speed was 220 mm/s [8].

From the foregoing, most macro-micro precision positioning systems only have one or two degrees of freedom and can not be used in vacuum environment such as scanning electron microscope (SEM). Then it is necessary and significant to investigate a kind of macro-micro precision positioning 3-DOF system for application in vacuum environment. There are many types of mechanisms have been developed as macro positioner, such as 3RRR (3 degrees of freedom, each branch consists of three rotating pairs (R)), 4RRR (4 degrees of freedom, each branch consists of three rotating pairs (R)) and 3PRR planar parallel mechanism [9]. The micro positioner also has attracted many researchers' interest and has been commercialized for a long time [10]. A kind of macro-micro precision positioning system with 3PRR planar parallel mechanism and piezoelectric micro stage is introduced and investigated in this paper. Firstly the experimental setup is shown and introduced in Sect. 2. Then analysis and experiment of loading effect on positioning accuracy is demonstrated in Sect. 3. At last, the experiment of macro-micro positioning system is conducted to show the improvement in positioning accuracy and travel range in Sect. 4. In the paper, laser interferometer is used as feedback control measurement, which means that only linear motion tracking but not circular motion tracking can be controlled. Therefore, the visual inspection and multiple sensors combination technique should be investigated to apply for the macro-micro positioning system in future work.

2   Introduction of the Experimental Setup

From Fig. 1, it can be seen that the experimental setup consists of macro positioner, micro positioner, laser interferometer, interference reflector, reflector and computer. The macro positioner is 3PRR planar parallel platform shown in Fig. 2(a), which is composed of static platform, mobile platform, sliding pair, revolving pair and grating scale. The sliding pair is driven by ultrasonic linear motor which can be used in vacuum environment. The macro positioner has three degrees of freedom as shown in Fig. 2(a), which are X axis translational freedom, Y axis translational freedom and rotation freedom around Z axis, respectively. Figure 3 shows the schematic diagram of macro-micro precision positioning system.

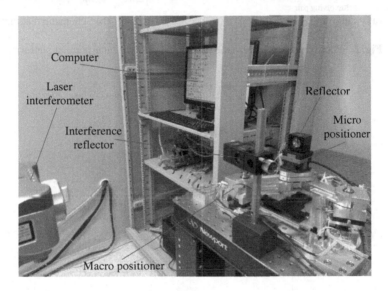

Fig. 1. Photograph of macro-micro precision positioning system experimental setup.

3PRR planar parallel positioning platform includes mechanical part and electric control part. Mechanical part is composed of three same kinematic chain, with each chain including ultrasonic linear motor, connecting rod, motor bracket and linear guide. The mobile platform is supported by three connecting rods of kinematic chains. The electric control part consists of linear grating encoder, photoelectric limit switch, motor driver, computer and motion control card.

The micro positioner is piezoelectric micro stage of PI company, which has compact size of 44 × 44 × 44 mm, weight of 320 g and XYZ strain gauge sensors. The travel range is 100 × 100 × 100 μm, and resolution is 0.2 nm.

Laser interferometer is used to measure the movement of reflector fixed on the micro positioner. In order to minimize the Abbe error and cosine error as far as possible, laser interferometer should be carefully fixed. The micro positioner is fixed on the mobile platform of macro positioner, which can realize the macro-micro superimposed motion.

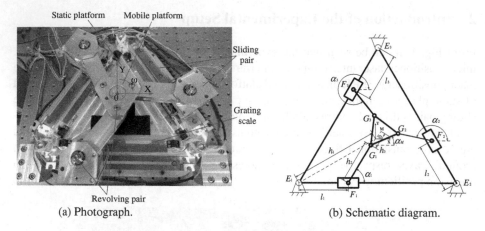

(a) Photograph.   (b) Schematic diagram.

Fig. 2. Photograph and schematic diagram of 3PRR planar parallel platform.

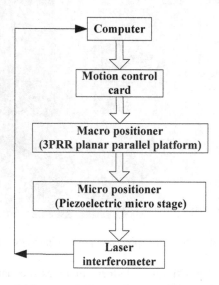

Fig. 3. Schematic diagram of macro-micro precision positioning system.

During the experiment, it is important to make the directions of motion of the macro positioner and micro positioner collinear.

Schematic diagram of 3PRR planar parallel mechanism is shown in Fig. 2(b). The closed-loop vector method is used to analyze the kinematic chain.

Because the sum of the closed loop vector is zero, then the kinematic constraint relation can be expressed as

$$\boldsymbol{E_iF_i} + \boldsymbol{F_iG_i} + \boldsymbol{G_iM} + \boldsymbol{MO} + \boldsymbol{OE_i} = \boldsymbol{0}, (i = 1, 2, 3) \tag{1}$$

where the meaning of E, F, G, M and O can be seen in Fig. 2(b), E is the fixed point, F is the first rotating pair, G is the second rotating pair, M is the geometric center of mobile platform, and O is the geometric center of static platform.

The projection along the X axis and Y axis of Eq. (1) can be expressed as

$$\begin{cases} l_i \cos(\dfrac{2\pi(i-1)}{3}) + h_2 \cos \alpha_i + h_3 \cos(\dfrac{\pi(4i-3)}{6} + \alpha_M) = X_M - X_{Ei} \\ l_i \sin(\dfrac{2\pi(i-1)}{3}) + h_2 \sin \alpha_i + h_3 \sin(\dfrac{\pi(4i-3)}{6} + \alpha_M) = Y_M - Y_{Ei} \end{cases}, \qquad (2)$$

where X_{Ei} and Y_{Ei} is the coordinate of point E_i, and X_M, Y_M, α_M is X axis translational freedom, Y axis translational freedom and rotation freedom around Z axis, respectively.

By solving Eq. (2), driving parameters l_i and α_i can be achieved and driving voltage of ultrasonic linear motor can be got for trajectory planning, which is also known as inverse kinematics. In order to make the paper concise, the expressions of l_i and α_i are omitted.

3   Experiment of Loading Effect

The dynamic model describes the transfer relationship of the motion of planar parallel mechanism and driving joint force and includes two main methods such as positive dynamics and inverse dynamics. Inverse dynamics is that one can use the known motion of planar parallel mechanism to calculate the driving joint force. In the mechanism design phase, inverse dynamics has important significance in the motor selection, energy consumption analysis and trajectory planning. In the design stage of mechanism motion control, the inverse dynamics is the basis of mechanism dynamics control. The Lagrange method can be used to model the inverse dynamics of 3PRR planar parallel positioning platform and deduce the dynamics equations. However, without considering factors such as deformation and friction, this method only represents the ideal modeling of 3PRR planar parallel platform. It is hard to get the relationship between different loads and positioning accuracy. Therefore, experimental research on the relationship between different loads and positioning accuracy is necessary.

From the analysis above, one can see that before conducting the experiment of macro-micro precision positioning system, it is necessary to investigate the loading effect on the positioning accuracy of 3PRR planar parallel platform. The weights of 92 g and 200 g are chosen to conduct the experiment. In order to have a scientific comparative result, four groups of experiments of different loads have been designed and conducted. These different loads are 0 g, 92 g, 292 g and 400 g, respectively. Figure 4 shows the photograph of different weights and loading experimental setup, in which the weight is used to replace the micro positioner as load in Fig. 1.

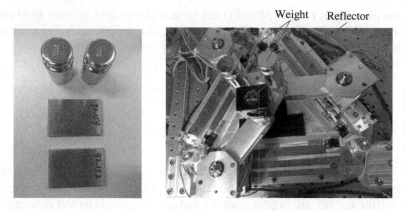

Fig. 4. Photograph of different weights and loading experiment.

In order to ensure experiment repeatable and accurate, the motion of 3PRR planar parallel platform is designed to move in straight line step by step from 0 mm to 12 mm and from 12 mm to 0 mm. Every loading experiment is designed to finish five linear motions to verify the repeatability positioning accuracy as Fig. 5 shows. Figure 5 only shows the loading experiment of 92 g load, and other experiments of 0 g, 292 g and 400 g can be showed in the same way. Figure 6 shows zoom in on the displacement of 2 mm of Fig. 5, in which the red point represents stable positioning point after the positioning motion has been finished. From Fig. 6, one can see that the positioning error will be adjusted and decreased in 0.5 s before the positioning motion is finished. The conventional PID feedback control is used to adjust the positioning error with parameter settings as $K_P = 25, K_I = 0.5, K_D = 50$.

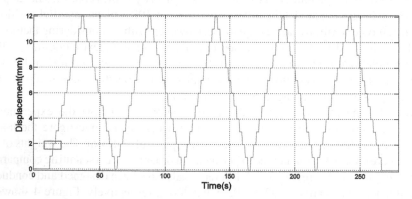

Fig. 5. Five linear motions step by step from 0 mm to 12 mm and from 12 mm to 0 mm of 92 g.

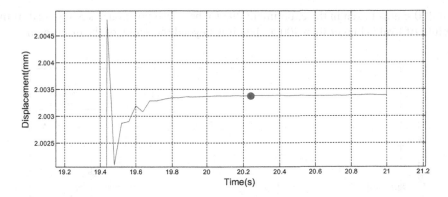

Fig. 6. Zoom in on the displacement of 2 mm of Fig. 4.

The absolute value of positioning error of different loads is shown in Fig. 7, from which on can see that the error from 0 mm to 12 mm becomes bigger and bigger due to error accumulation and from 12 mm to 0 mm becomes smaller and smaller due to repeatability positioning accuracy. It can also be seen that the absolute value of positioning error of no load is the smallest compared with other experiments of 92 g, 292 g and 400 g load. With the weight increasing, the absolute value of positioning error will become a little bigger than that of no load experiment, but the error is still much less than the travel range of the micro positioner.

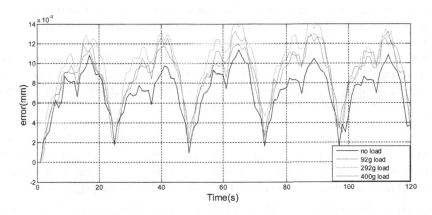

Fig. 7. Absolute value of positioning error of different loads.

Figure 8 shows the standard deviation from displacement 0 mm to 12 mm, from which it can be seen that the standard deviation of no load is also the smallest compared with other experiments of 92 g, 292 g and 400 g. Therefore, from the experiment of loading effect, on can draw a conclusion that both the absolute value of positioning error and standard deviation of no load is the smallest compared with other experiments of 92 g, 292 g and 400 g load, but the maximum error of 14 μm is still in the micro positioner's travel range of 100 μm. It should be noted that only loads of 0 g, 92 g, 292 g and 400 g are chosen in the experiment due to the micro positioner's 320 g load. If the weight of load is bigger than 400 g, then it is beyond the scope of the discussion.

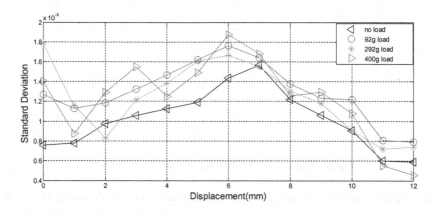

Fig. 8. Standard deviation from displacement 0 mm to 12 mm of different loads.

4   Experiment of Macro-Micro Precision Positioning System

To validate the higher positioning accuracy of the macro-micro precision positioning system compared with the macro positioner, it is necessary to conduct an experiment of macro-micro precision positioning system. The motion of macro-micro precision positioning system is designed to move in straight line step by step from 0 mm to 12 mm and from 12 mm to 0 mm, which is almost the same as the experiment of loading effect of Sect. 3. The only difference is that the micro positioner will be used to compensate the positioning error of macro positioner step by step, which will greatly improve the positioning accuracy of system. The experimental setup of macro-micro precision positioning system is shown and illustrated in Fig. 1.

The flow chart of experiment of macro-micro precision positioning system is shown in Fig. 9, from which the experimental process can be concluded by the following steps. Firstly the experiment is prepared and displacement input s_i is set to start the motion of macro positioner. After the motion of macro positioner is finished, displacement s_j measured by laser interferometer is used to compare with displacement input s_i. If the value of $|s_i - s_j|$ is smaller than 0.20 μm, then the micro positioner keeps still and another displacement input s_i will be set. If the value of $|s_i - s_j|$ is bigger than 0.20 μm, then the micro positioner begins moving and compensating error. After that, another displacement input s_i also will be set. After the linear motion from 0 mm to 12 mm and from 12 mm to 0 mm has finished, the experimental process is over.

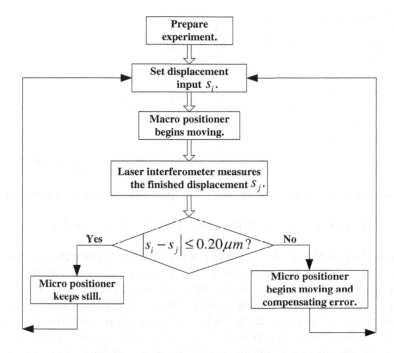

Fig. 9. Flow chart of experiment of macro-micro precision positioning system.

Figure 10 shows the linear motion step by step from 0 mm to 12 mm and from 12 mm to 0 mm of macro-micro precision positioning system. Zoom in on the displacement of 2 mm and 9 mm of Fig. 10 is shown in Fig. 11, from which one can see that the micro positioner can largely compensate the positioning error of macro positioner, which will greatly improve the positioning accuracy of system. In Fig. 11(a), it can be seen that the positioning error of 2 mm has decreased from 4.30 μm to 0.20 μm. In Fig. 11(b), it can be seen that the positioning error of 9 mm has decreased from 2.25 μm to 0.25 μm.

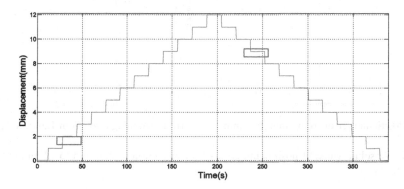

Fig. 10. Linear motion step by step from 0 mm to 12 mm and from 12 mm to 0 mm of macro-micro precision positioning system.

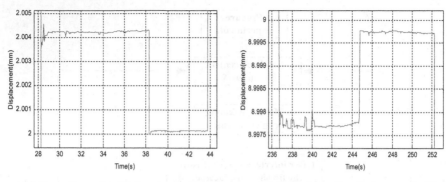

(a) Zoom in on the displacement of 2mm.   (b) Zoom in on the displacement of 9mm.

Fig. 11. Zoom in on the displacement of 2 mm and 9 mm of Fig. 9.

Figure 12 shows the absolute value of positioning error of linear motion from 0 mm to 12 mm and from 12 mm to 0 mm. In Fig. 12(a), one can see that the maximum positioning error of linear motion from 0 mm to 12 mm has decreased from 4.33 μm to 0.20 μm. The minimum positioning error of linear motion from 0 mm to 12 mm has decreased from 0.66 μm to 0.02 μm. In Fig. 12(b), it can be seen that the maximum positioning error of linear motion from 12 mm to 0 mm has decreased from 5.74 μm to 0.27 μm. The minimum positioning error of linear motion from 12 mm to 0 mm has decreased from 0.26 μm to 0.01 μm. Therefore, one can draw a conclusion that the macro-micro precision positioning system can greatly improve the positioning accuracy and also can enlarge the travel range.

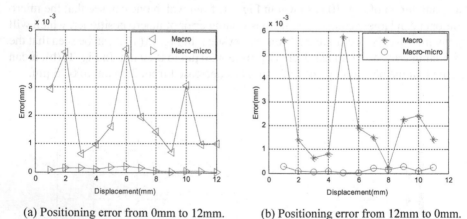

(a) Positioning error from 0mm to 12mm.   (b) Positioning error from 12mm to 0mm.

Fig. 12. Absolute value of positioning error of linear motion from 0 mm to 12 mm and from 12 mm to 0 mm.

5 Conclusions

A kind of 3-DOF macro-micro precision positioning system with 3PRR planar parallel mechanism and PI nanocube-piezo stage is introduced and investigated in this paper. Laser interferometer is used to measure the linear motion displacement of macro-micro precision positioning system for feedback control. Firstly the experimental setup is shown and introduced in Sect. 2, which consists of macro positioner, micro positioner, laser interferometer, interference reflector, reflector and computer. Then modeling analysis and experiment of loading effect is demonstrated in Sect. 3, from which one can see that there is some loading effect on positioning accuracy and standard variance of macro positioner, but the positioning error is much less than the travel range of micro positioner. At last, the experiment of macro-micro precision positioning system is conducted to show the improvement in positioning accuracy and travel range and experimental results show good performance.

Acknowledgment. The authors gratefully acknowledge these support agencies such as the National Natural Science Foundation of China (Grant No. U1501247, U1609206), the Science and Technology Planning Project of Guangdong Province (Grant No. 2014B090917001), the Natural Science Foundation of Guangdong Province (Grant No. S2013030013355) and the Scientific and Technological Project of Guangzhou (Grant No. 2015090330001).

References

1. Choi, S.B., Han, S.S.: Position Control System Using ER Clutch and Piezoactuator. Smart Structures and Materials 2003: Smart Structures and Integrated Systems, vol. 5056 (2003)
2. Liau, L.C.K., Chen, B.S.C.: Process optimization of gold stud bump manufacturing using artificial neural networks. Expert Syst. Appl. **29**, 264–271 (2005)
3. Ang, X.F., Zhang, G.G., Wei, J.: Temperature and pressure dependence in thermo compression gold stud bonding. Thin Solid Films **504**, 379–383 (2006)
4. Sharon, A., Hardt, D.: Enhancement of robot accuracy using endpoint feedback and a macro-micro manipulator system. In: American Control Conference, pp. 1836–1845 (1984)
5. Hongzhong, L., Bingheng, L., Yucheng, D., Hangsong, L., Le, Y.: Study of ultra precision positioning system and linearity compensation. J. Xi'an Jiaotong Univ. **37**(3), (2003)
6. Liu, C.H., Jywe, W.Y., Jeng, Y.R.: Design and control of a long-traveling nano-positioning stage. Precis. Eng. **34**(3), 497–506 (2010)
7. Juhász, L., Maas, J., Borovac, B.: Parameter identification and hysteresis compensation of embedded piezoelectric stack actuators. Mechatronics **21**(1), 329–338 (2011)
8. Shinno, H., Yoshioka, H., Sawano, H.: A newly developed long range positioning table system with a sub-nanometer resolution. CIRP Ann. **60**(1), 403–406 (2011)
9. Jiasi, M.: Research of the planar 3PRR parallel positioning system under the micro/nano operating environment. South China University of Technology (2016)
10. Wang, R., Zhang, X.: A planar 3-DOF nanopositioning platform with large magnification. Precis. Eng. **46**, 221–231 (2016)

Development of Control System
for the Assembly Equipments of Spacer
Bar Based on PLC

Hong He[1(✉)], Congji Li[1], Xiaoqin Li[1], Zegang Wang[1],
and Peng Xu[1,2]

[1] Beijing University of Chemical Technology, Beijing 100029, China
hehong@mail.buct.edu.cn
[2] Siping Power Line Hardware Co. Ltd., Siping 136001, China

Abstract. At present spacer bar cable clamp used for high voltage transmission line has been assembled manually at home, which lead to low production efficiency and unstable product quality. In order to achieve the high quality requirement of spacer bar cable clamp, a coordinated control system for spacer bar assembly equipments based on S7-200 PLC was developed. The composition of the control system was determined according to spacer bar assembly process and the hardware and the software included in the control system were designed. The new automatic control system greatly could improve the assembly efficiency and product quality, substituted for the hand labor.

Keywords: Spacer bar · PLC · Automatic assembly · Stamping riveting

1 Introduction

Due to the structure particularity of spacer bar, which is used in high voltage transmission line, the procedure of multiple fastening bolts riveting on punch for a spacer bar assembly can't realize continuous automatic production, the electric power tool manufacturer still rely on manual operations. Some unexpected failure could be caused by manual operation, for example riveting not tight would result in the nut off; riveting too tight would lead to damage the nut. As a result, this would lead to the riveting quality very unstable and high labor intensity, high labor costs, and the low production efficiency [1–3]. With the development of economy and the progress of science and technology, the process of mechanical assembly is also improving. Thus the automated assembly line is also an inevitable trend for mechanical assembly [4–13]. In view of this, the goal of this project was to improve the current situation of the actual operation for the spacer bar assembly, develop automatic fastening bolts riveting system based on PLC controlling and achieve automatic production of multiple fastening bolts riveting. This system could accelerate work efficiency greatly, improve product quality, and achieve higher manufacturing precision. At the same time, it could also reduce the cost and shorten the manufacturing cycle, complete the automatic riveting of the spacer bolt. It was of great significance for the processing of the spacer bar.

Y. Huang et al. (Eds.): ICIRA 2017, Part III, LNAI 10464, pp. 788–795, 2017.
DOI: 10.1007/978-3-319-65298-6_70

2 Assembly Process of Spacer Bar

The spacer bar assembly equipment mainly included the conveyer belt, manipulator, the rotary table, the limiter, the punch and other mechanical electrical equipment, as shown in Fig. 1. The process was as follows: First, the manipulator picked up the spacer bar and transferred it from the conveyor belt to the rotary table. Meanwhile the photoelectric sensor on the rotary table would detect the position of the spacer bar to perceive whether it is right on the turntable. If the photoelectric sensor light is turned on, the manipulator would place the work piece on the turntable, and the spacer bar would be clamped and fixed by the clamping device on the turntable. Then the rotating turntable started to rotate until it went to the limit position controlled by the limiter. At that moment the punch received the signal from limiter to rivet the spacer bar. After the spacer bar was punched, the turntable would rotate a certain angle and stop at a specified position controlled by the limit switch. At the same time the punch would receive the signal which signifies the second turns to rivet the spacer bar. After finishing six times such riveting action on spacer bar in turn, the counter would compute the punching times to six, and the punch would stop stamping, the manipulator would receive the signal from control system to pick up spacer again and put it back on the conveyor belt, next return to its original position, getting ready for the next piece. The process flow diagram is shown in Fig. 2.

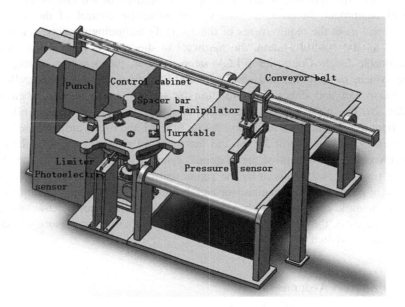

Fig. 1. The spacer bar assembly equipment

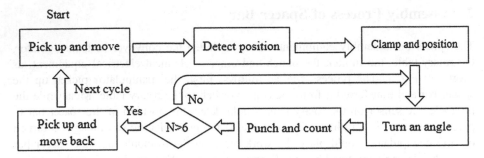

Fig. 2. Process flow of spacer bar assembly

3 The Hardware Design of the Control System for the Spacer Bar Assembly

3.1 Control System

The spacer bar assembly system could complete conveying, the rotating positioning and stamping riveting automatically. It integrated the automation technology, sensor technology, hydraulic transmission technology. The hardware system mainly included conveyor belt, manipulator, rotary working-table, limiter, punch and other auxiliary electrical equipment. In order to achieve the automatic control of the spacer bar assembly and meet the requirements of the assembly technology, safety, stability and economy for the control system, the hardware of the control system is as follows: Programmable Logic Controller (PLC), stepper motor, piezoelectric switch, photo-electric switch, limit switch and so on. S7-200 PLC was selected as controller. The rotary movement and direction of rotation for stepper motor and workbench was regulated by the pulse signal and direction signal sent by PLC controller. The manipulator movement, such as up and down, forward and backward, grab and release was driven by hydraulic cylinder. So did the stamping riveting motion of the punch. The reciprocating movement of the hydraulic cylinder was controlled by the electro-magnetic reversing valve driven by the signal sent by the on-off relay on the punch. The extreme position of the manipulator and the punch was limited by the limit switch. The rotation angle of rotary workbench was detected by the photoelectric switch and the identification of the work piece was completed by piezoelectric sensors. The control system frame is shown in Fig. 3.

3.2 I/O Address Assignment

Input/output module (I/O module) is the information exchange medium between CPU and site [14–18]. The input module is used to collect the input signal from the key switch and various sensors; while the output module may implement the program to drive the workload, such as contactors, solenoid valves and other equipment. The I/O address of the PLC is allocated in accordance with the control requirements and the number of input and output pairs for the spacer bar assembly process. The details of the

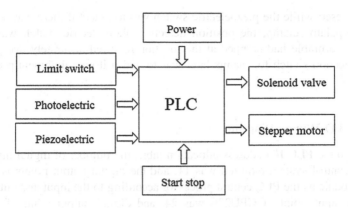

Fig. 3. Control system frame

address assignment is shown in Table 1. In order to improve the degree of automation and productivity, the limit switch, piezoelectric switch and photoelectric switch was used mainly to cooperate with PLC to complete some detection. Therein the limit switch of manipulator was to detect whether manipulator had risen or fallen to the right place; the limit switch on the punch was to check whether the stamping was done or the

Table 1. I/O address assignment.

Input address	Function	Output address	Function
I 0.0	Start button	Q0.0	Pulse signal output of rotary stepper motor
I 0.1	Stop button	Q0.1	Directional signal output of rotary stepper motor
I 0.2	Lower limit switch of manipulator	Q1.0	Solenoid valve of limiter YA1
I 0.3	Upper limit switch of manipulator	Q1.1	Downward solenoid valve of manipulator YA2
I 0.4	Left limit switch of manipulator	Q1.2	Upward solenoid valve of manipulator YA3
I 0.5	Right limit switch of manipulator	Q1.3	Right shift solenoid valve of manipulator YA4
I 0.6	Upper limit switch of punch	Q1.4	Left shift solenoid valve of manipulator YA5
I 0.7	Lower limit switch of punch	Q1.5	Stamping solenoid valve of punch YA6
I 1.0	Work piece detection sensor of manipulator	Q1.6	Reset solenoid valve of punch YA7
I 1.1	Position detection photoelectric switch of turntable	Q1.7	Solenoid valve of manipulator YA8
I 1.2	Position detection photoelectric switch of manipulator		

punch was reset; while the piezoelectric switch was to detect if there was a workpiece in the manipulator clamp; the position detection photoelectric switch was to check whether the turntable had stopped at the location required accurately every time; the position detection switch for manipulator was to find if it was alignment position with the turntable.

3.3 PLC I/O Wiring Diagram

According to the PLC I/O address allocation table, the number of digital inputs points which the control system required was 11, and the digital output points is 10. Select CPU226 module as the PLC central processor according to the input and output points. The digital input point of CPU226 was 24, and digital output point of it was 16. According to the I/O address allocation and PLC model, the PLC I/O wiring diagram is as shown in Fig. 4.

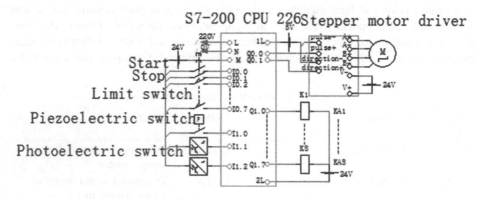

Fig. 4. PLC I/O Wiring diagram

4 Software Design of Control System

4.1 Action Cycle Graph

The action cycle of the control system is shown in Fig. 5. It was mainly divided into three parts which were the manipulator transfers the work piece from the conveyor belt to the turntable, the turntable rotated one turn for six stops and so did the stamping riveting for the punch, at last the manipulator moved the finished work piece from the turntable to the conveyor belt and then returned to its origin.

4.2 Program Flow Chart

The program for the control system, that is the PLC program mainly included the main program, the initialization subroutine, the manipulator conveying the work piece subroutine, the punch stamping-riveting subroutine, and the manipulator removing the

Action / Agency	Manipulator handling workpiece						Punch riveting (6 times)				Manipulator withdrawal					
	Falling	Grab	Rising	Move	Decline	Release	Backward	Forward	Punch	Reset	Falling	Grab	Rising	Move	Falling	Release
Manipulator																
Limiter																
Punch																
Stepper motor																

Fig. 5. Action cycle graph

work piece subroutine. The main program was executed to call subroutine and complete the automatic operation, the main program flow chart and PLC ladder diagram can be seen in Fig. 6. The initialization subroutine was used to initialize the system parameters, clear the data and reset the executive work piece. The manipulator conveying the work piece subroutine was applied to carry the work piece to the turntable. The punch stamping-riveting subroutine was to rivet the work piece, and the flow chart of it can be seen in Fig. 7. The manipulator removing the work piece subroutine was to move the finished work piece from the turntable back to the conveyor belt.

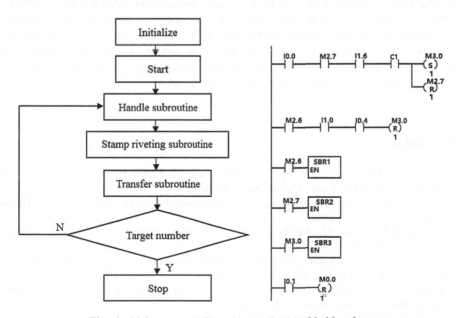

Fig. 6. Main program flow chart and part of ladder chart

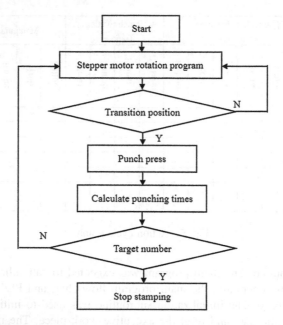

Fig. 7. The flow chart of punch stamping-riveting subroutine

5 Concluding Remarks

By means of the combination of hydraulic system and electrical control system, the modular design method was adopted to achieve the automatic linkage of conveying, stamping riveting of multiple fastening bolt of spacer bar assembly, the automatic rotation control and the accurate positioning of the turntable. The new control system based on PLC improved the situation of manual spacer bar clamps assembly in domestic production, increased automation and productivity greatly and reduced the production cost and enhanced the quality of bolt riveting.

References

1. Niu, H., Zhu, K., Sun, N., Wang, L.: Design and application of new type rotary clamp for 1000 KV spacer bar. Power Constr. **30**(2), 85–88 (2009)
2. La, Z.: Design and realization of machining control system based on PLC. Hunan University, Changsha (2004)
3. Hagedorn, P., Kraus, M.: On the performance of spacer dampers in bundled conductors. ETEP **3**(4), 304–311 (1993)
4. Lin, F., He, Y.: Design and realization of automatic assembly machine control system based on PLC. Equip. Manufact. Technol. **4**, 47–50 (2012)
5. Cohen, Y.: A technique for integrated modeling of manual and automatic assembly. J. Manufact. Technol. Manage. **26**(2), 164–181 (2015)
6. Zhou, B.: Bit based on S7-200PLC. Prod. Technol. Forum **9**, 58–60 (2014)

7. Chen, K.: Talking about the design method of PLC control system. Sci. Technol. Inf. China **20**, 116–118 (2009)
8. Luan, H., Hu, G., Chen, M.: Design and research of control system of wheel - mounted automatic assembly machine based on. Inf. Technol. **37**(6), 145–147 (2008)
9. Li, G.: Design and study of electroplating driving control system based on PLC. Appl. Mech. Mater. **727–728**, 641–644 (2015)
10. Wang, M.: Automatic gear assembly machine. CN 203156300 U.03-07 (2013)
11. Gao, P., Meng, J.: Electrical control basics and programmable controller application tutorial. Xi'an Univ. Electron. Sci. Technol. Press **22**, 210–221 (2007). Xian
12. Pan, H., Li, H.: A fully automated assembly machine. CN203197559U -09-18 (2013)
13. Zhang, P., Liao, W., Liu, C.: Automatic assembly of parts based on assembly characteristics. Electr. Technol. Autom. **05**, 93–96 (2004)
14. Yuan, H.: Program design of PLC control stepping motor. Mechan. Eng. Autom. **4**, 173–180 (2016)
15. Gu, J., Tang, M.: Design of position control system based on S7-200PLC and stepping motor. Precis. Manufact. Autom. **4**, 41–44 (2016)
16. Terada, Y., Murata, S.: Automatic modular assembly system and its distributed control. Int. J. Rob. Res. **4**, 445–453 (2008)
17. Xu, X., Dong, J.: Design of automatic assembly machine for single chip microcomputer. Mechanical **31**(8), 44–45 (2004)
18. Chen, R.: Design and Analysis of a New Type of Automatic Assembly Machine. Hefei University of Technology, Hefei (2015)

HADAMARD Transform Sample Matrix Used in Compressed Sensing Super-Resolution Imaging

Mei Ye[✉], Hunian Ye, and Guangwei Yan

Huazhong University of Science and Technology,
1037 Luoyu Road, Wuhan, China
yemei02@hust.edu.cn

Abstract. To realize super-resolution optical and terahertz imaging by the sub-wavelength hole arrays with extraordinary optical transmission (EOT) performance, the related sampling template structure is studied. In this paper, Hadamard matrix, cyclic S matrix and matrix chosen by random matrix selector are defined as a small probability matrix. The sub-wavelength coding imaging template based on the structure of this small probability matrix have extraordinary optical transmission (EOT) performance, which is the basis for achieving super-resolution imaging. It is showed that, in the case of array detector, high signal-to-noise ratio (SNR) improvement (exceeded 22) can be obtained only by sampling one frame using the imaging template designed by 3 order cyclic S matrix. For single detector, the SNR improvement using various order of cyclic S matrix designed imaging template is higher than N a quadratic root obtained by Hadamard matrix in classical Hadamard transform optical imaging method, and much higher than half a quadratic root of N by classical cyclic S matrix.

Keywords: Compressed sensing · Hadamard transform · Imaging · Sampling template

1 Introduction

The construction of sampling matrices in compressive sensing (CS) is a key constituent part of compressive sensing theory. Candes's Restricted Isometry Property (RIP) aimed at the construction of sampling matrices in CS creates the necessary condition of the construction of sampling matrices [1]. The door of rapid development stage of CS has been opened by RIP. Gaussian random matrix is good enough to meet RIP condition, and can be used as the sampling matrix in CS. But the sampling template on the random matrix is difficult to realize, so people turn to various formal deterministic matrices to grantee realizability of sampling matrices [2]. In the literatures [3–10] various formal deterministic matrices have been investigated. In this paper Hadamard matrix, cyclic S matrix and selected random matrix has studied as the construction of sampling matrices.

Y. Huang et al. (Eds.): ICIRA 2017, Part III, LNAI 10464, pp. 796–807, 2017.
DOI: 10.1007/978-3-319-65298-6_71

1.1   Hadamard Matrix

The Sylvester Hadamard matrix which has 2^n order is very important in the technology of Hadamard transform. It is constituted by the following formula:

$$[H_2^n] = [H_2] \otimes [H_2^{n-1}] = \begin{bmatrix} H_2^{n-1} & H_2^{n-1} \\ H_2^{n-1} & -H_2^{n-1} \end{bmatrix} \tag{1}$$

In which, $H_2 = \begin{bmatrix} 1 & 1 \\ 1 & -1 \end{bmatrix}$

1.2   Cyclic S Matrix

The cyclic S matrix of m sequence structure is constructed by the theory of linear feedback shift register. The sequence $a_i, a_{i+1}, \ldots, a_{i+2^p-2}$ owning the length of $2^p - 1$ is called pseudo-random sequence. They constitute a collection of $pa_m = (a_i, a_{i+1}, \ldots, a_{i+2^p-2})(i = 0, 1, \ldots, 2^p - 2)$. Take the pseudo-random sequence pa_m as the first row of the matrix, and construct n ($n = 2^p - 1$) order cyclic S matrix as follows:

$$S_n = \begin{bmatrix} a_0 & a_1 & \cdots & a_{n-1} \\ a_1 & a_2 & \cdots & a_0 \\ \vdots & \vdots & & \vdots \\ a_{n-1} & a_0 & \cdots & a_{n-2} \end{bmatrix} \tag{2}$$

Most cyclic S matrices can be folded into a two-dimensional [11].

1.3   The Sampling Template of Super Resolution Imaging

To realize super resolution imaging, the sub-wavelength hole arrays following the principle of a certain periodicity need to be made on the surface of the metal film. For array detector, the required measurements y for the recovery algorithm can be obtained by sampling one frame or several frames of image with a fixed template. However, in the case of single detector, y can be achieved by moving the template or transform the template for a certain amount of times. The foldable cycle S matrix can match the moving sampling mode of single detector. The Hadamard matrix can match the transformable sampling mode of single detector or two detectors. The selected binary random matrix can match the transformable sampling mode of single detector.

The coding matrix such as Hadamard matrix, cyclic S matrix is based on the theory of combination weighing. The Hadamard matrix is composed of +1, −1, and +1, 0 constitute the cyclic S matrix (+1, 0 and −1 denote transmittance, opacity and reflections respectively). The basic feature of cyclic S matrix is that each line of matrix can be obtained by a left or right moving of the previous line. The theoretical signal-to-noise ratio (SNR) improvement of Hadamard matrix called optimum chemical balance weighing design is $\sqrt{n}$ (n is the order of matrix), and $\sqrt{n/2}$ for cyclic S matrix called best spring dynamometer weighing design [12]. This kind of matrix can construct the periodic sub-wavelength hole arrays.

The detecting size of single detector is different due to the different folding sizes of imaging template. Table 1 lists the corresponding and folding situation of the different order cyclic S matrix.

Table 1. The element number of template corresponding to single detector

The style of matrix	The element number	The folded size
3 order cyclic S	3	3×1
7 order cyclic S	7	7×1
15 order cyclic S	15	5×3
35 order cyclic S	35	7×5
63 order cyclic S	63	9×7
255 order cyclic S	255	51×5
511 order cyclic S	511	73×7
1023 order cyclic S	1023	33×31

Note: The element number of template corresponding to single detector should be 3, 15, 35, 63 and 1023 which can be folded into a square matrix.

The some sampling template based on the cyclic S matrix is also generated by the cyclic rule of the cyclic S matrix line by line. So this template to array detector can be generated according to 256×256 array or 1024×1024 array.

In Fig. 1 shown the extraordinary optical transmission (EOT) [13–16] performance of the sampling template based on the 3 order cyclic S matrix. The size of hole (white dot, etched in Au film, tungsten lamp illumination) is 100 nm and manufactured by FIB.

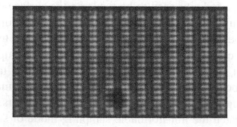

Fig. 1. Visible-light image of the extraordinary optical transmission (EOT) of the sampling template based on the 3 order cyclic S matrix.

2   Compressed Sensing Imaging

2.1   Compressed Sensing (CS)

Compressed sensing which breaks through the classical law of Nyquist sample, aims to recover a sparse image signal x from under sampled indirect measurements [17].

$$y = Ax \in X^{nxN}(n \leq N) \tag{3}$$

Then, x is expressed as $x = \Psi s$ with another sparse transform base $\Psi = \{\Psi_1, \Psi_2, \ldots, \Psi_N\}$. To solve the equations under defined namely NP-hard problem [18, 19], s can be estimated by convex optimization min $\|s\|_1$:

$$\arg\min\|s\|_1 \; s.t. \; y = \Lambda\Psi s \tag{4}$$

So, the main research contents in CS are about how to construct a sample matrix A and solve Eq. 4 accurately or the most approximately. Candes, Romberg and Tao et al. demonstrated that, the sensing matrix $\Theta = A\Psi$ satisfying the conditions of restricted isometry property (RIP) [20] is the guarantee of signal recovery:

$$(1 - \delta)\|s\|_2^2 \leq \|A\Psi s\|_2^2 \leq (1 + \delta)\|s\|_2^2 \tag{5}$$

In actuality, it is difficult to use RIP to instruct the design of the sample matrix. The equivalent condition of RIP presented by Baraniuk [21] is that the sample matrix A is irrelevant to sparse transform base Ψ and Gaussian random matrix has little coherence of most of the sparse transform base Ψ in order to satisfy the constraints preferably, which is usually used as sampling matrix in CS.

2.2   The Inevitability of Small Probability Events

In the classical probability statistics, a small probability event can be defined as an event owning low probability ($p < 0.05$). It is almost impossible to happen in one trial. Assuming that, the probability of a small probability event A is ($\varepsilon < 0.05$), then $\bar{A}(1 - \varepsilon)$. B can be defined as the event that A doesn't happen at all in n trials, so the probability of B is $(1 - \varepsilon)^n$. Therefore, the probability of $\bar{B}$ occurring at least once in n trails is as follows:

$$P = 1 - (1 - \varepsilon)^n \tag{6}$$

$P \to 1$, as $n \to \infty$, thus $\bar{B}$ namely the occurring of small probability event A will always be carried out in the constant independent trails. The transformation theorem of random events [22] presented by Ke-qin Zhao indicates that, if the event A occurred at the K-th test for the first time in the frequency-type random trial E (n is large enough), the event $\bar{A}$ will definitely happen at the $(K + m) - th$. Equation 6 attests that a small probability event of almost impossible occurrence is bound to happen, which aligns with the transformation theorem of random events.

There exists a small probability matrix in the random sampling matrix to obtain a better recovery image.

2.3   The Comparing of the Recovery Performance of 1023 Order Cyclic S Matrix and Other Orders Cyclic S Matrices in the Corresponding Position of the Target Image

Using 1023 order cyclic S matrix to compare with other orders cyclic S matrix in the corresponding position of the recovery image, we find that the recovery performance of 1023 order cyclic S matrix is better than other orders cyclic S matrices under the same measurement conditions. The recovery images of various order cyclic S matrix are shown in Fig. 2, which have an ideal quality of recovery from the intuitionistic view of human (See Table 2).

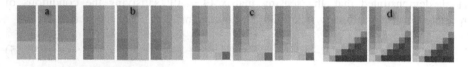

Fig. 2.   (a) Followed by the original image, the recovery image in the region of 3 × 1 and the recovery image of 1023 order cyclic S matrix in corresponding position; (b) Followed by the original image, the recovery image in the region of 5 × 3 and the recovery image of 1023 order cyclic S matrix in corresponding position; (c) Followed by the original image, the recovery image in the region of 7 × 5 and the recovery image of 1023 order cyclic S matrix in corresponding position; (d) Followed by the original image, the recovery image in the region of 9 × 7 and the recovery image of 1023 order cyclic S matrix in corresponding position.

Table 2.   The recovery performance of 1023 order and the other orders cyclic S matrices in the corresponding position

The style of matrix	MSE	RE	PSNR (dB)	M_{dB}
1023(3) order cyclic S	1.7080e-7	6.2270e-4	67.6752	115.3611
1023(15) order cyclic S	2.3654e-7	7.3172e-4	66.2609	113.8742
1023(35) order cyclic S	2.4723e-7	7.0459e-4	66.0689	113.6568
1023(63) order cyclic S	3.0744e-7	8.4457e-4	65.1224	112.7144

In Table 2, the region of 3 × 1, 5 × 3, 7 × 5, 9 × 7 is obtained in LENA image and the recovery image of various orders cyclic S matrices is obtained by OMP with the sampling rate of 100%. The mean square error (MSE)/The relative error (RE)/The peak signal to noise ratio (PSNR)/The improvement of SNR (M_{dB}).

2.4   The Selected Binary Random Matrix

The selected binary random matrix is a kind of small probability matrix. The random matrix selector with MATLAB software is designed. The function of the selector is to search a better binary random structure and obtain a preferable recovery by generating various random structures.

As is shown in Fig. 3, a random matrix selector module whose function is to select the random binary matrix using in CS is designed by MATLAB.

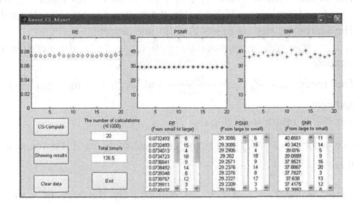

Fig. 3. The random matrix selector module.

In the experiments, the random matrix selector is used to generate thousands of different binary random matrices, and compute the PSNR of the image reconstructed by each sampling matrix. Owning the best performance, the first 18 matrices are work out. Based on the result above, the random matrix selector can be thought as an excellent method to search some preferable properties of sampling matrix.

3   Two Kinds of Recovery Algorithms and the Sampling Mode

Using the gradient projection for sparse reconstruction (GPSR) [23] in convex optimization algorithm and the orthogonal matching pursuit (OMP) [24] in greedy algorithm, it can be recovered that the sparse image signal x respectively.

The main thought of GPSR is searching along the feasible direction of descent from the initial feasible point and obtaining the new feasible point to decrease the objective function. When the iterative initial point is within the feasible region, one can achieve convergence threshold searching along the negative gradient direction.

3.1 The Difference Between the Imaging Modes of GPSR and OMP

GPSR. The measurements $y = A * x$ are obtained by means of matrix dot product (multiplying the corresponding element) in the program of MATLAB. So each element in the sampling matrix corresponds with each pixel of the original image one to one, which determines the sampling mode of the array detector in the super-resolution imaging experiment. The measurements y can be obtained by sampling one frame or several frames of image. The iterative process of GPSR is as follows:

*1 The initial feasible point $x_1 \in R^n$, the iterative threshold $\varepsilon \geq 0$ and $d_1 = -g_1, k = 1$ are given. If $\|g_1\| \leq \varepsilon$, the iterative process is stopped;

*2 The searching step α_k meets Armijo line search;

*3 $x_{k+1} = x_k + \alpha_k d_k, g_{k+1} = g(x_{k+1})$, if $\|g_{k+1}\| \leq \varepsilon$, the iterative process is stopped;

*4 The parameters $\beta_{k+1} = -\frac{\|g_{k+1}\|^2}{d_k^T g_k}$ and $d_{k+1} = -\theta_{k+1} g_{k+1} + \beta_{k+1} d_k$ are updated;

*5 $k = k+1$, return to step 2.

In the iterative process, $g_k = \nabla f(x_k)$ is the Gradient of x_k, $\alpha_k \geq 0$ is the searching step factor and $\theta_k = 1 - \frac{g_k^T d_{k-1}}{d_{k-1}^T g_{k-1}}$.

OMP. The measurements $y = A \times x$ are obtained by the way of standard matrix multiplication in the program of MATLAB. For the convenience of computing, the two-dimensional original image x is expanded into a one-dimensional column vector signal $\vec{x}$. The sampling matrix can be expressed as the set of row vector $A = (\alpha_1, \alpha_2, \ldots, \alpha_i)'$. Then, the measurements can be expressed as $y_i = [\alpha_i, \vec{x}]$ where the $[\alpha_i, \vec{x}]$ is the inner product of α_i and $\vec{x}$. The process above determines the sampling mode of the single detector in the super-resolution imaging experiment. The measurements y can be obtained only by the cyclic moving sampling with one imaging template based on cyclic S matrix.

15 order cyclic S matrix is selected as example. A brief description of the sampling process is shown in Figs. 4 and 5: The row vectors ①, ②, ③ etc., in matrix A correspond with the two-dimensional imaging coding template ①, ②, ③ etc. The single detector lies in the initial position corresponding with the two-dimensional imaging coding template ①. Then, it moves to ② and complete the second sampling. The measurements y is obtained by this way of recycling moving coding sampling.

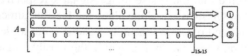

Fig. 4. The coding array for 15 order cyclic S matrix under the compressed sensing imaging method.

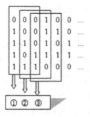

Fig. 5. Two-dimensional coding array based on 15 order cyclic S matrix (The folded size 5×3).

Figures 4 and 5 show the schematic diagram of cyclic moving sampling for single detector using 15 order cyclic S matrix as the sampling matrix.

The iterative process of OMP is as follows:

*1 The residual $r_0 = y$, the recovery image $x_0 = 0$, the index set $\Gamma^0 = \Theta$ and the number of iterations are initialized;

*2 The inner product $g^n = \Theta^T r^{n-1}$ is calculated;

*3 The element of absolute value of the largest element is found out and $k = \underset{i \in \{1,2,\dots,N\}}{\text{argmax}} |g^n[i]|$;

*4 The atoms set $\Theta_\Gamma = \Theta_{\Gamma^{n-1}} \cup \{\varphi_k\}$ and the new index set $\Gamma^n = \Gamma^{n-1} \cup \{k\}$ are updated;

*5 The approximate solution of signal $x^n = (\Theta_{\Gamma^n}^T \Theta_{\Gamma^n})^{-1} \Theta_{\Gamma^n}^T y$ is solved by the least Square method;

*6 The residual $r_0 = y - \phi x^n$ is updated;

*7 $n = n + 1$, if the condition of the iteration stopping is satisfied, the $\widehat{x} = x^n$, $r = r^n$ is exported. Otherwise the iterative process returns to step 2.

Compared with GPSR, the OMP can obtain a recovery image of higher precision. However, it has a higher complexity and a time-consuming feature due to the more sampling values. The column vector used to represent column atoms is obtained in the measurement matrix by the way of multiple iterations and gradually choosing, which is the main idea of OMP. The chosen atoms satisfying the principle of optimum matching atom have the maximum correlation with the current residual vector during each iteration. Then, the residual vector is constantly updated by subtracting the relevant part from measurement matrix until the iterations reach to K sparsity.

3.2   Study in the Recovery Performance of Small Probability Matrix with GPSR

Using the selected binary random matrix, Hadamard matrix and various order of cyclic S matrix as the sampling matrix in GPSR, the mean square error (MSE), the relative error (RE), the peak signal to noise ratio (PSNR), the SNR of image and the improvement of SNR (M_{dB}) are calculated. Results are shown in Table 3. The 3 order cyclic S matrix has the best recovery performance. As benchmark for the selected

Table 3. The recovery performance of random matrix and the small probability matrix under GPSR

The style of matrix	MSE	RE	PSNR (dB)	SNR	M_{dB}
The binary random	0.0012	0.0732	29.3113	38.1066	20.4797
Hadamard (+)	0.0011	0.0695	29.7626	36.9532	20.1172
Hadamard (−)	0.0010	0.0685	29.8951	36.3498	20.3107
3 order cyclic S	4.205e-4	0.0439	33.7626	33.0866	22.3908
7 order cyclic S	7.829e-4	0.0599	31.0631	33.7697	20.7678
15 order cyclic S	8.710e-4	0.0631	30.6001	33.1245	20.7020
35 order cyclic S	9.175e-4	0.0648	30.3741	35.5370	20.6341
63 order cyclic S	0.0010	0.0680	29.9560	35.5000	20.2910
255 order cyclic S	0.0010	0.0684	29.9096	38.5351	20.2231
511 order cyclic S	0.0010	0.0689	29.8459	34.8527	20.1733
1023 order cyclic S	9.958e-4	0.0675	30.0184	35.1509	20.3942

binary random matrix (RE <7.6%), these matrices above are desirable. The four image obtained by using the selected binary random matrix, Hadamard (+) matrix, Hadamard (−) matrix and 3 order cyclic S matrix are shown in Fig. 6. They all have an ideal quality of recovery from the intuitionistic view of human.

For Hadamard (+), +1 is retained and −1 is replaced by 0. For Hadamard (−), −1 is retained and +1 is replaced by 0.

Fig. 6. The four Lena image obtained by using the selected binary random matrix, Hadamard (+) matrix, Hadamard (−) matrix and 3 order cyclic S matrix. Followed by the recovery image of [a] binary random matrix, [b] Hadamard (+) matrix, [c] Hadamard (−) matrix, [d] 3 order cyclic S matrix.

3.3 Study in the Recovery Performance of Small Probability Matrix with OMP

As is known, the nature of OMP is using single detector to obtain the recovery image. The region of template corresponding with single detector is the approximate square. To reduce alignment errors, only the cyclic S matrix which can be folded into approximately square is studied, such as 3, 15, 35, 63, 1023 order cyclic S matrix (the difference between length and width in folded matrix is 2). Due to the long time consuming and the moving sampling mode only the corresponding part of x to be the

Table 4. The recovery performance of various orders cyclic S matrices under OMP

The style of matrix	MSE	RE	PSNR (dB)	M_{dB}
3 order cyclic S	8.7336e-4	0.0445	30.5881	27.1100
15 order cyclic S	2.5490e-6	0.0024	55.9364	69.2423
35 order cyclic S	1.4068e-5	0.0053	48.5176	70.0456
63 order cyclic S	2.8811e-7	0.0008	65.4044	91.3183
1023 order cyclic S	2.9350e-7	0.0011	65.3235	112.938

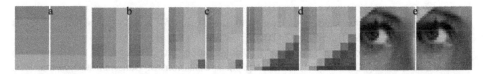

Fig. 7. Five divisions of Lena image. (a) Followed by the original image and the recovery image in the region of 3×1; (b) Followed by the original image and the recovery image in the region of 5×3; (c) Followed by the original image and the recovery image in the region of 7×5; (d) Followed by the original image and the recovery image in the region of 7×9; (e) Followed by the original image and the recovery image in the region of 33×31.

target image is selected. As is shown in Table 4 the 63, 1023 order cyclic S matrix can obtain a perfect recovery image than others. The corresponding recovery image is as shown in Fig. 7, which almost have no difference with the target image.

Study in the recovery performance of 1023 order cyclic S matrix with OMP. The 1023 order cyclic S matrix is used as sampling matrix. When the sampling rate reaches to 100%, the improvement of SNR <113> is higher than <32> obtained by Hadamard matrix in classical Hadamard transform optical imaging method, and <16> obtained by classical cyclic S matrix.

In Table 4, the region of 3×1, 5×3, 7×5, 9×7 is obtained in Lena image and the recovery results of various orders cyclic S matrices are obtained by OMP with the sampling rate of 100%.

4 Results

It is well demonstrated that a concept of small probability matrix from the CS sampling matrix with the inevitability theory of small probability event. For array detector, the selected binary random matrix, Hadamard matrix, cyclic S matrix can be used as the template of super resolution imaging by studying the recovery performance of them. The improvements of SNR among them are higher than 20, while it is preferable of the recovery performance using various order of cyclic S matrix designed imaging template for single detector. These matrices provide an important reference for achieving the wide-field super diffraction limit resolution optical and terahertz imaging with CS methods.

Acknowledgment. This research was partly supported by the National Natural Science Foundation of China "NSAF" Joint Fund Grants U1230109 and the National Natural Science Foundation of China Fund Grants 30627001, 60672058. This research was also supported by State Key Laboratory of Digital Manufacturing Equipment & Technology of HUST and advanced manufacturing testing center (School of Mechanical Science and Engineering of HUST).

References

1. Candes, E.J.: The restricted isometry and its implication for compressed sensing. Comtes Rendus Math. **346**(9/10), 589–592 (2008)
2. Monajemi, H., Jafarpour, S., Gavish, M., Collaboration, S.C., Donoho, D.L.: Deterministic matrices matching the compressed sensing phase transitions of Gaussian random matrices. Proc. Nat. Acad. Sci. USA **110**(4), 1181 (2012)
3. Zeng, L., Zhang, X., Chen, L., et al.: Deterministic construction of Toeplitzed structurally chaotic matrix for compressed sensing. Circ. Syst. Sign. Process. **34**(3), 797–813 (2015)
4. Nam, Y.Y., Na, Z.: Deterministic construction of real-valued ternary sensing matrices using optical orthogonal codes. IEEE Sign. Process. Lett. **20**(11), 1106–1109 (2013)
5. Calde, R.B.R., Howa, R.D.S., Jafa, R.P.S.: Construction of a large class of deterministic sensing matrices that satisfy a statistical isometry property. IEEE J. Sel. Top. Sign. Process. **4**(2), 358–374 (2010)
6. Amini, A., Montaze, R.H.V., Ma, R.V.F.: Matrices with small coherence using parry block codes. IEEE Trans. Sign. Process. **60**(1), 172–181 (2012)
7. Applebaum, L., Howa, R.D.S., Sea, R.L.S., et al.: Chirp sensing codes: deterministic compressed sensing measurements for fast recovery. Appl. Comput. Harmonic Anal. **26**(2), 283–290 (2009)
8. Li, S.X., Gao, F., Ge, G.N., et al.: Deterministic construction of compressed sensing matrices via algebraic curves. IEEE Trans. Inf. Theor. **58**(8), 5035–5041 (2012)
9. Ge, X., Xia, S.T.: LDPC codes based on Berlekamp-Justesen codes with large stopping distances In: Proceedings of the 2006 IEEE Information Theory Workshop, Piscataway, NJ, pp. 214–218 (2006)
10. Dimakis, A.G., Smarandache, R., Vontobel, P.O.: LDPC codes for compressed sensing. IEEE Trans. Inf. Theor. **58**(5), 3093–3114 (2010)
11. Gottlieb, P.: A television scanning scheme for a detector-noise-limited system. IEEE Trans. Inf. Theor. IT **14**(3), 428–433 (1968)
12. Ye, H., Ye, M., Yang, X.: Hadamard Transform Optical Imaging. HUST Publisher, Wuhan (2012). (in Chinese)
13. Genet, C., Ebbesen, T.W.: Light in tiny holes. Nature **445**(4), 39–46 (2007)
14. Ebbesen, T.W., Lezec, H.F., Thio, T., Wolff, P.A.: Extraordinary optical transmission through sub-wavelength hole arrays. Nature **391**(6668), 667–669 (1998)
15. Lalanne, P., Rodier, J.C., Hugonin, J.P.: Surface plasmons of metallic surfaces perforated by nanohole arrays. J. Opt. A: Pure Appl. Opt. **7**(8), 422–426 (2005)
16. Barnes, W.L., Dereux, A., Ebbesen, T.W.: Surface plasmon subwavelength optics. Nature **424**, 824–830 (2003)
17. Donoho, D.L.: Compressed sensing. IEEE Trans. Inf. Theor. **52**(4), 1289–1306 (2006)
18. Davis, G.: Adaptive nonlinear approximations. Ph.D. dissertation, New York University, New York (1994)

19. Natarajan, B.K.: Sparse approximate solutions to linear systems. SIAM J. Comput. **24**(2), 227–234 (1995)
20. Candes, E.J., Tao, T.: Decoding by linear programming. IEEE Trans. Inf. Theor. **51**(12), 4203–4215 (2005)
21. Baraniuk, R.G.: A lecture on compressive sensing. IEEE Sign. Process. Mag. **24**(4), 118–121 (2007)
22. Zhao, S.F., Zhao, K.Q.: Frequency-type contact probability and random events transformation theorem. CAAI Trans. Intell. Syst. **9**, 53–59 (2014). (in Chinese)
23. Figueiredo, M.A.T., Nowak, R.D., Wright, S.J.: Gradient projection for sparse reconstruction: application to compressed sensing and other inverse problems. IEEE Sel. Top. Sign. Process. **1**(4), 586–597 (2007)
24. Tropp, J.A., Gilbert, A.C.: Signal recovery from random measurements via orthogonal matching pursuit. IEEE Trans. Inf. Theor. **53**(12), 4655–4666 (2007)

Numerical Simulation of Forming Process Conditions and Wall Thickness for Balloon

Xuelei Fu[1], Hong He[1(✉)], and Wenchang Wang[1,2]

[1] Beijing University of Chemical Technology, Beijing 100029, China
hehong@mail.buct.edu.cn
[2] Lepu Medical Technology (Beijing) Co., Ltd., Beijing 102200, China

Abstract. The forming of balloons used in medical treatment is a kind of "black box art". When a new balloon is being developed, the process parameters and tube dimension are usually determined by a method of trial and error. This method is inefficient in current rapid development of computer technology. Numerical simulation is expected to replace the experiments and experience to guide the development of the new products. In this study, the moulding of the balloon was simulated by a finite element method and the results obtained from the simulation agreed with that of the experiments under the same actual process parameters. Therefore the numerical simulation used is feasible for the process of balloons forming. The effect of process parameters on the wall thickness of balloon was analyzed based on orthogonal design method. The results showed that the effect of first stretch rate on the wall thickness of the balloon was the most significant compared with other process parameters. A regression model of the relationship between wall thickness and the process parameters was established, which could be used to guide the selection of production process parameters.

Keywords: Balloon forming process · Numerical simulation · Orthogonal experiment · Regression model

1 Introduction

The balloon plays an important role in angioplasty with the functions of vascular pre-expansion, shaping and stent delivery [1–5]. The human body's blood vessels are very complex and the location for the arterial blockage is not the same. So the treatment of cardiovascular disease has a variety of specifications on the balloon size, which is a challenge for the moulding of balloons. When a new balloon is developed, the tube size, mold size, process parameters and other factors need to be considered, the experimental method is very low efficiency. How to determine the process parameters of the new balloon forming quickly is of great help to improve the market competitiveness of enterprises.

Numerical simulation for balloon forming process is difficult because the temperature, tensile speed, inflation pressure and other process parameters are variable in the forming process. At present, few reports on the numerical simulation of balloon formation at home and abroad can be found. Only Menary and Armstrong [6] have used

© Springer International Publishing AG 2017
Y. Huang et al. (Eds.): ICIRA 2017, Part III, LNAI 10464, pp. 808–818, 2017.
DOI: 10.1007/978-3-319-65298-6_72

the finite element software to simulate the first phase of the balloon formation and verified that the numerical simulation can be used in the balloon forming process by comparing forming process shoot in high speed camera with the simulation.

In this study, the finite element software was used to simulate the first stretching process at low blowing pressure and the secondary stretching process at high blowing pressure. In order to short the production development and improve production efficiency and product quality, the wall thickness of the balloon was taken as the research object, and the relation of wall thickness with process parameters was expected to be set up.

2   Balloon Forming Equipment and Process

The manufacturing process for balloons requires two kinds of equipment respectively. They are the Pipe drawing machine and Balloon forming machine (see Fig. 1). the Pipe drawing machine is used to make the tube into a parison, it can improve the stability of balloon molding. Balloon molding machine is controlled by the computer program, it can achieve synchronization among heating, axial stretching and internal pipe pressurization. Another advantage for balloon molding machine is its mold can be designed flexibly according to the needs of manufacturers.

(a) Pipe drawing machine   (b) Balloon forming machine

Fig. 1. Balloon forming machine

Figure 2(a) is the flow chart of the balloon molding process. First, the tube is stretched by using the pipe drawing machine to produce the parison. Then, the parison is put into the mold of the balloon forming machine whose process parameters such as temperature, drawing speed and inflation pressure are set. Finally, check the balloon quality after the process is completed. The forming procedure for the balloon forming machine is as follows: The first stage is to increase the mold temperature and to a certain blowing pressure at the same time. When the temperature reaches the set temperature, the parison is axially stretched. The second stage is to maintain the stability for a period of time after the first stage. Then increase the pressure with the secondary axial stretching. The third stage is to hold high pressure for some time, then cooling and demoulding. The evolution of the balloon in the forming process is shown in Fig. 2(b).

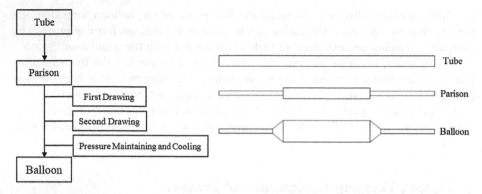

(a) The flow chart of the balloon molding process   **(b)** The evolution of the balloon

Fig. 2. Balloon forming process diagram.

3   Numerical Simulation

In this study, the size of a common PTA balloon for numerical simulation study is 9 × 40 mm and the material is PA12.

3.1   Geometry

The components involved in balloon molding are consisted of clamps, mold in the balloon molding machine. Clamps and mold can be considered to be excessively stiff relative to the tube. They were therefore modelled using an analytical rigid surface. Parison has two parts, one is the stretched part and the other is the unstretched part. Part of stretched section does not participate in the formation of the balloon, so the length of the stretched part can be appropriately reduced when building a geometric model. The structure and dimensions of the parison and target balloon are shown in Figs. 3 and 4, respectively.

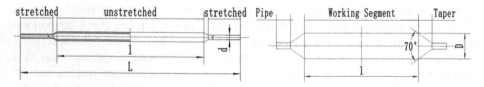

Fig. 3. Parison structure and dimensions   **Fig. 4.** Target balloon structure and dimensions

3.2 Meshing

Parsion was undergone a large deformation at biaxial directions in the formation process. Also the other dimensions are much larger than that of the wall thickness and the final shape of the balloon has the thin-walled structure, so the shell element is used for the meshing [7, 8]. In the meshing, the spacing between elements in the stretched part is lower than the unstretched part (see Fig. 3). The total number of mesh is 12240, as shown in Fig. 5.

Fig. 5. Meshing of the parison

3.3 Material Behaviour

PA12 is a semi-crystalline plastic. After stretching, the crystallinity increases, its glass transition temperature increases [9]. According to Literature [10–16], the yeoh model [17] was used to express the mechanical behavior of the material in the unstretched material. The Yeoh model was expressed as follows:

$$W = C_{10}(I_1 - 3) + C_{20}(I_1 - 3)^2 + C_{30}(I_1 - 3)^3 \tag{1}$$

Where W is the density of strain energy function, I_1 is the first principal strain invariant, C_{10}, C_{20}, C_{30} are model coefficient. The Yeoh model has a good prediction ability for a variety of deformation modes, and has a good expression ability for the medium and large deformation range. In the case of small deformation, C_{10} represents the initial shear modulus; When C_{20} is negative, which can reflect the material softening phenomenon in the secondary deformation; When C_{30} is positive, which can describe material hardening in the case of large deformation. The biaxial tensile data of PA12 at the temperature 110 °C is from the literature [6]. The material parameters of the yeoh model were fitted in the software, the fitting results were $C_{10} = 5.04808$, $C_{20} = -0.0162656$, and $C_{30} = 0.00445334$, the fitting results were shown in Fig. 6. In the process of balloon formation, the stretched part of parison does not participate in the molding, a linear elastic material was used, elastic modulus is 200 MPa, and Poisson's ratio is 0.33.

3.4 Process Conditions

Table 1 shows the actual parameters of the balloon formation.

According to the balloon forming parameters of Table 1, the chart of ideal time and the balloon forming process is drawn, as shown in Fig. 7. Use the "Table" command in the software to define the speed of the clamp stretching and the inflation pressure during the molding process. Isothermal numerical simulation was used because the first stretching and the second stretching were carried out under isothermal conditions.

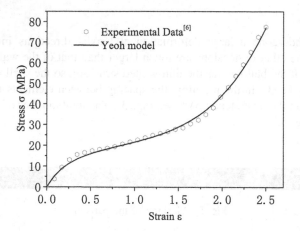

Fig. 6. Fitting curve of stress-strain and experimental data of PA12

Table 1. Balloon molding parameters

Stages parameters	1	2	3	4	5
Temperature (°C)	25–110	110	110	110	110–135-25
Pressure (MPa)	2.6	2.6	2.6	2.6–4	4
Stretching speed (mm/s)	0	70	0	10	0
Times(s)		0.5	7	0.5	

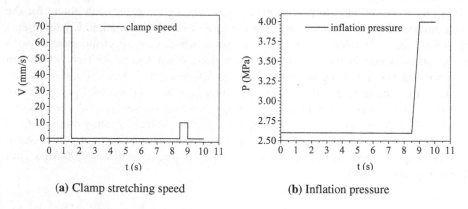

(a) Clamp stretching speed   (b) Inflation pressure

Fig. 7. Relationship between molding process and time.

4   Results and Discussion

In order to verify the feasibility of this numerical simulation, both the balloon forming process and the final wall thickness distribution for the work segment were compared to that of the actual situation.

Balloon molding process is unable to be observed directly because the mold is opaque. Observation and recordation this process could be carried out by opening the mold after the completion of each stage. Figure 8 shows the process of the numerical simulation and the experiment. When the first stretching was completed, the balloon had been basically formed. From the numerical simulation, it could be seen that only the corner part near the mold had not yet formed. After the second stretching, the balloon has been fully formed under the high pressure and the axial tension. Figure 8 shows that the balloon forming process described in the numerical simulation (left) is fully consistent with the actual process (right).

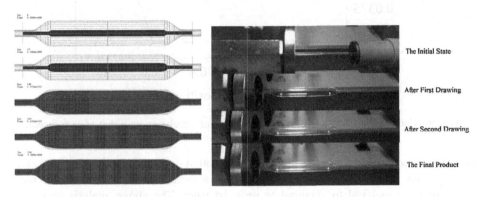

The Initial State

After First Drawing

After Second Drawing

The Final Product

Fig. 8. Comparison between numerical simulation and experimental process of balloon molding

Balloon wall thickness is very thin, ordinary measuring instruments cannot meet the requirements to measure it. The axial wall thickness of the balloon was measured by a Peacock G2-205 thickness gauge. Figure 9 is the comparison of wall thickness distribution of numerical simulation with the experiment. As can be seen from the Fig. 9, the change of the overall thickness simulated has the same trend as that in the experiment, but there is little difference in size. The wall thickness for 80% length of working segment changes gently. The actual wall thickness value of this part is smaller than the simulated wall thickness. The simulated wall thickness of this segment was 0.03812 mm, the experimental wall thickness is of this segment was 0.03683 mm, the difference between them was 3.5%. In engineering, the error less than five percent is acceptable, So the simulated wall thickness could be used to predict the actual wall thickness. There were two reasons for the actual wall thickness value of this part smaller than the simulated wall thickness. On the one hand, after the first stretching and the second stretching, the material will crystallize and retract in the subsequent high temperature holding and cooling stages, which results in a smaller thickness value. On the other hand, the material model may differ from the actual one, which also results in a difference between the actual value and the numerical simulation value. The wall thickness for the left 20% length of working segment increases, see Fig. 9, because this part is near the cone of the mold and the mold cone would result in a decrease in radial tensile ratio. The actual wall thickness simulated near the cone is too large because the

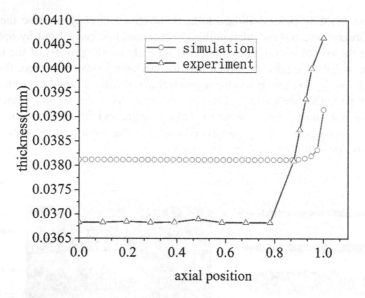

Fig. 9. Comparison of wall thickness distribution of simulation with that of experiment.

constitutive equation used in the numerical simulation does not take the viscous factors into account. In the actual processing, PA12 has a certain viscosity, and the stickiness will cause the material to accumulate near the taper. The above analysis and comparison shows that the numerical simulation is feasible.

5 Relationship Between Balloon Wall Thickness and Process Parameters

In the actual production, the balloon wall thickness refers to the thickness of the following three points, the middle point and both ends of balloon work segment. The wall thickness at the middle point is generally smaller than the wall thickness at both ends, where is most likely to burst when used. The study of the relationship between the thickness at the middle point and the molding process parameters is of great significance to the actual production guidance.

5.1 Orthogonal Experiment Design

According to the actual processing parameters and range, orthogonal experiments were carried out on the four factors, first stretching speed (V), low blowing pressure (P), second stretching speed (S) and high blowing pressure (H). Each factor was set to four levels, uniformly-spaced between the levels was selected according to the actual process range. For 4 factors 4 levels, select L_{16} Taguchi Orthographic Table as an experimental scheme [18, 19], as shown in Table 2. The results of orthogonal experiment are shown in Table 3. From these results, it was found that the first stretching

Table 2. Process factors and levels table.

Level	Process parameters			
	V(mm/s)	P(MPa)	S(mm/s)	H(MPa)
1	50	2.4	8	3.8
2	60	2.6	10	4
3	70	2.8	12	4.2
4	80	3	14	4.4

Table 3. Orthogonal experimental results.

Serial number	Process parameters				Thickness (mm)
	V(mm/s)	P (MPa)	S(mm/s)	H (MPa)	
1	50	2.4	8	3.8	0.0399225
2	50	2.6	10	4	0.0396598
3	50	2.8	12	4.2	0.0395479
4	50	3	14	4.4	0.0393219
5	60	2.4	10	4.2	0.0388971
6	60	2.6	8	4.4	0.0389441
7	60	2.8	14	3.8	0.038734
8	60	3	12	4	0.0386725
9	70	2.4	12	4.4	0.0381125
10	70	2.6	14	4.2	0.0381266
11	70	2.8	8	4	0.0383324
12	70	3	10	3.8	0.0382686
13	80	2.4	14	4	0.0377213
14	80	2.6	12	3.8	0.0377182
15	80	2.8	10	4.4	0.0374783
16	80	3	8	4.2	0.0375064
Range	0.008028	0.000884	0.000802	0.000786	

speed is the most significant influence on the thickness compared with the other factors. This is consistent with the actual forming results, and this phenomenon can be explained from a qualitative point of view. Since the size of tube and mold were defined and the radial draw ratio of the balloon could be deduced, the wall thickness was largely determined by the axial tensile ratio. The effect of the first stretching speed on the axial tensile ratio is the largest, so the first stretching speed has the greatest impact on wall thickness.

5.2   Regression Model for Wall Thickness

The regression model of wall thickness with process parameters was established based on above results simulated. Assuming that the variables of the regression model are the first stretching speed (V), low blowing pressure (P), the second stretching speed (S) and high blowing pressure (H) and the objective function is wall thickness f (x). Because process parameters affect each other and the relationship between each parameter and wall thickness is nonlinear, the objective function can be defined as follows:

$$f(x) = rV^q P^y S^t H^k \tag{2}$$

Where r, q, y, t, k are the model coefficients.

With the help of the mathematical analysis software MATLAB, the least square method [20] was applied to the nonlinear fitting of the formula (2). The model coefficients were obtained, r = 0.0669, q = −0.10957, y = −0.02556, t = −0.00916, k = −0.03473.

Then the final regression model became as follows:

$$f(x) = 0.06694 V^{-0.1095} P^{-0.02556} S^{-0.00916} H^{-0.03473} \tag{3}$$

The equation was tested for significance, $F = 5.19 > F_{0.05}(4,11) = 3.36$. The result showed that the regression effect of fitting model was significant, which was consistent with F test. The correlation coefficient R = 0.9978, very close to 1, which showed that there was a strong linear relationship between the balloon wall thickness and the process parameters. Figure 10 shows that the fitted curve by using formula (3) was consistent with the simulation results. Therefore the regression model can be used to express the relationship between wall thickness and process parameters for balloon forming.

In the actual production, the regression model can be used to estimate the wall thickness of the balloon. If the difference between the wall thickness of the balloon and the target value is larger, give priority to adjust the first stretching speed so as to

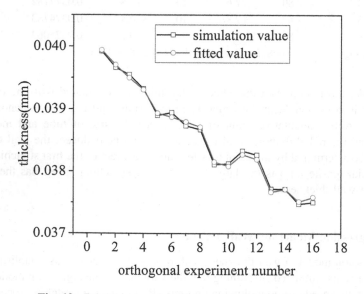

Fig. 10. Regression model values versus calculated values.

achieve the target wall thickness. Fine adjustment for the process parameters could be done, such as low blow pressure, secondary drawing speed and high blowing pressure when the difference with ideal wall thickness was little.

6 Conclusions

Numerical calculation for the forming process of the balloon was carried out and the feasibility of such simulation was verified by comparing with the experimental data. It showed that the numerical computation method can be used to simulate the process of balloon forming.

On the basis of simulation results from orthogonal experiment, it was found that the first stretching speed is the main technological factor to determine the wall thickness forming of the balloon.

Based on the computation results, the regression model of the wall thickness with the process parameters was established. It can be used to quickly optimize the combination of process parameters of the stretching and inflation stages of balloon forming within a given parameters range.

At present research about the balloon numerical simulation is isothermal. The effects of temperature and material crystallization could be considered in future research.

Acknowledgements. The authors thank the support from CHEMCLOUDCOMPUTING@ BUCT.

References

1. Xu, L., Liu, Y., Lu, C., et al.: State of aggregation changes of Nylon-12 balloon in molding process. Polymer Bull. **7**, 104–110 (2012)
2. Wang, Z., Wang, S., Wu, J., et al.: Balloon dilatation and stent placement method for treatment of Budd Chiari syndrome. Chinese Med. J. **2**, 97–99 (1995)
3. Ling, J., Xie, R., Xu, L., et al.: Experience and long-term outcome of percutaneous balloon pulmonary valvuloplasty. Chin. J. Cardiol. **5**, 6–8 (2003)
4. Yang, H., Shi, X.: Research progress of medical balloon. In: The First National Conference on Interventional Medical Engineering, Shandong, Weihai, China (2007)
5. Shichong, X., Biao, G.: Application of nylon 12 in medical equipment. New Chem. Mater. **2**, 183–184 (2014)
6. Menary, G.H., Armstrong, C.G.: Experimental study and numerical modelling of injection stretch blow moulding of angioplasty balloons. Plast., Rubber Compos. **35**(8), 348–354 (2006)
7. Lv, C.: The optimization of process conditions and preform structure parameters for PET bottles. Doctoral dissertation. Zhejiang University (2014)
8. Jie, L., Yaomin, H.: Finite element analysis of superplastic bulging process. J. Plast. Eng. **15** (3), 56–60 (2008)
9. Zhihan, P., Zupei, S.: Handbook of Plastics Industry: Polyamide. Chemical Industry Press, Beijing (2001)

10. Duan, Y., Saigal, A., Greif, R., et al.: A uniform phenomenological constitutive model for glassy and semicrystalline polymers. Polymer Eng. Sci. **41**(8), 1322–1328 (2001)
11. Schang, O., Billon, N., Muracciole, J.M., et al.: Mechanical behavior of a ductile polyamide 12 during impact. Polymer Eng. Sci. **36**(4), 541–550 (1996)
12. Zhang, K.: Numerical simulation of breadthwise stretching process of biaxially oriented plastics film. Doctoral dissertation. Nanchang University (2006)
13. Schmidt, L.R., Carley, J.F.: Biaxial stretching of heat-softened plastic sheets using an inflation technique. Int. J. Eng. Sci. **13**(6), 563–578 (1975)
14. Delorenzi, H.G., Nied, H.F.: Blow molding and thermoforming of plastics: finite element modeling. Comput. Struct. **26**(1–2), 197–206 (1987)
15. Erwin, L., Pollock, M.A., Gonzalez, H.: Blowing of oriented PET bottles: predictions of free blown size and shape. Polymer Eng. Sci. **23**(15), 826–829 (2004)
16. Martin, L., Stracovsky, D., Laroche, D., et al.: Modeling and experimental validation of the stretch blow molding of PET. In: Annual Technical Conference of SPE, pp. 14–22 (1999)
17. Yeoh, O.H.: Some forms of the strain energy function for rubber. Rubber Chem. Technol. **66**(5), 754–771 (1993)
18. Yong, H.U., Bolin, H., Hong, Y.: Numerical simulation of semi-solid die casting process of magnesium matrix composite. Chin. J. Nonferrous Met. **20**(7), 1260–1266 (2010)
19. Wu, Y., Wu, A.: Taguchi Methods for Robust Design. Mech. Eng. **5**, 78 (2001)
20. Gharaibeh, K.M.: Simulation of nonlinear systems in MATLAB ® nonlinear distortion in wireless systems: modeling and simulation with MATLAB, pp. 221–278 (2011)

Tri-Dexel Model Based Geometric Simulation of Multi-axis Additive Manufacturing

Shanshan He[1], Xiongzhi Zeng[1], Changya Yan[1(✉)], Hu Gong[2], and Chen-Han Lee[1]

[1] Manufacturing Intelligence Engineering Research Center,
Wuhan Institute of Technology, Wuhan, China
yanchangya@hotmail.com
[2] Tianjin University, Tianjin, China

Abstract. This paper presents a novel geometric simulation technique for multi-axis Additive Manufacturing (AM). In the proposed methodology, additive swept volume elements, which are represented with Tri-dexel models, are formulated for updating the virtual additive material. Triangular meshes are extracted from the Tri-dexel models using Marching Cube Algorithm for visualization. With the proposed methodology, either three-axis or five-axis AM tool paths of sculpture surfaces can be simulated before actually manufactured. Workpieces after additive manufacturing can be used as blanks for further Subtractive Manufacturing(SM). We developed a geometric simulation software based on the multi-axis AM simulation algorithm, and we carried out an actual three-axis AM experiment to compare with the simulation results. Computer implementation and practical example demonstrate the feasibility of the proposed multi-axis AM simulation method.

Keywords: Geometric simulation · Multi-axis additive manufacturing · Additive swept volume · Tri-dexel model

1 Introduction

Additive Manufacturing (AM) [1], also known as 3D printing or metal deposition, refers to create a three-dimensional object in which successive layers of material are formed, provides a novel and productive method for producing freeform parts relative to Subtractive Manufacturing (SM, also known as machining or cutting). There have been many AM technologies according to different material feeding method [1–3], such as Selective Laser Melting (SLM), Selective Laser Sintering (SLS) and Electron Beam Melting (EBM) based on powder bed, and Laser Engineered Net Shaping (LENS), Direct Metal Deposition (DMD) and Selective Laser Cladding (SLC) with simultaneous material delivering.

In recent years, Hybrid Manufacturing technologies in which AM methods are incorporated with SM methods in a traditional machine tool or industrial robot environment have been a trend. Before actual manufacturing, geometric simulation uses virtual manufacturing to simulate and verify the actual processing, which can reduce error probability of programs, save costs and improve efficiency in actual manufacturing.

© Springer International Publishing AG 2017
Y. Huang et al. (Eds.): ICIRA 2017, Part III, LNAI 10464, pp. 819–830, 2017.
DOI: 10.1007/978-3-319-65298-6_73

Geometric simulation of multi-axis SM is a mature technology, but there are fewer researches about geometric simulation of AM, especially for five-axis AM. Most AM geometric simulation solutions adapt the similar swept volume discretion method as SM methods.

Theoretically, the virtual sculpting systems have two types. One is derived from surface-based deformable geometric object modeling, which may be vertex-based, spline-based, particle based, FEM (finite element modeling)-based, etc. The other one is based on swept volume models, e.g., Voxel or Dexel models.

For geometric simulation solutions of SM, Dexel model, first proposed by Tim Van Hook in [4] in 1986, is used for swept volume discretion. Hui [5] proposed an algorithm for sweeping a 3D object in image space used for simulating the material removal process in an NC machining operation. The proposed method does not suit for five-axis machining, and using a Dexel model alone could not achieve very good modeling accuracy. Zhu et al. [6] gives a detail description of Dexel model updating processes for virtual sculpting. They gave detailed description of Dexel-based tool swept volume and some interaction methods. Mullel et al. [7] proposed multi-Dexel volume for representation of a solid and used the method in milling simulation. Ren et al. [8] proposed a virtual prototyping and manufacturing planning method by using Tri-dexel models and haptic force feedback. Ren et al. [9] presented a detailed algorithm for eliminating inconsistency in a Tri-dexel volumetric models and reconstructing a water-tight polyhedral surface model from the Tri-dexel volumetric model.

For geometric simulation solutions of AM, Choi and Chan [10] proposed two new simulation methodologies, namely the Dexel-based and the layer-based fabrication approaches to simulate the powder-based and the laminated sheet-based Rapid Prototyping (RP) processes. Gao et al. [11] proposed a method of Tri-dexel model of polyhedrons and its application in RP simulation. The methods are only suit for three-axis RP simulation.

Although there are few literatures about geometric simulation of multi-axis AM, two commercial software products have the geometric simulation function for multi-axis AM. NX software [12] provides support for DMG MORI Lasertec Hybrid machines with new hybrid-manufacturing technologies. Nevertheless, this module is bound with hardware. SKM software [13] has the function of tool simulation and mechanism simulation to confirm tool paths in AM processing but lack of material growth visualization.

In this study, we propose a geometric simulation technique for multi-axis AM based on LENS technology. The proposed AM geometric simulation method uses Tri-dexel model to represent the swept volume and triangular meshes for material growth visualization. The reminder of this paper is organized as follows. Section 2 reviews some related works about geometric simulation of multi-axis AM. Section 3 describes the Tri-dexel models of additive swept volume. Section 4 gives the flowchart of Tri-dexel model based geometric simulation algorithm. Section 5 describes two simulation examples and an actual experiment. At last, Sect. 6 is the conclusion.

2   Review of Related Works

This section reviews some basic algorithms relate to Tri-dexel model based AM geometric simulation algorithm. First of all, Sect. 2.1 introduces some multi-axis AM tool path generation methods. Then Sect. 2.2 describes data structures of Dexel model and Tri-dexel model. At last, Sect. 2.3 reviews Boolean operation used in AM simulation.

2.1   Multi-axis Additive Manufacturing Tool Path Generation

In AM, the quality of deposition depends on the choice of AM tool paths. AM tool paths need to fulfill some requirements such as avoiding collapse and deposition voids [14], saving time and cost, axes constraint and dimensional accuracy [15].

There are different tool path patterns for different parts and applications. (1) For three-axis AM tool path generation, contour-parallel offsetting pattern and zigzag pattern are two common patterns for general parts. The contour-parallel offsetting pattern results in better deposition for circular geometries and the zigzag pattern can avoid deposition voids [14]. Furthermore, staggered zigzag pattern is used for avoiding collapse and regional zigzag pattern is used for parts with inner voids. (2) For five-axis AM tool path generation, spiral pattern and regional zigzag pattern can be used, we should noting that the tool axes should be planed following surface feature. (3) For some special parts, for example, thin-wall parts, follow pattern and helical pattern can be used either for three-axis or five-axis AM tool path generation. Figure 1 shows some AM tool path patterns.

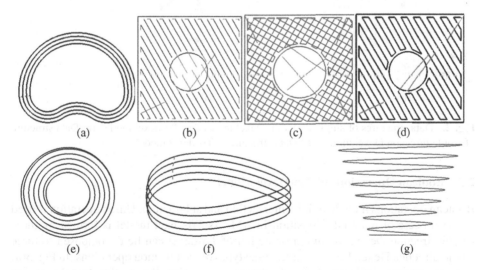

Fig. 1. AM tool path patterns. (a) Contour-parallel offsetting pattern. (b) Zigzag pattern. (c) Staggered zigzag pattern. (d) Regional zigzag pattern. (e) Spiral pattern. (f) Follow pattern. (g) Helical pattern.

2.2 Tri-Dexel Model

Dexel model uses a series of small cubes in a specific direction (usually the Z-axis direction) to represent a volume. The single direction Dexel model has deviation in expressing a complex shape. Comparing with the single direction DEXEL model, three direction DEXEL model could express a solid more precisely [11]. Tri-dexel rays are extracted in three directions. By Tri-dexel rays, the entity's bounding box is divided into a series of voxels. The neighbor voxels with same attributes can be merged into a big voxel to save memory.

Figure 2 shows data structures of a Dexel model and a Tri-dexel model [9]. The Dexel model is defined relative to an origin point O with a grid interval δ. As shown in Fig. 2(a), each Dexel segment $D_{i,j,k}$ consists of two node $\xi_{i,j,k}^{l}$ and $\xi_{i,j,k}^{u}$ as the lower and the upper ends of a Dexel segment. A Tri-dexel model is defined relative to three direction Dexel(Dexel-X, Dexel-Y, Dexel-Z), as shown in Fig. 2(b). N_x, N_y, N_z are the numbers of grid cells along the three edges of a 3D-grid.

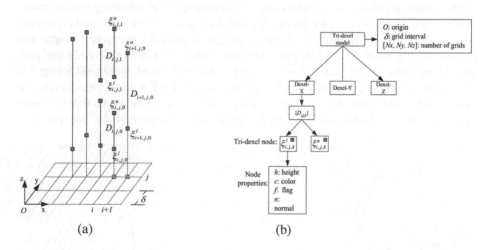

(a)   (b)

Fig. 2. Data structures of single direction Dexel model and Tri-dexel model. (a) Data structure of single direction Dexel model. (b) Data structure of Tri-dexel model

2.3 Boolean Operations of Tri-Dexel Model

Boolean operations include union, intersection and subtraction. Union operation is used in the process of material deposition in AM. As Tri-dexel model is formed by three single direction Dexel, the union operation of Tri-dexel can be decomposed to three single direction Dexel. Figure 3 gives three types of Dexel union operations. In Fig. 3(a) and (b),, the union of Dexel A, B is C. the union operation of more than two Dexels can transform to union of two Dexels. For example, in Fig. 3(c), the union of A1, A2 with B is the union result of A1, B then union with A2.

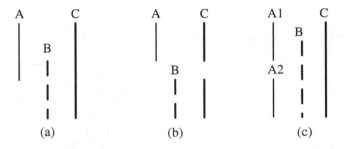

Fig. 3. Dexel union operations

For material deposition in AM, the whole additive swept volume is the union of all micro swept volume. Current swept volume after deposition is the union of current micro volume with deposited swept volume. The additive swept volume can be expressed as Tri-dexel models. The union operation of swept volume is based on Tri-dexel models which is linear, precise, and fast. In Sect. 3, we will introduce the method to generate the Tri-dexel model of additive swept volume. In Tri-dexel model, the Marching Cube Algorithm [16] could be easily used to extract triangles for visualization.

3   Tri-Dexel Models Generation of Additive Swept Volume

In this section, we will introduce a method of building additive volume and Tri-dexel models of swept volume. The whole swept volume are formed of many micro swept volume. The swept volume in this section is based on micro swept volume between two GOTO points.

3.1   Additive Swept Volume Model

In LENS processing, powders are deposited with the movement of laser tool. In our algorithm, the deposited volume between two GOTO points P_iP_{i+1} can be simplified as a cuboid, which the length is the distance of P_iP_{i+1}, the width is the laser spot diameter, and the height is the deposition thickness. As shown in Fig. 4(a).

The cuboid model is only suit for straight line type tool paths (such as Fig. 1(b)). There may be gaps at the corner of two segments. So we change the cuboid model to cylindrical model to get optimization results. The deposited shape between two GOTO points P_iP_{i+1} can be simplified as cylinder swept volume. As shown in Fig. 4(b).

After building the swept volume model, the micro swept volume can be transformed to Tri-dexel model, the following sections introduce Tri-dexel model of cylinder-shaped swept volume.

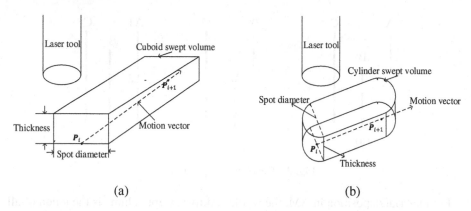

(a)   (b)

Fig. 4. Additive swept volume model. (a) Cuboid swept volume. (b) Cylinder swept volume

3.2 Tri-Dexel Model of Additive Swept Volume

To generate Tri-dexel models of additive swept volume, the first step is to compute the minimum bounding box of swept volume (as shown in Fig. 5(a)). Then the bounding box is divided into grids by grid interval (as shown in as shown in Fig. 5(b)). The grid vertexes can form all the rays to get the Tri-dexel model.

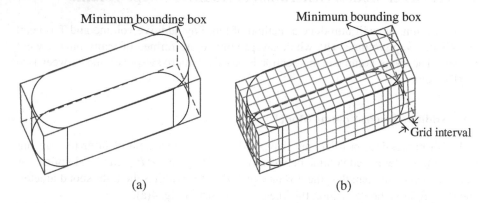

(a)   (b)

Fig. 5. Bounding box and Tri-dexel model of swept volume. (a) Minimum bounding box. (b) Tri-dexel model building.

The grid interval will determine the simulation accuracy. The more intense, the higher the accuracy of the simulation, but the computing time and memory space will rise sharply. So the ideal Tri-Dexel model of discrete space should be as large as possible under the premise of satisfying simulation error.

The cylinder-shaped swept volume is formed by some regular surfaces, as shown in Fig. 5(a): The left side and right side are cylindrical surfaces; the front side, back side, top side and bottom side are planes. The Tri-dexel swept volume is formed by computing the interaction of the rays and regular surfaces. The interaction algorithms is simple and can refer to [8].

4   Tri-Dexel Model Based Geometric Simulation Algorithm

As shown in Fig. 6, the flowchart of Tri-dexel model-based geometric simulation algorithm includes six main steps.

(0)   For the input data, the input tool paths are generated by any kind tool path pattern introduced in Sect. 2.1. The input geometry models is used to check the validity of input tool paths and compare the simulation result model with the original model.

(1)   In step 1, the micro swept volume and minimum bounding box is built for each segment formed by two GOTO points. The method is introduced in Sects. 3.1 and 3.2.

(2)   In step 2, the micro swept volume is transformed to Tri-dexel model. The method is introduced in Sect. 3.2.

(3)   In step 3, the micro swept volume and deposited material model are expressed in Tri-dexel models. Union operation of Tri-dexel models introduced in Sect. 2.3 is used for computing new deposited material models.

(4)   In step4, triangles meshes could be extracted by Marching Cube Algorithm [16]. Before extracting, the deposited material model expressed by Tri-dexel model should be transform to Voxel model because the Voxel model could express complex inner attributes.

(5)   The last step is the visualization of triangles meshes using display engine.

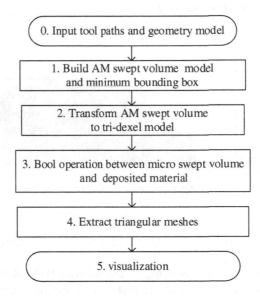

Fig. 6. Flowchart of Tri-dexel model based geometric simulation algorithm

5   Simulation and Experiments

5.1   Software Implementation and Simulation

A software framework is built to implement the multi-axis AM geometric simulation algorithm. The software is developed based on NX Open platform. The AM tool paths in this study are generated by software HybridCAM (the software has got China's compute software copyright registration certificate). The development of HybridCAM and simulation in this study is based on NX 9.0.

In this study, we use two parts to demonstrate the simulation process. The two parts come from real parts in aviation industry. The size of two parts are shown in Table 1, and the CAD models of two parts are shown in Fig. 7.

Table 1. AM geometry simulation configurations

Part name	Size(mm)	Tool path pattern	Processing parameters(mm)
Vane	Length: 53.3 Width: 4.8 Height: 100	Three-axis helical	Laser spot diameter: 1 Layer thickness: 2
Impeller	Diameter: 348.0 Height: 130.4 Blade thickness: 2.4	Five-axis follow	Laser spot diameter: 2 Layer thickness: 2

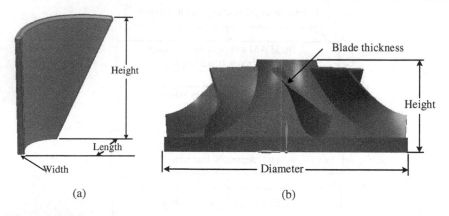

(a)   (b)

Fig. 7.   CAD models of two parts. (a) CAD model of "vane". (b) CAD model of "impeller".

Before the AM simulation processing, we use software HybridCAM to generate tool paths of the two parts. Tool paths of the first part "Vane" is generated by three-axis helical pattern, and tool paths of the second part "Impeller" use five-axis follow pattern. Processing parameters in Table 1 give two parameters in tool path generation and simulation. Figure 8(a) shows the generating tool paths of the part "Vane". Figure 8(b) shows the generating tool paths of one blade in the part "Impeller".

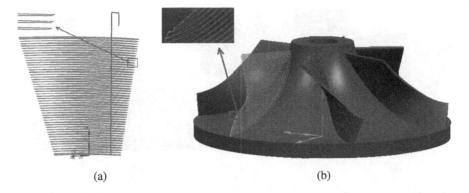

Fig. 8. Tool paths of two parts. (a) Tool paths of "vane". (b) Tool paths of "impeller".

In the AM simulation processing, the deposition models are first demonstrated by Tri-dexel models, and then transformed to triangular meshes for visualization. Figure 9(a) and (b) show Tri-dexel models of the two parts generated in the deposition processing. In the two models, grid interval for computing Tri-dexel model of swept volume is set to 1 mm. From partial magnification in Fig. 9, we can see that the Tri-dexel models are smooth.

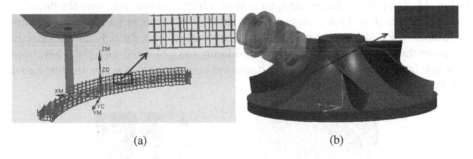

Fig. 9. Tri-dexel models of two parts. (a) Tri-dexel model of "vane". (b) Tri-dexel model of "impeller"

After the AM simulation processing, the deposition models are visualized by triangular meshes. As shown in Fig. 10. In the figures, we can clear see the layered deposition material and the deposition model is basically consistent with the CAD models. In Fig. 10, we can see the layer is smooth. We also notice that the laser tool axes are always perpendicular to the surface. In the two deposition models, there are some ladder shapes obviously on boundary but not obviously inside. The smaller the layer thickness, the smoother the deposition model. Comparing with the Dexel method in reference [11], the deposition results in this study are better with less ladder shapes and more overall smoothness.

(a)   (b)

Fig. 10. Deposition models of two parts. (a) Deposition model of "vane". (b) Deposition model of "impeller".

5.2 Additive Manufacturing Experiment

To evaluate that the AM simulation results are consistent with actual manufacturing results, we did the actual experiment of part "vane". The experiment of part "impeller" was not realized due to limitations of experimental equipment.

The experimental equipment is modified by a three-axis machining tool. As shown in Fig. 11. The powder feeder equipment (Fig. 11(a)) is connecting with the three-axis AM equipment (Fig. 11(b)). Actual AM deposition experiment parameters are shown in Table 2. The tool paths are generated similar with Fig. 8(a), the only difference is that the layer thickness is set to 0.3 mm. Considering the generated tool paths have more than 300 layers, the tool paths are not displayed here, and we only manufactured about 100 layers considering efficiency.

(a)   (b)

Fig. 11. Additive Manufacturing system. (a) Powder feeder equipment. (b) Three-axis AM equipment.

Table 2. Actual AM deposition experiment parameters

Part name	Powder material	Laser scanning speed	Laser power	Power feed rate
Vane	Stainless steel	1500 mm/s	1500 W	1r/s

The deposition model is shown in Fig. 12. We can see that the deposition model is intact, no collapse, no distortion and deformation. The actual manufacturing results are consistent with simulation results. The outer surface covers an oxide layer which can be knocked out. To get more smooth and high precision part, the deposition model can be used as blank for furthermore finishing.

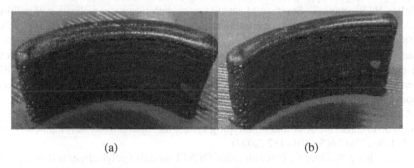

(a)   (b)

Fig. 12. Deposition model. (a) Front view. (b) Side view.

6 Conclusion and Future Work

This paper presents a novel geometric simulation technique for multi-axis Additive Manufacturing. The proposed method can be used for three-axis and five-axis AM processing geometric simulation. The cylinder-shaped additive swept volume model can simulate the deposition models lively, and simulation examples and actual manufacturing experiment demonstrate the feasibility of the proposed simulation method.

Currently we are working on an algorithm to simulate the AM processing more lively using new swept models. In addition, we plan to do more actual experiments using industrial robot to further verify and improve our algorithms.

Acknowledgement. The authors gratefully acknowledge National Natural Science Foundation of China(51575386). The authors would also like to thank the experimental supporting from Wuhan HuaGong Laser Engineering Co., Ltd.

References

1. Bikas, H., Stavropoulos, P., Chryssolouris, G.: Additive manufacturing methods and modelling approaches: a critical review. Int. J. Adv. Manuf. Technol. **83**(1), 389–405 (2016)

2. Griffith, M.L., Keicher, D.M., Atwood, C.L.: Free form fabrication of metallic components using laser engineered net shaping (LENS{trademark}). Office of Scientific & Technical Information Technical Reports (1996)
3. Thompson, S.M., Bian, L., Shamsaei, N., et al.: An overview of direct laser deposition for additive manufacturing; part I: transport phenomena, modeling and diagnostics. Add. Manuf. **8**, 36–62 (2015)
4. Van Hook, T.: Real-time shaded NC milling display. In: 13th Annual Conference on Computer Graphics and Interactive Techniques, p. 15. ACM, New York (1986)
5. Hui, K.C.: Solid sweeping in image space—application in NC simulation. Visual Comput. **10**(6), 306–316 (1994)
6. Zhu, W., Lee, Y.S.: Dexel-based force-torque rendering and volume updating for 5-DOF haptic product prototyping and virtual sculpting. Comput. Ind. **55**(2), 125–145 (2004)
7. Muller, H., Surmann, T., Stautner, M., et al.: Online sculpting and visualization of multi-dexel volumes. In: Eighth ACM Symposium on Solid Modeling and Applications, pp. 258–261 (2003)
8. Ren, Y., Lai-Yuen, S.K., Lee, Y.S.: Virtual prototyping and manufacturing planning by using tri-dexel models and haptic force feedback. Virtual Phys. Prototyping **1**(1), 3–18 (2006)
9. Ren, Y., Zhu, W., Lee, Y.S.: Feature conservation and conversion of tri-dexel volumetric models to polyhedral surface models for product prototyping. Comput.-Aided Des. Appl. **5**(6), 932–941 (2013)
10. Choi, S.H., Chan, A.: A virtual prototyping system for rapid product development. Comput. Aided Des. Cad **36**(5), 401–412 (2004)
11. Gao, X., Zhang, S., Hou, Z.: Three direction DEXEL model of polyhedrons and its application. In: International Conference on Natural Computation, vol. 5, 145–149 (2007)
12. NX Homepage, http://www.plm.automation.siemens.com/en_us/products/nx/for-manufacturing/cam/hybrid-additive-manufacturing.shtml#lightview-close
13. SKM Homepage, http://www.skm-informatik.com/skm/home.html
14. Ren, L., Sparks, T., Ruan, J., et al.: Integrated process planning for a multiaxis hybrid manufacturing system. J. Manuf. Sci. Eng. **132**(2), 237–247 (2010)
15. Gao, W., Zhang, Y., Ramanujan, D., et al.: The status, challenges, and future of additive manufacturing in engineering. Comput.-Aided Des. **69**(C), 65–89 (2015)
16. Lorensen, W.E., Cline, H.E.: Marching cubes: a high resolution 3D surface construction algorithm. In: Conference on Computer Graphics & Interactive Techniques, vol. 21(4), pp. 163–169 (1987)

Development of Rubber Aging Life Prediction Software

Hong He[✉], Kai Liu, Xuelei Fu, and Kehan Ye

Beijing University of Chemical Technology, Beijing 100029, China
hehong@mail.buct.edu.cn

Abstract. Raw rubber or rubber items are vulnerable to heat, oxygen, light and other factors during the processing, storing or using, due to exposure to the natural environment or a particular working environment, and easy to undergo physical or chemical change, such as the softening, sticky for crude rubber, and cracking, mildewy and brittle for rubber products which would degrade the material properties or even make it unusable. The loss of property caused by such rubber aging is up to hundreds of millions of dollars annually. Therefore how to accurately predict the rubber aging life and how to choose the appropriate test method are especially important for the selection of suitable rubber and reduction of the cost of existing aging experiment. Based on the summarizing and comparison of various kinds of rubber fatigue aging theories and experiment methods, a predicting rubber aging life software with VC++ programming language was developed. The software structure mainly included the theoretical prediction module, the commonly used rubber material data module, user guidance module. It could be used to predict rubber fatigue life, optimize the product material and help the manufacturing process more efficient implementation.

Keywords: Rubber aging · Life prediction · Programming software · VC++

1 Introduction

As is known to all, rubber products have been widely used in all aspects of industry and life [1, 2], for example, the automotive industry, military industry and health care industry, etc. since the date of its birth. Because they have unique properties such as elasticity, insulation, water and air impermeability. However, rubber has the characteristics of easy aging. In the process of processing, storing and use of rubber and its items, the physical and chemical properties and mechanical properties of rubber are gradually deteriorated due to the combined effects of internal and external factors and ultimately their use value are lost. This change is called rubber aging. The aging surface of rubber is usually characterized by cracking, sticking, hardening, softening, powdering, discoloration, mildew, and material performance decline accordingly, as a result, the use of rubber will be affected. If the fatigue aging life of rubber can be predicted accurately, preferred rubber types and replacement frequency of rubber products can be determined under the premise of safe production and use to maximize the effectiveness and make full use of resources. Therefore, predicting the service life of rubber, or determining the insurance period of rubber products is of practical value and significance.

© Springer International Publishing AG 2017
Y. Huang et al. (Eds.): ICIRA 2017, Part III, LNAI 10464, pp. 831–842, 2017.
DOI: 10.1007/978-3-319-65298-6_74

The commonly used experimental method for detecting aging life is time-consuming and some test conditions are not easy to get. So in recent years, many effort have been made to predict rubber aging life and develop "computer aging oven [3]".

There are many types of rubber and common theoretical life prediction models. The accuracy of the prediction results are different for rubber types and working conditions. Concerning the several commonly used aging prediction models [4], P-T-t three variables function model method was suitable for applying properties parameters from the test of rubber compression permanent deformation, elongation and compression stress relaxation etc.; S curve model could be applied to the test curves fitting and aging life predicting of various linear viscoelastic mechanical properties of rubber materials; Aging damage factor model was appropriate for the thermal oxidative aging damage of fiber reinforced rubber-based sealing materials with high confidence; Different from the model expressed the aging law in only single performance, The strain energy fractional factor model was used to characterize the aging index by the ratio of the two factors of the fracture strength and elongation at break before and after aging; In the Monte Carlo model and BP artificial neural network model, the method of computer simulation was used to analyze the material fatigue life. Based on the above summarization and analysis, this study focused on the developing of a rubber aging life predicting software which integrated the existing aging prediction models mentioned and included a database of rubber materials, which would help users to optimize the material choice of rubber items under different working conditions.

The software developed for predicting rubber fatigue life could be regarded as a computer aging oven. The purpose for the development was to accurately and quickly predict the aging life of types of rubber products and the use of time that user concerned about and maximize the production efficiency and development of rubber industry.

2   Software Features

The rubber fatigue life usually is predicted by the combination of theoretical model and experiment. Therefore, the software structure was designed to be divided into three modules: theory, experiment and material data and user help module, seen Fig. 1. The function of each module was as follows: (1) Aging prediction module. This module integrated the existing rubber aging prediction methods and theoretical models mentioned. It included three sub-options, experimental data curve model prediction, computer simulation model prediction, theoretical analytical model prediction. It was convenient for user to choose prediction method independently and compare the results of different prediction methods and models. (2) Database module. This module comprised commonly used rubber material parameters, such as elastic modulus, Poisson's ratio, etc. (3) User help module. This module involved the instructions for using the software, the theories and knowledge for each prediction method. It provided help for new user to learn the software operation and understand results.

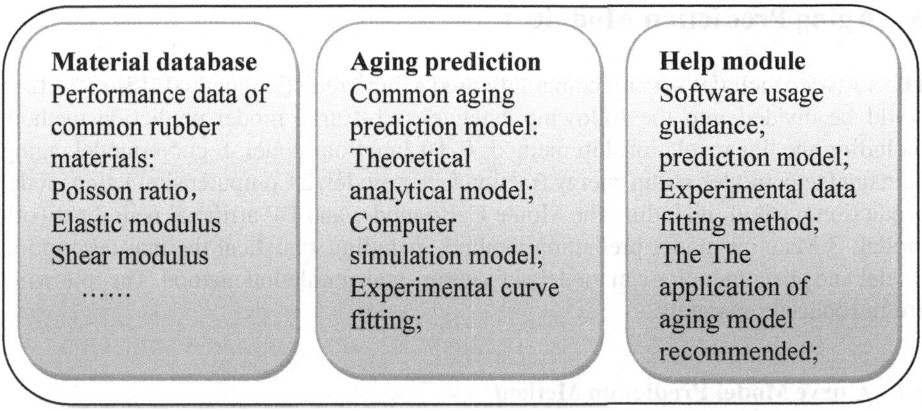

Material database	Aging prediction	Help module
Performance data of common rubber materials: Poisson ratio, Elastic modulus Shear modulus	Common aging prediction model: Theoretical analytical model; Computer simulation model; Experimental curve fitting;	Software usage guidance; prediction model; Experimental data fitting method; The The application of aging model recommended;

Fig. 1. The modules included in the software

The software was written in VC++ language, and the interface was designed by MFC class library. Using a single document mode and multi-window, it was convenient for the main program to call the subroutine. The database module was linked with the prediction module to get the prediction model required data handily. And the prediction module could be used independently because there was not necessary connection and complementary between the formulas in prediction module. User can choose the application formula by themselves or recommended in the software. The linkage of each module refers to Fig. 2.

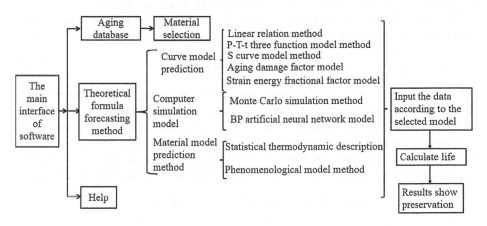

Fig. 2. The frame of the software

Software requirements were as follows: the system is Windows XP or above version, 32 bit operating system, the minimum storage space for more than 50 M. The accuracy of calculation results was more than 10^{-6}.

3   Aging Prediction Module

The software includes several commonly used aging prediction methods [5–18], which could be divided into the following categories: 1 Curve model prediction method, including the linear relationship method, P-T-t function model, S curve model, aging damage factor model, strain energy fraction factor model; 2 Computer simulation model prediction method, including the Monte Carlo model and BP artificial neural network model; 3 Material model prediction method, including statistical thermal description model and phenomenological model; 4 Experimental simulation method. The following are introduced separately.

3.1   Curve Model Prediction Method

The formula of this kind of prediction model were given in the form of analytic formula and there were more models. Here the aging damage factor model is introduced as an example. The aging damage factor model mentioned was mainly applied to the fiber reinforced rubber-based sealing material, and the leakage rate of the rubber material seal could be calculated by the aging factor. This method is simple and the scope of application was relatively narrow. the prediction accuracy is not too high. The degree of material aging damage was represented by the aging damage factor f in this model. According to the Arrhenius equation of chemical reaction kinetics, the relationship between aging damage factor and aging temperature, aging time was deduced and the model of thermal oxidative aging damage was established. This model was suitable to predict the fatigue life of fiber reinforced rubber-based sealing materials.

The aging damage factor f was defined as the product of the change of density and transverse tensile strength:

$$f = (\rho_2/\rho_1) \cdot (S_2/S_1) \tag{1}$$

Where, ρ_1 and ρ_2 are the material density before and after aging respectively; S_1 and S_2 are the transverse tensile strength before and after aging respectively.

The thermal oxidative aging of rubber is a slow process for chemical change, Therefore, the Arrhenius equation of chemical reaction kinetics was combined with the change of physical and mechanical properties of polymer materials with time, a three element dynamic model of damage factor f, time t and temperature T was established:

$$f = D \cdot exp(-k\,t^n) = D \cdot exp[-A \cdot exp(B/T)t^n] \tag{2}$$

Where, D, A, B, n, k are constant, obtained from regression of the experimental curve.

In order to link the dynamic aging damage factor with rubber fatigue life, the initial aging damage factor was set to 1, the maximum allowable leakage rate L_v was as the failure criterion, then the damage factor f_v in critical failure state could be got according to L_v and initial leakage rate L_0, where α is exponential factor:

$$L_v/L_0 = f_v^\alpha \tag{3}$$

The steps of using this model to predict the aging time is as following: First accelerated aging tests were conducted, the corresponding aging properties change values ρ and S at corresponding temperature were got. Then the damage factor was gained according to the formula (3); Using the formula (2), the parameters in the formula were regressed to obtain the function relation f-t; Finally, the critical aging damage factor was applied to the formula (2) to gain the fatigue life of the time t. The flow chart was shown in Fig. 3.

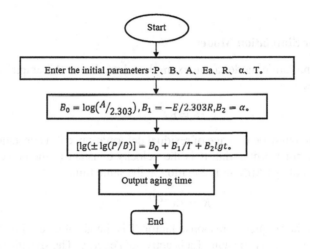

Fig. 3. Flow chart of aging damage factor model

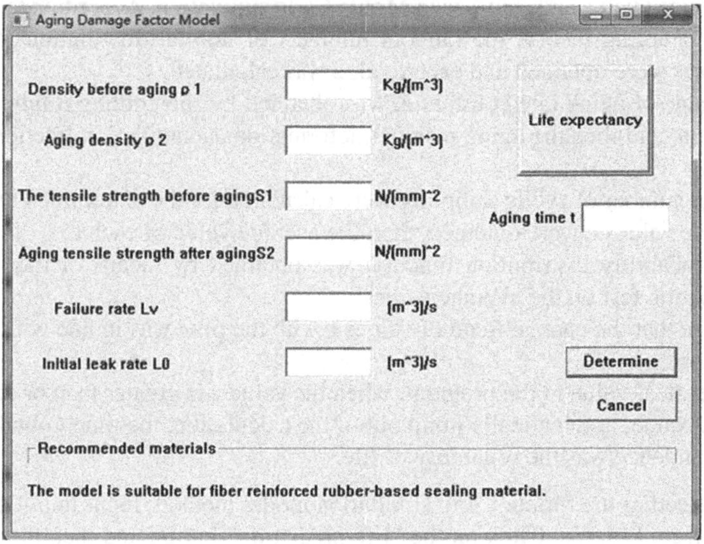

Fig. 4. Window for aging damage factor model

In the primary software interface, click "aging prediction method", open submenu, select the "aging damage factor model", the window of this model shown in Fig. 4, input initial parameters ρ_1, ρ_2, S_1, S_2, L_v, L_0, click the "life expectancy" button, the results of fatigue prediction under the corresponding conditions were found.

The operation of other models included in the curve model prediction method is similar to that of the aging damage factor model prediction method. The input parameters of different models are different and the interface settings for parameters input are different.

3.2 Computer Simulation Model

Monte Carlo simulation model as an example is introduced. An empirical formula for the aging process:

$$\varepsilon = Kt^c \tag{4}$$

Where, ε is cumulative percentage of residual deformation, c is a constant related to the aging resistance for the material, K is the velocity constant in the process of residual deformation and obeys Arrhenius equation at the same time:

$$K = Ae^{-B/T} \tag{5}$$

Where, A is the frequency response factor, T is the absolute environment temperature, $B = Ea/R$, R is a gas constant, Ea is activation energy. The specific steps of Monte Carlo simulation method are as follows:

(1) Determine t value and step size of Monte Carlo simulation. According to the Monte Carlo sampling theory, the random numbers of normal distribution for random variables were obtained and each t value was calculated.
(2) The value of aging rate constant K was obtained by substituting it into Arrhenius equation, and then the aging residual deformation accumulation fraction could be gotten.
(3) A large number of cyclic sampling were calculated for the determined t values and multiple values ε were obtained, then the average value of each ε.
(4) The probability distribution function was obtained by means of the population distribution test on the average value.
(5) Analyze that the change trend of values ε with the time was in line with the actual situation.
(6) Set a critical value in the program, when the value ε is greater than or equal to the critical value, automatically jump out of the calculation, the time t obtained under the confidence was the reliability of life.

After selecting the Monte Carlo simulation model method, input initial parameters at the interface, I, q, t, s. Click on the "life prediction" button and then the prediction results could can gotten. The computer program flow chart for the model is shown in Fig. 5. This method ensures the randomness nature of each parameter, and overcomes the deficiency of the traditional dynamic model. It could only be as a reference for the rubber aging life prediction because the reliability of it needs to be further confirmed.

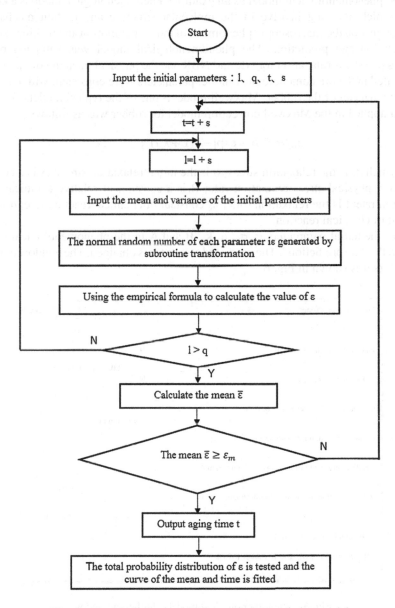

Fig. 5. The block diagram of Monte Carlo simulation model

3.3 Prediction Method of Material Model

Take the phenomenological model as an example. Phenomenological mode is a kind of model which does not involve in the molecular structure and motion mechanism, focusing only on the macroscopic phenomena of molecular motion and making relevant interpretation and prediction. The phenomenological model was validated by the compressive stress relaxation for NR and NBR and the physical meaning of parameters represented in the phenomenological model parameters were consistent with the actual aging mechanism of the rubber. Maxwell model is one of the typical models. The core formula applied in the Maxwell correction model for rubber was as follows:

$$\sigma_0/\sigma = B \cdot \exp[-A(1 + k_c t)^\alpha] \tag{6}$$

In which σ_0 is the relaxation stress; σ is the initial relaxation stress; A is a constant of relative physical stress relaxation rate; B is a constant of relative chemical stress relaxation rate; t is the relaxation time; k_c constant of chemical reaction rate; α is type constant of chemical reaction.

Input the initial parameters σ, σ_0, α, k_c, B and A in the phenomenological model. Click on the "life prediction". The predicted results t was obtained. The window of model parameters was shown in Fig. 6.

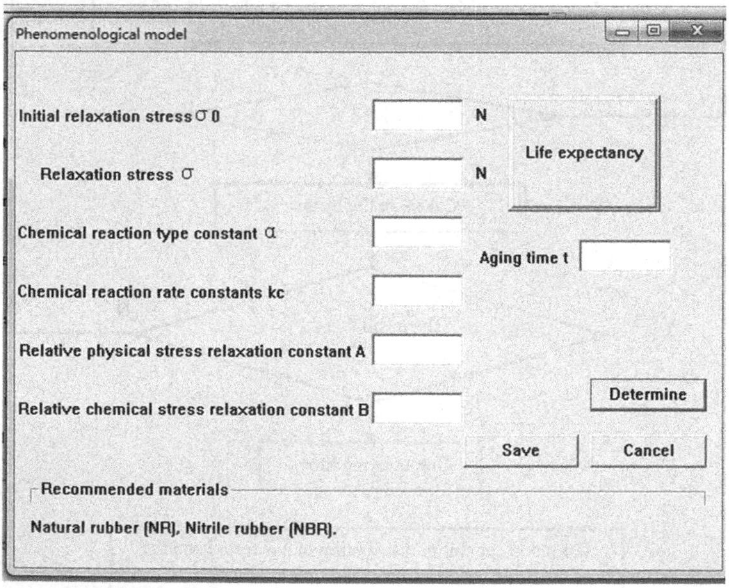

Fig. 6. Phenomenological model calculation window

4 Database

In order to facilitate the user to input parameters, the common types of rubber and related physical quantities were summarized in the database. The physical properties of each material included elastic modulus, shear modulus, poisson's ratio, Arrhenius activation energy and so on. Users check the corresponding experimental materials in the database interface according to materials used, the interface as shown in Fig. 7, call the material constants in database apply them to the corresponding model, eliminating the need for a user to find manual and manual input these material parameters. It is convenient and time-saving.

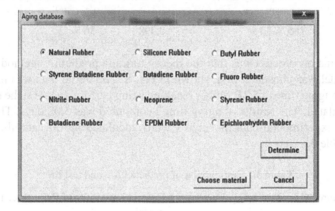

Fig. 7. Material selection interface and data interface for aging materials database

There are some interfaces left in the software database, which can be used to supplement the new material data in the future. Database data could be shared between the calculation models. The new results could also be stored in the database, which would be convenient and quick invoked for results comparison when needed.

5 An Example of Aging Life Prediction

Accelerated aging test of styrene butadiene rubber (SBR). Oven temperature was set to 80 °C. The tensile stress (M) and the other available model parameters was determined in the process of aging of rubber. Here the tensile stress is taken as an aging parameter, the aging data given as shown in Table 1. Called the database and the required parameters were given in Table 2. Aging performance index P- temperature T- time, P-T-t model was applied.

Table 1. Accelerated aging test data for SBR rubber [19]

Temperature T/°C	T/day	M/M$_0$
80	5	1.463
80	7	1.532
80	14	1.698
80	28	1.921

Table 2. Experimental parameters of SBR rubber in P-T-t prediction model

P Aging property	A Frequency factor constant	α constant	Ea Arrhenius activation energy/ kJ·mol^{-1}
M/M$_0$	78576.73	0.398	36.5

The calculation process was that the theory formula prediction method P-T-t three function model was chosen at first, shown in Fig. 8. The initial data was input, shown in Fig. 9. The aging time of SBR rubber reaching aging performance P at the temperature 80 was calculated. The results of aging time t computed was 5.08 days. The predicted life of other experimental data in Table 1 were calculated in turn, and the results are shown in Table 3.

Table 3. Comparison of predicted life and test life

Test temperature/°C	Test aging time/day	Calculation results of P-T-t three function model
80	5	5.08
80	7	6.77
80	14	13.84
80	28	29.18

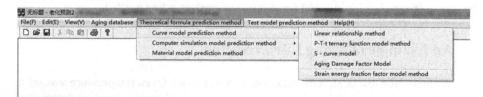

Fig. 8. Method selection window for aging prediction

From Table 3, compared with the actual test data, the error rate of life prediction was below 5% when the tensile stress was as the performance index. From that the prediction accuracy is very good.

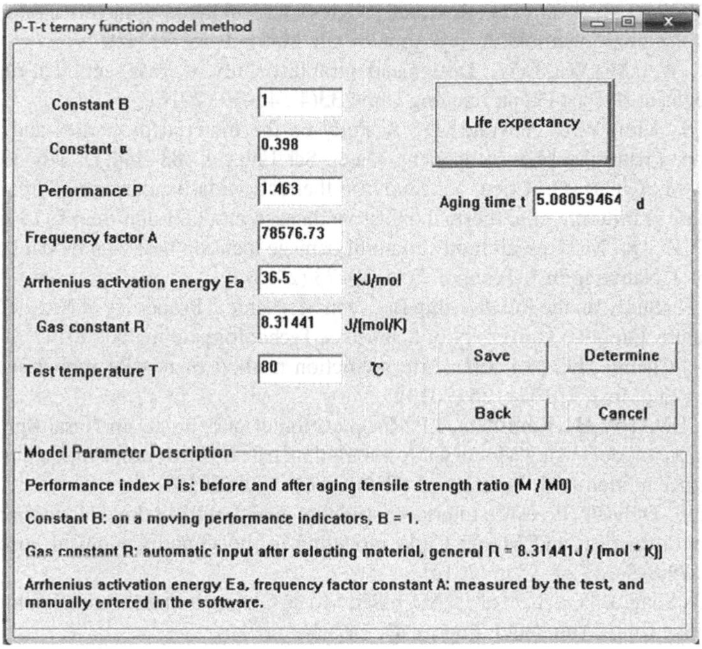

Fig. 9. P-T-t three function model input window

6 Conclusion

The software developed integrated the existing variety of aging life prediction methods of rubber materials. It can predict aging life using several theoretical models at one time and the results from different methods in predicting the same material were only a mutual reference. In addition, a commonly used rubber materials database was designed in the software to catch multiple material data at the same time and avoid the user manually checking material table and inputting data. Some space was left for the database to be improved constantly. For the results stored in the database, they could be used as a reference for future prediction model selection. The software also has a space for future development of the rubber material fatigue theory and experimental technology to make it more flexible on the human-computer interaction.

References

1. Li, Y.: Correlation between accelerated oven aging and indoor natural aging of nitrile vulcanizates. Spec. Rubber Prod. **22**(4), 51–56 (2001)
2. Re, N.-R., Li, H., Sun, B.G., et al.: Research report on tire aging of American highway traffic safety administration. China Rubber **31**(11), 8–12 (2015)
3. Qu, M.Z.: Research method of rubber life prediction. Sci. Technol. Innov. Appl. **21**, 4–5 (2012)

4. Liu, X.D., Xie, J., Fen, Z., et al.: Research progress of accelerated aging test and life prediction method for rubber materials. Aging Appl. Synth. Mater. **1**, 69–73 (2014)
5. Yang, D., Wei, M., Wu, D.W.: Design and durability study of cross section of rubber seal of shield segment. J. East China Jiaotong Univ. **33**(4), 45–49 (2016)
6. Woo, C.S., Kim, W.D., Kwon, J.D.: A study on the material properties and fatigue life prediction of natural rubber component. Mater. Sci. Eng., A **483–484**(1), 376–381 (2008)
7. Yuan, L.-M., Gu, B.-Q., Chen, Y.: Study on thermal oxidative damage of fiber reinforced rubber base sealing material thermal oxidative damage model. Lubr. Seal **1**, 78–80 (2006)
8. Cheng, X.Z., Qu, M.: Research on structural damage location based on modal strain energy relativity. J. Nanyang Inst. Technol. **7**(4), 70–73 (2015)
9. Zhao, X.K.: Study on the Relationship Between Mechanical Properties of Nitride Rubber and Temperature. Qingdao University of Science & Technology, Qingdao (2014)
10. Han, J.L., Cheng, H.G., Li, J.H.: Life prediction method of missile rubber seal. J. Naval Aeronaut. Eng. Inst. **2**, 172–176 (2013)
11. Guo, C., Doub, W.H., Kauffman, J.F.: Propagation of uncertainty in Nasal Spray in Vitro, performance models using Monte Carlo simulation: part II. Error propagation during product performance modeling. J. Pharm. Sci. **99**(8), 3572–3578 (2010)
12. Dazas, F.B., Delville, E., et al.: Interlayer structure model of tri-hydrated low-charge smectite by X-ray diffraction and Monte Carlo modeling in the Grand Canonical ensemble. Am. Mineral. **99**(8–9), 1724–1735 (2014)
13. Wang, D., Song, J., Yu, S., et al.: CAD-based Monte Carlo automatic modeling method based on primitive solid. Ann. Nucl. Energy **87**, 162–166 (2016)
14. Fang, Q.H., Lian, Y.X., Zhao, G.L., et al.: Prediction model of rubber aging based on BP artificial neural network. Aging Appl. Synth. Mater. **32**(2), 27–30 (2003)
15. Xie, H.X.: Analysis of water environmental quality using BP artificial neural network in Weihe River Baoji segment. Appl. Mech. Mater. **401–403**(3), 2147–2150 (2013)
16. Li, L., Qiu, P., Xing, S.B., et al.: Research on corrosion rate prediction of aluminum alloys in typical domestic areas based on BP artificial neural network. Adv. Mater. Res. **652–654**, 1088–1091 (2013)
17. Yin, W., Jin, X.X., Tong, G.: Finite element analysis and simulation of two common rubber constitutive models. J. Shanghai Dianji Univ. **13**(4), 215–218 (2010)
18. Xiao, X., Zhao, Y.F., Xu, W., et al.: Research progress of accelerated aging experiment and life evaluation model for rubber materials. Aerosp. Mater. Technol. **37**(1), 6–10 (2007)
19. Li, Y.J.: Reconsideration of rubber aging literature data. Plast. Ind. **43**(9), 515–530 (1997)

A Reliability Based Maintenance Policy of the Assembly System Considering the Dependence of Fixtures Elements Across the Stations

Shiming Zhang, Feixiang Ji, and Yinhua Liu[✉]

School of Mechanical Engineering,
University of Shanghai for Science and Technology, Shanghai, China
liuyinhua@usst.edu.cn

Abstract. In the mass production assembly process, the fixture system's reliability is vital for products' quality. The failure of the fixture system depends not only on the assembly operation times, but also upon the system degradation which is caused by the original manufacturing accuracy of fixtures, assembly forces, qualities of subassemblies from upstream assembly stations etc. In this paper, we propose a dynamic preventive maintenance policy for the fixture components based on the system reliability model of a multi-station assembly process. The proposed reliability model not only considers the degrading fixture components and other factors in the station, but also the dependence of process factors across the stations. Based on the reliability model and a given cost, an optimization method for the manufacturing tolerances and a dynamic maintenance schedule of locating pins are presented. At last, a body side assembly case is given to illustrate the proposed method.

Keywords: Preventive maintenance · Reliability modeling · Assembly system · Dependence

1 Introduction

The dimensional quality of the auto body is very important to the vehicle performance and appearance. The assembly system plays a key role on the auto body quality, thus the reliability of the assembly system is paid great attention to by both the researchers and manufacturers [1–3]. In the literature [4], Jin proposed a single station reliability model for the auto body assembly system, and afterwards this model was developed for multi-station system reliability modelling of assembly system. In real assembly process, the locating capability of fixtures has large effect on the product quality and system reliability. However, the above methods did not provide specific reasons of locating deviations and some factors that lead to locating errors are not included in the proposed reliability model. Besides, correlations of fixture component [5, 6] and the effect of part/subassembly quality on the assembly system reliability are not taken into account. In order to solve the problems, a new locating deviation model considering the original manufacturing tolerance, wearing *etc.* in each assembly operation is proposed.

© Springer International Publishing AG 2017
Y. Huang et al. (Eds.): ICIRA 2017, Part III, LNAI 10464, pp. 843–854, 2017.
DOI: 10.1007/978-3-319-65298-6_75

The dependence of fixture components in and across the assembly station is considered and the Copula function theory [7, 8] is used to describe the correlation. Based on the reliability model, a maintenance optimization method is proposed to prevent failures in the assembly systems and ensure the assembly quality.

2 Reliability Modeling of the Assembly System

2.1 Modeling of Locating Deviation

The locating accuracy of the system component is not only depend on the component state, but also related to the quality of the incoming part. In a single station, there are generally five factors that influence the locating accuracy of the fixture component and they are:

1. The wearing amount of fixture component;
2. The manufacturing accuracy of the circular degree of the fixture pins;
3. The install error of the component;
4. The manufacturing accuracy of the circular degree of the locating holes;
5. The manufacturing accuracy of the position degree of the locating holes;

Based on the above analysis, the locating deviation can be presented as follows:

$$\Delta x = (T_p + I_p + w_p + T_{H1} + T_{H2} + g)\cos\alpha$$
$$\Delta z = (T_p + I_p + w_p + T_{H1} + T_{H2} + g)\sin\alpha \tag{1}$$

$$\Delta x = (T_p + I_p + w_p + T_{H1} + T_{H2} + g)e\cos\beta$$
$$\Delta z = -(T_p + I_p + w_p + T_{H1} + T_{H2} + g)e\sin\beta \tag{2}$$

where T_P is the designed circular degree tolerance of the locating pin, T_H is the designed manufacturing tolerance of the incoming part it contains T_{H1}, T_{H2} and T_H is the circular degree tolerance, T_{H2} is the position degree tolerance, g is the fit clearance between the locating pin and the locating hole, I_p is the install error of the fixture, α and β are the contact angle and it will be shown in Fig. 1, e is a factor which is based on the contact point of the locating slot and the two way locating pins, Δx and Δz is the corresponding locating deviation of the locating pin.

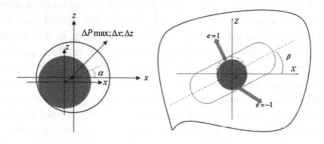

Fig. 1. The description of α, β, e

2.2   Modeling of the Wear of System Components

The wear of the locating pins is not just based on the assembly times but also based on the wear rate of the locating pins correlating with the locating accuracy, because the different locating errors correspond to a different interference fitting, which is corresponding to the frictional force. The locating pin is not just manufactured by a single material, actually each locating pin has a paint-coat and the paint-coat has a good mechanical property when the thin paint-coat is worn out the wear rate of the locating pins will increase fast so we can define the wear rate of each assembly time as:

$$
\begin{aligned}
\Delta w_l &= f_w(w_{l-1}, u_{1l}, \sigma_{1l}, u_{2l}, \sigma_{2l}) \\
w_l &= w_{l-1} + \Delta w_l
\end{aligned}
\tag{3}
$$

where u_{1l}, σ_{1l} are the vectors to describe the circular degree error and the install error of the fixture and each of the vector is formed by both the two aspects, u_{2l}, σ_{2l} are the vectors to describe the circular degree error and the position degree error of the locating holes and each of the vector is formed by both the two aspects. l is the born assembly numbers, Δw_l is the wear rate of the l-th assembly, w_l is the total wear amount.

Based on the wearing amount w_l, the assembly deviations of key product characteristics can be calculated. Based on the engineering knowledge, the wear rate of each assembly operation is not a certain value and the wear rate in the l-th operation follows a normal distribution and the two parameters of the distribution are shown as follows:

$$
\Delta w_l \sim N(\mu_{\Delta w_l}, \sigma_{\Delta w_l}^2)
\tag{4}
$$

$$
\begin{aligned}
\mu_{\Delta w_l} &= f_{\mu_l}(w_{l-1}, u_{1l}, \sigma_{1l}, u_{2l}, \sigma_{2l}) \\
\sigma_{\Delta w_l}^2 &= f_{\sigma_l^2}(w_{l-1}, u_{1l}, \sigma_{1l}, u_{2l}, \sigma_{2l})
\end{aligned}
\tag{5}
$$

Based on the empirical formula given by Jin [4], a definitive structure is given to describe the distribution of:

$$
\begin{aligned}
\mu_{\Delta w_l} =& [\phi_1(u_{1l} + u_{2l})^T + \phi_2(\sigma_{1l} + \sigma_{2l})^T + \phi_3 w_{l-1}]10^{-6} \\
&+ [\phi_4(u_{1l} + u_{2l})^T + \phi_5(\sigma_{1l} + \sigma_{2l})^T + \phi_6 w_{l-1}]10^{-6}\exp(-10^{-3}l)\ \text{mm} \\
\sigma_{\Delta w_l}^2 =& [\phi_7(u_{1l} + u_{2l})^T + \phi_8(\sigma_{1l} + \sigma_{2l})^T + \phi_9 w_{l-1}]10^{-9}\text{mm}^2
\end{aligned}
\tag{6}
$$

where ϕ is a coefficient, which is based on the form of the locator of the fixture and the shape of the parts.

During the long time manufacturing of locating pins, incoming parts and the install of the locating pins, the mean shifts were assumed to be zero, so u_{1l}, σ_{1l} and u_{2l}, σ_{2l} can be presented as follows:

$$
u_{1l} = [0\ \ 0]^T; u_{2l} = [0\ \ 0]^T; \sigma_{1l} = [I_{Pl}\ \ T_{pl}]^T; \sigma_{2l} = [T_{H1l}\ \ T_{H1l}]^T
\tag{7}
$$

The locating deviations of the all locating pins are denoted by $\{\Delta_{x1}\Delta_{z1} \cdots \Delta_{xi}$ $\Delta_{zi} \cdots \Delta_{xn} \Delta_{zn}\}^T$. i is the i-th locating pin. η_{xi} and η_{zi} are the corresponding threshold of the ith locating deviations. Then we can use the locating errors to calculate the corresponding reliability of single pair of locating pins if the locating deviation is beyond the threshold:

$$R_{fi} = (1 - \Delta_{xi}/\eta_{xi} + 1 - \Delta_{zi}/\eta_{zi})/2 \tag{8}$$

2.3   Modeling of the Fixture Components Reliability

To the in-station system components, the incoming part quality and the wear state of each locating pin can influence the locating accuracy of the in-station system components and they are not individual because the locating error of a locating pin can influence the locating accuracy of the other locating pins in this station and we can see it in Fig. 2. In the figure, the locating performance of the two way locating pin is not good, so the part in the point of the locating slot did not located will that caused a rotate of the part because of the designed fit clearance, so even the four way locating pin have a good locate ability but in the point of the four way locating pin still have a bad locating performance. The reliability of the system components, which reflect the locating performance of the correspond components, in this station can influence the product quality in this station, which is the incoming part as the next station and the quality of the part can influence the reliability of the system components in the next station, so to the between station system components, we can find that the product quality between each station can transmit this correlation. So divide the system components as in-station and between stations. As in the theory of Copula function there is an index to describe the degree of this correlation between each component, so we can use Copula function to describe the correlation between system components.

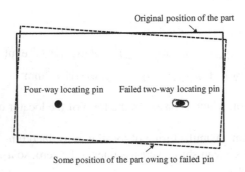

Fig. 2. The correlation between the locating pins in-station

According to the analysis, we can use the index in the Copula function to describe the correlation degree, which is affected mostly by the product quality and the system components state. The index can be shown as a function as following:

$$a = \begin{cases} f_{a1}(w,u,\sigma) \\ f_{a2}(u,\sigma) \end{cases} \tag{9}$$

where a is the Copula index and according to the theory of Copula the value range is $(0, 2]$. When the value of the Copula index is increasing the corresponding correlation degree is increasing.

We can assume that a follow an exponential variable, so we can give a definitive structure for a and that is as followings:

$$f_{a1}(w,u,\sigma) = 1 - \exp(\Re_1 w + \Re_2|u| + \Re_3\sigma); f_{a2}(u,\sigma) = 1 - \exp(\Re_4|u| + \Re_5\sigma) \tag{10}$$

where: $\Re$ is a coefficient and it's a negative value, which is based on the form of the locator of the fixture and the shape of the parts.

According to the theory of Copula function we know

$$\begin{aligned} R_{f_1 f_2} &= \mathrm{Pr}\,ob(w_1 \le \eta_1, w_2 \le \eta_2) = \hat{C}(R_{f_1}, R_{f_2}) \\ &= R_{f_1} + R_{f_2} - 1 + C(1 - R_{f_1}, 1 - R_{f_2}) \end{aligned} \tag{11}$$

$$C(1 - R_{f_1}, 1 - R_{f_2}) = \exp\{-[\sum_{i=1}^{2}(-\ln(1 - R_{f_i}))^\wedge(1/a)]^\wedge a\} \tag{12}$$

Take the function (12) to function (11) we can get the next function.

$$R_{f_1 f_2} = R_{f_1} + R_{f_2} - 1 + \exp\{-[\sum_{i=1}^{2}(-\ln(1 - R_{f_i}))^\wedge(1/a)]^\wedge a\} \tag{13}$$

So in the same way we can get function (14)

$$R_{f_1 f_2 f_3} = R_{f_1 f_2} + R_{f_3} - 1 + C(1 - R_{f_1 f_2}, 1 - R_{f_3}) \tag{14}$$

According to the logic of above we can get the system components reliability as followings

$$\begin{aligned} R_f &= R_{f_1 f_2 \cdots f_\lambda} = R_{f_1 f_2 \cdots f_{\lambda-1}} + R_{f_\lambda} - 1 + C(1 - R_{f_1 f_2 \cdots f_{\lambda-1}}, 1 - R_{f_\lambda}) \\ &= R_{f_1 f_2 \cdots f_{\lambda-1}} + R_{f_\lambda} - 1 + \exp\{-[(-\ln(1 - R_{f_1 f_2 \cdots f_{\lambda-1}}))^\wedge(1/a) \\ &\quad + (-\ln(1 - R_{f_\lambda}))^\wedge(1/a)]^\wedge a\} \end{aligned} \tag{15}$$

where λ is the total number of locating pins.

3 Effect of Product Quality on System Reliability

3.1 Assembly Deviation Modeling

In the multi-station assembly system the sheet mental part in the last station born the load, locate, assembly and then unload to go to the next station, during this process the deviation source contain the quality deviation of the incoming part, the fixture locating deviation and relocate deviation as show in Fig. 3. To describe this process we can use the state space model to get the final product quality of the assembly line.

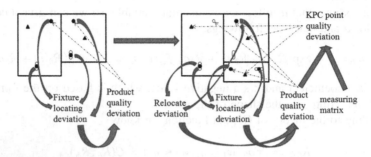

Fig. 3. The accumulate of the deviation flow diagram for state space model

In this figure it shows a two station assembly system, in the first station the deviation source are fixture locating deviation matrix and based on the theory of the state space model we can get the sensitive degree matrix of the fixture then combine the fixture locating deviation matrix and the sensitive degree matrix and add some white noise we get the product quality deviation of this station. The subassembly in this station goes to the next station, and in the following station some locating pins have relocating deviation to the assembly product. According to the propagation in all the stations, the final body in white quality is measured with a observation vector. This deviation propagation process can be modelled with a state space model which is shown:

$$D(1) = B(1)U(1) + \omega(1)$$
$$D(j) = A(j-1)D(j-1) + B(j)U(j) + \gamma(j) \qquad (16)$$
$$Y_j = C(j)D(j)$$

where: j is the assembly station index, $\mathbf{D}$ is the matrix of part deviation, $\mathbf{U}$ is the fixture locating deviation matrix, $\mathbf{A}$ is the relocate matrix, $\mathbf{B}$ is the sensitive degree matrix of the fixture, which is based on the locating position of the fixture, $\mathbf{C}$ is the measuring matrix, $\mathbf{Y}$ is the product quality deviation matrix, ω and γ are the white noise matrix.

3.2   Reliability Modeling Concerning Product Quality

In the engineering field, the coordinate measurement machine or white light scanning machine is used to measure the KPCs in the part and then the measurements are used to evaluate the product quality. In this paper we built the state space model to obtain the KPCs' deviation of the product. Because we want to build a reliability model related to the product quality, that is to say, all KPCs deviations of the product. The large amount of data of KPCs have mean values, so the mean values are used to describe the quality of the product. Each KPCs deviation mean has its corresponding thresholds η_Y. If the mean is beyond the threshold, the corresponding reliability is set to zero otherwise we can define the reliability index concerning product quality as follows:

$$R_{q\kappa} = 1 - Y_m(\kappa, 1)/\eta Y_m(\kappa, 1) \tag{17}$$

where m is station number, $Y_m(\kappa, 1)$ are the mean of some KPCs and η_{Y_m} is the corresponding threshold.

According the above reliability analysis considering the effect of product quality, the reliability on product quality in different stations is shown:

$$R(q) = \sum_{\kappa=1}^{n} K_\kappa R_{q\kappa} \tag{18}$$

where n is the number of KPC, K is a weight value to describe the importance of the different KPC reliability to the product quality reliability. According to the definition of the system reliability we can know that the system reliability of the auto-body assembly system is as followings:

$$R = R(f)R(q) = R_{f_1 f_2 \cdots f_\lambda} \sum_{\kappa=1}^{n} K_\kappa R_{q\kappa} \tag{19}$$

4   Application of System Reliability

4.1   Optimization of the Manufacturing Cost Allocation

To design and run a product line the manufacturing cost must be considered, and the manufacturing cost of the incoming parts and locating pins is based on the designed tolerance, so according to the analysis the cost of the incoming parts and the locating pins can be shown as followings:

$$C_p = \frac{\varphi_p}{T_p}; C_f = \frac{\varphi_f}{T_f} = \frac{\varphi_{f1}}{T_{H1}} + \frac{\varphi_{f2}}{T_{H2}} \tag{20}$$

where: φ is a coefficient that weigh the manufacturing cost to a designed tolerance, $C_\bullet$ is the manufacturing cost, T_f contains two parts T_{H1}, T_{H2}.

Taguchi proposed the concept of quality lost function, in this paper the corresponding quality lost cost function can be shown as followings:

$$C_L = \sum_{\kappa=1}^{n} K_\kappa M_\kappa \left(\frac{1}{C_P or C_{PK}}\right)^2 \tag{21}$$

where: M is a coefficient that weighs the quality lost cost, C_L is the quality lost cost. So the total manufacturing cost of this line is

$$C_T = C_L + C_p + C_f = \varphi_p/T_p + \varphi_{f1}/T_{H1} + \varphi_{f2}/T_{H2} + \sum_{\kappa=1}^{n} K_\kappa M_\kappa (1/C_P or C_{PK})^{\wedge}2 \tag{22}$$

When we design a assembly system product line the total manufacturing cost must be given, so if the total manufacturing cost must be in C_{UP} then different manufacturing cost allocate correspond to a different system reliability decline process, and among those decline process we can fine a best process so the correspond manufacturing cost allocation is the optimized manufacturing cost allocation. The optimization function can be shown as followings:

$$\mu_\Delta^* = \arg \quad \max\{R\} \\ s.t. \quad C_T \le C_{UP} \tag{23}$$

According to the analysis, the root cause of the system components reliability decline is the increase of the locating deviation and the main root cause of the decline of product quality reliability is also the increase of the locating deviation. After the study of the C_T we can find that to the same C_T by the different manufacture cost allocation we can get a different $T_P + T_{H1} + T_{H2}$ and that corresponding to a different $\Delta x, \Delta z$. So according to this analysis the optimization function can be put forward:

$$\mu_\Delta^* = \arg \quad \min T_p + T_{H1} + T_{H2} \\ s.t. \quad C_T \le C_{UP} \tag{24}$$

4.2   Maintenance Optimization of the Assembly System

In the industrial the root cause that to take a maintenance for the product line is the products' quality of the manufacturing system can't meet the current level. The aim to make a preventative maintenance is to solve the problem before the bad performance of the manufacturing system happened. In the industrial the index of C_p, C_{pk} really have the ability to describe the current product performance and there are also be used in the correspond field, but those indexes can not reflect the total root cause of the bad performance manufacturing system and they are the followings: (1) the system components arc in a good state, but the incoming parts are unqualified; (2) the incoming parts are qualified, but the fixture components are worn out; (3) the incoming parts and the system components are both unqualified. The incoming parts in the above three points are not only indicate the raw materials of the assembly line, but also indicate the subassemblies in the assembly process, for example the product of the first station is also a part of the incoming parts of the second station. Compared the system reliability of the manufacturing system to the traditional C_p, C_{pk} it has the advantage to find the total root cause of the bad performance of the line because the factor can reflect all the above three points are totally integrated in the system reliability so it is a comprehensive index to describe the state of both product quality and system components. Based on the analysis use the system reliability to monitor the manufacturing system is very scientific and the procedure that used to do the preventative maintenance is shown as followings:

Step1: Compare the current system reliability value, which include the system components reliability, product quality reliability and the system reliability, to the correspond set system reliability threshold, if the values are all bigger than the threshold then let the assembly line continue product, if the system quality reliability is less than the set value then go to step2, if the system components reliability is less than the set value then go to step3, if the system reliability of the line or one or both the system quality reliability and the system components reliability are less than the set value then go to step 4;

Step2: In this step the assembly line has been in a bad state because of the bad quality performance, so improve the quality of the incoming parts, then go to step 5;

Step3: In this step the assembly line has been in a bad state because of the bad system components performance, so renewal the failure system components, then go to step 5;

Step4: In this step the assembly line has been in a bad state because of the bad quality performance and the bad system components performance, so improve the quality of the incoming parts and renewal the failure system components, then go to step 5;

Step5: No maintenance actions. The assembly system will work continuously;

In conclusion, according to the reliability threshold, once the reliability index satisfies the following equations, the maintenance actions are performed.

$$R(q) < R(q)^*; R_f < R_f^*$$
$$R < R^* \quad or \quad one \quad or \quad both \quad R(q) < R(q)^* and \ R_f < R_f^* \tag{25}$$

5 Case Study

In the Fig. 4 a simple 2D side wall inner assembly case with three assembly stations and one measurement station were used to illustrate the proposed method. Table 1 shows the KPC coordinates and Table 2 shows the corresponding locating pins positions.

In the model, were set $I_P = 0.048$ mm; $g = 0.01$ mm; $C_{UP} = 5000$ RMB.

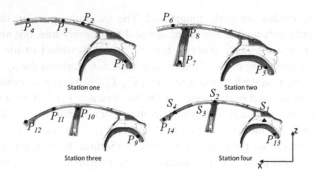

Fig. 4. Side wall assembly process

Table 1. KPCs'coordinate.

Unit mm	S_1	S_2	S_3	S_4
X	3655	4997	5093	6482
Z	1818	2271	1839	1894

Table 2. the coordinate of the locating pins.

Unit mm	P_1&P_5&P_9&P_{13}	P_2	P_2	P_4&P_6	P_7	P_8&P_{10}	P_{11}	P_{12}&P_{14}
X	3505	4198	4703	5287	5165	5061	5893	6654
Z	1409	2206	2235	2274	1453	2192	2240	1809

The corresponding cost function including manufacturing cost of the incoming parts and the locating pins also are shown as follows:

$$C_T = C_L + C_p + C_f = \frac{\varphi_p}{T_p} + \frac{\varphi_{f1}}{T_{H1}} + \frac{\varphi_{f2}}{T_{H2}} + \sum_{\kappa=1}^{n} K_\kappa M_\kappa \left(\frac{1}{C_p \, or \, C_{PK}}\right)^2$$

$$= \frac{4}{T_p} + \frac{200}{T_{H1}} + \frac{96}{T_{H2}} + \sum_{\kappa=1}^{n} K_\kappa M_\kappa \left(\frac{1}{C_p \, or \, C_{PK}}\right)^2$$

According to the Eq. (24) we get the optimized tolerance allocation results as follows:

Parameter	T_P	T_{H1}	T_{H2}
Optimized value (mm)	0.02	0.07	0.49

According to the function (19) the optimized system reliability curve as shown in the Fig. 5:

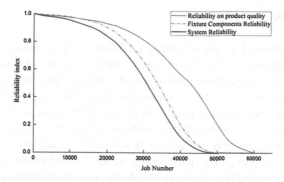

Fig. 5. Declining process of the system reliability of the optimized manufacturing allocation

In this paper the reliability thresholds was set as:

$$R(q)^* = 0.85 \, ; \, R_f^* = 0.85 \, ; \, R^* = 0.725$$

According to the function (25), the preventative maintenance procedure was shown in Fig. 6. As we can see in the figure, the 1st maintenance was triggered when R_f was less than 0.85, the 2nd maintenance was triggered when the reliability on product quality was less than 0.85 and the 3rd maintenance was triggered when the system reliability was less than 0.725. If the product quality and its dependence on the fixture components, the system reliability would be overestimated and the corresponding time interval between maintenance cycles based on the fixture component reliability would be longer. As a result, the dimension variations of auto bodies would be increased though the maintenance times were less. In real assembly shop, the maintenance cost for the fixture system is much more less than the dimension quality loss. Therefore, the proposed system reliability evaluation method considering the product quality and its dependence on the fixture components across stations is more effective than the traditional reliability model only considering the fixture component.

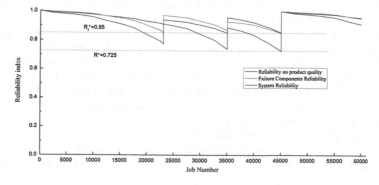

Fig. 6. Preventative maintenance of the fixture system

6   Conclusions

This paper has presented a new reliability modeling method considering the correlations of fixture components in and across the assembly stations. The method reveals the production capability for qualified products by integrating more process-related factors, such as wearing process of fixture components, manufacturing tolerance of locating pins, stamping holes and product quality in the assembly process. On the basis of the reliability modeling, the manufacturing costs of different locating pins concerning materials and coating process are allocated optimally to achieve maximum system reliability. Besides, a preventive maintenance method based on the reliability evaluation is proposed by setting the reliability thresholds and the maintenance actions should be triggered once the reliability declined to the threshold values. It should be pointed out that the reliability modeling on product quality is based on the variation propagation method of state space model. The variation models considering flexible deforming of sheet metal parts can be studied further to improve the accuracy of the system reliability modeling.

Acknowledgements. This research is supported by National Natural Science Foundation of China (No. 51405299), Natural Science Foundation of Shanghai (No. 14ZR1428700) and Open Fund of Shanghai Key Laboratory of Digital Manufacture for Thin-walled Structures.

References

1. He, Y., He, Z., Wang, L., et al.: Reliability modeling and optimization strategy for manufacturing system based on RQR chain. Math. Probl. Eng. **15**, 1–13 (2015)
2. Srikanth, D., Kulkarni, M.S.: Quality, reliability and maintenance (Q, R & M) issues in reconfigurable manufacturing systems (RMS). Appl. Mech. Mater. **110–116**, 1442–1446 (2011)
3. Chen, Y., Jin, J., Shi, J.: Integration of dimensional quality and locator reliability in design and evaluation of multi-station body-in-white assembly processes. IIE Trans. **36**(9), 827–839 (2004)
4. Jin, J., Chen, Y.: Quality and reliability information integration for design evaluation of fixture system reliability. Qual. Reliabil. Eng. Int. **17**(5), 355–372 (2001)
5. Tao, F., Mou, P., Jia, C., et al.: The gears system reliability analysis with correlation among the components of gear box in self-propelled gun. Appl. Mech. Mater. **526**, 230–235 (2014)
6. Yu, Y., Hao, Z., Wang, S., et al.: Correlation analysis for screening key parameters for passive system reliability analysis. Ann. Nucl. Energy **77**, 23–29 (2014)
7. Tang, J., He, P., Cheng, S.: Copula reliability calculation model for K/N(F) systems considering failure correlation. Appl. Mech. Mater. **446–447**, 1045–1051 (2013)
8. Han, W., Zhou, J.: Fatigue reliability copula model of structural systems with fatigue life correlation. Adv. Mater. Res. **199–200**, 548–554 (2011)
9. Ceglarek, D., et al.: Variation Reduction for Assembly: Methodologies and Case Studies Analysis. Technical report of the "2 mm" Program, University of Michigan, Ann Arbor (1993)
10. Liu, J., Jin, J., Shi, J.: State space modeling for 3-D variation propagation in rigid-body multistage assembly processes. IEEE Trans. Autom. Sci. Eng. **7**(2), 274–290 (2010)

Efficient Cutter-Freeform Surfaces Projection Method for Five-Axis Tool Path Computation

Xiyan Li[1], Chen-Han Lee[1,2(✉)], Pengcheng Hu[1], Yanyi Yang[1], and Fangzhao Yang[1]

[1] Huazhong University of Science and Technology, Wuhan, China
chenhanlee@hotmail.com
[2] Wuhan Institute of Technology, Wuhan, China

Abstract. The complex surfaces are usually approximated as Z-maps, point based models or facet models to simplify the computation for cutter positions. However, such surface approximation error may result in the conflict between the accuracy of cutter positions and the computational efficiency, and C1 discontinuity for tool paths and tool orientations. In this paper, an efficient cutter-freeform surfaces projection (CFSP) method based on the cutter-facet model projection (CFMP) method is proposed for five-axis tool path computation to solve the above issues. Motivated by the Newton-Raphson algorithm, the tangent plane-based search method is developed to find the accurate cutter contact points on surfaces. The golden section-based search algorithm is further carried out to deal with the convergence oscillation problem. Simulation on the part models validates that the algorithm is effective to improve the machining accuracy.

Keywords: Machining accuracy · Five-axis tool path computation · Cutter-freeform surfaces projection · Tangent plane-based search

1 Introduction

Five-axis machining can be widely exploited with full advantages in the manufacturing of parts with freeform surfaces. With two additional degrees of freedom, five-axis machining provides the benefits of less number of fixture, better tool accessibility, improved quality and efficiency, and shorter cycle of new product development compared with three-axis machining.

Accuracy of the cutter contact points gains more significance in five-axis tool path computation because the simultaneous five-axis motion increases the machine volumetric error. Error in the tool position computation is introduced when the complex surfaces are approximated as Z-maps, point based models or polyhedral model with triangular facets to reduce the computational complexity. The Z-map method computes the interference-free CLs from a grid data set. The precision of machining is dependent on the density of the grid data. A large memory space to be allocated for the grid data and large computation are needed [1]. P. Gray et al. [2] discussed a point-based method using a G buffer (essentially a modified Z buffer). Since a polygonal approximation of a surface requires fewer triangles than the points of an SPS for equivalent accuracy [3].

© Springer International Publishing AG 2017
Y. Huang et al. (Eds.): ICIRA 2017, Part III, LNAI 10464, pp. 855–866, 2017.
DOI: 10.1007/978-3-319-65298-6_76

The triangle mesh is a more common representation for tessellated models [4, 5]. In the process of tool path computation, the cutter contact point CC_{facet} is defined on a facet rather than on the original surface CC_{ori} (see Fig. 1). Therefore, the approximation error ε_{CC} exists because of the discretization of the freeform surfaces. A recursive algorithm is required to subdivide the surfaces and check the maximum distance of a set of facets from the surface. If the distance exceeds the bound, the surface patch is further subdivided. The accuracy of the algorithm depends on the density of the discretized facets. If high machining accuracy is required, a recursive algorithm will be very computationally expensive.

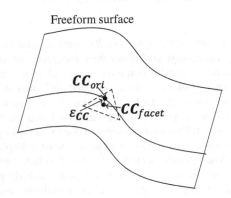

Fig. 1. Surface approximation error produced by discretization.

C1 continuous tool paths and tool orientations cannot be ensured when the cutter contact (CC) points are on the facet model. For tool paths on the facet model, the adjacent line segments are subjected to smoothing with C1 continuous smooth line segment to improve the machining efficiency. Sudden or drastic change of tool orientations may cause harmful results, such as streaks on the part surfaces and harshness on the tool for tool positions on facet model. If the cutter contact points are on the freeform surfaces, tool paths and tool orientations will be C1 continuous.

The principle of the projection method [6] is described (see Fig. 2). The original cutter positions are determined by the drive points on the pre-planned drive path. The cutter is located at a given initial position far away from the parts, then virtually moves along a projection direction towards the parts maintaining a constant tool axis orientation, and finally stops at the first cutter contact point (touching tangentially) with the part surface. Since the first cutter contact point in the direction of projection is selected, this algorithm can ensure gouge-free between the cutter and the part surface.

The cutter-facet model projection (CFMP) method along the tool axis direction [7–10] and along an arbitrary direction has been researched [11]. Facet models are used to approximate the freeform surfaces, then facets in the shadow of the cutter along the projection direction are selected to be projected for each drive cutter position, and finally the cutter contact points on facet model are calculated. Thus, the conflict between the accuracy of the cutter positions and computational efficiency exists, and C1 discontinuity for tool paths and tool orientations is caused.

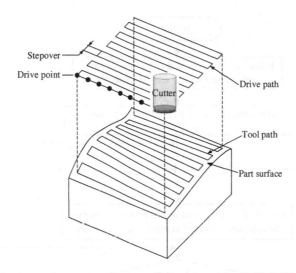

Fig. 2. Surface approximation error produced by discretization.

In this paper, an efficient cutter-freeform surfaces projection (CFSP) method, projecting onto the freeform surfaces instead of the facet models, is developed to compute precise five-axis tool paths. Initial cutter contact points are determined first by the CFMP method. Then the tangent plane-based search algorithm, similar to the Newton-Raphson algorithm, is proposed to find the cutter contact points on the freeform surfaces. To eliminate the convergence oscillation problem, the golden section-based algorithm is presented. In addition, simulation on the part models with several surfaces is implemented.

This paper is arranged as below: In Sect. 2, the overview of the implementation of the CFSP is presented. Section 3 illustrates the details of the tangent plane-based and the golden section-based search algorithms. In Sect. 4, simulation on part models are described. Conclusions and future work are provided in Sect. 5.

2 Overview

To solve the conflict between the accuracy of the cutter contact points and the computational efficiency, and the issue of C1 discontinuity for the tool paths and tool orientations, the CFSP method is proposed based on the cutter-facet model projection (CFMP) method [11].

The specific process to determine the cutter contact points on the freeform surfaces with compound surfaces is listed (see Fig. 3). For the ith drive point on the pre-determined drive path, the cutter contact point on a facet, the initial point for the CFSP method, is calculated first with the CFMP method. Then, the CFSP method is carried out to compute the projection distance and the projected point on the jth freeform surface. The projected point corresponding to the shortest projection distance is selected as the cutter contact point for the ith drive point. Finally the precise tool paths with cutter contact points on the freeform surfaces can be generated.

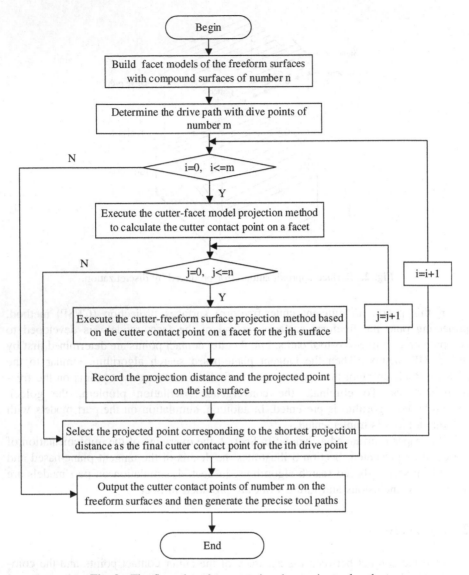

Fig. 3. The flow chart for computing the precise tool paths.

3   Details of the CFSP Method

Newton-Raphson algorithm requires an initial guess for the location of the root. This guess is improved by looking at the tangent of the function at this point and finding the root of the tangent function. Motivated by the Newton-Raphson algorithm, the initial cutter contact points on facets are computed with the CFMP method [11] and the tangent plane-based CFSP method is developed to compute the cutter contact points on freeform surfaces, which is presented in Sect. 3.1. However, it may oscillate between two cutter contact points, i.e., never converge towards the correct cutter contact point, even if the initial point is reasonable. To solve the convergence oscillation in finding the correct points on the freeform surfaces, a golden section-based search algorithm adjusting the search step is presented, which is presented in Sect. 3.2.

3.1 Tangent Plane-Based Search Algorithm

The Newton-Raphson algorithm for finding the root of function is shown in Fig. 4(a). The tangent plane-based CFSP method shown in Fig. 4(b) mainly includes the following steps: (1) Select the cutter contact point P_0 on a facet as the initial point; (2) Calculate the shortest distance D_0 between the cutter contact point P_0 and the surface, and obtain the corresponding point Q_0 on the surface; (3) Build the tangent plane of the point Q_0; (4) Execute the cutter-tangent plane projection and obtain the point P_1 on the tangent plane; (5) Calculate the shortest distance D_1 between the cutter contact point P_1 and the surface, and obtain the corresponding point Q_1 on the surface; (6) Repeat the iteration until the cutter contact point on the surface is found.

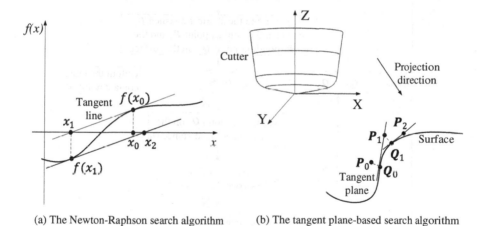

(a) The Newton-Raphson search algorithm   (b) The tangent plane-based search algorithm

Fig. 4. The tangent plane-based CFSP method.

The specific procedures of the tangent plane-based CFSP method are listed (see Fig. 5).

The cutter-tangent plane projection (see Fig. 6) includes the following steps: (1) Build the workpiece coordinate system (WCS) and the cutter coordinate system (CCS); (2) Transform the normal vector N_T of a facet from WCS to CCS; (3) Compute the cutter offset vector O in CCS and the CC_0 point on the cutter; (4) Intersect the line determined by the CC_0 point and the projection direction with the tangent plane in CCS, and obtain the intersection point; (5) Transform the intersection point from CCS to WCS and obtain the projected point P_1.

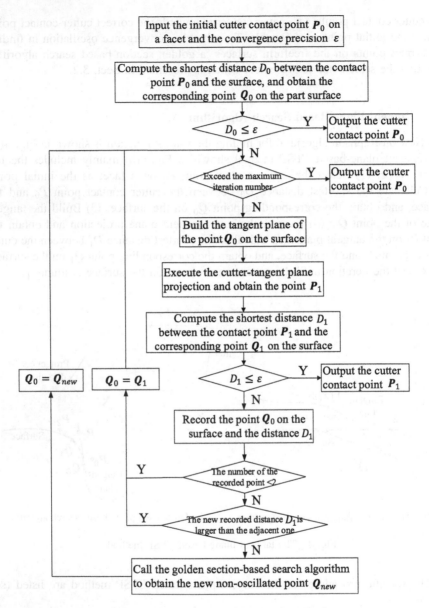

Fig. 5. The flow chart of the CFSP method.

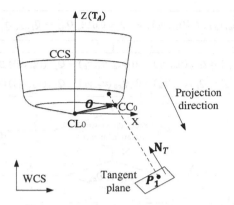

Fig. 6. The cutter-tangent plane projection.

3.2 Golden Section-Based Search Algorithm

The golden section-based search algorithm will be activated to adjust the search step when the two adjacent projected points oscillate. Motivated by the golden section algorithm which aims at finding the minimum, the golden section-based search algorithm is developed to seek the non-oscillated contact point corresponding to the minimum distance.

The two oscillated points are defined as Q_a and Q_b. The distance $D_a = f(Q_a)$ and $D_b = f(Q_b)$ can be calculated by executing the procedures (3), (4), and (5) in Sect. 3.1. To guarantee the search points Q_{a1} and Q_{a2} between the two oscillated points Q_a and Q_b on the freeform surface, the one-to-one mapping relationship between the two-dimension uv parametric space and the three-dimension WCS are built (see Fig. 7). The detailed procedures are given (see Fig. 8).

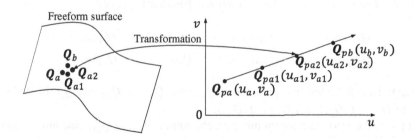

Fig. 7. The mapping relationship between the parametric space and the WCS.

Step1: Compute the corresponding points $Q_{pa}(u_a, v_a)$ and $Q_{pb}(u_b, v_b)$ in two-dimension uv parametric space from the two oscillated points Q_a and Q_b in WCS;

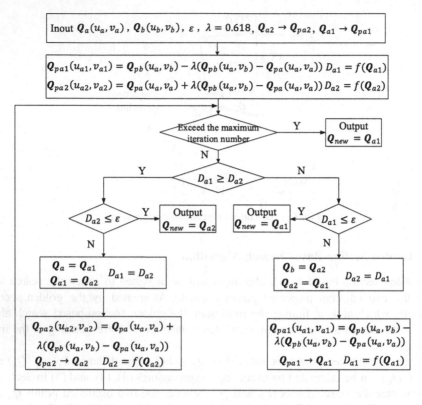

Fig. 8. The flow chart of the golden section-based search algorithm.

Step2: Calculate the two parameters Q_{pa1} and Q_{pa2} ($\lambda = 0.618$) between points $Q_{pa}(u_a, v_a)$ and $Q_{pb}(u_b, v_b)$ in two-dimension parametric space;

$$Q_{pa1}(u_{a1}, v_{a1}) = Q_{pb}(u_a, v_b) - \lambda\big(Q_{pb}(u_a, v_b) - Q_{pa}(u_a, v_a)\big)$$
$$Q_{pa2}(u_{a2}, v_{a2}) = Q_{pa}(u_a, v_a) + \lambda\big(Q_{pb}(u_a, v_b) - Q_{pa}(u_a, v_a)\big)$$

Step3: Transform the two parameters Q_{pa1} and Q_{pa2} to Q_{a1} and Q_{a2} in WCS, and compute $D_{a1} = f(Q_{a1})$ and $D_{a2} = f(Q_{a2})$;

Step4: If the max iteration number reach, output $Q_{new} = Q_{a1}$, else turn to step5;

Step5: If $D_{a1} \geq D_{a2}$ satisfy, turn to the step6, else turn to step7;

Step6: If $D_{a2} \leq \varepsilon$ satisfy, output $Q_{new} = Q_{a2}$; else make $Q_a = Q_{a1}$, $Q_{a1} = Q_{a2}$, $D_{a1} = D_{a2}$, $Q_{pa2}(u_{a2}, v_{a2}) = Q_{pa}(u_a, v_a) + \lambda\big(Q_{pb}(u_a, v_b) - Q_{pa}(u_a, v_a)\big)$, transform the parameter Q_{pa2} to Q_{a2}, compute $D_{a2} = f(Q_{a2})$, and turn to the step4;

Step7: If $D_{a1} \leq \varepsilon$ satisfy, output $Q_{new} = Q_{a1}$, else make $Q_b = Q_{a2}$, $Q_{a2} = Q_{a1}$, $D_{a2} = D_{a1}$, $Q_{pa1}(u_{a2}, v_{a2}) = Q_{pb}(u_a, v_a) - \lambda\big(Q_{pb}(u_a, v_b) - Q_{pa}(u_a, v_a)\big)$, transform the parameter Q_{pa1} to Q_{a1}, compute $D_{a1} = f(Q_{a1})$, and turn to the step4;

The convergence oscillation problem occurring in the tangent plane-based CFSP method can be solved with the golden section-based search algorithm and the correct cutter contact point on the freeform surface can be obtained.

4   Implementations and Simulation

To validate the effectiveness of the proposed CFSP method in five-axis tool path computation, simulation is carried out on two parts with several surfaces. The proposed method is implemented in c++ with Visual Studio 2010, and incorporated as the NC Blade software system, which was developed at Huazhong University of Science and Technology, China, for automatic planning and programming for five-axis machining.

The part 1 with compound surfaces and the part 2 with trimmed surfaces are tested (see Fig. 9).

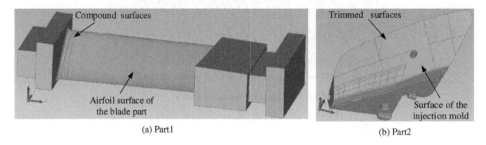

(a) Part1   (b) Part2

Fig. 9. Two parts of the blade and the injection mold.

The toroidal cutter D20R1.6 is used to cutting the part1 with the lead angle 15°. Tool paths both with the CFMP method and the CFSP method are generated in NC blade shown in Fig. 10(a). The quantitative change of the machining error is given in Fig. 10(b). For the CFMP method, only two cutter contact points can achieve the accuracy of 10^{-4} mm, 46.9% (4912/10455) of the cutter contact points cannot achieve the accuracy of 10^{-2} mm, and the maximum machining error is 0.10182 mm.

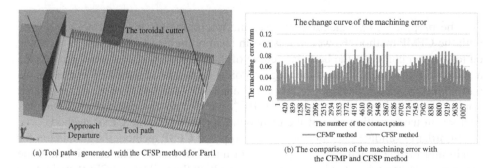

(a) Tool paths generated with the CFSP method for Part1

(b) The comparison of the machining error with the CFMP and CFSP method

Fig. 10. Tool paths and the comparison of the machining error for part 1.

By using the CFSP method, 99.6% (10415/10455) of the cutter contact points can achieve the accuracy of 10^{-4} mm, 99.9% (10447/10455) of the cutter contact points can achieve the accuracy of 10^{-2} mm, and the maximum machining error is 0.013134 mm.

Tool paths for the part2 are generated with the ball-end cutter D6R3 by using the CFMP and CFSP method shown in Fig. 11(a). It shows that 86.9% (1170/1346) of the cutter contact points can achieve the accuracy of 10^{-4} mm with the maximum machining error 0.003667 mm by using the CFSP method and only 3.49% (47/1346) of the cutter contact points can achieve the accuracy of 10^{-4} mm with the maximum machining error 0.026705 mm by using the CFMP method from Fig. 11(b).

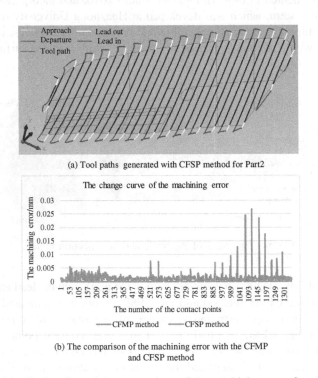

(a) Tool paths generated with CFSP method for Part2

(b) The comparison of the machining error with the CFMP and CFSP method

Fig. 11. Tool paths and the comparison of the machining error for part 2.

The proposed CFSP method improves the machining accuracy greatly, compared with the CFMP method. However, it is worth mentioning that 13.1% (176/1346) of the cutter contact points on the boundary curve of the trimmed surface cannot achieve the accuracy of 10^{-4} mm because the tangent points cannot be searched on curve for the part 2 with the CFSP method. For the part 1, about 4% (40/10455) of the cutter contact points cannot achieve the accuracy of 10^{-4} mm due to the discontinuous tangent vectors on the transition area of the compound surfaces with the CFSP method.

In order to validate the efficiency of the proposed CFSP method, the simulation for machining the Part1 and Part 2 is carried out under two cases. The case 1 is that the machining accuracy is set to be 0.1 mm for using both the CFMP and CFSP methods to generate tool paths. The case 2 is that the machining accuracy is set to be 10^{-4} mm for using the CFMP method, and 0.1 mm for using the CFSP method to generate tool paths. The compared results by using the CFMP and CFSP methods to generate tool paths under the two cases are shown in Table 1. The average computational time with the CFSP method increased about 37.26% compared to that with the CFMP method, while improving the machining accuracy from 0.1 mm (Part1) and 0.02 mm (Part2) to 10^{-4} mm in the first case. The results from the second case show that the average computational time with the CFSP method reduced about 23.64% compared to that with the CFMP method when achieving the same final machining accuracy of 10^{-4} mm.

The proposed CFSP method in this paper is more efficient than the CFMP method when high precision is required in the five-axis machining. The main reason is that the larger density of the facet models is required to ensure high machining accuracy when using the CFMP method, which leads to more computational time in building facet models and computing the cutter contact points than the CFSP method.

Table 1. Performance comparison of two methods for two tested parts under two cases.

Part	Number of tool paths	Number of CC points	Time/s (CFMP in case 1)	Time/s (CFSP in case 1 and case 2)	Time/s (CFMP in case 2)
Part1	5	1025	30.327	44.403	50.375
	10	1875	53.897	82.171	102.385
	20	3575	84.625	126.827	167.918
	30	5315	136.571	206.389	282.341
	40	7055	165.446	284.09	396.873
	50	8755	207.898	351.169	497.957
	60	10455	260.794	419.704	602.694
Part2	6	230	9.941	15.667	17.782
	11	440	19.704	30.166	38.552
	21	842	38.623	61.912	80.857
	33	1346	62.375	108.033	142.128
	42	1723	80.127	135.494	179.665
	51	2086	97.464	158.866	214.945
	64	2647	123.412	197.336	274.099

5 Conclusions and Future Work

This paper presents an efficient CFSP method to improve the machining accuracy. Based on the CFMP method for finding the initial iteration point, the tangent plane-based and golden section-based algorithms are proved effective to search the cutter contact points on the freeform surfaces. This CFSP method can solve the conflict between the accuracy of the cutter contact points and the computational efficiency, and can ensure C1 continuity for tool paths and tool orientations. The simulation results indicate that the machining accuracy can achieve 10^{-4} mm.

We plan to further extend the current work in the following three aspects. Cutting experiments to confirm the simulation need to be done. The cutter-curve projection problem should be addressed as a remedy for the proposed CFSP method to deal with the isolated curve or compound surfaces with discontinuous tangent vector on the transition area. The golden section-based search method for solving the convergence oscillation problem can be improved to speed up the computation.

Acknowledgements. The authors gratefully acknowledge the support of the National Science and Technology Major Project of the Ministry of Science and Technology of China (2013ZX 04007-041). We also acknowledge the help from Lixiong Gan, Fan Yang, and Dr. Changya Yan.

References

1. Gray, P.J., Bedi, S., Ismail, F.: Arc-intersect method for 5-axis tool positioning. CAD Comput. Aided Des. **37**(7), 663–674 (2005)
2. Gray, P.J., Ismail, F., Bedi, S.: Graphics-assisted rolling ball method for 5-axis surface machining. CAD Comput. Aided Des. **36**(7), 653–663 (2004)
3. Li, S.X., Jerard, R.B.: 5-axis machining of sculptured surfaces with a flat-end cutter. Comput. Des. **26**(3), 165–178 (1994)
4. Yau, H.T., Chuang, C.M., Lee, Y.S.: Numerical control machining of triangulated sculptured surfaces in a stereo lithography format with a generalized cutter. Int. J. Prod. Res. **42**(13), 2573–2598 (2004)
5. Duvedi, R.K., Bedi, S., Batish, A., Mann, S.: The edge-torus tangency problem in multipoint machining of triangulated surface models. Int. J. Adv. Manuf. Technol. **82**(9–12), 1959–1972 (2016)
6. Hansen, A., Arbab, F.: Fixed-axis tool positioning with built-in global interference checking for NC path generation. IEEE J. Robot. Autom. **4**(6), 610–621 (1988)
7. Hwang, J.S., Chang, T.-C.: Three-axis machining of compound surfaces using flat and filleted endmills. Comput. Des. **30**(8), 641–647 (1998)
8. Duvedi, R.K., Bedi, S., Batish, A., Mann, S.: A multipoint method for 5-axis machining of triangulated surface models. CAD Comput. Aided Des. **52**, 17–26 (2014)
9. Duvedi, R.K., Bedi, S., Batish, A., Mann, S.: Numeric implementation of drop and tilt method of 5-axis tool positioning for machining of triangulated surfaces. Int. J. Adv. Manuf. Technol. **78**(9–12), 1677–1690 (2015)
10. Kiswanto, G., Lauwers, B., Kruth, J.P.: Gouging elimination through tool lifting in tool path generation for five-axis milling based on faceted models. Int. J. Adv. Manuf. Technol. **32**(3–4), 293–309 (2007)
11. Li, X., Lee, C.-H., Hu, P., Sun, Y.: Efficient cutter-facet model projection methods and applications to five-axis tool path computation. CAD Comput. Aided Des. **38** (2017) (Submitted for publication), CAD-D-17-00021

Optimization of Milling Process Parameters Based on Real Coded Self-adaptive Genetic Algorithm and Grey Relation Analysis

Shasha Zeng[1](✉) and Lei Yuan[2]

[1] Hubei Key Laboratory of Waterjet Theory and New Technology,
School of Power and Mechanical Engineering, Wuhan University,
Wuhan 430072, China
zengshasha316@163.com
[2] Mechanical and Electrical Engineering College,
Hainan University, Haikou 570228, China

Abstract. In this paper, a method to optimize the milling process parameters based on the real-coded self-adaptive genetic algorithm (RAGA) and Grey relational analysis (GRA) is proposed. Experiments have been designed with four input milling process parameters at four different levels. The RAGA coupled with GRA has been applied for solving the proposed optimization problem to achieve the desired machined surface quality characteristics. Simulation experiments give the optimal parametric combination. Furthermore, experiments for the machined surface topography with the initial and optimal combination of milling process parameters are implemented and the results verify the feasibility of the proposed method.

Keywords: Process parameter optimization · Real coded self-adaptive genetic algorithm · Grey relational analysis · Surface topography

1 Introduction

Milling is the primary operation in most of the production processes in the industry. One of the essential criterions for selecting a proper milling process is the functional performance of machined surface [1]. Taking CNC milling for example, the surface roughness is the critical quality index for machined surface [2]. The corresponding performance is closely related to several machining parameters, such as the spindle speed, the feed per tooth, the axial depth of cut, and the cutter radius [3]. Surface topography of machined surface is significant for their functional performance. Among many parameters to characterize the surface topography, surface roughness is one of the most important parameters for evaluating the technological quality of a product. Accordingly, how to enhance the quality of the machined surface has been become an essential issue in the field of optimizing milling process parameters.

The RAGA and GRA are general methods that can be used to solve the complex problems for engineering design. Results demonstrate that RAGA can find optimal solutions [4]. Gong et al. [5] combined the RAGA and cumulative prospect theory to

© Springer International Publishing AG 2017
Y. Huang et al. (Eds.): ICIRA 2017, Part III, LNAI 10464, pp. 867–876, 2017.
DOI: 10.1007/978-3-319-65298-6_77

solve the portfolio choice problem. The computational results showed that the improved algorithm was more effective in realizing the global optimization and promoting evolution efficiency. Lee et al. [6] used RAGA to solve several benchmark optimization problems. This outcome of the study clearly demonstrated the effectiveness and robustness of the RAGA. Subbaraj et al. [7] introduced RAGA to solve the combined heat and power economic dispatch (CHPED) problem. The results showed that RAGA solved the CHPED problem efficiently. Subbaraj et al. [8] used the Taguchi-RAGA to solve the economic dispatch problem with valve-point loading. The result showed that the Taguchi-RAGA was very competitive in the field of solution quality, handling constraints and computation time. Abbas et al. [9] presented the adaptive real-coded genetic algorithm to identify the Volterra-system. The error between the identified nonlinear system and the Volterra model was reduced. Oyama et al. [10] obtained better performance in wing design by using the RAGA. The proposed method was to resolve the continuous search-space of an optimization problem. Deng [11, 12] proposed the Grey system theory, which was proven to be useful for dealing with the incomplete and uncertain information. Grey relational analysis (GRA) was adopted to combine multiple-quality parameters into one numerical score, and to determine the optimal setting for machine parameters by ranking these scores [13]. The optimization of many properties can be converted by Grey relational theory into a single grade value [14]. The RAGA-GRA method is proposed for the optimization of milling process parameters in this study.

2   Milling Process Parameters Optimization

2.1   Grey Relational Analysis (GRA)

In the Grey relational analysis, the first step is the generation of Grey relational, that is, normalize the S/N ratio ranging from zero to one and calculate the Grey relational coefficient to express the correlation between the desired and actual S/N ratios. In this study, the normalized values of surface roughness, corresponding to the lower-the-better characteristics criterion, can be expressed as

$$x_i(k) = \frac{\max y_i(k) - y_i(k)}{\max y_i(k) - \min y_i(k)} \tag{1}$$

where $x_i(k)$ is the value after the Grey relational generation, $\min y_i(k)$ and $\max y_i(k)$ are the smallest value and the largest value of $y_i(k)$ for the kth response, respectively. The Grey relational coefficient $\xi_i(k)$ can be formulated as

$$\xi_i(k) = \frac{\Delta_{\min} + \psi \Delta_{\max}}{\Delta_{0i}(k) + \psi \Delta_{\max}} \tag{2}$$

where $\Delta_{0i}(k) = \|x_0(k) - x_i(k)\|$ is the absolute value of the difference between the ideal sequence $x_0(k)$ and $x_i(k)$, ψ is the distinguishing coefficient satisfying $0 \le \psi \le 1$.

$\Delta_{\min}$ and $\Delta_{\max}$ are the smallest value and the largest value of Δ_{0i}, respectively. By averaging the Grey relational coefficients, the Grey relational grade can be given as

$$\gamma_i = \frac{1}{n}\sum_{k=1}^{n}\xi_i(k) \qquad (3)$$

where n is the number of process responses. The evaluation of overall performance characteristic is depended on the Grey relational grade. Using this method, the optimization of multiple performance characteristics can be converted to optimize a single Grey relational grade, since the optimal combination of process parameters is achieved corresponding to the highest Grey relational grade.

2.2 Real-Coded Self-adaptive Genetic Algorithm (RAGA)

Before executing the RAGA process, the population size, maximum generation number, crossover probability, mutation probability, fitness function, and range of each parameter must be assigned for the RAGA by expert knowledge and numerical experience [15]. In this paper, the population size is set at forty, and the maximum generation is twenty. The Grey relational grade is used as the fitness function. The reproduction procedure adopts the roulette wheel selection to pick chromosomes into the mating pool. Therefore, the probability of the jth chromosome entering into the mating pool is determined by the following equation.

$$fit\ ratio_j = \frac{fit\ value_j}{\sum_{i=1}^{k} fit\ value_j} \qquad (4)$$

where k is the population size.

The RAGA is able to adjust its crossover and mutation through contemporary adaptation of the maximum and minimum values, to accelerate evolutional speed and enlarge searching scope. The adaptation crossover and mutation can be used with the following equations [15].

$$C_c = C_{c0} \times \left[1 + \alpha \frac{\left(F_{avg}\right)^{m_c}}{\left(F_{\max} - F_{\min}\right)^{m_c} + \left(F_{avg}\right)^{m_c}}\right] \qquad (5)$$

$$C_m = C_{m0} \times \left[1 + \beta \frac{\left(F_{avg}\right)^{m_c}}{\left(F_{\max} - F_{\min}\right)^{m_c} + \left(F_{avg}\right)^{m_c}}\right] \qquad (6)$$

$$C_{new} = \beta F_{\max} + (1 - \beta) F_{\min} \qquad (7)$$

where C_c is the crossover, C_{c0} is the initial crossover, $\alpha = 0.3$, C_m is the mutation, C_{m0} is the initial mutation, $m_c = 2$, $F_{\max}$ is the fitness maximum, $F_{\min}$ is the fitness minimum, F_{avg} is the fitness average, $\beta = 0.2$, C_{new} is the gene after mutation.

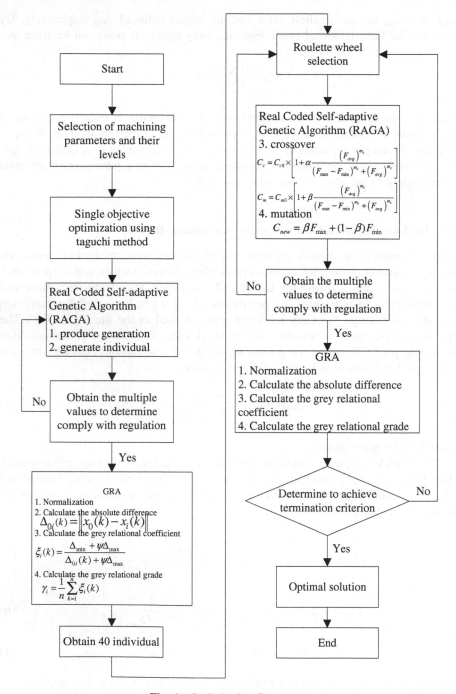

Fig. 1. Optimization flow chart.

2.3   Combined RAGA and GRA Methods

In this paper, the RAGA and GRA are combined for the optimization of milling process parameters. The regulation value of surface roughness can be used as the target value. The GRA is used to integrate multiple quality control values into a single one, which is the Grey relational grade, and is the fitness function of the RAGA. The crossover and mutation rates of the RAGA change in the next generation according to the obtained fitness function. If the average fitness becomes big, the crossover probability also becomes big, and vice versa. The whole process will do loop until achieving the stop condition or the optimal solution. In this study, the steps are repeated to search for the optimal solution until the end of the maximum generation number. The optimization flow chart can be illustrated in Fig. 1.

3   Verification

3.1   Experimental Design and Results

Various milling process parameters affect the machined surface topography and the surface roughness, which are indexes of importance for evaluating the machining quality. In the experiment, four cutting process parameters at four different levels each have been taken into consideration, namely, the cutter radius, the spindle speed, the feed per tooth, and the cutter helix angle. Process parameters with their symbols and values at different levels are listed in Table 1.

Table 1. Milling process parameters and their levels.

Symbol	Milling process parameters	Level 1	Level 2	Level 3	Level 4
A	Cutter radius (mm)	3	4[a]	5	6
B	Spindle speed (rpm)	7500	10000[a]	12500	15000
C	Feed per tooth (mm/tooth)	0.05	0.1[a]	0.2	0.25
D	Cutter helix angle (degree)	30	35[a]	40	45

[a]represents the initial milling process parameter settings

In order to compare the reliability, the Taguchi-Grey method is firstly performed with this case. The design matrix is selected according to the Taguchi's orthogonal array design, which consists of 16 sets of coded conditions. The surface roughness is usually quantified by the vertical deviations of a real surface from its ideal form. The commonly used three surface roughness parameters are Ra, Ry and Rz. Ra represents the arithmetical mean deviation of the profile. Ry represents the maximum height of the profile. Rz represents the mean roughness depth. Here, only Ra is considered in both down milling process and up milling process. The analysis of variance (ANOVA) results for the roughness of down milling process are listed in Table 2. The ANOVA results for the roughness of up milling process are listed in Table 3.

By using the Taguchi-GRA optimization, a set of optimization parameters can be found in Table 2 for down milling and Table 3 for up milling. As the result of Taguchi

Table 2. ANOVA results for roughness of down milling process.

Factor	S/N ratio (dB)				DOF	Sum of square	Variance	Contribution
	Level 1	Level 2	Level 3	Level 4				
A	−3.8845	−3.4980	−2.9895	**−2.6861**	3	0.0257	0.0086	4.81%
B	−3.3639	−3.4255	**−3.1030**	−3.1216	3	0.0034	0.0011	0.64%
C	**−0.5291**	−2.9859	−4.8354	−5.5539	3	0.5055	0.1685	94.51%
D	−3.2609	−3.3027	**−3.2180**	−3.2281	3	0.0002	0.0001	0.04%

The best combination is $A_4B_3C_1D_3$

Table 3. ANOVA results for roughness of up milling process.

Factor	S/N ratio (dB)				DOF	Sum of square	Variance	Contribution
	Level 1	Level 2	Level 3	Level 4				
A	−2.4319	−1.8250	−1.4699	**−1.1977**	3	0.00297	0.0099	7.48%
B	**−1.4200**	−1.5014	−1.9751	−1.9953	3	0.0099	0.0033	2.51%
C	**−0.0676**	−0.7499	−2.2831	−4.3878	3	0.3546	0.1182	89.38%
D	**−1.5527**	−1.6402	−1.8929	−1.7941	3	0.0025	0.0008	0.63%

The best combination is $A_4B_1C_1D_1$

method is not the best result, the result is used to setting feasible domain of the RAGA and GRA, as shown in Table 4. The comparisons of the two optimization methods for down milling process and up milling process are shown in Tables 5 and 6. The comparison results show that the RAGA-GRA method can find a better combination of milling process parameters than the Taguchi-GRA method.

Table 4. Feasible domain of the RAGA-GRA.

Milling process parameters	Upper-level	Low-level
Cutter radius (mm)	6	2
Spindle speed (rpm)	15000	7500
Feed per tooth (mm/tooth)	0.20	0.04
Cutter helix angle (degree)	50	30

Table 5. Comparison of the two optimization methods for down milling process.

	Initial	Taguchi-GRA	RAGA-GRA
Cutter radius (mm)	4	6	6
Spindle speed (rpm)	10000	12500	12000
Feed per tooth (mm/tooth)	0.10	0.05	0.04
Cutter helix angle (degree)	35	40	40
Surface roughness(μm)	1.60	0.40	0.28

Table 6. Comparison of the two optimization methods for up milling process.

	Initial	Taguchi-GRA	RAGA-GRA
Cutter radius (mm)	4	6	6
Spindle speed (rpm)	10000	7500	12000
Feed per tooth (mm/tooth)	0.10	0.05	0.04
Cutter helix angle (degree)	35	30	35
Surface roughness (μm)	7.6	6.2	6.0

3.2 Comparison of Surface Topography

With the derived optimal milling process parameters, this section mainly concentrates on the topography of the machined surface in the case of both optimal milling process parameter settings and initial milling process parameter settings for verifying the foregoing evaluated optimal parameters. In each case, both down milling process and

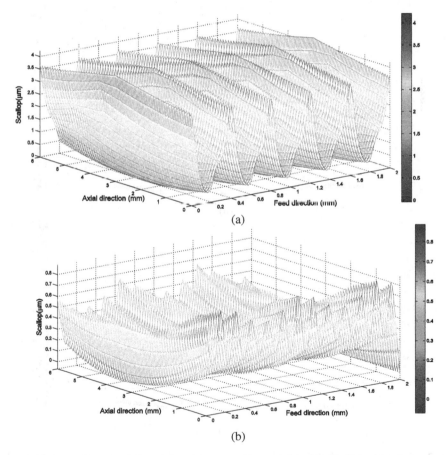

Fig. 2. Surface topography of down milling process. (a) Initial parameter settings. (b) RAGA-GRA parameter settings.

up milling process are studied. The results are shown in Figs. 2 and 3. Figure 2(a) and (b) show the surface topography of down milling process for the initial and the RAGA-GRA parameter settings, respectively. Figure 3(a) and (b) show the surface topography of up milling process for the initial and the RAGA-GRA parameter settings, respectively.

Some conclusions can be drawn by comparing and analyzing Figs. 2 and 3. The results show that the values of the machined surface roughness in the case of down milling process are much smaller than that of in the case of up milling process. The quality characteristic of the machined surface with the RAGA-GRA parameter settings is better than that of the machined surface with the initial parameter settings by comparing Fig. 2(a) and (b) as well as Fig. 3(a) and (b). Furthermore, the values of the machined surface roughness with the RAGA-GRA parameter settings are obviously smaller than that of with the initial parameter settings. This further demonstrates the RAGA-GRA method is feasible in optimization of milling process parameters.

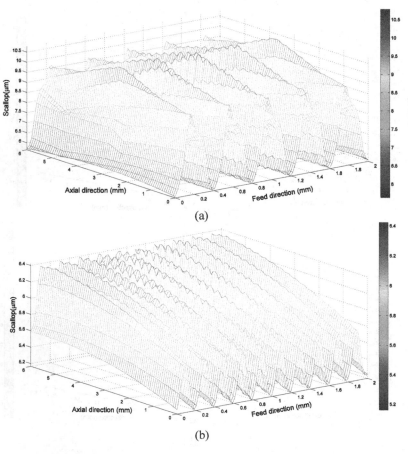

(a)

(b)

Fig. 3. Surface topography of up milling process. (a) Initial parameter settings. (b) RAGA-GRA parameter settings.

4 Conclusions

The present study concentrates on the application of real-coded self-adaptive genetic algorithm coupled with Grey relational analysis for solving the optimization of milling process parameters. The detailed methodology of the RAGA-GRA is applied for evaluating the optimal combination of milling process parameters so as to achieve the desired quality characteristics of the machined surface. Furthermore, the experiments of machined surface topography are implemented in the case of both down milling process and up milling process with the initial and optimal parameter settings. The experimental results indicate that the machined surface quality is improved after using the RAGA-GRA method, which further verify the feasibility of the proposed method.

Acknowledgements. This work is supported by the National Natural Science Foundation of China (Grant No. 51505343) and the Postdoctoral Science Foundation of China (Grant No. 2015M572192).

References

1. Tönshoff, H.K., Arendt, C., Amor, R.B.: Cutting of hardened steel. CIRP Ann. Manufact. Technol. **49**(2), 547–566 (2000)
2. Benardos, P.G., Vosniakos, G.C.: Prediction of surface roughness in CNC face milling using neural networks and taguchi's design of experiments. Robot. Comput. Integr. Manufact. **18**(5), 343–354 (2002)
3. Zhang, J.Z., Chen, J.C., Kirby, E.D.: Surface roughness optimization in an end-milling operation using the Taguchi design method. J. Mater. Process. Technol. **184**(1), 233–239 (2007)
4. Sabanayagam, A., Kumar, V.A., Raju, S., Kumar, N.S.: Optimization of interconnects for signal integrity using self-adaptation in real-parameter genetic algorithms with simulated binary cross over (SBX). In: 10th International Conference on Electromagnetic Interference & Compatibility, pp. 123–129. IEEE, India (2008)
5. Gong, C., Xu, C., Wang, J.: An efficient adaptive real coded genetic algorithm to solve the portfolio choice problem under cumulative prospect theory. Comput. Econ. 1–26 (2017). doi:10.1007/s10614-017-9669-5,
6. Lee, L.H., Fan, Y.: An adaptive real-coded genetic algorithm. Appl. Artif. Intell. **16**(6), 457–486 (2002)
7. Subbaraj, P., Rajnarayanan, P.N.: Optimal reactive power dispatch using self-adaptive real coded genetic algorithm. Electr. Power Syst. Res. **79**(2), 374–381 (2009)
8. Subbaraj, P., Rengaraj, R., Salivahanan, S.: Enhancement of self-adaptive real-coded genetic algorithm using Taguchi method for economic dispatch problem. Appl. Soft Comput. **11**(1), 83–92 (2011)
9. Abbas, H.M., Bayoumi, M.M.: Volterra-system identification using adaptive real-coded genetic algorithm. IEEE Trans. Syst. Man Cybern. Part A Syst. Hum. **36**(4), 671–684 (2006)
10. Oyama, A., Obayashi, S., Nakahashi, K.: Wing design using real-coded adaptive range genetic algorithm. In: 1999 IEEE International Conference on Systems, Man, and Cybernetics, pp. 475–480. IEEE, Japan (1999)
11. Deng, J.L.: Introduction to Grey system theory. J. Grey Syst. **1**(1), 1–24 (1989)

12. Deng, J.L.: A Course on Grey System Theory. Huazhong University of Science and Technology Press, Wuhan (1990)
13. Tzeng, C.J., Lin, Y.H., Yang, Y.K., Jeng, M.C.: Optimization of turning operations with multiple performance characteristics using the Taguchi method and Grey relational analysis. J. Mater. Process. Technol. **209**(6), 2753–2759 (2009)
14. Yang, Y.S., Huang, W.: A Grey-fuzzy Taguchi approach for optimizing multi-objective properties of zirconium-containing diamond-like carbon coatings. Expert Syst. Appl. **39**(1), 743–750 (2012)
15. Wang, L., Tang, D.: An improved adaptive genetic algorithm based on hormone modulation mechanism for job-shop scheduling problem. Expert Syst. Appl. **38**(6), 7243–7250 (2011)

HybridCAM: Tool Path Generation Software
for Hybrid Manufacturing

Xiongzhi Zeng, Changya Yan[✉], Juan Yu, Shanshan He, and Chen-Han Lee

Manufacturing Intelligence Engineering Research Center, Wuhan Institute of Technology,
Wuhan, China
yanchangya@hotmail.com

Abstract. In this paper, we demonstrate a Hybrid Manufacturing software-HybridCAM. HybridCAM provides a variety of tool path generation methods for sculpture surfaces. The tool path patterns include contour-parallel pattern, zigzag pattern, and helical pattern. HybridCAM can input multiple format of models, such as IGS, STL, and STP, and can be applied to multi-axis Additive Manufacturing (AM) and multi-axis Subtractive Manufacturing (SM). We present several tool path patterns for different parts and different applications generated by HybridCAM. An actual experiment was carried out to verity the feasibility and practicality of the software.

Keywords: Tool path generation · Additive manufacturing · Hybrid manufacturing · HybridCAM

1 Introduction

Subtractive Manufacturing (SM, known as machining or cutting) is a traditional high speed and high accuracy technology used in industrial manufacturing [1]. For complex parts, SM tool paths generation is difficult because it has to satisfy many requirements such as tool axes planning, without overcut and undercut, and collision avoidance. Additive Manufacturing (AM) is a novel technology to fabricate parts by adding material [2–4]. AM method has advantages of high material utilization and fit for complex part manufacturing, but has low surface precision. Hybrid Manufacturing integrates advantages of AM and SM to improve the efficiency and quality of manufacturing.

AM technology can be classified into two types by material feeding methods [5–7]: (1) Powder Beds Processes (PBP), including Selective Laser Melting (SLM), Selective Laser Sintering (SLS) and Electron Beam Melting (EBM); (2) Simultaneous Material Delivering (SMD), including Laser Engineered Net Shaping (LENS), Direct Metal Deposition (DMD) and Selective Laser Cladding (SLC). PBP is limited to layer deposition, while SMD is applied to complex parts generation with multi-axis tool paths.

In AM, deposition quality of parts depend on the choice of tool path patterns [8]. AM tool paths need to meet some requirements such as avoiding collapse and deposition voids. Contour-parallel pattern and zigzag pattern are two common patterns used in AM [9–13] while are not fit for some special parts, for example, thin-wall parts and blade in impeller. In SM, various multi-axis tool path patterns have been provided for various

© Springer International Publishing AG 2017
Y. Huang et al. (Eds.): ICIRA 2017, Part III, LNAI 10464, pp. 877–889, 2017.
DOI: 10.1007/978-3-319-65298-6_78

applications. At the same time, there also need abundant tool path patterns for special parts.

There are some achievements in the research of Hybrid Manufacturing. Ruan et al. [14] uses an adaptive slicing method to perform process planning and tool path generation. The method is limit to STL model. Lan Ren et al. [8] develops a Hybrid Manufacturing system which uses an adaptive trajectory planning method for tool path planning. This method combines with the contour-parallel and zigzag to avoid the deposition voids while doesn't provide five-axis AM tool path patterns.

In this paper, we introduce a software HybridCAM which integrates multi-axis AM module and multi-axis SM module. HybridCAM is developed by our team. HybridCAM provides various tool path patterns and supports for multiple format models, such as IGS, STL and STP. In HybridCAM, the AM module is developed based on SMD technology and the SM module uses machining module of NX 9.0. This paper focus on introducing AM module of HybridCAM. The reminder of this paper is organized as follows. In Sect. 2, we introduce the architecture of HybridCAM. In Sect. 3, we present several three-axis AM tool path patterns and five-axis AM tool path patterns. Section 4 presents AM tool paths of two actual parts generated by HybridCAM and demonstrates an actual AM experiment to verity the feasibility and practicality of the software. At last, Sect. 5 concludes this paper.

2 Architecture Overview

We use NX Open to develop the software HybridCAM. NX Open is a development language tool based on NX. The users can integrate their application to NX by NX Open easily. As shown in Fig. 1, NX Open supports for programming languages such as C++, .NET and JAVA and provides many tools such as Common API, Journaling and classic APIs.

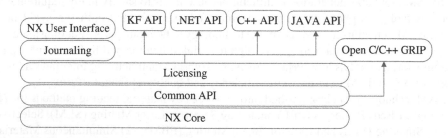

Fig. 1. NX Open framework

HybridCAM is developed base on NX Open by C++ language. As shown in Fig. 2, the architecture of HybridCAM includes four levels: Framework Layer, Core Layer, Application Layer and Interface Layer. The Framework Layer is supported by NX Open and the other three layers are provided by HybridCAM.

(1) The Framework Layer is the basic platform and consists of NX Geometric Engine, Mathematical Engine, Graphics Engine and the corresponding interface provided by C++ API.

(2) The Core Layer is the algorithm library for AM and includes many algorithm functions. This layer provides process planning and computes tool paths for AM.

(3) The Application Layer contains three-axis and five-axis AM units and provide multiple tool path patterns. Those tool path patterns will be described in detail in Sect. 3.

(4) The Interface Layer is the graphical user interface (GUI) and allow users input parameters. This layer provides graphic display function and shows tool paths to users.

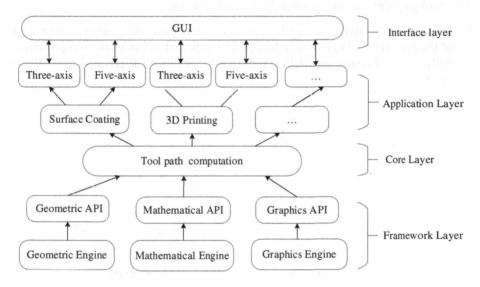

Fig. 2. Architecture of HybridCAM

3   Tool Path Patterns

HybridCAM has three-axis and five-axis AM unit, each unit has two subunits. We classify subunits according to shape of parts: (1) thin-wall part, (2) general part. AM has wide application field, faces many challenges. Single or simple tool path patterns cannot meet requirements of manufacturing. In order to meet different applications, AM also need multiple tool patterns.

3.1   Three-axis Tool Path Patterns

As shown in Fig. 3, three-axis tool path patterns include follow and helical of thin-wall parts; contour-parallel, zigzag, and regional zigzag of general parts.

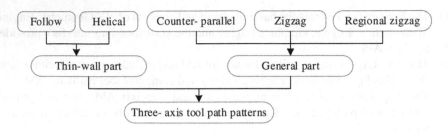

Fig. 3. Three-axis tool path patterns

Tool path patterns of thin-wall parts

Two tool path patterns are developed for thin-wall parts.

1. **Helical.** This pattern generates helical curve as tool path according to contour shape of the part. Tool path generated through helical pattern doesn't have non-processing path, has good continuity. Helical pattern is fit for thin blade processing. As shown in Fig. 4.

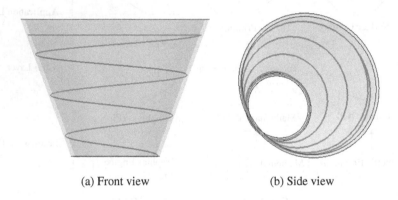

(a) Front view   (b) Side view

Fig. 4. Three -axis helical pattern

2. **Follow.** This pattern slices part depended on layer thickness and generates tool path by offsetting contour curve of each layer. Each layer only processing one ring. As shown in Fig. 5.

Tool Path Patterns of General Parts

HybridCAM provides three tool path patterns for general parts:

1. **Contour-parallel.** Tool path is obtained by equidistant offset contour curve. The internal and external boundary are processed at first, interior area later filled. Start point of each layer can be separated, separation distance also can be set. As shown in Fig. 6.
2. **Zigzag.** First, this pattern generates boundary tool paths based on the contour curves of the part, then uses the zig zag path to fill interior area. This pattern can improve the processing efficiency. As shown in Fig. 7.

3. **Regional zigzag.** This path pattern can identify holes or cavities of part, sub regional calculates path. It processes the outer and inner boundary at first, then fills the interior

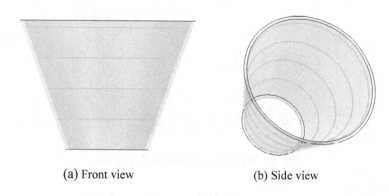

(a) Front view   (b) Side view

Fig. 5. Three-axis follow pattern

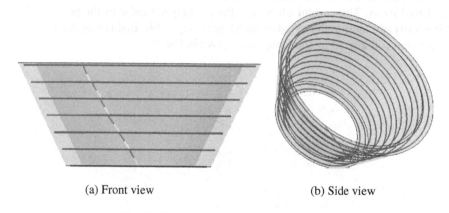

(a) Front view   (b) Side view

Fig. 6. Three-axis contour-parallel pattern

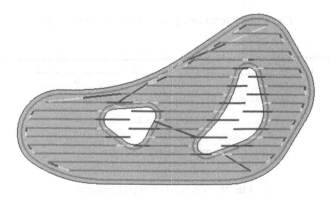

Fig. 7. Three-axis zigzag pattern

of the sub-area. Regional zigzag's non-processing path is less and the processing efficiency is higher than zigzag. As shown in Fig. 8.

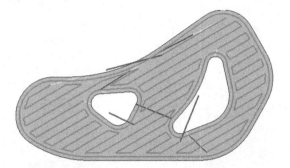

Fig. 8.   Three-axis regional zigzag pattern

Zigzag and regional zigzag patterns in HybridCAM have their own advantages. These tool path patterns can set the angle between the path and contour, while providing staggered tool paths. The current fill tool paths are perpendicular to the previous layer's filling tool paths and starting point of each layer is separable. In this way, the two patterns can prevent collapse from appearing. As shown in Fig. 9.

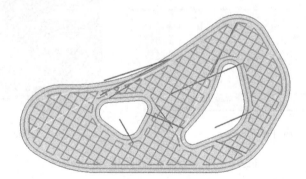

Fig. 9.   Staggered tool paths between layer and layer

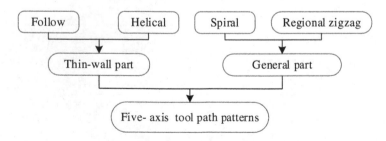

Fig. 10.   Five-axis tool path patterns

3.2 Five-Axis Tool Path Patterns

Five-axis AM is strict for the spindle of the laser tool, the laser beam should be perpendicular to the substrate. In order to avoid collisions, HybridCAM provides partial adjustment of the tool axis. Five-axis tool path patterns are shown in Fig. 10.

(1) Follow and helical tool path patterns are similar to that in three-axis AM. The difference is that in five axis, the direction of spindle change with the substrate. These tool path patterns are shown in Figs. 11 and 12.

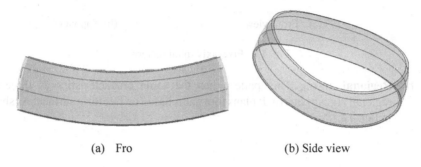

(a)   Fro   (b) Side view

Fig. 11. Five-axis follow pattern

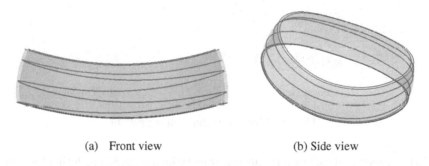

(a)   Front view   (b) Side view

Fig. 12. Five-axis helical pattern

(2) Spiral and regional zigzag of general parts.
  (a) **Spiral.** This pattern generates spiral curve as tool path in each layer after slicing the part. Spiral pattern has continuous tool path, no non-processing tool path, with high processing efficiency. As shown in Fig. 13.

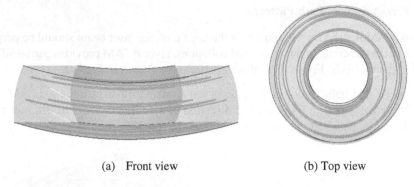

(a) Front view   (b) Top view

Fig. 13. Five-axis spiral pattern

(b) **Regional zigzag** This pattern has the same characteristics as three-axis regional zigzag pattern. It plans tool paths based on freeform surface. As shown in Fig. 14.

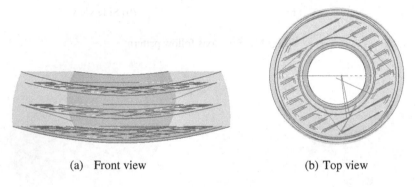

(a) Front view   (b) Top view

Fig. 14. Five-axis regional zigzag pattern

In this paper, tool path patterns are integrated with the machining technology. In tool paths planning, the boundary of the part deposits at first to ensure accuracy of parts dimension. The provided tool path patterns consider the continuity of processing, reduce the non-processing path and laser on or off frequency. When filling, staggered tool paths between layer and layer avoid deposition void, improve the compactness of parts.

3.3 Hybrid Manufacturing Programming

AM module is embedded in the NX CAM. When the user program, they can call all functions provided by NX CAM directly. NX CAM includes Mill_Planar, Mill_Contour, Mill_Multi-Axis, Mill_Multi-Blade, Bill_Rotary, Hole_Making etc. All functions of NX CAM can meet user's various machining requirements. AM tool paths and SM tool paths can be continuously output to ensure the continuity of manufacturing process. Hybrid Manufacturing build-sequence is shown in Fig. 15.

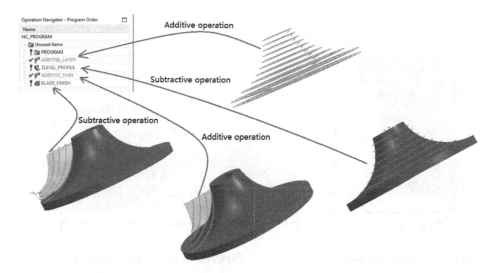

Fig. 15. Hybrid Manufacturing build-sequence

4   Applications and Discussion

In this section, we use two applications to verify the feasibility and practicality of HybridCAM. AM does not need to rely on the traditional machining tools, can manufacture many complex structural parts which are difficult to be manufactured by SM, and shortens the processing cycle [15]. For complex shape parts, such as the vane and impeller.

4.1   Tool Path Generation

The two parts come from actual parts in aviation industry. The size of two parts are shown in the Table 1, and the CAD model of the two parts are shown in Fig. 16. Tool paths of the vane is generated by three-axis helical, and tool paths of the impeller use five-axis follow. Processing parameters are shown in the Table 1. Tool paths for vane is too dense to observe, only part of it is shown in Fig. 17. Tool paths for impeller is shown in Fig. 18.

Table 1. AM tool path generation configurations

Part name	Size (mm)	Tool path pattern	Processing parameters (mm)
Vane	Length: 53.3	Three-axis helical	Laser spot diameter: 1.5
	Width: 3		Layer thickness: 0.3
	Height: 100		
Impeller	Diameter: 348.0	Five-axis follow	Laser spot diameter: 2
	Height: 130.4		Layer thickness: 2
	Blade thickness: 2.4		

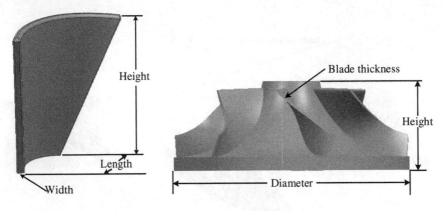

(a) CAD model of part vane   (b) CAD model of part impeller

Fig. 16. CAD models of two parts

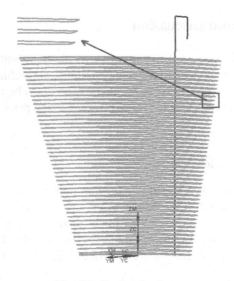

Fig. 17. Tool path of vane

Based on the two applications, it is found that the parameters needed for tool path computing are less by using HybridCAM. The software also provides parameters such as safe plane height, Lead in-Lead out, feed rate etc. In addition, AM unique laser start-stop, powder feed start-stop external parameters are also provided to user to set.

Fig. 18. Tool path of impeller

4.2 Machining Verification

To evaluate that tool paths generated by HybridCAM whether can be used, we manufactured the vane mentioned before. The Fig. 19 shows the three-axis AM system, the powder feeder equipment (Fig. 19(a)) is connecting with the three-axis AM equipment (Fig. 19(b)).

(a)   Powder feeder equipment   (b) Three-axis AM equipment

Fig. 19. Additive Manufacturing system

In the AM processing, some process parameters such as material, laser scanning speed need to be set beforehand. We list the processing parameters in the Table 2. Forming vane as shown in Fig. 20. The vine has smooth surface, no collapse, uniform wall thickness, no distortion.

Table 2. Actual AM deposition experiment parameters

Part name	Powder material	Laser scanning speed	Laser power	Power feed rate
Vane	Stainless steel	1500 mm/s	1500 W	1r/s

Fig. 20. Forming part

5   Conclusions

In this paper, we present the software HybridCAM used for Hybrid Manufacturing tool path generation. HybridCAM not only has common tool path patterns but also helical for thin-wall parts. Through the analysis of two application cases, the test results show that the software can reach the design requirements. HybridCAM can be carried out actual processing, confirmed the feasibility and practicality of this software.

At present, HybridCAM needs further development and improvement. The tool path patterns should be further expanded to meet the requirements of the special workpiece. Although AM and SM have been studied for many years, the integrated process planning still has a long way to go.

Acknowledgement. The authors gratefully acknowledge National Natural Science Foundation of China (51575386). The authors would also like to thank the experimental supporting from Wuhan HuaGong Laser Engineering Co., Ltd.

References

1. Fessler, J., Merz, R., Nickel, A., Prinz, F.: Laser deposition of metals for shape deposition manufacturing. In: Solid Freeform Fabrication Symposium, pp. 117–124 (1996)
2. Lu, B.-H., Li, D.-C.: Development of the additive manufacturing (3D printing) technology. Mach. Build. Autom. **42**, 1–4 (2013)
3. Turichin, G.A., Somonov, V.V., Babkin, K.D., Zemlyakov, E.V., Klimova, O.G.: High-speed direct laser deposition: technology. Equip. Mater. **125**, 012009 (2016)
4. Gao, W., Zhang, Y., Ramanujan, D., Ramani, K., Chen, Y., Williams, C.B., et al.: The status, challenges, and future of additive manufacturing in engineering. Comput. Aided Des. **69**, 65–89 (2015)
5. Bikas, H., Stavropoulos, P., Chryssolouris, G.: Additive manufacturing methods and modelling approaches: a critical review. Int. J. Adv. Manuf. Technol. **83**, 389–405 (2016)
6. Griffith, M.L., Keicher, D.M., Atwood, C.L.: Free form fabrication of metallic components using laser engineered net shaping (LENS{trademark}). Office of Scientific & Technical Information Technical Reports (1996)

7. Thompson, S.M., Bian, L., Shamsaei, N., Yadollahi, A.: An overview of direct laser deposition for additive manufacturing; part I: transport phenomena, modeling and diagnostics. Addit. Manuf. **8**, 36–62 (2015)
8. Ren, L., Sparks, T., Ruan, J., Liou, F.: Integrated process planning for a multiaxis hybrid manufacturing system. J. Manuf. Sci. Eng. **132**, 237–247 (2010)
9. Kao, J.H., Prinz, F.B.: Optimal Motion Planning for Deposition in Layered Manufacturing (2001)
10. Eiamsa-Ard, K., Liou, F.W., Ren, L., Choset, H.: Spiral-like path planning without gap for material deposition processes. In: ASME 2006 International Design Engineering Technical Conferences and Computers and Information in Engineering Conference, pp. 983–991 (2006)
11. Foroozmehr, E., Kovacevic, R.: Effect of path planning on the laser powder deposition process: thermal and structural evaluation. Int. J. Adv. Manuf. Technol. **51**, 659–669 (2010)
12. Tabernero, I., Calleja, A., Lamikiz, A.: Lacalle LNLD.: optimal parameters for 5-axis laser cladding ☆. Procedia Eng. **63**, 45–52 (2013)
13. Calleja, A., Tabernero, I., Fernández, A., Celaya, A., Lamikiz, A.: Lacalle LNLD.: improvement of strategies and parameters for multi-axis laser cladding operations. Opt. Lasers Eng. **56**, 113–120 (2014)
14. Ruan, J.: Automatic process planning and toolpath generation of a multiaxis hybrid manufacturing system. J. Manuf. Process. **7**, 57–68 (2005)
15. Kobryn, P.A., Moore, E.H., Semiatin, S.L.: Effect of laser power and traverse speed on microstructure, porosity, and build height in laser-deposited Ti-6Al-4 V. Scripta Mater. **43**, 299–305 (2000)

Author Index

Printed in the United States
By Bookmasters